THE FIRST RESULTS FROM SOHO

THE FIRST RESULTS FROM SOHO

Edited by

B. FLECK

ESA Space Science Department, ESTEC, Noordwijk, The Netherlands and NASA/GSFC, Greenbelt, MD, U.S.A.

and

Z. ŠVESTKA

SRON Space Research, Utrecht, The Netherlands and CASS, U.C. San Diego, La Jolla, CA, U.S.A.

Reprinted from Solar Physics, Volume 170, No. 1, 1997 and
Volume 175, No. 2, 1997

SPRINGER-SCIENCE+BUSINESS MEDIA, B.V.

A C.I.P. Catalogue record for this book is available from the Library of Congress.

DOI 10.1007/978-94-011-5236-5

Printed on acid-free paper

TABLE OF CONTENTS

THE FIRST RESULTS FROM SOHO
Part 1

THE FIRST RESULTS FROM SOHO
Part 2*

TABLE OF CONTENTS

TABLE OF CONTENTS

FIRST RESULTS FROM VIRGO, THE EXPERIMENT FOR HELIOSEISMOLOGY AND SOLAR IRRADIANCE MONITORING ON SOHO

CLAUS FRÖHLICH[1], BO N. ANDERSEN[2], THIERRY APPOURCHAUX[3], GABRIELLE BERTHOMIEU[4], DOMINIQUE A. CROMMELYNCK[5], VICENTE DOMINGO[3], ALAIN FICHOT[5], WOLFGANG FINSTERLE[1], MARIA F. GÓMEZ[6], DOUGLAS GOUGH[7,8], ANTONIO JIMÉNEZ[6], TORBEN LEIFSEN[9], MARC LOMBAERTS[5], JUDIT M. PAP[10,11], JANINE PROVOST[4], TEODORO ROCA CORTÉS[6], JOSÉ ROMERO[1], HANSJÖRG ROTH[1], TAKASHI SEKII[7], UDO TELLJOHANN[3], THIERRY TOUTAIN[4] and CHRISTOPH WEHRLI[1]

[1] *Physikalisch-Meteorologisches Observatorium Davos*
World Radiation Center, CH-7260 Davos Dorf, Switzerland
[2] *Norwegian Space Centre, N-0309 Oslo 3, Norway*
[3] *Space Science Department, ESTEC, NL-2200 AG Noordwijk, the Netherlands*
[4] *Departement Cassini, URA CNRS 1362, Observatoire de la Cote d'Azur, F-06304 Nice Cedex 4, France*
[5] *Institut Royal Météorologique de Belgique, B-1180 Bruxelles, Belgium*
[6] *Instituto de Astrofísica de Canarias,*
Universidad de La Laguna, E-38071 La Laguna, Tenerife, Spain
[7] *Institute of Astronomy, University of Cambridge, Cambridge CB3 0HA, U.K.*
[8] *Department of Applied Mathematics and Theoretical Physics,*
University of Cambridge, Cambridge CB3 9EW, U.K.
[9] *Institute of Theoreretical Astrophysics, University of Oslo, N-0315 Oslo*
[10] *Department of Physics and Astronomy, Division of Astronomy and Astrophysics*
University of California, Los Angeles, CA90095-1562, U.S.A.
[11] *Jet Propulsion Laboratory, California Institute of Technology,*
Pasadena, CA 91109, U.S.A.

(Received 21 October, 1996; in revised form 30 November, 1996)

Abstract. First results from the VIRGO experiment (Variability of solar IRradiance and Gravity Oscillations) on the ESA/NASA Mission SOHO (Solar and Heliospheric Observatory) are reported. The observations started mid-January 1996 for the radiometers and sunphotometers and near the end of March for the luminosity oscillation imager. The performance of all the instruments is very good, and the time series of the first 4–6 months are evaluated in terms of solar irradiance variability, solar background noise characteristics and *p*-mode oscillations. The solar irradiance is modulated by the passage of active regions across the disk, but not all of the modulation is straightforwardly explained in terms of sunspot flux blocking and facular enhancement. Helioseismic inversions of the observed *p*-mode frequencies are more-or-less in agreement with the latest standard solar models. The comparison of VIRGO results with earlier ones shows evidence that magnetic activity plays a significant role in the dynamics of the oscillations beyond its modulation of the resonant frequencies. Moreover, by comparing the amplitudes of different components of *p*-mode multiplets, each of which are influenced differently by spatial inhomogeneity, we have found that activity enhances excitation.

Solar Physics **170:** 1–25, 1997.

1. Introduction

The aim of the VIRGO investigation (Variability of solar IRradiance and Gravity Oscillations) on SOHO (SOlar and Heliospheric Observatory) is to determine the characteristics of pressure and internal gravity oscillations by observing irradiance and radiance variations, to measure the solar total and spectral irradiance and to quantify their variability over periods of days to the duration of the mission.

VIRGO contains two different active-cavity radiometers for monitoring the solar 'constant' (DIARAD and PMO6-V), two three-channel sunphotometers (SPM) for the measurement of the spectral irradiance at 402, 500, and 862 nm with a bandwidth of 5 nm, and a low-resolution imager (Luminosity Oscillation Imager, LOI) with 12 'scientific' and 4 guiding pixels, for measuring the radiance distribution over the solar disk at 500 nm. The instrumentation has been described in detail by Fröhlich *et al.* (1995). In addition, the observed in-flight performance and operational aspects of the irradiance observations are described by Fröhlich *et al.* (1997), and those of the LOI by Appourchaux *et al.* (1997).

SOHO was successfully launched on 2 December, 1995. VIRGO was switched on a few days later. The covers were kept closed for most of the time, but were released in order to outgas the experiment thoroughly during the following 7 weeks without having the optical surfaces exposed to the Sun. On Christmas Eve all covers were opened for a short period of time in order to test the performance of the instruments in a 'first light' experiment. Real observations with the radiometers started mid-January, and observations with the sunphotometers at the end of January 1996. At that time it was realized that the LOI cover could not be opened. A concerted effort, as described by Appourchaux *et al.* (1997), was needed, and finally on 27 March the cover was successfully opened. The LOI has been operating continuously since then. The performance of all the instruments is very good.

The products of the VIRGO experiment are the following observational and derived data:

– continuous high-precision, high-stability and high-accuracy measurements of the solar total and spectral irradiance and spectral radiance and their variation;

– continuous measurements of the solar polar and equatorial diameter;

– frequencies, amplitudes and phases of oscillation modes in the frequency range of <0.1 μHz to 8 mHz.

The observational data products are time series which are corrected for all *a priori* known effects such as temperature, pointing, instrument–Sun distance and relative velocity (for level 1), and inferred effects such as degradation (for level-2 data products). The present scientific evaluation is carried out on preliminary level-2 data, which may change in the future owing to improvements in the production of level-2 data from level-1 data. Such datasets from the first 4 – 6 months of operation are the basis for the results we present here.

The following sections describe contributions to the study of total and spectral irradiance variability, namely the influence of active regions on irradiance variability and the characteristics of the solar background 'noise'. For helioseismology, frequencies and linewidths of the low-degree p modes ($l = 0 \ldots 7$) in the frequency range from 2150 to 3850 μHz ($n = 13 \ldots 27$) are presented, together with a preliminary interpretation from inversions. Furthermore, the temporal amplitude variation of p-modes is examined.

2. Solar Irradiance Variability

2.1. Total and Spectral Solar Irradiance and Radiance Measurements

The performance of the VIRGO radiometers and the procedures for the data evaluation are described by Fröhlich *et al.* (1997). The radiometers are operated in active mode, and have a basic sampling cadence of 3 min for the DIARAD and of 1 min for the PMO6-V. Originally the latter was 2 min; the change is due to the fact that the PMO6-V shutters ceased to operate shortly after the start of the measurements. The PMO6-V radiometer is now operated quasi-continuously, with a reference value obtained every 8 hours by closing the cover (not the shutter) for about 6 min. This extends the accessible frequencies of variation into the p-mode range, thereby allowing p modes to be studied also with total irradiance data. The continuously exposed DIARAD left channel shows no measurable degradation up to August 1996. This conclusion was obtained from several comparisons with the right DIARAD channel which is operated about every 60th day for 30 min. In contrast, the operational PMO6-V radiometer degrades at a rate of ≈ 2.6 ppm day^{-1} relative to DIARAD. It is interesting to note that radiometers of both types (DIARAD and PMO6-V) on EURECA showed a degradation similar to that of the VIRGO PMO6-V, if referred to the same exposure times, whereas on SOHO the DIARAD behaves quite differently. There are also some small differences in the time series of both radiometers which are not readily explained. An investigation is under way.

The absolute irradiance values are deduced from the radiometric characterization of each radiometer obtained before launch. In space the two radiometers differ by 1.5 W m^{-2} or 0.11% with the PMO6-V reading lower; this difference is within the estimated absolute accuracy of each radiometer of $\pm 0.15\%$. The VIRGO solar irradiance values as shown in Figure 1 are DIARAD values decreased by 0.75 W m^{-2} to account for the difference between both type of radiometers. The absolute values will be checked against measurements with similar radiometers on a planned stratospheric balloon flight in spring 1997 and later with SOLCON onboard the NASA Hitchhiker program (Crommelynck *et al.*, 1996). Moreover, they will be complemented by a detailed study of the time series of simultaneous measurements of VIRGO and ACRIM-II on UARS.

The spectral irradiance data from the SPM are used to study the solar variability and solar oscillations; results from the former are presented in this section, those

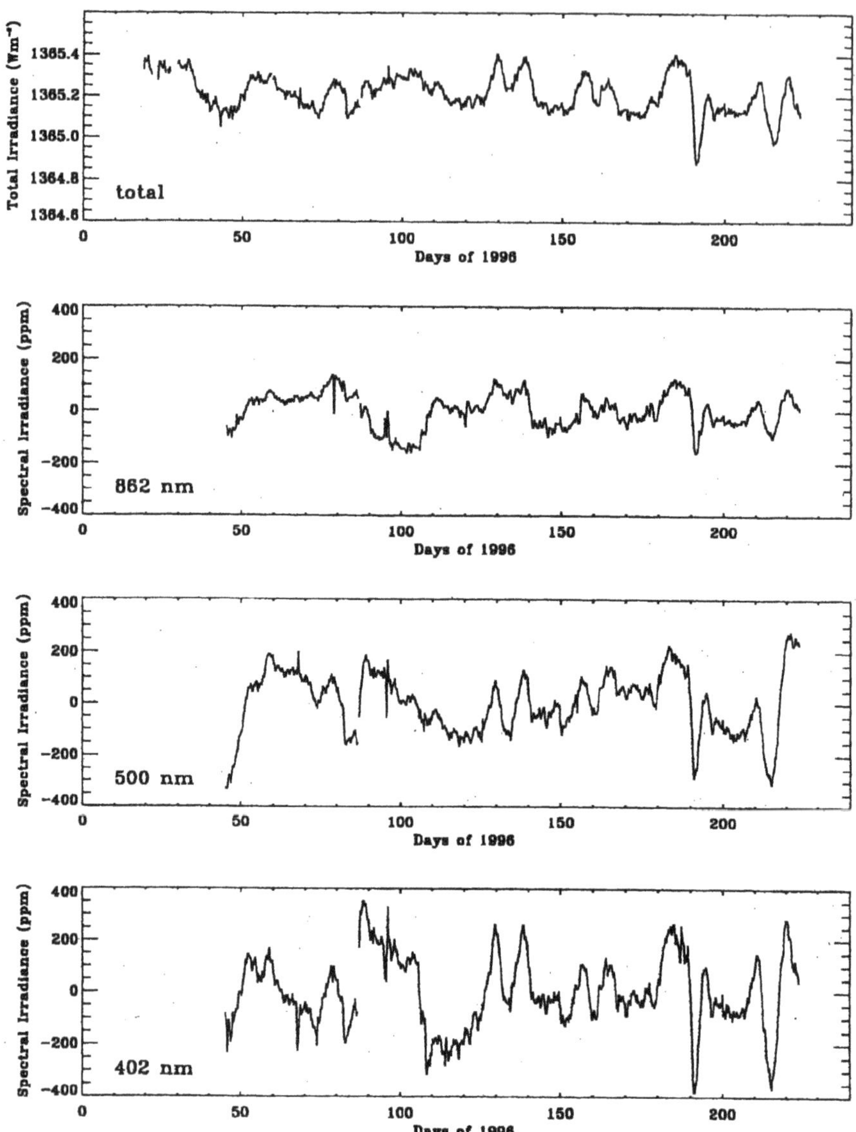

Figure 1. Six-hours averages of total solar irradiance in W m^{-2} and detrended spectral irradiance at 862 nm, 500 nm, and 402 nm in ppm observed by VIRGO.

from the latter in Section 3. The performance of the SPM and the procedures for the data evaluation are described by Fröhlich *et al.* (1997). The sensitivity of the main instrument has slowly degraded with rates of about -52 ppm day^{-1} for the 862 nm, -310 ppm day^{-1} for the 500 nm and -650 ppm day-1 for the 402 nm channel.

These rates are determined assuming a constant solar irradiance over the period of evaluation. They are an order-of-magnitude lower than what was experienced by similar SPM instruments on EURECA or PHOBOS; more stable filters and strict cleanliness control during construction and in the spacecraft have indeed paid off. The spare instrument has been exposed only 4 times for 18-min intervals, and shows no net trend; this is compatible with the observations of the total solar irradiance, and indicates that the assumption of constancy is reasonable. The detrended time series are shown in Figure 1 together with the total solar irradiance.

During the 'first light' experiment all three irradiances were about 2.9% below the values of the spectrum of Neckel and Labs (1984). While this is still within the combined errors of the NBS-1973 spectral irradiance scale of the lamps used and the reference spectrum of Neckel and Labs (1984), the lower values of all channels hint at a systematic error during calibration. The absolute value will be checked during the stratospheric balloon flight planned for 1997 by comparing with a similar SPM calibrated before and after the flight, by means of the recently developed, very accurate radiometric method (Friedrich, Fischer, and Stock, 1996) instead of with standard lamps.

The variations seen in the time series of spectral and total irradiance are caused by a superposition of random and periodic phenomena. Variability caused by individual phenomena at the photosphere is modulated further by the solar rotation, leaving a very complicated mixture of temporal and spatial effects. Although the main objective of the LOI is the study of solar oscillations (cf., Section 3), its spatial resolution allows us to deconvolve the effects of different spatial and temporal influence of photospheric variations on the solar irradiance. The performance of the LOI and the procedures for the data evaluation are described by Appourchaux *et al.* (1997). The signals from the 16 pixels are corrected to fluxes measured at 1 AU, using the orbital data to take into account the real shape of the pixels, the limb darkening at 500 nm, and the in-flight size of the solar image. In addition, for comparing the LOI integrated flux with the SPM, four conversion factors are derived for each type of pixel. These factors take into account the pixel shape, the limb darkening and the insensitive lanes between the pixels. Since the variation of these factors is slow, they are calculated only once per day. Using the four SPM conversion factors we compared over the period of simultaneous measurements the LOI integrated flux with the one from the SPM green channel. The LOI sensitivity decreases much faster with time than the SPM 500 nm channel. At the beginning of the exposure it amounted to about 1000 ppm day^{-1}, decreasing to 500 ppm day^{-1} after 100 days. The stronger degradation is probably due partly to the fact that the LOI entrance filter sees more solar wind particles because of the rather large viewing angle to space, and is partly a result of having different coatings on the front surfaces. After correction for degradation, the LOI integrated flux compares quite well with the SPM, as shown by Appourchaux *et al.* (1997). Moreover, there is no indication of relative sensitivity changes between different pixels. Thus, the

LOI is well suited to distinguish between spatial and temporal variability in the SPM and radiometer signals.

2.2. SOLAR NOISE

The continuous measurements allow us to produce for the first time alias-free power spectra of the variance of total and spectral irradiance over a wide range of frequencies corresponding to periods from 2 min to the duration of observation. This not only permits studies of the Sun with helioseismology, but also allows us to study the variability itself. In Figure 2 the power density spectra of the total and spectral irradiance are shown. The most important features are:

– the p-mode signal stands above the solar background noise with a signal-to-noise at the peak of about 40 : 1;

– there is no visible excess in the solar noise in the p-mode region;

– in and above the p modes, the solar noise decreases with frequency as about ν^{-4}; this value depends weakly on the wavelength of the observations, the blue and green channels being the steepest;

– from 10 to 100 μHz the solar noise decreases with frequency as ν^{-2};

– from 100 to 800 μHz the solar noise spectrum is quite flat; this was not expected, and it is a very encouraging finding because this is the range where g modes are expected;

– a distinct hump in the power spectrum is seen at about 1.2 mHz in all spectra;

– fitting the spectra with function types proposed by Harvey (1985) gives time constants of the three regimes of the solar noise to be about 66, 480, and 40 000 s; this indicates that there are no structures visible with time constants in the mesogranulation regime; the shortest time constant could possibly be attributed to network bright points;

– at low frequencies peaks exist corresponding to periods of about 27, 9, and 7 days which are related to activity features crossing the visible disk (cf., Figure 9).

If one compares the total irradiance spectrum with the one deduced from orbital mean data of SMM-ACRIM in 1984/85 (Fröhlich *et al.* , 1991) one notices that the power in the latter between about 40 μHz and the Nyquist frequency of 86 μHz is up to a factor of 3 higher than the one observed by the VIRGO radiometers. This is an artifact from folding back high-frequency noise due to only partial coverage during an orbit, and demonstrates that continuous time series are a necessary prerequisite for unbiased power spectra.

Another new result is shown in Figure 3 where the ratio of the variance of the different spectral channels to that of total irradiance is plotted. It shows a complicated dependence of these ratios upon frequency. In the frequency range of granulation and meso-granulation they are more or less constant, then there is around 10 μHz a region with about equal ratios for all colours. At periods of a few days a broad hump is observed which is most important for the blue and about equal for the red and green. At the lowest frequencies the variance in the red is less than 20% of the total.

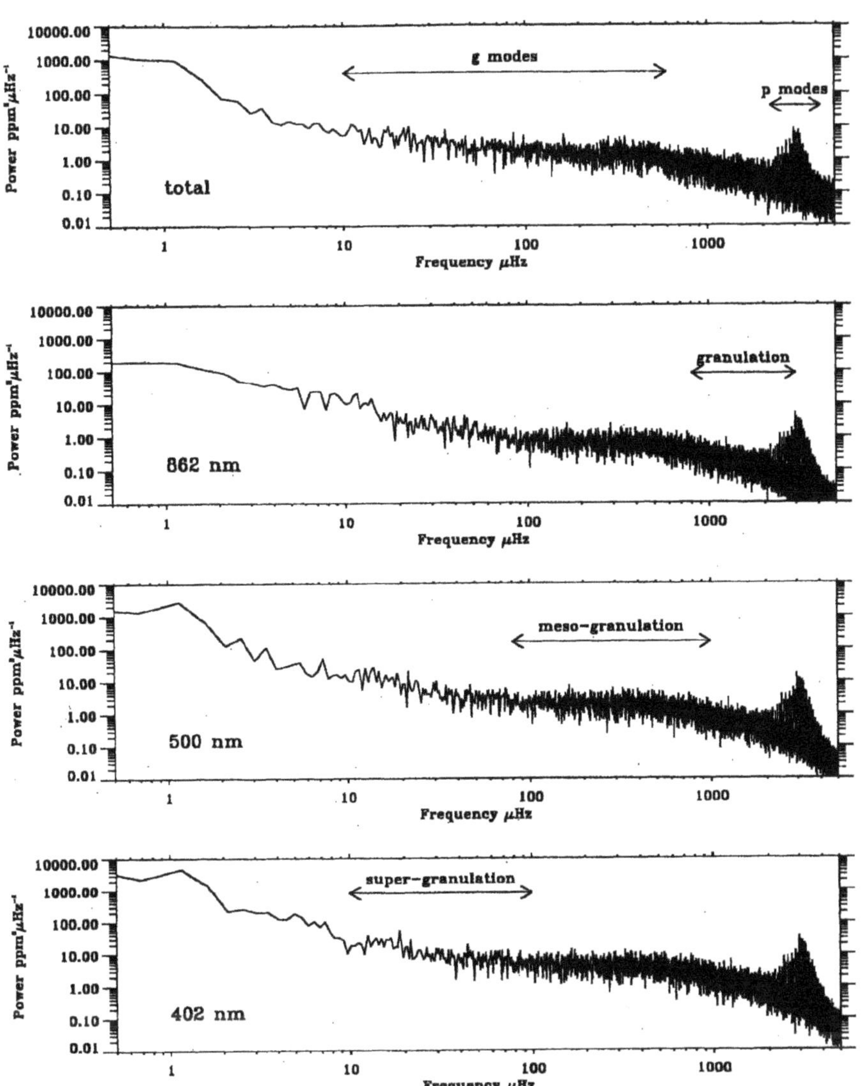

Figure 2. Power spectra of total and spectral irradiance time series for the period from January 28 until August 10, 1996.

2.3. INFLUENCE OF ACTIVE REGIONS

The time scales associated with the lifetime of magnetically active regions have revealed extremely interesting variations in total solar irradiance. Analyses based on former observations of total solar irradiance showed that the most striking events in the short-term irradiance changes are the sunspot-related temporary dips (Willson

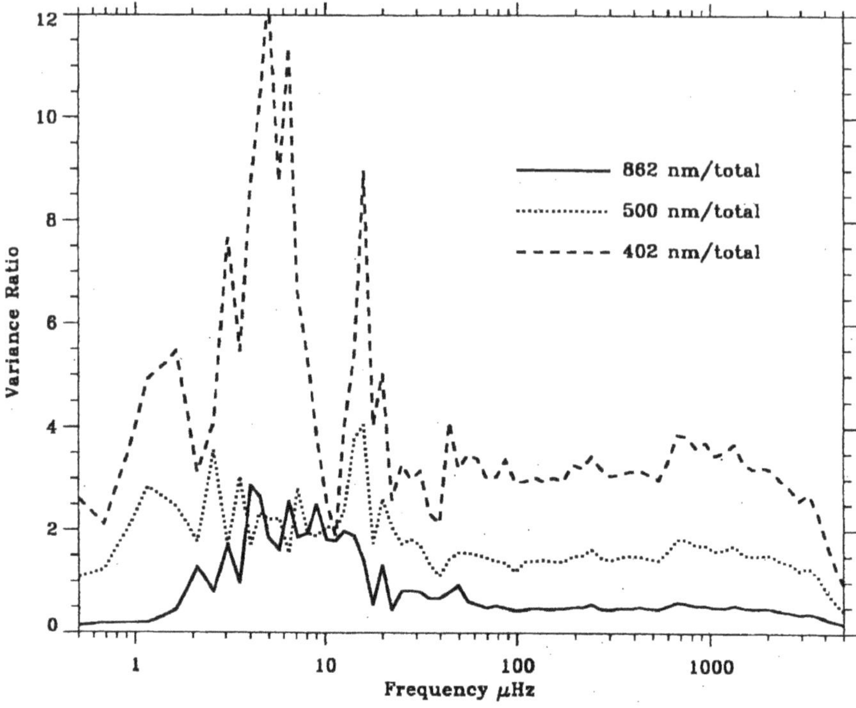

Figure 3. Ratio of the power spectra from Figure 3 of the spectral channels to the one of total irradiance.

et al., 1981). The fundamental question is whether the observed short-term changes are real luminosity changes or whether they represent a simple redistribution of the solar irradiance by sunspots and active region's faculae (Kuhn, 1996). To distinguish between these possibilities and to understand the underlying physical mechanisms of solar variability, it is important to study the spectral distribution of the variations observed in the entire solar flux. The VIRGO experiment provides the first high-precision observations of solar spectral irradiance at 402, 500, and 862 nm in parallel with the total solar irradiance observations. These observations permit the study of simultaneous temporal changes in the near-UV, visible, and infrared ranges, and to determine the spectral distribution of total solar irradiance variations, as is quite evident from Figure 1. The most prominent passages of small active regions have their central meridian transit around days 120, 147, 182, and 202. Here we discuss the first one (occurrence of sunspot group NOAA 7962 in early May 1996).

At the top of Figure 4 we show magnetograms from Mt Wilson to illustrate the active region, and below we show time series of the total and spectral irradiance and of the four southern pixels of the LOI. The active region is causing significant

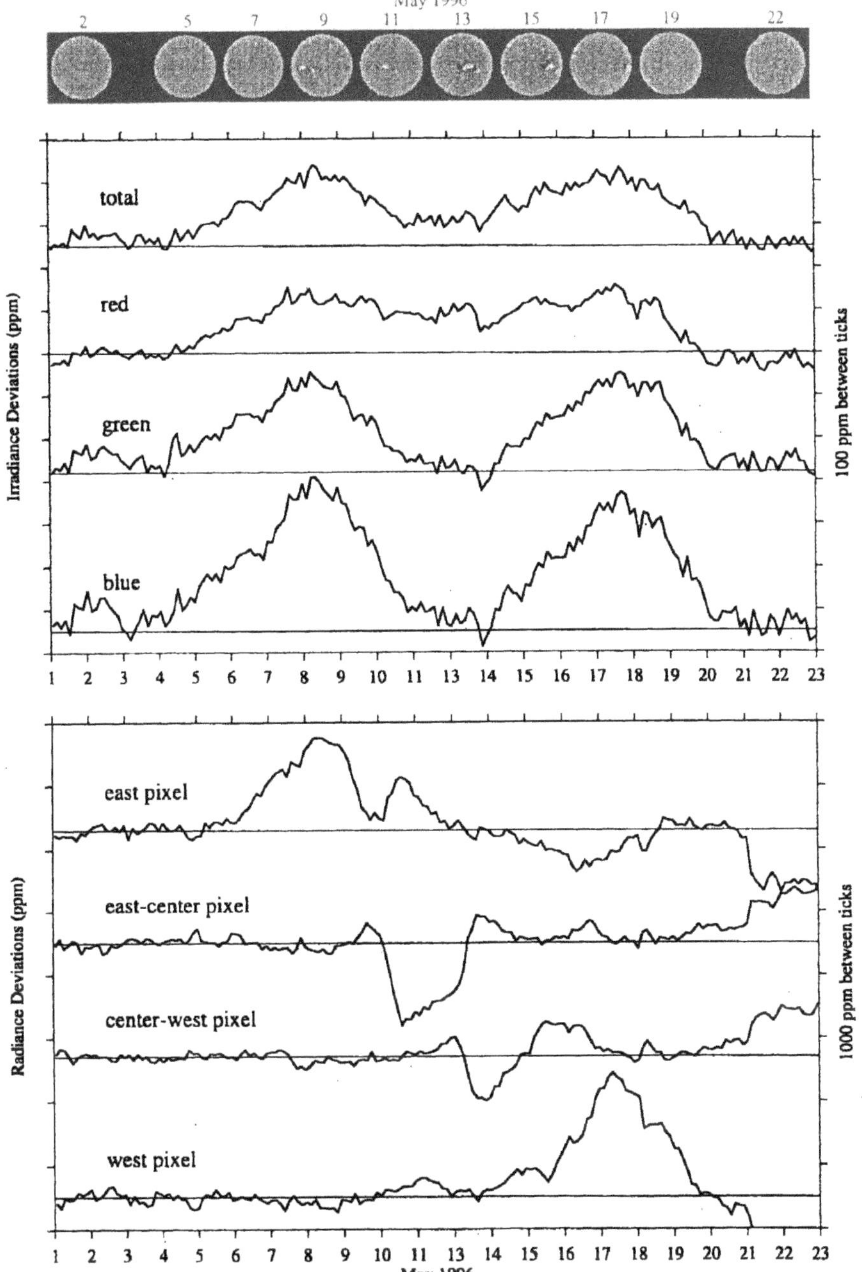

Figure 4. Passage of the active region NOAA 7962 in May 1996 shown by the Mt Wilson magnetograms in the Fe I line (*top panel*, black and white correspond to ±5 G), the time series of total and spectral irradiance (*middle panel*) and of the four LOI pixels south of the equator (*bottom panel*).

changes in the spectral irradiance in the blue and green channels and the LOI, and much less in total irradiance and in the near IR. This particular active region appeared close to the limb in the southern part of the visible solar disk on 6 May (day 127) and contained a very small sunspot which started to vanish around 14 May, just after the central meridian transit. Very close to this region, a smaller active region without sunspots was also visible.

The LOI east pixel intensity value started to rise around 6 May with the appearance of the faculae in the active region rotating onto the east limb of the solar disk with a primary peak three days later. In parallel with the LOI east pixel intensity increase, both total solar and spectral irradiance in all the 3 wavelengths started to increase. When referring the intensities to the level of the quiet Sun before and after the passage, the largest increase in irradiance within a 4-day of period was measured in the blue channel: it was about 300 ppm, and is to be compared to about 200 ppm for the green channel and about 100 ppm in both the red channel and the total irradiance. It is interesting to note that at the meridian transit only the blue and green irradiances reached the quiet-Sun level, whereas the red channel and the total irradiance decreased by only about 40% and 60% of the total increase respectively. The photometric deficit of the spot group as determined for total irradiance by the San Fernando Observatory was around 50 ppm (*Solar-Geophysical Data*, 1996) at the maximum sunspot influence (around May 12–13). Taking the wavelength dependence of the spot contrast into account one would expect 45, 65, and 75 ppm for red, green and blue channels respectively. This accounts for only about half of the observed decrease in total irradiance by the sunspot darkening. The other half must be due to the facular contribution which is smaller at disk center. Photometric observations of faculae show a limb brightening and the facular contrast at the disk centre should be zero or even negative. The behaviour close to the limb can account for the strong peak of the irradiance at central meridian distances of about $\pm 60°$. The fact that only a partial decrease at the centre is observed indicates that the faculae are still seen in the spectral and total irradiance, or that the faculae extend over a much larger area than the magnetogram suggests. Moreover, the centre-to-limb increase depends strongly on wavelength, and must be substantially smaller in the red and total irradiance than in the green and blue.

The LOI radiance values confirm this general behaviour. The east pixel shows mainly the increase due to faculae; on the crossover to the next pixel the radiance first increases due the faculae in front of the active region and then it suddenly decreases due to the dark spot. The effect of the 65 ppm in the green sun-as-a-star signal is greater by a factor of about 12, owing to the smaller area of the pixel, yielding a decrease down to about -1100 ppm; then the spot slowly vanishes, indicated by the steady increase of the radiance with time. The change-over to the next pixel shows again an increase and then the decrease due to the now smaller spot. At the central crossover there is still an increase, indicating that indeed the faculae are still bright close to the centre. This whole sequence shows very interesting details of how active regions influence the irradiance. Yet more detailed studies will definitively reveal clues for the understanding of irradiance and radiance variability.

3. Solar Oscillations

The analysis is based on the power spectra calculated from the time series of the SPM channels and the 12 LOI pixels. The time series are detrended by piecewise fits of quadratic polynomials for the SPM and with a curve obtained by triangular smoothing with a full width of one day for the LOI. The power spectra are calculated from the Fourier transform of apodized time series by squaring the complex amplitude. As an example the p-mode spectra of the total, red, green and blue irradiance are shown in Figure 5. Note the high signal-to-noise ratio for the SPM channels and the somewhat lower one for the total irradiance. This is due to the sampling duty cycle of only 31% for the PMO6-V observations compared to the SPM with 94%. The ratios of the increasing amplitudes observed in the red, total, green and blue are $\approx$ 0.4:0.5:0.7:1.0. Another obvious feature is the relative increase of the amplitudes of the higher degree modes from red to blue as shown by the ratio of the $l = 2$ to the $l = 0$ (e.g. around 3030 μHz) or the $l = 3$ to the $l = 1$ (e.g. around 2950 μHz), albeit less obvious due to the low visibility of the $l = 3$ mode. This is due to the increasing center-to-limb variation towards the blue which weights the central part of the disk more and more and increases the visibility of the higher degrees relative to lower ones.

3.1. Data Reduction

From the SPM spectra the p-mode parameters are derived by fitting Lorentzians using a maximum-likelihood method as described in Toutain and Fröhlich (1992). The only difference is that the mode amplitudes in a given (l, m) multiplet are no longer constrained to be the same. For the LOI a given (m, ν) diagram is fitted simultaneously using a maximum-likelihood method as described in Appourchaux *et al.* (1995). Moreover, the crosstalk between the $2l + 1$ components of a multiplet as well as the aliases from higher-degree modes are appropriately taken into account (Appourchaux *et al.* , 1997).

3.2. Frequencies of p Modes

The most important parameters of the p-mode characteristics are the frequencies. Four such sets are available in VIRGO. As the degeneracy splitting is not resolved in the low-degree $l = 1, 2$ modes determined from global observations of the sun the frequencies of SPM spectra can be biased. The frequencies of the uninfluenced $l = 0$ modes determined from the SPM channels are in good agreement among themselves and with the ones determined from the LOI as shown in Table I. There might be, however, a systematic decrease of the frequencies from the red

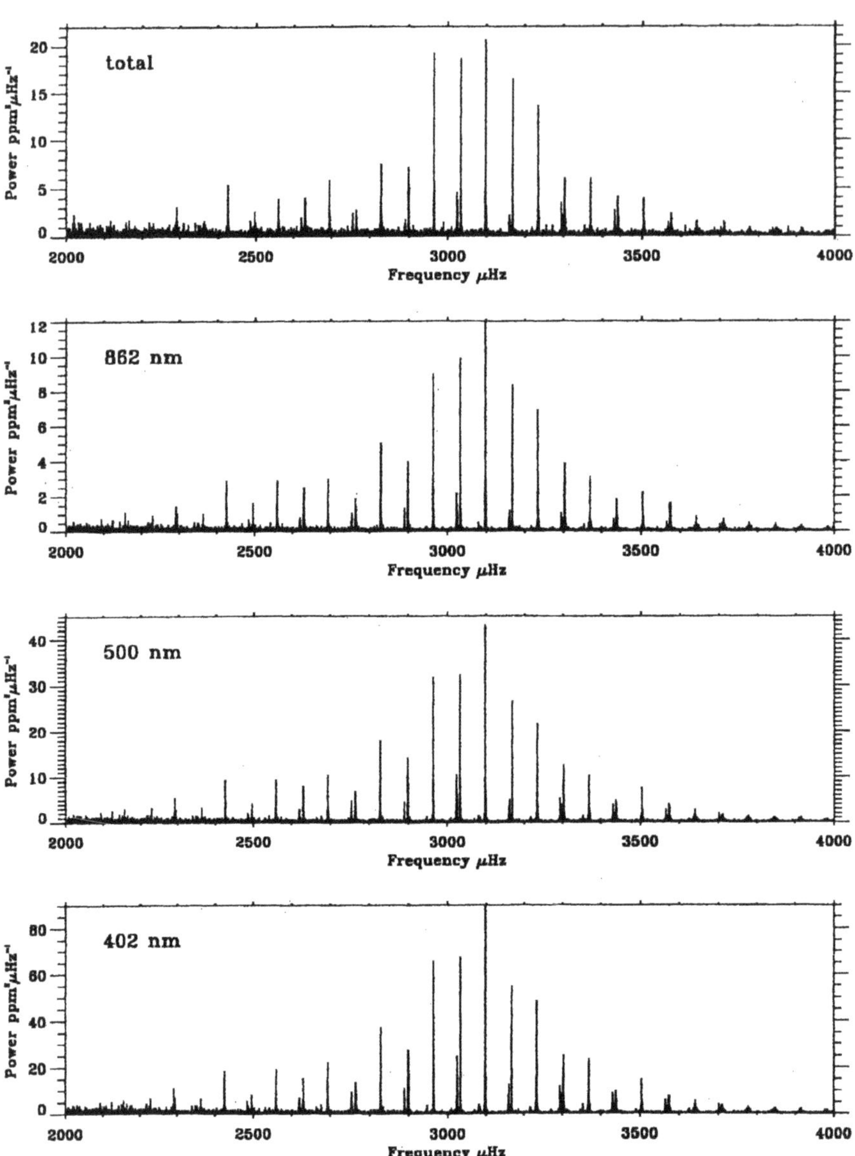

Figure 5. *p*-mode spectra from about 200 days of observation of the total, red, green and blue irradiance.

to the blue. This effect cannot be the influence of excitation noise as the same realizations are observed in all three colours. It could be due to an effect in the solar atmosphere changing the asymmetry of the line profile with height which

Table I

Comparison of the $l = 0$ frequencies determined from the LOI and the three SPM channels

n	LOI	Red	Green	Blue
15	2228.34±0.086	2228.499±0.097	2228.434±0.118	2228.440±0.113
16	2362.96±0.120	2363.108±0.094	2363.107±0.095	2363.106±0.095
17	2496.14±0.088	2496.233±0.083	2496.219±0.088	2496.238±0.085
18	2630.03±0.151	2630.014±0.088	2629.948±0.089	2629.996±0.094
19	2764.43±0.156	2764.367±0.117	2764.326±0.115	2764.311±0.119
20	2898.95±0.097	2899.028±0.082	2898.983±0.087	2898.970±0.078
21	3033.75±0.085	3033.683±0.060	3033.662±0.058	3033.668±0.058
22	3168.67±0.085	3168.610±0.072	3168.559±0.066	3168.556±0.067
23	3303.21±0.114	3303.232±0.094	3303.205±0.091	3303.228±0.092
24	3438.95±0.160	3438.863±0.126	3438.854±0.132	3438.860±0.133
25	3574.62±0.291	3574.855±0.219	3574.630±0.225	3574.719±0.227

may then be interpreted as a frequency shift because it is analysed assuming a symmetric profile.

In order to have an internally consistent data set we present as VIRGO first-result frequencies those determined from LOI. These frequencies are listed in Table II. They can be compared to those determined by the IPHIR experiment (Toutain and Fröhlich, 1992) during the second half of 1988. Figure 6 shows the result of this comparison, and it is interesting to note that the frequencies of the $l = 1$ modes are systematically higher, with a slight increase towards higher frequencies which could well be due to differences in the level of solar activity.

3.3. Linewidths and Amplitudes

Linewidths are determined by the same fitting procedure from both the SPM green channel and the LOI. The results are shown in Figure 7. Note the pronounced dip around the peak of the 5-min oscillations. This dip is a resonance effect with convection which decreases the damping of the modes, and was predicted by Gough (1980; cf., Balmforth, 1992). It is interesting that the IPHIR data did not show this dip, which is suggesting that the resonance effect may be decreased by increasing solar activity.

Detailed studies of the time variation of mode amplitudes for solar p modes are essential for understanding both the excitation mechanisms and the response of the solar surface to the perturbations caused by the modes. In this context it is crucial to understand that the properties of the solar surface acts as a filter between the oscillations and the observer. These studies require continuous data sets.

Wavelet analysis is an appropriate tool for investigating time-dependent phenomena. With an optimal selection of parameters this method also provides a better combined frequency and time resolution than is achieved by conventional Fourier

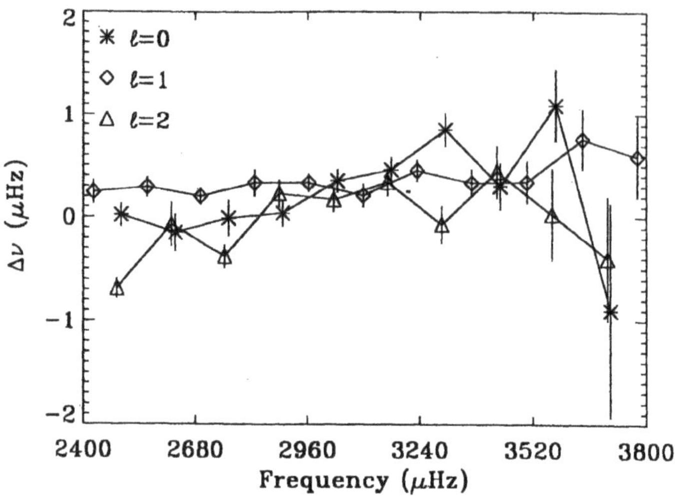

Figure 6. Comparison of the VIRGO frequencies with the ones determined by IPHIR.

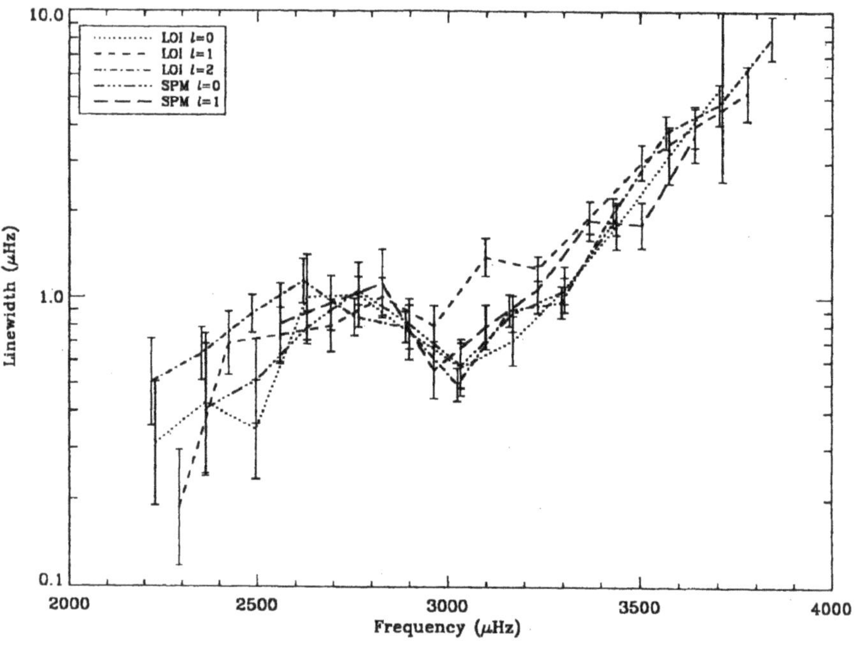

Figure 7. Linewidth determined from the SPM green channel and LOI for $l = 0$–2.

analysis. They have previously been successfully used on IPHIR data (Baudin, Gabriel, and Gibert, 1994; Leifsen *et al.*, 1995). Here we present results from a study of the amplitudes of the p modes from the SPM blue channel and the LOI.

Table II
List of frequencies of solar p modes determined from VIRGO.

order	$l=0$	$l=1$	$l=2$	$l=3$
13	–	–	–	2137.79±0.07
14	–	–	2217.63±0.08	2273.47±0.06
15	2228.34±0.09	2292.17±0.05	2352.29±0.07	2407.73±0.09
16	2362.96±0.12	2425.68±0.10	2486.05±0.07	2541.70±0.06
17	2496.14±0.09	2559.29±0.07	2619.69±0.08	2676.23±0.05
18	2630.03±0.15	2693.32±0.07	2754.44±0.07	2811.49±0.06
19	2764.43±0.16	2828.19±0.08	2889.59±0.05	2946.97±0.04
20	2898.95±0.10	2963.32±0.08	3024.75±0.04	3082.31±0.04
21	3033.75±0.09	3098.16±0.11	3159.83±0.07	3217.74±0.05
22	3168.67±0.09	3233.21±0.04	3295.25±0.06	3353.58±0.07
23	3303.21±0.12	3368.64±0.11	3430.76±0.11	3489.65±0.08
24	3438.95±0.16	3504.39±0.17	3567.02±0.15	3626.35±0.12
25	3574.62±0.29	3640.35±0.24	3703.41±0.22	3763.01±0.18
26	3711.45±0.98	3777.07±0.18	3840.84±0.31	-
27	3847.28±1.57	–	–	–

Order	$l=4$	$l=5$	$l=6$	$l=7$
13	2188.43±0.35	2235.42±0.02	2280.08±0.07	2322.40±0.09
14	2324.31±0.05	2371.23±0.07	2415.59±0.06	2458.25±0.07
15	2458.53±0.05	2506.00±0.06	2551.12±0.05	2594.27±0.06
16	2593.16±0.08	2641.25±0.04	2687.08±0.05	2731.08±0.09
17	2728.51±0.09	2777.24±0.07	2823.81±0.05	2868.40±0.05
18	2864.28±0.09	2913.51±0.04	2960.62±0.04	3005.54±0.04
19	3000.13±0.05	3049.84±0.05	3097.09±0.05	3142.77±0.04
20	3135.96±0.05	3186.26±0.05	3234.17±0.06	3279.76±0.05
21	3271.74±0.06	3322.67±0.10	3370.91±0.11	3417.35±0.07
22	3408.14±0.09	3459.74±0.12	3508.80±0.12	3554.93±0.15
23	3544.67±0.10	3596.35±0.17	3646.05±0.14	-
24	3682.25±0.16	3734.72±0.17	-	-

The results from the LOI data on the $l = 1$ indicate no correlation between the different rotationally split components of each radial order. This is to be expected from our view that the oscillations are excited randomly by the turbulence in the upper boundary layer of the convection zone, despite the fact that each individual excitation event applies equal impulses to eastward and westward propagating modes with like values of l and $|m|$. The results with high frequency resolution for the $l = 0, n = 16 - 26$ modes from the SPM blue channel are shown in Figure 8. From the data we see no clear and consistent correlation of the excitation of the different modes. This too is as we expect (Chang, 1995). A prominent phenomenon is the substantial variation of the central frequency position of the modes. Com-

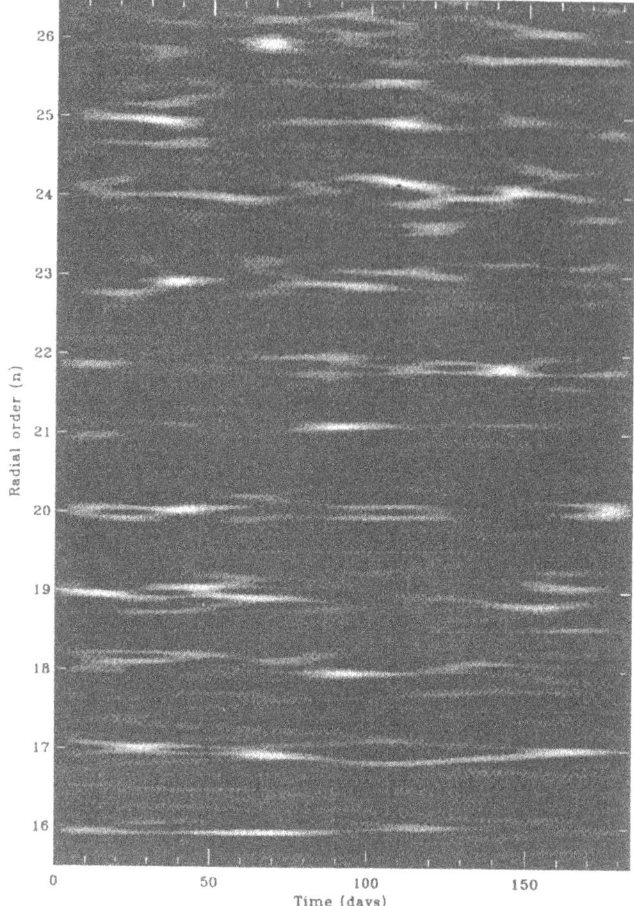

Figure 8. Wavelet analysis of $l = 0$, $n = 16-26$ from the SPM blue channel. The time axis starts on the January 29, 1996. The frequency coverage for each radial order is 4 μHz.

plicated structures with shifts and splittings of the modes are clearly seen. This is a result of phase wandering due to the excitation, and is manifest as a power variation in different regions of the mean Lorentzian profile of the modes. It shows the danger in fitting Lorentzians to the power spectra of modes obtained from short time strings of data. Such a fitting will give spurious frequency shifts. The required length of the time series to achieve a reasonable accuracy of the Lorentzian fitting may be so long that solar cycle variations could be masked.

The time dependence of the amplitude variation of the different modes have been studied by calculating the power spectra. For this, wavelet parameters giving a time resolution of about 10 hours were used. The possible effect of interference from the nearby $l = 2$ modes has been filtered out. The results are given in the upper panel of Figure 9. The dominant peak is seen at approximately the position of the solar rotation and should be compared with the lower panel showing the periodicities in the irradiance signal. Within the current resolution some of the minor peaks are also common. The periodicities are also seen in the power spectra of the individual modes, but there the amplitudes of the different peaks vary strongly between the modes. From these results one cannot infer a latitudinal variation of the excitation. With current understanding of these mechanisms this would not create a discernible signal in the modulation. The explanation of the modulation stems from the variation of the solar surface structure, i.e. with active regions, which modulate what we observe. The response of the surface to the perturbations of the oscillations is different in active and quiet regions of the Sun. Further observations are needed to see whether the periodicities in the amplitudes correspond to the mean solar rotation or to the rotation of the active regions.

Another interesting property of the oscillations concerns the relative amplitudes of the components of the multiplets. Although the datasets are perhaps not yet long enough to secure reliable statistics, there does appear to be a systematic variation in the power, as is illustrated in Figure 10. Such a distribution is not inconsistent with preferential excitation of the oscillations in regions of enhanced activity. In particular, if we adopt a latitudinal variation of the excitation rate that has maxima at midlatitudes, we can account for the general trends in the observed amplitudes if the maximum excess rate of energy input into oscillations is about 50% of the uniform background. The best fit to the data of a smooth symmetric function with single maxima in each hemisphere has the maxima at about $\pm 40°$ latitude, which is similar to though somewhat greater than the mean latitudes of the sunspot belts early in the cycle. Nevertheless, the broad agreement does suggest that magnetic activity plays a role in exciting the oscillations, in addition to modifying their frequencies (cf., Kuhn and Libbrecht, 1991). The difference between the locations of the maxima in the excitation function that we have deduced and the locations of the sunspots could perhaps be a result of inadequate statistics, or it could perhaps be an artifact of having misjudged the instrumental sensitivity to different modes by ignoring the magnetic perturbation to the eigenfunctions which causes the apparent decrease in power which Bogdan *et al.* (1993) have observed to occur in active regions. Alternatively, it could simply be that the regions of enhanced excitation extend beyond the observed regions of activity, as we were led to suspect from the variations in the radiance measurements discussed in Section 3.2.

3.4. Inversions of the VIRGO Frequencies

p-mode frequencies from Table II have been inverted to investigate the structure of the solar core. A total of 92 modes with $0 \leq l \leq 7$ were used in the frequency range 2.1 mHz $\lesssim \nu \lesssim$ 3.8mHz. The mode with the shallowest inner turning point penetrates to $r/R_\odot \simeq 0.3$, and our investigation concentrates on the layers below these depths,

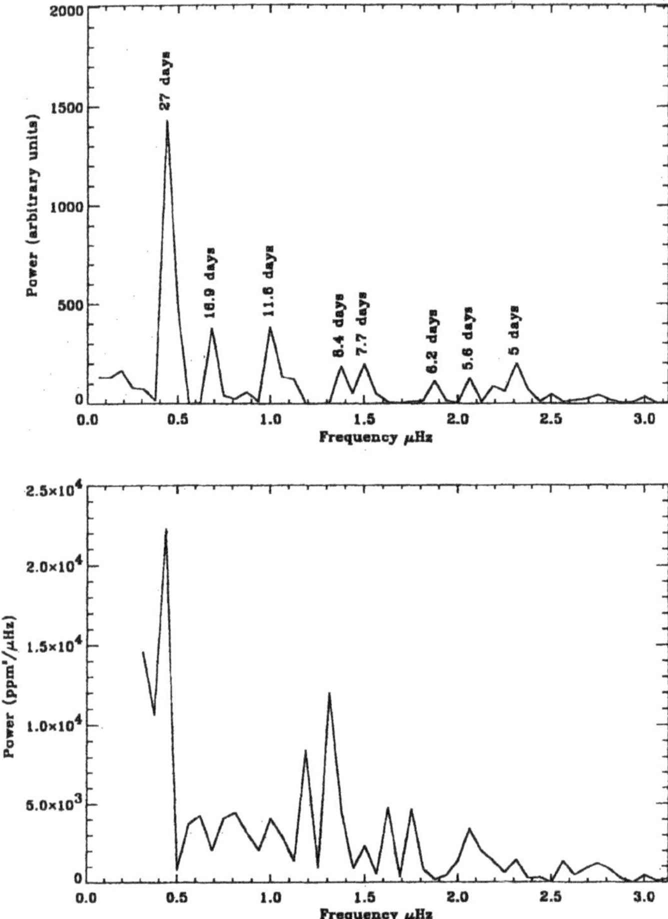

Figure 9. Power spectrum of the average amplitude variation for the modes of Figure 8 in the very low-frequency range (*upper panel*). The most prominent peaks are labelled with the corresponding period in days. For comparison the power spectrum of the SPM blue channel is shown in the *lower panel*.

since outside these regions the datasets do not provide much information about the solar structure.

The standard strategy of linearization was adopted (e.g., Gough, 1996): we use a theoretical model of the Sun as a reference, and then consider the difference between the Sun and the model. We attempt to determine how the Sun differs from the reference model by considering the difference between the observed frequencies

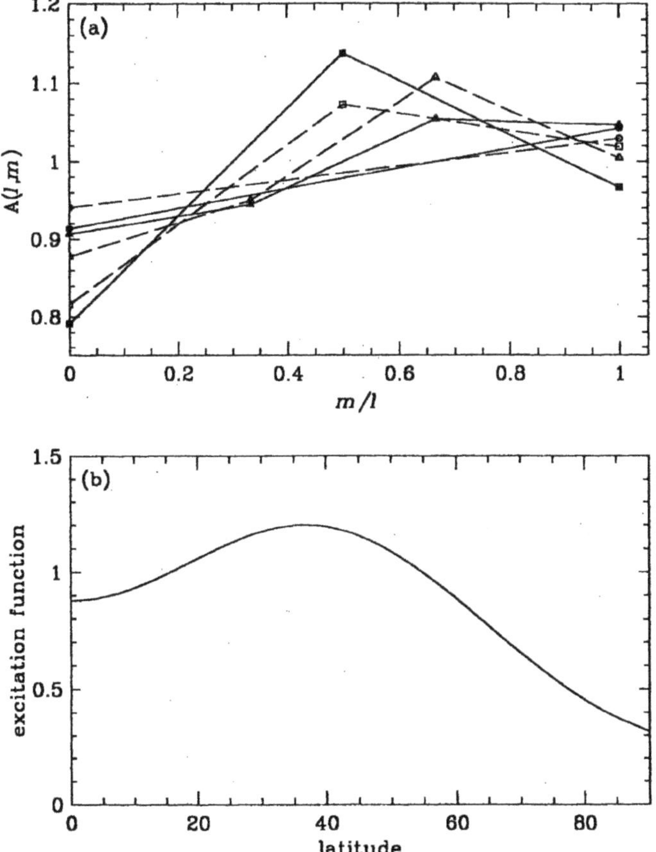

Figure 10. (a) Mean relative amplitudes of the components of dipole (circles), quadrupole (squares) and octupole (triangles) multiplets, plotted against $|m|/l$. The amplitudes of each multiplet are normalized such that the mean power per mode is unity. Filled symbols joined by continuous lines are derived from observations by LOI; the corresponding open symbols joined by dashed lines are theoretical values computed assuming the latitudinal variation of energy input illustrated in (b).

and those of the reference model. The relative frequency difference is equated to a linear functional of the difference in structures:

$$\frac{\delta\nu_i}{\nu_i} \equiv \frac{\nu_{\mathrm{obs},i} - \nu_i}{\nu_i} = \int \left[K^i_{u,\gamma_1}(r)\frac{\delta u}{u} + K^i_{\gamma_1,u}(r)\frac{\delta\gamma_1}{\gamma_1} \right] \mathrm{d}r + \frac{F(\nu_i)}{I_i}, \tag{1}$$

where ν_i and $\nu_{\mathrm{obs},i}$ are respectively the frequencies of the model and the observed frequencies of the mode i. and K^i_{u,γ_1} and $K^i_{\gamma_1,u}$ are the kernels for the squared isothermal sound speed u and the adiabatic exponent γ_1, which are our choice of inversion variables. The two functions δu and $\delta\gamma_1$ are the differences between the values of u and and the ones of γ_1 in the Sun and in the model. We present below inversions for δu. The term $F(\nu_i)/I_i$, where $F(\nu_i)$ is an unknown function of

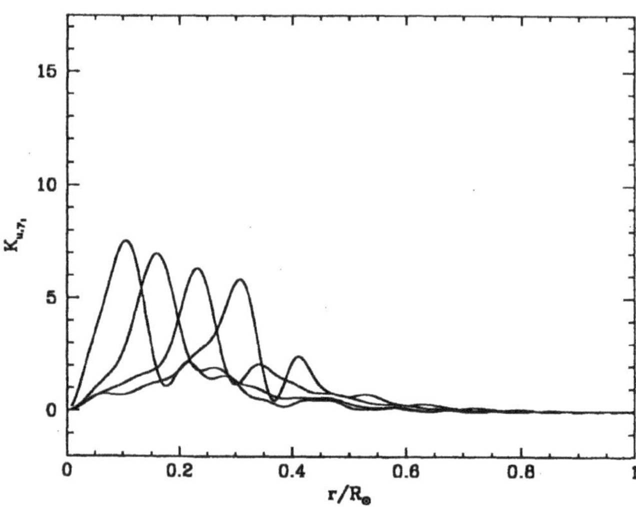

Figure 11. Averaging kernels for squared isothermal sound speed u obtained from 92 LOI frequencies.

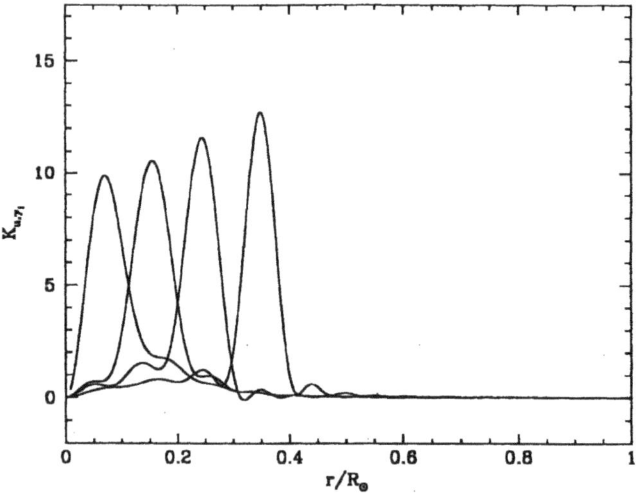

Figure 12. Averaging kernels for squared isothermal sound speed u obtained from 1105 combined LOI+LOWL frequencies.

frequency (which we represent as an expansion in terms of Legendre polynomials) and I_i is the mode inertia (normalization is such that the vertical component of displacement eigenfunction is unity at the surface), is added to represent all the uncertainties that arise from our lack of knowledge of the subsurface layers (for detailed discussion see Gough, 1995).

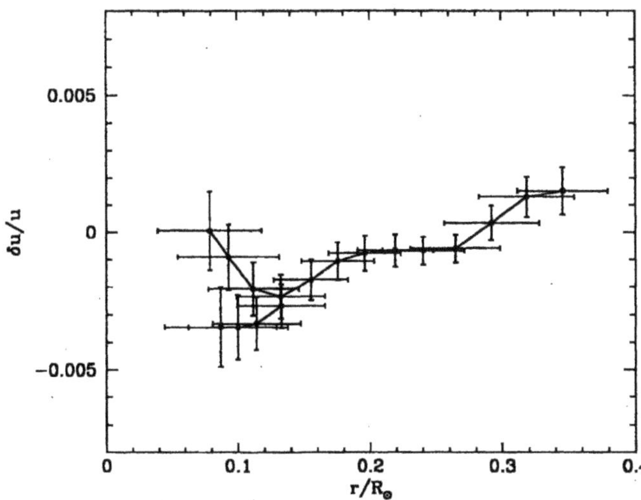

Figure 13. The relative difference in the squared isothermal sound speed between the Sun and the reference model, inferred from the LOI+LOWL frequencies. The lower curve represented by 4 points was obtained by excluding $l = 0, n = 15$ from the inversion; at greater values of $r/R_\odot$ the removal of the datum hardly influences the inversions.

To carry out the inversions we used optimally localized averaging (e.g., Gough, 1996; also Däppen *et al.*, 1991); to estimate δu at radius r_0, we construct a linear combination of the relative differences,

$$\sum c_i(r_0)\frac{\delta\nu_i}{\nu_i} = \int \left(\sum c_i K^i_{u,\gamma_1}\right) \tfrac{\delta u}{u}\mathrm{d}r + \int \left(\sum c_i K^i_{\gamma_1,u}\right) \tfrac{\delta\gamma_1}{\gamma_1}\mathrm{d}r + \\ + \sum c_i \tfrac{F(\nu_i)}{I_i} \tag{2}$$

in such a way that it provides an estimate of a localized average of δu in the vicinity of $r = r_0$. Specifically, we choose the coefficients c_i that minimize

$$S \equiv \int K_{u,\gamma_1}(r, r_0)^2(r - r_0)^2\mathrm{d}r + \alpha \int K_{\gamma_1,u}(r, r_0)^2\mathrm{d}r + \beta \sum_i c_i^2\sigma_i^2 \tag{3}$$

subject to

$$\int K_{u,\gamma_1}\mathrm{d}r = 1 \tag{4}$$

and

$$\sum_i c_i \frac{P_\lambda(\nu_i)}{I_i} = 0 \ , \lambda = 0, \ldots, \lambda_{\max} \ , \tag{5}$$

where

$$K_{u,\gamma_1}(r, r_0) \equiv \sum_i c_i K^i_{u,\gamma_1}(r) \ , \tag{6}$$

$$K_{\gamma_1,u}(r, r_0) \equiv \sum_i c_i K^i_{\gamma_1,u}(r) \quad (7)$$

are the averaging kernel for u and the kernel determining the contamination of the inference by $\delta\gamma_1$, and $P_\lambda(\nu_i)$ is a Legendre polynomial.

The first term in S encourages localization of the averaging kernel, which has unit modulus imposed by the first constraint; the second suppresses the amplitude of the contaminating kernel, and the third suppresses error magnification. Thus the parameters α and β are used to balance the weights between resolution, contamination from $\delta\gamma_1$ and error. The purpose of the second set of the constraints is to eliminate the contribution from the unknown surface layers, with a resolution determined by the value of $\lambda_{\max}$ in the Legendre decomposition of $F(\nu_i)$.

A selection of the averaging kernels we have obtained is shown in Figure 11. Owing to the small number of modes, the degree of localization is not very good. To obtain better localization, it is essential to supplement the dataset with higher-degree modes. We used the LOWL dataset (Tomczyk, Schou, and Thompson, 1996) for this purpose. From the LOWL 2-year average dataset, 1013 modes with $l \geq 8$ in the frequency range 1.5 mHz $\leq \nu \leq$ 3.5 mHz were chosen and added to 92 LOI frequencies. The improved averaging kernels obtained from the combined dataset are shown in Figure 12, which illustrates the importance of having higher-degree modes. Figure 13 shows the result of the inversion, together with a result of a second inversion, where we excluded a single mode ($l = 0, n = 15$) because a problem with that mode was recognized during the analysis. We note how sensitive our inference of the deepest part of the Sun can be to just a single mode. We also add a further word of caution, because there seem to be systematic differences between the LOWL and LOI frequencies for $l < 8$, which may come about from the different intervals of observation coupled with a systematic difference between the outcome of Lorentz fitting to velocity and radiance data or similar to the difference between the colours from an atmospheric effect. It is therefore incumbent upon us to carry out the frequency analysis with the greatest care before any definitive conclusion be drawn. From the current results alone there is no compelling evidence that the Sun is substantially different from the reference model.

4. Conclusions

All instruments within the VIRGO package work very well, and interesting new results have been gathered from our preliminary investigation of the first 4–6 months of observations. The very quiet environment of SOHO, together with the continuity of the observations, enables us to obtain time series of a quality never previously achieved. Our first results confirm broadly what has been found from previous observations: firstly, that the solar irradiance is modulated by the passage of active regions across the disk, but that not all of the modulation is straightforwardly explained in terms of sunspot flux blocking and facular enhancement, and secondly, that helioseismic inversions are at present more-or-less in agreement with the latest standard solar models. However, we have in addition some new observations providing hints that profound solar activity is more extensive than its superficial manifestation. This has been obtained partly from a comparison of coarsely resolved measures of radiance with the total irradiance. We also have evidence that magnetic activity plays a significant role in the dynamics of the oscillations beyond its modulation of the resonant frequencies. Pallé, Régulo, and Roca Cortés (1990) and Elsworth *et al.* (1993) have already shown that the mean linewidths vary with the solar cycle. Comparison of our VIRGO data with IPHIR data reveals that the variation depends on frequency, in a manner that is suggestive of a modulation in the mechanism by which the oscillations interact with the turbulence in the convection zone. Moreover, by comparing the amplitudes of different components of p-mode multiplets, each of which is influenced differently by spatial inhomogeneity, we have found evidence that activity enhances excitation. This appears to be contrary to the temporal variation reported by Pallé *et al.* (1990) and Elsworth *et al.* (1993), who found that on average mode amplitudes decrease as the global level of activity rises. Further analysis is required to ascertain whether this disaccord is an indication that the excitation process is substantially more complicated than one might suspect or whether the discrepancy is simply a result of the modification of the oscillation eigenfunctions in the solar atmosphere by the magnetic field. Whatever is the case, it is evident that our first results have already raised interesting issues concerning the role played by activity in the global dynamics of the Sun.

Acknowledgements

VIRGO is a co-operative effort of many individual scientists and engineers at several institutes in Europe and USA. The VIRGO team is led by Claus Fröhlich (PMOD/WRC, CH) and includes the following co-investigators (in alphabetical order): Bo Andersen (NSC, N), Thierry Appourchaux (SSD-ESTEC NL), Gabrielle Berthomieu (OCAN, F), Dominique Crommelynck (IRMB, B), Werner Däppen (USC, CA USA), Vicente Domingo (SSD/GSFC, USA), Alain Fichot (IRMB, B), Douglas Gough (IoA & DAMTP, GB), Todd Hoeksema (CSSA, Ca USA), Antonio Jiménez (IAC, E), Judit Pap (UCLA, Ca USA), Janine Provost (OCAN, F), Teodoro Roca Cortés (IAC, E), José Romero (PMOD/WRC, CH), Thierry Toutain (OCAN, F), Christoph Wehrli (PMOD/WRC, CH), Richard Willson (Columbia University, NY USA). The VIRGO technical team lead by Hansjörg Roth (PMOD/WRC, CH) was responsible for the successful realization of the experiment and the following indivuals were involved at PMOD/WRC, Davos, Switzerland: Thomas Fröhlich, Adolf Geisseler, Silvio Koller, Urs Schütz, Jules Wyss; at SSD/ESTEC, Noordwijk, The Netherlands: Thierry Beaufort, Jos Fleur, Samuel Lévêque, Didier Martin, Udo Telljohann; at IRMB, Bruxelles, Belgium: André Chevalier, Christian Conscience, Marc Lombaerts, Pierre Malcorps and at IAC, Tenerife, Spain: Maria Gómez, José

Herreros. Without the continuous and concerted efforts of the team this experiment would never have reached the success it demonstrates today.

The VIRGO team has been supported by several national and international funding agencies which are gratefully acknowledged: the PMOD/WRC by the Swiss National Science Foundation under grants 2.860-0.88, 20-28779.90, 20-33941.92, 20-40589.94 and PRODEX, the IRMB by the Fonds de la Recherche Fondamentale Collection d'initiative ministerielle and PRODEX, the SSD/ESA by their annual funds from the Science Directorate, the IAC by the CICYT through PNIE under grants ESP88-0354 and ESP90-0969, the OCA by the CNES and CNRS. Individual contributions to the project have been supported by grants from PPARC of the UK to D. O. Gough, by UCLA under a contract with NASA to J. M. Pap.

Thanks are extended to S. Tomczyk and J. Schou for providing the LOWL data through www.hao.ucar.edu/public/research/mlso/LowL/lowl.html and to the Mt.Wilson Observatory for the magnetograms retrieved from the SOHO archive. We have also received help with the inversions from Umin Lee (Tohoku University, Japan).

References

Appourchaux, T., Toutain, T., Telljohan, U., Jiménez, A., and Andersen, B. N.: 1995, *Astron. Astrophys.* **294**, L13.

Appourchaux, T., Andersen, B. N., Fröhlich, C., Jiménez, A., Telljohan, U., and Wehrli, C.: 1997, *Solar Phys.* **170**, 27 (this issue).

Balmforth, N. J.: 1992, *Monthly Notices Roy. Astron. Soc.* **255**, 603.

Baudin, F., Gabriel, A., and Gibert, D.: 1994, *Astron. Astrophys.* **285**, L29.

Bogdan, T. J., Brown, T. M., Lites, B. W., and Thomas, J. H.: 1993, *Astrophys. J.* **406**, 723.

Chang, H.-Y.: 1995, *Ph.D. dissertation*, University of Cambridge,, Cambridge.

Crommelynck, D., Fichot, A., Domingo, V., Lee R. III: 1996, *Geophys. Res. Letters* **23**, 2293.

Däppen, W., Gough, D. O., Kosovichev, A. G., and Thompson, M. J.: 1991, in D. O. Gough and J. Toomre (eds.), *Challenges to Theories of the Structure of Moderate-Mass Stars*, Springer-Verlag, Berlin. p. 111.

Elsworth, Y., Howe, R., Isaak, G. R., McLeod, C. P., Miller, B. A., New, R., Speake, C. C., and Wheeler, S. J.: 1993, *Monthly Notices Roy. Astron. Soc.* **265**, 888.

Friedrich, R., Fischer, J., and Stock, M.: 1996, *Metrologia* **32**, 509.

Fröhlich, C., Foukal, P. V., Hickey, J. R., Hudson, H. S., Willson, R. C.: 1991, in C. P. Sonnett, M. S. Giampapa, and M. S. Matthews (eds.), *The Sun in Time*, The University of Arizona Press, Tucson. p. 11.

Fröhlich, C., Romero, J., Roth, H., Wehrli, C., Andersen, B. N., Appourchaux, T., Domingo, V., Telljohann, U., Berthomieu, B., Delache, P., Provost, J., Toutain, T., Crommelynck, D., Chevalier, A., Fichot, A., Däppen, W., Gough, D. O., Hoeksema, T., Jiménez, Gómez, M., Herreros, J., Roca-Cortés, T., Jones, A. R., Pap, J., and Willson, R. C.: 1995, *Solar Phys.* **162**, 101.

Fröhlich, C., Crommelynck, D., Chevalier, A., Fichot, A., Finsterle, W., Jiménez, A., Romero, J., Roth, H., and Wehrli, C.: 1997, in preparation.

Gough, D. O.: 1980, in H. A. Hill and W. A. Dziembowski (eds.), *Nonradial and Nonlinear Stellar Pulsation, Lecture Notes in Physics* **125**, 273.

Gough, D. O.: 1995, in B. Battrick (ed.), *The 4th SOHO Workshop, Helioseismology*, **SP 376**, 49.

Gough, D. O.: 1996, in T. Roca Cortés and F. Sánchez (eds.), *The Structure of the Sun*, Cambridge University Press, Cambridge, p. 141.

Harvey, J.: 1985, in E.Rolfe and B. Battrick (eds.), *Future Missions in Solar, Heliospheric and Space Plasma Physics* **SP 235** 199.
Kuhn, J. R.: 1996, in T. Roca-Cortes and F. Sánchez (eds.), *The Structure of the Sun*, Cambridge University Press, Cambridge, p. 231.
Kuhn, J. R. and Libbrecht, K. G.: 1991, *Astrophys. J.* **381**, L35.
Leifsen, T., Hanssen, A., Andersen, B. N. and Toutain, T.: 1995, *ASP Conf. Ser.* **76**, 520.
Neckel, H. and Labs, D.: 1984, *Solar Phys.* **90**, 205.
Pallé, P. L., Régulo, C., and Roca Cortés, T.: 1990, in Y. Osaki and H. Shibahashi (eds.), *Progress of Seismology of the Sun and Stars, Lecture Notes in Physics* **367**, 189.
Solar-Geophysical Data: 1996, *Solar-Geophysical Data: Comprehensive Reports* No. 624 (II), 24.
Tomczyk, S., Schou, J., and Thompson, M. J.: 1996, *Astrophys. J.* **448**, L57.
Toutain, Th. and Fröhlich, C.: 1992, *Astron. Astrophys.* **257**, 287.
Willson, R. C., Gulkis, S., Janssen, M., Hudson, H. S., Chapman, G. A.: 1981, *Science* **211**, 700.

IN-FLIGHT PERFORMANCE OF THE VIRGO LUMINOSITY OSCILLATIONS IMAGER ABOARD SOHO

THIERRY APPOURCHAUX
Space Science Department, European Space Research and Technology Center, 2200 AG Noordwijk, The Netherlands

BO N. ANDERSEN
Norwegian Space Centre, 0309 Oslo 3, Norway

CLAUS FRÖHLICH
Physikalisch-Meteorologisches Observatorium Davos, World Radiation Center, 7260 Davos Dorf, Switzerland

ANTONIO JIMÉNEZ
Instituto de Astrofísica de Canarias, Universidad de La Laguna, 38071 La Laguna, Tenerife, Spain

UDO TELLJOHANN
Space Science Department, European Space Research and Technology Center, 2200 AG Noordwijk, The Netherlands

CHRISTOPH WEHRLI
Physikalisch-Meteorologisches Observatorium Davos, World Radiation Center, 7260 Davos Dorf, Switzerland

(Received 17 October, 1996; in revised form 4 December, 1996)

Abstract. The Luminosity Oscillations Imager (LOI) is a part of the VIRGO instrument aboard the Solar and Heliospheric Observatory (SOHO). The scientific objective of the LOI experiment is to identify and characterize pressure and internal gravity oscillations of the Sun by observing the radiance variations. The LOI is a low-resolution imager with 12 pixels, for the measurement of the radiance distribution over the solar disk at 500 nm. The low resolution capability of the instrument allows the identification of individual azimuthal orders for $l = 0$ to 7, without suffering the mixing that affects integrated solar disk instruments. The performance, calibrations and instrumental effects of the LOI are described together with the procedures for extracting the solar *p* modes.

1. Introduction

The European Space Agency (ESA) embarked in 1985 on a collaborative mission with the National Aeronautics and Space Administration, the so-called Solar and Heliospheric Observatory (SOHO). On board this spacecraft are 3 helioseismology instruments: Michelson Doppler Imager/Solar Oscillation Instrument, Global Oscillations at Low Frequency, both in velocity, and VIRGO (Variability of solar Irradiance and Gravity Oscillations). They respectively observe the Sun, in velocity and in intensity, on a detector with 1000×1000 pixels, in velocity of the Sun as a star, and in intensity at three different wavelengths.

VIRGO aims specifically at detecting the gravity modes of the Sun. These modes are extremely important, as they propagate mainly in the core and the radiative zone. They would considerably improve our knowledge of the deep interior of the Sun,

Solar Physics **170:** 27–41, 1997.

which is presently known only by inverting the frequencies of the pressure modes, which have a smaller resolving power than the gravity modes in the core. The VIRGO experiment (Fröhlich *et al.*, 1995) consists of several instruments. One of these, the Luminosity Oscillations Imager (LOI) was designed and built by the Space Science Department of ESA. Its scientific objective is to detect the p and g modes of the Sun and to study the radiance variability due to solar active regions and other phenomena.

Here we discuss the performance and operation of the LOI and present some first results as an illustration. First we describe how the instrument cover was opened; it is a story of perseverance that has to be told once and for all. Next we briefly describe the instrument and show how it performs and how the data are analyzed for extracting reliable p-mode parameters.

2. The Opening of the Cover

On December 2, 1995, the SOHO satellite was launched by an Atlas-Centaur rocket from Cape Canaveral. A few days later, VIRGO was powered on, and the LOI remained in calibration mode until December 24, 1995 when its cover was opened for 5 min for a 'first light' experiment. Then the cover was closed to continue the outgassing of the instrument. On January 17, 1996, we tried to open the cover in order to start the regular operation of the LOI. Unfortunately, the cover remained closed, even after several attempts.

During this operation the LOI could see the Sun for about 1–2 s which was sufficient to lock the guider onto the Sun. The cover is operated by a 90-degree motor which is switched to the open position by a nominal current pulse length of 710 ms. The only explanation we had was that the cover was bouncing back to the closed position and/or that the length of the impulse sent to the motor was too short to keep it in the open position.

At this stage, the inability to open the cover concerned the whole SOHO community and a team was set up to resolve the problem. A model of the cover was developed in collaboration with the Mechanics Section of the Mechanical Systems Department of the European Space Research and Technology Centre. The first idea to open the cover by reducing even more the length of the pulse sent to the motor was confirmed. The pulse length to be used was determined empirically from the extensive modeling of the cover movement together with actual measurements on the spare in Davos. The major question was how to perform the same operation on an experiment without computer and thus without the necessary flexibility. The only way was to send a command for opening the cover, and then to switch-off VIRGO with a delay according to the pulse length wanted. To achieve this, MATRA/MARCONI (main contractor of SOHO), SAAB (on board computer) and the SOHO project devised a patch that would allow the accurate timing of the switch-off of VIRGO. One week before the real execution with the spacecraft, a rehearsal was performed at the SAAB premises using the VIRGO flight spare and a prototype of the SOHO onboard computer. The timing of the switch-off worked very well and allowed a variation in the pulse length from 70–700 ms with a precision of a few ms.

On March 27, 1996, the two most likely pulse lengths were sent to the spacecraft and the second one with 480 ms length worked. Clearly, this was a relief to all of us who had been working for more than 10 years on this project. The opening of the LOI cover was a concerted effort by many people as listed in the acknowledgments. The success was mainly due to the highly motivated co-operation between the different parties in order to allow LOI to play its important role for the unambiguous identification of low degree solar oscillations.

3. Description of the LOI

The design of the LOI instrument is extensively described in Appourchaux *et al.* (1995a). Briefly, it consists of a Ritchey–Chrétien telescope ($f = 1300$ mm) imaging the Sun through a 5 nm bandpass interference filter at 500 nm. The image is projected on a photodiode array detector built to our specifications (Appourchaux, Martin, and Telljohann, 1992). The detector has 12 scientific pixels and 4 guiding pixels (Figure 1). The shape of the scientific pixels are trimmed to detect low-degree modes for $l < 7$ (Appourchaux and Andersen, 1990). The guiding pixels are 4 quadrants of an annulus with an equivalent inner and outer radius of 0.95 and 1.05 solar radius, respectively. The error signals for the guiding servo are the differences between two opposite pixels. The error signals are fed back to two piezoelectric actuators that move the secondary mirror around its vertex. The actuators provide both the coarse and fine pointing; the total range is $\pm 7.5'$ with a $0.2''$ resolution; the 3 db bandwidth of the servo control loop is at 10 Hz. The current of each scientific pixel is amplified and digitised using Voltage-to-Frequency Converters (VFCs). The current of each guiding pixel is amplified and digitised by the common data acquisition system of VIRGO. The 12 scientific pixels are read simultaneously, while the 4 guiding pixels are read sequentially. The sampling time of the instrument is 60 s. The integration time of the scientific pixels is the same as the sampling time. The integration time of the guiding pixels is about 3 s distributed over the 1-min period.

The LOI data products are of two kinds: the level 1 and the level 2 data. The level-1 data are those that have been converted from raw counts (level 0 data) to physical units; they include the calibrations and contain all the corrections known *a priori* for instrument-related effects, such as the influence of temperature variations. The signals are also reduced to 1 AU distance and to zero radial velocity. The level-2 data are the level-1 data corrected for degradation. The data products at either level are the following:

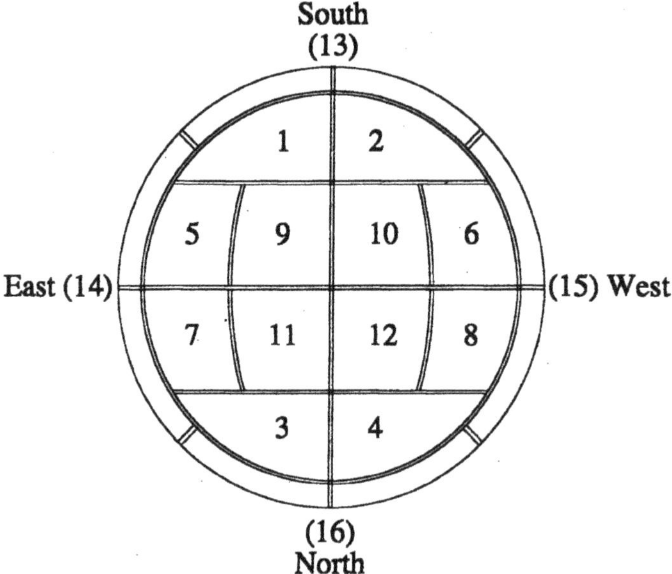

Figure 1. Configuration of the in-flight detector of the LOI with the correct orientation.

– 12 radiances of the LOI scientific pixels every minute,
– 4 radiances of the LOI guiding pixels every minute,
– 2 diameters (N/S and E/W) every minute,
– 4 VIRGO sunphotometers (SPM) flux corrections every day.

The first year's level-2 data will be made available to the scientific community one year after completion; for more detail on the data dissemination, see Fröhlich *et al.*(1996). From the level-2 data products the following data are derived for the scientific evaluation:

– time series of an l, m mode for p-mode analysis ($l < 9$);
– p-mode characteristics such as frequencies, splittings, amplitudes and linewidths;
– same as the above but for the g modes if they are found.

The level 0 and level 1 data will be made available to the scientific community only on request by a specific proposal to the VIRGO Principal Investigator. The proposal will be discussed and eventually approved by the VIRGO team.

4. Instrument Performances

4.1. Operations

The LOI was switched on December 6, 1995 and began, on this date, calibration measurements with the cover closed. These measurements entailed applying reference voltages to the amplifiers of the 16 pixels. The calibration measurements continued until the opening of the cover on March 27, 1996. A few minutes after the opening of the cover, the guiding servo loop was activated and the LOI was in its nominal observational mode where all 16 pixels are acquired every minute. On April 17, 1996, the calibration voltages were switched off, allowing a better determination of the offset voltages.

In addition to the normal operations of the LOI, the instrument took part in various useful joint payload operations: J3 (for guiding calibrations) and the roll operations. J3 occurred from April 3 to 5, 1996. During J3, the spacecraft scanned the solar rotation axis by ± 5 arc min (north–south) and the solar equator (east–west) by ± 20 arc min. During the north–south scan the LOI active guiding was activated in the range -10 to 0 arc min, and switched off outside this range. Similarly, during the east–west scan the guiding was activated in the range ± 5 arc min. The roll maneuver was performed on May 20, 1996. During the roll operations the spacecraft was rotated by 90° around the line of sight for about 20 hours.

Other regular spacecraft operations, such as the momentum management perturb the normal operations of the LOI; but due to the large guiding range, they can be handled without any additional specific operation of the LOI. Nevertheless, the data quality during such short periods is somewhat degraded.

4.2. In-Flight Performance

4.2.1. *Dark Current Noise*

During the 4 months that the cover was closed, we had ample time to characterize the noise of the electrical chain. This was derived using the so-called calibration mode described above. Figure 2 shows the power spectrum of an equivalent $l = 0$ mode together with a typical real solar spectrum. A similar spectrum was measured from the ground (Fröhlich *et al.*, 1996) for a single pixel, confirming an adequate performance of the instrument in space. The noise in the p mode range, mainly due to solar granulation, is about 0.1 ppm^2 μHz^{-1} at 3000 μHz which is about a factor 4 lower than what was observed from the ground by Appourchaux *et al.* (1995b). This higher noise is obviously due to the Earth's atmosphere.

4.2.2. *Guiding Calibrations and Performance*

The J3 operations permitted calibration of the guiding when the servo loop was activated. This was useful to make sure that the piezoelectric actuators were performing as expected. It helped also to measure an offset towards the south pole already observed by all the SOHO instruments. This offset was due to the spacecraft and amounted to 3 arc min for the LOI. A repointing occurred on April 17, 1996 that put the solar image close to the LOI optical axis. Figure 3 shows the calibration of the guiding, i.e., the high voltage applied to the actuators as a function of the offset angle. The sensitivity is about 1.6 arc sec V^{-1}. The performance is the same as measured on the ground. Figure 4 shows the guiding stability of the instrument, which is better than 0.01 arc sec at 1 σ. It is interesting to note that higher noise in the E–W direction remained when the spacecraft was rotated by 90°, which

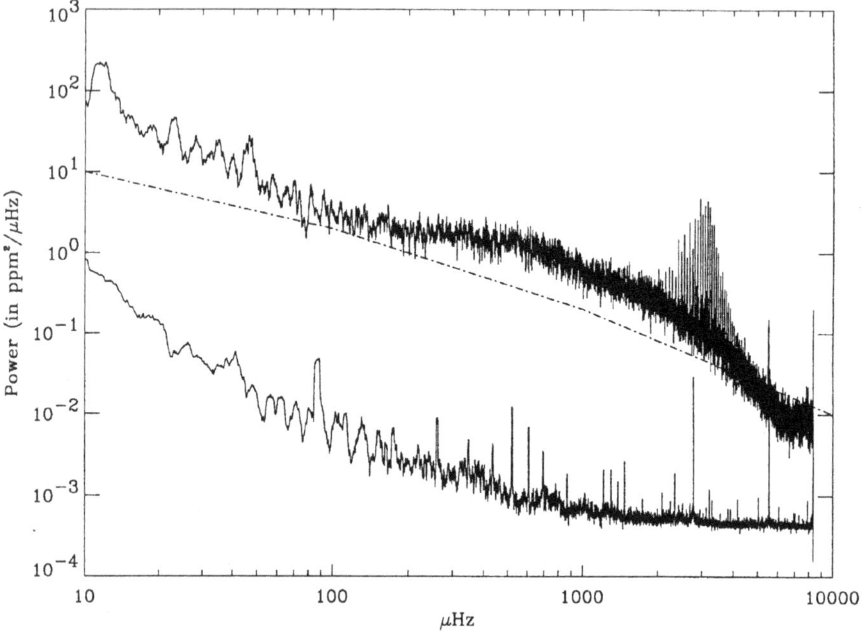

Figure 2. Top: LOI full-power spectrum for an $l = 0$, showing the 5-min oscillation modes. The spectrum becomes flatter around 1 mHz and then rises again; this is mostly due to the granulation noise and a Sun at its minimum of activity. Dashed is the noise as simulated by Andersen *et al.* (1994). *Bottom*: this is the equivalent $l = 0$ noise obtained during the calibration mode of the LOI (with the cover closed). The spectrum shows the noise due only to the electronic chain. The noise level in the *p*-mode range is about two orders of magnitude less than the one of the granulation noise. Several spikes are present which are due to changing calibration modes (at regular intervals) of another VIRGO instrument. A large spike at 3 min is also present (at 5555 μHz and its 'harmonics' at 2778 μHz). This is due to the basic acquisition block rate of VIRGO, which is 3 min.

indicates that the noise is from the pointing system of the spacecraft and not a solar feature.

4.2.3. *Pixel Sensitivity Calibrations*

The J3 operations were also used to perform a flat field calibration. The idea is derived from Kuhn, Lin, and Loranz (1991): a flat field of a multi-element detector (such as a charge coupled device or a photodiode array) can be derived by moving an image with a known intensity distribution on the detector. We tried to use a very similar technique for flat fielding the LOI pixels.

As an example, Figure 5 shows how the intensity varied on the pixels as a function of the offset angle. The measured profile is then compared with a theoretical profile, taking into account the limb darkening at 500 nm (Neckel and Labs, 1994) and the LOI pixel shapes. Since the active guiding was off (a fixed voltage was applied to the actuators) the position of the Sun was derived by maximizing the correlation between the measured and theoretical profiles (Figure 6).

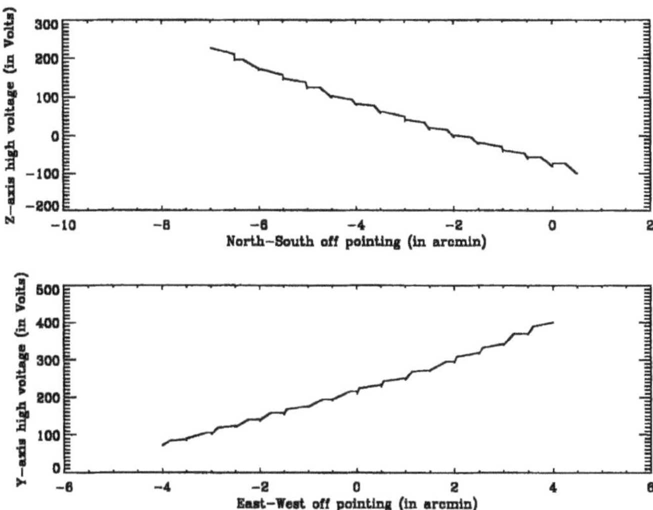

Figure 3. High voltage as a function of the offset in arc min for the east–west axis (*bottom*), and the north–south axis (*top*). The variations observed in voltage for a given offset angle are due to the creep in the piezoelectric material of the actuators.

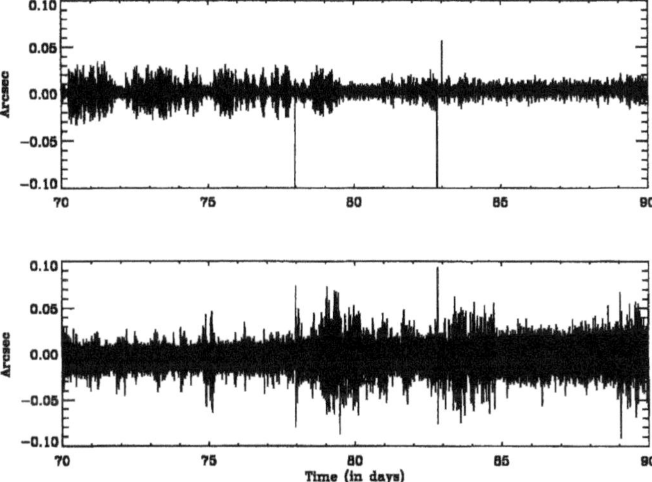

Figure 4. Guiding stability for the north–south and east–west direction as a function of time. The east–west guiding pixels are noisier than those of north–south; this could be due to a combination of instrumental and spacecraft pointing noise. It is probably not of solar origin.

Unfortunately, the precision in the sensitivity obtained using this technique was not better than 1%. It did not work because the departure from the theoretical profiles was about a few %, mainly due to the effect of the point spread function

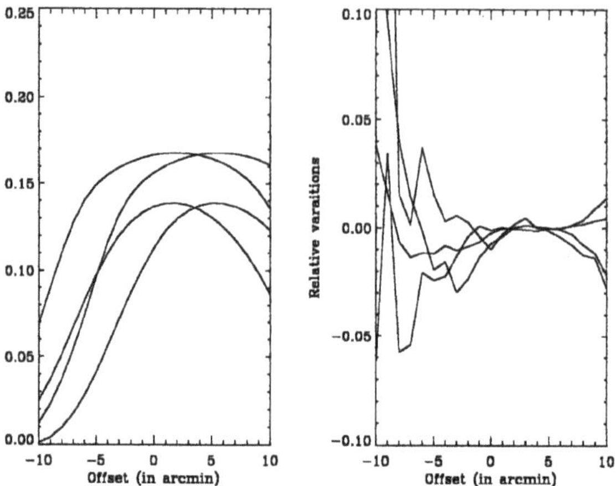

Figure 5. *Left*: theoretical east–west pixel intensity as a function of the offset angle during the north–south scan. The lack of symmetry between the 4 profiles is due to the offset of the Sun during the J3 operations; the active guiding was off and a fixed voltage was applied to each actuator. *Right*: relative difference between the measured profile and the theoretical ones for these pixels.

(PSF) of the telescope that needs to be taken into account. The PSF of the instrument varies a lot, especially when the image is not centered on axis; the variation of the PSF in the field can lead to variations in the intensity falling on the pixels of a few %. The effect of a varying PSF was also measured when the active guiding was on. For example on Figure 7, the pixel intensity varies with the offset angle.

These variations are calibrated for each pixel and used for producing the level-1 data.

4.2.4. *Solar Diameters*

The guiding pixels of the LOI are also used to derive an equivalent solar diameter. We used the limb-darkening function at 500 nm to derive the ratio of the guiding pixels to the central pixels. This ratio is almost a linear function of the size of the solar image, but it is insensitive to the decrease in flux which is due to the degradation. The diameters derived, for the north–south and east–west directions, are then converted to 1 AU using the orbit data of SOHO. Figure 8 shows the solar diameters for June 1996. At that time, the Sun's apparent diameter was still decreasing. The opposite trend is still visible in Figure 8, showing that the calibration has a systematic error of about 5%. The seasonal variation of the Sun's apparent diameter will be used to correct for this systematic error.

The noise in a 0.3-Hz bandwidth in the diameters is in a 0.3-Hz bandwidth about 0.1 arc sec and 0.2 arc sec peak-to-peak for the north–south and east–west directions, respectively.

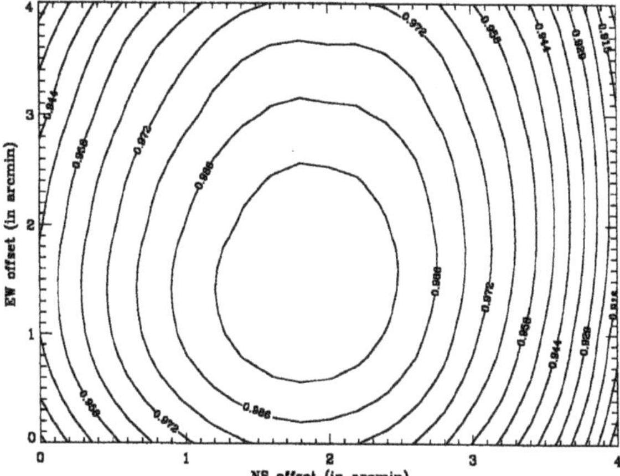

Figure 6. Correlation map between the measured profiles and theoretical profiles for the east–west pixels as a function of the north–south and east–west offset angles. The scan was north–south, which explains why the minimum is better defined in this direction. The maximum is very close to 1. Its location gives the offset of the solar image with respect to the center of the detector during the J3. Similar correlation maps have been obtained for the other pixels and the other east–west scan.

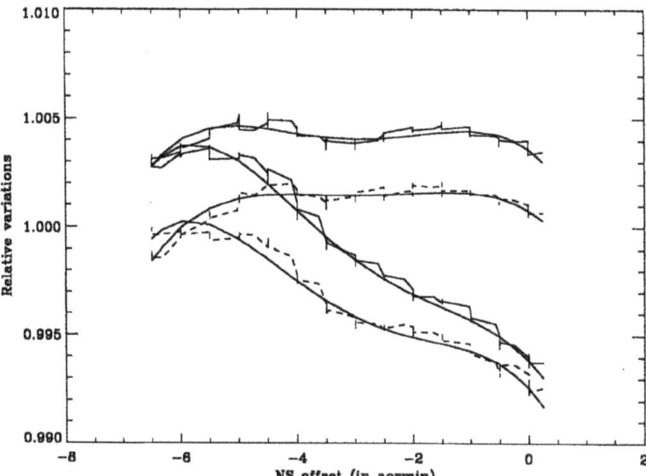

Figure 7. Variation of the intensity of the 4 east–west pixels as a function of the offset angle when the guiding is active and during the north–south scan; solid line for the pixels 6 and 7, dashed line for the pixels 5 and 8. The smooth lines are fifth-order polynomial fits.

4.2.5. *Solar Image Distortion*

Although most of the performance characteristics are excellent, the LOI is still very sensitive to slight distortions in the shape of the solar image. The distortion affects both the size of the image and its circular shape.

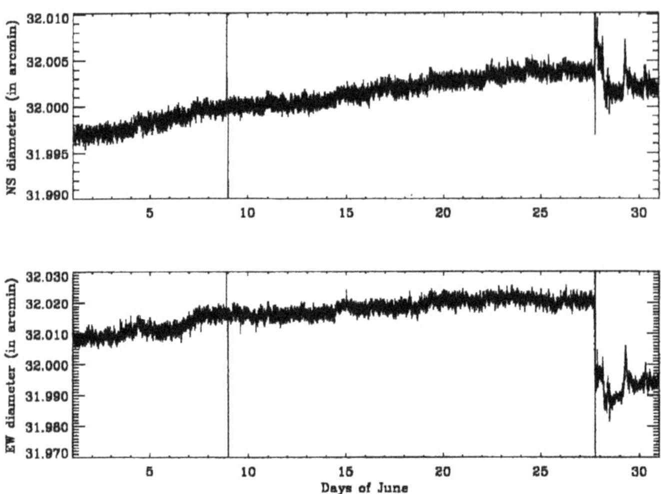

Figure 8. The 1 AU diameters for the north–south (*top*) and east–west (*bottom*) as a function of time for the month of June. The slight trend remaining is due the imperfect correction for a decreasing apparent solar diameter; the SOHO spacecraft was moving away from the Sun. The typical precision over one day is less than 1 arc sec.

In addition, this distortion has a 1-day periodicity creating, mainly in the $m = \pm l$ spectra, harmonics below 200 μHz, making the detection of the g modes more difficult. It has been empirically found that the temperature of the DC-DC converter of the high voltage can be used for compensating the distortion. Using the guiding pixels, we can derive that the magnitude of the distortion of the solar image is about 0.4 arc sec and 1.3 arc sec for a temperature change of 1 °C on the DC-DC for the north–south and east–west diameters, respectively. The magnitude of the correction is about 900 ppm / °C, 1550 ppm / °C and <400 ppm / °C, for the pixels 1–4, 5–8, and 9–12, respectively. The most affected pixels are, of course, the ones closest to the limb. The linear correction given above can reduce the amplitude of these harmonics by about a factor 3. This attenuation is not sufficient to facilitate the detection of the g modes in this range. A nonlinear correction has yet to be derived that would provide better attenuation of the harmonics and would cope with larger temperature variations occurring after the spacecraft's momentum management. An example of undercorrection can be seen in Figure 8 after the momentum management of 27 June, 1996.

At the time of writing, the reason for the distortion is not yet understood. However, since it has been observed that the distortion redistributes the flux among the pixels (i.e., the distortion disappears when summing up all the 16 pixels), there is great confidence that the effect will be explained. We used the linear corrections mentioned above for producing the level-1 data.

5. Scientific Data Analysis

5.1. Level 0 to Level 1 Processing

After unpacketizing the data, the level-0 data are built from the 3 delayed transmissions of the VIRGO data; this is done at the VIRGO data center (VDC). The data are then converted from counts to engineering values using the calibration performed on the ground. The offset calibrations were refined in flight after the calibration mode switch-off. The 16 pixels are then converted to flux measured at 1 AU, using the SOHO–Sun distance derived from the orbital data. The correction takes into account the real shape of the pixels, the limb darkening at 500 nm, and the in-flight size of the solar image at 1 AU. The precision of the correction better than 0.1% is for all the scientific pixels. The pixels are also converted to zero velocity using the orbital data. These level-1 conversions are done at the VDC.

In addition, for comparing the LOI integrated flux with the SPM, 4 conversion factors are derived. These factors take into account the pixel shapes, the limb darkening and the non sensitive tracks on the detector. Since the variations are slow, these factors are generated once for a whole day. This is also done at the VDC.

The 16 pixels are corrected for the variations in intensity using the attitude data of the spacecraft. Finally, the distortion effect is corrected, using the temperature of the DC-DC, by applying the same coefficient for all pixels having the same shape. These latter corrections are yet to be implemented at the VDC.

5.2. Comparison with the Green Channel of the SPM

Using the 4 SPM conversion factors, we compared the LOI integrated flux with the SPM. Figure 9 shows the result of the comparison. The LOI flux decreases much faster than that of the SPM. The degradation affects all the pixels in the same manner. The stronger degradation is probably due either to the fact that the LOI entrance filter has a much larger viewing angle for the solar wind than the SPM, or to the higher temperature of the LOI filter (about 50 °C), and possibly also to the fact that the filters were produced by different manufacturers. After correction for degradation, the LOI integrated flux compared quite well with the SPM. Especially the decrease in flux due to the emergence and the passage of a sunspot is very well correlated.

5.3. *p*-Mode Analysis

After the production of the level-1 data, the 12 scientific pixels are ready to be used for extracting the *p* modes. Each pixel is detrended using a triangular smoothing with a full width of 1 day, and then the residuals are converted to relative values. For extracting a given degree, the 12 pixels are combined using so-called optimal filters. The optimal filters were derived by Appourchaux and Andersen (1990). Since these

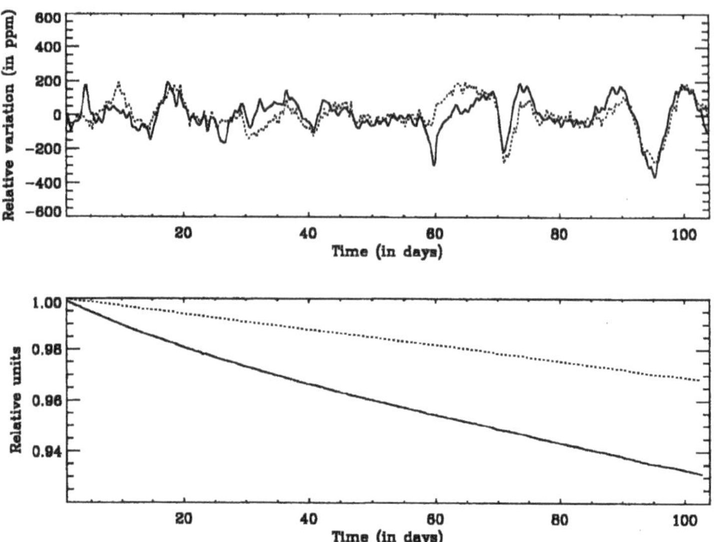

Figure 9. Comparison of the LOI (solid) and SPM (dotted) green channel, both viewing the Sun as a star. The time series starts on May 1, 1996. *Top*: relative variations. The agreement is quite good but there are still unexplained differences which need more detailed analysis. Both signals show very nicely the emergence and passage of an active region on day 70 and 90 respectively. *Bottom*: absolute variations normalized to the first datum.

filters are complex, they allow each m to be separated in an l, n multiplet. These filters were successfully used by Appourchaux *et al.* (1995b, c) and Rabello-Soares *et al.* (1996). The optimal filters were computed weekly, using the real size of the solar image (calibrated in flight) and the orientation of the Sun (only the B angle as P is maintained to zero by the spacecraft attitude control).

The spectra obtained have been used to derive the p-mode parameters. A given (m, ν) diagram is fitted using the maximum-likelihood method as described by Appourchaux *et al.* (1995b). As outlined in that paper, the different $2l+1$ component spectra of a given l, n multiplet are correlated with each other because of the imperfect isolation of the modes (see also Rabello-Soares, 1996). In principle, the full covariance matrix should be used to fit not the power spectra, but the amplitude or Fourier spectra (Schou, 1992). An additional complication is taken into account in the fitting of the spectra, higher-degree modes leaking into the lower-degree modes, and *vice versa*. For example, in the $l = 1$ signal we can detect $l = 6$ and *vice versa*; the $l = 4$ signal is contaminated by $l = 7$ (Figure 10) and *vice versa*, the $l = 5$ by $l = 8$ and *vice versa*. The full leakage between the $2l + 1$ modes of a multiplet and between the mode of the higher (or lower) degree has been computed using the computed optimal filters in a manner similar to that of Rabello-Soares (1996). This is an improvement of the procedure adopted by Appourchaux *et al.* (1995b), who regarded the mode leakage matrix not to be known *a priori* and

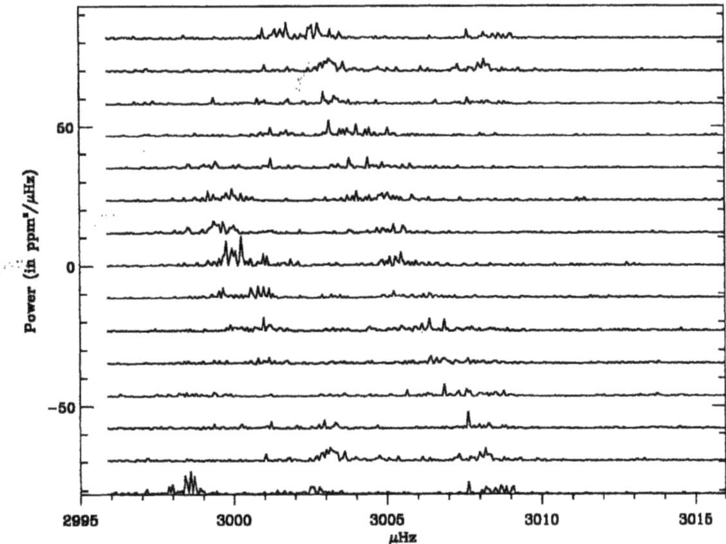

Figure 10. The m, ν diagram for the $l = 7$, $n = 18$ multiplet; $m = -7$ is at the top. The resolution is about 84 nHz. It shows an example of the effect of the leakage of the $l = 4$, $n = 19$ into the $l = 7$. We see clearly the splitting of the $l = 7$ mode and also the effect of the leakage of unwanted modes with different values of m. The splitting of the $l = 4$ is not visible because the leakage is more complex: the $l = 4$, $m = +4$ mode is partially transmitted by the filter for $l = 7$, $m = -7$, and the $l = 4$, $m = -4$ is transmitted by the $l = 7$, $m = +7$.

fitted it to the data. The amplitudes of all the modes in the spectrum, together with the frequencies of the target and leaked modes, were fitted simultaneously. The frequencies of the unwanted leaked modes were sometimes fixed using the frequencies obtained from BBSO (Libbrecht, Woodard, and Kaufman, 1990). This simplification did not influence the frequencies of the target modes substantially, yet it increased the rate of convergence of the iterations in the fitting. In this way the bias in the frequency estimates due to spatial mode leakage was reduced.

As far as the rotational splitting is concerned, we believe that the estimates are biased for ($l \leq 3$) because we do not fit the amplitude. In addition, the inaccuracy in the knowledge of the leakage matrix will lead to a bias in the splitting, the sign of which depends on the under- or over-estimation of the leakage. This effect is more important for low-degree modes ($l \leq 3$). Fortunately, fitting the amplitude spectra is easier to implement for these degrees because the higher degree aliases are either very weak or non-existent. In any case, the sources of systematic errorsin the estimate of the low-degree splitting have to be carefully assessed before drawing any conclusion on, e.g., the internal rotation of the Sun. From this point of view, longer time series will improve the situation.

6. Conclusion

In retrospect, only a few changes would have improved the instrument's performance. For example, we could have had more pixels; this would have reduced the leakage from the higher (or lower) degree. We could have read the guiding pixels using also individual VFCs; this would have reduced the high frequency noise by a factor of about 4–5.

Up to now, the LOI has been performing extremely well. There are also great expectations from the measurement of the diameters. Nevertheless, a lot of work remains to be done to understand the cause of the distortion of the solar image, especially if we want to detect the g modes. The data allow us to measure the p modes for $l \leq 7$ with a very high precision, as shown in the first scientific result paper by Fröhlich *et al.* (1997).

Acknowledgements

Those responsible for the LOI (ThA) are eternally thankful to Trevor Sanderson of the Solar and Heliospheric Science Division of ESA. He was the one who proposed to replace the 'non-intelligence' of VIRGO by a well timed 'pulling of its plug'. We are also grateful to our colleagues of the Mechanics Section of ESTEC without whom the model of the cover could have not been validated: P. Coste, M. Eiden, M. York, and M. Verin. The latter performed several tests on the motor in the Mechanical and Thermal Laboratory of ESTEC. Sincere thanks also to F. Dufrechou and P. Temporelli of MATRA-MARCONI-SPACE for the general management of the problem, to A.Lago and P. Roos of SAAB for the implementation of the software patch and to P. Strada of the SOHO project for coordinating the software patch efforts. Last but not least, we are thankful to H. J. Roth for his faithful dedication to the VIRGO instrument and all tests performed on the spare.

The building of the LOI would not have been possible without the contribution of the following ESTEC personnel: J. Fleur, R. Scheper, A. Smit for the mechanical designs; J. Blommers, J. Postema, M. J. Kikkert, K. F. Frumau, for the machining and the assembling; K. Bouman, A. Fransen, J. Heida, L. Smit for the electronics; T. Beaufort and D. Martin for the electronic design; S. Lévêque for the optical and detector tests; additional contributions were given by G. Gourmelon for the spectrophotometry, by B. Johlander for various irradiations tests, by J. M. Guyt for checking the contamination, by A. Zwaal for the cleanliness levels and by P. Brault for the bakings; we also thank R. Czichy for loaning us the Zygo interferometer and D. Doyle for the technical assistance on this interferometer. The various parts of the LOI instrument could not have been built without the help of the following individuals in industry: T. E. Hansen at Ame for the detector, C. Shannon and T. Hicks at Queensgate for the piezoelectric actuators, and P. Robert at SESO for the telescope.

Last but not least the LOI would never have had a place in space without the efforts of the whole VIRGO team which are gratefully acknowledged.

One of the authors (ThA) is very grateful to Richard Marsden for checking the English.

References

Andersen, B. N., Leifsen, T., and Toutain, T.: 1994, *Solar Phys.* **152**, 247.

Appourchaux, T. and Andersen, B. N.: 1990, *Solar Phys.* **128**, 91.

Appourchaux, T., Martin, D., and Telljohann: 1992, *SPIE* **1679**, 200.

Appourchaux, T., Telljohann, U., Martin, D., Lévêque, S., and Fleur, J.: 1995a, V. Domingo and T. Hoeksema (eds.), 'The Luminosity Oscillations Imager', *Fourth SOHO Workshop on Helioseismology*, ESA SP-376, p. 359.

Appourchaux, T., Toutain, T., Telljohann, U., Jiménez, A., Rabello-Soares, M.C., Andersen, B. N., and Jones, A. R.: 1995b, *Astron. Astrophys.* **294**, L13.

Appourchaux, T., Toutain, T., Telljohann, U., Jiménez, A., Rabello-Soares, M. C., Andersen, B. N., and Jones, A.R.: 1995c, in V. Domingo and T. Hoeksema (eds.), 'Results from the Luminosity Oscillations Imager', *Fourth SOHO Workshop on Helioseismology*, ESA SP-376, p. 265.

Fröhlich, C., Romero, J., Roth, H., Wehrli, C., Andersen, B.N., Appourchaux, T., Domingo, V., Telljohann, U., Berthomieu, B., Delache, P., Provost, J., Toutain, T., Crommelynck, D., Chevalier, A., Fichot, A., Däppen, W., Gough, D.O., Hoeksema, T., Jiménez, Gómez, M., Herreros, J., Roca-Cortés, T., Jones, A. R., Pap, J., and Willson, R. C.: 1995, *Solar Phys.* **162**, 101.

Fröhlich, C., Roth, H., Wehrli, C., Andersen, B.N., Appourchaux, T., Domingo, V., Berthomieu, B., Provost, J., Toutain, T., Crommelynck, D., Chevalier, A., Gough, D.O., Jiménez, Gómez, M., Herreros, J., Roca-Cortés, T. and Pap, J.: 1997, *Solar Phys.* **170**, 1 (this issue).

Kuhn, J. R., Lin, H., and Loranz, D.: 1991, *Publ. Astron. Soc. Pacific* **103**, 1097.

Neckel, H. and Labs, D.: 1994, *Solar Phys.* **153**, 91.

Libbrecht, K. G., Woodard, M. F., and Kaufman, J. M.: 1990, *Astrophys. J. Suppl.* **74**, 1129.

Rabello-Soares, M. C.: 1996, 'Helioseismological Study of the Solar Interior', Ph.D. Thesis, Universidad de La Laguna, La Laguna, Tenerife, Spain.

Rabello-Soares, M. C., Roca-Cortes, T., Jiménez, A., Appourchaux, T., and Eff-Darwich, A.: 1996, *Astrophys. J.*, submitted.

Schou, J.: 1992, 'On the Analysis of Helioseismic Data', Ph.D. Thesis, Aarhus University, Aarhus, Denmark.

STRUCTURE AND ROTATION OF THE SOLAR INTERIOR: INITIAL RESULTS FROM THE MDI MEDIUM-L PROGRAM

A. G. KOSOVICHEV, J. SCHOU, P. H. SCHERRER, R. S. BOGART, R. I. BUSH, J. T. HOEKSEMA, J. ALOISE, L. BACON, A. BURNETTE, C. DE FOREST, P. M. GILES, K. LEIBRAND, R. NIGAM, M. RUBIN, K. SCOTT and S. D. WILLIAMS
W.W. Hansen Experimental Physics Laboratory, Stanford University, Stanford, CA 94305, U.S.A.

SARBANI BASU and J. CHRISTENSEN-DALSGAARD
Theoretical Astrophysics Center, Danish National Research Foundation, and Institute of Physics and Astronomy, Aarhus University, DK-8000 Aarhus C, Denmark

W. DÄPPEN and E. J. RHODES, JR.
Department of Physics and Astronomy, University of Southern California, Los Angeles, CA 90089, U.S.A.

T. L. DUVALL, JR.
Laboratory for Astronomy and Solar Physics, NASA Goddard Space Flight Center, Greenbelt, MD 20771, U.S.A.

R. HOWE and M. J. THOMPSON
Astronomy Unit, Queen Mary and Westfield College, London, E1 4NS, U.K.

D. O. GOUGH and T. SEKII
Institute of Astronomy, and Department of Applied Mathematics and Theoretical Physics, Madingley Road, Cambridge, CB3 0HA, U.K.

J. TOOMRE
JILA, University of Colorado, CO 80309, U.S.A.

T. D. TARBELL, A. M. TITLE, D. MATHUR, M. MORRISON, J. L. R. SABA, C. J. WOLFSON and I. ZAYER
Lockheed-Martin Advanced Technology Center 91-30/252, 3251 Hanover St., Palo Alto, CA 94304, U.S.A.

P. N. MILFORD
Parallel Rules, Inc. 41 Manzanita Ave., Los Gatos, CA 95030, U.S.A.

(Received 23 October, 1996; in revised form 13 November, 1996)

Abstract. The medium-l program of the Michelson Doppler Imager instrument on board SOHO provides continuous observations of oscillation modes of angular degree, l, from 0 to $\sim$ 300. The data for the program are partly processed on board because only about 3% of MDI observations can be transmitted continuously to the ground. The on-board data processing, the main component of which is Gaussian-weighted binning, has been optimized to reduce the negative influence of spatial aliasing of the high-degree oscillation modes. The data processing is completed in a data analysis pipeline at the SOI Stanford Support Center to determine the mean multiplet frequencies and splitting coefficients.

The initial results show that the noise in the medium-l oscillation power spectrum is substantially lower than in ground-based measurements. This enables us to detect lower amplitude modes and, thus, to extend the range of measured mode frequencies. This is important for inferring the Sun's internal structure and rotation. The MDI observations also reveal the asymmetry of oscillation spectral lines. The line asymmetries agree with the theory of mode excitation by acoustic sources localized in the upper convective boundary layer. The sound-speed profile inferred from the mean frequencies gives evidence for a sharp variation at the edge of the energy-generating core. The results also confirm the previous finding by the GONG (Gough *et al.*, 1996) that, in a thin layer just beneath the convection zone, helium appears to be less abundant than predicted by

Solar Physics **170:** 43–61, 1997.

theory. Inverting the multiplet frequency splittings from MDI, we detect significant rotational shear in this thin layer. This layer is likely to be the place where the solar dynamo operates. In order to understand how the Sun works, it is extremely important to observe the evolution of this transition layer throughout the 11-year activity cycle.

1. Introduction

Solar oscillations observed at the photosphere as fluctuations of Doppler velocity or intensity are usually described in terms of normal modes that are identified by three integers: angular degree l, angular order m, and radial order n. The frequencies of the normal modes depend on structures and flows in the regions where the modes propagate. These regions are different for different modes. In Figure 1 we show the spatial eigenfunction of a typical mode that propagates half way through the Sun in depth and at relatively low latitudes. The surface structure of the modes is represented by spherical harmonics, $Y_l^m(\theta, \phi)$. Because different modes sample different regions inside the Sun, by observing many modes one can, in principle, map the solar interior.

The principal goal of helioseismology is to infer the structure and dynamics of the Sun's interior from the oscillation frequencies of normal modes. The resolving power of these inferences depends on the accuracy of the frequency measurements, and on the frequency and angular degree bandwidths. Until now the frequencies of solar oscillations have been reliably measured only in a band around 3 mHz, from approximately 1.5 to 4.5 mHz, where the mode amplitudes peak. These oscillations are acoustic (p) modes of low l and high n, or of high l and low n, and also fundamental (f) modes of high l. Modes of low l and low n have not been observed from the ground because their amplitude is very small, much less than 1 cm s^{-1}. Thus, one of the prime tasks of the MDI space experiment on board SOHO is to detect low-order modes. This requires long-term, continuous, stable observations free of atmospheric disturbances.

MDI has three basic helioseismology programs: medium-l, low-l, and dynamics (Scherrer *et al.*, 1995). The regions of the Sun covered by these programs are shown in Figure 2. The first two programs have been run with almost no interruptions since 18 April, 1996. The medium-l data are spatial averages of the full-disk Doppler velocity out to 90% of the disk's radius measured each minute. This results in 26 000 bins of approximately 10 arc sec resolution that provide sensitivity to solar p modes up to $l = 300$. The low-l observables are velocity and continuum intensity images summed into 180 bins, with the intent of detecting oscillations up to $l = 20$. The dynamics program provides 1024×1024 images of the whole disk, thus covering all of the p modes up to $l = 1500$. However, the dynamics program can run continuously for only 2 months each year when the high-rate telemetry channel is available. Therefore, the frequency resolution of the dynamics data is somewhat limited. Also, the dynamics program cannot monitor the evolution of the solar structure and dynamics continuously. Therefore, it was important to extend the sensitivity of the medium-l program to as high a degree as possible.

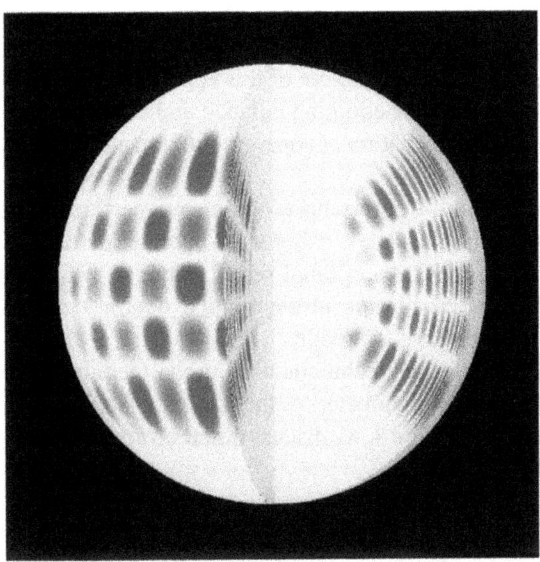

Figure 1. The spatial structure of the solar mode of angular degree $l = 20$, angular order $m = 16$, and radial order $n = 14$. Red and blue show element displacements of opposite sign. The frequency of this mode determined from the MDI data is 2935.88 ± 0.02 μHz.

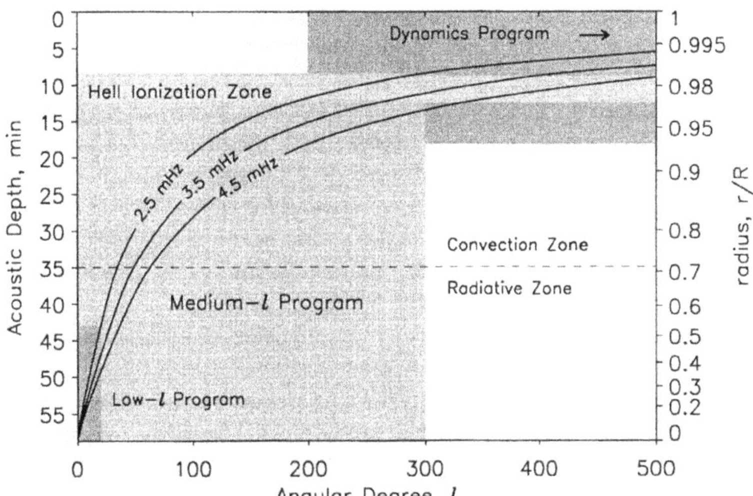

Figure 2. Acoustic depth and radius of penetration of solar modes of frequency of 2.5, 3.5, and 4.5 mHz as a function of angular degree, l (black curves); and the areas covered by the MDI medium-l, low-l, and dynamics programs (shaded regions). The lower boundary of the convection zone is shown by the dashed line. The light-gray horizontal strip shows the location of the zone of the second ionization of helium.

The target for the medium-l program is to observe the modes up to $l = 300$, because these modes have their inner reflection points in the He II ionization zone, and are, therefore, particularly sensitive to the structure of this zone. Studying the He II ionization zone is important for measuring the helium abundance and calibrating the equation of state, and also for understanding the nature of supergranulation and the large-scale dynamics of the convection zone.

To optimize the medium-l observing program, we performed simulations of several vector-weighted binning schemes, using a 20-hour series of full-disk Dopplergrams obtained on 25 January, 1996. An optimal set of Gaussian weights has been found that substantially reduces spatial aliasing in the angular degree range from 0 to 300 while preserving most of the power of the modes.

In Section 2, we describe the technique used in the medium-l program, the on-board data processing, and the basic parameters of the program. Analysis of the medium-l data is presented in Section 3. In Section 4, we discuss initial inversion results for the sound-speed profile and rotation.

2. Medium-l Program Technique

The medium-l data are transmitted through the low-rate telemetry channel (5 kilobits per second). The bit rate limits the spatial resolution of the observations. Prior to transmission the Doppler velocity is calculated from the original 1024×1024-pixel CCD filtergrams and binned on board the spacecraft. The binning can be performed in several ways depending on the scientific objectives of the program. For instance, Kosovichev (1992) considered spatially non-uniform binning schemes that maximize the range of the angular degree and maintain sensitivity to all the modes in this range. Alternatively, Gough (1992) suggested focusing on zonal and sectoral modes only, pushing the upper limit to $l = 1500$. These and other possibilities will be explored in the course of the mission. However, the initial goal of the program is to obtain the most reliable measurements of modes of moderate l that provide information about most of the interior of the Sun.

In observations performed with relatively low spatial resolution, the medium-l data can be corrupted because of spatial aliasing. Spatial aliasing contaminates normal mode peaks in the power spectrum with the signal from high-degree modes. This poses significant problems for measuring parameters of the modal lines. Therefore, we have adopted the Gaussian vector binning scheme suggested by Milford and Scherrer (1992). This reduces the spatial aliasing by averaging the data with Gaussian weights centered on a regular grid (say, of 4×4- or 5×5-pixel spacing) on the CCD with a half-width approximately equal to the spacing of the grid. Such a 2-D Gaussian filter also reduces solar noise. We have evaluated Gaussian-weighted binning using a 20-hour long time series of one-minute full disk 1024×1024-pixel Doppler images obtained during the MDI commissioning and initial calibration observations.

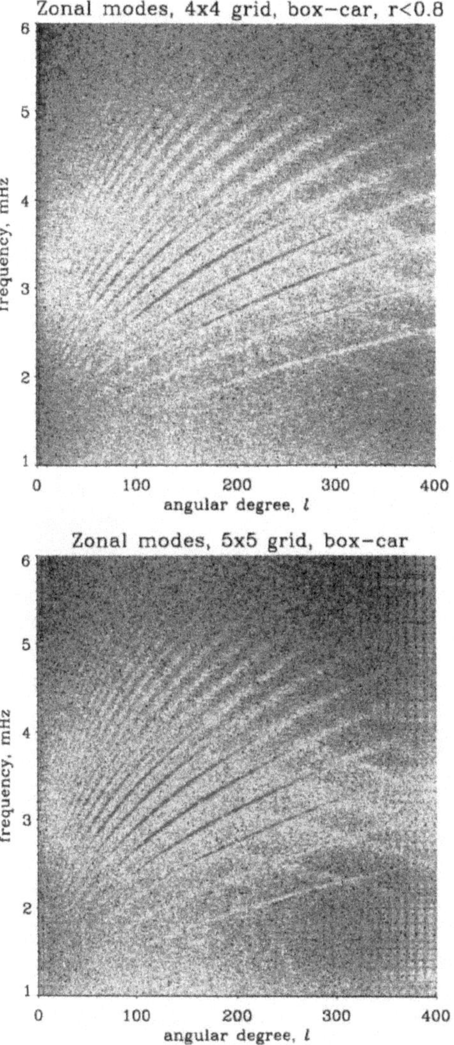

Figure 3. Power spectra of zonal modes computed using 20 hours of MDI full-resolution data box-car averaged onto a grid spacing of 4 × 4 pixels, covering the central part of the solar disk ($r < 0.8R$) (*top*) and for a 5 × 5 grid spacing covering 90% of the disk (*bottom*).

2.1. GAUSSIAN VECTOR BINNING

For a vector binning scheme, the 2D filter function for a bin centered at the CCD pixel with coordinates (i_0, j_0) takes the form

$$W(i_0, j_0; i, j) = u(i_0; i)v(j_0; j), \tag{1}$$

Table I

Shot-noise reduction coefficient, C_n.

	a=2.5	a=3.5	**a=4**	a=4.5	a=5.4	a=6
C_n	0.160	0.114	**0.100**	0.089	0.074	0.066

where $u(i_0; i)$ and $v(j_0; j)$ are functions of the CCD column number, i, and row number, j, respectively. While, in general, these functions can be arbitrary, the dyadic form of W is essential for the on-board processing.

Following Milford and Scherrer (1992), we have represented u and v in the Gaussian form, i.e.,

$$u(i_0; i) = W_0 \exp\left[-(i - i_0)^2/a^2\right], \tag{2}$$

and similarly for $v(j_0; j)$. Here, a is the characteristic width of the Gaussians, and W_0 is a scaling factor. The centers (i_0, j_0) of the bins were uniformly distributed on a 5 × 5-pixel rectangular grid in the area within 0.98 solar radii of disk center. The total number of bins was 29804. Results were evaluated for a=2.5, 3.5, 4, 4.5, 5.4 and 6. For comparison we also computed oscillation power spectra for 5 × 5-pixel box-car averages over the full-disk and for 4 × 4-pixel box-car averages in the central area restricted to $r < 0.8R$. Each case matches the telemetry bandwidth limit.

The data binning reduces the shot noise that determines the precision required in the on-board accumulations. We have estimated the coefficient of reduction of the shot noise, C_n, relative to the single pixel value as

$$C_n = \frac{\sqrt{\sum_{i,j} W(i,j)^2}}{\sum_{i,j} W(i,j)}. \tag{3}$$

The coefficients for the various values of a are shown in Table I.

2.2. Power Spectra of the Spatially Binned Data

Power spectra of the binned data were computed using two different techniques: the standard method of remapping the images onto a rectangular grid of heliographic coordinates, and a method of direct projection on eigenmode masks (Kosovichev, 1992). The results obtained by both techniques are consistent with each other.

The power spectra of zonal modes ($m = 0$) computed with 4 × 4 and 5 × 5 box-car averaging are shown in Figure 3. Obviously, the data sampled on the finer spatial grid resolve p modes at higher l better. Also, the relative amplitudes of the aliased modes crossing the original mode ridges are substantially smaller. However, observations with 4 × 4-pixel averaging can only cover a smaller portion of the solar disk, increasing the spatial leakage from adjacent modes. This makes the mode ridges broader and complicates identification of individual oscillation mode frequencies. Therefore, for the initial observations, we require a 5 × 5-pixel grid which covers 90% of the full disk.

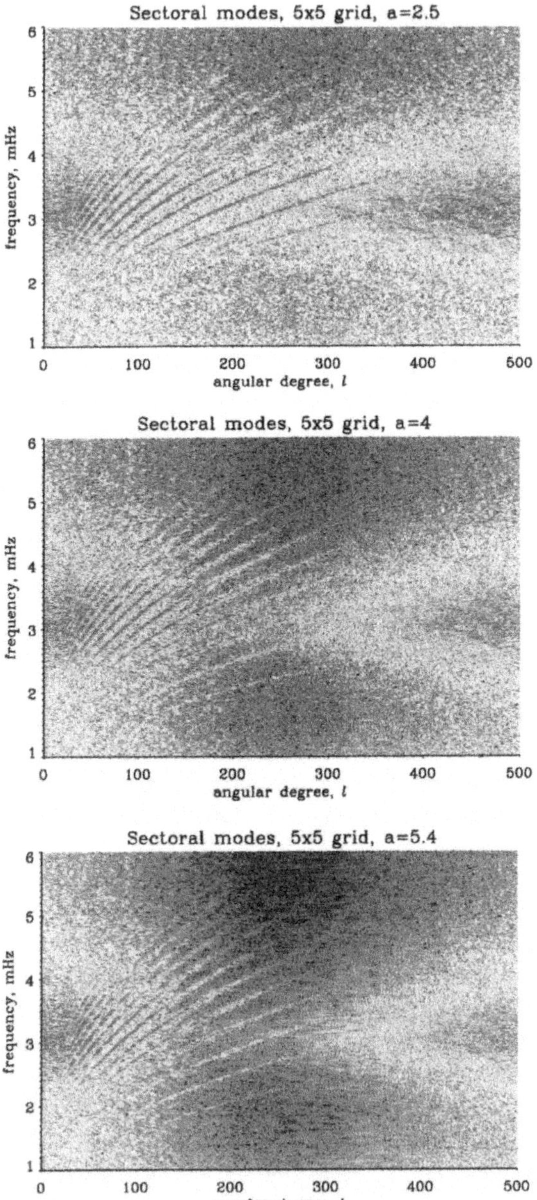

Figure 4. Power spectra of the sectoral modes obtained from 20 hours of the MDI full resolution spectra using the Gaussian spatial filter of $a = 2.5$ (*top*), 4 (*middle*), and 5.4 (*bottom*).

Figure 4 shows the power spectra of the sectoral modes ($m = l$) computed with Gaussian weighting with widths a=2.5, 4, and 5.4. As expected, increasing a reduces the amplitudes of the aliased ridges, but it also reduces the amplitude of the original modes at higher l. For $a > 4$, the amplitude of the 'true' modes in the range $l = 200 - 300$ becomes too small. It seems that $a = 4$ represents a reasonable tradeoff between anti-aliasing and high-l cut-off. In this case, the aliasing is substantially reduced up to $l = 300$.

Therefore, for the initial medium-l program observations, we have chosen to use the Gaussian vector-weighted data binning on a rectangular grid of 5×5 pixels, in the form of Equations (1) and (2) with $a = 4$. The shot noise of the binned data is reduced to 0.1 of the single-pixel value (Table I). If the noise per pixel per 60 s is 12 m s^{-1} then the noise per bin is 1.2 m s^{-1}. The binned images are generated on board once per minute.

3. Analysis of the Medium-l Data

3.1. Power Spectra

A standard helioseismic data analysis procedure (e.g., Hill *et al.*, 1996) has been applied to obtain the oscillation power spectra from the medium-l data. 60 days of data have been processed. The images were apodized with a cosine bell between $0.85R$ and $0.89R$ and remapped onto a grid with 768 points around the equator and 128 intervals between the equator and pole. After this, they were Fourier transformed in longitude and multiplied by the associated Legendre functions $P_l^m(\theta)$ in colatitude θ to do the spherical harmonic decomposition.

The power spectra of the sectoral and zonal modes, and also the m-averaged power spectrum, obtained from 10 days of the data are shown in Figure 5. The ridges in the diagrams correspond to modes of different radial order n. The lowest weak ridge is the f mode. At low l, the ridges are not resolved in the plot of the $l - \nu$ diagrams because the spectrum becomes too dense.

The aliasing effect is strongest for zonal ($m = 0$) and sectoral ($m = l$) modes because these modes have the smallest spatial scale. It is clearly seen in the power spectra of these modes for $l > 220$. In the m-averaged spectrum (Figure 5, bottom), the effect is much weaker. It is important that aliasing of the fundamental modes, f, and first-order acoustic modes, p_1, be negligible up to $l = 300$. This makes it possible to measure frequencies of individual modes along these two lowest ridges in the $l - \nu$ diagram. For higher p-mode overtones, individual modes are not resolved at $l > 220$ because the frequency difference between the modes of adjacent l becomes smaller than the line widths of the modes. In this case, instead of measuring individual peaks, we plan to use 'ridge-fitting' techniques.

The medium-l program also provides good sensitivity to low-degree modes (Figure 6), thus covering almost the whole range of the solar oscillation spectrum, in which individual p modes can be resolved.

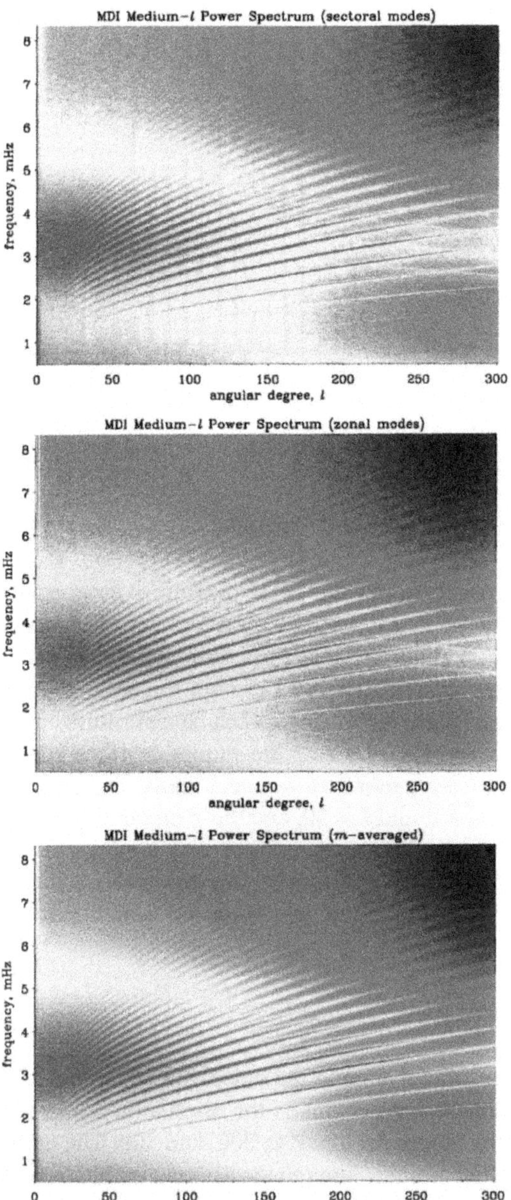

Figure 5. Power spectra ($l - \nu$ diagrams) obtained from 10 days of the MDI medium-l data for the sectoral ($l = m$) modes (*top*), zonal ($m = 0$) modes (*middle*), and for the modes averaged over m (*bottom*).

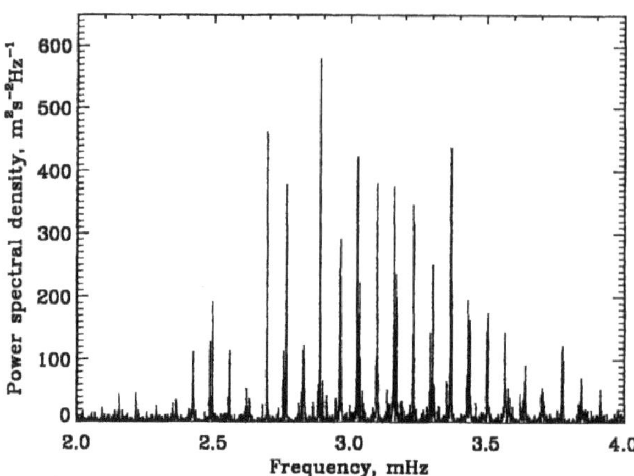

Figure 6. Power spectrum of $l = 0$, 1, and 2 modes obtained from 10 days of the MDI data.

3.2. Resolution of Low-Frequency Modes

An important difference between the MDI medium-l program and the ground-based networks is the ability to detect the low-n low-frequency modes that carry substantial independent information about the solar structure and dynamics. The lifetimes of these modes are much longer than the lifetimes of high-frequency modes. Therefore, in principle, the frequencies of these low-n modes can be determined more accurately than the frequencies of the other modes. However, the amplitude of the low-n modes is very small, so long stable time series are required to detect them. Two examples of low-frequency modes: the p mode of $n = 1$ and $l = 50$, and the f mode of $l = 95$, are shown in Figures 7 and 8. Parameters of these modes have been measured for the first time. With longer time series, we should be able to observe more low-frequency modes, with even lower amplitudes. So far, the smallest mode amplitude we have been able to detect was about 1 mm s^{-1}.

3.3. Line Asymmetry

Because of the significant reduction in the noise, the medium-l data have revealed interesting characteristics of the line profiles of the oscillation power spectra. Figure 9 shows the power spectra of the modes for $l = 100, 200$, and 300. The most interesting feature is the asymmetry of the line profiles. Though the asymmetry has been noticed in the ground-based data (Duvall *et al.*, 1993), frequencies of solar modes are usually determined by assuming that the line profile is symmetric and can be fitted by a Lorentzian, which would be the case if the solar p modes were damped simple harmonic oscillators excited by a stochastic source. However, this leads to systematic errors in the determination of frequencies (Hill *et al.*, 1996; Abrams and Kumar, 1996). Several authors have studied this problem theoretically and have found that there is an inherent asymmetry whenever the

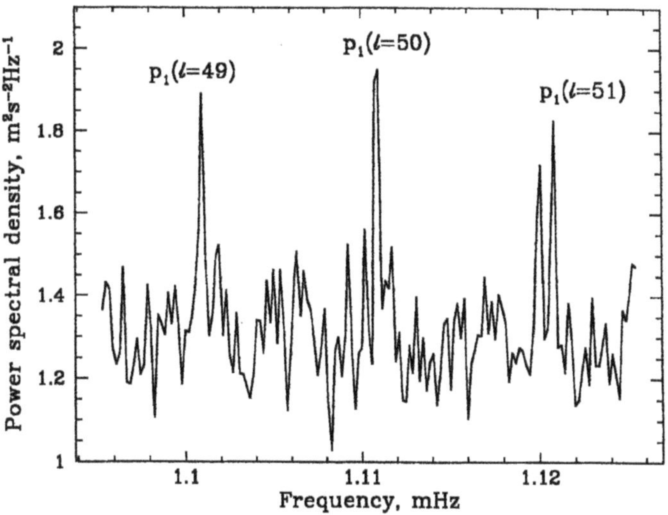

Figure 7. Power spectrum of p_1 mode of $l = 50$ obtained from 2 months of the MDI medium-l data. Modes of adjacent l leak into the spectrum because of imperfect spatial filtering.

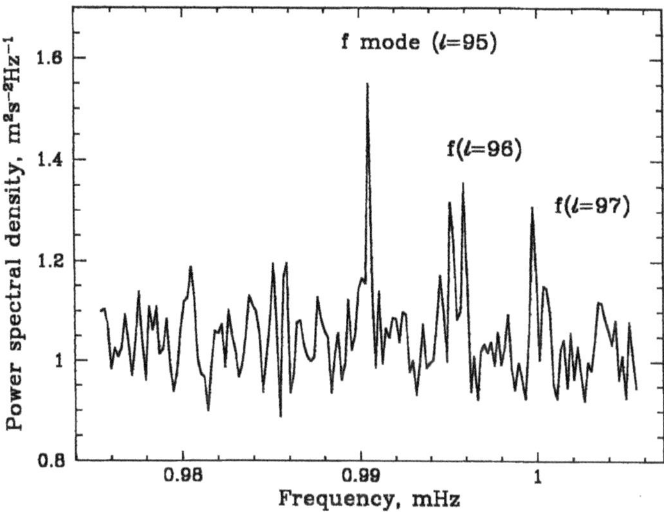

Figure 8. Power spectrum of f mode of $l = 95$ obtained from 2 months of the MDI data.

waves are excited by a localized source (Gabriel, 1992, 1993, and 1995; Kumar *et al.* 1994; Roxbourgh and Vorontsov, 1995). Physically, the asymmetry is an effect of interference between an outward direct wave from the source and a corresponding inward wave that passes through the region of wave propagation (Duvall *et al.*, 1993). Figure 10 shows a theoretical power spectrum of p modes of $l = 200$ obtained by Nigam and Kosovichev

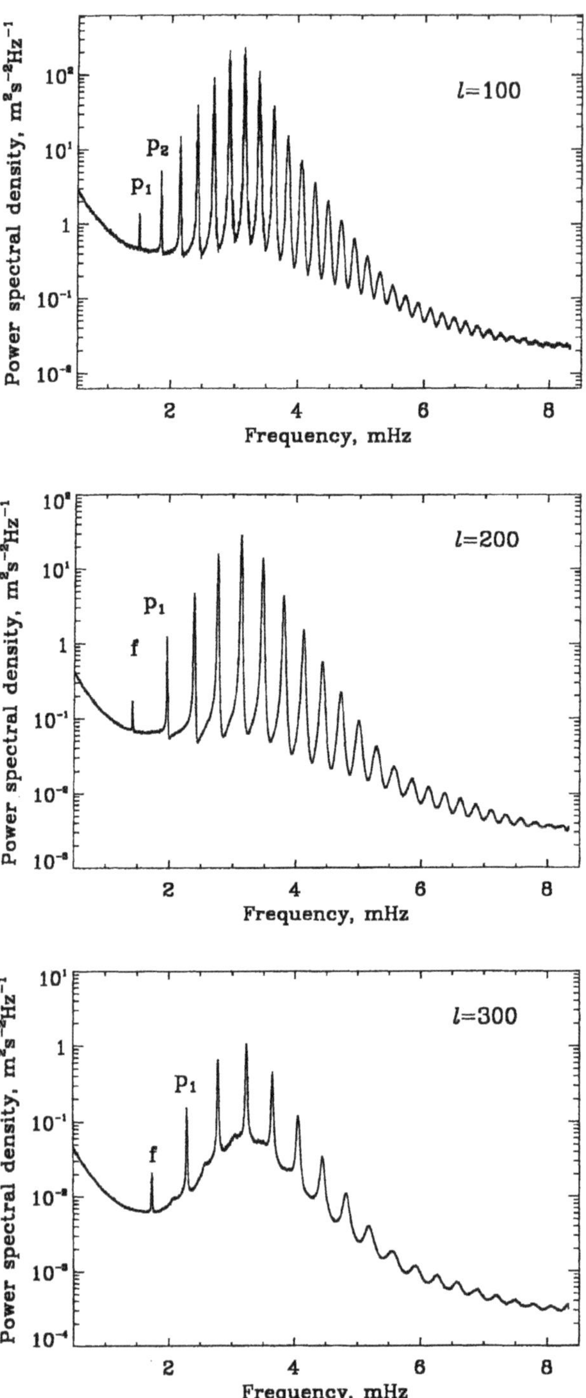

Figure 9. Power spectra obtained from 10 days of the MDI medium-l data of modes of $l = 100$ (*top*), $l = 200$ (*middle*), and $l = 300$ (*bottom*). Small bumps between the main peaks in the $l = 300$ spectrum are spatial aliases (see Section 3.2).

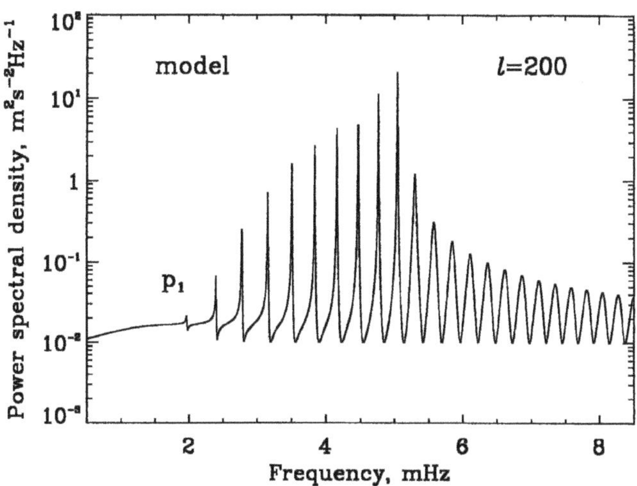

Figure 10. Theoretical power spectrum of $l = 200$ mode for a model with a source of the uniform spectral density localized 80 km beneath the photosphere. The background noise is assumed to be uniform with the power spectral density 10^{-2} m^2s^{-2} Hz^{-1}.

(1996). This model is in good qualitative agreement with the observations. The degree of the asymmetry depends on the relative locations of the acoustic sources and the upper reflection layer of the modes. This opens the prospect of using the observations of the line profiles of solar modes to test theories of excitation of solar and stellar oscillations and of their interaction with turbulent convection.

3.4. Frequency Measurements

The frequencies of the solar oscillations have been measured as described by Schou (1992). The frequency splitting of mode multiplets due to rotation and asphericity is approximated using a polynomial expansion similar to that of Duvall, Harvey and Pomerantz (1986):

$$\nu_{nlm} = \nu_{nl} + \sum_{k=1}^{6} a_{nl}^{(k)} \mathcal{P}_k^{(l)}(m), \tag{4}$$

where ν_{nl} is the mean frequency of a mode multiplet, and $\mathcal{P}_k^{(l)}(m)$ are orthogonal polynomials of degree k defined by

$$\mathcal{P}_k^{(l)}(l) = l, \quad \text{and} \quad \sum_{m=-l}^{l} \mathcal{P}_i^{(l)}(m)\mathcal{P}_j^{(l)}(m) = 0 \quad \text{for} \quad i \neq j. \tag{5}$$

The polynomials, $\mathcal{P}_k^{(l)}(m)$, can be expressed in terms of the Clebsch–Gordan coefficients, C_{k0lm}^{lm},

$$\mathcal{P}_k^{(l)}(m) = \frac{l\sqrt{(2l-k)!(2l+k+1)!}}{(2l)!\sqrt{2l+1}} C_{k0lm}^{lm}. \tag{6}$$

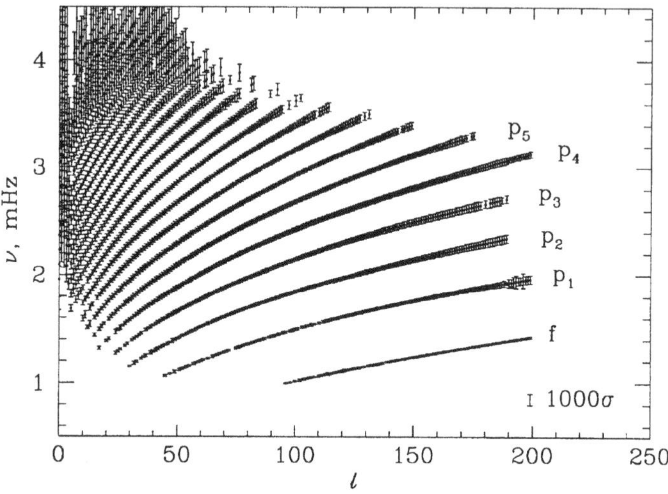

Figure 11. Mean frequencies of the mode multiplets obtained from 2 months of the medium-l data. The error bars show the standard errors multiplied by 1000.

Since

$$\sum_{m=-l}^{l} \mathcal{P}_k^{(l)}(m) = 0 \quad \text{for} \quad k > 0, \tag{7}$$

the mean frequencies, ν_{nl}, depend only on the spherically-symmetric component of the solar structure. Frequency splitting coefficients, $a_{nl}^{(k)}$, for even k depend on the aspherical component of the structure, while the coefficients for odd k measure the rotation rate (e.g., Gough, 1993).

For the initial frequency measurements, we approximated the line profiles by Lorentzians and used a maximum likelihood method to determine the parameters of the Lorentzians, taking into account possible overlaps of lines. The effect of the line asymmetry was not taken into account in these initial measurements. The mean frequencies of mode multiplets are shown in Figure 11. The error bars in this figure indicate estimated errors multiplied by 1000.

4. Initial Inversion Results

4.1. Radial Stratification

We have determined the spherically-symmetric structure of the Sun by using the optimally localized averaging techniques (e.g., Gough, 1985; Pijpers and Thompson, 1992) to invert the mean frequencies of split mode multiplets, ν_{nl}. Figure 12 shows the relative difference between the square of the sound speed in the Sun and a reference solar model. The

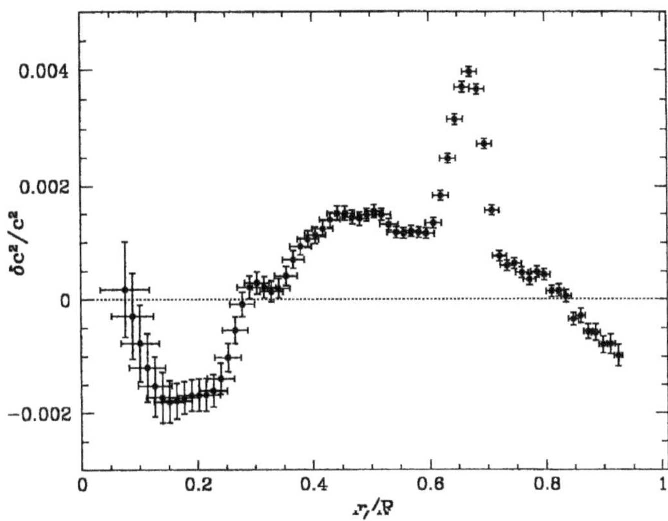

Figure 12. Relative differences between the squared sound speed in the Sun and a standard solar model as inferred from 2 months of the MDI data. The horizontal bars show the spatial resolution, and the vertical bars are error estimates.

reference model was model S of Christensen-Dalsgaard *et al.* (1996). This model was a standard evolutionary model computed using the most recent information on nuclear reaction rates (Bahcall and Pinsonneault, 1995) and radiative opacity and the equation of state (Rogers and Iglesias, 1996). The gravitational settling and diffusion of helium and heavier elements were taken into account following the theory by Michaud and Proffitt (1993).

The inversion results show that the maximum difference in the square of the sound speed between the model and the Sun is only 0.4%. Nevertheless, this difference is very important for understanding solar evolution and physical processes inside the Sun. Two features of the sound-speed profile are particularly notable. The first is a narrow peak centered at 0.67 R, just beneath the convection zone. This peak was previously detected in the LOWL (Basu *et al.*, 1996) and GONG data (Gough *et al.*, 1996) and is most likely due to a deficit of helium in this narrow region. The deficit of helium decreases the mean molecular weight and thus increases the sound speed. The deficit of helium could result from additional mixing of the material in the layer with the surrounding plasma if turbulence is generated in this layer because of rotational shear. Indeed, as we show in the next section, there is a strong radial gradient of the rotation rate in this layer.

Another interesting feature is the sharp decrease of the sound speed compared to the model at the boundary of the energy-generating core, at 0.25 R. This variation is not explained by the theory. Kosovichev and Fedorova (1991) noticed a possible sharp increase of density at the edge of the core when inverting the BBSO data (Libbrecht, Woodard, and Kaufman, 1990) if the spatial resolution of the inversion was increased at the expense of substantially increased errors of the estimates. The evidence from the MDI data is more convincing. It is quite possible that the drop in the sound speed results from an overabundance of helium at the edge of the solar core.

The steep increase of the sound speed towards the solar center can be explained if helium is less abundant than in the standard solar model. This indicates that the helium abundance profile seems to be more flat in the solar core than it is in the standard solar model. Gough *et al.* (1996) came to a similar conclusion from the GONG data. In principle, this variation of the helium abundance could arise from errors in the rates of the nuclear reactions or element diffusion. It could also be explained by uncertainties in the equation of state and in radiative opacity. Helioseismology is unable to distinguish among the possibilities without additional assumptions about the physical processes. However, if the transition at the edge of the core is really as sharp as we have found from the initial MDI data, then it strongly suggests material redistribution by macroscopic motions in the core, possibly induced by the instability of ^{3}He burning, as first suggested by Dilke and Gough (1972). Detailed studies of this phenomenon are very important for understanding stellar evolution. The deviations from the standard model of the spherically averaged structure of the solar core are fairly small; too small to solve the solar neutrino problem directly. However, if an instability really occurred, as predicted by theory (Christensen-Dalsgaard, Dilke, and Gough, 1974; Boury *et al.*, 1975; Shibahashi, Osaki and Unno, 1975; Kosovichev and Severny, 1985; Merryfield, Toomre and Gough, 1991), then the basic assumptions of solar evolution theory about the thermal balance in the core must be questioned. This is certainly important for understanding the nature of the solar neutrino problem and for using the Sun as a laboratory for particle physics.

4.2. Internal Rotation

The internal rotation rate at three latitudes, 0 (equator), 30 and 60 deg, is shown in Figure 13. The rotation is inferred from the odd $a_{nl}^{(k)}$ coefficients of Equation (3) using a regularized least-squares technique (e.g., Tikhonov and Arsenin, 1977). The inversion results confirm the previous findings that latitudinal differential rotation occurs only in the convection zone, that the radiative interior rotates almost rigidly, and that there is a thin shear layer near the surface (Goode *et al.*, 1991; Thompson *et al.*, 1996). The most interesting feature is that the transition layer (tachocline) between the radiative and convection zone is mostly located in the radiative zone, at least at the equator where it is also fairly thin, certainly less than 0.1 R. The layer seems to be wider at high latitudes. Previously, the parameters of the tachocline were estimated by Kosovichev (1996) and by Charbonneau *et al.* (1996) by fitting an analytical model to the data from the BBSO (Woodard and Libbrecht, 1993), and from the LOWL instrument (Tomczyk *et al.*, 1995).

The sharp radial gradient of the angular velocity beneath the convection zone strongly suggests that the sharp narrow peak of the sound speed at 0.67 R seen in the structure inversion (Figure 12) is indeed due to rotationally-induced turbulent mixing in the tachocline (Spiegel and Zahn, 1992). The location and the width of the tachocline are also consistent with the requirement of recent dynamo theories of the solar cycle (Rüdiger and Brandenburg, 1995; Weiss, 1994). It will be extremely important to observe the variations

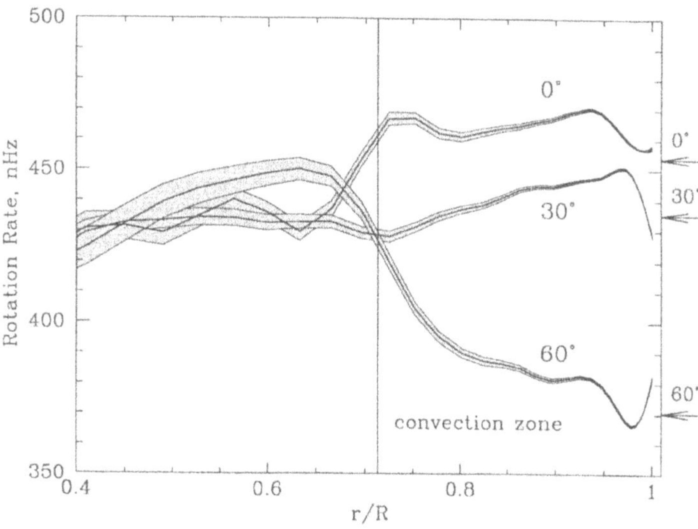

Figure 13. Solar rotation rate inferred from 2 months of MDI medium-l data as a function of radius at three latitudes, 0°, 30°, and 60°. The formal errors are indicated by the shaded regions. The arrows indicate the Doppler rotation rate directly measured on the surface.

of the angular velocity in the tachocline through the solar cycle in order to understand the mechanism of solar activity.

5. Conclusions and Perspectives

The MDI medium-l program is designed for continuous monitoring of the structure and dynamics of the Sun's interior. The initial results show that the noise in the medium-l oscillation power spectrum is substantially lower than in the ground-based measurements. This enables us to resolve lower amplitude modes and, thus, to extend the range of measured mode frequencies. This is important for inferring the Sun's internal structure and rotation. The MDI observations clearly show the asymmetry of oscillation line profiles in agreement with the theory of mode excitation by acoustic sources localized in the upper convective boundary layer. The sound-speed profile inferred from the mean frequencies gives evidence for a sharp variation at the edge of the energy-generating core, which is not accounted for by the standard evolution theory. The results also confirm the previous finding from GONG data that helium appears to be less abundant than predicted by theory in a thin layer just beneath the convection zone (Gough *et al.*, 1996). Inverting the multiplet frequency splittings from MDI, we have found significant rotational shear in this thin layer. This shear flow probably generates turbulence that mixes the plasma in the upper radiative zone. This layer ('tachocline') is likely to be the place where the solar dynamo operates.

The first results of the medium-l program suggest the initial binning scheme should be maintained for the first year of SOHO observations to obtain the time series sufficient to resolve low-frequency p modes. After the first year, it may be interesting to explore other possibilities, such as decreasing the remaining aliasing for modes of $l = 200$–300 to allow identification of a greater number of modes, or using an inhomogeneous binning scheme to focus on some particular groups of modes. Certainly, the decision should be taken after consideration of the performance of the other SOHO and ground-based helioseismic programs. However, the principal goal of the medium-l program will remain the same – continuous measurement of the solar structure and dynamics as they change through the solar cycle. This is crucial for understanding the mechanisms for solar variability.

Acknowledgements

We thank John Leibacher for commenting, as referee, on the original submitted version of this paper.

The authors acknowledge many years of effort by the engineering and support staff of the MDI development team at the Lockheed Palo Alto Research Laboratory (now Lockheed-Martin) and the SOI development team at Stanford University. SOHO is a project of international cooperation between ESA and NASA. This research is supported by the SOI-MDI NASA contract NAG5-3077 at Stanford University.

References

Abrams, D. and Kumar, P.: 1996, *Astrophys. J.* **472**, 882.
Bahcall, J. N. and Pinsonneault, M. H.: 1995, *Rev. Mod. Phys.* **67**, 781.
Basu, S., Christensen-Dalsgaard, J., Schou, J., Thompson, M. J., and Tomczyk, S.: 1996, *Bull. Astr. Soc. India* **24**, 147.
Boury, A., Gabriel, M., Noels, A., Scuflaire, R., and Ledoux, P.: 1975, *Astron. Astrophys.* **41**, 279.
Brandenburg, A.: 1994, in M. R. E. Proctor and A. D. Gilbert (eds.), *Lectures on Solar and Planetary Dynamo*, Cambridge University Press, Cambridge, p. 177.
Charbonneau, P., Christensen-Dalsgaard, J., Henning, R., Schou, J., Thompson, M. J., and Tomczyk, S.: 1996, in J. Provost and F. X. Schmider (eds.), 'Sounding Solar and Stellar Interiors', *Proc. IAU Symp.* **181**, in press.
Christensen-Dalsgaard, J., Dilke, F. W. W., and Gough, D. O.: 1974, *Monthly Notices Roy. Astron. Soc.* **169**, 429.
Christensen-Dalsgaard, J., Gough, D. O., and Thompson, M. J.: 1991, *Astrophys. J.* **378**, 413.
Christensen-Daalsgard, J., Däppen, W., and the GONG Team: 1996, *Science* **272**, 1286.
Dilke, F. W. W. and Gough, D. O.: 1972, *Nature* **240**, 262.
Duvall, T. L., Harvey, J. W., and Pomerantz, M. A.: 1986, *Nature* **321**, 500.
Duvall, T. L., Jr., Jefferies, S. M., Harvey, J. W., Osaki, Y., and Pomerantz, M. A.: 1993, *Astrophys. J.* **410**, 829.
Gabriel, M.: 1992, *Astron. Astrophys.* **265**, 771.
Gabriel, M.: 1993, *Astron. Astrophys.* **274**, 935.
Gabriel, M.: 1995, *Astron. Astrophys.* **299**, 245.
Goode, P. R., Dziembowski, W. A., Korzennik, S. G., and Rhodes, E.J.: 1991, *Astrophys. J.* **367**, 649.
Gough, D. O.: 1985, *Solar Phys.* **100**, 65.
Gough, D. O.: 1992, *On POX Strategy*, SOI-TN-099, Stanford University.

Gough, D. O.: 1993, in J. P. Zahn and J. Zinn-Justin (eds.), *Astrophysical Fluid Dynamics*, Elsevier, Amsterdam, p. 399.
Gough, D. O., Kosovichev, A. G., and the GONG Team: 1996, *Science* **272**, 1296.
Hill, F., Stark, P. B., Stebbins, R. T., and the GONG Team: 1996, *Science* **272**, 1292.
Kosovichev, A. G. and Fedorova, A. V.: 1991, *Astron. Zh.* **68**, 1051.
Kosovichev, A. G. and Severny, A. B.: 1985, *Izv. Krymsk. Astrofiz. Obs.* **72**, 188.
Kosovichev, A. G.: 1992, *Optimal Masks for Structure Program*, SOI-TN-096, Stanford University.
Kumar, P., Fardal, M. A., Jefferies, S. M., Duvall, T. L., Jr., Harvey, J. W., and Pomerantz, M. A.: 1994, *Astrophys. J.* **422**, L29.
Libbrecht, K. G., Woodard, M. F., and Kaufman, J. M.: 1990, *Astrophys. J. Suppl.* **74**, 1129.
Merryfield, W. J., Toomre, J., and Gough, D. O.: 1991, *Astrophys. J.* **367**, 658.
Milford, P. N. and Scherrer, P. H.: 1992, *Outline of Gaussian Mask Mode*, SOI-TN-058, Stanford University.
Michaud, G. and Proffitt, C. R.: 1993, in A. Baglin and W. Weiss (eds.), *Inside the Stars*, ASP Conf. Series, Vol. 40, San Francisco, p. 246.
Nigam, R. and Kosovichev, A. G.: 1996, in J. Provost and F. X. Schmider (eds.), 'Sounding Solar and Stellar Interiors', *Proc. IAU Symp.* **181**, in press.
Pijpers, F. P. and Thompson, M. J.: 1992, *Astron. Astrophys.* **262**, L33.
Rogers, F. J. and Iglesias, C. A.: 1996, *Astrophys. J.* **456**, 902.
Roxbourgh, I. W. and Vorontsov, S. V.: 1995, *Monthly Notices Roy. Astron. Soc.* **272**, 850.
Rüdiger, G. and Brandenburg, A.: 1995, *Astron. Astrophys.* **296**, 557.
Saio, H.: 1980, *Astrophys. J.* **240**, 685.
Scherrer, P. H., Bogart, R. S., Bush, R. I., Hoeksema, J. T., Kosovichev, A. G., Schou, J., Rosenberg, W., Springer, L., Trabell, T. D., Title, A., Wolfson, C. J., Zayer, I., and the MDI Engineering Team: 1995, *Solar Phys.* **162**, 129.
Schou, J.: 1992, 'On the Analysis of Helioseismic Data', Thesis, Aarhus University.
Shibahashi, H., Osaki, Y., and Unno, W.: 1975, *Publ. Astron. Soc. Japan* **27**, 401.
Spiegel, E. A. and Zahn, J.-P.: 1992, *Astron. Astrophys.* **265**, 106.
Thompson, M. J., Toomre, J., and the GONG Team: 1996, *Science* **272**, 1300.
Tikhonov, A. N. and Arsenin, V. Y.: 1977, *Solution of Ill-posed Problems*, Winston, Washington, DC.
Tomczyk, S., Streander, K., Card, G., Elmore, D., Hull, H., and Cacciani, A.: 1995, *Solar Phys.* **159**, 1.
Weiss, N. O.: 1994, in M. R. E. Proctor and A. D. Gilbert (eds.), *Lectures on Solar and Planetary Dynamo*, Cambridge University Press, Cambridge, p. 59.
Woodard, M. F. and Libbrecht, K. G.: 1993, *Science* **260**, 1778.

TIME-DISTANCE HELIOSEISMOLOGY WITH THE MDI INSTRUMENT: INITIAL RESULTS

T. L. DUVALL JR.
Laboratory for Astronomy and Solar Physics, NASA Goddard Space Flight Center, Greenbelt, MD 20771, U.S.A.

A. G. KOSOVICHEV, P. H. SCHERRER, R. S. BOGART, R. I. BUSH, C. DE FOREST, J. T. HOEKSEMA and J. SCHOU
W. W. Hansen Experimental Physics Laboratory, Stanford University, Stanford, CA 94305, U.S.A.

J. L. R. SABA, T. D. TARBELL, A. M. TITLE and C. J. WOLFSON
Lockheed-Martin Advanced Technology Center, 91-30/252, 3251 Hanover St., Palo Alto, CA 94304, U.S.A.

P. N. MILFORD
Parallel Rules, Inc., 41 Manzanita Ave., Los Gatos, CA 95030, U.S.A.

(Received 28 October, 1996; in revised form 20 November, 1996)

Abstract. In time-distance helioseismology, the travel time of acoustic waves is measured between various points on the solar surface. To some approximation, the waves can be considered to follow ray paths that depend only on a mean solar model, with the curvature of the ray paths being caused by the increasing sound speed with depth below the surface. The travel time is affected by various inhomogeneities along the ray path, including flows, temperature inhomogeneities, and magnetic fields. By measuring a large number of times between different locations and using an inversion method, it is possible to construct 3-dimensional maps of the subsurface inhomogeneities.

The SOI/MDI experiment on SOHO has several unique capabilities for time-distance helioseismology. The great stability of the images observed without benefit of an intervening atmosphere is quite striking. It has made it possible for us to detect the travel time for separations of points as small as 2.4 Mm in the high-resolution mode of MDI (0.6 arc sec pixel^{-1}). This has enabled the detection of the supergranulation flow. Coupled with the inversion technique, we can now study the 3-dimensional evolution of the flows near the solar surface.

1. Introduction

Three strong features of the MDI instrument on SOHO (Scherrer *et al.*, 1995) are the stability coming from the absence of an intervening atmosphere between the telescope and the Sun, an orbit in continuous sunlight enabling long sequences of 'around-the-clock' observing, and the relatively high resolution compared to some groundbased helioseismology experiments. In this paper, we describe work that fully exploits these features of MDI and which points to a future in which we will be able to follow in detail the subsurface convection.

We are using a new technique coming to be known as time-distance helioseismology (Duvall *et al.*, 1993, 1996a, b; Kosovichev, 1996; Kosovichev and Duvall, 1996; D'Silva, 1996; D'Silva *et al.*, 1996), in which the time $t(x_1, x_2)$ for acoustic waves to travel between different surface locations (x_1, x_2) is measured from a cross-correlation technique. In the first approximation, the time $t(x_1, x_2)$ depends on inhomogeneities along a geometric ray path connecting the surface locations. If the temperature is locally high, the waves will traverse the region more quickly leading to a shorter travel time. If there is a flow with a component in the direction of the ray path, the flow will tend to speed up the wave in the direction of the flow, while slowing it down in the reciprocal direction. In this case, we are led to the very distinctive signature of the travel time being different for waves traveling in opposite directions between the same points. We have thought of no other effect which can yield this signature.

A magnetic field in the region of wave travel can also lead to travel time anisotropy (Duvall *et al.*, 1996a). This is somewhat harder to isolate, as the distinctive signature of the field is that waves travelling along the field direction have a different wave speed than waves travelling perpendicular to the field. For some field geometries, it is possible to set up pairs of rays that intersect at right angles to look for this type of signature, but it is more difficult as the two rays have different paths except at the location where they intersect, and so other inhomogeneities can confuse the issue. The general form of rays below the surface is shown in Figure 5. In this paper the effect of magnetic anisotropy is not considered, because we have studied a very quiet region.

2. Observations

For this study we have used 8.5 hours of Doppler images from January 27, 1996 taken in the high-resolution mode of MDI, which has 0.6 arcsec pixel^{-1}. To study convection near the surface, it is necessary to use fairly short time intervals in order that the evolution not be too great during the observing interval. On the other hand, it is necessary to observe for some length of time in order to get a statistical sample of waves. The 8.5 hours is a compromise between these two competing requirements, with the supergranule lifetime of about 1 day putting an upper limit on how long to observe. Another point is that we need to observe for longer than the travel time, since the correlation that we are detecting arises from the same wave traveling from one location to the other through the subsurface layers. The travel times used in the present study are for surface point separations in the range 6-30 Mm, and are in the range 17–34 min.

An example of a Doppler image is shown in Figure 1. There is considerable high spatial frequency information in the image, as measurements of the point-spread function of the instrument would predict (Scherrer *et al.*, 1995). And in fact, with 0.6 arc sec pixel^{-1} we are somewhat oversampling the p-mode signal. With this sampling the spatial Nyquist frequency is near spherical harmonic degree $l = 5000$. But the fundamental mode, the lowest frequency mode in the p-mode range, crosses the peak acoustic cutoff frequency (~ 5.3 mHz) at degree $l = 2700$. Since the peak of the power envelope is near temporal frequency 3.2 mHz, there is very little power

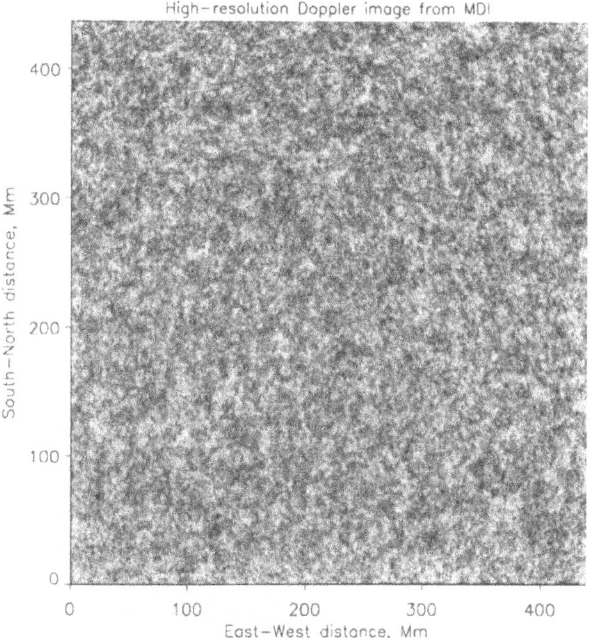

Figure 1. A MDI Dopplergram for the high-resolution field on 27 January, 1996. White is for receding velocity, black for approaching. The solar rotation has been removed. The image is dominated by the p-mode oscillations and granulation. The standard deviation of the velocity values is 244 m s^{-1}.

due to oscillations in the outer half (in spatial frequency) of the power spectrum. We have used this information explicitly in some later experiments in which the images are binned 2 × 2 on board the SOHO spacecraft to reduce the telemetry required.

The mean power spectrum of the 8.5 hours of data is shown in Figure 2. The spectrum is smooth, as it has been averaged over azimuth. The spatial frequency response has been flattened to remove the effects of the point-spread function. The excellent quality of the data is apparent. One striking thing is the large width of the f mode, which increases to $\sim$1 mHz (full-width at half maximum) in width at $l = 2500$. It seems likely that much of this width could be caused by motions at the surface causing the waves to be Doppler shifted. Using the standard Doppler-shift relation, $v = \Delta\omega/k$, we find a flow velocity, v, of 0.9 km s^{-1} for a frequency shift, $\Delta\omega/2\pi$, of 0.5 mHz (one half of full-width at half maximum), and for a horizontal wave number, $k = \sqrt{l(l+1)}/R_{\odot}$. This is of the order of surface motions from convection, as determined by correlation tracking or as we find later in this paper by the time-distance technique.

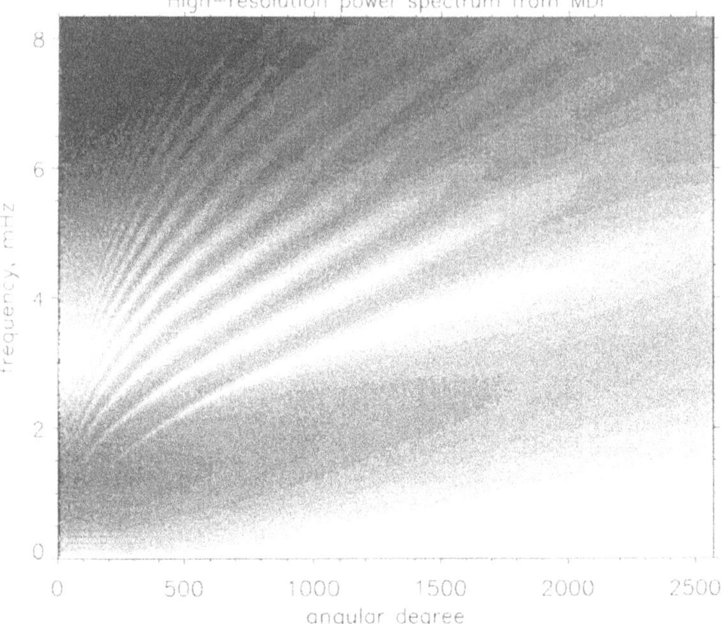

Figure 2. A mean power spectrum for the 8.5 hour interval studied on 27 January, 1996. The solar rotation was removed before the power spectrum was computed. The raw data was binned 2 × 2 before the analysis, so the spatial Nyquist frequency shown is half of the maximum achievable, although we get almost all the oscillation signal with the present analysis. The power spectrum is smooth because it has been averaged over wave direction.

3. Analysis

To measure the travel times between different locations, we calculate the temporal cross-correlation function between the data at one location and the data within an annulus at some great-circle distance from the point. We look at both positive and negative lags of the correlation function, as this tells us in which direction the waves are travelling. For example, a signal at location 1 first and later at location 2 will lead to a lag signal of one sign while a signal first at location 2 will lead to a correlation at the opposite sign of lag.

The cross-correlation function, averaged over a number of origins, is shown in Figure 3, along with a curve showing the time versus distance for a simple theory. If we imagine generating a pulse of acoustic energy at the surface that propagates along the ray paths to the distant location, we might expect to see a pulse at the distant location. But in fact we see something that is much broader and really looks like a wave packet with a period of 5 min, and a width in time of about 1/bandwidth of our oscillation power spectrum, or $(1\ \text{mHz})^{-1} = 17$ min. This makes sense, because the pulse that we generated at the surface had an infinite

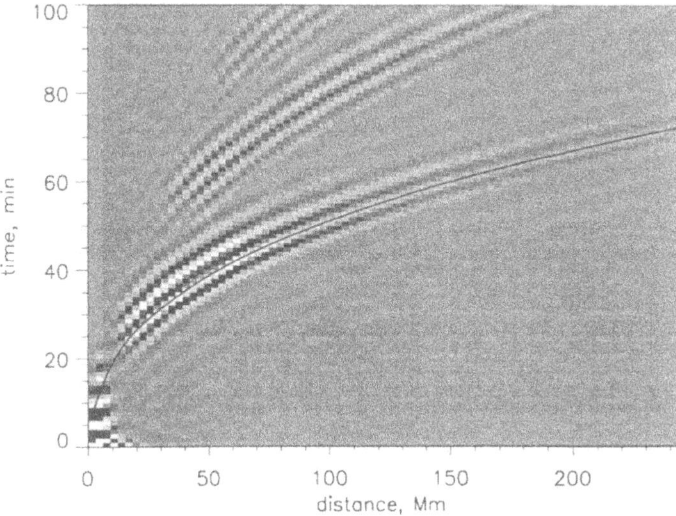

Figure 3. A mean cross-correlation function for the data. The solid line is a theoretical plot of the time for waves to travel along a ray path for the specified great-circle distance. The grayscale picture is the cross-correlation function. The first, second and third bounces are visible near 30 min, 60 min, and 90 min The fine structure in each of the ridges is caused by the finite bandpass of the oscillations.

bandwidth and so could make a zero-width feature in the correlation function. As shown earlier by Duvall *et al.* (1993), the correlation function that we observe is the Fourier transform of the power spectrum (e.g., Figure 2). The remaining features in Figure 3 correspond to the multiple bounces. If there is a correlation observed at time t and at distance d, there will be similar signals at $n \times t$ and $n \times d$, where n is integral. In the figure, the second and third bounces are seen.

For distances shorter than 10 Mm in Figure 3, we see a gap in the signal level and then some features parallel to the abscissa, which appear quite different to the wave packet type of feature present at larger distances. We hypothesize that the falloff of signal below 10 Mm distance is caused by the relative paucity of high spatial-frequency signal in the raw signal (note that the spectrum in Figure 2 has had the high spatial power corrected, while this has not been done to the data in Figure 3). We also hypothesize that the large signal in the neighborhood of the origin is caused by long wavelength features. To get around these problems and to separate adequately the first and second skip at short distances, we have used cross-correlations that have been filtered in phase speed (ω/k) with a fairly narrow bandpass. An example of one of these cross-correlations, covering the distance range 4.7-8.3 Mm, is shown in Figure 4. Our problems with lack of time-distance signal at short distances and spurious signals would appear to be solved.

In some earlier work (Duvall *et al.*, 1993), the correlation function was rectified by using the analytic signal formalism, and times were measured from this signal,

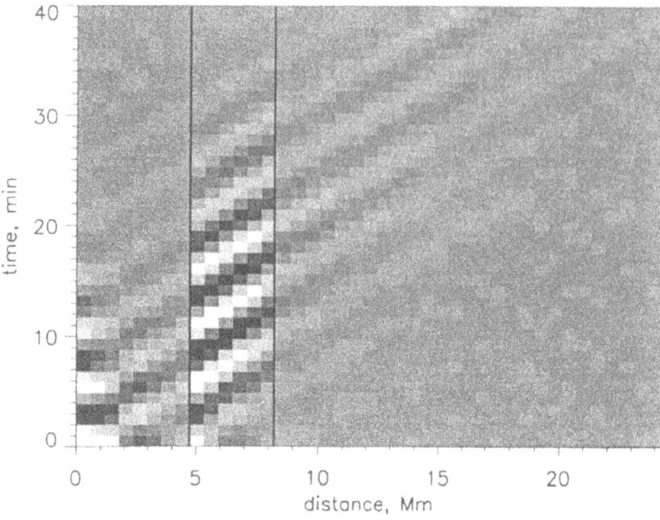

Figure 4. Cross-correlation function after bandpass filtering in phase velocity. The signal in the target distance range is enhanced.

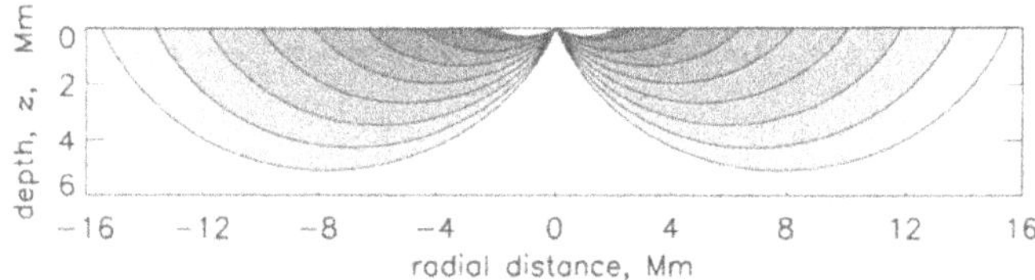

Figure 5. The rays going from the center to the different annuli. The ray paths are curved because of the increasing sound speed with depth. The sound speed increases from 7 km s^{-1} near the surface to 35 km s^{-1} at 10 Mm depth. The horizontal separation between surface points is approximately π times the depth. The measured travel times are sensitive to the component of velocity along the particular ray path.

which is the envelope of the correlation function. Later it was found that the higher frequency structure in the correlation function is actually a more sensitive measure of subsurface inhomogeneities (Duvall *et al.*, 1996a). In that work, the location of one of the fine-structure peaks was determined by measuring the zero crossing of the instantaneous phase of the correlation function. For the present work, this has been further refined (Kosovichev and Duvall, 1996). A 20-min interval in the neighborhood of the peak in the correlation function is fit to a Gaussian wave packet, with independent parameters for the location of the envelope, the location of one of the fine-structure peaks, the amplitude and width of the Gaussian, and the frequency. All the results shown are from the location of the fine-structure peak.

In Figure 5, we show some examples of raypaths below the surface. The ray paths are curved because of the increasing sound speed with depth in the Sun.

A useful rule of thumb is roughly that the horizontal distance between surface reflections is π multiplied by the depth of the turning point of the ray. Also the length along the ray path is four times the depth. These relations are exactly true for a polytrope and approximately true for a real solar model.

Following Duvall *et al.* (1996a), we compute the mean cross-correlation for waves travelling both out from the center to the annuli and the reverse. The difference of the times measured from these cross-correlations is a measure of divergence of the flow. It could be due to either a downflow near the central location or to an average horizontal outflow (or inflow) near the annulus. In addition to this divergence signal, a mean time is computed for the inward and outward waves.

To get more directional information about flows, we divide the annulus surrounding a point into quadrants centered on the four directions north, south, east, west (Duvall *et al.*, 1996b). The signals from the four quadrants are cross-correlated separately with the signal from the central point. The positive and negative temporal lags of these four correlation functions (corresponding to the two directions of propagation) are separated, yielding 8 correlations in all. Pairs of these correspond to each of the four cardinal directions, e.g., waves going from the eastern quadrant to the central point and from the central point to the western quadrant. The average correlation function for each of the four directions is computed from these pairs. The travel times are measured from the first skip from these correlations. Signals proportional to a flow are obtained by taking the difference of eastward and westward times (and likewise the northward and southward times).

In Figure 6 we show the four different sets of times computed for each range of distance: (i) outward–inward, (ii) west–east, (iii) north–south, and (iv) mean time. These 32 images are the input to the inversion described below.

In general, the signals for smaller annuli should be more sensitive to near-surface inhomogeneities. The cellular pattern with the scale size of supergranulation is most apparent in the outward–inward signal. Of course, the rays focus at the center points at the surface, but also reach the surface in the annuli. One interesting thing to notice is the apparent large-scale patterns visible in the signals from the larger annuli. We cannot distinguish directly whether these patterns are due to large scale signals near the surface that are being picked by the annular filter or they are due to larger scale flows near the bottom of the ray paths visible because the rays for the larger annuli penetrate deeper. In principle the inversion is able to distinguish between these options, but it would be useful to have another way to see this. Maybe in future work the signal from the second bounce could be used to make this distinction.

To check that we are seeing something with a basis in reality, we have used the west–east and north–south signals for the smallest annulus to make an apparent horizontal velocity to compare with the mean Doppler image over the 8.5-hour period. The west–east and north–south times are taken as the components of a horizontal vector and the line-of-sight component of this vector is taken. The times are calibrated using the mean sound speed over the depth range of the rays from a solar model. The comparison between these two is shown in Figure 7. The very

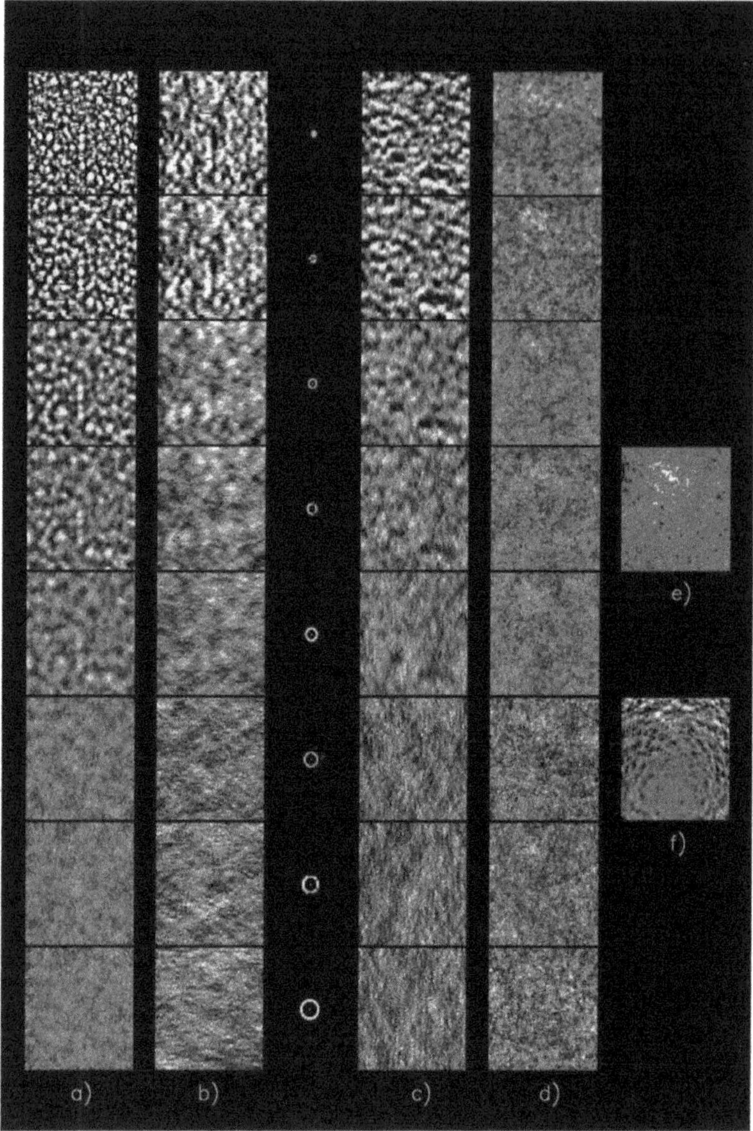

Figure 6. Maps of the times measured and input to the inversion procedure. The 8 pictures vertically are for the 8 different annulus sizes. The sizes of the annuli are shown in the central column, smallest at the top and largest at the bottom. The horizontal size of each image is 370 Mm. (a) Time for outward-going waves minus inward-going waves with white displayed as a negative signal. The r.m.s. signal in the top image is 0.2 min while in the bottom one is 0.04 min. (b) Westward times minus eastward times. The r.m.s. signal in the top image is 0.35 min while in the bottom image is 0.19 min. (c) Northward times minus southward times. The magnitudes are similar to b). (d) The variation in mean time for the average of inward and outward times with a negative signal displayed as white. The r.m.s. of the top image is 0.05 min while for the bottom image it is 0.08 min. A correlation of this signal with the location of the magnetic features can be seen, in agreement with results of Duvall *et al.* (1996a, b). (e) The magnetic field in this region, as observed with the MDI instrument. (f) The mean Doppler signal observed for the 8.5 hour interval. This shows mainly the horizontal motions of supergranules. Near the center of the disk (lower center of picture) the signal from horizontal flows disappears because it is perpendicular to the line of sight. The images in a single column are all on the same scale.

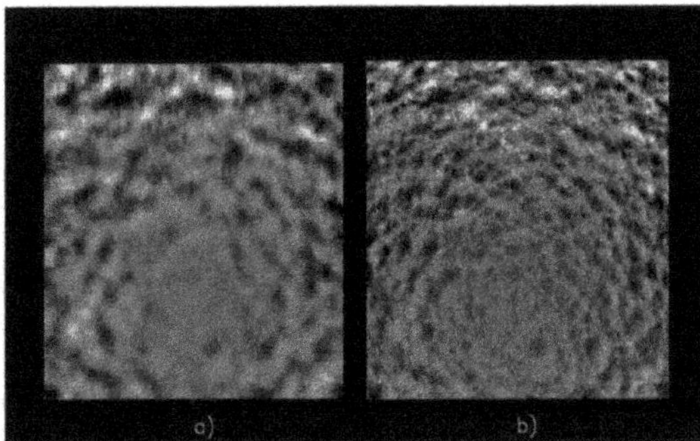

Figure 7. A comparison between a simulated Doppler image using the images from the top line of Figures 6(b) and 6(c) and the mean Doppler image for the 8.5 hours (Figure 6(f)). (a) The simulated image. The east–west and north–south times are combined and then projected onto the line of sight. (b) The mean Doppler image for the 8.5 hours.

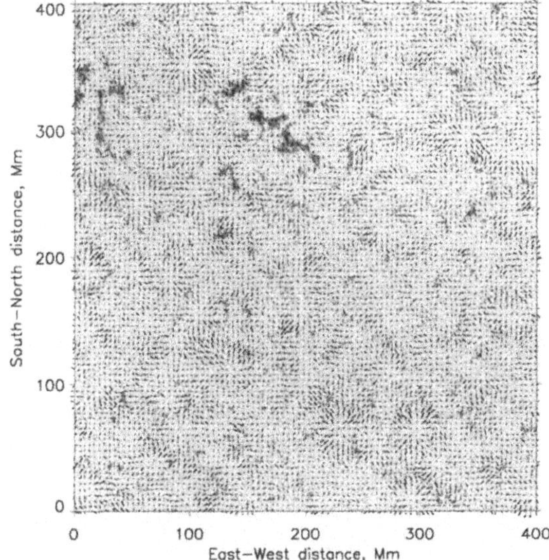

Figure 8. A comparison between the absolute value of magnetic field (in color) and the simulated surface velocity from the time-distance pictures (arrows). The longest arrow is for 1 km s^{-1} velocity.

high correlation is apparent. The correlation coefficient between these two images is 0.74. The signals disappear near the disk center because the supergranulation velocity is predominantly horizontal which is perpendicular to our line of sight at disk center.

Another check on our understanding is shown in Figure 8. The same components of east–west and north– south velocity used in Figure 7 are combined to show horizontal velocity vectors. The vectors are overlaid on an image of the absolute value of magnetic field. We see that the magnetic field tends to lie at the boundaries

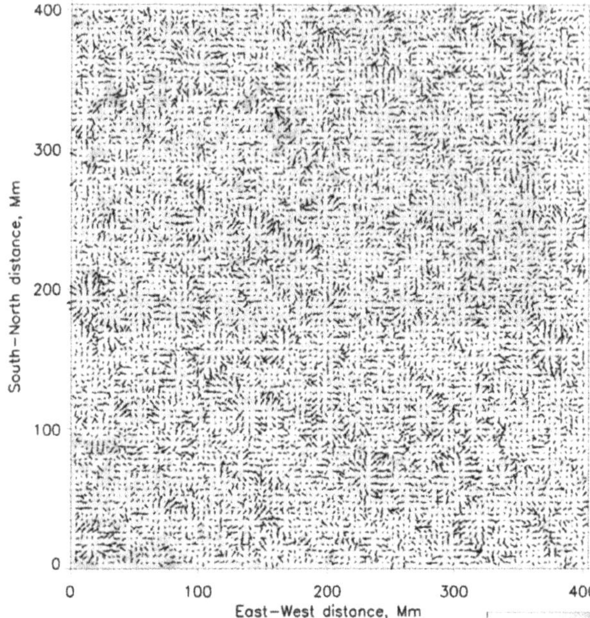

Figure 9. The horizontal flows near the surface from the inversion procedure are shown as vectors. The vectors are overlaid on the inverted temperature structure shown in blue and red for hotter and cooler.

of the cellular outflows, in agreement with our general picture of the magnetic network.

The time – distance maps are used in a ray-theory inversion (Kosovichev, 1996; Kosovichev and Duvall, 1996). The region of the Sun is separated into a 3-D grid of inhomogeneities with the same horizontal sampling interval as the input images (4.3 Mm) and an equal increment in depth covering the same depth range as the input data. The number of grid points in depth is taken to be the same as the number of input images (in this case, 8). Each 3-D grid point has four independent variables, the three components of flow velocity and a sound speed inhomogeneity. No attempt is made to satisfy any conservation equations, such as the continuity equation.

In Figure 9, we show the inversion results for the surface layer. The high correlation with the mean Doppler image is apparent. One advantage of studying near-surface regions is that we do have extra information for comparison at the surface that we would not have for the deep interior.

In Figure 10 we present a vertical cut showing the subsurface flows and sound speed inhomogeneities. It would appear that the pattern of horizontal motions at the surface only persists to a few Megameters in depth. But it is probably still premature to draw strong conclusions because of the newness of the method.

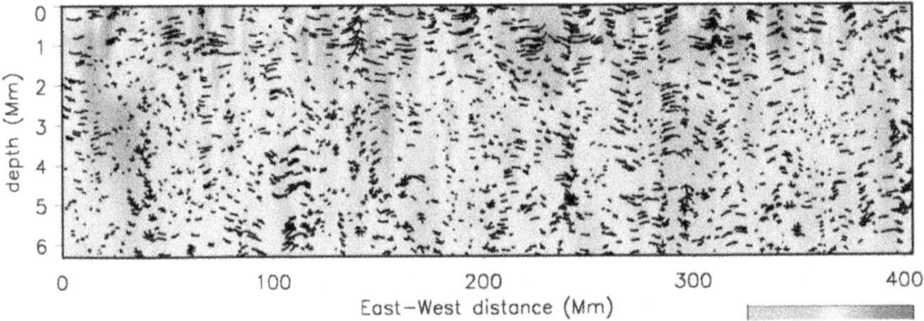

Figure 10. A vertical cut showing the component of flow in the plane. The flow is again shown as vectors and is overlaid on the temperature inhomogeneities.

4. Conclusions

We have shown that there is detailed information about the subsurface inhomogeneities contained in the helioseismology data. We have also shown it should be possible to extract this information and have shown some first-order attempts to do so. We would appear to have an exciting new window into the solar interior.

Acknowledgements

The authors acknowledge many years of effort by the engineering and support staff of the MDI development team at the Lockheed Palo Alto Research Laboratory (now Lockheed-Martin Advanced Technology Center) and the SOI development team at Stanford University. SOHO is a project of international cooperation between ESA and NASA. This research is supported by the SOI-MDI NASA contract NAG5-3077 at Stanford University.

References

D'Silva, S.: 1996, *Astrophys. J.* **469**, 964.
D'Silva, S., Duvall, T. L., Jr., Jefferies, S. M., and Harvey, J. W.: 1996, *Astrophys. J.* **471**, 1030.
Duvall, T. L., Jr., Jefferies, S. M., Harvey, J. W., and Pomerantz, M. A.: 1993, *Nature* **362**, 430.
Duvall, T. L., Jr., D'Silva, S., Jefferies, S. M., Harvey, J. W., and Schou, J.: 1996a, *Nature* **379**, 235.
Duvall, T. L., Jr., Kosovichev, A. G., Scherrer, P. H., and Milford, P. N.: 1996b, *Bull. Am. Astron. Soc.* **188**, 898.
Kosovichev, A. G.: 1996, *Astrophys. J.* **461**, L55.
Kosovichev, A. G. and Duvall, T. L., Jr.: 1996, in J. Christensen-Dalsgaard and F. Pijpers (eds.), 'Solar Convection and Oscillations and their Relationship', *Proc. of SCORe'96 Workshop*, Aarhus, Denmark, May 27–31, 1996, Kluwer Academic Publishers, Dordrecht, Holland, in press.
Scherrer, P. H., Bogart, R. S., Bush, R. I., Hoeksema, J. T., Kosovichev, A. G., Schou, J., Rosenberg, W., Springer, L., Tarbell, T. D., Title, A., Wolfson, C. J., Zayer, I., and the MDI Engineering Team: 1995, *Solar Phys.* **162**, 129.

FIRST RESULTS OF THE SUMER TELESCOPE AND SPECTROMETER ON SOHO

I. *Spectra and Spectroradiometry*

K. WILHELM
Max-Planck-Institut für Aeronomie, D-37189 Katlenburg-Lindau, Germany

P. LEMAIRE
Institut d'Astrophysique Spatiale, Unité Mixte CNRS – Université, Paris XI, Bat 121, F-91405 Orsay, France

W. CURDT, U. SCHÜHLE and E. MARSCH
Max-Planck-Institut für Aeronomie, D-37189 Katlenburg-Lindau, Germany

A. I. POLAND, S. D. JORDAN and R. J. THOMAS
NASA/Goddard Space Flight Center (GSFC), Greenbelt, Maryland, U.S.A.

D. M. HASSLER
High Altitude Observatory/NCAR, Boulder, Colorado, U.S.A.

M. C. E. HUBER
ESA, Space Science Department, ESTEC, NL-2200 AG Noordwijk, The Netherlands

J.-C. VIAL
Institut d'Astrophysique Spatiale, Unité Mixte CNRS – Université, Paris XI, Bat 121, F-91405 Orsay, France

M. KÜHNE
Physikalisch-Technische Bundesanstalt, D-10587 Berlin, Germany

O. H. W. SIEGMUND
The University of California, SSL, Berkeley, California, U.S.A.

A. GABRIEL
Institut d'Astrophysique Spatiale, Unité Mixte CNRS – Université, Paris XI, Bat 121, F-91405 Orsay, France

J. G. TIMOTHY
The University of New Brunswick, Fredericton NB, Canada E3B 5A3

M. GREWING
Astronomisches Institut, D-72076 Tübingen, Germany

U. FELDMAN
Naval Research Laboratory, Washington, D.C., U.S.A.

J. HOLLANDT
Physikalisch-Technische Bundesanstalt, D-10587 Berlin, Germany

P. BREKKE
The University of OSLO, ITA, Oslo, Norway

(Received 6 August, 1996; in revised form 17 September, 1996)

Solar Physics **170:** 75–104, 1997.

Abstract. SUMER – the Solar Ultraviolet Measurements of the Emitted Radiation instrument on the Solar and Heliospheric Observatory (SOHO) – observed its first light on January 24, 1996, and subsequently obtained a detailed spectrum with detector B in the wavelength range from 660 to 1490 Å (in first order) inside and above the limb in the north polar coronal hole. Using detector A of the instrument, this range was later extended to 1610 Å. The second-order spectra of detectors A and B cover 330 to 805 Å and are superimposed on the first-order spectra. Many more features and areas of the Sun and their spectra have been observed since, including coronal holes, polar plumes and active regions. The atoms and ions emitting this radiation exist at temperatures below 2×10^6 K and are thus ideally suited to investigate the solar transition region where the temperature increases from chromospheric to coronal values. SUMER can also be operated in a manner such that it makes images or spectroheliograms of different sizes in selected spectral lines. A detailed line profile with spectral resolution elements between 22 and 45 mÅ is produced for each line at each spatial location along the slit. From the line width, intensity and wavelength position we are able to deduce temperature, density, and velocity of the emitting atoms and ions for each emission line and spatial element in the spectroheliogram. Because of the high spectral resolution and low noise of SUMER, we have been able to detect faint lines not previously observed and, in addition, to determine their spectral profiles. SUMER has already recorded over 2000 extreme ultraviolet emission lines and many identifications have been made on the disk and in the corona.

1. Introduction

The Sun, the central star of our planetary system, has always been of fundamental importance to the Earth and consequently to mankind. It is said that Utu, the Sumerian Sun God, inspired Enmerkar, the lord of Uruk and Kulaba, to invent writing of cuneiform characters on clay. We called our extreme ultraviolet (EUV) solar telescope/spectrometer SUMER – Solar Ultraviolet Measurements of Emitted Radiation – in recognition of the ancient Mesopotamian civilization, which held the Sun in such high regard.

In the late 1960s and early 1970s, a group at Culham Laboratory carried out a series of rocket flights to investigate the physics of the solar chromosphere and transition region covering the 300 – 2950 Å wavelength range with a spectral resolution of about 0.4 Å (Burton, Ridgeley, and Wilson, 1967; Burton and Ridgeley, 1970; Gabriel *et al.*, 1971). As a result of these rocket flights, many intrinsically intense solar lines were identified; however, because of insufficient spectral resolution, line shapes could not be obtained.

Additional observations of the EUV solar spectrum in the range from 280 to 1340 Å were obtained with the Harvard College Observatory (HCO) spectroheliometer S055 on *Skylab* ATM (Apollo Telescope Mount) (Reeves, Huber and Timothy, 1977) and smaller instruments flown earlier on Orbiting Solar Observatories OSO-4 and OSO-6 (cf., Huber *et al.*, 1973). Although the spectral and spatial resolutions of the S055 instrument were only 1.6 Å and 5×5 arc sec^2, respectively, its photoelectric recording permitted the study of time variations (Vernazza *et al.*, 1975; Withbroe *et al.*, 1976) and resulted in a well-established spectroradiometric measurement of solar radiance in this wavelength range (Reeves *et al.*, 1977).

The Naval Research Laboratory (NRL) photographic objective-grating imaging spectrometer S082A on *Skylab* ATM covered the spectrum from 171 to 630 Å with nominally high spectral and spatial resolutions (Tousey *et al.*, 1977). However, the interlaced spectral and spatial imaging did not allow the extraction of reliable line profiles.

The *Skylab* ATM NRL/S082B EUV spectrograph was designed to operate in the 970–3940 Å range (Bartoe *et al.*, 1977). This instrument had a spectral resolution of approximately 50 mÅ at 1500 Å and a 2 × 60 arc sec^2 angular resolution. The optical surfaces were coated with aluminium and a thin layer of MgF_2 which made the instrument very efficient at wavelengths longer than 1150 Å, but very inefficient below 1100 Å. The solar spectrum in the 1150–2000 Å range contains a large number of bright lines, mainly, from plasmas of the chromosphere and the lower-transition region, i.e., at electron temperatures $T_e < 2.5 \times 10^5$ K. However, no line from the upper transition region ($2.5 \times 10^5 < T_e < 1 \times 10^6$ K) was observed and only a small number of forbidden lines originating at coronal temperatures appeared in the spectra. During its operation, the instrument recorded spectra from disk features and from regions located slightly above the limb. Because of the fairly low efficiency of the photographic plates used, few if any spectra were obtained from regions that extended more than 20 arc sec above the limb. Results from the instrument were published in a large number of scientific papers. An extensive list of references to the results is given in the book 'The Solar Transition Region' (Mariska, 1992).

The NRL/High Resolution Telescope and Spectrograph (HRTS) which also operated at wavelengths longer than 1150 Å combined both high spectral resolution (50 mÅ) and high spatial resolution (1 arc sec) (Bartoe and Brueckner, 1975; Brueckner *et al.*, 1986). It was launched repeatedly on suborbital rockets and also on the Spacelab-2 mission, and provided important information on lower transition-region properties.

A University of Colorado 1.2 m EUV sounding rocket spectrometer obtained high spectral resolution EUV line profiles in the wavelength range 610–630 Å, providing observations of systematic radial flows in the transition region and corona, including early evidence for blueshifts or net radial outflows in coronal holes (Rottman, Orrall and Klimchuk, 1982; Hassler *et al.*, 1990; Hassler, Rottman, and Orrall, 1991).

Until the launch of SOHO, only limited spectral line profiles were available for the disk or features above the limb between 450 and 1100 Å. Exceptions are the profiles obtained for various solar features in H I Lα and Lβ (cf., Gouttebroze *et al.*, 1978; Lemaire *et al.*, 1981; Vial, 1982) and profiles of O VI 1032 Å recorded with high spectral resolution by the Laboratoire de Physique Stellaire et Planétaire (LPSP) polychromator on OSO-8 (Vial *et al.*, 1980). The situation changed dramatically when spectra from SUMER became available, and high spatial and spectral resolution spectroheliograms in this wavelength range were obtained routinely on the disk and in the corona.

SUMER is part of SOHO's payload, i.e., of the ESA/NASA Solar and Heliospheric Observatory. SOHO was launched on December 2, 1995 by an Atlas II A

Centaur into a transfer trajectory to the first Lagrange point L1. It was injected into a halo orbit around L1 on February 14, 1996, where, in constant view of the Sun, it accompanies the Earth at a sunward distance of 1.5×10^6 km.

Two of the main scientific goals of SOHO are to understand coronal heating and to determine the mechanism(s) for acceleration of the solar wind. Of particular importance to the coronal heating problem is the rôle of the transition region between the 'million degree' corona and the much cooler chromosphere, in which temperatures are typically two orders of magnitude lower. The temperature gradient in the transition region must be very large, as observations have already shown it to be quite thin, in some places less than one thousand kilometres. Thus the transition region is the site of rapid energy flow by conduction down as well as mass flow both up and down, with a small net upward component necessary to balance the loss due to the solar wind. In addition, the energy that heats the corona must pass through the transition region from below. Because of the dynamic nature of the transition region, and the small spatial scale over which physical conditions are changing, this critical region has not yet been adequately studied, leaving a serious gap in information necessary to study both coronal heating and the flow of mass to the solar wind.

In this paper and its companion (Lemaire *et al.*, 1997) we present the first results of SUMER obtained during the first months of the SOHO mission. Emphasis is put on the scientific achievements, but some of the technical performance characteristics are also discussed in order to put the observations into their right perspective.

SUMER offers for the first time high spatial, spectral and temporal resolution over a large wavelength range ideal for transition region studies. Moreover, this instrument is scheduled to operate for an extended period. In these two papers we show that observations already obtained give promise of yielding models for the transition region and many of the physical processes occurring there that will fill in many gaps in our knowledge. This knowledge in turn should lead directly to better models for the coronal heating process(es) and mass flow to the solar wind.

Most of the observations discussed here have been made during the technical commissioning phase of SOHO and SUMER, and only a limited amount of scientific data analysis could be done before this first presentation. It is felt, however, that an early publication is justified for at least two reasons:

(1) The inflight performance characteristics of SUMER should be demonstrated by examples. This will allow the scientific community to consider and appreciate the SUMER capabilities of doing extensive studies of the solar atmosphere and to propose suitable observations to be made during the expected lifetime of SOHO and SUMER.

(2) Although, in general, SUMER performs very well in nearly every respect, there are a few areas in which the inflight performance of SUMER deviates from expectations. For the design of elaborate observation programmes it is important to know these differences. Their impact on potential observations, and strategies to minimize their effects, therefore, have to be described here.

In this Paper I, we mainly concentrate on the spectroscopic capabilities of the instrument and in Paper II (Lemaire *et al.*, 1997) we describe its imaging aspects. It is, however, important to point out that the combination of good spectral and spatial resolution is essential for the study of the solar atmosphere and its dynamics. The optical, mechanical as well as the electronics and the software aspects of SUMER have already been presented by Wilhelm *et al.* (1995a, b). They will not be repeated unless reference is made to a specific item. The instrument descriptions given earlier reflect the status at the end of the development, calibration and test phases, and performance predictions given at the time were based on that status. This report updates the earlier instrument descriptions and covers the observation period until June 3, 1996.

2. The SUMER Spectrometer

2.1. Grating Focus Adjustment

Focusing of the SUMER instrument under operational conditions is done in two separate stages. First, the spectrometer entrance slit is adjusted to the position of best telescope focus. Since this also changes the separation between slit and collimating mirror, a second adjustment inside the spectrometer is then required, which is accomplished by an offset of the grating position to optimize final imaging on the detector. Both settings are periodically checked and optimized if necessary.

The adjustment of the slit to the best telescope focus will be described in Paper II. In the grating focus procedure, a 1×1 arc sec^2 pinhole is used as spectrometer entrance aperture and the spherical-concave grating is moved through various axial offset locations on either side of its nominal position for each of nine strong solar emission lines that cover the full first-order spectral range. The spatial widths of all resulting images recorded by detector A (cf., Section 2.3) are then measured, and quadratic fits give the location of best grating offset for each of the nine wavelengths. In Figure 1, we give as an example the result at 1032 Å. These offsets do not vary with wavelength, as can be seen in Figure 2, so that a weighted mean of the entire set can be used to determine the overall best offset value to within 100 μm. This corresponds to an image blur of less than 0.1 arc sec in the spatial direction, and less than 4 mÅ (first order) in the spectral direction, due to inaccuracies in positioning of the grating. In addition, Figure 1 and the corresponding measurements at other wavelengths demonstrate that the SUMER spectrometer can image a 1 arc sec structure in the focal plane of the telescope with a width of better than 2.3 px (px: pixel, picture element) or 2.3 arc sec full width at half maximum (FWHM) on the detector over nearly all of its spectral range. The spatial resolution performance is detailed in Paper II.

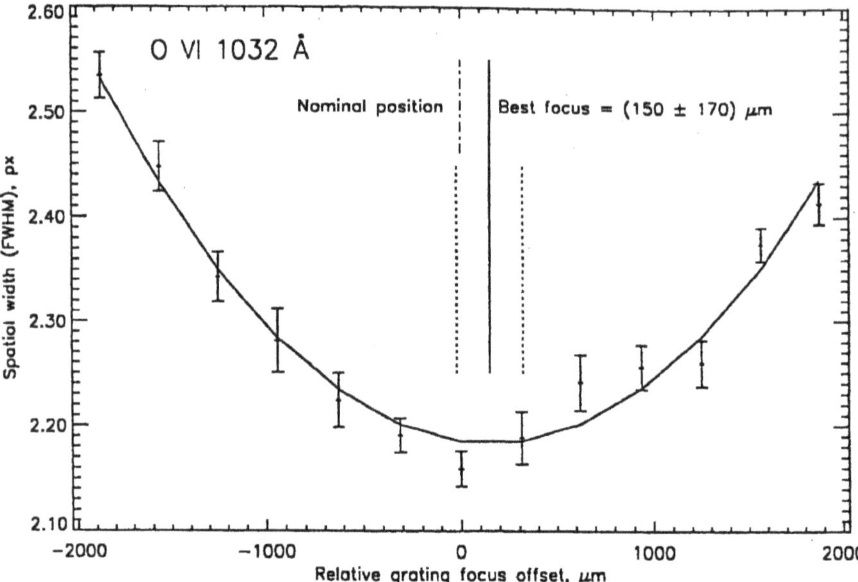

Figure 1. Determination of the best focus of the SUMER spectrometer at a wavelength of 1032 Å. The nominal position indicates the pre-launch configuration. One pixel corresponds to an angular size of 1.01 arc sec at 1032 Å. All errors shown are 1σ values.

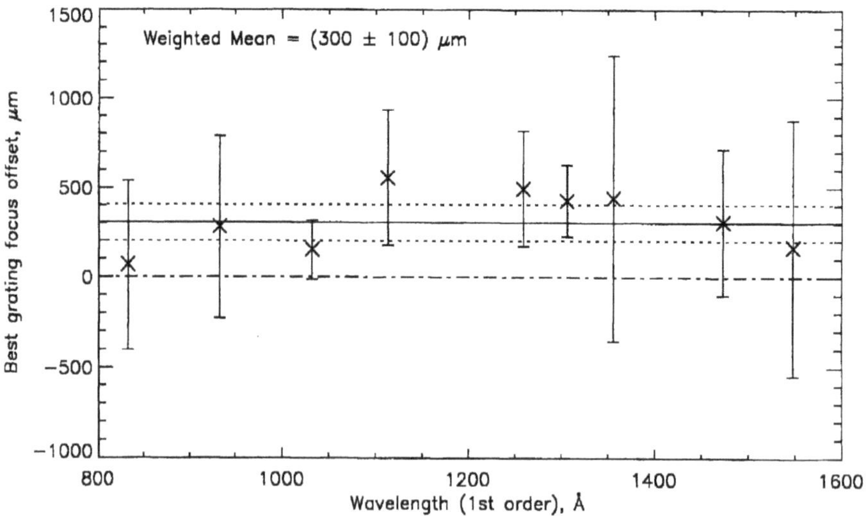

Figure 2. Summary of the first grating focus offset measurements at nine wavelengths. The error bars show 1σ values. The offset (weighted mean) so determined gave the adjustment needed for the grating focus mechanism. The difference between the best focus positions on the ground and in-flight was 300 μm.

2.2. Wavelength Selection

The wavelength calibration of the spectrometer was performed before launch with laboratory calibration devices. For wavelengths above 1100 Å, the spectrum of a Pt – Ne-lamp was used, and for shorter wavelengths, single reference lines were employed, which were emitted from a differentially-pumped, open-discharge source operating on noble gases. However, the change of the in-flight-operation temperature of the instrument (cf., Section 2.7) made it necessary to revise the parameters of this calibration. Making use of many solar lines and the perfectly stable thermal conditions at L1, we have restored the onboard wavelength calibration such that a spectral line position commanded is now accurately placed within a 25 px window. Consequently, any EUV line in the SUMER range can be routinely observed inside this smallest spectral format provided by the instrument software (corresponding to approximately 1.1 Å in first order).

2.3. Detectors

Both detectors A and B have been turned on during the commissioning phase, and their performance with and without UV illumination has been verified during the first days of their operation. The operating parameters have been found to be unchanged with respect to the values used during ground tests, and full spectral scans have been performed with both detectors, which established for the first time the wavelength range capabilities of the SUMER spectrometer under flight conditions. Only one detector can be switched on at a time. Detector A has mainly been used since the initial commissioning, and detector B is kept as a backup. During the first four months of continuous operation (with short interruptions during spacecraft manoeuvers) a few anomalies have been found, but the performance generally conforms to the descriptions in our previous publications (Siegmund *et al.*, 1994, 1995; Wilhelm *et al.*, 1995a). The main limitation found was a reduction of local gain when bright solar lines are observed. With lines such as C III 977 Å or O VI 1032 Å, the counting rates at certain detector locations can be much higher than 10 count s^{-1} px^{-1}, leading to a local reduction in the microchannel plate (MCP) gain and, therefore, loss of dynamic range and radiometric calibration for these particular lines. In addition, long exposures of such bright lines at the same pixel location on the detector will leave an imprint of reduced gain after a certain amount of total counts (total charge) has been extracted. Both effects, the reduction of detector gain when the local counting rate is high, and the overall reduction of gain through progressive extraction of charge by the total number of counts, are well known for MCPs and are discussed below.

Because the capability of the MCP to process events is finite, the gain per detected photon gradually decreases as the event rate in any spectral line exceeds a certain level (Siegmund *et al.*, 1994). This instantaneous (and recoverable) gain loss becomes a problem when the rate is so high that the reduction in gain results

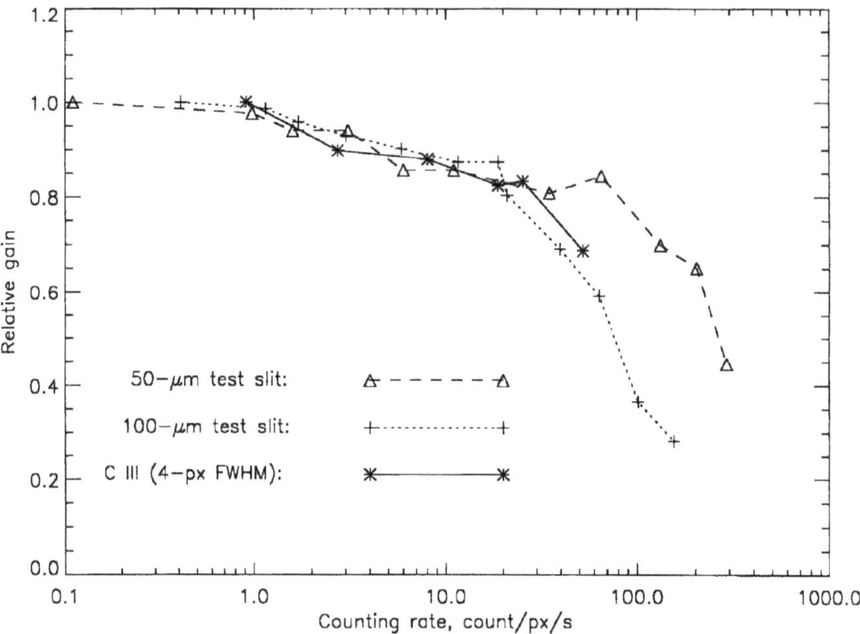

Figure 3. Laboratory and in-flight measurements demonstrating the effect of local gain depression of the MCP of the SUMER detectors. The relative gain is shown versus the counting rate with the test slit width as parameter for the laboratory data. The 100 μm slit corresponds to a line width of four pixels and is thus representative of the C III line at 977 Å used for the in-flight data.

in events falling below the lower level threshold and being lost. In experiments with the SUMER testbed detector in the laboratory and with the flight model on board SOHO, we have quantified the gain decrease as a function of counting rate and spectral line width. The results are plotted in Figure 3 and give an indication of the local gain depression effect. There are several ways to ensure that the local counting rate is not exceedingly high: besides a careful selection of targets on the Sun, a reduced slit size of 0.3×120 arc sec^2 can be chosen, and lines can be placed on the, usually less sensitive, bare part of the detector. When planning observation programmes for SUMER, a software simulation scheme must therefore be used to perform photometric evaluations and predict counting rates to be expected under various solar conditions.

The progressive (and non-recoverable) reduction of gain as the overall charge extraction increases is due to outgassing of the microchannel walls and chemical changes in the MCP active layers (Siegmund, 1989). The effect may be controlled by a commonly used 'scrubbing' process, where a certain amount of charge is extracted to stabilize the MCP performance. The SUMER data processing unit keeps in its memory the total number of counts per pixel location (binned by a factor of 2 in the spectral range and 20 in the spatial dimension) so that, at any time, the history of the count distribution on the whole sensitive area of both detectors can be evaluated, even if sections of the detector area have not been read out in the science data channel. The operational significance is that the observations are scheduled so that each detector as a whole is on the average 'evenly' exposed by placing spectral lines at different positions. In Figure 4, we show such a 'history memory' image for detector A, integrated up to day 130 after the start of detector operation. Based on the distribution obtained, we have, for instance, decided to engage the asymmetric positions of the short slits more often and to use, in particular, the narrow slit in its upper (slit No. 6, large spatial pixel number) or lower (slit No. 8) position for active region studies. Occasionally the MCP operating high voltage must be raised to compensate for gain losses (cf., Figure 6(b)). The latter can immediately be recognized by monitoring the pulse-height distribution of the MCP output in the data stream. In this manner the overall radiometric calibration can be restored and conserved (cf., Section 2.5). It is anticipated that this progressive gain reduction will eventually cease as the MCPs stabilize. This is indeed suggested by the measurements shown in Figure 6(b).

Only the overall sensitivity of the instrument has been radiometrically calibrated. The sensitivity of the detector active area is, however, nonuniform on scales of about 20 px or less. In particular, when compression schemes are used in order to reduce the telemetry load, it is recommended to compensate for these nonuniformities by employing an onboard flat-field correction routine. The flat-field data are being accumulated during a three-hour exposure in the H I Lyman continuum between 860 and 900 Å while the spectrometer grating is defocused. This provides a deep, although not entirely 'flat', exposure from which the SUMER processor extracts all those features which are smaller than 20 px. This flat-field image is available on board and is also transmitted to the ground. The small-scale variations amount to as much as 50%. This includes a pixel-to-pixel variation of approximately 20% in the spatial dimension caused by an analogue-to-digital converter differential non-linearity (electronic bin-width variation).

It is necessary to apply a flat-field correction either on board or on the ground to every image meant for intensity measurements. The variations reduce slightly with time as the detector is being 'scrubbed' (see above), and therefore the flat-field data need to be updated quite frequently. The option of a specific flat-field approach should also be mentioned. This requires a raster scan of the line(s) to be observed in a quiet-Sun region and the availability of a statistically homogeneous observation area (to be checked by inspecting the data obtained).

The fringe fields in the detector MCP-anode gap lead to a geometric distortion of the image of 'inverse cushion' type which is most pronounced at the extremities of the image. We have produced artificial 'rectangular grid' images, using averaged stable solar spectral lines and continuum data, from which the distortion can be deduced and then applied to subsequent images to correct for this shortcoming. In Figure 5, we show a full detector spectrum centred near 920 Å, which has been flat-field and geometric-distortion corrected. After this correction the continuum

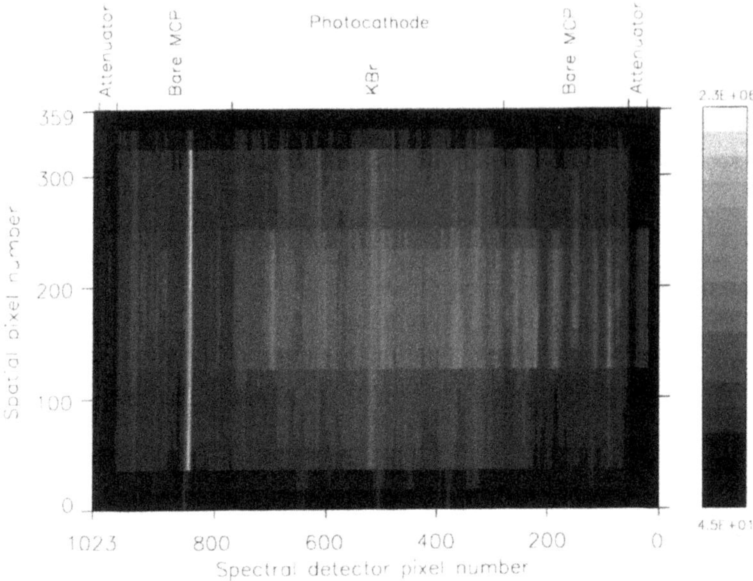

Figure 4. Diagram showing the 'history memory' of detector A after 130 days of operation. The KBr section of the photocathode from spectral detector pixel number 280 to 770 is clearly visible and so is the central position of the short slits. The traces of a slit position adjustment along the spatial dimension can also be seen. The right attenuator covers pixels 23 to 56 and the left one 982 to 1016. Pixels below 23 and above 1016 are dark and cannot be used. The deep exposure at pixel number 850 stems from the first full-Sun image in C III 977 Å. The total number of counts (including the ground test exposure of 7.2×10^9) of this detector at this day was 1.07×10^{11}. The total counts of detector B, collected during the ground tests and early commissioning activities, amount to 4.25×10^9 at this stage.

appears to be smooth in the dispersion direction and the lines are straight to within one spectral pixel. The intensity variations along the spatial direction reflect true brightness changes related to the solar network structure intersected by the slit. Members of the Lyman series, but also C II, N IV, and S VI lines, are very prominent in this segment of the SUMER spectrum (cf., Table I).

2.4. Background and Scatter Performance of the Spectrometer

The contribution of background counts to the total signal is extremely small. This virtue of the detector allows us to make deep exposures and observe very faint spectral lines. The background counting rate of detector A had been measured during ground tests and a value of approximately 2×10^{-5} count s^{-1} px^{-1} was determined. After several months of operation, it could be established that this value has not changed significantly. When the SUMER aperture door is closed, only visible light is admitted to the instrument (cf., Section 2.5). With long exposures of three-hour duration at several wavelengths, we could verify that the sensitivity

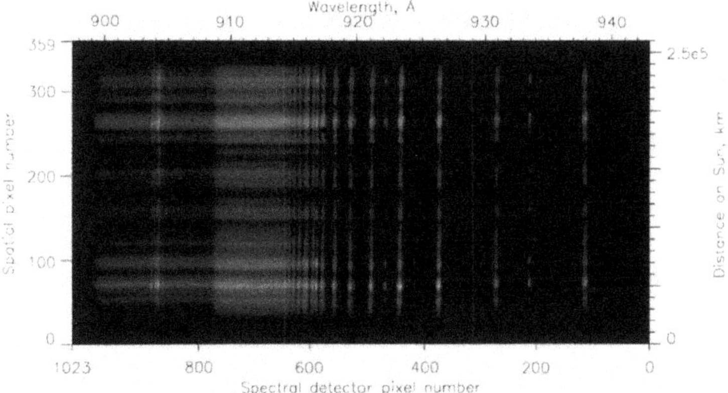

Figure 5. The H I Lyman edge and the high members of the Lyman series (up to 20) are depicted on the full area of detector A with 1024 spectral and 360 spatial pixels. Only 300 px are illuminated by the slit in the spatial dimension. Before transmission to ground a 2 to 1 byte data compression had been applied. It then takes about 5 min to transmit such a full frame detector spectrum of approximately 45 Å (first order). Smaller windows can be selected for higher temporal resolution requirements. The data have been de-compressed on the ground and the diagram has been corrected for geometric distortions and for flat-field variations of the detector. Note that the KBr section of the detector is more sensitive in this wavelength range by a factor of about 2. This causes the apparent edge at pixel number 770 and an attenuation of the lines at pixel numbers smaller than 280. The largest detector defect can also be seen near the pixel coordinates (575, 250).

to visible light is negligible and that the background signal is very uniform across the detector area at a rate of 3.5×10^{-5} count s^{-1} px^{-1}.

With respect to the spectrometer scatter characteristics, a preliminary evaluation of H I Lα observations on May 6, 1996 confirmed our ground test results (Wilhelm *et al.*, 1995a). With H I Lα on the attenuator, we found a mean scatter contribution of 3.1×10^{-9} px^{-1} of the total line counting rate (normalized to KBr without attenuator). This has to be compared with a figure of 2×10^{-9} px^{-1} obtained during ground tests taking into account that the illumination of the attenuator must be considered to be a worst case scenario as far as scattering is concerned.

2.5. Radiometric Calibration and its Medium-Term Stability

A careful radiometric calibration of the SUMER flight model based on a secondary-source standard traceable to an electron-storage ring primary standard had been accomplished before launch (Hollandt *et al.*, 1996a, b). The stability of such a radiometric calibration has been a major concern from the beginning of the project, because of the experiences with previous solar UV missions, which suffered from serious sensitivity losses during their operations under solar irradiation. In particular, normal-incidence UV optical components are very susceptible to this effect (Huber *et al.*, 1973; Reeves *et al.*, 1977; Lemaire, 1991). The degradation is attributed to contamination by organic substances, which are constantly outgassing from parts and materials used in the instrument construction and are subsequently photoactivated and deposited on irradiated surfaces through polymerization. The process, although now well understood, is very difficult to prevent in complex optical instruments for space research. A very stringent cleanliness control programme was therefore set into effect from the start of the design phase of SUMER (Schühle, 1993), including special design features of the optical housing, material selection specifications as well as cleaning and clean-room procedures during assembly and tests.

As major differences to previous programmes we cite the telescope design (which avoids illumination of any reflecting surface by more than one solar constant), and the extremely limited usage of organic materials. Any organic compound was subjected to a thorough examination of its outgassing behaviour under vacuum, before being approved for use in the SUMER instrument. In addition, flight hardware was intensively cleaned and baked out during assembly. Also included in SUMER is an electrostatic solar-wind deflector, because it had been demonstrated in tests that solar-wind protons would degrade the optical performance of the primary silicon-carbide telescope mirror and, in addition, contribute to the activation of hydrocarbons and thus increase the polymerization process. These measures constituted a substantial part of the technical investment for the production of the instrument.

The entrance door played an important rôle in the cleanliness concept of SUMER as well. Until launch, SUMER was purged by the dry nitrogen (GN_2) distribution system of SOHO. The door was set ajar on mission day 7 to facilitate the removal of any adsorbed and occluded gases. The internal optical compartment was then held at a temperature close to 30 °C for 27 days before UV light was admitted to the telescope mirror. Through a window in the door, which was opaque to UV light, this mirror was illuminated by visible and infrared light from the first day of the mission and was thus held at a temperature approximately 25 K above its environment. This provision also helped in avoiding any condensation on this critical item. Note that under operational conditions the mirror is approximately 55 K warmer than the bulk of the instrument. Both detectors were initially commissioned before the door was completely open and dark images were obtained in this configuration. The door has also been closed during all spacecraft manoeuvres.

The sensitivity of the instrument has been monitored from the moment when first UV light was impinging on the telescope and the other optical elements including the detectors. After the entrance aperture had been fully opened, the counting rates for well-known solar spectral lines were measured for quiet solar conditions and found to be just as predicted by the calibration made on the ground and previous measurements (Vernazza and Reeves, 1978).

During several months of almost uninterrupted operation, at regular time intervals, long exposures of three-hours duration have been made to derive data for the flat-field correction function of detector A. This also resulted in a good tracking of the calibration during this period: it provided deep exposures in the H I Lyman continuum near 880 Å at quiet regions near Sun centre.

All measurements have been performed by exactly the same observation sequence. Although even the quiet Sun is highly variable at these wavelengths, the three-hour average over a 40 Å bandpass of the H I Lyman continuum with no prominent spectral lines is a good indicator for the sensitivity of the instrument. To transfer the calibration to other wavelengths, a full spectral scan is also made on a regular basis in quiet Sun areas.

The counting rates observed during the flat-field exposures are presented in Figure 6 together with other trend data concerning detector A. Based on the solid curve in panel (a), we can state that the counting rates (total detector) of the flat-field exposures have not varied by more than $\pm 15\%$ of the mean value and no trend can be detected during the first 110 days of operation. The variations observed thus have to be attributed to changes of the solar brightness in this band, which are by no means surprising (cf., Lean, 1991). This result, being much better than anticipated, is a great success and justifies all the efforts made for the cleanliness of the instrument and the pre-launch calibration. It will have to be confirmed by calibration rockets and compared with other relevant SOHO instruments.

We have also determined two specific features of the instrument:

(1) The spectrometer efficiency ratio in first and second orders for detector A could be measured to be $r_g = 1.66$ near 790 Å, by looking at the same lines (O IV) in either order at opposite ends of the spectrum and comparing the total counts in the line spectra, i.e., $r_g = I_{1st}/I_{2nd}$.

(2) The ground calibration was accomplished with a beam of UV light smaller than the SUMER aperture (Hollandt *et al.*, 1996b). Although both azimuth and elevation scans were performed, it could not be established beyond any doubt that the nominal SUMER aperture was indeed available and no vignetting was present. Solar H I Lα observations with attenuator now allow us to verify that there is indeed no vignetting in any section of the SUMER optical path. The design of the attenuator (Wilhelm *et al.*, 1995a) would lead to a signal modulation on the detector if the full aperture were not available. We do not find such a modulation.

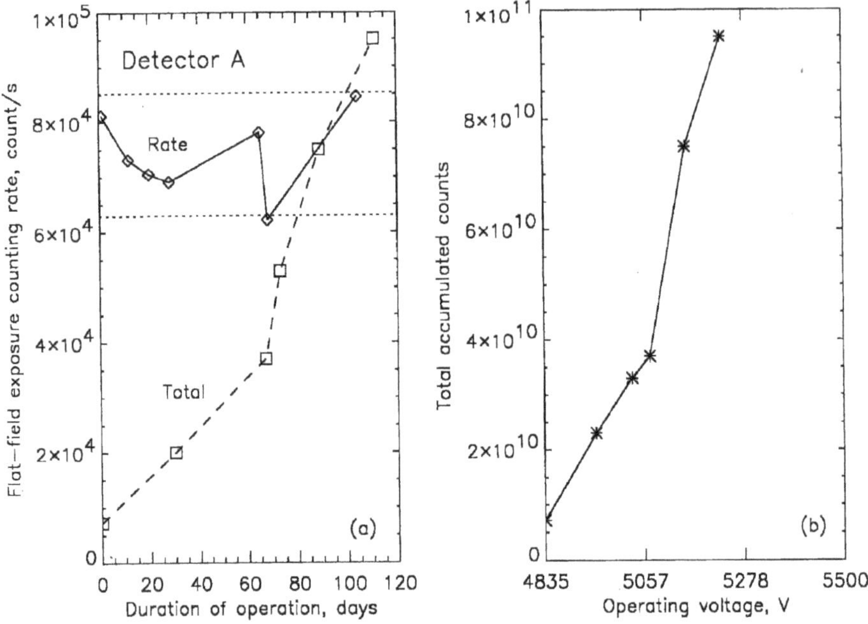

Figure 6. Radiometric and detector trend information is plotted versus the in-flight operation time of detector A in (a). The solid curve gives the counting rate of the whole detector area for several flat-field exposures. Also indicated are the levels of the mean value $\pm 15\%$. The dashed curve shows the total accumulated counts of the detector (calculated from history memory data) on the same scale as (b). During ground tests 7.2×10^9 counts were accumulated. The plot in the right panel (b) indicates the high-voltage levels required to restore the pulse-height distribution (cf., Section 2.3) after the accumulation of the corresponding total count levels. Operating voltages up to 5500 V will be available.

2.6. Spectral Resolution

Many extremely narrow spectral lines could be observed in the solar spectra by SUMER. Most of them are emitted by neutral species in the cool chromosphere. The spectral resolution performance of SUMER can be deduced from the profiles of such lines. The narrowest lines detected up to now (O I 929.517 Å and S I 1300.91 Å) had a width of 2 px (FWHM) or 86 mÅ in first order with the 1 arc sec slits. This corresponds to a Doppler width of $\Delta\lambda_D = 52$ mÅ (first order). If – in the most unfavourable case – we assume that the latter value corresponds to the instrumental line width, $\Delta\lambda_I$, then its influence can be predicted to lie below 10% for all wider lines with $\Delta\lambda_D > 115$ mÅ. Note also that H I Ly 19 and Ly 20 ($\Delta\lambda = 219$ mÅ; cf., Table I) as well as the C II lines at 903.9616 and 904.1416 Å ($\Delta\lambda = 180$ mÅ) can be resolved without difficulty.

SUMER's spectral resolution is thus adequate to provide line profile information even on lines that are emitted in non-turbulent chromospheric regions. The shift of

2 km s^{-1} induced by the solar rotation is easily detected in most narrow lines. This is very encouraging and demonstrates the capability of the instrument for making reliable measurements of chromospheric and transition region bulk velocities and oscillations.

2.7. Thermal control system

The instrument was switched on after the launch of SOHO for the first time on December 9, 1995 (with the exception of the high-voltage circuits). Soon it became apparent that the thermal control system, designed to maintain the SUMER optical bench at a nominal temperature of 20 °C, suffered from an unexpectedly high performance of the flight-model, multi-layer insulation (MLI) of the telescope/spectrometer unit. (For cleanliness reasons, both the thermal-balance and thermal-vacuum tests were conducted with the engineering model of the MLI.) Not only was the equilibrium temperature several degrees above the nominal control temperature, but also a gradient of several degrees developed along the optical axis of the instrument. Together with the fact that the pointing and alignment of SUMER varied by nearly 25 arc sec for a gradient of 1 K along the optical bench, this thermal problem was a serious threat to the ability of our instrument to do high-resolution measurements. By reducing the internal power consumption to the absolute minimum, by shading a portion of the front shield of SUMER with the help of the movable entrance door, and by increasing the control temperature from 20 to 23 °C, it was possible to bring SUMER into a stable thermal configuration even at perihelion. Since the beginning of March 1996, the temperature sensors of the telescope/spectrometer compartment are controlled to within ± 0.05 K, and no further drift of the optical system caused by thermal effect has been observed so far.

2.8. Rear slit camera (RSC)

While the entrance door with its optical window was still in its partially open position (cf., Section 2.5), the first observations with the Rear Slit Camera (RSC) were performed. An alignment error (that had not been eliminated owing to schedule pressures during the final ground tests) caused the slit image in visible light to miss the photosensor of the RSC by 0.8 mm. However, it was found that a yet unexplained internal reflection could be used to image at least the 1×120 arc sec^2 slit in its offset position (slit No. 3), albeit with reduced sensitivity. By employing the longest sampling time of 206 ms available, determinations of the solar limb and observations of sunspots for alignment purposes, as well as focus checks of the telescope, can still be accomplished as planned (cf., Paper II).

3. SUMER Spectra of Solar Features

Spectral catalogues of different solar features have been derived from scans over the entire wavelength range of SUMER. They include emissions from neutrals and ions of various elements in the temperature range from $T_e = 1 \times 10^4$ to 2×10^6 K, i.e., emission lines and continua emitted from the lower chromosphere to the corona. The broad wavelength coverage of SUMER provides important new plasma diagnostics and means to deduce essential physical parameters of the solar atmosphere. The catalogues can also be used for planning purposes: to determine adequate integration times, identify possible blends, and to select proper data extraction windows in future SUMER studies.

The wavelength ranges covered are 660 to 1610 Å in first order and 330 to 805 Å in second order. (Below 500 Å the normal-incidence optics are very inefficient, yet the line with the shortest wavelength identified so far is Ne VII at 465.219 Å.) The spectra presented in this and the following sections are averaged along a portion of slit, while being pointed to typical solar features. The entire spectral range covered is shown in Figure 7 as a composite spectrum using data of detectors A and B. Note that first- and second-order spectra are superimposed. The full spectral range was put together from a number of exposures each covering approximately 20 Å in first order on the KBr coated, and therefore most sensitive, part of the detector (with the exception of wavelength ranges near H I Lα and 1610 Å). No radiometric calibration of the data has been applied in this figure, as the separation of the first- and second-order spectra (making use of the different responses of the KBr and bare portions of the MCPs to short and long wavelengths) is a complicated and time-consuming process. For each identified line, however, the radiometric calibration of SUMER (Wilhelm *et al.*, 1995a; Hollandt *et al.*, 1996b) provides a provisional estimate, which is expected to be quite accurate, given the calibration stability demonstrated in Section 2.5. The panels in Figure 8 show a selection of detailed spectra for various solar regions and demonstrate the resolving power of the instrument. The spectra were observed in a quiet-Sun region, near the limb, and in a coronal streamer 2 arc min above the limb. Note that the H I Lyman series of the limb spectrum shows self-absorption for members up to at least Ly 12. It should also be pointed out that many narrow and cool O I lines in the disk and limb spectra are essentially absent in the streamer spectrum.

3.1. Line Identification

Extensive line lists derived from spectra obtained by S082B and by HRTS were published mostly for the 1175–2000 Å range (Sandlin *et al.*, 1986; Cohen, Feldman, and Doschek, 1978). The line lists for the wavelength range between 600 and 1175 Å derived from high-resolution spectra are more scarce and in most cases contain only the prominent lines. The best of the line lists for this range are those of Burton and Ridgeley (1970) and of Feldman and Doschek (1991).

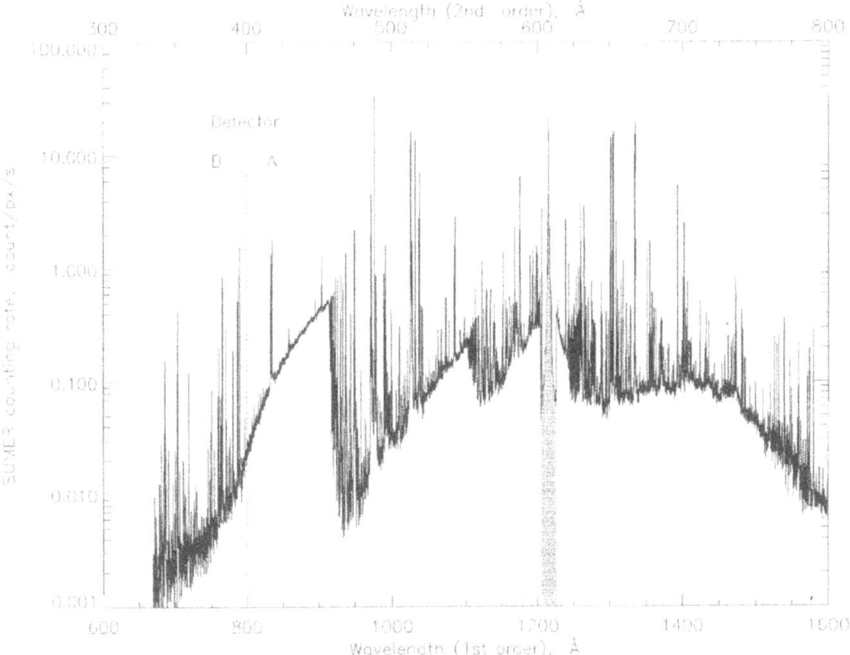

Figure 7. Solar spectrum covering the range from 660 to 1600 Å in first order with the second-order spectrum from 330 to 800 Å superimposed. The range from 660 to 780 Å represents a limb spectrum obtained with detector B and the 0.3 arc sec slit. The long-wavelength portion was observed with detector A and with a slit of 1 arc sec near Sun centre in a quiet region. SUMER counting rates are given for the KBr photocathode, except for the neighbourhood of H I Lα and near the upper and lower limits, where observations are only possible on the bare MCP. The H I Lα line itself is, in addition, reduced by a factor of 10 by the attenuator. No adjustments have been performed and actual counting rates are provided for a 1 arc sec slit (the 0.3 arc sec slit measurements have been normalized to 1 arc sec).

The spectra acquired by SUMER are a significant improvement over those recorded in the past: they are much richer than previously published spectra and contain lines emitted by neutrals and many ions from a large number of elements. Since the Sun is fairly quiet during the current phase of the solar cycle, the recorded lines are those emitted from ions abundant in plasma with $T_e < 1.5 \times 10^6$ K. It is expected that as we move towards solar maximum in the coming six years, lines formed at higher temperatures will appear in the SUMER spectrum as well. Preliminary 'working line lists' from spectra recorded inside the limb and above the limb were compiled. The list covering the disk observations contains wavelengths and peak intensities of over 500 spectral lines in the 680 – 1175 Å wavelength range. The list with lines above the limb indicates many more lines, of which only a fraction could be identified so far. To identify the rest of the lines, a significant amount of work needs to be done and more basic atomic physics data may be

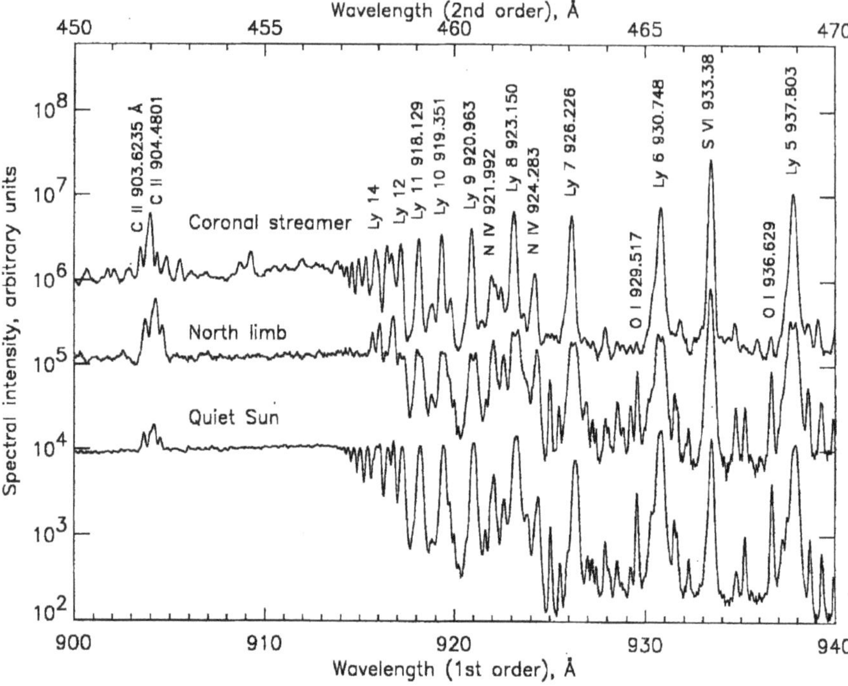

Figure 8. High-resolution spectra from 900 to 940 Å obtained during observations of a coronal streamer, a polar coronal hole (north limb) and the quiet Sun near the centre of the disk (cf., Table I).

required. Table I provides samples of lines present in a 40 Å section centred at 923 Å. In this table we include, in addition to the lines shown in Figure 8, some lines identified in more active regions (e.g., Ne VII 465.22 Å observed in second order). The wavelengths $\lambda_{\rm lit}$ are taken from Kelly (1987). The wavelengths $\lambda_{\rm obs}$ are obtained from SUMER observations. Since not all geometric and flat-field corrections could be performed before measuring these wavelengths, a more accurate determination will be attempted later. Special mention should be made of the high members of the H I Lyman and He II Balmer series and the fact that the ratio between the corresponding H I Ly and He II Ba lines is temperature sensitive (Feldman, 1995). It varies in our observations between 2.5 in active regions above the limb, 5.0 in a plume, 7 to 10 on the disk in quiet and moderately active regions, and 25 in prominences.

3.2. FORBIDDEN LINES

Coronal forbidden lines, first recorded during solar eclipses, allowed Grotrian (1939) and Edlén (1943) to establish for the first time the temperature of the solar corona. Ever since, such lines have been used in the diagnostics of low-

Table I
Line identifications in the range between 903.6235 and 942.538 Å

$\lambda_{\rm lit}$, Å	Line	$\lambda_{\rm obs}$, Å	Transition			Remarks
903.6235	C II	903.63	$2s^22p\,^2P_{1/2}$	–	$2s2p^2\,^2P_{3/2}$	
903.9616	C II	903.97	$2s^22p\,^2P_{1/2}$	–	$2s2p^2\,^2P_{1/2}$	
904.1416	C II	904.15	$2s^22p\,^2P_{3/2}$	–	$2s2p^2\,^2P_{3/2}$	
904.4801	C II	904.48	$2s^22p\,^2P_{3/2}$	–	$2s2p^2\,^2P_{1/2}$	
–	H I Ly 20	913.87	$1s\,^2S$	–	$21p\,^2P$	
–	H I Ly 19	914.09	$1s\,^2S$	–	$20p\,^2P$	
–	H I Ly 18	914.33	$1s\,^2S$	–	$19p\,^2P$	
914.576	H I Ly 17	914.60	$1s\,^2S$	–	$18p\,^2P$	
914.919	H I Ly 16	914.94	$1s\,^2S$	–	$17p\,^2P$	
915.329	H I Ly 15	915.35	$1s\,^2S$	–	$16p\,^2P$	
915.612	N II	915.59	$2s^22p^2\,^3P_0$	–	$2s2p^3\,^3P_1$	
915.824	H I Ly 14	915.84	$1s\,^2S$	–	$15p\,^2P$	
915.962	N II		$2s^22p^2\,^3P_1$	–	$2s2p^3\,^3P_0$	
916.012	N II	916.00	$2s^22p^2\,^3P_1$	–	$2s2p^3\,^3P_2$	blend
916.429	H I Ly 13	916.44	$1s\,^2S$	–	$14p\,^2P$	
916.701	N II		$2s^22p^2\,^3P_2$	–	$2s2p^3\,^3P_2$	
916.710	N II	916.69	$2s^22p^2\,^3P_2$	–	$2s2p^3\,^3P_1$	blend
917.181	H I Ly 12	917.18	$1s\,^2S$	–	$13p\,^2P$	
–	He II Ba 22		$2s\,^2S$	–	$24p\,^2P$	
–	He II Ba 22	917.74	$2p\,^2P$	–	$24d\,^2D$	blend
918.129	H I Ly 11	918.13	$1s\,^2S$	–	$12p\,^2P$	
918.724	O I	918.73	$2s^22p^4\,^3P_0$	–	$2s^22p^312d\,^3D_1$	
–	He II Ba 20		$2s\,^2S$	–	$22p\,^2P$	
–	He II Ba 20	919.01	$2p\,^2P$	–	$22d\,^2D$	blend
919.351	H I Ly 10	919.35	$1s\,^2S$	–	$11p\,^2P$	
919.658	O I	919.69	$2s^22p^4\,^3P_2$	–	$2s^22p^310d\,^3D_3$	
–	He II Ba 19		$2s\,^2S$	–	$21p\,^2P$	
–	He II Ba 19	919.68	$2p\,^2P$	–	$21d\,^2D$	blend
919.908	O I	919.91	$2s^22p^4\,^3P_2$	–	$2s^22p^311s\,^3S_1$	
919.971	O I	919.97	$2s^22p^4\,^3P_0$	–	$2s^22p^311d\,^3D_1$	
–	He II Ba 18		$2s\,^2S$	–	$20p\,^2P$	
–	He II Ba 18	920.62	$2p\,^2P$	–	$20d\,^2D$	blend
920.963	H I Ly 9	920.92	$1s\,^2S$	–	$10p\,^2P$	
–	He II Ba 17		$2s\,^2S$	–	$19p\,^2P$	
-	He II Ba 17	921.57	$2p\,^2P$	–	$19d\,^2D$	blend
921.860	O I	921.86	$2s^22p^4\,^3P_2$	–	$2s^22p^39d\,^3D_3$	
921.992	N IV		$2s2p\,^3P_1$	–	$2p^2\,^3P_2$	
922.0081	O I		$2s^22p^4\,^1D_2$	–	$2s^22p^33d\,^1F_3$	
922.0727	O I	922.01	$2s^22p^4\,^1D_2$	–	$2s^22p^33d\,^1D_2$	blend
922.46	O I		$2s^22p^4\,^1D_2$	–	$2s^22p^33d\,^1P_1$	
922.519	N IV	922.51	$2s2p\,^3P_0$	–	$2p^2\,^3P_1$	blend
–	He II Ba 16		$2s\,^2S$	–	$18p\,^2P$	
–	He II Ba 16	922.77	$2p\,^2P$	–	$18d\,^2D$	blend

Table I
Continued

λ_{lit}, Å	Line	λ_{obs}, Å	Transition			Remarks
923.057	N IV		$2s2p\ ^3P_1$	–	$2p^2\ ^3P_1$	
923.150	H I Ly 8		$1s\ ^2S$	–	$9p\ ^2P$	
923.220	N IV	923.15	$2s2p\ ^3P_2$	–	$2p^2\ ^3P_2$	blend
923.549	O I	923.68	$2s^22p^4\ ^3P_1$	–	$2s^22p^310s\ ^3S_1$	
923.675	N IV		$2s2p\ ^3P_1$	–	$2p^2\ ^3P_0$	
923.790	O I	923.60	$2s^22p^4\ ^3P_0$	–	$2s^22p^39d\ ^3D_1$	blend
924.283	N IV	924.27	$2s2p\ ^3P_2$	–	$2p^2\ ^3P_1$	
924.952	O I	924.97	$2s^22p^4\ ^3P_2$	–	$2s^22p^38d\ ^3D_3$	
925.442	O I	925.45	$2s^22p^4\ ^3P_2$	–	$2s^22p^39s\ ^3S_1$	
–	He II Ba 14		$2s\ ^2S$	–	$16p\ ^2P$	
–	He II Ba 14	925.84	$2p\ ^2P$	–	$16d\ ^2D$	blend
926.226	H I Ly 7		$1s\ ^2S$	–	$8p\ ^2P$	
926.295	O I	926.23	$2s^22p^4\ ^3P_1$	–	$2s^22p^38d\ ^3D_2$	blend
926.809	O I	926.84	$2s^22p^4\ ^3P_1$	–	$2s^22p^39s\ ^3S_1$	
926.903	O I	926.91	$2s^22p^4\ ^3P_0$	–	$2s^22p^38d\ ^3D_1$	
927.178	Fe II	927.20	$3d^6(a^5D)4s\ ^6D_{7/2}$	–	24°	
927.394	O I	927.41	$2s^22p^4\ ^3P_0$	–	$2s^22p^39s\ ^3S_1$	
–	He II Ba 13		$2s\ ^2S$	–	$15p\ ^2P$	
–	He II Ba 13	927.85	$2p\ ^2P$	–	$15d\ ^2D$	blend
928.004	Fe III	928.06	$3d^6\ ^1F_3$	–	$3d^5(b^2F)4p\ ^1F_3$	
928.474	Fe III	928.48	$3d^6\ ^3G_5$	–	$3d^5(a^2I)4p\ ^3H_6$	
929.163	Fe III	929.17	$3d^6\ ^3G_4$	–	$3d(a^2I)4p\ ^3H_5$	
929.5168	O I	929.56	$2s^22p^4\ ^3P_2$	–	$2s^22p^37d\ ^3D_3$	
930.086	Fe III	930.06	$3d^6\ ^3G_3$	–	$3d(a^2I)4p\ ^3H_4$	
930.2566	O I		$2s^22p^4\ ^3P_2$	–	$2s^22p^38s\ ^3S_1$	
–	He II Ba 12		$2s\ ^2S$	–	$14p\ ^2P$	
–	He II Ba 12	930.33	$2p\ ^2P$	–	$14d\ ^2D$	blend
930.442/2	Ne VII	465.19	$2s^2\ ^1S_0$	–	$2s2p\ ^1P_1$	
930.748	H I Ly 6		$1s\ ^2S$	–	$7p\ ^2P$	
930.8862	O I	930.75	$2s^22p^4\ ^3P_1$	–	$2s^22p^37d\ ^3D_2$	blend
931.4820	O I	931.50	$2s^22p^4\ ^3P_0$	–	$2s^22p^37d\ ^3D_1$	
931.6282	O I	931.64	$2s^22p^4\ ^3P_1$	–	$2s^22p^38s\ ^3S_1$	
932.0537	Ar II	932.00	$2s^22p^5\ ^2P_{1/2}$	–	$2s2p^6\ ^2S_{1/2}$	
932.2249	O I	932.23	$2s^22p^4\ ^3P_0$	–	$2s^22p^38s\ ^3S_1$	
933.38	S VI	933.39	$3s\ ^2S_{1/2}$	–	$3p\ ^2P_{3/2}$	
934.703	Fe III	934.69	$3d^6\ ^3P_2$	–	$3d^5(a^4P)4p\ ^3S_1$	
935.1930	O I	935.19	$2s^22p^4\ ^1D_2$	–	$2s^22p^34s\ ^1DA_2$	
936.6295	O I	936.66	$2s^22p^4\ ^3P_2$	–	$2s^22p^36d\ ^3D_3$	
–	He II Ba 10		$2s\ ^2S$	–	$12p\ ^2P$	
–	He II Ba 10	937.41	$2p\ ^2P$	–	$12d\ ^2D$	blend
937.803	H I Ly 5		$1s\ ^2S$	–	$6p\ ^2P$	
937.8405	O I		$2s^22p^4\ ^3P_2$	–	$2s^22p^37s\ ^3S_1$	

Table I
Continued

λ_{lit}, Å	Line	λ_{obs}, Å	Transition			Remarks
938.0200	O I	937.80	$2s^2 2p^4\ ^3P_1$	–	$2s^2 2p^3 6d\ ^3D_2$	blend
938.6249	O I	938.62	$2s^2 2p^4\ ^3P_0$	–	$2s^2 2p^3 6d\ ^3D_1$	
939.2346	O I	939.27	$2s^2 2p^4\ ^3P_1$	–	$2s^2 2p^3 7s\ ^3S_1$	
939.8412	O I	939.86	$2s^2 2p^4\ ^3P_0$	–	$2s^2 2p^3 7s\ ^3S_1$	
942.363	Fe III	942.39	$3d^6\ ^1F_3$	–	$3d^5(a^2H)4p\ ^1G_4$	
942.490	He II Ba 9		$2s\ ^2S$	–	$11p\ ^2P$	
942.538	He II Ba 9	942.56	$2p\ ^2P$	–	$11d\ ^2D$	blend

Table II
Forbidden lines newly observed by SUMER in a coronal streamer 2 arc min above the east limb and their relative peak intensities

Line	λ_{cal}, Å	Intensity
Ar XII	1054.5	500
Al VII	1053.94	50
Si VII	1049.23	200
Ar XII	1018.9	1000
Si IX	950.13	9400
Si VIII	949.24	4000
Si VIII	944.38	9400
S IX	871.73	1600
S X	787.38	1000
S XI	782.91	840

density plasmas in general and in coronal studies in particular. The streamer spectra recorded by SUMER contain forbidden lines with excellent diagnostic properties. Many of the forbidden lines present in the SUMER wavelength range belong to highly ionized abundant solar ions having $2s^2\ 2p^k$ and $3s^2\ 3p^k$ (where $k = 1$ to 5) ground state configurations. Some of the forbidden lines were previously measured in spectra emitted by the solar corona and by Tokamak plasmas. However, many other forbidden lines have been observed for the first time in the SUMER spectra of a coronal streamer. Preliminary identifications and the observed relative intensities of the more intense among the newly observed forbidden lines are given in Table II. The wavelengths λ_{cal} and λ_{obs} in this and the following two tables are cited after Kaufman and Sugar (1986).

Some of the forbidden lines originating within the ground configuration of N-like ions are expected to be quite intense in streamer spectra, provided the ions are sufficiently abundant in the plasma. Four transitions are shown in Table III

Table III

N I iso-electronic sequence and their peak intensity in a streamer spectrum

Transition	Line	λ_{obs}, Å	λ_{cal}, Å	Intensity
$2s^2\,2p^3\,{}^4S_{3/2} - {}^2P_{3/2}$	Mg VI	1190.07	1190.074	100
	Al VII	1053.94	1054.08	50
	Si VIII	944.38	944.38	9400
	Ar XII		648.93	?
$2s^2\,2p^3\,{}^4S_{3/2} - {}^2P_{1/2}$	Mg VI	1191.62	1191.611	500
	Si VIII	949.23	949.24	4000
	S X	787.38	787.56	1000
$2s^2\,2p^3\,{}^4S_{3/2} - {}^2D_{5/2}$	Si VIII	1440.50	1440.497	200
	S X	1196.24	1196.245	10000
	Ar XII		1018.6	1000
$2s^2\,2p^3\,{}^4S_{3/2} - {}^2D_{3/2}$	Si VIII	1445.75	1445.753	1100
	S X	1212.96	1212.970	20000
	Ar XII		1054.9	500

with their relative intensities as seen by SUMER in the same coronal streamer. The N-like ions Mg^{+5}, Si^{+7}, S^{+9}, Ar^{+11}, and Ca^{+13} have their peak abundance at electron temperatures of 5×10^5, 8×10^5, 1.3×10^6, 2×10^6, and 3×10^6 K, respectively. The forbidden lines of Si^{+7}, S^{+9} are most intense among the N-like ions in the streamer spectrum, while the Ca XIV lines are barely visible or perhaps not detectable at all, indicating a streamer temperature that peaks near 1×10^6 K. It is most likely that the streamer contains little if any plasmas at temperatures higher than $T_e = 2 \times 10^6$ K.

Intensity ratios of lines emitted by the above-mentioned ions are sensitive to electron densities in the range between 10^8 and 10^{11} cm^{-3} (Feldman *et al.*, 1978; Dwivedi, 1994; Mason and Monsignori-Fossi, 1994). Relative peak intensities of the Si VIII and S X density-sensitive forbidden lines present in a SUMER spectrum are given in Table IV for two streamers at positions 30 arc sec and 2 arc min above the west and east limb. A more thorough determination will be required to determine the total line intensities. But, since the line widths are not significantly different, the peak intensities provide a good measure for calculating line ratios. The electron density derived from the line ratios is also given in the table. As can be seen, the electron density in the streamers in both cases is about a factor of four higher at 30 arc sec than at 2 arc min above the limb. Detailed studies will be required to determine density variations in a systematic manner and to establish the error contributions.

Table IV
Electron densities in streamers from forbidden line ratios of the N I iso-electronic sequence

Location date	Line	λ_{cal}, Å	30 arc sec above limb		2 arc min above limb	
			Ratio	Density N_e, cm^{-3}	Ratio	Density N_e, cm^{-3}
West	Si VIII	1445.753	14	4×10^8	8	1×10^8
28 May, 1996	Si VIII	1440.497				
East	S X	1212.970	3.4	7×10^8	2.2	2×10^8
1 June, 1996	S X	1196.245				

3.3. Elemental Abundances

Elemental abundances in the solar corona were studied extensively in the past. Until the mid 1980s it was assumed that abundances in the upper solar atmosphere and photosphere were identical. Mass spectrometers measuring particles originating from the Sun, on the other hand, showed that elements with first ionization potential (FIP) of less than 10 eV are enriched in the solar wind by a factor of four relative to the photosphere. Meyer (1985) reviewed early measurements of abundances in the upper solar atmosphere and concluded that they were consistent with the solar-wind values obtained by mass spectrometers. In more recent studies it was discovered that elemental abundances in the upper solar atmosphere can be enriched by any number between 1 and 20 depending on the properties of the particular region. See Feldman (1992) for a review on elemental abundances in the solar atmosphere, and Geiss *et al.* (1995) for the latest solar-wind composition results obtained on board the Ulysses spacecraft.

Some issues regarding the enrichment of the upper solar atmosphere by elements with FIP less than 10 eV (low FIP) can be answered by SUMER observations. For example, are all low-FIP elements, independent of FIP, enriched by the same amount? In particular, are the most abundant low-FIP elements Si, Fe, and Mg which have FIPs of 8.1, 7.9, and 7.6 eV, respectively, enriched to the same degree as Ca with a 6.1 eV, Al with 6.0 eV, Na with 5.2 eV and K with a FIP of 4.3 eV? (Cf., Bürgi and Geiss, 1986; Marsch, von Steiger, and Bochsler, 1995.) Until recently the spectral information on Al, Na, and K was scarce. SUMER spectra for the first time provide ample information that will allow us to study this important question. Figure 9 displays spectra from a single detector reading centred at 630 Å that clearly shows lines of Fe XII ($3s^2 3p^3\ ^4S_{3/2} - 3s^2 3p^3\ ^2P_{3/2}$) 1242.00 Å, Mg X ($2s^2 S_{1/2} - 2p^2 P_{3/2}$) 624.95 Å, Al X ($2s^2\ ^1S_0 - 2s2p^3 P_1$) 637.99 Å and, possibly, K IX ($2s^2 S_{1/2} - 2p^2 P_{3/2}$) 636.32 Å blended by a first-order line.

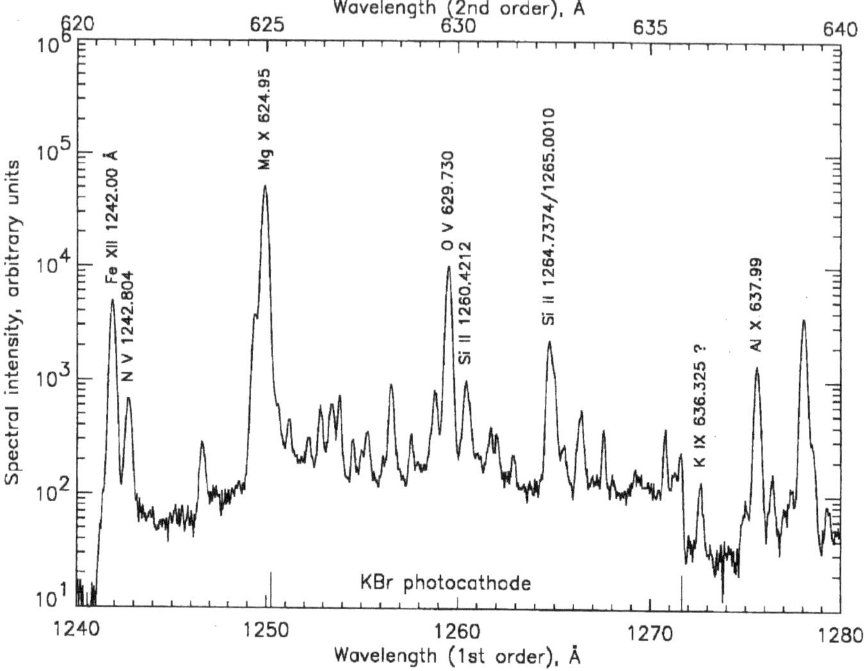

Figure 9. Spectrum centred at 630 Å (second order) taken with the 4 arc sec slit in a coronal streamer 2 arc min above the east limb on April 11, 1996. The spectrum has not been corrected for changes of the sensitivity due to the different photocathodes.

3.4. LINE PROFILE RESULTS

Line-profile studies of spectra recorded by S082B and HRTS provided extensive information on opacities, mass motions and on ion temperatures in a variety of lower-transition-region features ($T_e < 2.5 \times 10^5$ K). Forbidden lines detected by the same instruments provided limited information on the plasma conditions in some coronal regions. Examples of line profiles from the lower-transition region in a quiet solar region, a coronal hole, and in three active regions were reported by Feldman *et al.* (1976), Feldman and Doschek (1978), and Mariska (1992).

Early studies of line profiles in the EUV (from 170 to 805 Å) were published by Feldman and Behring (1974) and Doschek, Behring, and Feldman (1974), including line profiles of He I 537 Å and He I 584 Å. The He lines had widths in the range of 0.10–0.14 Å, perhaps broadened by opacity, while coronal lines had an excess broadening corresponding to turbulent velocities of approximately 30 km s^{-1}. High spectral resolution line profiles of Mg X 625 Å and O V 630 Å on the disk and above the limb were observed by Hassler *et al.* (1990). Increased broadening of the Mg X line profile with height above the limb was observed out to 1.2 solar radii (over 3.2 arc min).

More recently, the GSFC/Solar EUV Rocket Telescope and Spectrograph (SERTS) has measured the intensities and profiles of many lines in the range of 170–450 Å with good spatial and spectral resolutions for a variety of solar features. An extensive spectral atlas has been published from the results of this instrument (Thomas and Neupert, 1994). Studies using these data include element abundances and He II 304 Å line formation mechanisms (Jordan *et al.*, 1993). However, above 450 Å, studies of line profiles have been performed on very few additional EUV coronal lines over the years, until the SOHO mission. Now SUMER spectra provide for the first time detailed line-profile information in this important wavelength region. We present here, as examples of the type of analysis that can be done with SUMER, a dynamic event and a quiescent phenomenon.

3.4.1. *Explosive Event*

Among the many theoretical concepts of how the Sun heats its corona, reconnection processes of oppositely directed magnetic field lines play a prominent role. Small-scale reconnection events in the transition region have been observed by HRTS on many occasions (cf., Dere *et al.*, 1991; Kjeldseth-Moe and Cheng, 1994). SUMER with its high spatial, spectral and temporal resolution capabilities allows us to investigate such events in more detail. An example will be discussed here. It was observed near the centre of the Sun in the line O VI 1032 Å at approximately 3×10^5 K. The diagram in Figure 10 depicts in several frames a slit image of 1×110 arc sec^2, corresponding to a range of $700 \times 77\,000$ km^2 on the Sun, which is about 1/10 000 of its total area, and a spectral range of 1.76 Å. The line shown in the first frame is relatively undisturbed near the beginning of the event. In the following frames, a time series of the event is given in the same format with exposures every 6 s. The pronounced protrusions of the line images to shorter wavelengths, which developed with time, stem from the Doppler shifts, indicating a line-of-sight bulk-velocity component of the emitting ions of up to 120 km s^{-1} in addition to their thermal and turbulent velocities. The North-South dimension of the event (along the slit) is about 5000 km. As the axis of the protrusion is inclined with respect to the dispersion direction, it appears as if, in this particular case, the velocity vectors of the ions were at an angle with the line of sight. The distance of the apparent motion of the jet along the slit would be of the order of 4000 km. The duration of the growth phase for this event was about 30 s and, consequently, a velocity of 130 km s^{-1} was deduced. However, it is also possible that the high-velocity ions are not directly linked to the brightest region near the spatial pixel number 191. As the highest velocities were observed at pixel 197 near a relatively undisturbed location, this assumption does not seem to be very likely.

Many of these dynamic events, which occur at a rate of several thousands per minute on the whole Sun, have been and will be observed by SUMER during the mission time of SOHO. Their contribution to the heat and mass input into the

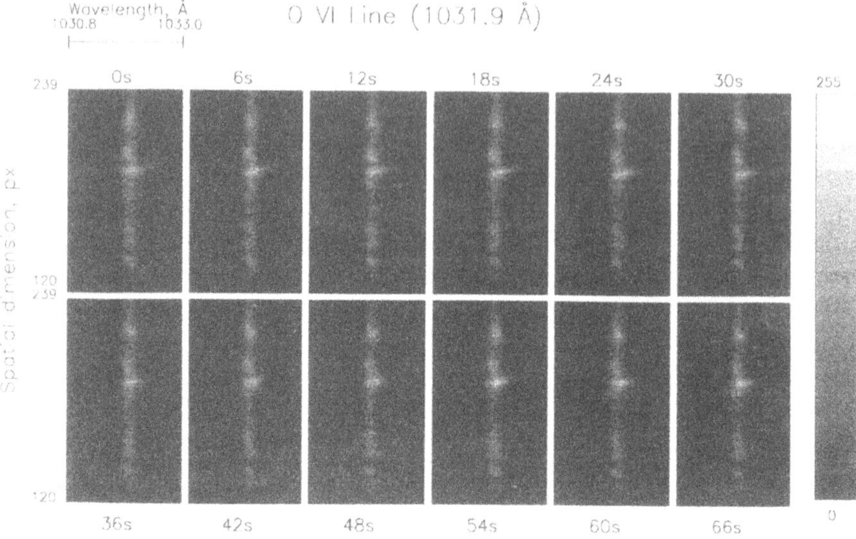

Figure 10. Explosive event seen in the line O VI 1032 Å on May 1, 1996 at 00:04 UT. Every 6 s the spectral line is shown along a slit length of about 110 px (corresponding to 110 arc sec).

corona will be studied, but other physical processes occurring on the Sun will also be investigated.

3.4.2. *Prominence*

Preliminary observations of prominences demonstrate that SUMER is a powerful tool to understand these phenomena. In Figure 11(a) we present an image of a solar prominence as seen by SUMER in He I 584 Å and in Figure 11(b) an area on the disk taken from the same full-Sun image. For each pixel in these images we have a high-resolution line profile. Examples of the average profiles in the prominence and on the disk are shown in Figure 12. The width is 120 mÅ (FWHM) in the prominence, and 150 mÅ (FWHM) on the disk. The total line intensity in the prominence is about 0.4 the intensity on the disk. Such a value (of the dilution factor) confirms that the helium emission of the prominence results from the scattering of the incident chromospheric radiation. Based on the 500 points measured, the prominence profile is about 20% narrower than that on the disk and is shifted by 40 mÅ towards the blue, corresponding to a line-of-sight velocity of 20 km s^{-1}. Such horizontal (line-of-sight) flows have been observed by, e.g., Mein (1977) and Vial *et al.* (1980) (for a review, see Schmieder (1990)). The ability to measure spectral lines with such precision will allow us to determine accurately temperatures and turbulence based on line widths, and bulk velocities based on line shifts in prominences and filaments.

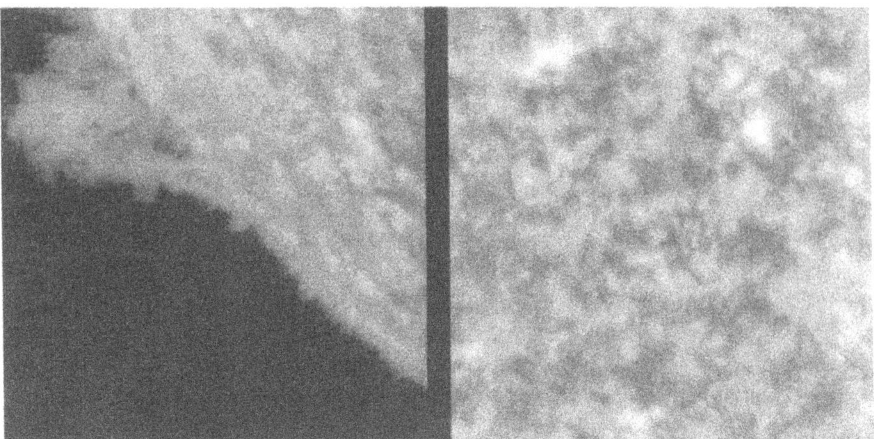

Figure 11. Two sections of the full Sun observed in He I 584 Å between March 2 and March 4, 1996. The limb with a prominence is shown on the left (a) and a region on the disk on the right (b). The areas depicted are 304 × 300 arc sec^2 each.

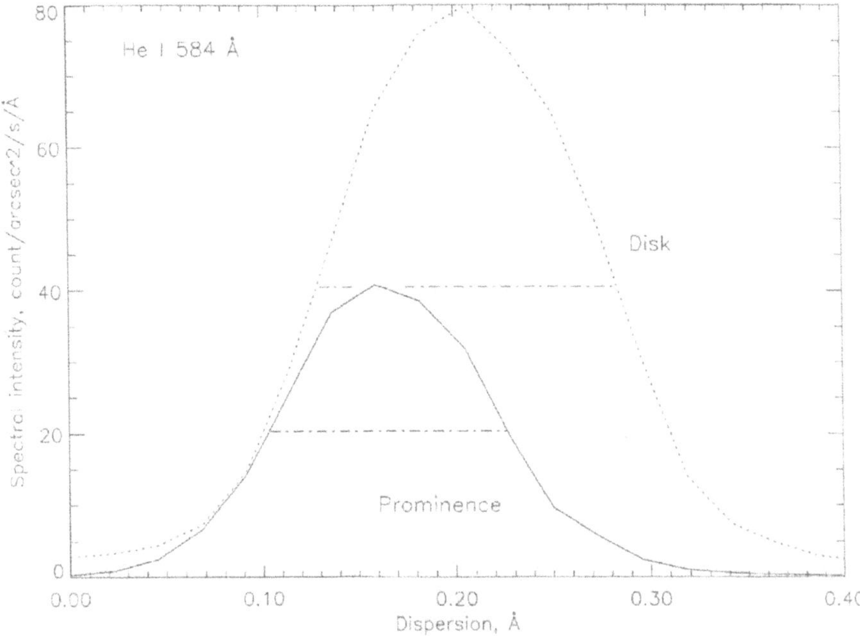

Figure 12. Examples of spectral shapes of the line He I 584 Å. The averages were taken along the slit over 50 px and across the slit over 10 positions (approximately 52 × 10 arc sec^2) in the prominence and on the disk.

4. Concluding Remarks

The commissioning phase of SUMER has been completed successfully. The performance characteristics of the instrument conform, with a few exceptions, to those specified before launch. Many significant scientific observations could be performed during the first months of the SOHO mission, examples of which are presented in this Paper I and in Paper II. In summary, we can state that SUMER is available for detailed observations of the solar atmosphere.

Acknowledgements

The SUMER project is financially supported by DARA, CNES, NASA and the ESA PRODEX programme (Swiss contribution). Additional financial support is being provided by the participating institutions along with general administrative assistance at various phases of the project. SUMER is part of SOHO, the Solar and Heliospheric Observatory, of ESA and NASA. The instrument development and/or commissioning has been carried out by a large dedicated team of engineers, scientists and technicians co-ordinated by a management support group. The authors would like to thank A. Arondel, S.J. Battel, L. Bemmann, H.-M. Bock, C. de Boer, R. Boucarut, P. Boutry, P. Bouyries, H.-J. Braun, I. Büttner, I. E. Dammasch, A. Dannenberg, W. Donakowski, D. Dumont, W. Engelhardt, H.-G. Engelmann, K. Eulig, A. Fischer, E. Frank, G. A. Gaines, D. Germerott, K. Gräbig, W. Grill, M.A. Gummin, J.W. Hamilton, H. Hartwig, E. Hertzberg, J. Hoberman, G. Hoch, J. Hull, D. Innes-Markiewicz, P. Jelinsky, P. Jensen, M. Jhabvala, H.G. Kellner, E. Keppler, R. Keski-Kuha, E. Korzac, T. Kucera, D. Leviton, J. Linant, Y. Longval, T. Magoncelli, C. Maurel, R. Meller, T. Moran, N. Mosquet, N. Neisen, W. Neumann, J. Osantowski, A. Pauly, W. Paustian, J. Platzer, K. Poser, R. Raffanti, W. Reich, T. Rodriguez-Bell, T. Saha, T. Sasseen, H. Schild, R. Schmidt, H. Schüddekopf, H. Specht, E. Steinmetz, J.M. Stock, U. Strohmeyer, W. Tappert, W.T. Thompson, S. Töpfer, J.L. Tom, G. Tomasch, B. Wand, L. Wang, B. Welsh and G. Wright for their contributions. We are particularly appreciative of the dedication of N. Morisset during the operation of the SUMER instrument.

Scientific advice was provided by J. Mariska and a team of Associate Scientists: J. Aboudarham, B. Aschenbach, W.I. Axford, J. Barnstedt, F. Bely-Dubau, V. Bommier, J.L. Culhane, J.P. Delaboudinière, C. Diesch, G. Doschek, J.G. Doyle, J. Dubau, B.N. Dwivedi, G. Einaudi, A. Fludra, B. Foing, P. Gouttebroze, R.A. Harrison, H.F. Haupt, E. Haug, W.-H. Ip, C. Jordan, F. Keenan, O. Kjeldseth-Moe, B. Kliem, F. Kneer, K. Lang, B. Leroy, O. v.d. Lühe, P. Maltby, H.E. Mason, R.W.P. McWhirter, P. Mein, F. Millier, J.H. Parkinson, E.R. Priest, A.K. Richter, H. Rosenbauer, S. Sahal, D. Samain, B. Schmieder, J.H.M.M. Schmitt, M. Schüssler, R. Schwenn, G.M. Simnett, G. Tondello, J. Trümper, V.M. Vasyliunas and O. Vilhu.

We acknowledge the contributions of our late colleagues I. Liede, B. Monsignori-Fossi and B.E. Patchett. We dedicate this paper to the memory of Christian Becker, whose initial project management efforts until his untimely death in 1991 were instrumental in arriving at the current performance of SUMER.

References

Bartoe, J.-D. F. and Brueckner, G. E.: 1975, *J. Opt. Soc. Am.* **65**, 13.
Bartoe, J.-D. F., Brueckner, G. E., Purcell, J. D., and Tousey, R.: 1977, *Appl. Opt.* **16**, 879.
Behring, W. E., Cohen, L., Feldman, U., and Doschek, G. A.: 1976, *Astrophysics J.* **203**, 521.
Brueckner, G. E., Bartoe, J.-D. F., Cook, J. W., Dere, K. P., and Socker, D.G.: 1986, *Adv. Space Res.* **6**, 263.
Bürgi, A. and Geiss, J.: 1986, *Solar Phys.* **103**, 347.
Burton, W. M. and Ridgeley, A.: 1970, *Solar Phys.* **14**, 3.
Burton, W. M., Ridgeley, A., and Wilson, R.: 1967, *Monthly Notices Roy. Astron. Soc.* **135**, 207.
Cohen, L., Feldman, U., and Doscheck, G. A.: 1978, *Astrophys. J. Suppl.* **37**, 393.
Dere, K. P., Bartoe, J.-D. F., Brueckner, G. E., Ewing, J., and Lund, P.: 1991, *J. Geophys. Res.* **96**, 9399.
Doschek, G. A., Behring, W. E., and Feldman, U.: 1974, *Astrophys. J.* **190**, L141.
Dwivedi, B. N.: 1994, *Space Sci. Rev.* **65**, 289.
Edlén, B.: 1943, *Z. Astrophys.* **22**, 30.
Feldman, U.: 1992, *Physica Scripta* **46**, 202.
Feldman, U.: 1995, *Comm. Atmospheric Mol. Phys.* **31**, 11.
Feldman, U. and Behring, W. E.: 1974, *Astrophys. J.* **189**, L45.
Feldman, U. and Doschek, G. A.: 1978, *Astrophys. J. Suppl.* **37**, 443.
Feldman, U. and Doschek, G. A.: 1991, *Astrophys. J. Suppl.* **75**, 925.
Feldman, U., Doschek, G. A., Van Hoosier, M. E., and Purcell, J. D.: 1976, *Astrophys. J. Suppl.* **31**, 445.
Feldman, U., Doschek, G. A., Mariska, J. T., Bahtia, A. K., and Mason, H. E.: 1978, *Astrophys. J.* **226**, 674.
Gabriel, A. H., Garton, W. R. S., Goldberg, L., Jones, T. J. L., Jordan, C., Morgan, F. J., Nicholls, R. W., Parkinson, W. J., Paxton, H. J. B., Reeves, E. M., Shenton, C. B., Speer, R. J., and Wilson, R.: 1971, *Astrophys. J.* **169**, 595.
Geiss, J., Gloeckler, G., von Steiger, R., Balsiger, H., Fisk, L. A., Galvin, A. B., Ipavich, F. M., Livi, S., McKenzie, F. J., Ogilvie, K. W., and Wilken, B.: 1995, *Science* **268**, 1033.
Gouttebroze, P., Lemaire, P., Vial, J.-C., and Artzner, G.: 1978, *Astrophys. J.* **225**, 655.
Grotian, W.: 1939, *Die Naturwissenschaften* **27**, 214.
Hassler, D. M., Rottman, G. J., Shoub, E. C., and Holzer, T. E.: 1990, *Astrophys. J.* **348**, L77.
Hassler, D. M., Rottman, G. J., and Orrall, F. Q.: 1991, *Astrophys. J.* **372**, 710.
Hollandt, J., Kühne, M., Huber, M. C. E., and Wende, B.: 1996a, *Astron. Astrophys. Suppl.*, **115**, 561.
Hollandt, J., Schühle, U., Paustian, W., Curdt, W., Kühne, M., Wende, B., and Wilhelm, K.: 1996b, *Appl. Opt.* **35**, 5125.
Huber, M. C. E., Dupree, A. K., Goldberg, L., Noyes, R. W., Parkinson, W. H., Reeves, E. M., and Withbroe, G. L.: 1973, *Astrophys. J.* **183**, 291.
Jordan, S. D., Thompson, W. T., Thomas, R. J., and Neupert, W. M.: 1993, *Astrophys. J.* **406**, 346.
Kaufman, V. and Sugar, J.: 1986, *J. Phys. Chem. Ref. Data* **15**, 321.
Kelly, R. L.: 1987, *J. Phys. Chem. Ref. Data* **16**, 1.
Kjeldseth-Moe, O. and Cheng, C. C.: 1994, *Space Sci. Rev.* **70**, 85.
Lean, J.: 1991, *Rev. Geophys.* **29**, 505.
Lemaire, P.: 1991, *ESA J.* **15**, 237.
Lemaire, P., Gouttebroze, P., Vial, J.-C., and Artzner, G.: 1981, *Astron. Astrophys.* **103**, 160.
Lemaire, P., Wilhelm, K., Curdt, W., Schühle, U., Marsch, E., Poland, A. I., Jordan, S. D., Thomas, R. J., Hassler, D. M., Vial, J.-C., Kühne, M., Huber, M. C. E., Siegmund, O. H. W., Gabriel, A., Timothy, J. G., and Grewing, M.: 1997, *Solar Phys.* **170**, 105 (this issue).
Mariska, J. T.: 1992, *Cambridge Astrophysics Series* **23**.
Marsch, E., von Steiger, R., and Bochsler, P.: 1995, *Astron. Astrophys.* **301**, 261.

Mason, H. E. and Monsignori-Fossi, B.: 1994, *Astron. Astrophys.* **6**, 123.
Mein, P.: 1977, *Solar Phys.* **54**, 45.
Meyer, J. P.: 1985, *Astrophys. J. Suppl.* **57**, 173.
Reeves, E. M., Huber, M. C. E., and Timothy, J. G.: 1977, *Appl. Opt.* **16**, 837.
Reeves, E. M., Timothy, J. G., Huber, M. C. E., and Withbroe, G. L.: 1977, *Appl. Opt.* **16**, 849.
Rottman, G. J., Orrall, F. Q., and Klimchuk, J. A.: 1982, *Astrophys. J.* **260**, 326.
Sandlin, G. D., Bartoe, J.-D. F., Brueckner, G. E., Tousey, R., and Van Hoosier, M. E.: 1986, *Astrophys. J. Suppl.* **61**, 801.
Schmieder, B.: 1990, in V. Ruždjak and E. Tandberg-Hanssen (eds.), 'Dynamics of Quiescent Prominences', *IAU Colloq.* **117**, 85.
Schühle, U.: 1993, in E. H. Silver and S. M. Kahn (eds.), *UV and X-Ray Spectroscopy of Laboratory and Astrophysical Plasmas, Proc. of the 10th International Colloquium*, **373**.
Siegmund, O. H. W.: 1989, *Proc. SPIE* **1072**, 111.
Siegmund, O. H. W., Gummin, M. A., Stock, J. M., Marsh, D., Raffanti, T., Sasseen, T., Tom, J., Welsh, B., Gaines, G. A., Jelinsky, P., and Hull, J.: 1994, *Proc. SPIE* **2280**, 89.
Siegmund, O. H. W., Gummin, M. A., Sasseen, T., Jelinsky, P., Gaines, G. A., Hull, J., Stock, J. M., Edgar, M., Welsh, B., Jelinsky, S., and Vallerga, J.: 1995, *Proc. SPIE* **2518**, 344.
Thomas, R. J. and Neupert, W. M.: 1994, *Astrophys. J. Suppl.* **91**, 461.
Tousey, R., Bartoe, J.-D. F., Brueckner, G. E., and Purcell, J. D.: 1977, *Appl. Opt.* **16**, 870.
Vernazza, J. E. and Reeves, E. M.: 1978, *Astrophys. J. Suppl.* **37**, 485.
Vernazza, J. E., Foukal, P. V., Huber, M. C. E., Noyes, R. W., Reeves, E. M., Schmahl, E. J., Timothy, J. G., and Withbroe, G. L.: 1975, *Astrophys. J.* **199**, L123.
Vial, J.-C.: 1982, *Astrophys. J.* **253**, 330.
Vial, J.-C., Lemaire, P., Artzner, G., and Gouttebroze, P.: 1980, *Solar Phys.* **68**, 187.
Wilhelm, K., Curdt, W., Marsch, E., Schühle, U., Lemaire, P., Gabriel, A., Vial, J.-C., Grewing, M., Huber, M. C. E., Jordan, S. D., Poland, A. I., Thomas, R. J., Kühne, M., Timothy, J. G., Hassler, D. M., and Siegmund, O. H. W.: 1995a, *Solar Phys.* **162**, 189.
Wilhelm, K., Curdt, W., Marsch, E., Schühle, U., Lemaire, P., Gabriel, A., Vial, J.-C., Grewing, M., Huber, M. C. E., Jordan, S. D., Poland, A. I., Thomas, R. J., Kühne, M., Timothy, J. G., Hassler, D. M., and Siegmund, O. H. W.: 1995b, *Proc. SPIE* **2517**, 2.
Withbroe, G. L., Jaffe, D. T., Foukal, P. V., Huber, M. C. E., Noyes, R. W., Reeves, E. M., Schmahl, E. J., Timothy, J. G., and Vernazza, J. E.: 1976, *Astrophys. J.* **203**, 528.

FIRST RESULTS OF THE SUMER TELESCOPE AND SPECTROMETER ON SOHO

II. *Imagery and Data Management*

P. LEMAIRE
Institut d'Astrophysique Spatiale, Unité Mixte CNRS – Université Paris XI, Bat 121, F-91405 Orsay, France

K. WILHELM, W. CURDT, U. SCHÜLE and E. MARSCH
Max-Planck-Institut für Aeronomie, D-37189 Katlenburg-Lindau, Germany

A. I. POLAND, S. D. JORDAN and R. J. THOMAS
NASA/Goddard Space Flight Center, Greenbelt, Maryland, U.S.A.

D. M. HASSLER
High Altitude Observatory/NCAR, Boulder, Colorado, U.S.A.

J. C. VIAL
Institut d'Astrophysique Spatiale, Unité Mixte CNRS – Université Paris XI, Bat 121, F-91405 Orsay, France

M. KÜHNE
Physikalisch-Technische Bundesanstalt, D-10587, Berlin, Germany

M. C. E. HUBER
ESA, Space Science Department, ESTEC, NL-2200 AG Noordwijk, The Netherlands

O. H. W. SIEGMUND
The University of California, SSL, Berkeley, California, U.S.A.

A. GABRIEL
Institut d'Astrophysique Spatiale, Unité Mixte CNRS – Université Paris XI, Bat 121, F-91405 Orsay, France

J. G. TIMOTHY
The University of New Brunswick, Fredericton NV, Canada E3B5A3

M. GREWING
Astronomisches Institut, D-72076 Tübingen, Germany

(Received 6 August, 1996; in revised form 19 September, 1996)

Abstract. SUMER – Solar Ultraviolet Measurements of Emitted Radiation – is not only an extreme ultraviolet (EUV) spectrometer capable of obtaining detailed spectra in the range from 500 to 1610 Å, but, using the telescope mechanisms, it also provides monochromatic images over the full solar disk and beyond, into the corona, with high spatial resolution. We report on some aspects of the observation programmes that have already led us to a new view of many aspects of the Sun, including quiet Sun, chromospheric and transition region network, coronal hole, polar plume, prominence and active region studies.

After an introduction, where we compare the SUMER imaging capabilities to previous experiments in our wavelength range, we describe the results of tests performed in order to characterize and optimize the telescope under operational conditions. We find the spatial resolution to be 1.2 arc sec across the slit and 2 arc sec (2 detector pixels) along the slit. Resolution and sensitivity are adequate to provide details on the structure, physical properties, and evolution of several solar features which we then present. Finally some information is given on the data availability and the data management system.

Solar Physics **170:** 105–122, 1997.

1. Introduction

This Paper II is the second of the two papers presenting the first results of the SUMER instrument on the ESA/NASA Solar and Heliospheric Observatory (SOHO). As outlined in Paper I (Wilhelm *et al.*, 1996), the goal of the SUMER investigation is to collect data in the wavelength range from below 500 to 1610 Å at any location on the solar disk and in the low corona. Observations in this wavelength range should be suitable to provide new insights into and constraints on coronal heating and solar-wind generation mechanisms.

In this paper, Paper II, we mainly concentrate on the imaging capabilities of the instrument while in the companion Paper I we describe its spectroscopic aspects. It is, however, important to point out that the combination of good spatial and spectral resolution is essential for the study of the solar atmosphere and its dynamics.

The first important step is to perform studies at very small spatial scales in order to locate and identify specific phenomena. This will be followed by combining all the information on a variety of structures, to estimate the global energy budget needed to understand coronal heating. The quality of the imaging capabilities of SUMER will play a key role in understanding the complexity of structures in the upper chromosphere, the transition region and the corona.

The very good stability of the SOHO spacecraft (a jitter smaller than 0.2 arc sec root-mean-square (r.m.s.) over several hours during small perturbations from experiments has been measured during tests in February 1996) gives us the ability to use the full SUMER resolution. In the spectral domain covered by SUMER, few instruments have provided high quality imaging together with good spectral resolution so that contributions from lines formed at different temperatures can be separated.

On the *Skylab* Apollo Telecope Mount (ATM) in 1973–1974, two EUV spectroheliograph/spectrometers had imaging capabilities: the Naval Research Laboratory (NRL) S082A spectroheliograph (Tousey *et al.*, 1977) and the Harvard College Observatory (HCO) EUV spectrometer/spectroheliometer S055 (Reeves *et al.*, 1977). The S082A instrument has provided series of full solar images in the range of 320 – 630 Å using a slitless spectrometer with a 2–3 arc sec resolution on a photographic emulsion. The S055 spectrometer, with a spectral resolution of 1.6 Å, was able to register simultaneously seven lines in the 280 – 1340 Å range and to image a field of 5×5 arc min^2 by scanning in a boustrophedonic pattern with a spatial resolution of 5×5 arc sec^2 in horizontal and vertical steps of 2.5 or 5 arc sec, respectively. The wealth of data collected by these two instruments and other instruments on *Skylab*/ATM has established the base of our present knowledge of upper solar atmospheric structures.

Following *Skylab*/ATM were the Orbiting Solar Observatory 8 (OSO-8) with its Laboratory for Atmospheric and Space Physics (LASP) Ultraviolet Spectrometer (UVS) (Bruner *et al.*, 1977) and its Laboratoire de Physique Stellaire et Planétaire (LPSP) Multichannel Spectrometer (Bonnet *et al.*, 1978), and the Solar Maximum Mission's (SMM) Ultraviolet Spectrometer and Polarimeter (UVSP) (Miller *et al.*,

1981). These instruments provided good imaging with 3 arc sec or better resolution in a few UV lines in the 1200 – 4000 Å range, but they were hampered by rapid sensitivity losses.

On several occasions the NRL/High Resolution Telescope and Spectrometer (HRTS) was able to obtain high-resolution images during rocket flights. However, only the *Spacelab 2* mission (Brueckner *et al.*, 1986) with a duration of 5 days has provided a good statistical distribution of fine structures over the solar disk with a 1 – 2 arc sec spatial resolution in the 1200 – 1700 Å wavelength range on photographic emulsion.

Before the development of sophisticated multilayer coatings deposited on mirrors, only the LPSP/Transition Region Camera rocket telescope (Bonnet *et al.*, 1980) was able to image the Sun in several UV wavelength bands (1216 – 2200 Å) with 1 arc sec resolution on photographic emulsion. Since then, the MSSTA rocket (Hoover *et al.*, 1991; Walker, Hoover, and Barbee, 1992), with several wavelengths in the range 500 – 1600 Å, has obtained better than 1 arc sec resolution images on photographic emulsion.

After a discussion of the SUMER telescope characteristics, we present some results obtained during the first months of operation on solar features such as network, plage, prominence and coronal hole. We then provide some information on the data processing and availability.

2. The SUMER Telescope

The SUMER telescope has being described in an earlier paper (Wilhelm *et al.*, 1995).

We recall the relevant pre-flight characteristics here:

– a single surface, off-axis parabola as telescope mirror,

– a good image quality with a full width at half-maximum (FWHM) of less than 1 arc sec and a sharp decrease of the point spread function (PSF), reaching 0.001 of the peak at 3.5 arc sec,

– a low scattering to permit observations up to half a solar radius above the solar limb,

– scanning mechanisms capable of rotating the parabola around its focus in order to maintain the nominal angular resolution within a field of 48×48 arc min^2 with a geometric limitation of 64×64 arc min^2, with step sizes being multiples of 0.38 arc sec,

– slits used at the telescope focus: 0.3×120, 1×120, 1×300, or 4×300 arc sec^2. Also available is a 1×1 arc sec hole for special applications, – the slit can be moved by the slit focussing mechanism to select the best telescope focus.

The in-flight determination of the telescope performance and its maintenance are continuous tasks that require regular checks and updates of telescope parameters.

We describe the results obtained from some of the tests on telescope focussing, angular resolution and telescope scattering in the following sections.

2.1. Telescope Focussing

The SUMER structure is made out of aluminum and, as such, is very sensitive to changes of the thermal equilibrium of the instrument. As noted in Paper I, the thermal temperature of the whole instrument had to be increased by 3 K. We cannot measure the temperature of the silicon-carbide (SiC) telescope mirror, but there is no reason why it should not be close to its design value of 75 ± 5 °C. To keep the image quality, the telescope has to be brought to a focus in-flight. Using the Rear Slit Camera (RSC) (Wilhelm *et al.*, 1995) we are able to determine the shape of the solar limb as it appears in the visible continuum near 6000 Å within a band of 760 Å (FWHM). We then have the capability of adjusting the distance between the telescope and the slit by moving the slit with the help of the focussing mechanism (this, in turn, requires a re-adjustment of the spectrometer focus).

In Figure 1 we present the variation of the inverse slope of the limb as a function of the telescope focus. The best focus can be determined with an accuracy of tens of micrometres (30 μm out of focus corresponds to an image blur of 0.5 arc sec). We regularly check the telescope focus. It has been very stable within a few dozen micrometres, since the temperature is accurately controlled at 23°C.

2.2. Image Quality

The in-flight angular resolution of the telescope has been estimated by measuring the smallest width of a solar structure that has high contrast along the slit, and while the telescope is scanning perpendicularly to the long slit dimension (as illustrated in Figures 2 and 3). It can be seen that the smallest structure detected has a width of 2 detector pixels (or 2 arc sec) (FWHM) along the slit. By use of the smallest slit width (0.3 arc sec) and the smallest scanning steps (0.38 arc sec), the width of the smallest structure measured to date was 1.2 arc sec (FWHM).

Another very important parameter in evaluating the performance of the telescope is the evaluation of the sharpness of the PSF wings, which results from the low-frequency contribution of telescope mirror surface imperfections (Saha, Leviton, and Glenn, 1996). An example of the SUMER image sharpness obtained in the S VI 933 Å line is shown in Figure 4. The wings of the bright point have no extension in the adjacent network perpendicular to the slit, while two adjacent solar features are detected at different slit positions during the telescope scan.

2.3. Telescope Scatter Characteristics

Observations above the solar limb of lines which are bright on the disk (e.g., Lyman series, O VI lines) are difficult because of the scattered light from the extended solar disk. During the laboratory tests it has been shown (Saha, Leviton,

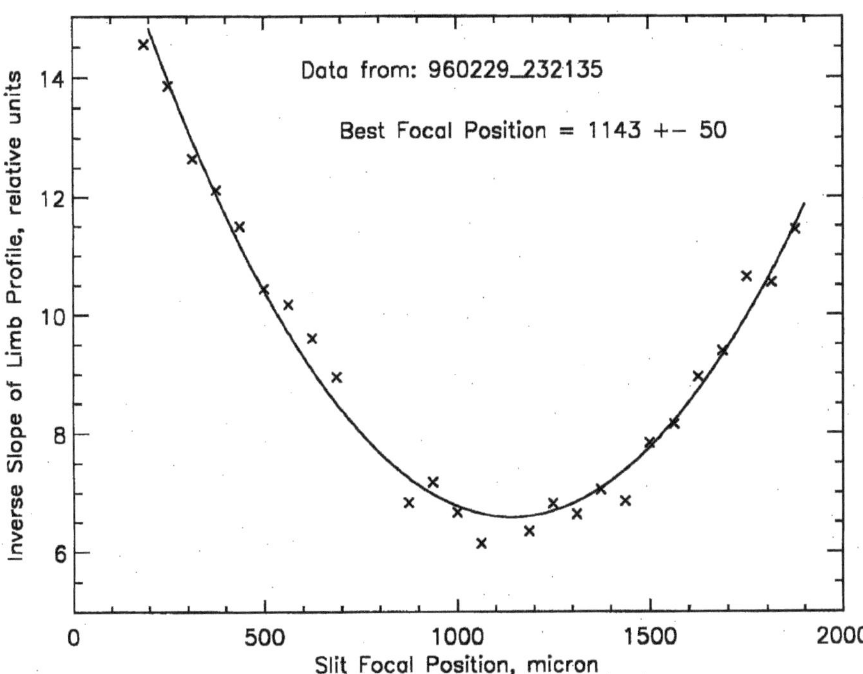

Figure 1. Determination of the telescope best focus by use of the Rear Slit Camera and the 1 × 120 arc sec^2 slit in its asymmetric position. The best focus of the telescope is attained at the minimum of the inverse slope of the solar limb.

and Glenn, 1996) that the scattered-light contribution of a point source focussed by the SUMER telescope mirror at 1236 Å was below 10^{-10} at an angular separation of 50 arc min.

The determination of the scattered-light characteristics of the telescope is very complicated. Here we can only give some preliminary results obtained from our early measurements:

– As one might expect, the level of scattered light seems to be insensitive to the azimuthal location around the solar disk. This result will have to be confirmed with more accuracy in using different sets of data taken at different positions above the solar disk.

– At this stage, the measurements made near the poles, where the coronal contribution is rather small over the solar-minimum polar holes, are more reliable than those taken near the equator where strong contributions from streamers interfere with the scattered light measurements.

– The level of scattered light measured at 844 Å (Lyman continuum) varies from 1.7×10^{-3} of the limb intensity at 1.5 arc min above the solar north limb to 7×10^{-4} at 11.0 arc min.

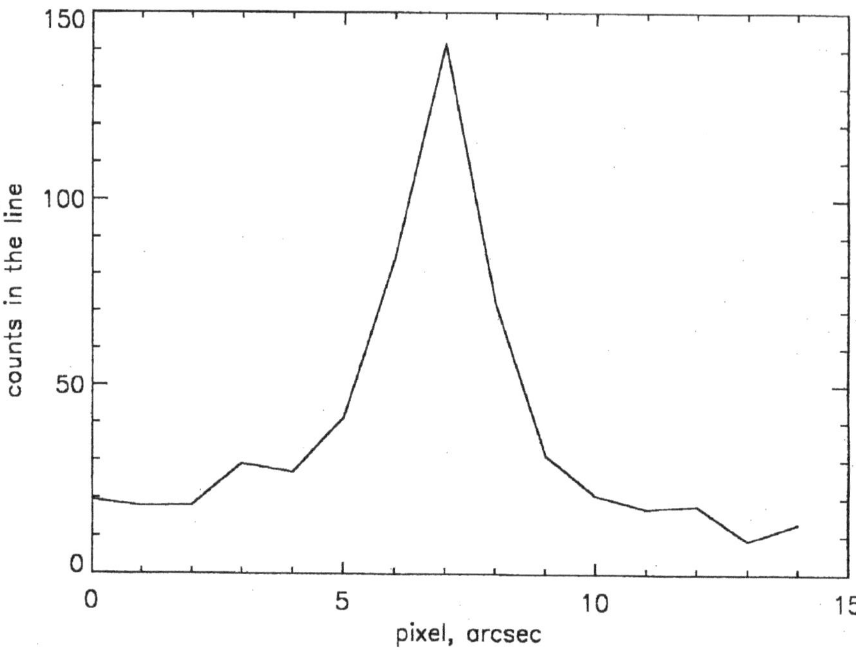

Figure 2. Smallest solar structure detected along the slit length (0.3×120 arc sec^2) in the O VI 1032 Å line. The width of 2 pixels (FWHM) on the detector corresponds to 2 arc sec.

To evaluate the impact of scattered solar light on the solar measurement above the limb we have plotted (Figure 5) the measured intensity variations, relative to the solar limb, above the limb in a few lines: H I Lβ, and O VI 1032 Å.

The strong intensity drop of H I Lβ above the limb reaches nearly the scattered light level at 2 arc min.

From there on, the relative intensity varies at the same rate as the scattered light, while being always above it.

The relative intensity variation of the O VI 1031.9 Å line in the coronal hole shows a variable slope on a logarithmic scale. The relative intensity is lower than the one published in Mariska and Withbroe (1978) up to 4.5 arc min above the limb, in a streamer, and seems to reach the scattered light level near 6 arc min.

3. First Results

Although SUMER was primarily designed to obtain spectra for the determination of temperatures, densities, and velocities in solar atmospheric features, it also has sophisticated imaging capabilities to identify the spatial features of structures or events where spectra are taken. The limitations are due to the step-by-step raster build-up of solar maps. Together with the photon count statistics and the limited

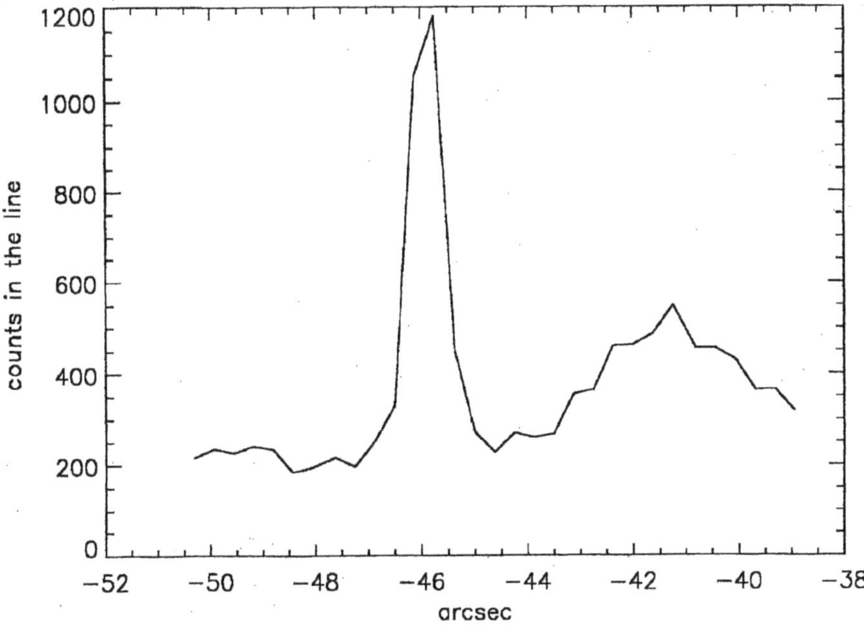

Figure 3. Smallest solar structure detected when the telescope mechanism is scanning perpendicularly to the slit (0.3 × 120 arc sec^2) by steps of 0.38 arc sec in the O VI 1032 Å line. The range of three steps (FWHM) corresponds to 1.14 arc sec.

telemetry rate, rastering can be a rather slow process. To demonstrate its high quality imaging capabilities, we present, in what follows, a selection of results obtained in network, plage, coronal hole, prominence, streamer, polar coronal holes, and plumes. From full-Sun observations, we derive limb-brightening curves and temporal evolutions of the network structure as illustrated.

3.1. Network and Plage

The importance of the chromospheric network and its extension to the base of the corona has been emphasized since the close relation between the brightness enhancement in the network and the ambient magnetic structure was established (Howard, 1959; Leighton, 1959). The network is thought to channel the energy transport from the upper chromosphere to the base of the solar corona (Gabriel, 1976). High resolution images of the network at different temperatures, combined with line spectra, will help to define better the real constraints on any heating mechanism. Only in this way will it be possible to evaluate the relative importance of conduction, convection, diffusion, wave energy transport, and radiation in the network energy balance and in transporting energy to the corona.

From images taken simultaneously in the S VI 933 Å ($T_e = 2 \times 10^5$ K) and H I Lϵ 937 Å ($T_e < 2 \times 10^4$ K) lines with a scan step size of 1.5 arc sec and a pixel size

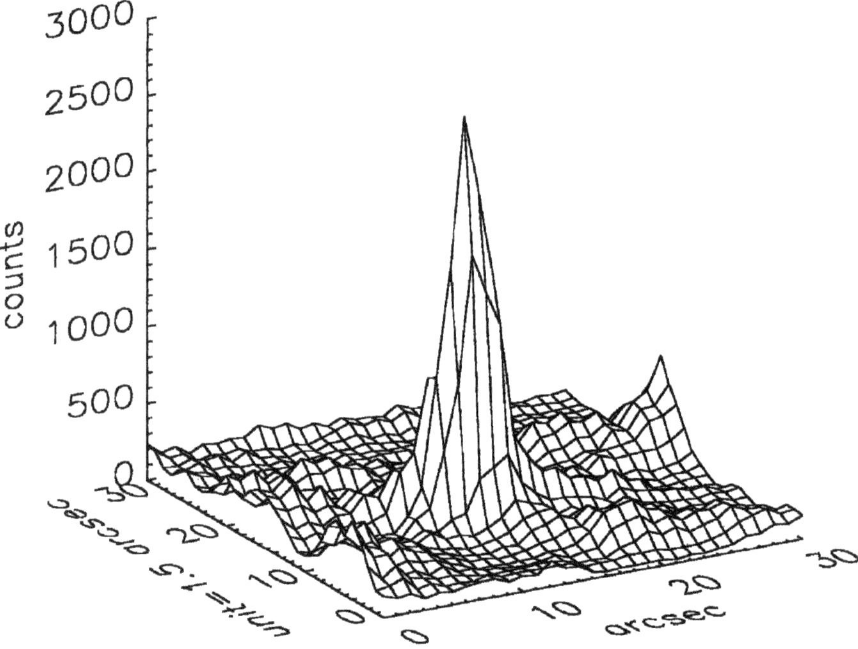

Figure 4. Two-dimensional contours of a bright point seen in the S VI 933 Å line that show the telescope image sharpness.

of 1 arc sec along the slit (Figure 6), we have established histograms of intensities shown in Figure 7. Use of the frequency distribution of the intensity in cells and networks has been developed by Skumanich, Smythe, and Frazier (1975). It has been used in the case of comparison between quiet and coronal-hole regions by Huber *et al.* (1974). Reeves (1976) has applied this technique in combination with the study of the shape of the structures in different lines spanning the 2.0×10^4 to 1.0×10^6 K temperature range to derive properties of the network with the HCO/ATM 5×5 arc sec^2 angular resolution. Here, although the structure of the network in the two temperatures is very similar, the histograms have different shapes. The S VI distribution shows a sharper peak than the Lϵ distribution that corresponds to a sharper separation in network and cell in the S VI line.

An important aspect of the network mapping is the detection of very bright network elements. Figure 8 shows a cut across one very bright network element. The bright network elements seem related to explosive events detected in the spectra. Their lifetime, possible recurrence, and connection with heating mechanisms are been studied in several programmes.

The low level of solar activity at this phase of the solar cycle, with very few plages, has directed our priority to quiet-Sun observation. Yet a few active regions have been mapped. Figure 9 displays a plage mapping in 3 simultaneous

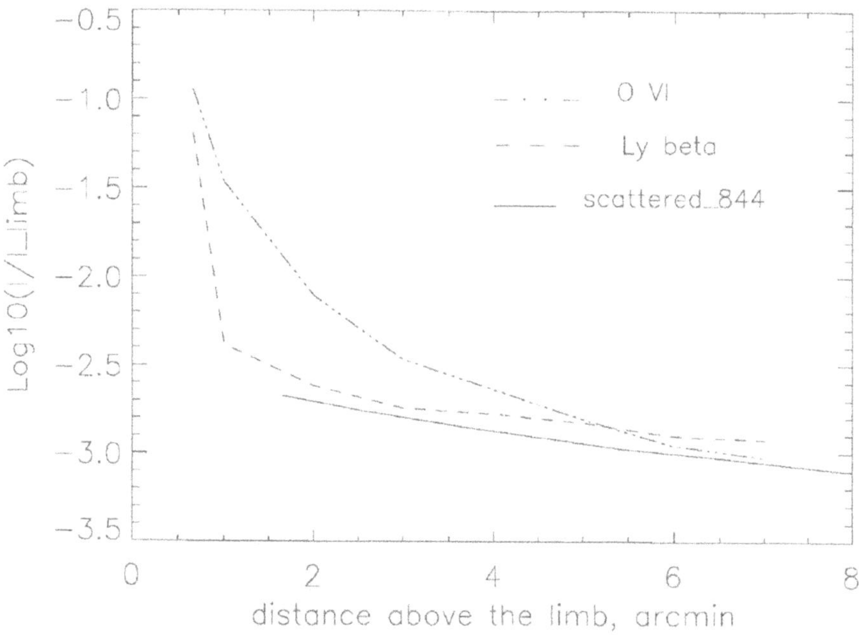

Figure 5. Relative variation of intensity above the limb in a coronal hole compared to the scattered-light level determined at 844 Å.

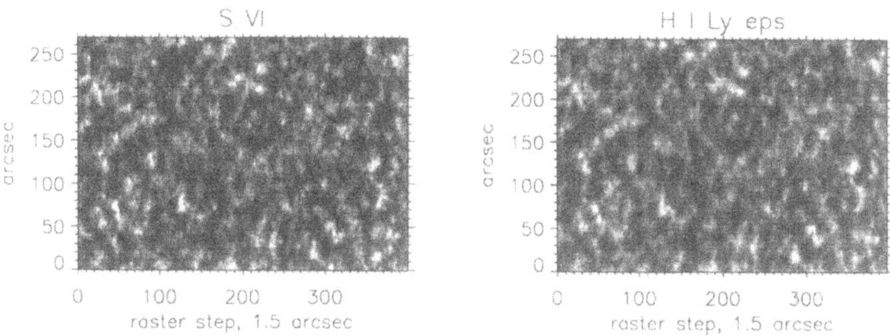

Figure 6. Network maps taken simultaneously at two wavelengths (S VI 933 Å and H I Lϵ) near disk center.

wavelengths (Si VI 1393.755 Å – 63 000 K, O VI 1401.16 Å – 160 000 K, and S VI 1406.06 Å – 100 000 K). The display in two different intensity scales allows us to see the loops' extension in the upper part and the bright cores in the lower part. The sharpest definition of the loops is given at the lowest temperature. Above 100 000 K the loops seems more diffuse than below, and some bright points in the three lines appear at different locations, which may indicate that the loops are different or, on the same loop the temperature varies along the loop.

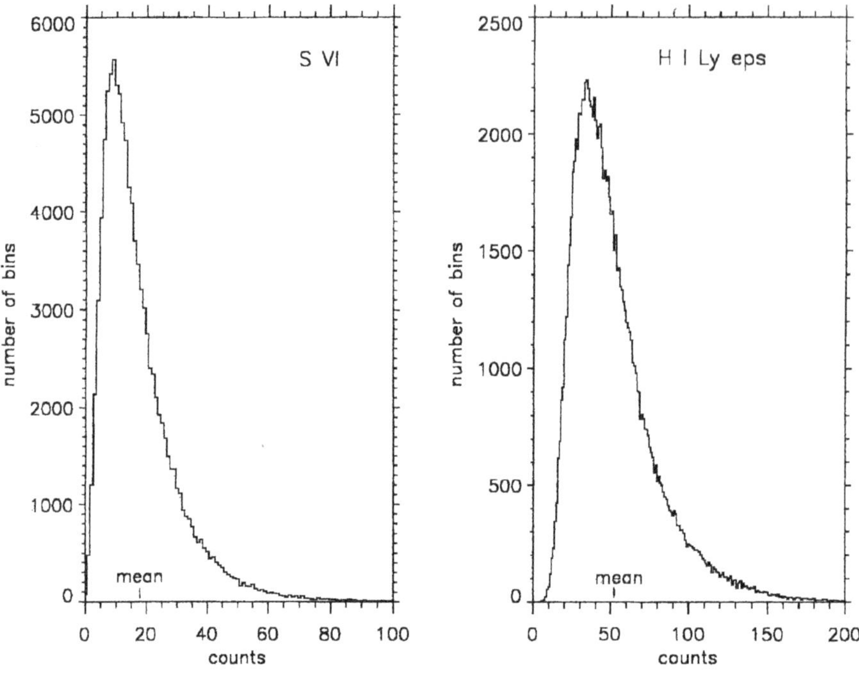

Figure 7. Frequency distribution of intensities within the network maps shown in Figure 6.

3.2. PROMINENCES, STREAMERS, CORONAL HOLES AND PLUMES

The period of minimum solar activity is characterized by many filaments and prominences which appear at rather high latitudes. Already several prominences have been mapped in different lines by SUMER; an example in the C III 977 Å line ($T_e < 7 \times 10^4$ K) is displayed in Figure 10. At the same time we have established a map of the flows that run along the prominence thread; velocities along the line of sight are measured up to 25 km s^{-1}. An accurate estimation based on the topology will enhance this value by a factor of as much as 2.

Two distinct features characterize the external corona during the solar activity minimum period and are also of great importance as far as the solar wind acceleration is concerned: the equatorial streamers and the polar coronal holes. In coordination with other SOHO instruments we have performed several observations to infer the shape and the evolution of these structures. An example of the distribution of intensities in coronal holes, taken at O VI 1032 Å is given in Figure 11. The polar plumes, which are a subset of polar coronal holes, seem to have their roots in some bright points seen on the disk in the polar hole. The time evolution of polar plume structures shows a slow evolution with the solar rotation

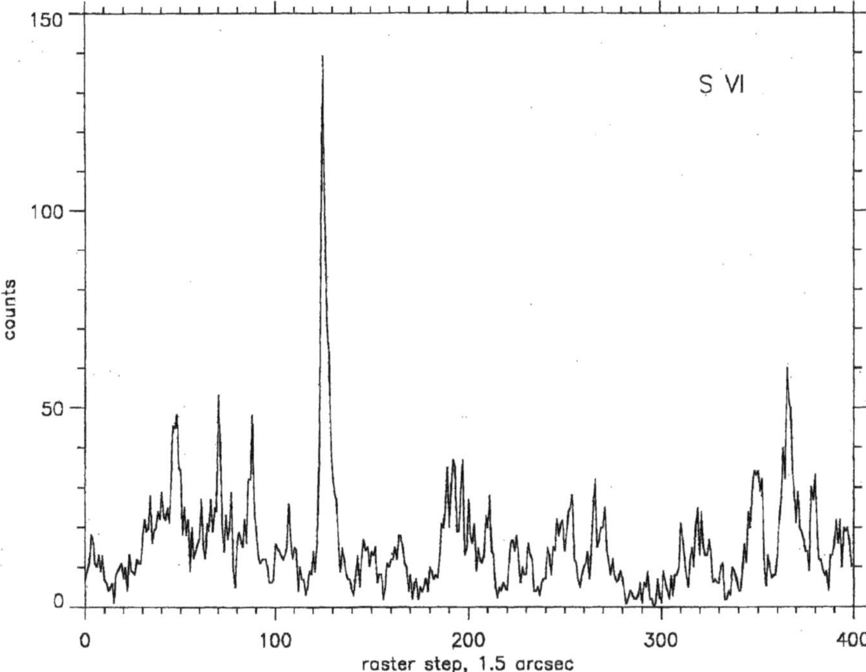

Figure 8. Variation of intensity along a cut across the quiet network and a bright network element from a map taken at S VI 933 Å line.

with intensity ratios of a factor of 2 between the plumes and the inter-plume lanes in the O VI line 1032 Å ($T_e = 3 \times 10^5$ K).

3.3. Full Sun and Limb Brightening

Although the SUMER instrument is not optimized to produce full-Sun images, because it takes a long time to build such a map in given spectral lines, it has through its spectroscopic capabilities the advantage of providing a simultaneous temperature coverage from lines selected within the same detector range or of obtaining line profiles at each point of the scan. This capability permits us to derive Doppler shifts at each location and leads us to produce velocity maps at one or more temperatures in the transition region.

Several solar maps in different wavelengths (temperatures) have been produced by now. In parallel, full-Sun maps in two wavelengths (S VI 933 Å and H I Lε 937 Å) have been generated on a regular basis (every 5 days) to follow the solar evolution.

A subset of these full-Sun maps is shown in Figure 12. An enlargement of a small area where there is an active region demonstrates the strong variation of the intensity in Figure 13. This example illustrates the extreme variation of energy dissipation

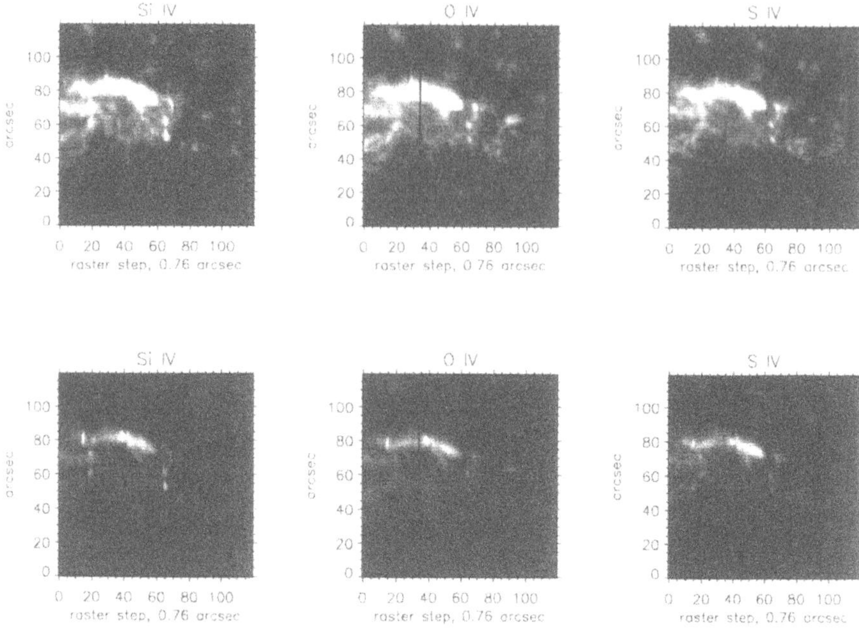

Figure 9. Plage mapped in Si VI 1393.755 Å, O VI 1401.16 Å, and S VI 1406.06 Å in two intensity scales to show the plage core (*lower panel*) and the loops' extension (*upper panel*).

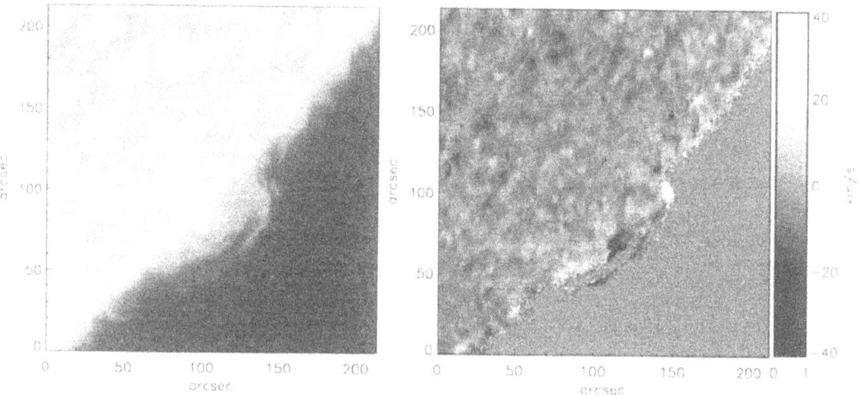

Figure 10. Intensity and velocity (with associated velocity scale) maps of the solar limb and a small prominence taken at C III 977 Å.

in local structures. From these images we can see a very strong limb brightening effect, which will require us to revise our models of the solar atmosphere. This effect is strongly related to the small-scale structure of the network that seems to be formed of small loops (Figure 14).

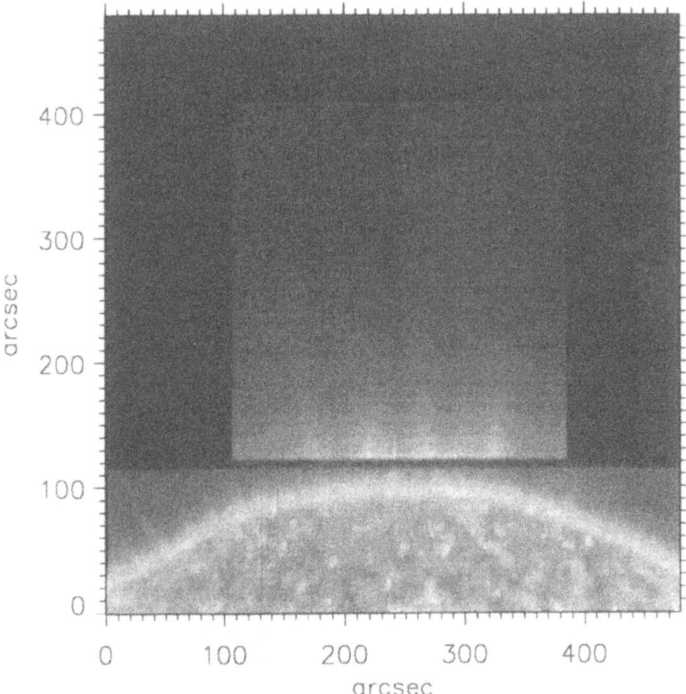

Figure 11. Map of the north polar coronal hole and its extension above the limb in O VI 1032 Å line.

3.4. TEMPORAL EVOLUTION

Temporal evolution in small-scale structures has been established at several temperatures in small areas. Figure 15 displays a temporal series of 8 small rasters (60×120 arc sec^2) taken every 341 s in O VI 1032 Å and H I Lβ lines in a quiet-Sun area near Sun centre. The variation of the network fine structure in shape and brightness is strongly enhanced in O VI line with emergence and disappearance of several small bright points. Time series of small area, combined with line profiles, provide the base to evaluate the local rate of energy dissipation.

4. The Data Processing and Availability

After data decompression and reformating Flexible Image Transport System (FITS) files are generated and put into the SUMER data base before being transferred to the SOHO archive data base. The first version of the data is uncalibrated and not corrected for detector distortions and, if not programmed during the observation, for flat field variations. A SUMER readout of FITS files with binary table extensions

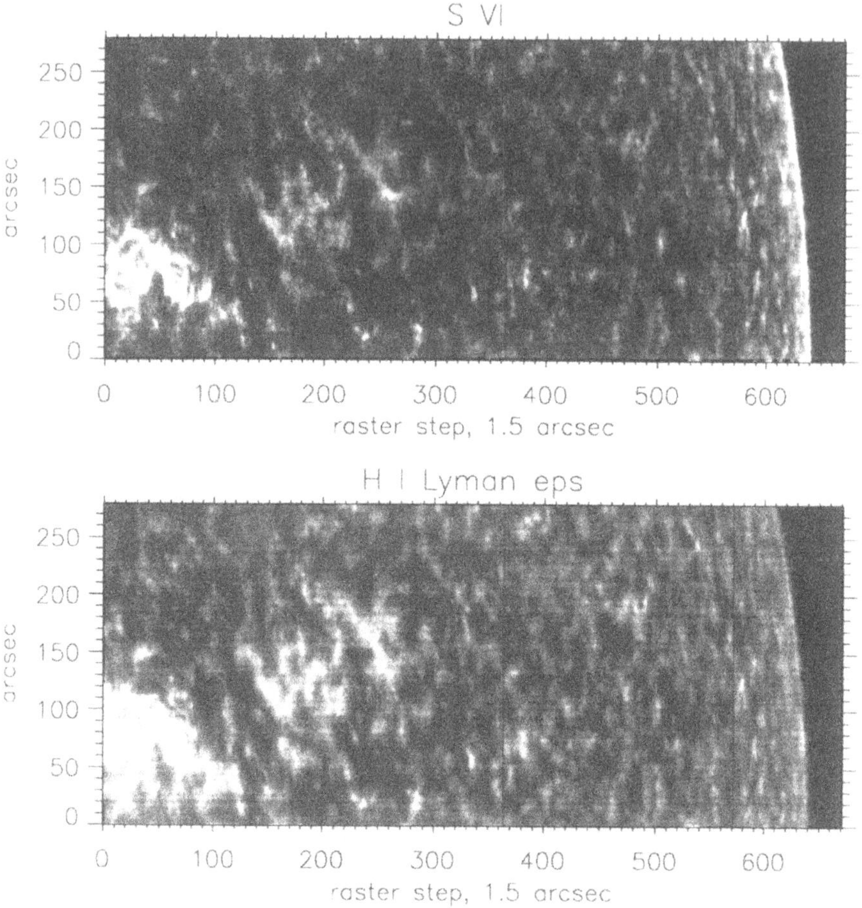

Figure 12. Subset of the full Sun in S VI 933 Å and H I Lε 937 Å lines.

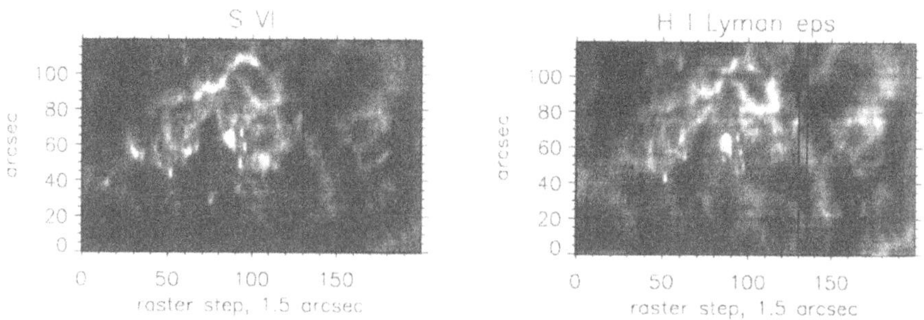

Figure 13. Subset of the full Sun in S VI 933 Å and H I Lε 937 Å lines: map of part of an active region (May 13, 1996, 02:03 UT).

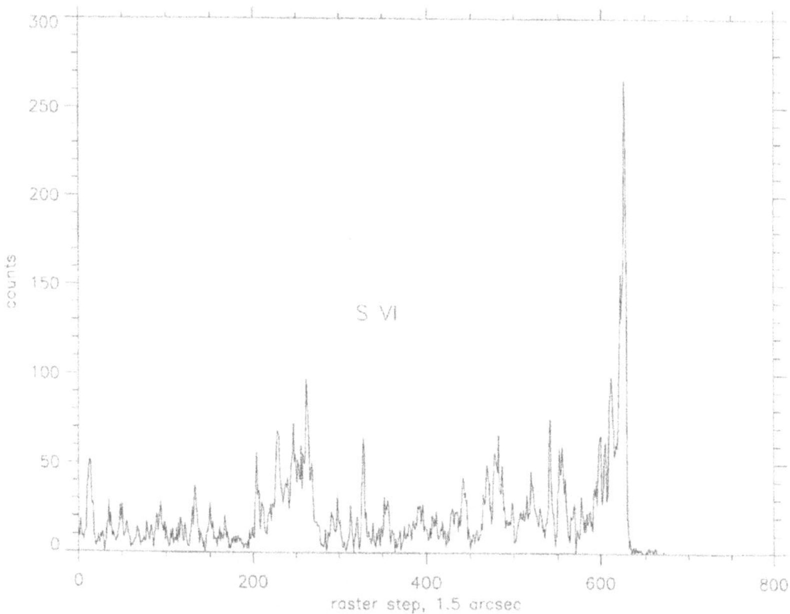

Figure 14. Centre-to-limb variation in S VI at a wavelength of 933 Å.

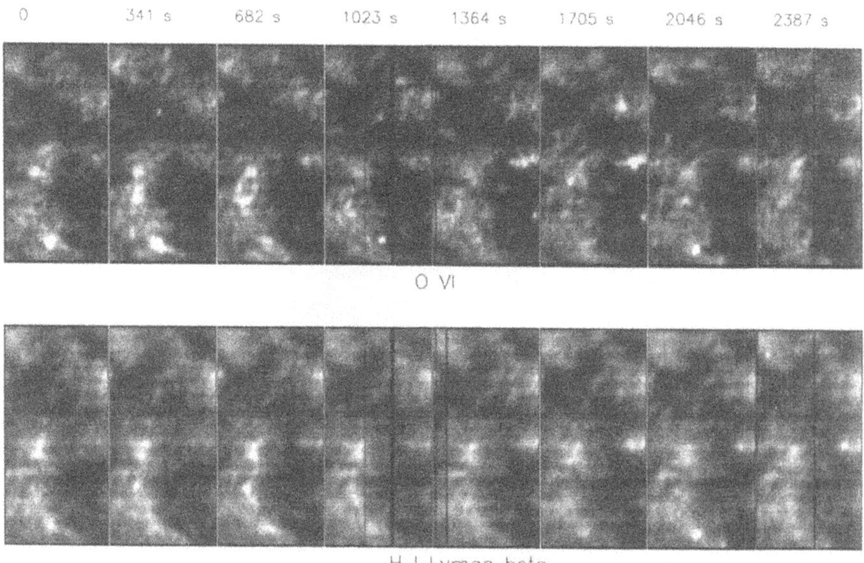

Figure 15. Temporal evolution of a small (60 × 120 arc sec^2) quiet area in O VI 1032 Å and H I Lβ lines.

is available with a growing set of data analysis programmes on the 'Sumop1' anonymous account; on the same account a catalogue can be used to identify the files. Within a few months the catalogue will be available on the Web and, for identified observers, the data corresponding to their programmes will be retrievable.

As agreed within the SUMER team, all data will be available to anyone one year after they have been taken. Given the complexity of the operation of the instrument, it is advisable to contact a SUMER team member during the analysis to make the best use of the data.

5. Conclusion

We have demonstrated the robust properties of SUMER in terms of spatial resolution, off-limb scattered light, and control of these properties against thermal perturbations. First results obtained in imaging various solar structures, either on the disk (network, plage) or at the limb (prominences, coronal holes, streamer, plumes) exhibit the power of combining high spectral resolution, multi-line (i.e., multi-temperature) spectroscopy with the best spatial resolution ever obtained on a solar free-flyer spacecraft. These capabilities are complemented by a data organization which provides easy access to a catalogue and to files in a standard (FITS) format and allows for sophisticated 'searches'.

Acknowledgements

The SUMER project is financially supported by DARA, CNES, NASA and ESA PRODEX programme (Swiss contribution). Additional financial support is being provided by the participating institutions along with general administrative assistance at various phases of the project. The instrument development and/or commissioning has been carried out by a large dedicated team of engineers, scientists, and technicians coordinated by a management support group. The authors would like to thank A. Arondel, S. J. Battel, L. Bemmann, H.-M. Bock, C. de Boer, R. Boucarut, P. Boutry, P. Bouyries, H.-J. Braun, I. Büttner, I. E. Dammasch, A. Dannenberg, W. Donakowski, D. Dumont, W. Engelhardt, H.-G. Engelmann, K. Eulig, A. Fischer,E. Frank, G. A. Gaines, D. Germerott, K. Gräbig, W. Grill, M. A. Gummin, J. W. Hamilton, H. Hartwig, E. Hertzberg, J. Hoberman, G. Hoch, J. Hull, D. Innes-Markiewicz, P. Jelinsky, M. Jhabvala, H. G. Kellner, E. Keppler, R. Keski-Kuha, E. Korzac, T. Kucera, D. Leviton, J. Linant, Y. Longval, T. Magoncelli, C. Maurel, R. Meller, T. Moran, N. Morisset, N. Mosquet, N. Neisen, W. Neumann, J. Osantowski, A. Pauly, W. Paustian, J. Platzer, K. Poser, R. Raffanti, W. Reich, T. Rodriguez-Bell, T. Saha, T. Sasseen, H. Schild, R. Schmidt, H. Schüddekopf, H. Specht, E. Steinmetz, J. M. Stock, U. Strohmeyer, W. Tappert,

W. Thompson, S. Töpfer, J. L. Tom, G. Tomasch, B. Wand, L. Wang, B. Welsh, and G. Wright for their contributions. Scientific advice was provided by J. Mariska and a team of Associate Scientists: J. Aboudarham, B. Aschenbach, W. I. Axford, J. Barnstedt, F. Bely-Dubau, V. Bommier, J. L. Culhane, J. P. Delaboudinière, C. Diesch, G. Doschek, J. G. Doyle, J. Dubau, B. Dwivedi, G. Einaudi, A. Fludra, B. Foing, P. Gouttebroze, R. A. Harrison, H. F. Haupt, E. Haug, J. Hollandt, W.-H. Ip, C. Jordan, F. Keenan, O. Kjeldseth-Moe, B. Kliem, F. Kneer, K. Lang, B. Leroy, O. v.d. Lühe, P. Maltby, H. E. Mason, R. W. P. McWhirter, P. Mein, F. Millier, J. H. Parkinson, E. R. Priest, A. K. Richter, H. Rosenbauer, S. Sahal, D. Samain, B. Schmieder, J. H. M. M. Schmitt, M. Schüssler, R. Schwenn, G. M. Simnett, G. Tondello, J. Trümper, V. M. Vasyliunas, and O. Vilhu.

We particularly acknowledge the contributions of our late colleagues C. Becker, I. Liede, B. Monsignori-Fossi, and B. E. Patchett.

References

Bonnet, R. M., Lemaire, P., Vial, J.-C., Artzner, G., Gouttebroze, P., Jouchoux, A., Leibacher, J. W., Skumanich, A., and Vidal-Madjar, A.: 1978, *Astrophys. J.* **221**, 1032.

Bonnet, R. M., Bruner, E. C., Jr, Acton, L. W., Brown, W. A., and Decaudin, M.: 1980, *Astrophys. J.* **237**, L45.

Brueckner, G. E., Bartoe, J.-D. F., Cook, J. W., Dere, K. P., and Socker, D. G.: 1986, *Adv. Space Res.* **6** (8), 263.

Bruner, E. C., Jr.: 1977, *Space Sci. Instrum.* **3**, 369.

Gabriel, A. H.: 1976, *Phil. Trans. Roy. Soc. London* **A281**, 339.

Hoover, R. B., Walker, A. B. C., Jr., Lindblom, J., Allen, M., O'Neal, R., and DeForest, C.: 1991, *SPIE* **1546** 175.

Howard, R.: 1959, *Astrophys. J.* **130**, 193.

Huber, M. C. E., Foukal, P. V., Noyes, R. W., Reeves, E. M., Schmahl, E. J., Timothy, J. G., Vernazza, J. E., and Withbroe, G. L.: 1974, *Astrophys. J.* **194**, L115.

Leighton, R. B.: 1959, *Astrophys. J.* **130**, 366.

Mariska, J. T. and Withbroe, G. L.: 1978, *Solar Phys.* **60**, 67.

Miller, M. S., Caruso, A. J., Woodgate, B. E., and Sterk, A. A.: 1981, *Appl. Optics* **20**, 3805.

Reeves, E. M.: 1976, *Solar Phys.* **46**, 53.

Reeves, E. M., Huber, M. C. E., and Timothy, J. G.: 1977, *Appl. Optics* **16**, 837.

Saha, T. T., Leviton, D. B., and Glenn, P.: 1996, *Appl. Optics* **35**, 1742.

Skumanich, A., Smythe, C., and Frazier, A. M.: 1975, *Astrophys. J.* **293**, 551.

Tousey, R., Bartoe, J.-D. F., Brueckner, G. E., and Purcell, J. D.: 1977, *Appl. Optics* **16**, 870.

Walker, A. B. C., Jr., Hoover, R. B., and Barbee, T. W., Jr.: 1992, *SPIE* **1546**, 500.

Wilhelm, K., Curdt, W., Marsch, E., Schühle, U., Lemaire, P., Gabriel, A. H., Vial, J.-C., Grewing, M., Huber, M. C. E., Jordan, S. D., Poland, A. I., Thomas, R. J., Kühne, M., Timothy, J. G., Hassler, D. M., and Siegmund, O. H. W.: 1995, *Solar Phys.* **162**, 189.

Wilhelm, K., Lemaire, P., Curdt, W., Schühle, U., Marsch, E., Poland, A. I., Jordan, S. D., Thomas, R. J., Hassler, D. M., Huber, M. C. E., Vial, J.-C., Kühne, M., Siegmund, O. H. W., Gabriel, A. H., Timothy, J. G., Grewing, M., Feldman, U., Hollandt, J., and Brekke, P.: 1997, *Solar Phys.* **170**, 75 (this issue).

HIGH-RESOLUTION OBSERVATIONS OF THE EXTREME ULTRAVIOLET SUN

R. A. HARRISON, A. FLUDRA, C. D. PIKE and J. PAYNE
Astrophysics Division, Rutherford Appleton Laboratory, Chilton, Didcot, Oxfordshire OX11 0QX, U.K.

W. T. THOMPSON and A. I. POLAND
NASA Goddard Space Flight Center, Code 682, Greenbelt, MD 20771, U.S.A.

E. R. BREEVELD, A. A. BREEVELD and J. L. CULHANE
Mullard Space Science Laboratory, University College London, Holmbury St. Mary, Dorking, Surrey RH5 6NT, U.K.

O. KJELDSETH-MOE
Institute of Theoretical Astrophysics, University of Oslo, PO Box 1029, Blindern, 0315 Oslo, Norway

M. C. E. HUBER
ESA, Space Science Department, ESTEC, P.O. Box 299, 2200 AG Noordwijk, The Netherlands

B. ASCHENBACH
Max-Planck Institut für Extraterrestrische Physik, D-85748 Garching, Germany

(Received 27 June, 1996; in revised form 27 November, 1996)

Abstract. This paper presents first results of the Coronal Diagnostic Spectrometer (CDS) recently launched aboard the Solar and Heliospheric Observatory (SOHO). CDS is a twin spectrometer, operating in the extreme ultraviolet range 151–785 Å. Thus, it can detect emission lines from trace elements in the corona and transition region which will be used to provide diagnostic information on the solar atmosphere. In this paper, we present early spectra and images, to illustrate the performance of the instrument and to pave the way for future studies.

1. Introduction

The Sun's extreme ultraviolet (EUV) spectrum, in the range about 150–1200 Å, is extremely rich in bound-bound and free-bound transitions, mainly from trace elements in the Sun's atmosphere. Observations of these transitions can be used to provide a unique facility for probing the corona and transition region. Emission line intensities, intensity ratios and line shapes can be used to obtain diagnostic information on plasmas in the temperature range 10^4 to over 10^6 K, allowing the derivation of characteristics such as temperature, density, abundance, and flows.

Despite the importance of observing this region of the Sun's spectrum, to understand the workings of its atmosphere, relevant observations in the EUV range are rare.

The first EUV observations of the Sun were made in the early 1960s. However, most observations have been made either integrated over the disc or in wide bands (integrated over wavelength). Spectral and spatial information are essential to identifying the processes which may drive such phenomena as coronal heating and

Solar Physics **170:** 123–141, 1997.

solar wind acceleration. In addition, an inspection of emission line wavelengths shows that in order to probe the hottest layers of the Sun's 'quiet' atmosphere (over 10^6 K) one needs to detect radiation from wavelengths well below 300 Å. This allows one to detect emission from sequences of highly ionised iron and magnesium in particular. Finally, since the Sun's atmosphere is such a transient environment, and the processes which we anticipate to be at work may occur over very short time scales, one must demand extremely short time resolutions.

Although several of the OSO satellite series, *Skylab*, and a few rocket and Shuttle flights carried EUV instrumentation, if one demands non-integrated sunlight, spectral information, temporal resolutions of order seconds or better, and observations extending across the EUV but including lines well below 300 Å, we find only one previously flown device which satisfies our requirements. This is the CHASE (Coronal Helium Abundance Spacelab Experiment, see Lang *et al.*, 1990). This experiment operated in the range 150–1335 Å, but it viewed only selected narrow wavelength bands centred on a limited series of prime emission lines. Furthermore, the spatial element was 15 arc sec, which is rather large when compared to known solar structures, and the duration of the observations was just a few days.

It is clear that there is a great potential for very significant advances in understanding the solar atmosphere with extended high resolution (spatial, spectral, temporal) observations of the EUV Sun. The Coronal Diagnostic Spectrometer (CDS) aboard the Solar and Heliospheric Observatory (SOHO) is designed to fulfill this requirement. SOHO was launched on December 2, 1995 and, after a cruise phase, was inserted into an orbit around the Sun–Earth L1 Lagrangian point on February 14, 1996. During the cruise phase and immediately after arrival at L1, the CDS instrument was commissioned. Its scientific operations started in March 1996 and, thus, it is the purpose of this paper to present some initial results and discuss the instrument performance.

It is not the purpose of this paper to present a complete analysis of the performance of the CDS instrument. This requires several months of in-flight analysis and monitoring, and is not appropriate material for this journal. However, it is timely to report on the initial observations of CDS; to display examples of the spectra and images being recorded. Thus, we announce the first set of emission line identifications for this instrument, for quiet-Sun observations, and we show images of a prominence, bright point and active region loop system. In many ways, this paper is a forerunner for in-depth CDS papers, providing reference material and paving the way for future work.

2. The CDS Instrument

CDS is a twin spectrometer operating in the range 151–785 Å. It has been described in detail by Harrison *et al.* (1995). Its principal characteristics are given in Table I. These are the characteristics immediately applicable to scientific considerations;

Table I
The principal characteristics of CDS

General	Spatial element	2–3 arc sec
	Pointing	Anywhere on disc and low corona
	Raster step sizes	E–W = 2.032 arc sec
		N–S = 1.016 arc sec
	Field of view	4 × 4 arc min
Normal incidence	Wavelength ranges	308–381 Å
		513–633 Å
	Prime slits	2 × 240, 4 × 240, 90 × 240 arc sec
Grazing incidence	Wavelength ranges	151–221 Å
		256–338 Å
		393–493 Å
		656–785 Å
	Prime slits	2 × 2, 4 × 4, 8 × 50 arc sec

further information on details such as mass, power, telemetry and geometry are given by Harrison *et al.* (1995).

The spatial element is defined by the point spread function of the CDS Wolter–Schwarzschild type 2 telescope and the performance of the detectors. The telescope feeds two spectrometers, only one of which is used at any point in time. The instantaneous field of view of CDS (without repointing) is 4 × 4 arc min.

The normal-incidence spectrometer (NIS) is stigmatic. It makes use of a toroidal grating system to image long, thin slits on a 2-dimensional detector. Thus, by rastering the Sun's image across the entrance slit, using a flat mirror, one can build up an image of a portion of the Sun in all of the lines which one chooses to telemeter to the ground. Of course, the amount of the full spectrum one would wish to return will depend on requirements on temporal resolution due to telemetry limitations.

The second spectrometer, the grazing-incidence spectrometer (GIS), is astigmatic. Thus, it uses a pin hole slit and the spectrum can be collected, simultaneously from four detectors mounted round the Rowland circle. Images in the lines one chooses to record can be built up by rastering with the scan mirror in one direction and rastering the slit itself in the other.

3. Early Spectra

3.1. Normal Incidence

Let us consider the NIS system first. Figure 1 shows the spectral 'image' as it appeared on the detector for an active region observation made on April 9, 1996.

Two portions of the 1024 × 1024 pixel detector array are shown. Each portion is, in effect, a plot of wavelength (abscissa) against space (ordinate). The wavelength

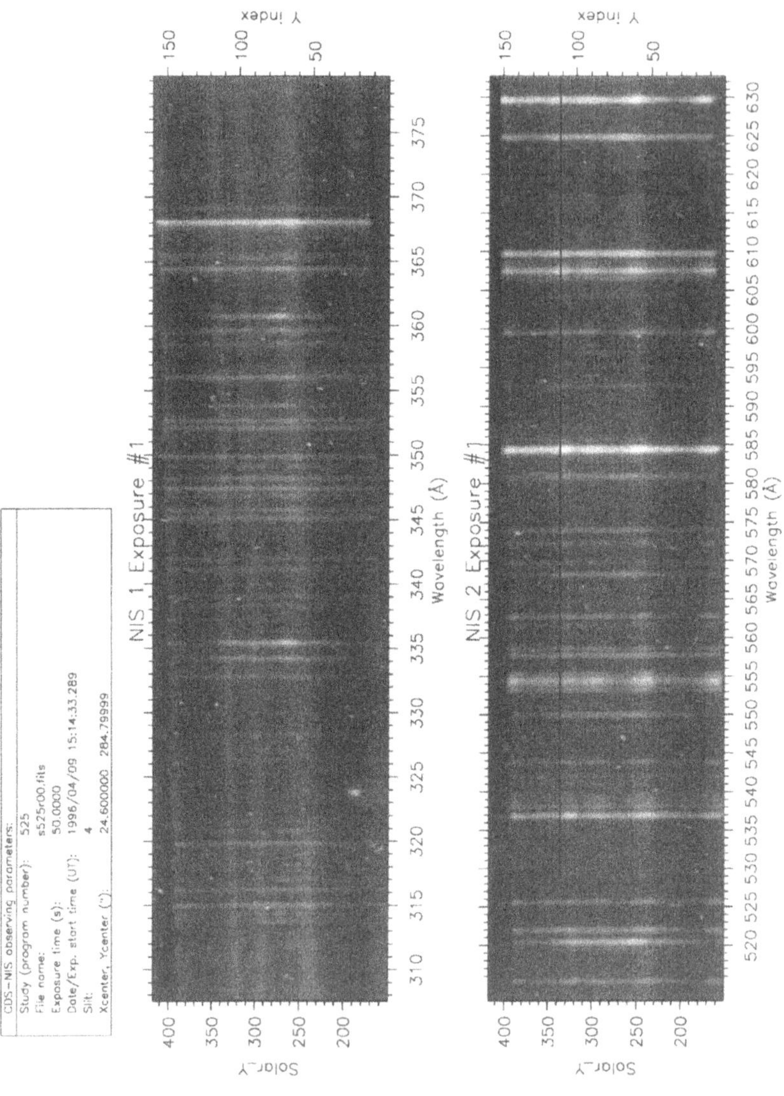

Figure 1. A NIS spectrum as it appears on the face of the detector. This observation, made on April 9, 1996 used a 2 × 240 arc sec slit directed toward an active region.

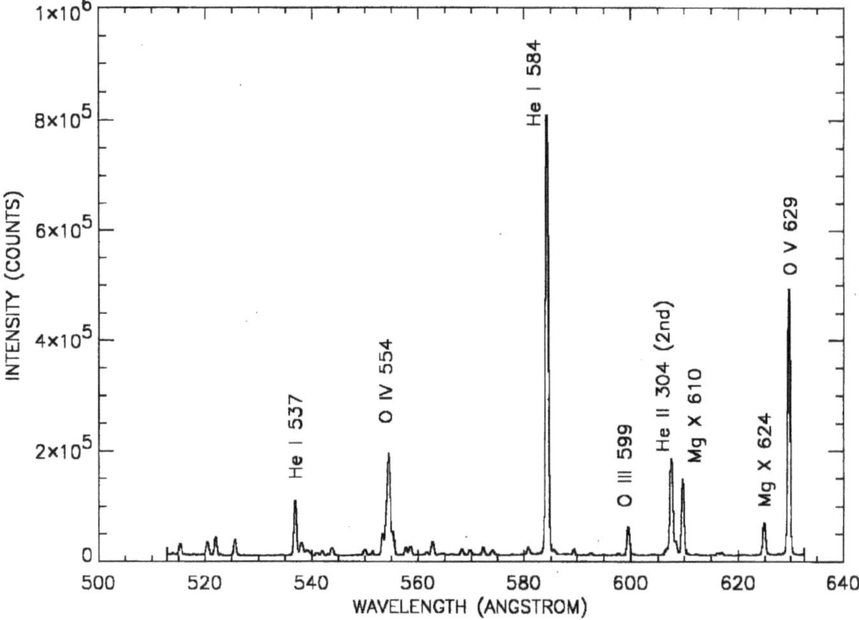

Figure 2. The brightest lines of the 513–633 Å range.

ranges are marked. Spatial measures are given in arcseconds from Sun centre (left scale) and pixels (right scale, each pixel being 1.68 arc sec). The observation is made using the 2 × 240 arc sec slit, so one sees a 2 × 240 arc sec image of the Sun dispersed in wavelength space, across the two ranges.

Clearly one observes many, many spectral emission lines. In addition the image shows evidence for cosmic rays interacting with the detector system, as bright dots or short tracks. This is a feature of all of the imaging instruments on SOHO and, in most instances, can be removed by software techniques. SOHO is located well outside the magnetosphere so such 'hits' were anticipated.

Since the NIS system uses a toroidal grating, the solar image is focussed along the slit. Thus, the structure which is clearly evident along the slit 'images' is solar in nature.

To discuss line identifications we reduce the 2 dimensional displays to a 1-dimensional plot of intensity.

Figures 2–4 show plots of intensity versus wavelength for NIS observation made on May 12, 1996. For this, an observation of a relatively quiet portion of the Sun was summed over a 20 × 240 arc sec region (to avoid a quiet-Sun measure dominated by a particular cell location). For this reason, the intensity is given for a region of 10 × 143 'pixels'. The exposure time for this particular observation was 50 s. So, the count-rate Per pixel can be calculated by dividing by 71 500. A count

Table II
Well-detected lines in the 513–633 Å range

Ion	Wavelength (Å)	Log T	Intensity (see text)
He I	516.60	4.3	d
Si XII	520.67	6.3	d
He I	522.20	4.3	d
O III	525.80	5.0	d
He I	537.03	4.3	b
Ne IV	543.80	5.2	d
Al XI	550.00	6.2	d
O IV	554.52	5.2	b
Ca X	557.76	5.8	d
Ne VI	558.59	5.6	d
Ne VI	562.83	5.6	d
Al XI	567.80	6.2	d
Ne V	569.20	5.5	d
Ne V	572.20	5.5	d
Ca X	574.00	5.8	d
Si XI	580.90	6.2	d
He I	584.33	4.3	a
O III	599.59	5.0	c
He II	303.78	4.7	b
Mg X/O IV	609.79	6.0/5.2	b
O II	616.60	4.5	d
Mg X	624.94	6.0	c
O V	629.73	5.4	a

value of 2×10^5 is equivalent to about 3 counts per second per pixel, where a pixel is of order 2×1.68 arc sec.

Figure 2 shows the brightest lines in the 513–633 Å range. It is dominated by the He I 584 Å line but also contains bright emission from O III, IV, and V, Mg X and He II (in second order). Figure 3 shows an expansion of this plot with the brighter lines extending out of the intensity range. A selection of well identified lines are marked. A summary of the bright lines in this range is given in Table II. However, one should realise that the available spectral emission lines will vary depending on the targets being observed. These data are for a 'modestly active' portion of quiet Sun.

Table II illustrates well the 'tools' of this first NIS wavelength range. Many more lines may be detected through long-duration exposures or in different solar features but this shows a selection of lines from a wide range of temperatures. Our primary purpose at this stage is to show that identifications have been made and to enable researchers to identify the emission lines required for detailed studies of particular

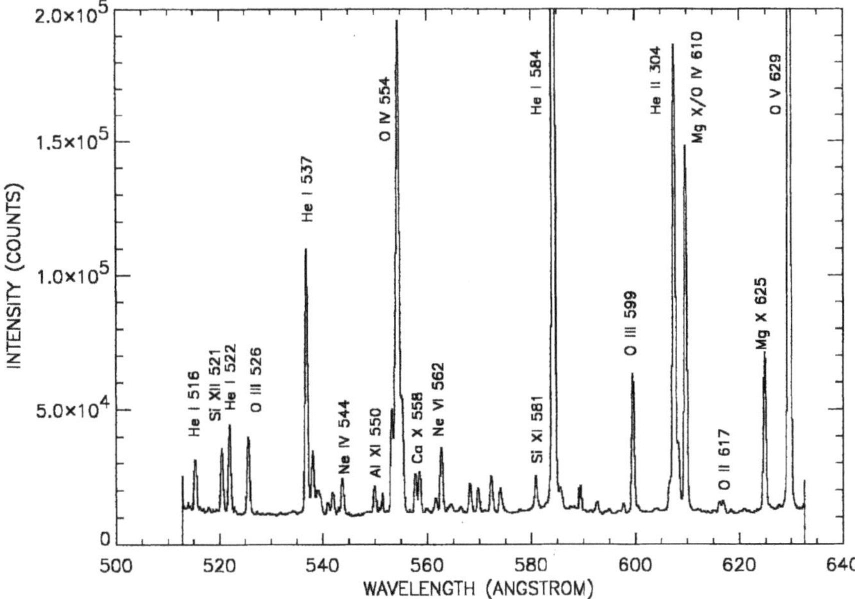

Figure 3. A plot of the 513–633 Å spectrum with an expanded intensity scale.

aspects of solar physics. Actual counts are not listed here since a full, calibrated spectral 'atlas' will be the subject of a future paper, with a proper consideration of the variations between solar features. Thus intensity is given on an arbitrary graded scale from 'a' to 'd' with divisions defined as (a) intensities greater than O IV 554 Å, (b) intensities between Mg X 625 Å and O IV, (c) intensities between He I 522 Å and Mg X, and (d) well-identified peaks at intensities less than this. Remember that this plot is from one observation and should be considered as an illustration of the specta observed. The temperatures for maximum ion abundance for all of the tables are taken from Arnaud and Rothenflug (1985).

It is important to note, also, that the wavelengths listed in Table II are not derived accurately from CDS data. Although a pixel to wavelength calibration has been applied, the locations to two decimal places are the product of past observation. Certainly the consistency of the CDS line locations with the accepted wavelength values will be the subject of future work.

The emission line structure seen in the shorter wavelength range of the NIS is shown in Figure 4. This region is more complex because of generaly lower intensities and the crowded nature of the iron ion spectrum in particular. In addition, since most lines are weaker than the longer wavelength range lines, the occasional cosmic-ray hit on the detector is more apparent. In Figure 4, the major cosmic-ray events are marked with a cross. As with Table II, we list the lines from this wavelength range in Table III.

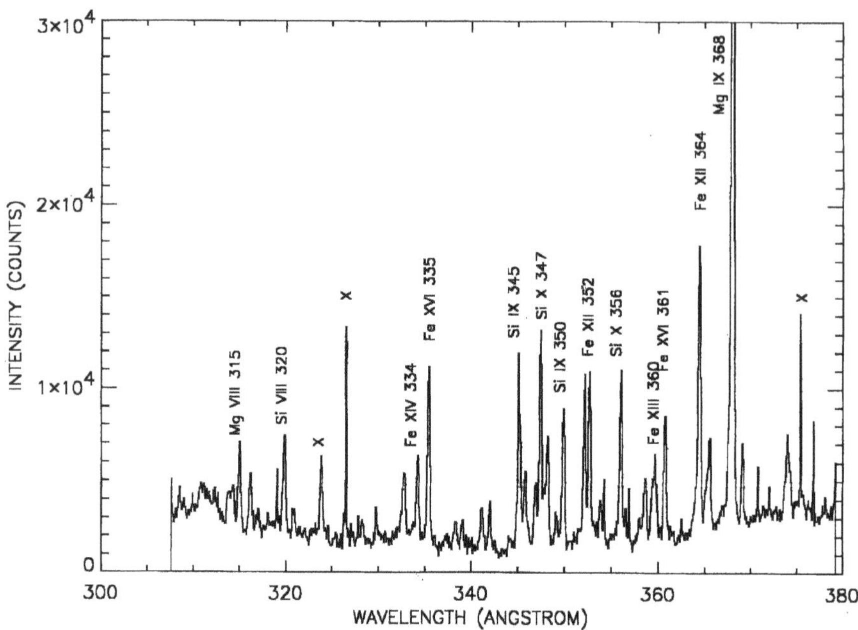

Figure 4. The NIS 308–381 Å range.

Again, we list intensities using an arbitrary scale defined as (a) intensities greater than that of the Fe XII 364 Å line, (b) intensities between the Fe XVI 335 and Fe XII 364 Å lines, and (c) other well detected lines.

Tables II and III demonstrate that the CDS NIS is able to detect the primary lines required by CDS for these two bands. The longer wavelength band contains an enormous spread of temperatures and well separated, bright lines, useful for dynamic studies throughout the corona and transition region and for analyses of structure and evolution at different temperature values. The shorter wavelength band is more limited in temperature range. It contains good coverage of the hottest plasmas (to $\log T = 6.3$) and is geared towards the production of, in particular, density diagnostics using ratios such as the Si X 347/356, Si IX 345/349 and Fe XIII 359/348 lines.

Note, again, that the spectra presented here are for one observation and should be regarded as representative of what CDS can detect rather than a definitive spectral 'atlas'. Before moving on to the GIS wavelength identifications, we should briefly examine the spectral resolution performance of the NIS. Figure 5 shows expanded wavelength plots around four of the NIS lines – He I 584 Å, O V 629 Å, Mg IX 368 Å, and Fe XVI 335 Å.

A quick examination of the four principal lines in these plots indicates line full width at half maximum intensity values of about 0.6 Å for the 584 and 629 Å lines,

Table III
Well-detected lines in the 308–381 Å range

Ion	Wavelength (Å)	Log T	Intensity (see text)
Mg VIII	315.02	5.9	c
Si VIII	316.21	5.9	c
Mg VIII	317.01	5.9	c
Si VIII	319.83	5.9	c
Al X	332.77	6.1	c
Fe XIV	334.17	6.2	c
Fe XVI	335.40	6.3	c
Si IX	345.13	6.0	b
Si X	347.40	6.1	b
Fe XIII	348.18	6.2	c
Si IX	349.87	6.0	c
Fe XII	352.11	6.2	c
Fe XI	352.67	6.1	c
Fe XIV	353.83	6.2	c
Si X	356.04	6.1	c
Fe XIII	359.64	6.9	c
Fe XVI	360.76	6.3	c
Fe XII	364.47	6.9	b
Mg IX	368.06	6.0	a
Fe XI	369.16	6.1	c

and 0.3–0.4 Å for the 368 Å and 336 Å lines. The line widths are a function of the natural line width (thermal and non-thermal) but they are dominated by the width of the image of the slit on the detector. In practice, the spectral resolution of CDS is dependent on our ability to separate lines and this can be done for fractions of a line width.

3.2. Grazing Incidence

The GIS system utilises 1-dimensional detectors which view four portions of the EUV spectrum between 151 and 785 Å. Figures 6–9 show examples of quiet-Sun spectra for each band in turn. Figures 6, 8, and 9 are produced from data taken on February 10, 1996. The GIS detector system for that particular observation was tuned to the observation of the He II 304 Å line in the second wavelength band and thus several of the other lines are subdued. As a result, for this particular range, we choose to show a more suitable example, in Figure 7, from an observation made on June 10, 1996. Figures 6, 8, and 9 were from a 36 exposure sequence of exposure time 50 s, for a 2 × 2 arc sec slit. The exposures have been summed. Thus, to obtain a count rate per pixel (i.e., per 2 × 2 arc sec of the Sun), the intensities must be divided by 50 × 36 = 1800. Similarly, for Figure 7, we show the sum of 225

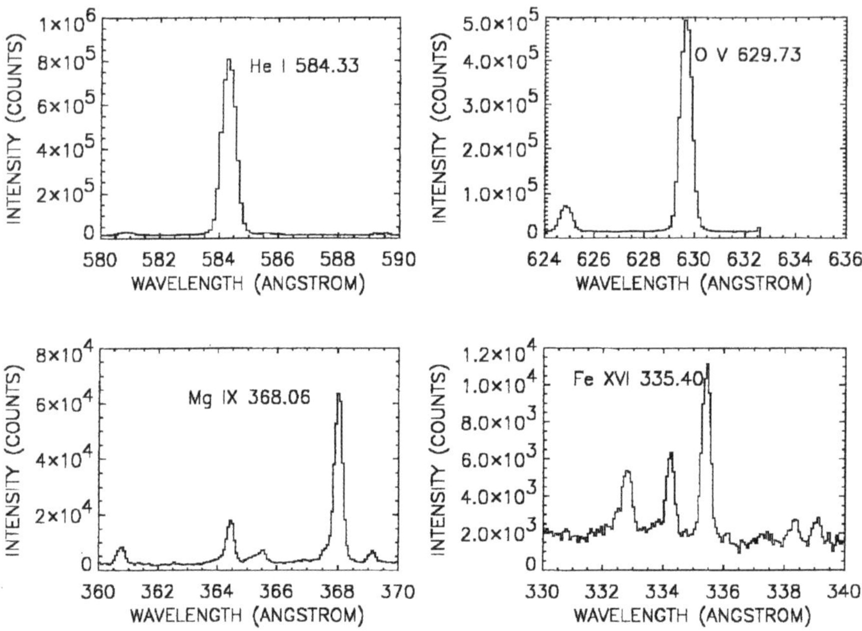

Figure 5. Four expanded wavelength plots to illustrate the line profiles from the NIS.

exposures of 50 s for the same slit, and a division by $50 \times 225 = 11\,250$ would be necessary. Remember that this is for quiet Sun and for the smallest slit!

Lines which can be easily identified from Figures 6–9 are tabulated in Tables IV and V. The tables do not contain all of the lines which can be seen since some of the weaker line identifications need to be confirmed. However, it is clear that we have a good range of well resolved lines with which to examine the solar atmosphere. Again, we use a rough scale for peak intensity, set at approximately (a) ten times the background level, (b) 5–10 times the background level, and (c) well-identified lines at less than this.

The first GIS band (151–221 Å), shown in Figure 6, contains a superb range of iron ion lines, from Fe IX to Fe XIII. Most of the ions are from the temperature range $\log T = 5.9–6.2$, so it offers a detailed analysis of the hottest layers of the corona. Many of the iron ions are useful for density diagnostics (see Harrison *et al.*, 1995, and references therein). It contains also the O VI emission which has long been advertised as a useful temperature diagnostic in the corona. One should note that these are early displays of the CDS spectra. The plots for Figures 6–9 contain a wavelength scale which was produced from early identifications of the principal emission lines. As further line identifications are confirmed, the wavelength scales will be refined. In general, the emission line locations are fairly accurate on these plots, though some lines (such as the Ne VIII 780 Å line on the last plot) are slightly displaced. In addition, some edge effects of the detector system, which have yet

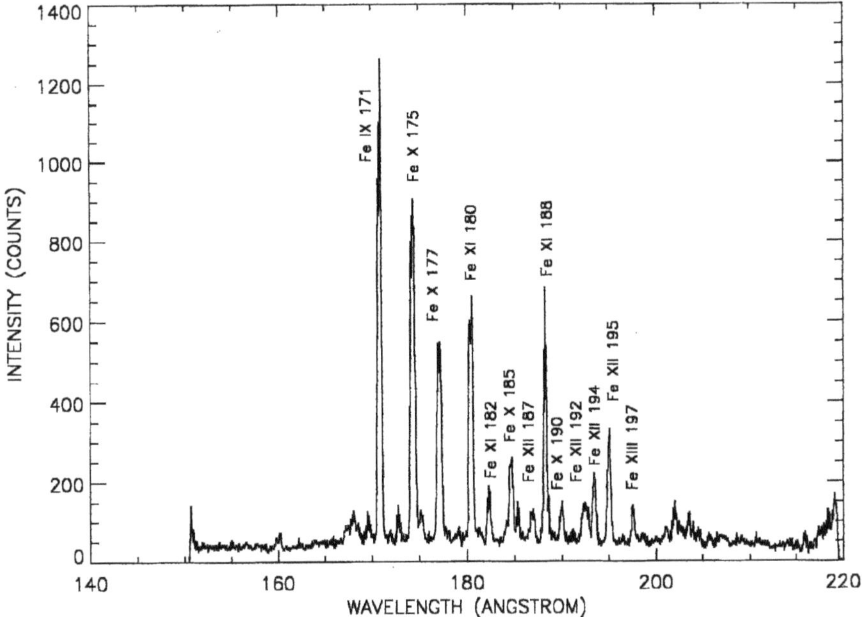

Figure 6. The GIS 151–221 Å band.

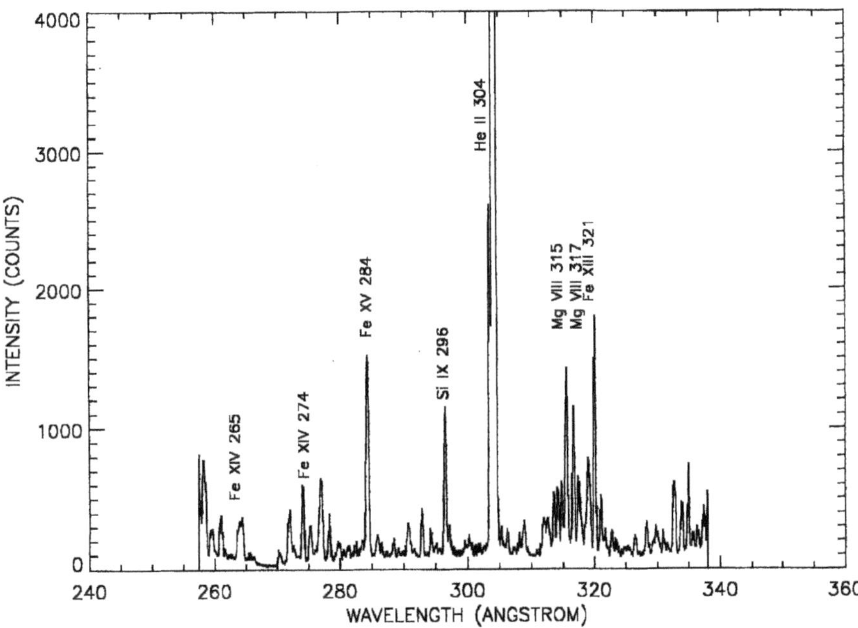

Figure 7. The GIS 256–338 Å band.

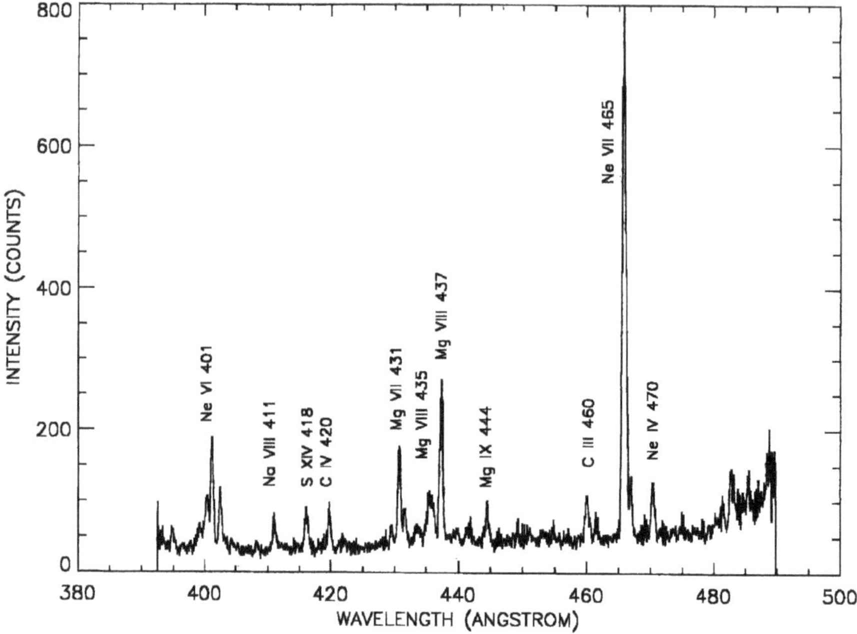

Figure 8. The GIS 393–493 Å band.

to be corrected are evident due to the increased emission at the extremes of the wavelength ranges. Again, such effects will be corrected as the mission progresses.

Figure 7 shows the second band (256–338 Å) which contains a range of lines from iron, silicon and magnesium as well as the He II 304 Å line. Again, some of these are density sensitive pairs (Harrison *et al.*, 1995) and, in particular, the Si VII and Mg VIII lines are important for observations just at the top of the transition region temperature range. The Mg VIII line at 315.02 Å is a density diagnostic with the Mg VIII line at 430.46 Å in the next wavelength band. This second band has a number of lines which we do not identify at this time, such as the components of the group between 310 and 320 Å. Such identifications need further, more detailed work.

The third band (393–493 Å), shown in Figure 8, contains emission lines from many cooler ions, such as C III and IV, Ne IV and VI and a range of Mg ions, and this range, along with the last band will provide useful tools for studying the transition region. In this band, the Mg lines offer some density diagnostic capabilities.

Figure 9 shows the last band (656–785 Å) which contains many of the coolest emission lines detected by CDS, including emission from O II and III, Ne I and N III,

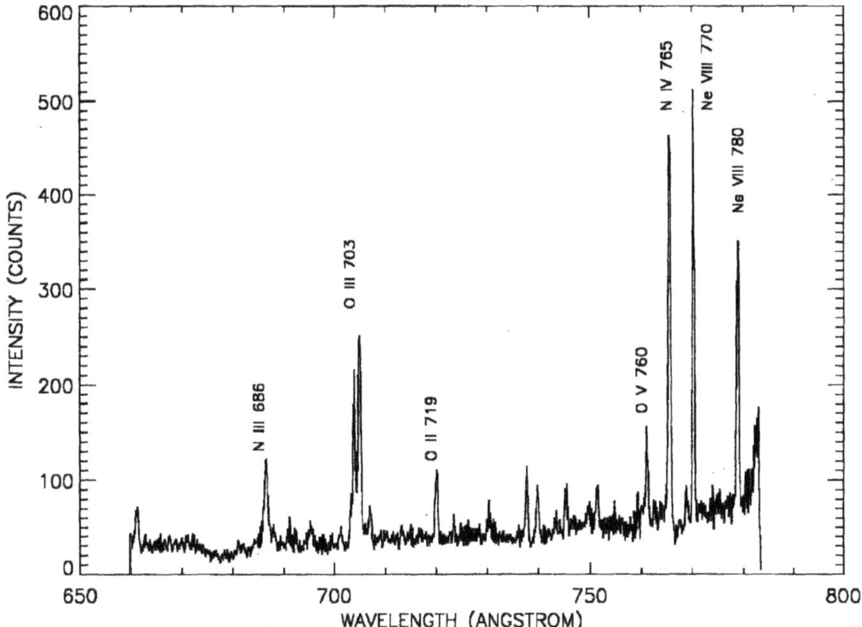

Figure 9. The GIS 656–785 Å band.

but also contains the important, bright Ne VIII and N IV lines which will be valuable lines for high temperature transition region studies.

It is clear that the GIS wavelength bands contain a rich array of emission line 'tools' for selecting and analysing emissions from the different temperature regimes of the solar atmosphere, with plenty of diagnostic ratios available. However, it should be noted that the GIS spectra shown here are representative of quiet-Sun conditions. Intensities and intensity ratios will vary depending on the solar feature being observed. Also, one should note that an intensity calibration has not been applied to these spectra.

Figure 10 shows expanded wavelength regions around 180 and 770 Å. It shows that for both wavelength bands the full width at half maximum intensity is of order 0.4–0.5 Å. Due to the influence of fixed patterning in the detector processing, some structure is found in the GIS line shapes. This effect will be reported on in full, but to avoid confusion, in this figure, we have applied a simple boxcar average over 3 pixels. As with the NIS, the line width is dominated by the instrumental effects such as the width of the slit image on the detector face.

Table IV
Well-detected lines in the GIS bands 151–221 Å and 256–338 Å

Ion	Wavelength (Å)	Log T	Intensity (see text)
Fe IX	171.07	5.9	a
O VI	173.08	5.5	c
Fe X	174.53	6.0	a
Fe X	177.24	6.0	b
Fe XI	180.40	6.1	b
Fe XI	182.17	6.1	c
Fe X	184.54	6.0	c
Fe XII	186.88	6.2	c
Fe XI	188.22	6.1	b
Fe X	190.04	6.0	c
Fe XII	192.39	6.2	c
Fe XII	193.51	6.2	c
Fe XII	195.12	6.2	b
Fe XIII	196.53	6.2	c
Fe XII/XIII	201.12	6.2	c
Si X	261.06	6.1	c
Fe XIV	264.79	6.2	c
Fe XVI	270.52	6.3	c
Si X	271.99	6.1	c
Fe XIV	274.20	6.2	c
Si VII	275.37	5.8	c
Si X	277.27	6.1	c
Fe XV	284.16	6.3	b
Fe XII	291.01	6.2	c
Si IX	292.80	6.0	c
Si IX	296.12	6.0	b
He II	303.78	4.7	a
Mg VIII	315.02	5.9	b
Mg VIII	317.01	5.9	c
Fe XIII	320.80	6.2	b

4. Images

CDS produces images in selected wavelengths by rastering in the solar east–west direction with a flat scan mirror, and in the solar north–south direction by physical motion of the slit. Figure 11 shows an example of a NIS images. It shows a 4 × 4 arc min area near the limb taken on April 14, 1996. The region is mapped, simultaneously in 11 selected emission lines. The coolest lines, He I 584 Å and 537 Å ($\log T = 4.3$) clearly show a prominence. The prominence feet are apparently separated by about 2 arc min, over 100 000 km, and the loop reaches some 60 000 km above the limb. The solar 'surface' shows the mottled

Table V
Well-detected lines in the GIS bands 393–493 and 656–785 Å

Ion	Wavelength (Å)	Log T	Intensity (see text)
Ne VI	401.14	5.6	b
Na VIII	411.16	5.9	c
Fe XV/S XIV	417.26/0.60	6.3/6.4	c
C IV	419.50	5.0	c
Mg VIII	430.46	5.9	c
Mg VII	431.33	5.8	c
Mg VIII	436.73	5.9	b
Mg VII	434.93	5.8	c
Mg IX	444.03	6.0	c
C III	459.50	4.8	c
Ne VII	465.22	5.7	a
Ne IV	469.80	5.2	c
N III	685.83	4.9	c
O III	702.98	5.0	b
O II	718.53	4.5	c
Ne I	735.90	4.2	c
Ne I	743.70	4.2	c
S IV	748.90	5.0	c
S IV	750.20	5.0	c
O V	760.40	5.4	c
N IV	765.14	5.2	b
Ne VIII	770.40	5.8	a
Ne VIII	780.30	5.8	b

supergranular structure. The O III 599 Å ($\log T = 5.0$), O V 629 Å ($\log T = 5.4$) and Ne VI 562 Å ($\log T = 5.6$) lines represent transition region temperatures and display similar supergranulation but little evidence of the prominence (it can just be seen in the Ne VI image). The coronal lines of Mg IX 368 Å ($\log T = 6.0$), and Mg X 624 Å ($\log T = 6.0$) show some structure on the disc, relating to the brighter portions of the supergranular patterns, but in general, the disc looks dark and we see a bright, narrow band of emission from the corona, lying above the limb. The brightest region is on the limb at the southern extreme (bottom) of the image and this is emphasized by the emission from the hottest lines, Fe XIV 334 Å ($\log T = 6.2$), Fe XVI 335 Å ($\log T = 6.3$) and Si XII 520 Å ($\log T = 6.3$), suggesting that this location, just south of the prominence footpoint contains plasma at 2 000 000 deg.

Images such as this really display the power of CDS. We can literally take the atmosphere to pieces and inspect the different temperature regimes. Inclusion of diagnostic pairs and time series allows a truly thorough examination of the processes at work and the resulting evolution.

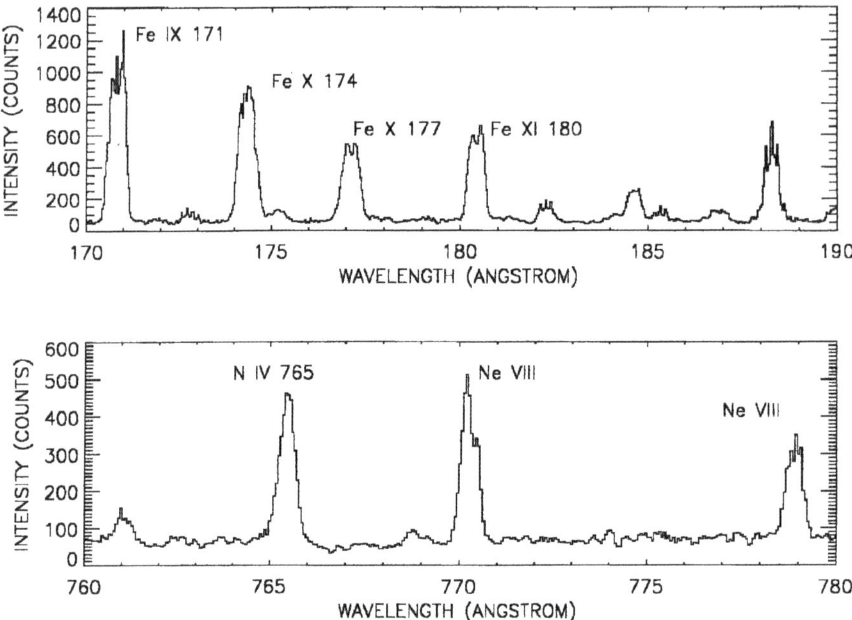

Figure 10. Two expanded wavelength plots to illustrate the line profile from the GIS.

Figure 12 shows two further observations, one of a bright point and one of an active region loop system. As with Figure 11, all of the images are made by rastering with a 2 × 240 arc sec slit over a 4 × 4 arc min area of the Sun. The top left image shows emission from the Ne VI line at 562 Å of an active region, and the eastern limb on June 19, 1996. This represents a temperature of $\log T = 5.6$. Loop structures can be seen quite clearly, rooted in the bright region just within the limb. One particularly bright loop leans northward, even though both feet are in the bright region. In addition, loop structures can be seen extending out of the north (top) and south (bottom) extremes of the image. The top right image, taken at the same time, is of Mg IX 368 Å light, i.e., one million degree plasma. The bright active region core seen in Mg IX encompasses the Ne VI bright region, and a system of Mg IX loops can be seen entending from the active region. The inclination and extent of one of the loops appears to coincide with the bright Ne VI loop which leans northwards. Observations of this kind, across the CDS wavelength ranges will allow a detailed analysis of the temperature and density structure, and their variations with time, of active region loops.

The lower two images show an observation taken on March 18, 1996 of a bright point. The O III 599 Å image on the left shows the $\log T = 5.0$ emission, with plenty of supergranular structure. The brightest patch is to the eastern (left) side of the image but there is evidence for numerous bright patches of scale size about 4–5 arc sec. The Mg IX 368 Å image, which is simultaneous in space and time,

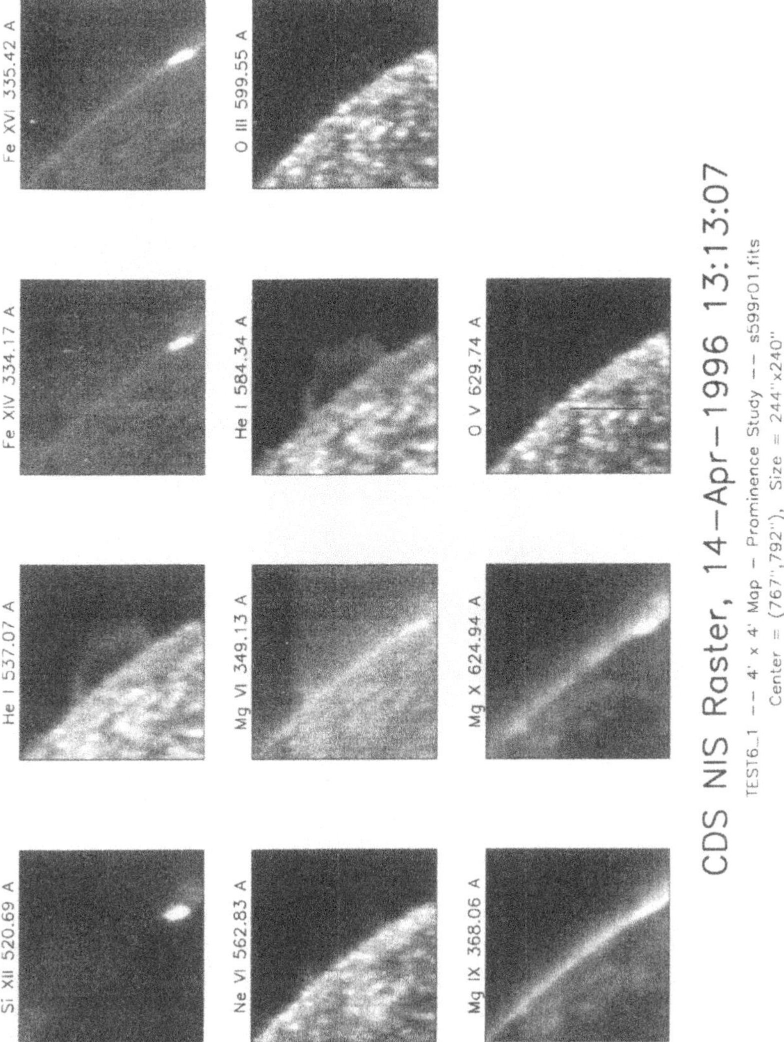

Figure 11. **A limb region mapped by CDS in 11 wavelength bands simultaneously, on April 4, 1996.**

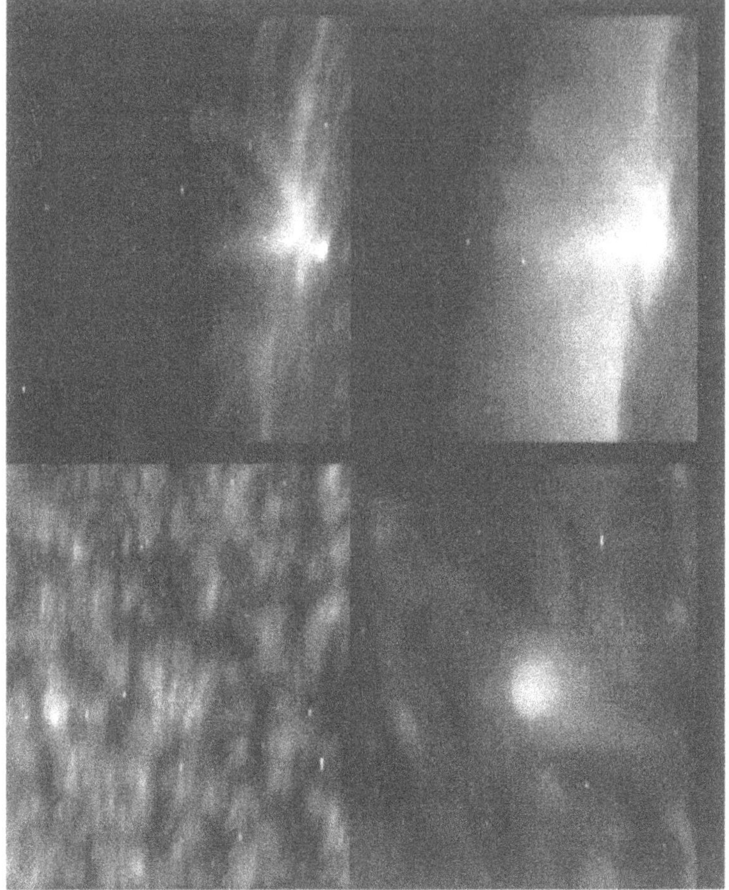

Figure 12. Top: an active region loop system in Ne VI 562 Å (*left*) and Mg IX 368 Å (*right*). *Bottom*: a bright point in O III Å (*left*) and Mg IX 368 Å (*right*).

tells a very different story. There is a very intense region, a bright point, located in the centre of the image. Some of the weaker structures map to the O III features, but in general the image is much less structured than the O III image and contains the bright point. There is little evidence for the existence of the bright point in the cooler emission even though the magnetic structure of the bright point must be rooted to the lower atmosphere. Again, multi-wavelength observations of this kind will allow us to examine the temperature profile of the bright point, allowing an understanding of how the emission from the transition region and corona evolve as the bright point develops, and hopefully providing some understanding of the nature of these features.

5. Discussion

The main messages of this work are the following:

(1) The first set of line identifications for the CDS EUV wavelength ranges has been given. This has identified an excellent range of well detected lines in the temperature range $\log T = 4.3$ to 6.3 with numerous density diagnostic pairs. In short, the tools for much useful diagnostic work have been identified.

(2) We have confirmed that the CDS instrument is also capable of producing high quality images of a number of solar phenomena in wavelength ranges which can be selected by the user.

This work should be regarded as a forerunner of many in depth studies using CDS data. It is hoped that these first results will be used by many, to provide information for the development of future studies using the instrument for example, by allowing useful line selections to be made. In this sense, this paper is opening a new era in solar coronal and transition region EUV studies.

Acknowledgements

The CDS instrument was built and is operated by an international consortium led by the Rutherford Appleton Laboratory (RAL). We would like to thank the many members of the hardware groups at RAL and the Mullard Space Science Laboratory in the UK, the Max Planck Institut für Extraterrestrische Physik and Physicalisch-Technische Bundesanstalt in Germany, the Goddard Space Flight Center in the U.S.A., and the University of Oslo, in Norway. SOHO is a project of international cooperation between ESA and NASA.

This paper is dedicated to the memory of Bruce E. Patchett who, as the original Principal Investigator for CDS, brought the instrument through the design phase, to successful selection in 1988. The ongoing success of CDS is a tribute to his foresight and skills.

References

Arnaud, M. and Rothenflug, R.: 1985, *Astron. Astrophys. Suppl.* **60**, 425.
Harrison, R. A., Sawyer, E. C., and 37 co-authors: 1995, *Solar Phys.* **162**, 233.
Lang, J., Mason, H. E., and McWhirter, R. W. P.: 1990, *Solar Phys.* **129**, 31.

APPLICATION OF SPECTROSCOPIC DIAGNOSTICS TO EARLY OBSERVATIONS WITH THE SOHO CORONAL DIAGNOSTIC SPECTROMETER

H. E. MASON and P. R. YOUNG
Department of Applied Mathematics and Theoretical Physics, Silver Street, Cambridge CB3 9EW, U.K.

C. D. PIKE, R. A. HARRISON and A. FLUDRA
Rutherford Appleton Laboratory, Chilton, Didcot, Oxfordshire OX11 OQX, U.K.

B. J. I. BROMAGE and G. DEL ZANNA
Centre for Astrophysics, University of Central Lancashire, Preston PR1 2HE, U.K.

(Received 4 July, 1996; in revised form 23 October, 1996)

Abstract. The Coronal Diagnostic Spectrometer (CDS) has as a scientific goal the determination of the physical parameters of the solar plasma using spectroscopic diagnostic techniques. Absolute intensities and intensity ratios of the EUV spectral emission lines can be used to obtain information on the electron density and temperature structure, element abundances, and dynamic nature of different features in the solar atmosphere. To ensure that these techniques are accurate it is necessary to interface solar analysis programs with the best available atomic data calculations. Progress is reported on this work in relation to CDS observations.

1. Introduction

The CDS instrument provides an opportunity to study the EUV (151–785 Å) wavelength range with good spectral, spatial, and temporal resolution. Details of the CDS instrument and science objectives can be found in Harrison *et al.* (1995). CDS has the capability of obtaining temperature and density estimates from diagnostic line ratios; emission measure distributions from sequences of ions and elemental abundance variations; flow patterns from Doppler shifts. These spectroscopic diagnostic techniques depend on an accurate knowledge of the relevant atomic physics processes, including radiative decay rates, electron collisional excitation, ionisation and recombination. The electron excitation rates required for the analysis of CDS spectra were the subject of an Atomic Data Assessment Workshop held in March 1992 (Lang, 1994). These atomic data are being continuously improved. The Iron Project, an international collaboration of atomic physicists, aims to calculate new electron excitation rates for all the iron ions (Hummer *et al.*, 1993). Several atomic databases and analysis packages have been developed and adapted for application to the SOHO data. The Atomic Data and Analysis Structure (ADAS) package (Summers *et al.*, 1996) provides an accurate method of atomic data analysis, storage and processing. ADAS solves the generalised collisional-radiative model for equilibrium or non-equilibrium conditions. In this paper we use the CHIANTI atomic database (Dere *et al.*, 1996) and associated software for initial comparisons

Solar Physics **170:** 143–161, 1997.

with the CDS spectra. A comprehensive review of spectroscopic diagnostics for solar and stellar plasmas in the EUV (100–2000 Å) was published by Mason and Monsignori Fossi (1994). Line shifts and broadenings give information about the dynamic nature of the solar and stellar atmospheres. The transition region spectra from the solar atmosphere are characterised by broadened line profiles, which put constraints on possible heating processes. Initial results from CDS on spectral line broadening and shifts are discussed in Brekke *et al.* (1997).

2. The CDS Instrument

The CDS incorporates two spectrometers. The grazing incidence spectrometer (GIS) has four microchannel plate plus spiral anode detectors which cover the wavelength ranges 151–221, 256–341, 393–492, and 659–785 Å. The normal incidence spectrometer (NIS) covers two wavelength ranges (307–379 and 513–633 Å) using a microchannel plate/CCD combination detector. Full details of the CDS optical and detector design are given in Harrison *et al.* (1995) with initial results and performance described in Harrison *et al.* (1997).

In order to carry out quantitative analyses of the CDS spectral data, it is necessary to determine the relative calibration. Sensitivities of the four grazing incidence spectrometer detectors and the two normal incidence detectors were determined from the pre-launch calibration data (Bromage *et al.*, 1996). Further calibration of the detectors will continue during flight.

3. Optically Thin Emission Lines

3.1. Spectral Line Intensities

We are concerned with atomic processes in a hot ($T > 2 \times 10^5$ K) and low density ($N_e < 10^{12}$ cm^{-3}) plasma. We assume that the spectral lines are optically thin, which is usually valid for the outer atmosphere of the Sun.

The emission line flux, $I(\lambda_{ij})$, at a distance d from the Sun, from a volume V for an optically thin spectral line of wavelength λ_{ij} is

$$I(\lambda_{ij}) = \frac{1}{4\pi d^2} \int_V N_j A_{ji} \, \mathrm{d}V \quad (\text{photons cm}^{-2}\ \text{s}^{-1}\ \text{st}^{-1}) , \tag{1}$$

where i, j are the lower and upper levels, A_{ji} is the spontaneous transition probability and N_j is the number density of the upper level j of the emitting ion. In low-density plasmas the collisional excitation processes are generally faster than ionisation and recombination time scales, therefore the collisional excitation is dominant over ionisation and recombination in populating the excited states. This allows the low-lying level populations to be treated separately from the ionisation and recombination processes.

In the coronal model approximation it is assumed that the upper level j of the ion X^{+m} is populated via electron collisional excitation from the ground state g and that the radiative decay overwhelms any other depopulation process. Thus

$$N_g(X^{+m})N_e C^e_{gj} = N_j(X^{+m})A_{jg} \,. \tag{2}$$

The electron collisional excitation rate coefficient, C^e_{gj}, for a Maxwellian electron velocity distribution with a temperature T_e (K), is given by

$$C^e_{gj} = \frac{8.63 \times 10^{-6}}{T_e^{1/2}} \frac{\Upsilon_{gj}(T_e)}{\omega_g} \exp\left(\frac{-\Delta E_{gj}}{kT_e}\right) , \tag{3}$$

where ΔE_{gj} is the energy difference between levels g and j; ω_g is the statistical weight of level g, k is the Boltzmann constant, and Υ_{gj} is the thermally-averaged collision strength.

The population of the level g can be expressed as

$$N_g(X^{+m}) = \frac{N_g(X^{+m})}{N(X^{+m})} \frac{N(X^{+m})}{N(X)} \frac{N(X)}{N(H)} \frac{N(H)}{N_e} N_e \,. \tag{4}$$

Here, $N(X^{+m})/N(X)$ is the ionisation ratio of the ion X^{+m} relative to the total number density of element X (Arnaud and Rothenflug, 1985; Arnaud and Raymond, 1992); $Ab(X) = N(X)/N(H)$ is the elemental abundance; $N(H)/N_e$ is the hydrogen abundance ($\simeq 0.8$). For allowed transitions (electric dipole), the radiative decay rate is very large (10^{10} s^{-1}) compared to the electron excitation rates, so the population of the upper level N_j is negligible in comparison with the ground level g, $[N_g(X^{+m})/N(X^{+m}) \simeq 1]$.

From the equations above, we find

$$I(\lambda_{ij}) = \frac{1}{4\pi d^2} Ab(X) \int_V G(T) N_e^2 \, \mathrm{d}V \quad (\text{photons cm}^{-2}\,\text{s}^{-1}\,\text{st}^{-1}) \,, \tag{5}$$

assuming $Ab(X)$ constant over the emitting volume and that the transition is not density dependent; the function $G(T)$, called the *contribution function*, contains all of the relevant atomic physics parameters and is strongly peaked in temperature.

Above we have considered a very simple system. If a radiative transition probability is small, other mechanisms can de-excite the upper level. For example, for forbidden or intersystem transitions, the electron de-excitation rate may become larger than the radiative decay rate as electron density increases. Eventually, at high enough densities, the populations of the upper and lower levels are in statistical equilibrium. A level which has a significant population relative to the ground level is called a *metastable level.* Other levels in the ion can be excited from these metastable levels. In practice the level populations are determined by solving the statistical equilibrium equations for a number of low-lying levels of the ion including all the important collisional and radiative excitation and de-excitation mechanisms.

3.2. Atomic Databases and Analysis Software

An extensive amount of atomic data are required for the analysis of CDS spectra. ADAS - an atomic database and software package is central to the CDS analyis software. It has been adapted from work carried out over the past decade for fusion research by Prof. H. P. Summers and co-workers (Summers *et al.*, 1996). It is an extensive atomic dataset and set of highly sophisticated routines which treats the ionisation and recombination processes together with radiative and collisional processes in the low lying metastable levels. This enables non-equilibrium processes to be calculated. The atomic data are assessed and stored, these include energy levels, radiative data, electron, and proton collisional rates, as well as ionisation and recombination processes. The analysis routines have been adapted to have an IDL (Interactive Data Language) interface for consistency with other solar software.

The CHIANTI atomic database and analysis software (Dere *et al.*, 1996) is the result of an international collaboration (U.S.A./Italy/U.K.) to provide a comprehensive dataset for ions of astrophysical interest seen in the wavelength region $\lambda > 50$ Å. There are three basic sets of files – the observed and theoretical energy levels, the wavelengths and radiative data for each transition in an ion and the electron collisional data for each transition. The latter are fitted using the Burgess and Tully (1992) approach. CHIANTI also includes a set of IDL routines to solve the statistical equilibrium equations; to provide theoretical line intensity ratios; to calculate CDS synthetic spectra and to determine the differential emission measure. CHIANTI has been incorporated into the CDS science analysis software. It does not have the sophistication of ADAS, does not calculate any ionisation or recombination rates and cannot treat non-equilibrium processes. CHIANTI is used in this paper for an initial analysis of the CDS spectra.

4. Temperature Distribution of CDS Observations

4.1. Emission Measure

The emission measure is a measure of the amount of plasma emitting the observed radiation as a function of temperature and it is the primary characteristic which any theoretical model should predict. Following Pottasch (1963), one can assume that the spectral line is emitted from a volume V, over a temperature range ΔT around the temperature $T_{\max}$ corresponding to the peak value of its contribution function, $G(T_{\max})$. Taking a constant value for the contribution function $\langle G(T)\rangle$ $(= 0.7G(T_{\max}))$ over this temperature range, the flux can be expressed as

$$I_{\lambda_{ij}} = \frac{1}{4\pi d^2} Ab(X) \langle EM \rangle \langle G(T) \rangle \,, \tag{6}$$

where $\langle G(T) \rangle = \int G(T) \, \mathrm{d}T/\Delta T$ and the average volume emission measure $\langle EM \rangle$ for that spectral line is defined as

$$\langle EM \rangle = \int_V N_e^2 \, \mathrm{d}V \,. \tag{7}$$

This method, with further refinements, can be used to derive the $\langle EM \rangle$ as a function of temperature using lines emitted for a wide range of values of T_{max}.

The average electron density $\langle N_e^2 \rangle = \langle EM \rangle / V$ can be crudely deduced assuming that the spectral line is emitted over a homogeneous, isothermal volume estimated from images at that temperature.

Figure 1 shows a typical example of how the structures in the solar atmosphere vary with the temperature of the emitting plasma. Six consecutive ionisation stages of magnesium emit lines in the NIS bandpasses, covering temperatures from 3×10^5 K (Mg V) to 1.5×10^6 K (Mg X), showing the change from fairly compact transition region structures to more diffuse, extended coronal features. The distribution of the line emission from the Mg ions as a function of temperature is shown in Figure 2. Only the Naval Research Laboratory's S082A instrument, on *Skylab*'s Apollo Telescope Mount, has been capable of observing images in such a wide range of ions before, but the images overlapped and for the weaker lines, such as those of Mg V–VII, separating these images was extremely difficult. With NIS there is no such problem, and such sequences of images can be readily formed not only for Mg, but also for the O, Ne, Si, and Fe ions.

The Mg V–VII images in Figure 1 show an excellent example of a transition region brightening, a compact intense feature that is often seen in UV data. They are most noticeable in active region observations and appear to coincide with the footpoints of coronal loops, although further investigation is required to confirm this. From the data we can extract the spectra from both the brightening, seen in the top right of the Mg V image, and the quiet region to the left of centre. These spectra are displayed in Figure 3, where one can clearly see that the Mg V and Mg VI lines are enhanced by factors of around 10 to 20, whereas the coronal lines are largely unaffected. Such examples of strong enhancements of transition region lines show how lines seen as weak in one spectrum may be very strong in another. Further CDS studies are planned to define the nature of these transition region brightenings. It would be advantageous to coordinate these observations with ground based vector magnetograms.

An important aspect of CDS NIS observations comes in the designing of the studies to be run by the instrument. Telemetry constraints make it necessary to preselect a number of wavelength windows. Being able to estimate the background in EUV spectra is crucial in fitting profiles to lines. In the 307–379 Å bandpass of NIS, we find the lines are packed quite tightly together and it is often necessary to

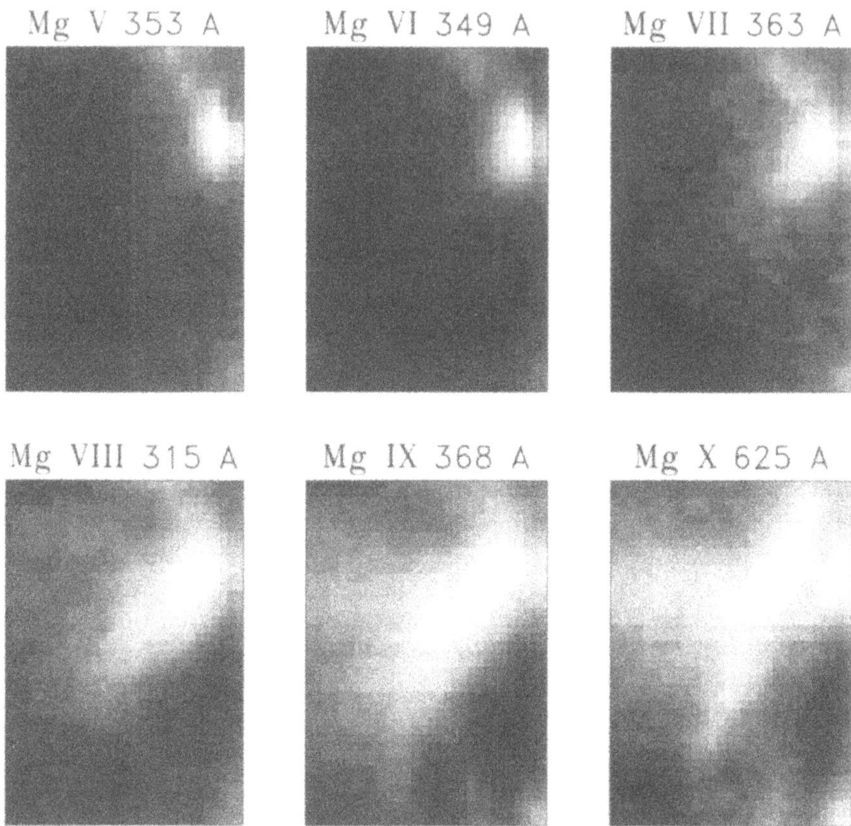

Figure 1. CDS NIS images from a range of Mg ions taken on 23 April, 1996.

go 1–2 Å away from a line to find a 'true' background. For this purpose, the study used to obtain the spectra in Figure 3 used fairly wide wavelength windows. Often there are minor blends which need to be treated cautiously. For example, Mg X 624.94 Å is a popular choice in NIS studies and, being formed at just over 10^6 K, is very prominent in virtually all coronal conditions, the left frame of Figure 4 shows a typical line profile. However, the right frame shows the line profile at another transition region brightening that took place in the newly-emerged active region of 6 June, 1996. The Mg X line is still very strong, but has been joined by three normally undetectable O IV lines at 624.62, 625.13, and 625.83 Å. Another problem associated with blending has been mentioned by Brekke *et al.* (1997) with regard the strong Fe XVI 335.40 Å line seen in active regions. In quiet conditions this line becomes weak, revealing a Mg VIII line at 335.25 Å which can prove hazardous if velocity shifts in the Fe XVI line are being studied.

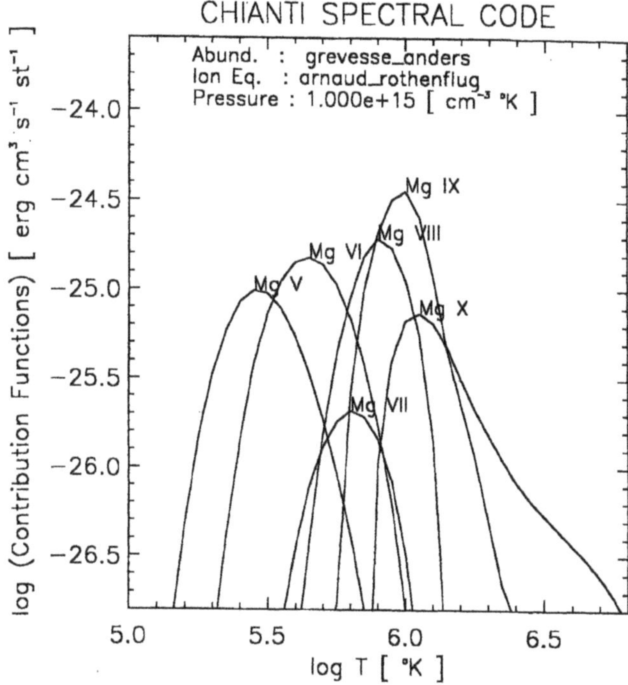

Figure 2. The contribution functions $(G(T)Ab(X)/4\pi)$ for various ionization stages of Mg: Mg V (353.09 Å), Mg VI (349.16 Å – multiplet), Mg VII (363.77 Å), Mg VIII (315.04 Å), Mg IX (368.07 Å), Mg X (624.94 Å).

As SOHO has been launched during the quiet phase of the solar cycle, observations of highly-ionised ion lines have been rare, a matter not helped by the fact that, beyond the Fe XVI doublet at 335.40 Å and 360.75 Å, there are not many strong lines representing the temperature region $6.4 < \log T < 7.0$. Perhaps the strongest and most frequently seen lines are the 291.98 and 320.56 Å lines of Ni XVIII (actually the same transitions as the Fe XVI lines), both of which can be observed in the second band of the GIS, while the latter can also be seen in the first band of the NIS. Figure 5 shows examples of the appearance of both lines in active region spectra. The GIS spectrum is from an active region observation on the 24 March, 1996, and for comparison data are shown from an active region observation from the 29 March, 1996, showing significantly lower high-temperature emission. The NIS data comes from the 6 June, 1996 observation of a recently emerged bipolar active region, the dashed line being from a cooler part of the same data-set. Flare lines from Fe XXI (335.9 Å) and Fe XXII (349.25 Å) may just be observable with CDS, however it should be noted that the wavelength for the former is uncertain, while the nearby Mg VI 349.16 Å line will make the latter difficult to analyse. These

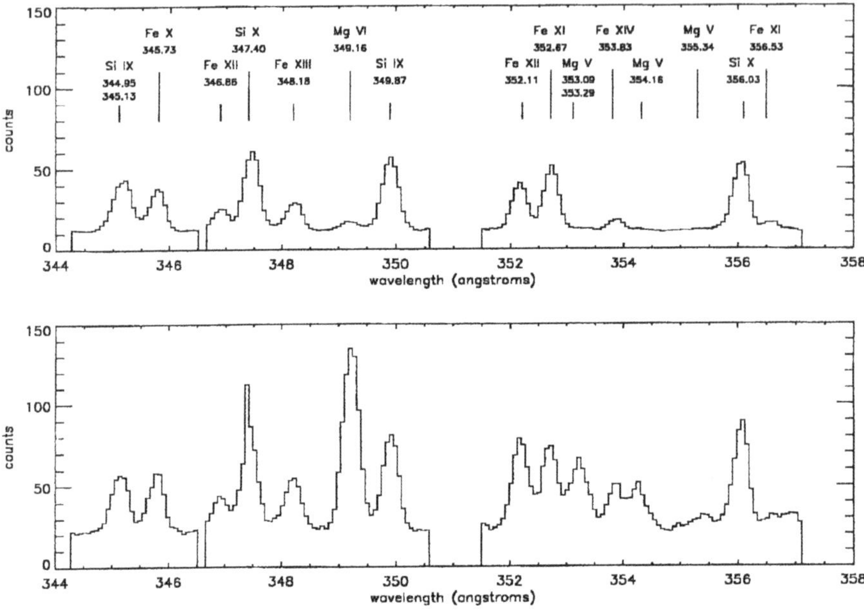

Figure 3. NIS spectral scans taken on 23 April, 1996 from a quiet region (*top*) and transition region brightening (*bottom*).

high-temperature lines should be useful in determining the peak temperatures of loop structures and transient brightenings that are seen in the *Yohkoh* X-ray images.

In Table I we give a selection of reasonably unblended lines in the NIS wavelength ranges from the O, Ne, Mg, Si, and Fe ions covering a wide temperature range. Some of the lines are multiplets of lines from the same ion. The temperatures listed are $T_{\max}$ from Arnaud and Rothenflug (1985) and, for the iron ions, Arnaud and Raymond (1992).

4.2. DIFFERENTIAL EMISSION MEASURE

More sophisticated inversion methods involve re-writing Equation (5) in the form

$$I(\lambda_{ij}) = \frac{1}{4\pi} Ab(X) \int\limits_{T} G(T)\phi(T)\,\mathrm{d}T \quad (\text{photons cm}^{-2}\ \text{s}^{-1}\ \text{st}^{-1})\,, \tag{8}$$

where the differential emission measure (DEM) function, for a plane parallel atmosphere, is defined as

$$\phi(T) = \frac{N_e^2}{d^2} A_s \frac{\mathrm{d}h}{\mathrm{d}T}\,, \tag{9}$$

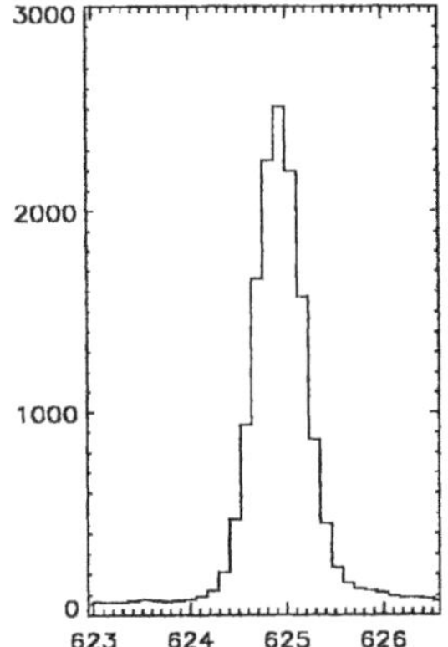

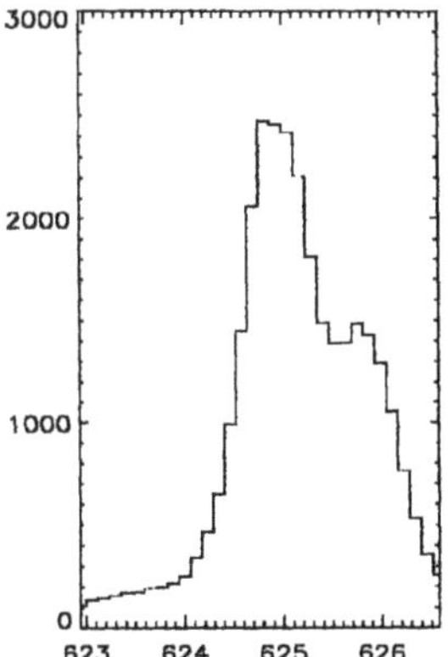

Figure 4. Two line profiles for the Mg X 624.94 Å line recorded by NIS; the one to the left showing typical coronal conditions, while that to the right shows a transition region brightening. The y-axis scale gives the counts recorded by NIS over a 50 s exposure time from a 2×4 arc sec^2 region.

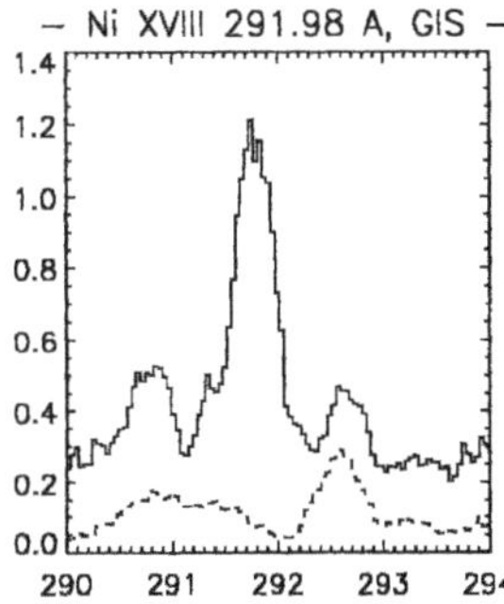

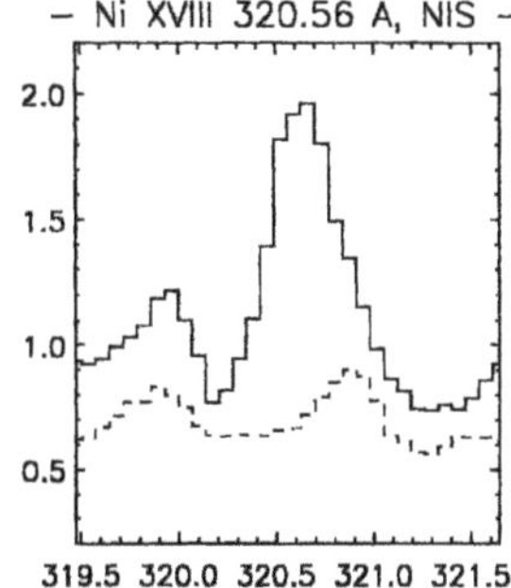

Figure 5. The Ni XVIII 291.98 Å, 320.56 Å lines seen in GIS (band 2) and NIS (band 1). The spectra both come from active regions, with the dashed lines showing spectra from a cooler active region.

where A_s is the projected area of the emitting volume.

The DEM relates to the amount of material in the temperature interval T and $T + \mathrm{d}T$ and the temperature gradient along the line of sight, h. The DEM is related to the conductive flux and gives information about the structure of the atmosphere. Often, it is more convenient to express the DEM on a $\mathrm{Log}_{10} T_e$ scale. To derive the DEM from the observations one must solve a set of integral equations, like Equation (8). The problem, given a set of observed spectral intensities, the values of the abundances and the $G(T)$ functions, is to invert the system of integral equations. The inversion problem itself is not simple and requires some assumptions about the nature of the solution. A series of workshops was sponsored by CDS in 1990–1991 to study differential emission measure techniques (Harrison and Thompson, 1992). It was found that most codes eventually gave consistent results, but that

Table I

A selection of useful lines in the NIS wavelength bands for a range of ions

log T	Oxygen		Neon		Magnesium		Silicon		Iron	
5.0	O III	599.59	–		–		–		–	
5.2	O IV	554.52	Ne IV	543.89	–		–		–	
5.4	–		–		Mg V	353.09	–		–	
5.5	O V	629.73	Ne V	572.33	–		–		–	
5.6	–		Ne VI	562.80	Mg VI	349.16	–		–	
5.7	–		Ne VII	561.73	–		–		–	
5.8	–		–		Mg VII	363.77	–		–	
5.9	–		–		Mg VIII	315.04	Si VIII	319.83	–	
6.0	–		–		Mg IX	368.06	Si IX	349.87	Fe X	345.72
6.1	–		–		Mg X	624.94	Si X	347.40	Fe XI	352.67
									Fe XII	364.47
6.2	–		–		–		Si XI	580.91	Fe XIII	348.18
6.3	–		–		–		Si XII	520.67	Fe XIV	334.17
									Fe XV	327.02
6.4	–		–		–		–		Fe XVI	360.75
6.9	–		–		–		–		Fe XIX	592.20

the DEM derived depends rather critically on the methods used to constrain the solution and the errors in the observed intensities and atomic data.

Figure 6 shows in the right hand plots the DEMs derived from three GIS observations chosen as representative of three types of different coronal emission – an active region, the quiet Sun and a coronal hole. The observations were obtained with the $2'' \times 2''$ slit on the February 28, March 4, and February 25, 1996, respectively. The Arcetri inversion technique (Monsignori Fossi and Landini, 1991) and the CHIANTI package were used.

The ground-based calibration for GIS was used as a starting point, but some variations in the sensitivity of the GI detectors as a function of wavelength have occurred since launch so these results must be considered preliminary. Further analysis will take into account the in-flight calibration which is being performed. In general, most of the spectral lines have a 30–40% agreement between the theoretical and observed intensities (see the left-hand plots of Figure 6). Adjustments were made to the elemental abundances to obtain consistency among spectral lines from different elements. The Grevesse and Anders (1989) abundances (with variations in C, N, O, S) were used to deduce the DEMs for the CH and QS observations, whilst the Feldman *et al.* (1992) abundances (modified in C, N, S) were used for the AR observation. The preliminary DEM analysis shown in Figure 6 confirms a differentiation in the elemental abundances of different regions. Further refinements could be made to the assumed pressure and elemental abundances to improve the fits.

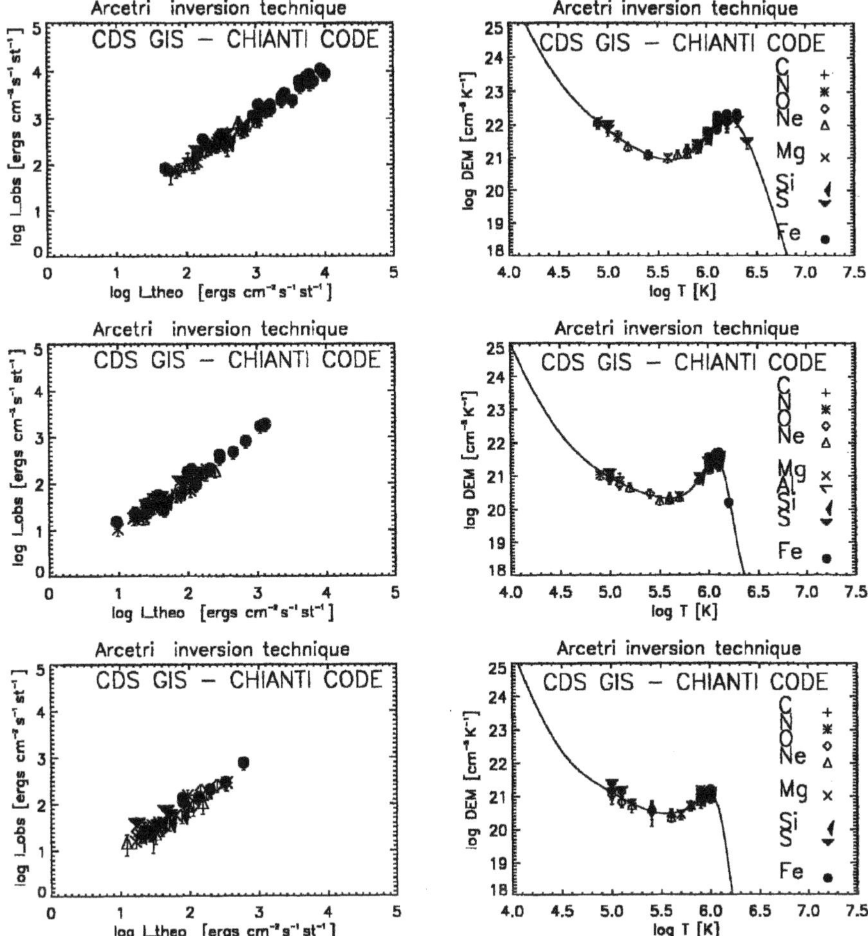

Figure 6. The derived DEMs (right-hand plots) and comparison of observed (Log I_{obs}) and theoretical (Log I_{theo}) intensities (left-hand plots) for an active region, the quiet Sun and a coronal hole (*top to bottom*).

More CDS studies are required to define the precise nature of the FIP differentiation for different solar regions.

There remain some anomalous lines which need further investigation. It should be noted that the well-known problem of the lack of consistency between the theoretical and observed intensity ratios for the Fe XV 284 Å and 417 Å lines still persists. (The Fe XV 417 Å line is blended with S XIV). Similar problems are present for various Fe XIV lines.

The low-temperature region (Log $T \sim 4.5$) is constrained by the continuum, which is seen at longer wavelengths.

The high-temperature region (Log $T \sim 6.5$–7) is constrained by the absence of high temperature lines. The highest temperature lines used are from Fe XV, Fe XVI, and S XIV. In the quiet-Sun and active region spectrum, the Fe XV and Fe XVI lines are observed and can be used to determine the drop in the DEM at high temperatures. We note that the DEM falls rapidly in the coronal hole spectrum. This is constrained by the lack of any Fe XIII ines – in particular the lines at 202 Å and 203.8 Å and any other higher Fe ionization stages.

In Figures 7 and 8 we show as examples, comparisons of the observed and synthetic spectrum for the active region (28 February, 1996) for the GIS 1 and GIS 3 wavelength bands. The GIS 1 wavelength band is dominated by a sequence of iron lines from Fe IX–Fe XIV. These are well reproduced by the CHIANTI synthetic spectra, bearing in mind that work is still in progress on the CDS wavelength and intensity calibrations. In the GIS 3 wavelength band, a number of second-order lines are evident in addition to the first-order lines.

There has been a great deal of discussion and controversy recently about variations in the elemental abundances in the solar atmosphere (Meyer, 1993). One approach to determining element abundances is to use the detailed shape of the $\langle EM \rangle$ or DEM distribution for ions from the same element and apply an iterative procedure to normalising the curves for different elements. Another procedure is to use the intensity ratios for individual spectral lines which have very similar $G(T)$ functions (Widing and Feldman, 1992). All the methods used depend rather critically on the assumed accuracy for the atomic data (Mason, 1992, 1995).

A determination of electron temperature can be obtained from the intensity ratio of two allowed lines excited from the ground level but with significantly different excitation energy (cf., Bely-Dubau, 1994). However, such spectral lines tend to be far apart in wavelength and it may be necessary to use lines from different instruments. Problems can arise owing to uncertainties in the relative calibration of the instruments. Some observing sequences have been carried out but quantitative analyses await in-flight calibrations.

5. Electron Density Diagnostic Ratios from CDS

The electron pressure ($N_e T_e$) is an important parameter in any theoretical model for the plasma. Experience from solar observations is that the plasma often exists in the form of unresolved filamentary structures, even down to the best spatial resolution which has yet been obtained. At one extreme is the solar transition region, where only a very small fraction of the observed emitting volume is actually filled with plasma. The situation is a little better in the corona, but even here the filamentary nature of the emission is evident. The determination of electron density from spectral line intensity ratios from the same ion, makes no assumption about the size of the emitting volume or the element abundance value. It therefore provides a powerful and important diagnostic for the solar plasma.

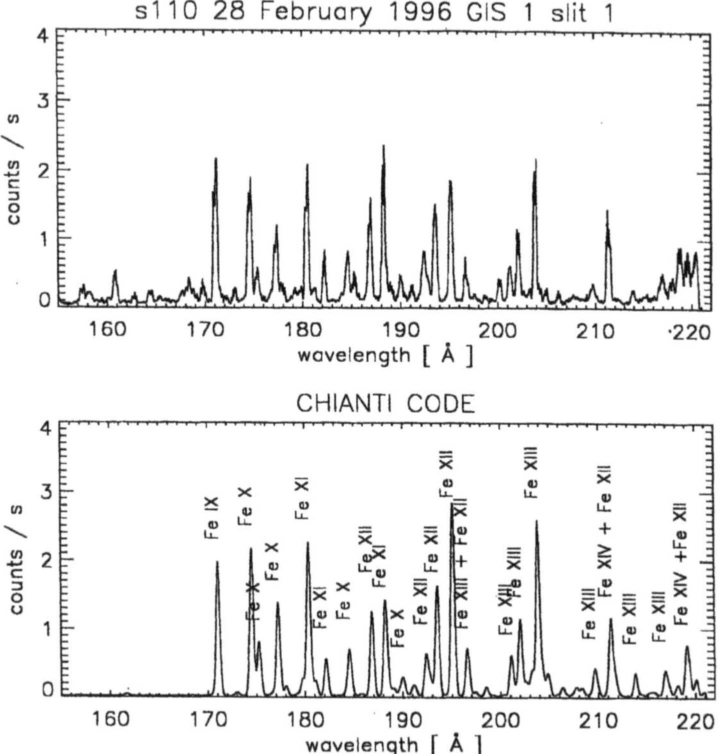

Figure 7. The observed (*top*) and synthetic (*bottom*) GIS 1 spectra of the active region observation.

Spectral lines may be grouped into different categories according to the behaviour of the upper level population: allowed lines collisionally excited from the ground level (the coronal model); forbidden or intersystem lines originating from a *metastable level* – *m*; allowed lines excited from a metastable level.

For forbidden and intersystem transitions the radiative decay rate is generally very small ($A_{mg} \simeq 10^0$–10^2 s^{-1}). Collisional de-excitation then becomes an important depopulating mechanism and may even be the dominant mechanism. Moreover, the population of the metastable level becomes comparable with the population of the ground level, then other excited levels (k) can be populated from this metastable level as well as from the ground level and the dependence of the intensity on electron density becomes

$$I(\lambda_{mk}) \simeq N_e^{\beta}\,; \quad 2 < \beta < 3\,.$$

This latter type of diagnostic is most frequently associated with coronal ions – the metastable levels being in the ground configuration. Since the intensities of different spectral lines within the same ion have a different dependence on the

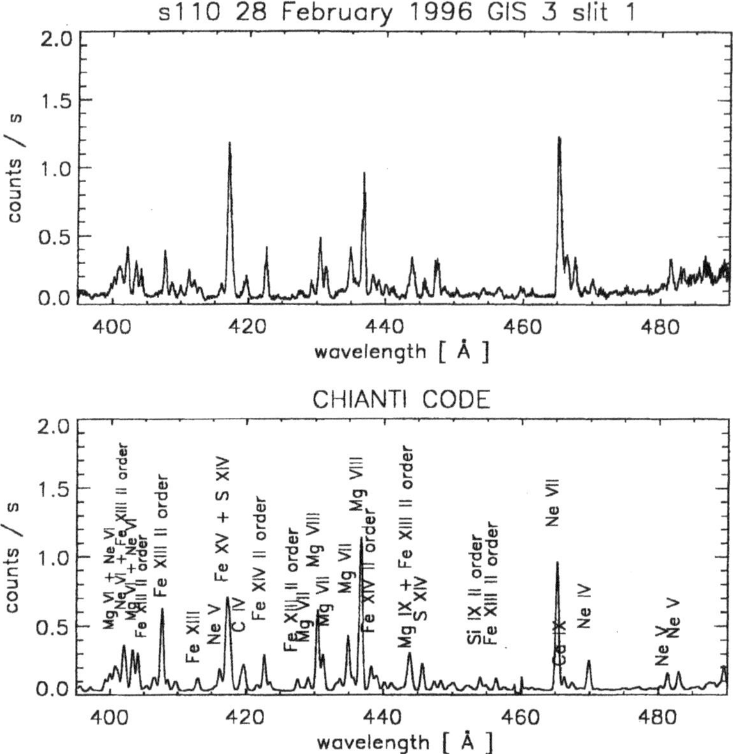

Figure 8. The observed (*top*) and synthetic (*bottom*) GIS 3 spectra of the active region observation.

Table II
Density-sensitive line ratios (NIS)

Ion	Ratio	$\log T_e$	Density sensitivity ($logN_e$)
Si IX	345.13/341.95	6.0	7.5–9.0
Si IX	349.87/345.13	6.0	7.5–9.5
Si X	356.03/347.40	6.1	8.0–10.0
Fe XII	338.26/364.47	6.1	8.5–12.0
Fe XIII	318.12/320.80	6.2	9.5–11.5
Fe XIII	359.64/348.18	6.2	8.5–10.5
Fe XIV	353.83/334.17	6.3	9.0–11.0

electron density, the ratio of the observed intensities allows a determination of the average electron density for the emitting region. The intensity ratio is independent of the ionisation ratio, helium abundance, element abundance, and emitting volume.

The 307–379 Å band of NIS contains several excellent density diagnostics, covering mainly coronal temperatures ($1-2 \times 10^6$ K). These are listed in Table II.

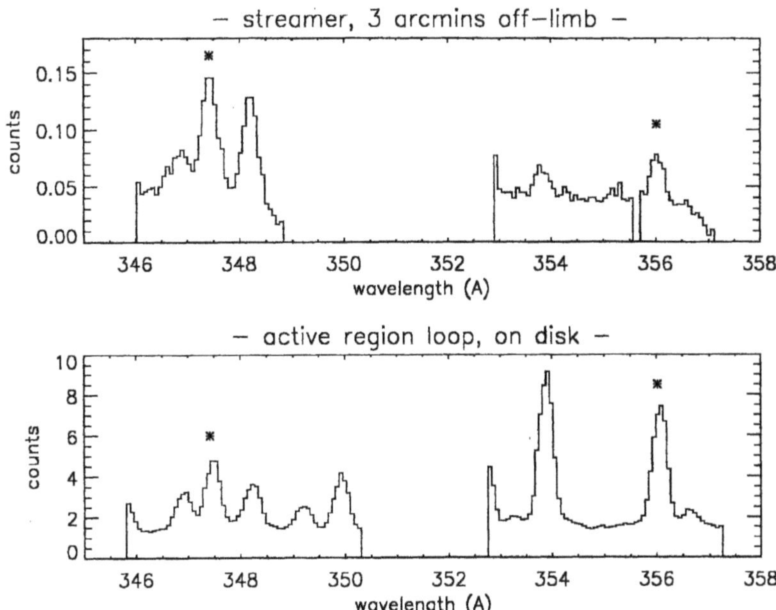

Figure 9. The upper spectrum was taken around 3 arc min above the east limb on 5 June, 1996, where there was a streamer. The lower spectrum is from an active region loop on 6 June, 1996. The counts have been normalised and represent the counts entering one 2 × 4 arc sec 'bin' in one second in both spectra. Stars mark the positions of the two Si X lines at 347.40 Å and 356.03 Å.

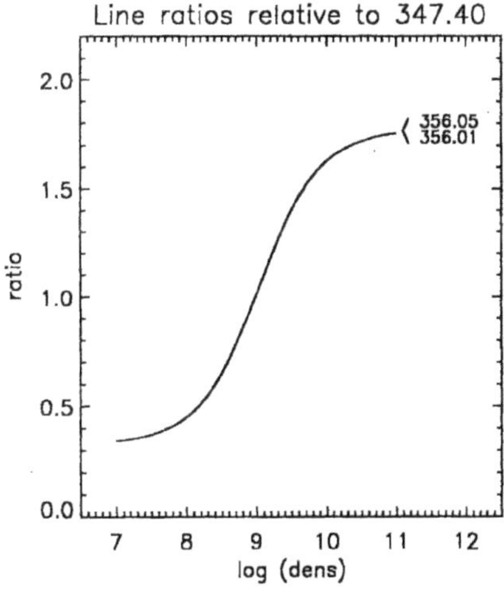

Figure 10. The Si X density-sensitive 356.03/347.40 ratio (the observed 356.03 Å line is a blend of two Si X lines).

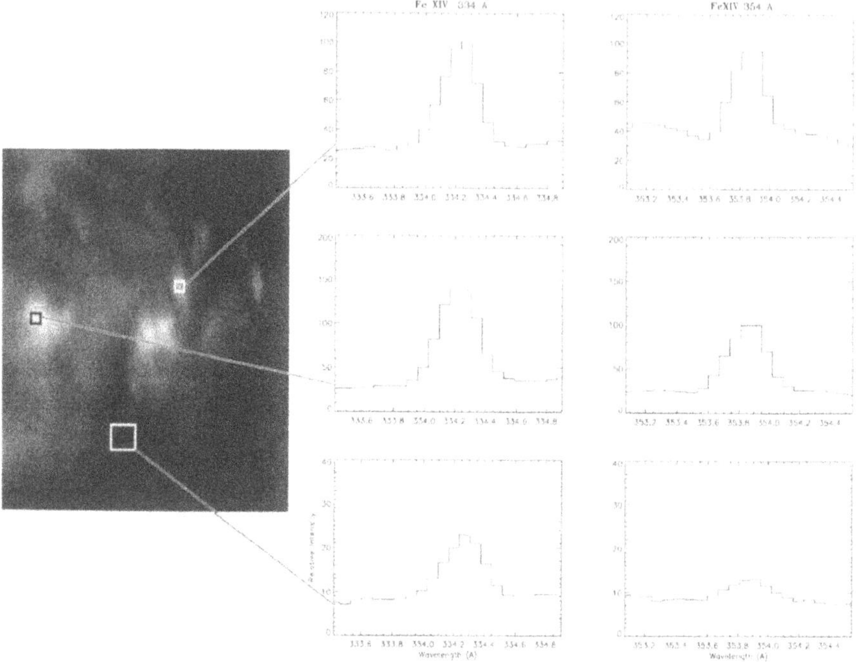

Figure 11. Fe XIV spectral lines for the three different regions of the raster, as shown. The observations were made on 13 May, 1996.

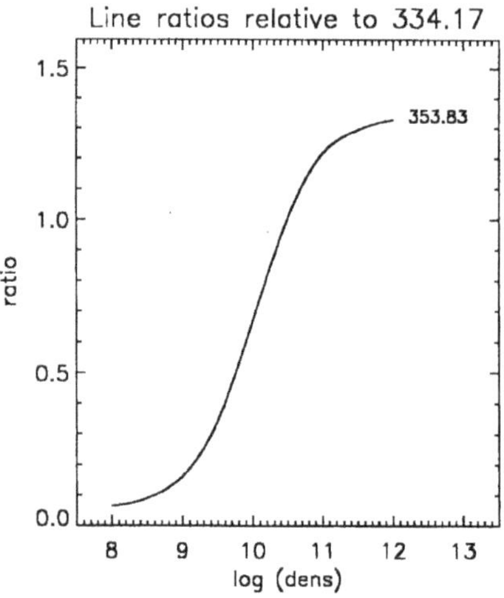

Figure 12. Fe XIV density-sensitive ratio.

Table III
Density-sensitive line ratios (GIS)

Ion	Ratio	GIS	Density sensitivity ($logN_e$)
Fe X	175.27/174.53	GIS 1	9.0–11.0
Fe XI	182.17/180.41	GIS 1	9.0–11.0
Fe XII	186.88/195.12	GIS 1	8.5–12.0
Fe XIII	203.83/202.04	GIS 1	8.5–10.5
S XII	299.50/288.41	GIS 2	9.0–11.0
Mg VIII	436.73/430.46	GIS 3	7.0–9.0

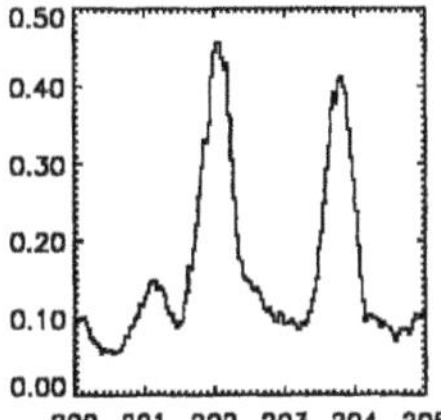

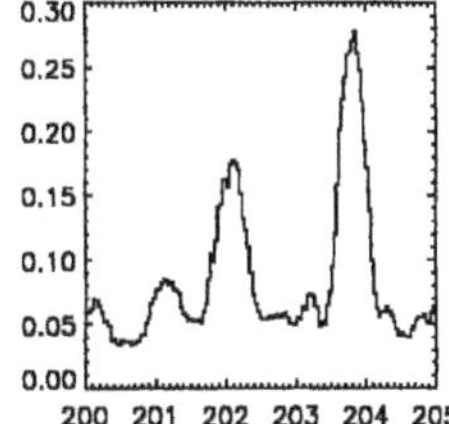

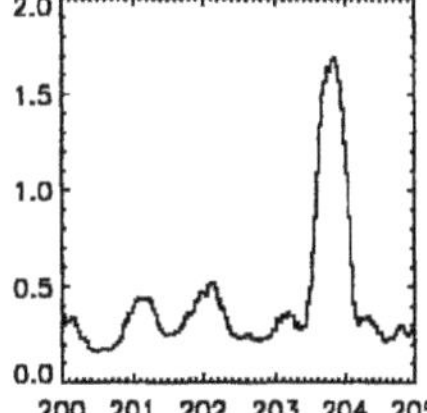

Figure 13. Three examples of the Fe XIII 202.04 and 203.83 Å lines. All three are from different parts of an active region. The lowest density can be seen to the left, while the highest density is to the right. The intensity scale represents the counts emitted by a 2 × 2 arc sec area per second. The left-hand spectrum was obtained on 29 March, the middle and right-hand spectra were obtained on 24 March, 1996.

We seek to show how two of the strongest and most useful of these diagnostics vary in different solar conditions.

Si X is formed at around 1.3×10^6 K and the 347.40 Å and 356.03 Å lines are a fairly prominent feature of most NIS spectra (see, for example, Figure 3). Around active regions the lines have similar intensities, suggesting densities close to 10^9 cm^{-3}, however in the hotter parts of active regions the 356.03 Å line can be found to be around two times more intense than 347.40 Å line. An example is shown in the lower portion of Figure 9. This is close to the high density limit of Si X (see Figure 10), implying a density $> 10^{10}$ cm^{-3}. By contrast, above the limb we can find the Si X ratio approaching the low density limit. The upper portion of Figure 9 shows a spectrum obtained on the 5 June around 3 arc min above the limb, showing the 356.03 Å far weaker than the 347.40 Å line. This corresponds to a density close to 10^8 cm^{-3}.

Fe XIV is formed at a higher temperature (around 1.8×10^6 K) and so the lines are not always prominent in the quiet Sun. For an active region observation from 13 May, 1996, we show three different locations in the raster and the corresponding 334.17 Å and 353.83 Å profiles (Figure 11). The variations in the ratio are clear, with the density varying from around 10^9 cm^{-3} in the quiet region to around $10^{10.5}$ cm^{-3} in the knot/footpoint near the centre of the raster (see Figure 12). Care

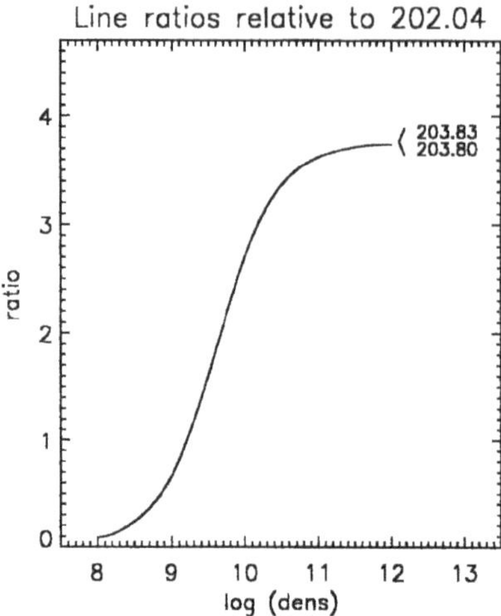

Figure 14. The theoretical variation of the Fe XIII 203.83/202.04 ratio with electron density. The 203.83 line is actually a blend of two lines at 203.80 and 203.83 Å.

must be taken when using this Fe XIV diagnostic ratio in very hot (flaring) regions, since the 353.83 Å line is blended with an Ar XVI line (formed at $\log T$ (K) ~ 6.6).

There are also some excellent density diagnostics within the GIS bandpasses and these are summarised in Table III. Figure 13 shows spectra for one of these ratios, namely Fe XIII 203.83/202.04. One can see that the ratio is varying from around 1 to 4, suggesting densities of between 10^9 and 10^{11} cm^{-3} (see Figure 14).

6. Conclusions

In this work, electron density and temperature distributions have been determined from the EUV observations made with the CDS instrument, for different solar features. This is very much a forerunner of future, more in-depth, analyses of solar plasma diagnostics, and confirms that such studies can be made. Thus, we have illustrated the power of CDS to probe the solar atmosphere.

We have established line lists which are much refined compared to pre-launch estimates. This will set us in a good position for future observation and analysis.

In short, our initial analyses of the CDS spectra promise a rich harvest of scientific results.

Acknowledgement

The support of PPARC is acknowledged in this work.

References

Arnaud, M. and Raymond, J.: 1992, *Astrophys. J.* **398**, 394.
Arnaud, M. and Rothenflug, R.: 1985, *Astron. Astrophys. Suppl. Ser.* **60**, 425.
Bely-Dubau, F.: 1994, *Adv. Space. Res.* **14**, 153.
Brekke, P., Kjeldseth-Moe, O., Brynildsen, N., Maltby, P., Haugan, S. V. H., Harrison, R. A., Thompson, W. T., and Pike, C. D.: 1997, *Solar Phys.* **170**, 163 (this issue).
Bromage, B. J. I., Breeveld, A. A., Kent, B. J., Pike, C. D., and Harrison, R. A.: 1996, UCLan Report CFA/96/09.
Burgess, A. and Tully, J. A.: 1992, *Astron. Astrophys.* **254**, 436.
Dere, K. P., Landi, E., Mason, H. E., Monsignori Foss, B. C., and Young, P. R.: 1996, *Astron. Astrophys.*, submitted.
Feldman, U., Mandelbaum, P., Seely, J. F., Doschek, G. A., and Gursky, H.: 1992, *Astrophys. J. Suppl.* **81**, 387.
Grevesse, N. and Anders, E.: 1989, in C. J. Waddington (ed.), *Cosmic Abundances of Matter*, AIP, New York, p. 1.
Harrison, R. A. and Thompson, A. M. (eds.): 1992, RAL-91-092.
Harrison, R. A., Sawyer, E. C., Carter, M. K., Cruise, A. M., Culter, R. M., Fludra, A., Hayes, R. W., Kent, B. J., Lang, J., Parker, D. J., Payne, J., Pike, C. D., Peskett, S. C., Richards, A. G., Culhane, J. L., Norman, K., Breeveld, A. A., Breeveld, E. R., Al Janabi, K. F., McCalden, A. J., Parkinson, J. H., Self, D. G., Thomas, P. D., Poland, A. I., Thomas, R. J., Thompson, W. T., Kjeldseth-Moe, O., Brekke, P., Karud, J., Maltby, P., Aschenbach, B., Bauninger, H., Kuhne, M., Hollandt, J., Siegmund, O. H. W., Huber, M. C. E., Gabriel, A. H., Mason, H. E., and Bromage, B. J. I.: 1995, *Solar Phys.* **162**, 233.
Harrison, R. A., Fludra, A., Pike, C. D., Payne, J., Thompson, W. T., Poland, A. I., Breeveld, E. R., Breeveld, A. A., Culhane, J. L., Kjeldseth-Moe, O., and Aschenbach, B.: 1997, *Solar Phys.* **170**, 123 (this issue).
Hummer, D. G., Berrington, K. A., Eissner, W., Pradhhan, A. K., Saraph, H. E., and Tully, J. A.: 1993, *Astron. Astrophys.* **279**, 298.
Lang, J. (ed.): 1994, *A.D.N.D.T.* **57**.
Mason, H. E.: 1992, *Proc. of the First SOHO Workshop*, Annapolis, Maryland, 25–28 August, 1992, ESA SP-348, p. 297.
Mason, H. E.: 1995, *Adv. Space Res.* **15**, 53.
Mason, H. E. and Monsignori Fossi, B. C.: 1994, *Astron. Astrophys. Rev.* **6**, 123.
Meyer, J.-P.: 1993, *Adv. Space Res.* **13** (9), 377.
Monsignori Fossi, B. C. and Landini, M.: 1991, *Adv. Space Res.* **11**, 281.
Pottasch, S. R.: 1963, *Astrophys. J.* **137**, 945.
Summers H. P., Brooks D. H., Hammond T. J., Lanzafame, A. C., and Lang, J.: 1996, RAL Technical report RAL-TR-96-017, March 1996.
Widing, K. G. and Feldman, U: 1992, in E. Marsch and R. Schwenn (eds.), *Proc. of the Solar Wind Seven Conference*, Goslar, Germany, 16–20 September, 1991, p. 405.

FLOWS AND DYNAMICS IN THE CORONA OBSERVED WITH THE CORONAL DIAGNOSTIC SPECTROMETER (CDS)

P. BREKKE, O. KJELDSETH-MOE, N. BRYNILDSEN, P. MALTBY and S. V. H. HAUGAN
Institute of Theoretical Astrophysics, University of Oslo, Oslo, Norway

R. A. HARRISON
Space Science Department, Rutherford Appleton Laboratory, Chilton, Didcot, Oxfordshire OX11 0QX, U.K.

W. T. THOMPSON
Applied Research Corporation, 8201 Corporate Drive, Landover, MD 20785, U.S.A.

C. D. PIKE
Space Science Department, Rutherford Appleton Laboratory, Chilton, Didcot, Oxfordshire OX11 0QX, U.K.

(Received 24 June, 1996; in revised form 15 November, 1996)

Abstract. EUV spectra obtained with the Coronal Diagnostic Spectrometer (CDS) on the Solar and Heliospheric Observatory (SOHO) show significant flows of plasma in active region loops, both at coronal and transition region temperatures. Wavelength shifts in the coronal lines Mg IX 368 Å and Mg X 624 Å corresponding to upflows in the plasma reaching velocities of 50 km s^{-1} have been observed in an active region. Smaller velocities are detected in the coronal lines Fe XVI 360 Å and Si XII 520 Å. Flows reaching 100 km s^{-1} are observed in spectral lines formed at transition region temperatures, i.e., O V 629 Å and O III 599 Å, demonstrating that both the transition region and the corona are clearly dynamic in nature. Some high velocity events show even higher velocities with line profiles corresponding to a velocity dispersion of 300–400 km s^{-1}. Even in the quiet Sun there are velocity fluctuations of 20 km s^{-1} in transition region lines.

Velocities of the magnitude presented in this paper have never previously been observed in coronal lines except in explosive events and flares. Thus, the preliminary results from the CDS spectrometer promise to put constraints on existing models of the flows and energy balance in the solar atmosphere. The present results are compared to previous attempts to observe flows in the corona.

1. Introduction

Our knowledge of the flow of mass and energy from beneath the solar surface to the solar wind remains incomplete despite several space missions devoted to understanding the structure and dynamics of the outer layers of the solar atmosphere. Most observations until now have reported on flows in the chromosphere and transition region at temperatures less than 2.5×10^5 K. For a review of the quiet-Sun transition region see Mariska (1992). In this paper we present the first evidence for an extension of the dynamic nature of the transition region into the corona itself and argue that it is the characteristics of the CDS instrument that make this detection possible.

In the hotter portions of the solar transition region there have only been a few attempts to measure Doppler shifts. Inside coronal holes systematic blue shifts of

Solar Physics **170:** 163–177, 1997.

a few km s^{-1} relative to the quiet Sun have been reported in the upper transition region and corona (Cushman and Rense, 1976; Rottman, Orrall and Klimchuk, 1982; Orrall, Rottman, and Klimchuk, 1983). However, absolute values are difficult to obtain owing to the lack of wavelength reference lines near the strong coronal lines. Thus, the blue shift may in reality be a smaller red shift in coronal holes than in the quiet Sun. CDS will enable us to determine if there are systematic differences in line shift between coronal holes and the quiet Sun.

Observations in lines from Ne VIII formed at 520 000 K suggest no systematic average shifts for the full Sun or in quiet regions (Hassler, Rottman, and Orrall, 1991; Mariska and Dowdy, 1992). Local variations were found and the latter investigation also reported Doppler shifts corresponding to downflow velocities of up to 70 km s^{-1} above an active region. Downflows of 15 km s^{-1} in Fe XII (1242 Å) were reported in a plage region by Brekke (1993) using data from the High Resolution Telescope and Spectrograph (HRTS). Brosius *et al.* (1996), using data from the Solar EUV Rocket Telescope and Spectrograph (SERTS) which has similarities to the normal incidence part of CDS, observed no large Doppler shifts above active regions during the 1993 flight. However, upflows of 25 ± 5 km s^{-1} were detected above a sunspot and smaller shifts were found at various locations in the quiet Sun. During the 1989 flight upflows were reported only in Mg IX above a large sunspot (Neupert *et al.*, 1992).

Substantial flows have previously been reported in the lower transition region. The CDS observations reported in this paper confirm the dynamical nature of the transition region and extend this picture also to coronal temperatures. Upflowing plasma is observed in active region loops, reaching velocities of 50 km s^{-1} in lines like Mg IX 368 Å and Mg X 624 Å formed at $T \sim 1$ MK. Smaller velocities are detected in the even hotter coronal lines Si XII 520 Å and Fe XVI 360 Å at ~1.9 MK and ~2.7 MK, respectively. We note that the flows do not necessarily occur in the brightest areas of the active regions. The flows occur in localized areas with spatial extent of typically 20 × 20 arc sec.

The purpose of this paper is to present samples of observations made with CDS during the early phase of the mission and to demonstrate that the coronal and transition region plasma is in a very dynamical state. A more detailed analysis is beyond the scope of this paper and some of the above results will be discussed in more detail in following papers.

2. Instrument and Observations

The Coronal Diagnostic Spectrometer (CDS) on the Solar and Heliospheric Observatory (SOHO) is a dual extreme ultraviolet (EUV) spectrometer covering the wavelength range 150–780 Å. The normal incidence spectrometer (NIS) gives stigmatic spectral images along a 240 arc sec long slit in two wavelength bands (308–381 and 513–633 Å). The astigmatic grazing incidence spectrometer (GIS) covers four wavelength bands in the range 150–780 Å. For more details of the instrument and the performance we refer the reader to Harrison *et al.* (1995, 1997, this issue).

The data presented in this paper were obtained with NIS. By rastering the instrument, i.e., moving the solar image perpendicular to the slit, the spectrometer builds up images of solar features in a series of spectral lines. Each image retains detailed spectral information and can be used to give maps of total emission, velocities, line widths, plasma densities etc. Using this technique CDS is able to distinguish differences in the emitting plasma on a finer graded temperature scale than for example *Yohkoh*. The total field of view (FOV) is 4 × 4 arcmin while in the raster mode. By moving the legs of the CDS instrument the FOV can be moved anywhere on the Sun and can be extended to 1.4 solar radius.

Figure 1 shows monochromatic images obtained in different lines, arranged such that the line formation temperature increases from the top left to the bottom right. The raster was obtained on May 11, 1996, 06:55 UT (CDS study number 2693). The FOV is 4 × 4 arc min and includes part of the active region NOAA 7962, located at a heliocentric angle of 20 deg. The panels show from top left; He I 584 Å ($T \sim 2.0 \times 10^4$ K), O V 629 Å ($T \sim 2.4 \times 10^5$ K), Mg IX 368 Å ($T \sim 9.5 \times 10^5$ K), Mg X 624 Å ($T \sim 1.1 \times 10^6$ K), Si XII 520 Å ($T \sim 1.9 \times 10^6$ K), Fe XVI 335 Å ($T \sim 2.7 \times 10^6$ K). These images demonstrate how the active region structure changes with temperature. In particular the CDS observations have demonstrated that even images obtained in Mg IX and Mg X looks quite different. Although the difference in line formation temperature between the Mg IX and Mg X lines is modest it is clear that the two plasmas are located in different loop structures.

For each single pixel element in the CDS raster images we can extract the entire line profile to derive line-of-sight velocities. Since the heliocentric angle is small we will refer to line-of-sight velocities as upflows and downflows, although we cannot exclude the presence of horizontal components. The vertical dashed lines in the O V and Mg X panels in Figure 1 represents the locations of the slit from where the measurements displayed in Figures 2, 5, and 7 have been extracted.

Flow velocities in the present data were found by measuring the wavelength position of Gaussian fits to the observed profiles, which provide a measure of the line position, line width, and line intensity. An absolute wavelength calibration has not yet been applied. Thus, all the reported velocities are measured relative to the average position of the line itself in a nearby quiet-Sun region where no obvious net flows are observed. Where the profile is strong the uncertainty in the velocity determination is estimated to 10 km s^{-1}. When deriving Doppler shifts we have avoided lines that are blended since this can seriously affect our results, as discussed below. In the following we give results obtained for different lines, beginning with O V 629 Å ($T \sim 2.4 \times 10^5$ K).

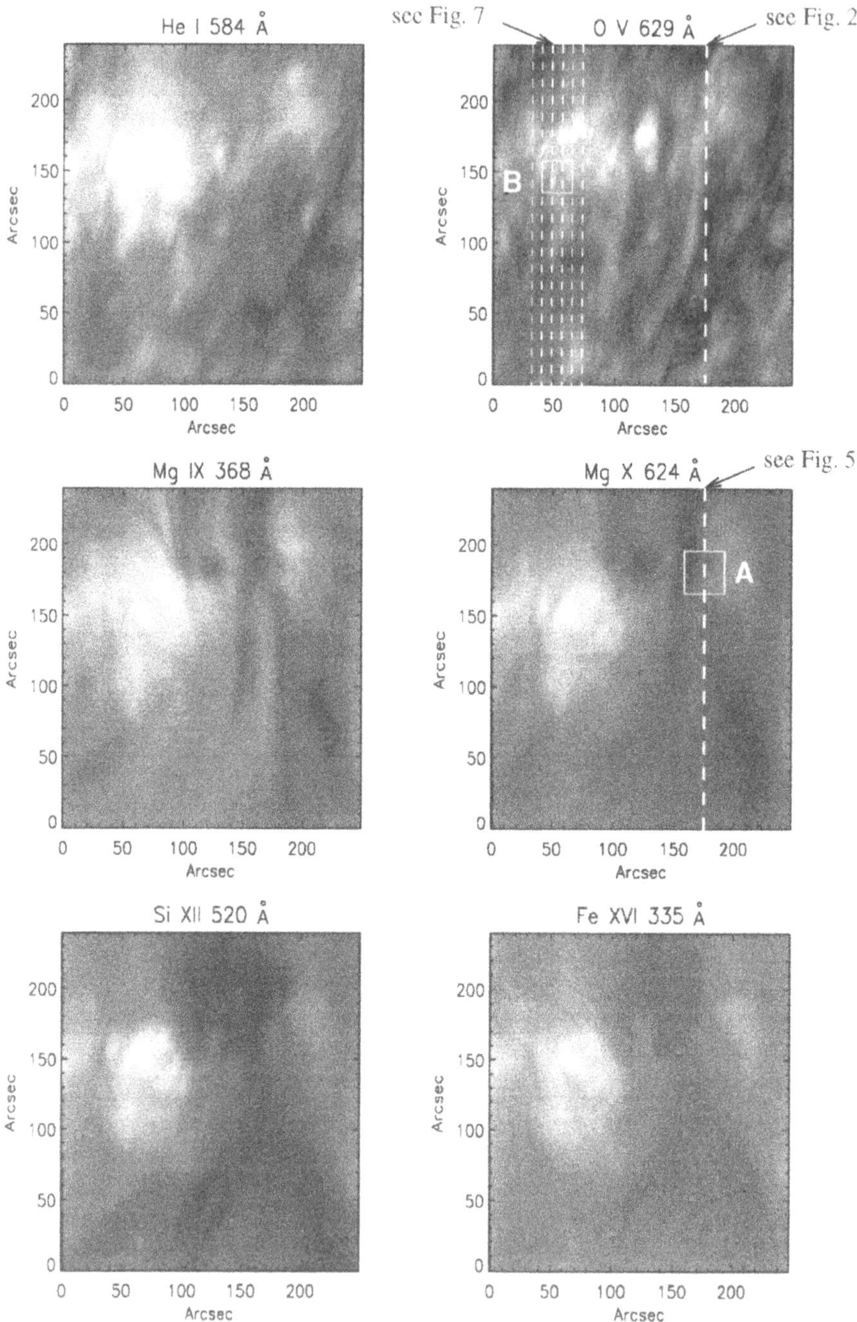

Figure 1. Monochromatic images obtained in different lines, ordered from top left with increasing line-formation temperature. The observations were made on May 11, 1996 (CDS study number 2693). The raster covers $4' \times 4'$ and includes part of the active region NOAA 7962. The vertical dashed lines in the O V and Mg X panels represent the locations of the slit from where the measurements displayed in Figures 2, 5, and 7 have been extracted.

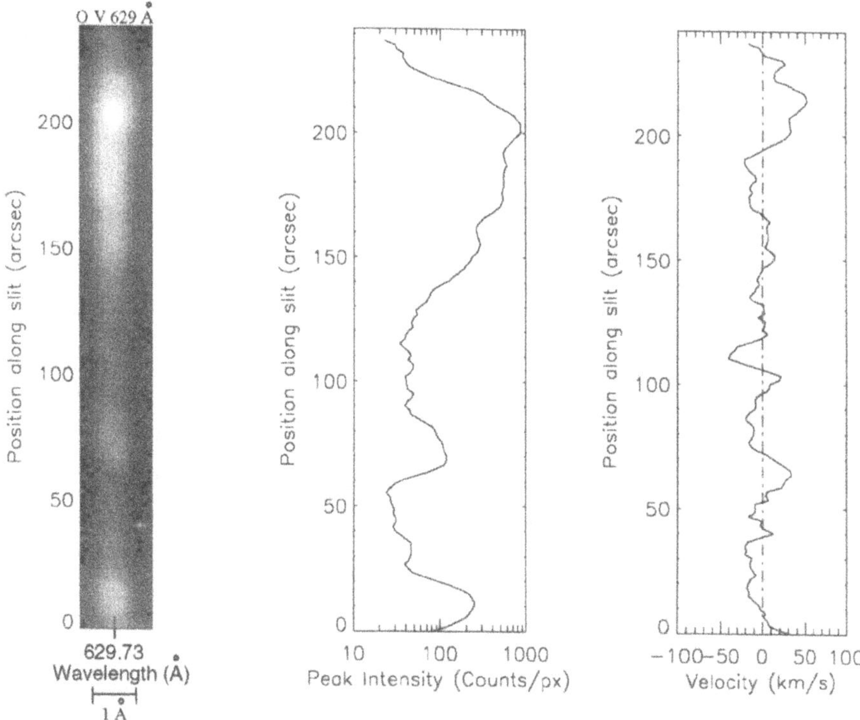

Figure 2. The left panel shows the spectrum of O v 629 Å (CDS study number 2693, exposure 85). The center panel gives the corresponding peak intensity, while the rightmost panel gives the line-of-sight velocity, both quantities are plotted as function of the position along the slit of the spectrograph. The slit position is marked in Figure 1.

2.1. O v 629 Å

Figure 2 gives one example of the derived line-of-sight velocities for O v 629 Å as function of position along the slit in the active region NOAA 7962. It is evident from Figure 2 that O v shows large variations in wavelength shifts along the NIS slit. We find that the observed velocity in O v 629 Å varies between ±50 km s^{-1} in this observing sequence. The actual position of the slit during this particular exposure is marked in Figure 1. Flow velocities reaching 70 km s^{-1} were observed in another, smaller active region on May 8 (CDS study number 2651).

Another example of high velocities is illustrated in Figure 3 which shows an off limb active region observation on July 27, 1996, 10:00 UT (CDS study 3819). This raster was obtained using the 4 arc sec wide slit and the slit was moved 4 arc sec between each exposure. The line profiles from three different spatial locations (A, B, and C) are displayed in the right panel together with a quiet-Sun disk center profile.

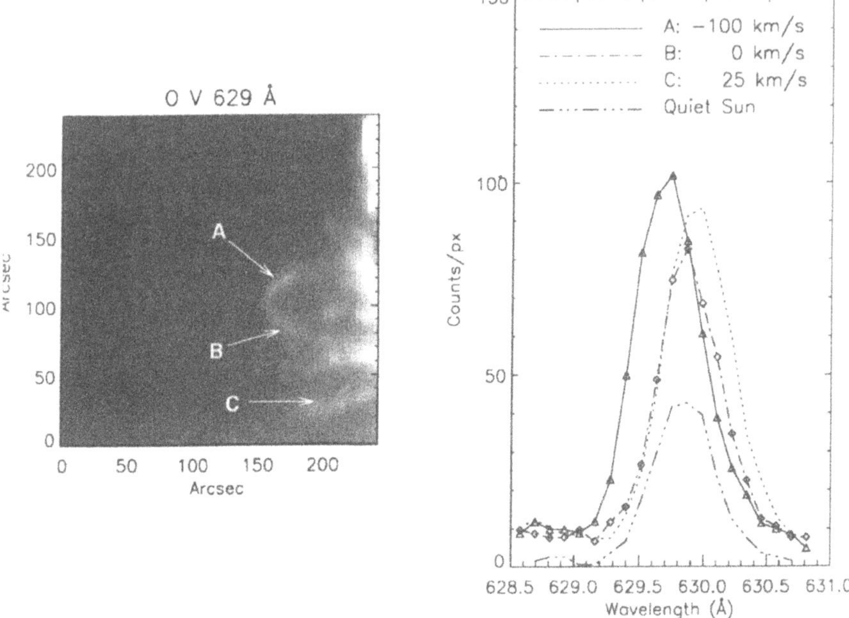

Figure 3. Active-region loop system above the east limb observed in O V on July 27 1996 10:00 UT (CDS study 3819). The line profiles from three different spatial locations (A, B, and C) are displayed in the right panel together with a quiet-Sun disk center profile.

The plasma at location A is moving towards the observer with a velocity of 100 km s^{-1} relative to the quiet Sun. It is difficult to see which foot point is the closer one from the current image. Thus, it is impossible to establish the orientation of the loop to tell if the plasma is moving upward or downward the leg of the loop. On the other side of the loop summit (location B) there is no evidence for significant flows. We find a red shift corresponding to 25 km s^{-1} in another loop (location C).

Figure 4 shows monochromatic images of the quiet Sun obtained at disk center as part of the daily CDS synoptic survey on September 24 1996 (CDS study 4908). In typical quiet-Sun regions we can clearly see the network structure in He I and O V while in the Mg IX image the network pattern is less apparent. The intensity ratio between the bright network elements and the cell interior is a factor 5–10 but reaches 15 in some of the localized bright areas seen in Figure 4. These number are considerable larger that reported by Reeves (1976). Based on *Skylab* data with 5 arc sec resolution he found the contrast between representative network elements and cell interior to be approximately 6 for emission lines formed near 2×10^5 K. The discrepancy between this result and the CDS observations can be explained by the higher resolution of CDS. Observations with even higher resolution, such

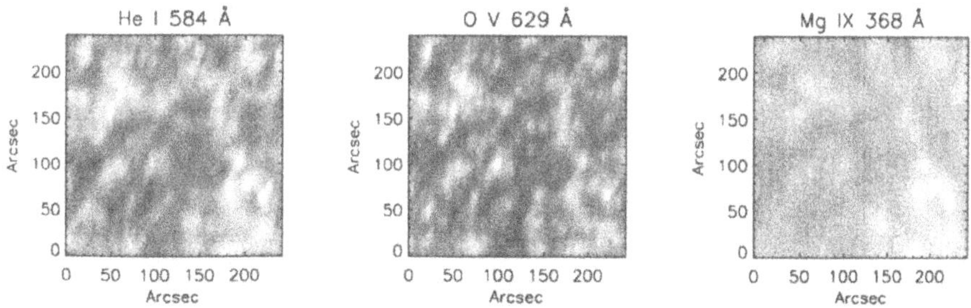

Figure 4. Monochromatic images of the quiet-Sun disk center obtained during the daily synoptic survey on September 24 1996 (CDS study 4908). The He I and O V images show the network structure while in the Mg IX image the network pattern is less apparent.

as HRTS, reveal even higher contrast ratios between the network and cell interior (Judge and Brekke, 1997).

The quiet-Sun spectra show velocity fluctuations with values around ± 20 km s^{-1}. Further investigation of an extensive set of observations is needed to study possible correlations between the magnitude of the flows and the brightness of the line.

2.2. Observed shifts in Mg X 624 Å and Mg IX 368 Å

In the active region NOAA 7962 we observe blue shifts in Mg X reaching 50 km s^{-1} in localized areas. This is illustrated by one example in Figure 5. The left panel shows the spectrum of the Mg X 624 Å line in one single slit position (representing one exposure or readout from the CCD). The slit location within the CDS FOV is marked in Figure 1. The panel in the middle shows the variation of the peak intensity along the slit, while the rightmost panel illustrates the variation of the line-of-sight velocity. The flow area extends over approximately 20 arc sec along the slit and 10 -15 arc sec perpendicular to the slit. The latter number was estimated by looking at all the exposures in this particular raster. The location of the upflowing material is marked as A in the Mg X image in Figure 1. In this particular case the strongest upflow is found between two brighter features where no significant Doppler shifts are apparent.

The Mg IX 368 Å line shows similar flow patterns as the Mg X line. However, measurements of large blue shifts in Mg IX 368.06 Å are complicated by a nearby Mg VII line at 367.67 Å which shows significant brightness in some parts of the active region. This is illustrated in Figure 6 in the left panels. Thus, the combination of large blue shifts in this Mg IX line and, at the same time, a bright Mg VII line makes the Mg IX line appear even more blue-shifted. As discussed below the Mg IX 368 Å line is not suitable for accurate velocity determinations. The Mg X line is blended with three, normally weak O IV lines at 624.62, 625.13, and 625.83 Å as discussed by Mason *et al.* (1997). These lines could in some region influence measurements of red shifts in the Mg X line.

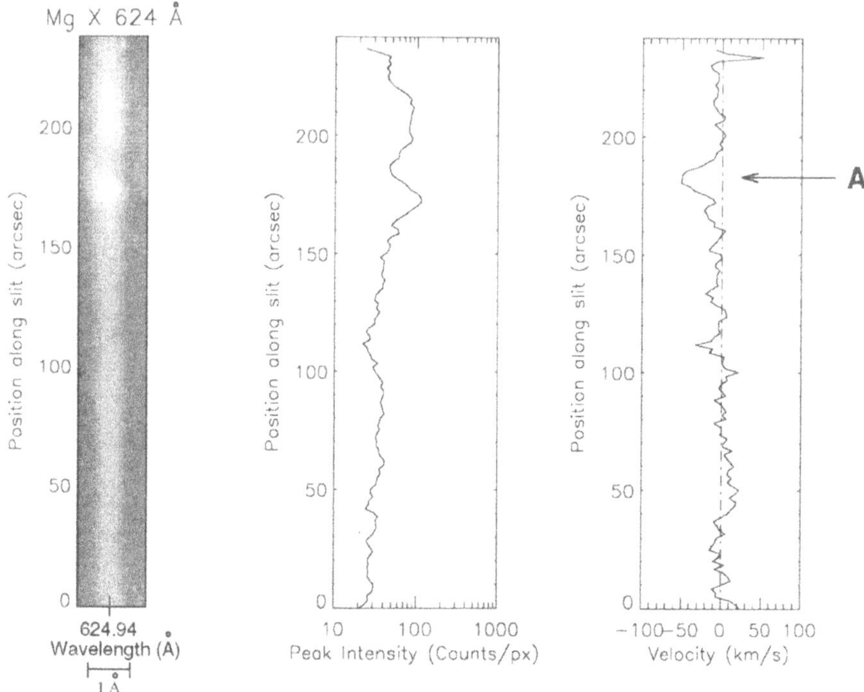

Figure 5. Spectrum, peak intensity and line-of-sight velocity as a function of position along the slit observed in Mg X 624 Å (CDS study number 2693, exposure 88). The large upflow at position 180 was located within the box marked A in Figure 1. The large spike in the velocity plot at position 234 is caused by a cosmic ray hit as can be seen in the leftmost panel.

2.3. Observed Shifts in Si XII and Fe XVI

Compared to the Mg X 624 Å and Mg IX 368 Å lines the hotter coronal lines Si XII 520 Å and Fe XVI 360 Å show smaller Doppler shifts. The large shifts observed in these lines are predominantly blue shifts with values around 25 km s^{-1}, in the CDS material used in this study. The above measurements are relative to the average position of the lines in the surrounding quiet-Sun regions. By browsing the CDS data base we find examples of even higher velocities in these hot coronal lines. Thus, a more detailed analysis of a larger set of observations is required to determine the properties of flows in active region coronal plasma.

Among the high-temperature lines the Fe XVI 335 Å was initially selected as the primary line for studying coronal mass motions. The blending of this line complicates the measurements and suggests that the weaker Fe XVI 360 Å line should be used instead. A short discussion on blended lines follows.

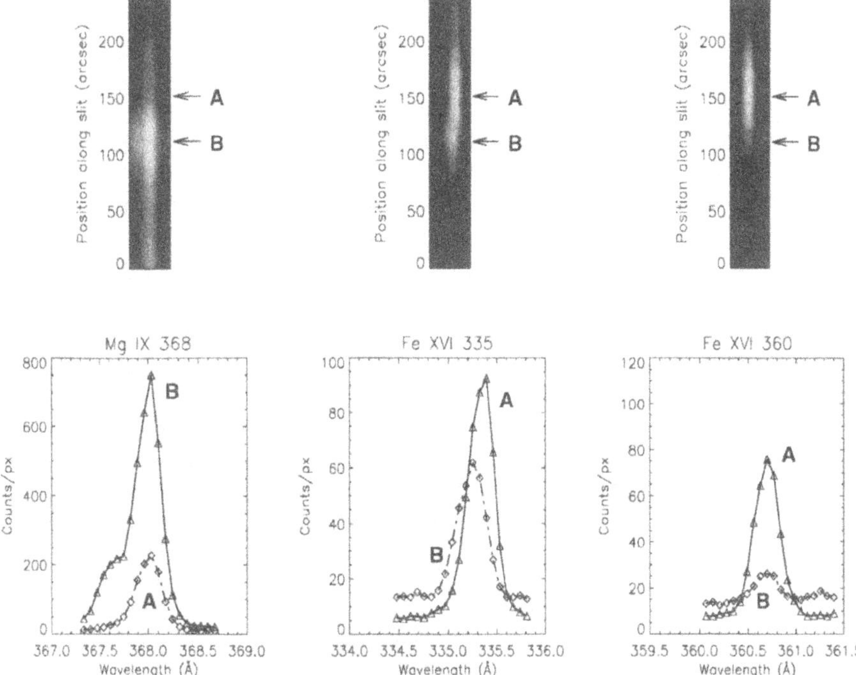

Figure 6. Line profiles from Mg IX 368, Fe XVI 335, and Fe XVI 360 Å demonstrating the problem with blends. The line profiles shown in the lower panel were extracted from the positions A and B. Both Mg IX 368 and Fe XVI 335 are clearly contaminated by blends. The observations were made on September 20, 1996 (CDS study number 4830).

2.4. Caution about Blended Lines

When determining Doppler shifts in emission lines one should be careful with lines that might be blended. Based on studies of the CDS material we find that some of the strong coronal lines listed by Harrison *et al.* (1995) as suitable lines for velocity diagnostic are clearly contaminated by blends which become very strong in active regions.

The Mg IX 368 Å and Fe XVI 335 Å lines are strong lines that are frequently used in CDS studies. Figure 6, upper panels, shows simultaneous observations of the lines Mg IX 368, Fe XVI 335 and Fe XVI 360 Å and is well suited to illustrate the importance of blends in the first two of these lines in some parts of the active region. The Mg IX 368.06 Å line is blended with Mg VII 367.67 Å which shows significant brightness in some parts of the active region, see line profile B in lower left panel in Figure 6. Evidently this complicates measurements of large blue shifts in the Mg IX 368 Å line.

Just by inspecting by eye the spectrum of the Fe XVI 335.40 Å line in Figure 6, top middle, it is obvious that in some parts of the active region this line is seriously contaminated by one or more blends. The Fe XVI 335.40 Å line is blended with a Mg VIII line with a theoretical wavelength of 335.25 Å (Martin and Zalubas 1980). It has not been observed or listed in earlier EUV line lists (e.g., Behring *et al.*, 1976; Dere, 1978; Thomas and Neupert, 1994). However, the present observations show that in some locations of the active region the line Mg VIII 335.25 Å line becomes strong enough to affect the velocity measurements, compare profiles B and A in Figure 6, middle panel. The Mg VIII 335 Å line, interestingly enough, appears to be bright in areas where both the Fe XVI lines, 335 Å and 360 Å, almost disappear. This gives a spurious result in that the Fe XVI 335 Å line seems to be strongly blue-shifted corresponding to a velocity of 100 km s^{-1} whereas no Doppler shift can be detected in the unblended Fe XVI line at 360 Å.

The above discussions of the Mg IX 368 Å and Fe XVI 335 Å lines suggest that these lines should be used with great care for velocity or density diagnostic, particularly in active regions where the contribution to these lines from blends may be stronger than the principal line. The contribution from the Mg VIII 335 Å line to the Fe XVI 335 Å line intensity can be estimated by also observing the Fe XVI 360 Å line.

3. Spatial and Temporal Variations

Figure 7 shows the O V line from ten adjacent raster locations, observed subsequently in the active region as illustrated in the O V panel in Figure 1. Note that the step increment between each slit position has been exaggerated in the illustration shown in Figure 1. The slit was moved 2 arc sec between each exposure and the integration time for each exposure was 20 s. Figure 7 clearly demonstrates the spatial variations seen on small scales. The profile broadens in one or two slit positions as illustrated by the event seen as a bright feature in frames 24 and 25. This feature looks similar to the so-called 'Explosive Events' first observed with HRTS (e.g., Brueckner and Bartoe, 1984). The line profiles extracted from exposure 24 at location A and B are shown in the left panel. Profile B represents an average plage profile while A represents the explosive event. The broad line profile corresponds to a velocity dispersion of 300–400 km s^{-1} and extends over an area of about 4 arc sec. This particular event also shows up strongly in O III 599 Å and Ne VI 562 Å. A weak signature can be observed in the lines from Mg VI 349 Å , Mg IX 368 Å, and Mg X 624 Å, while no trace of this event was observed in Si XII 520 Å or Fe XVI 335 Å.

In addition to large variations in plasma flow as a function of spatial position there is also clear evidence of rapid variations in both intensity and velocity as a function of time. This is best observed when the instrument slit is kept in the same location on the Sun while recording a series of exposures. Figure 8 illustrates this

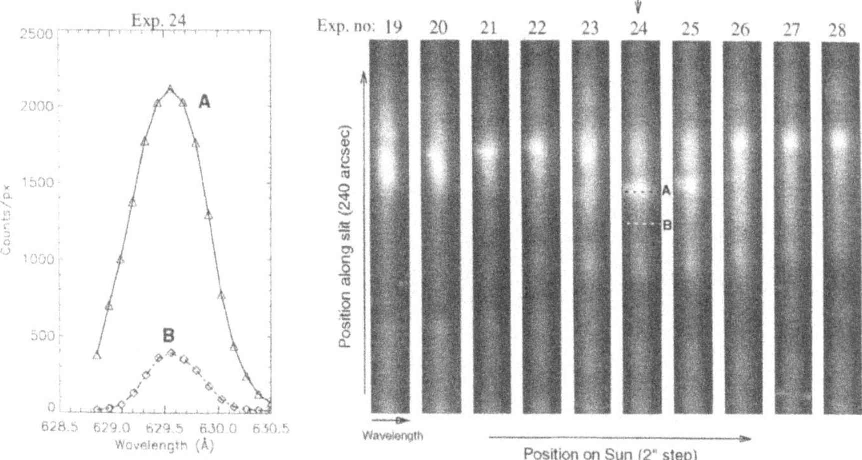

Figure 7. Line profiles from a selection of the O V raster demonstrating the large changes that occur over a few arcsec in both intensity and velocity. The step size between each exposure is 2 arc sec. The location of the high-velocity event seen in profile number 24 (or 6th as seen in the figure) is located inside the box market B in the O V image in Figure 1. The line profiles extracted from locations A and B are shown in the left panel. The observations were made on May 11, 1996 (CDS study number 2693).

point from a time sequence obtained on April 22, 1996 (CDS study 672). In this particular study 50 exposures were obtained on the same location in the quiet Sun. The 2 × 240 arc sec slit was used with an exposure time of 20 s.

The top panel shows to the right the spectra of O V 629 Å obtained between $t = 600$ s to the end of this sequence ($t = 980$ s). Only the top 130 arc sec of the slit is shown in this panel. The line profiles extracted from exposure $t = 720$ s at the three spatial locations A, B, and C are displayed to the left. Profiles A and B represent two different bright features while profile C represents a typical quite-Sun profile. The profiles A and B differ from C in that they are both much stronger and show significant Doppler shifts. The shifts measured in profiles A and B relative to C are approximately 80 km s^{-1} and -50 km s^{-1}, respectively.

The lower two panels of Figure 8 demonstrate the time evolution of the integrated line intensity of the O V line during the entire observing sequence. The middle panel gives the variation of the line intensity at each single slit position. The rectangular area in the upper right corner represents the data displayed in the top panel. The locations of the three positions A, B, and C are marked and the time variation of the integrated line intensity at these locations are show in the bottom panel.

It should be noted that CDS was not compensating for solar rotation during this particular time sequence. However, with a drift of 7 arc sec per hour the image of the Sun will drift approximately 1.8 arc sec during the 1000 s observing sequence. Thus, some of the variations seen in Figure 8 may be caused by features drifting

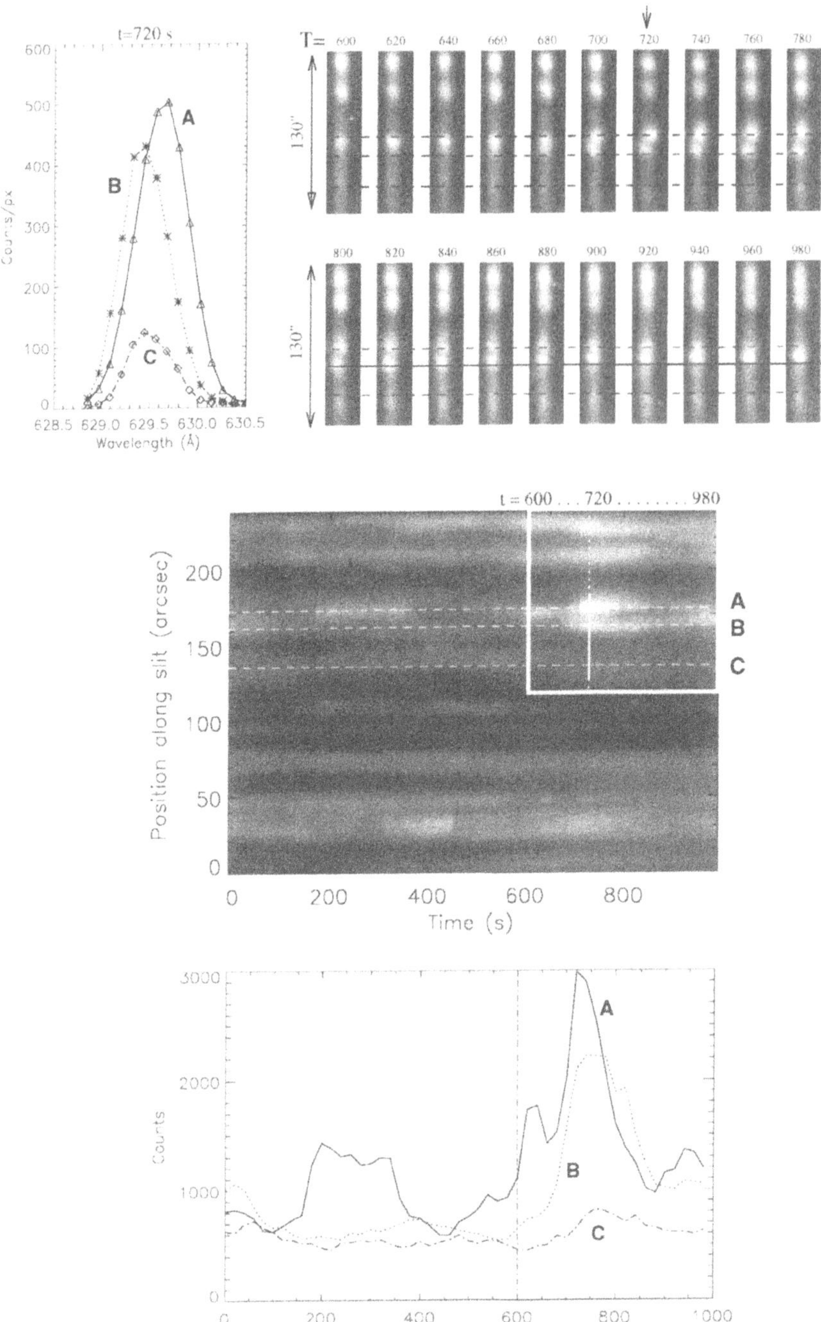

Figure 8. Time sequence of the O V 629 Å line observed on April 22, 1996 (CDS study 672). The top panels to the right show the spectra observed between $t = 600$ s and the end of this sequence ($t = 980$ s). The line profiles extracted from exposure $t = 720$ s at the three spatial locations A, B, and C are displayed to the left. The middle panel gives the variation of the line intensity at each single slit position during the entire sequence. The rectangular area in the upper right corner represents the data displayed in the top panel. The time variation of the integrated line intensity at locations A, B, and C are show in the bottom panel.

into or disappearing from the slit field of view. However, this effect can not explain the sudden brightening that takes place between $t = 700{-}840$ where the solar image drifted only about 0.25 arc sec or $\frac{1}{8}$ of the slit width.

The above estimate of time variations are supported by other CDS observations, obtained by operating CDS in the wide 90×240 arc sec slit mode. In this mode the CDS instrument is extremely fast and both slow and rapid time variations may be studied by following individual solar features. Long-lived features may be followed for hours without moving the instrument. Analysis of these observations show that localized brightenings can appear and disappear within time scales less than a minute.

It is clear from Figure 8 that the dynamical, as well as the density and temperature properties of the plasma change on short time scales. Variations in velocities and intensities are observed to occur on time scales from a few minutes down to a few tens of seconds based on a larger CDS material. This is much faster than expected based on the earlier observations (e.g., Rabin and Dowdy, 1992). Such rapid variations in the emission are important to understand since they may indicate that the solar atmosphere is in a state of non-equilibrium ionization, which occurs when the observed time scales become comparable to or shorter than the time it takes to ionize or recombine from one stage of ionization to another.

4. Concluding Remarks

The origin of flows in the transition region have been discussed in numerous papers, either as a unidirectional flow along the whole loop (e.g., Spadaro *et al.*, 1991) or as upflows and subsequent downflows along the same lines of force (Pneuman and Kopp, 1978). Bray *et al.* (1991) and Mariska (1992) have reviewed these models. Recently, Hansteen, Maltby, and Malagoli (1996) showed that MHD disturbances traveling from the corona towards the chromosphere along magnetic field lines often will have a nonlinear character and may explain (1) the observed multiple velocities within a single spatial resolution element (Kjeldseth-Moe *et al.*, 1988, 1993), (2) the overabundance of red shifts over blue shifts (Doschek, Feldman, and Bohlin, 1976; Bruner *et al.*, 1976), and (3) the wavelength shift being larger in active than in quiet regions (Lites *et al.*, 1976; Brueckner, 1981; Dere, 1982; Feldman, Cohen, and Doschek, 1982). The observations needed to distinguish between alternative interpretations include observations of lines formed at temperatures above 2.5×10^5 K and an improved knowledge about the temporal variations. Both quantities may be measured by the CDS instrument.

Based on a few series of observations with the CDS we have detected significant plasma flows in active region loops formed in the corona and the upper part of the transition region. Material moving upward with velocities reaching 50 km s^{-1} are observed in the coronal lines Mg IX 368 Å and Mg X 624 Å. Smaller velocities are detected in the coronal lines Fe XVI 360 Å and Si XII 520 Å in the data set discussed here. Velocities of the magnitude presented in this paper has never previously been observed in coronal lines except in explosive events and flares. Thus, the preliminary results from the CDS spectrometer promise to provide data needed to differentiate between alternative interpretations of the observations of the transition region lines and to put constraints on models of the energy balance in the solar atmosphere.

We point out that shifts observed in Fe XVI 335 Å is strongly affected by a nearby Mg VIII line and that this line should not be used for velocity or density diagnostic, at least not in active regions. This is sometimes also the case for the Mg IV 368 Å.

The velocities found in O V confirm and extend the previous results obtained with the HRTS instrument. Whereas the first rocket flights with HRTS showed high line-of-sight velocities, corresponding to supersonic speeds approaching 180 km s^{-1} in the second HRTS flight (Brueckner, 1981; Brekke *et al.*, 1992), more moderate speeds were reported from the Spacelab 2 observations near or above sunspots by Kjeldseth-Moe *et al.* (1988, 1993). The large Doppler shifts seen in the HRTS material are not restricted to sunspots and active regions but also appear in the quiet Sun with velocities ranging from 50–100 km s^{-1} (e.g., Brekke *et al.*, 1992). The observations presented here show that the high-speed flows occur in localized areas with spatial extent of typically 20 × 20 arc sec. We also note that the flows do not necessarily correspond to the brightest areas of the active regions. Taking the spatial extension of the high-speed regions into account it is not surprising that, for instance, SERTS did not observe these strong flows since their observations are based on one single slit position, while CDS moves the slit and covers a large part of the active region.

Acknowledgements

We would like to thank the many members of the main hardware groups at Rutherford Appleton Laboratory and Mullard Space Science Laboratory in the UK, Max-Planck-Institut für Extraterrestrische Physik and Physikalisch-Technische Bundesanstalt Berlin in Germany, Goddard Space Flight Center in the US, and Institute of Theoretical Astrophysics, Norway who made CDS such an excellent instrument. We would also like to thank Peter Young and Helen Mason at Cambridge for help searching the CDS data base. This work was supported by the The Research Council of Norway. SOHO is a project of international cooperation between ESA and NASA.

References

Behring, W. E., Cohen, L., Feldman, U., and Doschek, G. A.: 1976, *Astrophys. J.* **203**, 521.
Bray, R. J., Cram, L. E., Durrant, C. J., and Loghhead, R. E.: 1991, *Plasma Loops in the Solar Corona*, Cambridge University Press, Cambridge.
Brekke, P.: 1993, *Astrophys. J.* **408**, 735.
Brekke, P., Brynildsen, N., Kjeldseth-Moe, O., Maltby, P., and Brueckner, G. E.: 1992, *Proceedings of the 26th ESLAB Symposium on Study of the Solar-Terrestrial System*, ESA SP-346, p. 211.
Brosius, J. W., Davila, J. M., Thomas, R. J., Saba, J. L. R., Hara, H., and Monsignori-Fossi, B. C.: 1996, *Astrophys. J.*, submitted.
Brueckner, G. E.: 1981, in F. Q. Orrall (ed.), *Solar Active Regions*, Colorado Associated University Press, Boulder, p. 113.
Brueckner, G. E. and Bartoe, J.-D. F.: 1984, *Astrophys. J* **272**, 329.
Bruner, E. C., Jr., Chipman, E. G., Lites, B. W., Rottman, G. J., Shine, R. A., Athay, R. G., and White, O. R.: 1976, *Astrophys. J.* **210**, L97.
Cushman, G. W. and Rense, W. A.: 1976, *Astrophys. J* **207**, L61.
Dere, K. P.: 1978, *Astrophys. J.* **221**, 1062.
Dere, K. P.: 1982, *Solar Phys.* **77**, 77.
Doschek, G. A., Feldman, U., and Bohlin, J. D.: 1976, *Astrophys. J.* **205**, L177.
Feldman, U., Cohen, L., and Doschek, G. A.: 1982 *Astrophys. J.* **255**, 325.
Hansteen, V., Maltby, P., and Malagoli, A.: 1996, in B. Bentley and J. Mariska (eds.), *Yohkoh Conf. on Observations of Magnetic Reconnection in the Solar Atmosphere*, PASP Conference Series, in press.
Harrison, R. A. and 38 co-authors: 1995, *Solar Phys.* **162**, 233.
Harrison, R. A., Fludra, A., Pike, C. D., Payne, J., Thompson, W. T., Poland, A. I., Breeveld, E. R., Breefeld, A. A., Culhane, J. L., Kjeldseth-Moe, O., and Aschenbach, B.: 1997, *Solar Phys.* **170**, 143 (this issue).
Hassler, D. M., Rottman, G. J., and Orrall, F. Q.: 1991, *Astrophys. J.* **372**, 710.
Judge, P. G. and Brekke, P.: 1997, in preparation.
Kjeldseth-Moe, O., Brynildsen, N., Brekke, P., Engvold, O., Maltby, P., Bartoe, J-D. F., Brueckner, G. E., Cook, J. W., Dere, K. P., Socker, and D. G.: 1988, *Astrophys. J.* **334**, 1066.
Kjeldseth-Moe, O., Brynildsen, N., Brekke, P., Maltby, P., and Brueckner, G. E.: 1993, *Solar Phys.*, **145**, 257.
Lites, B. W., Bruner, E. C., Chipman, E. G., Shine, R. A., Rottman, G. J., White, O. R., and Athay, R. G.: 1976, *Astrophys. J.* **210**, L111.
Mariska, J. T.: 1992, *The Solar Transition Region*, Cambridge University Press, Cambridge.
Mariska, J. T. and Dowdy, J. F., Jr.: 1992, *Astrophys. J.* **401**, 754.
Martin, W. C. and Zalubas, R.: 1980, *J. Phys. Chem. Ref. Data* **9**, 1.
Mason, H. E., Young, P. R., Pike, C. D., Harrison, R. A., Fludra, A., Bromage, B. S. I., and Del Zanna, G.: 1997, *Solar Phys.* **170**, 143 (this issue).
Neupert, W. M., Brosius, J. W., Thomas, R. J., and Thompson, W. T.: 1992, *Astrophys. J.* **392**, L95.
Orrall, F. Q., Rottman, G. J., and Klimchuk, J. A.: 1983, *Astrophys. J.* **216**, L65.
Pneuman, G. W. and Kopp, R. A.: 1978, *Solar Phys.* **57**, 49.
Rabin, D. and Dowdy, J. F.: 1992, *Astrophys. J.* **398**, 665.
Reeves, E. M.: 1976, *Solar Phys.* **46**, 53.
Rottman, G. J., Orrall, F. Q., and Klimchuk, J. A.: 1982, *Astrophys. J.* **260**, 326.
Rottman, G. J., Hassler, D. M., Jones, M. D., and Orrall, F. Q.: 1990, *Astrophys. J.* **358**, 693.
Spadaro, D., Antiochos, S. K., and Mariska, J. T.: 1991, *Astrophys. J.* **382**, 338.
Thomas, R. J. and Neupert, W. M.: 1994, *Astrophys. J. Suppl.* **91**, 461.

FIRST ENERGETIC PARTICLE EVENTS OBSERVED BY THE ERNE INSTRUMENT

J. TORSTI, E. VALTONEN, L. G. KOCHAROV, R. VAINIO, E. RIIHONEN,
A. ANTTILA, T. LAITINEN, M. TEITTINEN and J. KUUSELA
Space Research Laboratory, Department of Physics, FIN-20014 Turku University, Finland

(Received 20 June, 1996; in final form 5 September, 1996)

Abstract. The energetic particle instrument ERNE on-board SOHO started its observations on December 15, 1995. The low-energy sensor of ERNE, LED, is capable of measuring particles in the energy range from 1 to 10 MeV $nucl^{-1}$. From the beginning of the year 1996 until May 22, 1996, LED-observations included four energetic particle events above threshold intensities.

An energetic particle event caused by a corotating interaction region that accelerated protons upto $\sim$10 MeV, was observed during January 20–25. Another similar particle event occured on May 6 –12. The events were separated by four solar rotation periods. They had similar time profiles, but the one in May had a harder spectrum and a lower intensity level. The ^{4}He-to-proton ratios were in accordance with the solar wind value.

Energetic particles observed during April 22–23 and May 14–17 were accelerated at the Sun. The first one was apparently an outcome from an active region observed on the west limb by telescopes on-board SOHO. Protons were detected at energies from 1 to $\sim$10 MeV. For this event, the ^{4}He-to-proton ratio in the range 1.5–5 MeV $nucl^{-1}$ was $\sim$3%. No ^{3}He ions were detected. The period of May 14 –15 was, in contrast, extremely ^{3}He-rich: it had a ^{3}He-to-proton ratio of 1.5 ± 0.6 and a ^{3}He-to-^{4}He ratio as high as $\sim$8. The period of May 14–17 comprised at least three individual, one-day-long events. The first two events were ^{3}He-rich, while the last one seemed to have a normal composition.

1. Introduction

Of the solar particle detectors on SOHO*, the ERNE instrument covers the highest energies, roughly from 1 to 100 MeV $nucl^{-1}$ for light elements and a few times more for heavier elements. The sensitivity of the ERNE sensors is high enough to detect even weak solar energetic particle (SEP) events and low-level particle fluxes originating from corotating interaction regions (CIR). Low-intensity SEP events enriched in ^{3}He may be detected during solar minimum. The ^{3}He-rich SEP events have been extensively studied for the past 25 years (e.g., reviews by Kocharov and Kocharov,1984; Reames, Meyer and von Rosenvinge,1994). In spite of many attempts, no reliable explanation has been found for the observed huge enrichments in ^{3}He.

ERNE was switched on December 7, 1995. Subsequently, the observations were started on December 15. In this paper, we present the first energetic particle events observed by the ERNE instrument from January through May, 1996.

* SOHO is a project of international co-operation between ESA and NASA.

Solar Physics **170:** 179–191, 1997.

2. Instrument Description

The ERNE instrument was designed for measuring the composition and energy spectra of particles encountered in interplanetary space, covering the energy range from 1 to 100 MeV $nucl^{-1}$ for protons and ^4He nuclei. The instrument consists of two detector telescopes, low-energy detector (LED) and high-energy detector (HED), and their individual electronics. A full description of the ERNE experiment was recently given by Torsti *et al.* (1995). A brief description of LED, concentrating on the features relevant to the data analysis and interpretation of the results of the present paper, is given below.

LED operates in the low-energy range of ERNE. In principle, it is a conventional particle telescope consisting of two layers of silicon detectors, D1 and D2. For acceptable events, simultaneous signals from both D1 and D2 are required. The determination of energies and identification of particles are based on the two signals, one obtained from the traversed D1 detector and the other one from D2, in which the particles stop. An anticoincidence detector under D2 ensures rejection of high energy particles passing through D2. For these particles, the total energies cannot be determined and the identification would be difficult. Particles above the energy range of LED are recorded by HED.

The novelty in LED is the division of the first detector layer into seven independent detectors in a honeycomb structure. This approach serves several purposes. First of all, by using several detectors, a large total area of the first detector layer is achieved, giving a large geometric factor. Six of the D1 detectors are located on the corners of a hexagon and one at the center. The planes of the corner detectors are tilted at an angle of 16.7° relative to the plane of the center detector. This keeps the energy thresholds of all the D1-D2 combinations the same and increases the field of view and acceptance of the sensor. The half angle of the total view cone of LED is about 32°.

Each of the seven detectors in the first layer has a mechanical collimator shielding the detector against the background caused by the high flux of lowest energy particles during solar flares. The path length variations of incident particles in the D1 detectors are limited by the collimator and by the distance between the D1 detectors and the upper surface of D2. For reaching a good mass resolution, minimizing the effect of path length variations in the fluctuations of the signals is very important. In LED the maximum path length variation is below 4%, which allows separation of ^{20}Ne from ^{22}Ne in a limited energy range.

Using separate detectors in the first layer leaves a free choice of the thickness of each individual detector. Thereby, the energy thresholds of various D1-D2 combinations can be different. In LED, detectors with different thicknesses are used. Two of the detectors have a nominal thickness of 20 microns and five a thickness of 80 microns. Taking into account the thin kapton foil above each detector, protecting them thermally and against sunlight, the low-energy limits of the telescopes consisting of a thin D1 and D2 or a thick D1 and D2 become about 1.2 MeV or

3 MeV, respectively, for protons. The total acceptance of the two telescopes with the lower energy threshold is 0.260 cm^2 sr and that of the five telescopes with the higher threshold 0.655 cm^2 sr.

Using detectors with two different thicknesses in the first layer allows an easy extension of the energy range, not only to lower but also to higher energies. This follows from the fact that the accuracy of particle identification worsens with the decreasing ratio of energy loss in D1 over residual energy in D2. Therefore, a combination of a thin D1 with a very thick D2 is impractical. A thicker D1, giving still reasonable signals for particles stopping at the bottom of D2, is needed to cover the high energy end. Selection of the events from one or the other D1-D2 combinations can then be done by the analysis software to achieve the best accuracy. In LED, the thickness of D2 is 1000 microns giving the upper energy limit of about 12 MeV for protons.

In the nominal observation mode of the instrument, the on-board analysis software is capable of identifying all hydrogen and helium isotopes. The program forms a 12-bin energy spectrum of each of these isotopes. During quiet times, however, the allocated telemetry rate allows all recorded particles to be transmitted to earth as pulse height data, as well. From this data, more accurate energy spectra can be constructed because we can freely choose the energy channels, and also study the instrumental background. The results of this paper are based on the utilization of the pulse height data.

In addition to the nominal observation mode, measurements in the so called maintenance mode can be used for analysis of weak low-energy particle fluxes. In the maintenance mode the instrument records all particles, but only a fixed number of them can be transmitted down. The reason is that in the absence of on-board analysis, the telemetry buffer of the instrument is filled in some fraction of the nominal one minute interval. The actual portion of the low-energy particles transmitted depends on the HED count rate. For this reason, the maintenance mode measurements can be used for studies of relative particle fluxes detected by LED during quiet time when the HED count rate is almost constant.

3. Observations

The LED observations of ERNE from the beginning of 1996 until May 22, 1996 include four periods with enhanced energetic particle fluxes. The first one (on January 20–25, 1996) was recorded while the instrument was in the maintenance mode, and the other three in the nominal observation mode. During the commissioning phase of SOHO, from launch until the end of March 1996, calibration runs and performance tests of the on-board scientific software of ERNE were carried out, which resulted in a few data gaps lasting several days. Therefore, it is possible that some other energetic particle events have also occurred, which are not included in the present analysis.

The first months of the SOHO flight, from December 1995 till May 1996 have been quiet, especially as far as energetic particles are concerned. For most of the active measuring time of the ERNE instrument, the LED intensity at energies 1.2–3 MeV $nucl^{-1}$ has been at its background level, which is well below 10^{-3} protons/(cm^2 s sr MeV). The estimated threshold intensity for a detectable proton event in LED at energies 1.2–3 MeV $nucl^{-1}$ is 3×10^{-3} protons/(cm^2 s sr MeV) with a 2-hour integration time. In this energy range, the thresholds for ^{4}He and ^{3}He events are 3×10^{-4} and 8×10^{-5} particles/(cm^2 s sr MeV $nucl^{-1}$), respectively.

Riihonen *et al.*(1996) studied the IMP-8 proton intensity observations above 1 MeV over the 5-year period 1984–88. The least active year in the production of 1–10 MeV protons was the year 1986. In 1986, the proton flux from 1 to 10 MeV was above 0.1, 1, and 10 (cm^2 s sr$)^{-1}$ for about 25%, 10%, and 5% of the IMP-8 observation time, respectively, when the satellite was in the solar wind environment outside the geomagnetic field. None of the three particle events of the present analysis observed in the nominal observation mode exceeded even the lowest threshold value of the analysis of Riihonen *et al.*(1996). For example, the maximum proton intensity observed on April 22, 1996, integrated over energies from 1 to 10 MeV, just exceeds the intensity level of 10^{-2} protons/(cm^2 s sr). The virtually undisturbed solar system during the observation period of this analysis provides an exceptional scene to monitor the intensity profiles of small events with minimal interplanetary disturbances.

3.1. The January 20–25, 1996 Event

The ERNE instrument was in the maintenance mode between January 6 and January 25. During this time, one energetic particle event was oserved in LED (Figure 1). On January 20, the count rates of protons and ^{4}He in the energy range from 1.3 to 3 MeV $nucl^{-1}$ rose simultaneously. The event lasted until January 25, and it had a more or less symmetric profile relative to the maximum on January 22. The energy spectra integrated over the event are presented as distribution functions (i.e., the number of particles per unit volume in velocity space) for protons and ^{4}He in Figure 2. The proton distribution function is well represented by an exponential in momentum, with characteristic momentum $p_0 = 6.6 \pm 0.6$ MeV c^{-1}. The helium distribution function is slightly softer with $p_0 = 5.3 \pm 1.4$ MeV c^{-1} per nucleon. The ^{4}He-to-proton ratio at energies 1.3–3 MeV $nucl^{-1}$ is $(3.8 \pm 0.2)\%$. No enhancement in ^{3}He intensity was observed.

3.2. The April 22–23, 1996 Event

The count rates of 1.5–5 MeV $nucl^{-1}$ protons and ^{4}He revealed an energetic particle event starting on April 22, 1996 (Figure 3). The proton spectrum extended up to energies above 10 MeV, as HED count rates of 12–18 MeV protons also

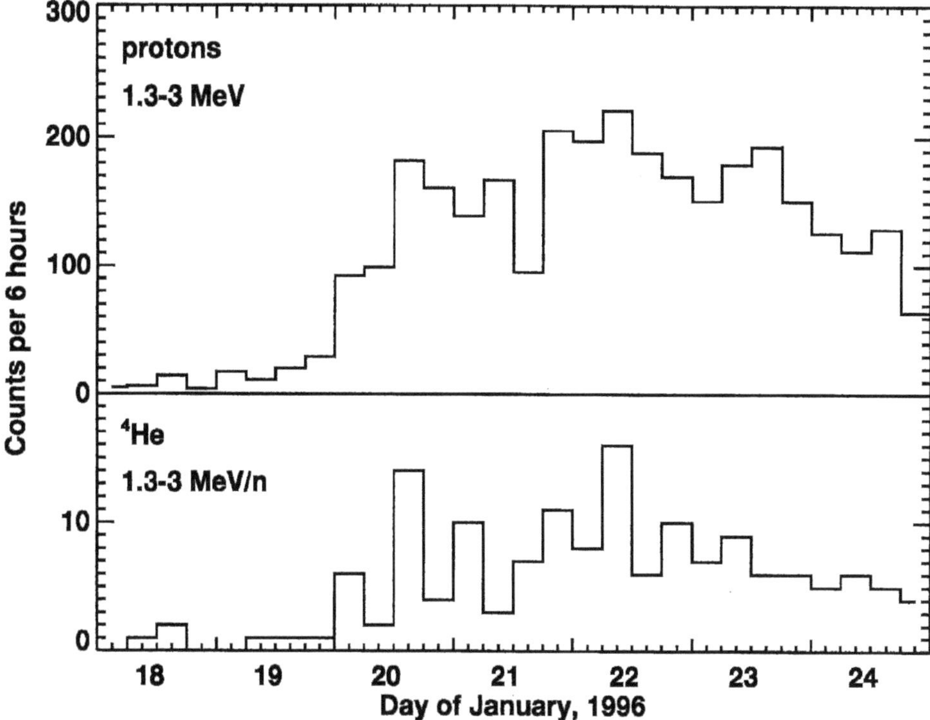

Figure 1. Six-hour count rates of protons and ^{4}He observed by ERNE in January 1996 in the energy range 1.3–3 MeV nucl^{-1}.

showed enhancement over the background levels by about 3σ. The ^{4}He-to-proton ratio at 1.5–5 MeV nucl^{-1} was $(3.4 \pm 1.5)\%$ when calculated from 09 UT April 22 to 06 UT April 23, which was the time interval when the proton intensity remained continuously above the 3σ-level. During the same interval, the proton spectrum exhibits a slope $\gamma = 1.9 \pm 0.5$ on the $\log j - \log E$ plot at energies 1.5–5 MeV. These results include a background subtraction, calculated over the quiet period of 00 UT April 17 – 00 UT April 20, 1996.

3.3. The May 6–12, 1996 Event

Starting on May 6, 1996, the proton count rates at 1.5–8 MeV showed an increase over the background level as presented in Figures 4 and 5. The increase was most clearly detected at relatively high energies, 3–8 MeV, whereas the increase at 1.5–3 MeV was comparatively weak. The hard spectrum of the event was also evident from the HED measurements, where the proton count rates at energies 12–16 MeV rose clearly above the background level on May 6. The intensity enhancements in LED remained above the background until May 13. The time profile of the event

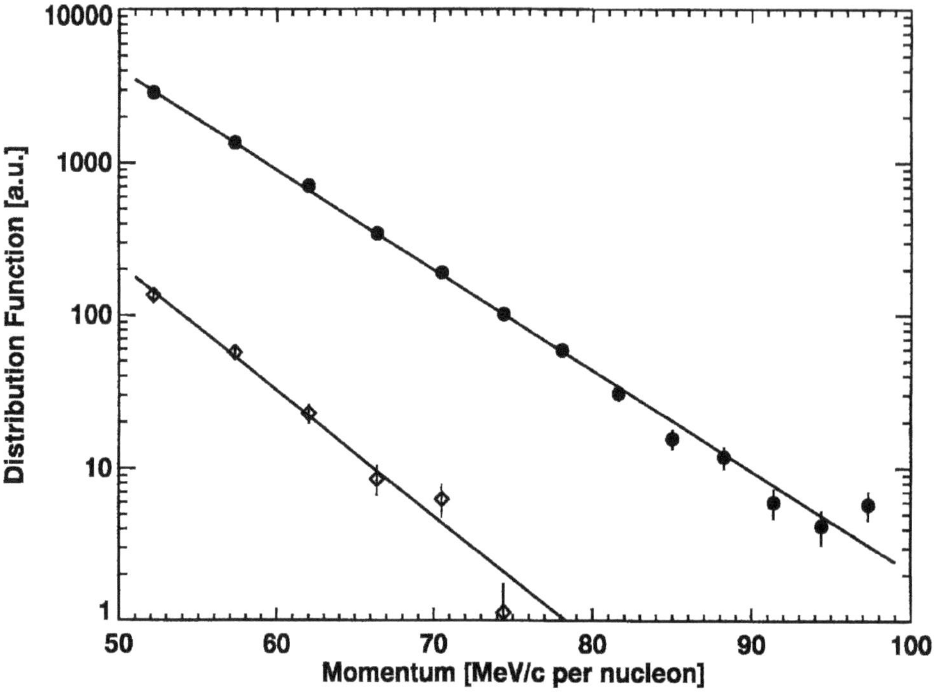

Figure 2. Proton (solid circles) and helium (open diamonds) distribution functions integrated over the time period of the January 1996 event. Solid lines represent the best-fit exponentials to the data.

was very flat and the maximum flux in the 3–8 MeV channel was observed between May 7–9. During the proton event, no statistically significant enhancement of ^{4}He was observed. The most probable value of the ^{4}He-to-proton ratio integrated over the time from May 6 to May 12 at 3–8 MeV nucl^{-1} was $(5^{+7}_{-5})\%$. The subtracted backgrounds of the proton and helium count rates have been calculated using the quiet period from 00 UT May 19 to 00 UT May 22, 1996.

3.4. The May 14–17, 1996 Event

Observations from May 13 through May 18, 1996 revealed solar energetic particle events enhanced in ^{3}He as shown in Figures 4 and 5. The search for ^{3}He enhancement was performed using the 1.5–3 MeV nucl^{-1} count rates in 4-hour intervals. An event was defined as ^{3}He-rich if the following two criteria were met: (i) the statistical error in the ^{3}He flux was less than 50%; (ii) the ^{3}He-to-^{4}He ratio was $> 10\%$. The search revealed two clear ^{3}He-rich events during May 14–15. About 16 hours before the first of these events, a less distinct increase was registered in the 1.5–3 MeV nucl^{-1} ^{3}He count rate. The ^{3}He–^{4}He ratio integrated from 00 UT May 14 through 00 UT May 16 in the energy range 3–8 MeV nucl^{-1} was as high

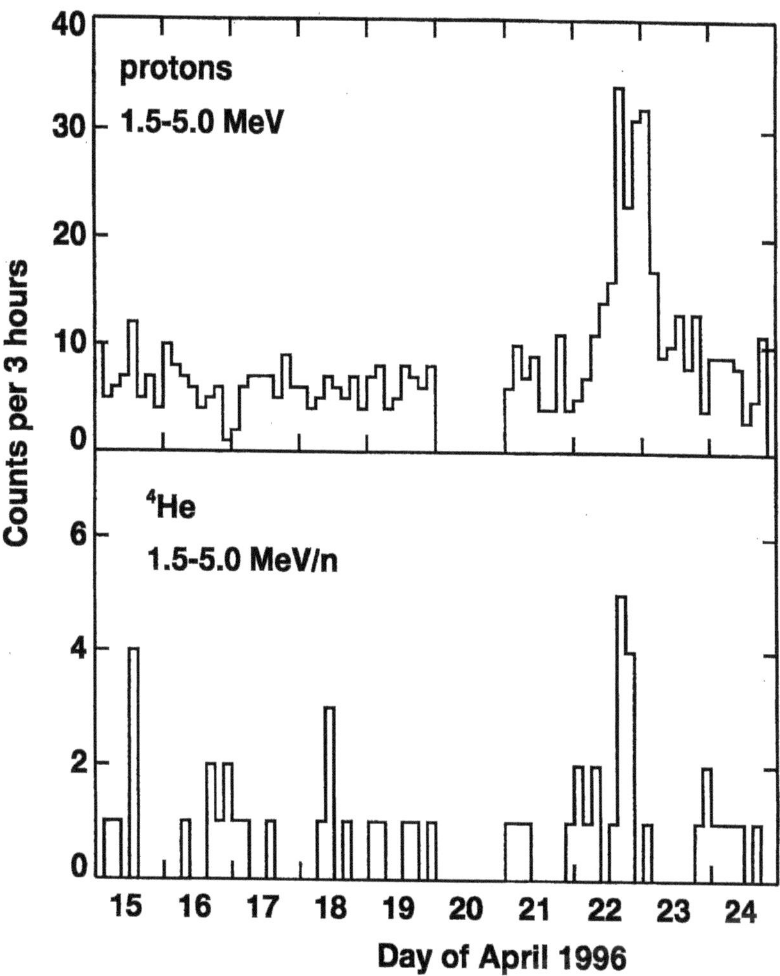

Figure 3. Three-hour count rates of protons and ^{4}He at energies 1.5–5 MeV nucl^{-1} observed by ERNE in April 1996. The quiet period backgrounds are not subtracted from these count rates.

as ~8. The ^{3}He-to-proton ratio for the same interval was deduced to be 1.5 ± 0.6. This value is among the highest values ever observed (cf., Figure 8 by Kocharov and Kocharov, 1984). For the interval from 00 UT May 14 through 00 UT May 16, the spectrum of ^{3}He was rather hard. In the 2–8 MeV nucl^{-1} range, it can be approximated by a power law in energy with a spectral index $\gamma = 3.4 \pm 0.8$. The fluence of ^{3}He was very small, $F(> 3\,\mathrm{MeV}) \approx 10^2\,\mathrm{cm}^{-2}\mathrm{sr}^{-1}$. This is in agreement with the general tendency of the ^{3}He-to-proton ratio to get higher values in weaker events (e.g., Kocharov and Kocharov, 1984). All the results are given with the quiet-period backgrounds subtracted. The backgrounds were obtained by

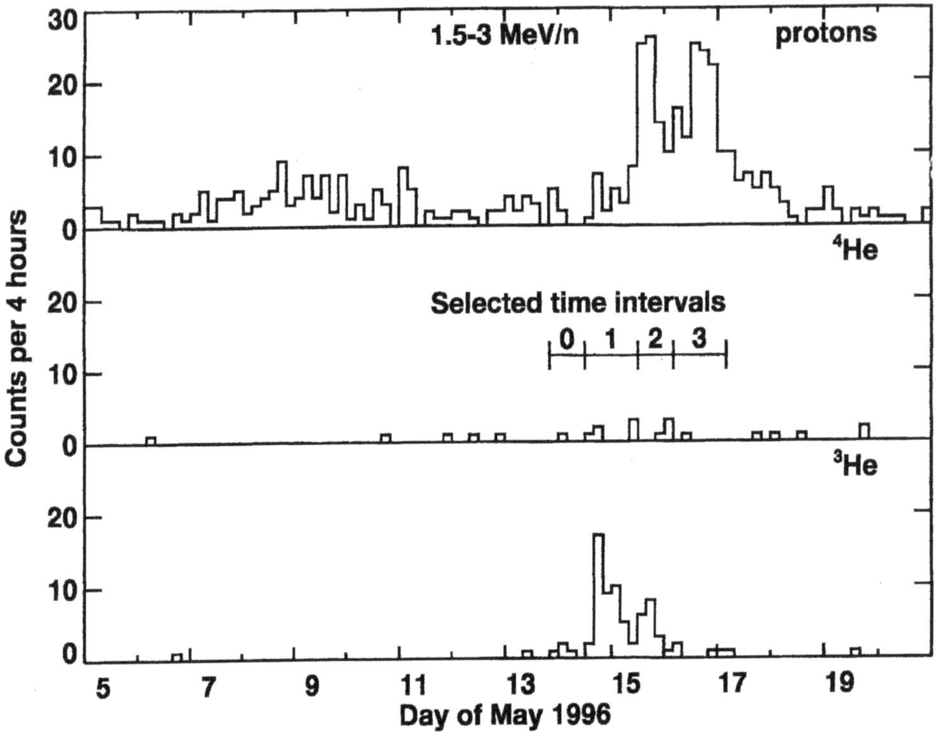

Figure 4. Four-hour count rates of protons, ^{4}He and ^{3}He in the energy range 1.5– 3 MeV nucl^{-1} observed by ERNE during May 5–20. The indicated time intervals were chosen for the analysis of ^{3}He enrichment. The quiet-period backgrounds are not subtracted from these count rates.

averaging corresponding count rates observed from 00 UT May 19 to 00 UT May 22.

Four time intervals were selected for more detailed analysis of the ^{3}He-rich period (the intervals are marked with bars 0, 1, 2, and 3 in Figure 4). During the interval 0 (20 UT May 13 – 12 UT May 14), a weak increase of ^{3}He count rate was registered in the lower-energy channel only (Figure 4). This increase would not attract considerable attention, if it was not followed by stronger increases observed during the next two days. In the 1.5–3 MeV nucl^{-1} range, the ^{3}He-to-proton ratio for the interval was measured to be ≈ 2 with very poor statistics. A weak 3–8 MeV proton increase was also observed during the interval 0 (Figure 5). The strongest ^{3}He-rich event was registered during the interval 1 (12 UT May 14 – 12 UT May 15). This event was clearly seen in both 1.5–3 MeV nucl^{-1} and 3–8 MeV nucl^{-1} ^{3}He channels. In the 1.5–3 MeV nucl^{-1} energy range, the ^{3}He-to-proton ratio was 2.4 ± 0.9. Next in time a ^{3}He-rich event was registered during the interval 2 (12 UT May 15–04 UT May 16). This event was weaker in ^{3}He, while the flux of

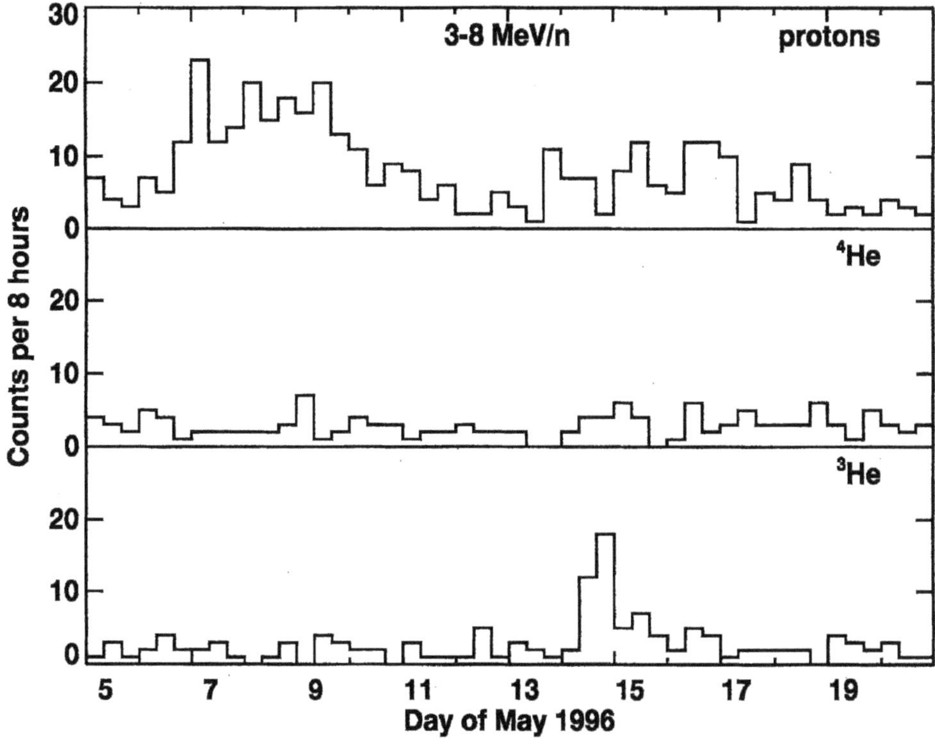

Figure 5. Same as in Figure 4, but in the energy range 3–8 MeV $nucl^{-1}$.

1.5–3 MeV protons was much higher compared with the previous interval 1. The ^{3}He-to-proton ratio in the 1.5–3 MeV $nucl^{-1}$ energy range was 0.26 ± 0.09. Finally, during the interval 3 (04 UT May 16 – 04 UT May 17), a proton event was recorded while there was no statistically significant accompanying ^{3}He enrichment.

4. Discussion

4.1. Particles Accelerated in Interplanetary Space

The January 20–25, 1996 energetic particle event occurred in a time period when several high-speed solar wind streams were detected by the SWE experiment on-board the WIND spacecraft. The first rapid rise in the solar wind speed from about 270 to 650 km s^{-1} took place on January 14–15. At that time, no energetic particle enhancement was observed by the ERNE sensors. The increase of particle fluxes took place on January 20, when another peak in the solar wind speed was observed by the SWE instrument. According to the measurements of the MFI experiment, also on-board WIND, the polarity of the interplanetary magnetic field (IMF) was mainly towards the Sun during the first high-speed stream. During the energetic particle event of January 20–25, it was mainly away from the Sun. This is the usual situation for observing energetic particles accelerated in a CIR: when the IMF is pointing outwards, the accelerated particles may be detected in the inner solar system (e.g. Kunow *et al.*,1991). The distribution functions of protons and ^{4}He are well represented by exponentials in momentum. Within the error limits, the characteristic momenta are $p_0 \approx 6$ MeV c^{-1} per nucleon for both ion species. This is consistent with previously observed typical values for corotating events (Gloeckler, Hovestadt and Fisk, 1979): $p_0 = 6.5$–11 MeV c^{-1} per nucleon. The observed time profile was characteristic for a corotating event (Kunow *et al.*, 1991). No velocity dispersion was observed. Also, the observed ^{4}He-to-proton ratio agrees well with the solar wind value. We thus conclude that the energetic particles of the event on January 20–25 were accelerated in a CIR beyond the orbit of SOHO.

The energetic particle event on May 6–12 occurred almost exactly four solar rotation periods later than the January event. The two events had a similar time profile in proton channels. The energetic particle composition during the May 6–12 event was probably close to that of the solar wind, but due to the low statistics this can not be deduced with a high certainty. The intensity level of the May 6–12 event was considerably lower than in January 20–25 at the energies 1.5–8 MeV. However, the proton spectrum observed on May 6–12 was hard in comparison with the January event: the count rates observed in the two energy channels of Figures 4 and 5 would require a characteristic momentum $p_0 \approx 12$ MeV c^{-1} for protons. The changes in the solar wind speed from May 1 to May 8 where not very dramatic, but since the WIND spacecraft entered the magnetosphere on May 8, its observations of the solar wind were not available during the latter half of the particle event. Hence, to conclude whether the May 6–12 event was an outcome from the same CIR that operated in January, further investigations of both the solar wind and the particle data are needed.

4.2. Solar Energetic Particle Events

We studied the time profiles of protons during April 22 in two energy channels, 1.2–3 MeV and 3–5 MeV, with a three-hour integration time. A comparison of the count rates before the intensity maximum at 15–18 UT indicates that the particles at high velocities arrived before the low-velocity particles. This effect may be interpreted as a sign of a solar source of the event. At the time of the event, there was an active region near the west limb observed by telescopes onboard the SOHO spacecraft. The energetic particle event on April 22 had a hard spectrum. The proton count rate at 1.5–5 MeV was above the background level for approximately one day. The ^{4}He-to-proton ratio for the event was about 3%, but most of the ^{4}He ions were recorded in coincidence with the first arriving protons during a few hours, so the composition seems to evolve in time. The decay times of the particle intensities were also quite fast: of the order of a few hours. These observations are difficult to explain with a single impulsive injection of particles from the Sun. A possible association with isolated optical/radio events may be a matter of further investigation.

During the 1996 May 14 ^{3}He-rich event, the first ^{3}He ions with an energy $\sim$6 MeV nucl^{-1} were detected by ERNE at $\sim$14 UT. Using WIND MFI data, we verified the IMF direction with respect to the view cone of LED. Around 14 UT on May 14, the magnetic field direction was favourable for the detection of particles with small pitch angles. For the first particles, the injection time is estimated to be $\sim$12 UT on May 14, 1996. Close to this time, a short ($\sim$5 min) microwave burst was registered in Bern (Arzner, 1996). The maximum flux of 3–8 GHz emission was detected at 11:35 UT. Type III radio bursts were registered at 8:14, 10:10, 11:33 –11:39 and 15:06 UT on May 14, 1996 (*Solar-Geophysical Data*, 1996). Thus, the only possible solar association for the ^{3}He-rich event is the radio burst at 11:33 –11:39 UT. Using Monte-Carlo simulations of the interplanetary transport, we estimated that the observed intensity–time profile is consistent with an impulsive injection of ^{3}He at 11:35 UT, if the radial mean free path of 2 MeV nucl^{-1} ions was $\sim$0.15 AU.

Reames and von Rosenvinge (1983) analyzed ^{3}He-rich events observed aboard ISEE-3 during 1978–1982 (see also Kahler *et al.*,1985 and references therein). Of the 67 events investigated, they found only three events which had ^{3}He/^{4}He ≥ 4 and ^{3}He spectral index $\gamma < 4.5$ at 1–3 MeV nucl^{-1}. In the case of the 1996 May 14 event, both an extremely high enrichment factor and a hard spectrum of ^{3}He were observed at 1–8 MeV nucl^{-1}. Note that usually in ^{3}He-rich events, such a hard spectrum does not extend above 2 MeV nucl^{-1} (Möbius *et al.*, 1980).

As a result of the fixed view cone of the ERNE detectors and the variability of the IMF direction, some increases or decreases in the ERNE count rates might result from scanning different parts of the angular distribution with no changes in the omnidirectional particle intensities. We have studied the magnetic field observations to validate the existence of separate events during May 14–17, 1996. The WIND MFI data suggest that the two maxima in the ^{3}He count rate (in intervals 1 and 2, Figure 4) represent actual events and can not be explained by scanning caused by rotations of the magnetic field. To reach this conclusion, we have calculated the fraction of time when the magnetic field was pointed inside the LED view cone. This fraction revealed no correlation with the observed particle count rates. For instance, the fraction was increasing during the interval from 20 UT May 14 to 12 UT May 15, while the ^{3}He count rate was decreasing. We can not exclude that the change of the magnetic field direction affected the intensity–time profiles, but it is unlikely that the variation in the IMF direction was the source of the two maxima in the observed ^{3}He count rate.

During the entire period of 1996 May 13–17, at least two ^{3}He-rich events were observed by the ERNE instrument. Appearance of such a group of ^{3}He-rich events is not a unique phenomenon, but rather a general property of events with ^{3}He enhancements. For instance, among the 132 ^{3}He-rich events observed between

1969 April and 1982 June, we have found 8 groups consisting of three or four events each. In every group, consecutive events were separated in time by not more than 3 days. Among the 66 events observed aboard ISEE-3 between 1978 October and 1982 June, there were 4 groups consisting of three events each and one group of four events. Additionally, 6 close pairs of events were observed during the same time interval. These numbers are extremely large to be explained as a statistical artifact. Hence, the tendency of ^{3}He-rich events to appear in groups seems clear. Some special magnetic structure at the Sun may be responsible for a series of ^{3}He-rich events to appear, while the total duration of a group observed (2–5 days) may also be limited by conditions of the magnetic connection between the solar source region and a spacecraft.

We used the solar wind measurements on-board the WIND spacecraft to trace the position of the root of the Earth-connected interplanetary magnetic field line. Two bright regions were observed by the telescopes on-board SOHO during May 13–17, 1996. Both regions were located in the western hemisphere, to the south of the solar equator. At the beginning of the ^{3}He-rich period, SOHO got an interplanetary magnetic connection with the leading (west) region and stayed well connected during May 14–15. For these reasons, one can propose that the region observed by the SOHO telescopes was the source of the ^{3}He-rich events recorded by the ERNE particle detectors during May 14–15, 1996. We feel that this issue deserves a more detailed study in the future.

Acknowledgements

We thank the referee for useful comments. We made use of the solar wind parameters from the WIND spacecraft by courtesy of Drs R. P. Lepping and K. W. Ogilvie. We thank Dr V. Melnikov for comments on the microwave data and Drs D. Reames and T. von Rosenvinge for providing the ISEE-3 ^{3}He Event List used in the discussion.

The Academy of Finland is thanked for financial support.

References

Arzner, K.: 1996, personal communication.
Gloeckler, G., Hovestadt, D., and Fisk, L.A.: 1979, *Astrophys. J.* **230**, L191.
Kahler, S., Reames, D. V., Sheeley, N. R., Howard, R. A., Koomen, M. J., and Michels, D. J.: 1985, *Astrophys. J.* **290**, 742.
Kocharov, L. G. and Kocharov, G. E.: 1984, *Space Sci. Rev.* **38**, 89.
Kunow, H., Wibberenz, G., Green, G., Müller-Mellin, R., and Kallenrode, M.-B.: 1991, in R. Schwenn and E. Marsch (eds.): *Physics of the Inner Heliosphere* **2**, Springer-Verlag, Berlin, p. 234.
Möbius, E., Hovestadt, D., Klecker, B., and Gloeckler, G.: 1980, *Astrophys. J.* **238**, 768.
Reames, D. V. and von Rosenvinge, T. T.: 1983, *Proc. 18th Int. Cosmic Ray Conf.* **4**, 48.
Reames, D. V., Meyer, J. P., and von Rosenvinge, T. T.: 1994, *Astrophys. J. Suppl.* **90**, 649.

Riihonen, E., Torsti, J., Anttila, A., Kuusela, J, Lumme, M., and Valtonen, E.: 1996, *Ann. Geophys.* **14**, 503.
Solar-Geophysical Data: 1996, No. 623, Part I.
Torsti, J., Valtonen, E., Lumme, M., Peltonen, P., Eronen, T., Louhola, M., Schultz, G., Teittinen, M., Ahola, K., Holmlund, C., Kelhä, V., Leppälä, K., Ruuska, P., Strömmer, E.: 1995, *Solar Phys.* **162**, 505.

ANOMALOUS COSMIC-RAY HELIUM, NITROGEN AND OXYGEN IN 1996

Measurements of the ERNE Instrument on-board SOHO

J. TORSTI, E. VALTONEN, A. ANTTILA, R. VAINIO, P. MÄKELÄ,
E. RIIHONEN and M. TEITTINEN
Space Reseach Laboratory, Department of Physics
FIN-20014 Turku University, Finland

(Received 20 June, 1996; in final form 5 September, 1996)

Abstract. The energy spectra of the anomalous components of helium, nitrogen and oxygen have been measured by the ERNE experiment on board the SOHO spacecraft. During February 28–April 30, 1996, the maximum intensity of anomalous helium was found to be 3.8×10^{-5} cm^{-2} sr^{-1} s^{-1} (MeV nucl^{-1})$^{-1}$ in the energy range 10–15 MeV nucl^{-1}. During the period January 26–April 30, 1996, the maximum oxygen intensity was 1.2×10^{-5} cm^{-2} sr^{-1} s^{-1} (MeV nucl^{-1})$^{-1}$ at 4–7 MeV nucl^{-1}, and the maximum nitrogen intensity 1.7×10^{-6} cm^{-2} sr^{-1} s^{-1} (MeV nucl^{-1})$^{-1}$ at 4–9 MeV nucl^{-1}. These peak intensities are at the same level as two solar cycles ago in 1977, but significantly higher than in 1986. This gives observational evidence for a 22-year solar modulation cycle. A noteworthy point is that the spectra of anomalous nitrogen and oxygen appear to be somewhat broader than in 1977.

1. Introduction

In 1972, during the solar quiet time, it was discovered for the first time that the differential intensity of helium below 60 MeV nucl^{-1} was essentially independent of kinetic energy (Garcia-Munoz *et al.*, 1973). This observation could not be explained by the solar modulation models prevailing at that time. The idea of interstellar neutral gas penetrating into the heliosphere as the origin of this so-called anomalous component of cosmic rays was proposed by Fisk, Kozlovsky, and Ramaty (1974). They suggested that neutral interstellar atoms with high first ionization potentials are swept into the heliosphere as the Sun travels through the local interstellar medium. Closer to the Sun the atoms are singly ionized by the UV radiation of the Sun and by charge exchange processes with the solar wind. These singly charged ions are then picked up by the solar wind and carried to the outer heliosphere where they are accelerated by some mechanism. Pesses, Jokipii, and Eichler (1981) extended this model by proposing that the accelaration up to the energies of tens of MeV nucl^{-1} could take place by diffusive shock acceleration in the heliospheric termination shock. These views of the origin and acceleration of the anomalous cosmic rays (ACR) are widely accepted today. Hence, the ACR observations give important knowledge about the local interstellar matter (Cummings and Stone, 1995).

Solar Physics **170:** 193–205, 1997.

However, interplanetary shocks and corotating interaction regions might contribute in the acceleration of particles (Giacalone and Jokipii, 1996; Gloeckler *et al.* 1994). It is also possible that all the ACR particles do not originate from the interstellar neutral gas, but also from other sources, such as giant planets (Luhmann, 1994), comets, or heliospheric dust grains.

During the years, evidence from direct measurements of the charge states of ACR has accumulated, showing that they indeed are singly ionized, as predicted by the theory of Fisk, Kozlovsky, and Ramaty (1974). For oxygen, Klecker *et al.* (1995) obtained an upper limit of 10% for ionization states higher than one. However, more recently Klecker *et al.* (1996) and Selesnick *et al.* (1996) have reported that at energies above 20 MeV $nucl^{-1}$ a significant fraction of ACR oxygen is doubly ionized, and that at energies above 30 MeV $nucl^{-1}$, charge states higher than one become predominant.

A subject closely connected to the anomalous cosmic rays is the solar modulation, because the variations in cosmic ray intensities are most clearly detectable in anomalous cosmic rays. The four basic physical mechanisms responsible for the solar modulation of cosmic rays are convection, diffusion, adiabatic deceleration/acceleration, and drift. The relative importance of these different mechanisms is not very well understood. The current modulation theories can be divided into two groups. In the first category are the theories which are based on the idea that the 11-year modulation results from stepwise pileup of distinct intensity decreases in the outer heliosphere. The second category includes theories in which gradient and curvature drifts are assumed to be the most important factors in the solar modulation of cosmic rays. These so-called drift models can explain very well large-scale variations in the level of modulation. Especially during low solar activity, global drifts are believed to be dominant (Potgieter *et al.*, 1993).

The present study is based on the observations of the ERNE experiment on board the SOHO spacecraft. The purpose of the study was to examine if the intensities of anomalous cosmic rays during the current solar activity minimum are equal to those during the 1977 minimum (i.e., two solar cycles ago), and how do they compare with the intensities in 1986, very near to the last minimum. Earlier observations have shown that there are considerable differences in the intensities of ACR during the 1977 minimum as compared to the minimum in 1965 and in 1986 (Evenson *et al.*, 1983, and references therein; Cummings, Stone, and Webber, 1984; Garcia-Munoz, Pyle, and Simpson, 1987; Cummings and Stone, 1987a).

2. Instrument Description

ERNE is a high-energy particle instrument consisting of two sensors. Observations with both of the ERNE sensors, low-energy detector (LED) and high-energy detector (HED), are used in this study. For oxygen, LED operates at energies from 2.5 MeV $nucl^{-1}$ up to about 30 MeV $nucl^{-1}$ and HED covers energies a further order of magnitude upwards. The ERNE instrument has been described in some detail in Torsti *et al.* (1995), and some features of LED are given in an adjacent paper (Torsti *et al.*, 1997). In the following, the main characteristics of the HED sensor are summarized.

As LED, HED is a particle telescope, where the particle identification and total energy determination are based on sampling the ionization losses in one or more transmission detectors along the track of the particle, and absorbing the residual energy of the stopping particle in one detector layer.

One-dimensional position sensitive strip detectors are used in the first four layers of the HED telescope. The position resolution in the direction perpendicular to the strips is 1.0 mm. A pair of such detectors was constructed with the strips of the closely located separate detectors perpendicular to each other. Two such pairs with a distance of 31.0 mm, each giving the positions of incident particles in two dimensions, are used to determine the trajectories. This information is used both in the on-board and on-ground analysis to correct for the path length variations in the observed energy losses.

A capacitive read-out method is used in the strip detectors. The charge produced by an incident particle is capacitively transferred to the two edge strips of the detectors and fed to the inputs of two charge sensitive pre-amplifiers. The sum of the two signals gives the total energy loss, and the ratio of the signal from one edge to the total signal gives the position.

The total thickness of the strip detectors is 1200 microns. These detectors are followed by a conventional silicon detector with a thickness of 1000 microns. The active area of each strip detector layer is about 46 cm^2 and that of the conventional silicon detector 52 cm^2. The last two layers of HED consist of scintillators. Photodiode read-outs of the signals are applied. This allows a very compact structure of the sensor. The material of the first scintillator layer is CsI(Tl) and the second one is BGO.

3. Particle Identification and Data Analysis

Due to telemetry limitations, hydrogen and helium with their isotopes are fully analysed on-board by the ERNE data processing unit. The data corresponding to ions heavier than helium, are stored as pulse heights. Due to the limited capabilities of the on-board identification algorithm, these ions are identified and analysed on the ground.

The algorithm used in the on-board particle identification is based on the power-law range–energy relationship and on the assumption that all ions are fully stripped (e.g., Goulding, 1979).

The analysis software converts the observed pulse heights to energy units. This is performed by assuming for the silicon detectors a linear amplification, independent of the particle species, and by applying element- and scintillator-dependent light production curves for the scintillators. The parameters used in the conversions are based on the calibrations before the flight at accelerator facilities, and on the actual flight data, particularly for the scintillators.

From the strip detector signals, corrected for nonlinearities, the coordinates of particles, and from these the angles of incidence, are calculated. Therefore, in HED the path length variations in the transmission detectors can be accurately taken into account.

After the real, corrected energy loss, residual energy and path length have been calculated, the value of the particle identification function can be determined. This value is used as the argument of a tabulated function giving the actual particle number corresponding to hydrogen, helium, or one of their isotopes.

Due to fluctuations in the recorded energies and due to inaccuracies of the identification algorithm, the particle number obtained is not always the real one. In the on-board analysis, the range of allowed particle numbers is divided into 25 parts, allowing each of the particles mentioned above to have five different values, a nominal proton corresponding to the value 3, and a nominal ^{4}He to the value 23. Thereby, a mass distribution for each ion is allowed.

The results of the analysed particles, which represent protons, deuterons, tritons, ^{3}He, or ^{4}He at a certain energy, are stored in a matrix. The particle numbers constitute one dimension of the matrix. The other dimension is the measured energy. The energy range is divided logarithmically into 12 parts. In addition to the matrix and the pulse heights of ions heavier than helium, a one-minute data buffer contains samples of the pulse heights of the light particles. The light ion pulse heights are used for calibration purposes and as complementary data for the analysed events.

On the ground a more accurate and complicated analysis of the heavy ions is carried out. First, the pulse heights of silicon detectors are converted to energies by applying the amplification functions based on calibrations. By using the strip detectors, the particle trajectory is constructed.

For particles stopping in the silicon layers, the analysis then proceeds with identification of the particle. This begins with the calculation of the ratio of the residual energy to the total energy. The total energy is then compared with tabulated theoretical values for particles with a mass number equal to twice the nuclear charge number, and with the calculated ratio of the residual and total energies taking into account the measured angle of incidence. The particle number is obtained by interpolating from the table.

For particles stopping in the scintillators, the analysis is a bit more complicated, because the light output is dependent on the particle type. An iterative approach is used by first making a guess of the particle number. The corresponding light output parameters are then used for the energy loss determination. Since we are not aiming for isotopic resolution in this analysis, we use tabulated values of the light output of the most abundant isotope of each element. When the energy loss has been determined, the particle number can be calculated by using the same method as for the silicon detectors. If the calculated value matches the guessed one, the iteration loop is terminated. Otherwise, the calculated particle number is used as a new guess.

To improve the data quality, a method based on the so-called energy dissipation pattern is used. By comparing the relative energy losses in different layers to the theoretical calculations, one can reject the particles which differ a pre-defined amount from the nominal behavior. For particles stopping in silicon layers, this can be done prior to any particle identification, and for particles stopping in scintillators, only one particle number has to be calculated to be able to use the light output tables.

4. Results and Discussion

The LED observations used in this study are from February 28 to April 30, 1996 for helium, and from January 26 to April 30, 1996 for the heavier elements. The HED observations are from the period May 13–21, 1996 for helium and from the period January 26–March 3, 1996 for heavier elements. The observed energy spectra of helium, oxygen, and nitrogen are shown in Figures 1, 2 and 3, respectively. The intensity error bars correspond to one sigma statistical errors. The energy ranges covered in the analysis are from 3.4 to 48 MeV $nucl^{-1}$ for He, from 3.1 to 65 MeV $nucl^{-1}$ for O, and from 4.3 to 78 MeV $nucl^{-1}$ for N.

4.1. Helium

In Figure 1, our helium spectrum is compared with the *Voyager* 1 and 2 observations during the 1977 solar activity minimum (Cummings, Stone, and Webber, 1984), and with the IMP and ISEE-3/ICE measurements in 1986 (Garcia-Munoz, Pyle, and Simpson, 1987; von Rosenvinge and Reames, 1987). The *Voyager* measurements are from a time period when the spacecraft were between 1.3 and 2.5 AU from the Sun. Therefore, the results can be directly compared with our measurements at 0.99 AU. Respectively, the IMP and ISEE-3/ICE observations are from 1 AU.

Cummings, Stone, and Webber (1984) used a quiet-time selection criterion requiring that the 4.5–7.8 MeV $nucl^{-1}$ He intensity should be below 7×10^{-5} particles /(cm^2 sr s (MeV $nucl^{-1}$)). Using a one-day integration time, this criterion is met for the whole time period of this analysis.

In Figure 1, the lowest four points between 3 and 6 MeV $nucl^{-1}$ in our helium spectrum are probably contaminated by solar particles accelerated in the solar atmosphere or in interplanetary space. In this energy range our intensities are about 30–50% lower than in 1977. This is one indication that the Sun has been exceptionally quiet during the first part of 1996, at least in producing energetic particles. On the other hand, during March 1–May 1, 1986 the helium intensity of von Rosenvinge and Reames (1987) at about 6 MeV $nucl^{-1}$ is about 30% of

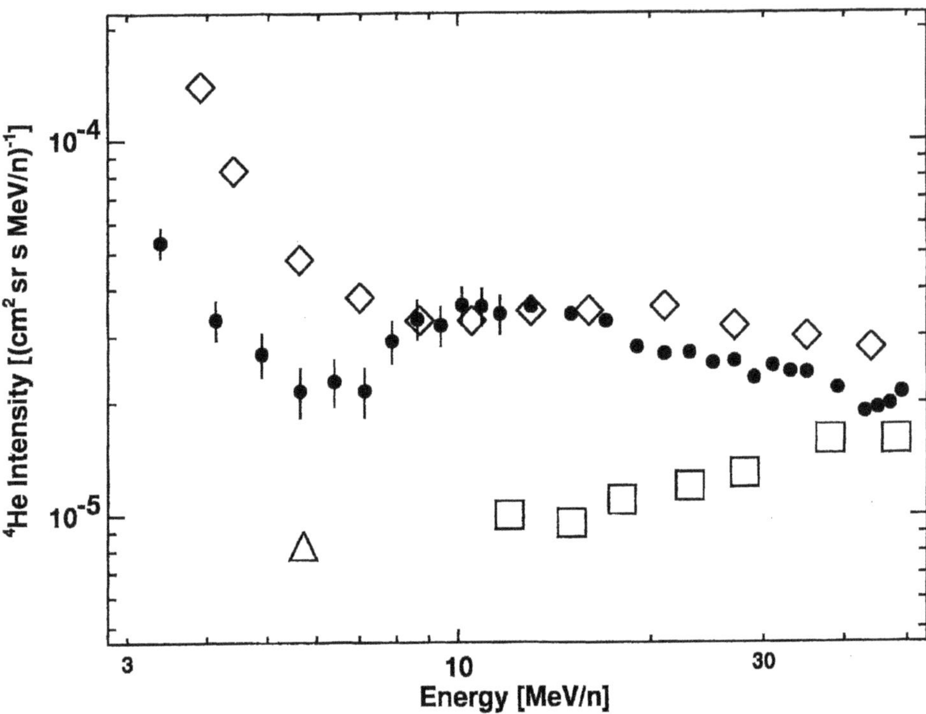

Figure 1. The energy spectrum of helium as observed by ERNE during a quiet time period from February 28 to April 30, 1996 (solid circles). For comparison are shown the spectra from *Voyager* 1 and *Voyager* 2 during September 1977–February 1978 (open diamonds, Cummings, Stone, and Webber 1984), from ISEE-3/ICE during March 1–May 1, 1986 (open triangle, von Rosenvinge and Reames, 1987), and from IMP 8 during July 19–October 8, 1986 (open squares, Garcia-Munoz, Pyle, and Simpson, 1987).

the intensity we obtained for the 1996 period. Direct comparison is, however, not straightforward because they have excluded some solar active days, but do not give the exact exclusion criterion.

According to our present results, helium has a rather flat spectrum from about 8 MeV nucl^{-1} upwards. This broad part of the spectrum is dominated by the anomalous helium. The intensity maximum, 3.8×10^{-5} particles/(cm^2 sr s (MeV nucl^{-1})), is within the statistical error limits at the same level as in 1977. The corresponding energy is in the range 10–15 MeV nucl^{-1}. In 1977 the maximum seems to be reached at somewhat higher energies. An explanation could be that the level of solar modulation was higher in 1977 than in 1996.

When comparing our spectra to the results obtained in 1986, significant differences can be seen. Our intensities of He between 10–30 MeV nucl^{-1} are by a factor of 2–4 higher than those measured by IMP 8 (Garcia-Munoz, Pyle, and Simpson, 1987) one solar cycle ago. A natural explanation could be that the anomalous helium intensity was still recovering in late 1986, as was the case for oxygen (Mewaldt *et al.*, 1993a).

4.2. Oxygen

The oxygen intensities in the range from 3.1 to 28 MeV nucl^{-1} (Figure 2, solid circles) represent the results obtained from LED. The data from 28 to 64 MeV nucl^{-1} (open circles) are based on the HED observations. During the observation period, the particles recorded by LED are almost purely anomalous cosmic rays. In the HED energy range, comparable contributions come from the galactic component and from the anomalous cosmic rays.

Oxygen has a clear maximum in its spectrum. The peak intensity is 1.2×10^{-5} particles/(cm^2 sr s (MeV nucl^{-1})) between 4–7 MeV nucl^{-1}. The present statistics of LED at low energies (637 particles from 2.5 to 15 MeV nucl^{-1}) does not allow a precise estimation of the peak intensty. The agreement with the results of Cummings, Stone, and Webber (1984) from the period September 1977 to February 1978 is, however, excellent. The minimum in the spectra in the range 40 –70 MeV nucl^{-1} was at the level of 4×10^{-7} particles/(cm^2 sr s (MeV nucl^{-1})) both in 1977 and in 1996. In both years, there is a 30-fold difference in the observed maxima and minima of the spectra. The only difference in the shapes between the 1977 and 1996 observations is that in 1996 there seems to be an excess of oxygen in the energy range from 10 MeV nucl^{-1} to 20 MeV nucl^{-1}.

Comparing the present results with the measurements in 1986, significant differences can be seen again. In Figure 2, IMP 8 results from the period July 25 to November 6, 1986 are presented (Cummings and Stone, 1987a). In the energy range between 5 and 20 MeV nucl^{-1} the anomalous oxygen intensities in 1996 are a factor of 3 to 7 higher than in 1986, as was the case in 1977. The spectrum in 1986 seems to be flatter than in 1977 or 1996. Also, although the spectrum of Cummings and Stone (1987a) extends down to about 4 MeV nucl^{-1}, no peak in the spectrum is visible.

The recovery of the ACR fluxes during the present solar minimum began in early 1992 (Mewaldt *et al.*, 1993b). At the time of the measurements by Klecker *et al.* (1995) with the HILT instrument on-board SAMPEX from October 1992 to May 1993, the ACR intensities had recovered already above the level in 1986, but were still a factor two lower than during the 1977 minimum. The oxygen intensities in 1996 in the energy range 5–28 MeV nucl^{-1} were a factor of 2.5 higher than during the 1992–1993 period as seen in Figure 2. The more recent observations made by SAMPEX and *Geotail* reveal a constant increase in the oxygen intensity in the energy range 10–30 MeV nucl^{-1}. Hasebe *et al.* (1994) and Cummings *et al.* (1995b) measured, at the end of the year 1993, oxygen intensities that were 30–50% lower than our measurements. Accordingly, SAMPEX measurements in 1994 show 10–20% lower oxygen intensities (Cummings *et al.*, 1995a). At higher

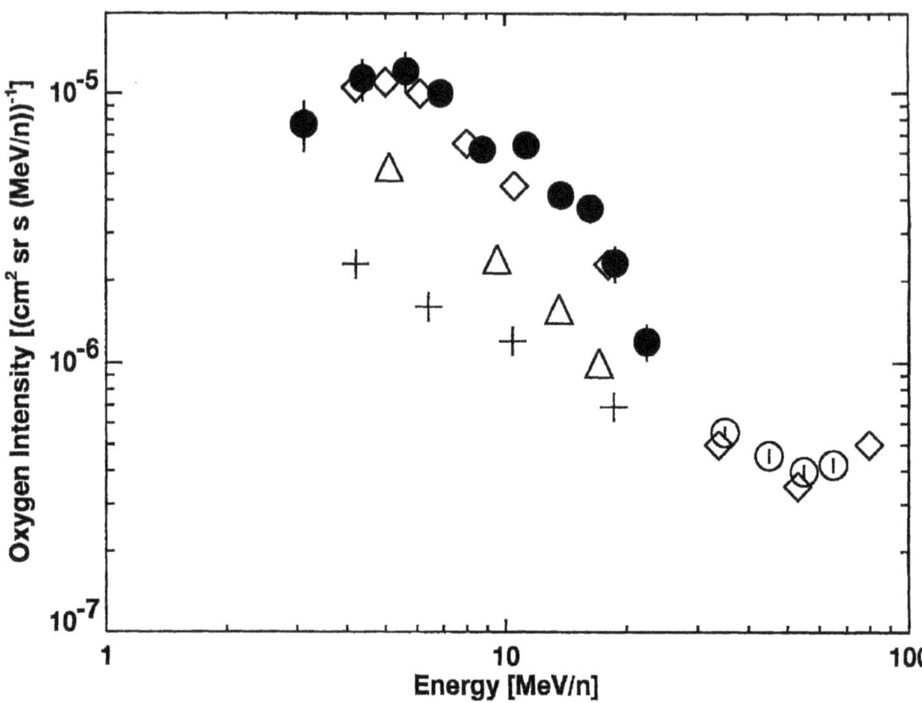

Figure 2. The energy spectrum of oxygen as observed by ERNE during a quiet time period from January 26 to April 30, 1996 (LED: solid circles; HED: open circles). For comparison are shown the spectra from *Voyager* 1 and *Voyager* 2 during September 1977–February 1978 (open diamonds, Cummings, Stone, and Webber, 1984), from IMP 8 during July 25–November 6, 1986 (crosses, Cummings and Stone, 1987a) and from SAMPEX during October 1992–May 1993 (open triangles, Klecker *et al.*, 1995).

energies near the turn-up of the cosmic ray component, Mewaldt *et al.* (1995) have measured 60% lower intensities from mid 1992 to the end of 1994.

4.3. Nitrogen

In the case of anomalous nitrogen our statistics during the period January 26 – April 30, 1996 consists of 134 nuclei in the LED energy range 2.8–25 MeV nucl^{-1}. The resulting intensities are in agreement with the *Voyager* measurements in 1977, as shown in Figure 3. A broad maximum is seen in the spectrum with a peak intensity of 1.7×10^{-6} particles/(cm^2 sr s (MeV nucl^{-1})) in the range 4–9 MeV nucl^{-1}. At energies between 10–50 MeV nucl^{-1} our spectrum tends to be flatter than in 1977. In the range covered by LED, the differences in intensities are within the limits of one and a half sigmas. The statistics of HED are much higher, and between 25 and 45 MeV nucl^{-1} (3 lowest energy points of HED) a significant deviation from the 1977 results can be observed. At the highest energies, 45–80 MeV nucl^{-1},

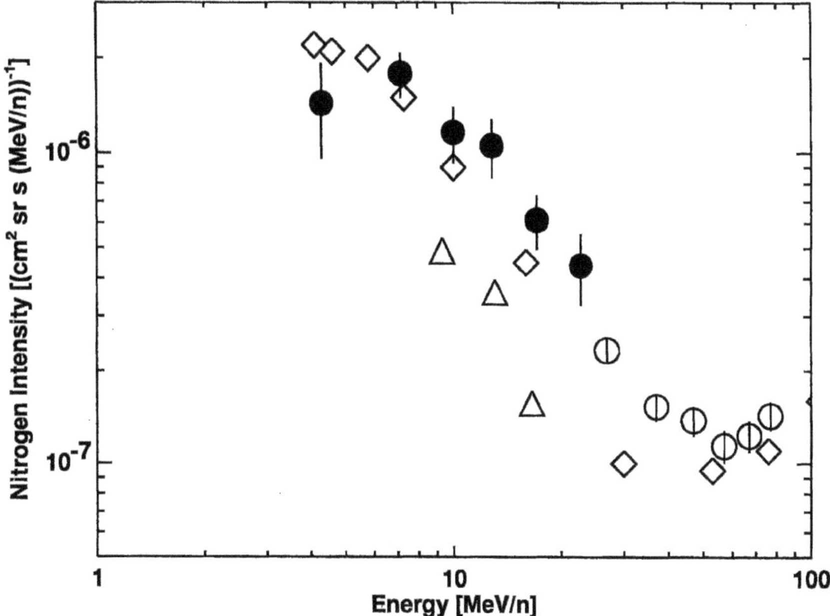

Figure 3. The energy spectrum of nitrogen as observed by ERNE during a quiet time period from January 26 to April 30, 1996 (LED: solid circles; HED: open circles). For comparison are shown the spectra from Voyager 1 and Voyager 2 during September 1977–February 1978 (open diamonds, Cummings, Stone, and Webber 1984) and from SAMPEX during October 1992–May 1993 (open triangles, Klecker *et al.*, 1995).

where galactic cosmic rays become dominant, our measurements again agree with those of Cummings, Stone, and Webber (1984). This indicates an excess in the anomalous component at the beginning of 1996 in the range 25–45 MeV nucl^{-1}, whereas the galactic component is at the same level as in 1977. The intensity enhancement of ACR nitrogen during the 3-year period since the end of 1992 has been 2.5-fold in the energy range 10–20 MeV nucl^{-1} as shown in Figure 3. At the end of 1993, nitrogen intensity levels measured by *Geotail* were 30–50% lower than our measurements in the energy range 10–40 MeV nucl^{-1} (Hasebe *et al.*, 1994). Their nitrogen energy spectrum below 30 MeV nucl^{-1} seems to be steeper than ours and the cosmic ray component turns up earlier, near 40 MeV nucl^{-1}. The enhancements of both oxygen and nitrogen fluxes have followed each other quite precisely.

4.4. Global Solar Magnetic Field Reversal

There are several alternative explanations for the differences in the intensity levels of anomalous cosmic rays from one solar cycle to another. Garcia-Munoz, Pyle, and Simpson (1987) showed that anomalous helium is particularly sensitive to variations in the solar modulation levels. According to their analysis, the low intensity of anomalous helium during the solar minimum in 1965 and observations in 1986, could be explained by the conventional modulation theories.

It is also possible that variations in the ACR intensity levels between consecutive solar cycles are due to the effects caused by the opposite solar magnetic field orientations (Pesses, Jokipii, and Eichler, 1981; Jokipii, 1986; see also a review by McKibben, 1987). When the solar magnetic field is pointing outward at the north pole ($qA > 0$), as is the case during the current solar minimum as well as in 1977, the anomalous cosmic rays accelerated in the polar regions of the solar wind termination shock have, due to drift effects, an easy access to the equatorial regions of the inner solar system. In 1965 and 1986 the magnetic field was pointing inwards ($qA < 0$). In this case, lower intensities of anomalous cosmic rays would be expected due to drift patterns.

A particular feature of the model of Jokipii (1986) is that it also predicts a change with different magnetic polarities in the energy at which the peak intensities of anomalous cosmic rays occur. For $qA < 0$, the peak intensities shift to higher energies and the spectra become narrower, as compared with $qA > 0$. In the anomalous helium and oxygen spectra of Garcia-Munoz, Pyle, and Simpson (1987) and Cummings and Stone (1987a) there is no evidence of maxima, which would allow comparison with our results. However, Cummings and Stone (1987b) have obtained a good agreement between the model of Jokipii (1986) and their measurements, by using the *Voyager* data from the years 1977–1978 and 1985–1986 at 1.8 and 19.5 AU, respectively.

Most drift models are based on the assumption that the solar magnetic field geometry follows the standard Archimedean spiral solution everywhere. It has been suggested (Jokipii and Kota, 1990) that the polar heliospheric magnetic field might deviate considerably from Parker's solution. This could cause significant changes in the transport of cosmic rays, which would be dependent on the polarity of the solar magnetic field. Thereby, particularly the anomalous cosmic-ray component could be affected.

4.5. The Present Minimum of Heliospheric Modulation

In order to verify whether the activity minimum of the present solar cycle has actually been reached, we have plotted in Figure 4 the intensity of anomalous oxygen as a function of time. The time span is from day 28 to day 125 of 1996 (January 28–May 4). This interval has been divided into 5.4-day subperiods. The black points represent the oxygen intensities in the energy range 25–35 MeV nucl^{-1} averaged over these subperiods. From these intensities 5-point averages were calculated to correspond to the average intensities over the 27-day solar rotation periods. It can be seen that during the first three solar rotation periods, days 28–108 (January 28 – April 19, 1996), the oxygen flux has stayed at the same level or slightly (2%) decreased. Instead, the last solar rotation period shows a 20%

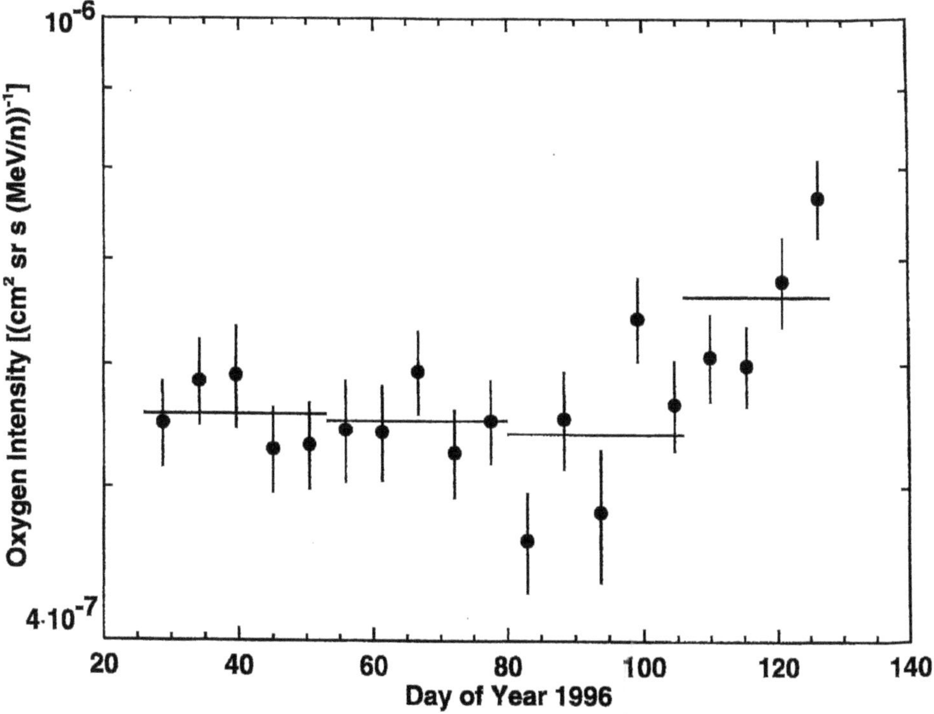

Figure 4. Time profiles of 5.4-day oxygen intensity averages in HED from day 28 to 125 of 1996 in the energy range 25–35 MeV $nucl^{-1}$ (solid points with statistical error limits). The full lines represent the averages of four 27-day periods at the beginning of 1996.

increase in the anomalous and galactic cosmic ray oxygen intensities. Therefore it seems that the present solar minimum has not yet been reached.

5. Conclusions

Our observations give evidence that the heliospheric particle environment is quite similar to that during the previous $qA > 0$ solar minimum (1977). This is distinctly different from the $qA < 0$ minimum of 1986 as seen in both the shapes of the energy spectra and more clearly in the intensities of anomalous cosmic rays. It seems likely that the direction of the solar magnetic field has an effect on the particle flow into the heliosphere, as predicted by the models of Pesses, Jokipii, and Eichler (1981) and Jokipii (1986).

The maximum intensities of anomalous cosmic ray helium, oxygen and nitrogen coincide with the 1977 measurements (Cummings, Stone, and Webber, 1984). However, the present spectra of oxygen and nitrogen are somewhat broader than they were in 1977.

The low-energy part (< 10 MeV nucl^{-1}) of helium intensity shows that during the first months of 1996 the solar production of energetic helium nuclei is on a much lower level than in 1977.

The cosmic-ray oxygen intensities in the interval 25–35 MeV nucl^{-1} were analyzed for days 28–128 of the year 1996. The flux experienced an average increase of 20% during the last 27-day period compared to the first 80 days, during which no intensity changes were observed. This is an indication that the heliopheric modulation might still be decreasing.

Acknowledgements

The Academy of Finland is thanked for its financial support.

References

Cummings, A. C. and Stone, E. C.: 1987a, *Proc. 20th Int. Cosmic Ray Conf., Moscow* **3**, 425.
Cummings, A. C. and Stone, E. C.: 1987b, *Proc. 20th Int. Cosmic Ray Conf., Moscow* **3**, 421.
Cummings, A. C. and Stone, E. C.: 1995, *Proc. 24th Int. Cosmic Ray Conf., Rome* **4**, 497.
Cummings, A.C., Stone, E.C., and Webber, W.R.: 1984, *Astrophys. J.* **287**, L99.
Cummings, A. C., Blake, J. B., Cummings, J. R., Fränz, M., Hovestadt, D., Klecker, B., Mason, G. M., Mazur, J. E., Mewaldt, R. A., Stone, E. C., and Webber, W. R.: 1995a, *Proc. 24th Int. Cosmic Ray Conf., Rome* **4**, 800.
Cummings, A. C., Mewaldt, R. A., Blake, J. B., Cummings, J. R., Fränz, M., Hovestadt, D., Klecker, B., Mason, G. M., Mazur, J. E., Stone, E. C., von Rosenvinge, T. T., and Webber, W. R.: 1995b, *Geophys. Res. Letters* **22**, 341.
Evenson, P., Garcia-Munoz, M., Meyer, P., Pyle, K. R., and Simpson, J. A.: 1983, *Astrophys. J.* **275**, L15.
Fisk, L. A., Kozlovsky, B., and Ramaty, R.: 1974, *Astrophys. J.* **190**, L35.
Garcia-Munoz, M., Pyle, K. R., and Simpson, J. A.: 1987, *Proc. 20th Int. Cosmic Ray Conf., Moscow* **3**, 438.
Garcia-Munoz, M., Mason, G. M., and Simpson, J. A.: 1973, *Astrophys. J.* **182**, L81.
Giacalone, J. and Jokipii, J.R.: 1996, *Proc. AGU 1996 Spring Meeting*, SH42C-2, S225.
Gloeckler, G., Geiss, J., Roelof, E.C., Fisk, L.A., Ipavich, F.M., Ogilvie, K.W., Lanzerotti, L.J., von Steiger, R., and Wilken, B.: 1994, *J. Geophys. Res.* **99**, 17637.
Goulding, F.S.: 1979, *Nucl. Instr. Meth.* **162**, 609.
Hasebe, N., Mishima, Y., Fujiki, K., Fujii, M., Kobayashi, M., Doke, T., Kikuchi, J., Hayashi, T., Shino, T., Ito, T., Takashima, T., Yanagimachi, T., Nakamoto, A., Murakami, H., Nagata, K., Kohno, T., Munakata, K., Kato, C., Yanagita, S., Kashiwagi, T., Maezawa, K., Muraki, Y., Nishida, A., Terasawa, T., and Wilken, B.: 1994, *Geophys. Res. Letters* **21**, 3027.
Jokipii, J. R.: 1986, *J. Geophys. Res.* **91**, 2929.
Jokipii, J. R. and Kota, J.: 1990, *Proc. 21st Int. Cosmic Ray Conf., Adelaide* **6**, 104.
Klecker, B,. McNab, M. C., Blake, J. B., Hamilton, D. C., Hovestadt, D., Kästle, H., Looper, M. D., Mason, G. M., Mazur, J. E., and Scholer, M.: 1995, *Astrophys. J.* **442**, L69.
Klecker, B., Hovestadt, D., Scholer, M., Blake, J. B., McNab, M. C., Looper, M. D., and Mason, G. M.: 1996, *Proc. AGU 1996 Spring Meeting*, SH42C-3, S225.
Luhmann, J. G.: 1994, *J. Geophys. Res.* **99**, 13285.
McKibben, R. B.: 1987, *Rev. Geophys.* **25**, 711.
Mewaldt, R. A., Cummings, A. C., Cummings, J. R., Stone, E. C., and von Rosenvinge, T. T.: 1993a, *Proc. 23rd Int. Cosmic Ray Conf., Calgary* **3**, 404.

Mewaldt, R. A., Cummings, A. C., Cummings, J. R., Stone, E. C., Klecker, B., Hovestadt, D., Scholer, M., Mason, G. M., Mazur, J. E., Hamilton, D. C., von Rosenvinge, T. T., and Blake, J. B.: 1993b, *Geophys. Res. Letters* **20**, 2263.
Mewaldt, R. A., Cummings, J. R., Leske, R. A., Selesnick, R. S., Stone, E. C., and von Rosenvinge, T. T.: 1995, *Proc. 24th Int. Cosmic Ray Conf., Rome* **4**, 477.
Pesses, M. E., Jokipii, J. R., and Eichler, D.: 1981, *Astrophys. J.* **246**, L85.
Potgieter, M. S., Le Roux, J. A., McDonald, F. B., and Burlaga, L. F.: 1993, *Proc. 23rd Int. Cosmic Ray Conf., Calgary* **3**, 525.
Selesnick, R. S., Mewaldt, R. A., Cummings, J. R., and Stone, E. C.: 1996, *Proc. AGU 1996 Spring Meeting*, SH42C-4, S225.
Torsti, J., Valtonen, E., Lumme, M., Peltonen, P., Eronen, T., Louhola, M., Riihonen, E., Schultz, G., Teittinen, M., Ahola, K., Holmlund, C., Kelhä, V., Leppälä, K., Ruuska, P., and Strömmer, E.: 1995, *Solar Phys.* **162**, 505.
Torsti, J., Valtonen, E., Kocharov, L., Vainio, R., Riihonen, E., Anttila, A., Laitinen, T., Teittinen, M., and Kuusela, J.: 1997, *Solar Phys.* **170**, 179 (this issue).
von Rosenvinge, T. T. and Reames, D. V.: 1987, *Proc. 20th Int. Cosmic Ray Conf., Moscow* **3**, 434.

PERFORMANCE AND EARLY RESULTS FROM THE GOLF INSTRUMENT FLOWN ON THE SOHO MISSION

A. H. GABRIEL[1], J. CHARRA[1], G. GREC[2], J.-M. ROBILLOT[3], T. ROCA CORTÉS[4], S. TURCK-CHIÈZE[5], R. ULRICH[6], S. BASU[7], F. BAUDIN[4,8], L. BERTELLO[6], P. BOUMIER[1], M. CHARRA[1], J. CHRISTENSEN-DALSGAARD[9], M. DECAUDIN[1], H. DZITKO[5], T. FOGLIZZO[5], E. FOSSAT[10], R. A. GARCÍA[5], J. M. HERREROS[4], M. LAZREK[2,11], P. L. PALLÉ[4], N. PÉTROU[5], C. RENAUD[2] and C. RÉGULO[4]

[1]*Institut d'Astrophysique Spatiale, CNRS/Université Paris XI, 91405 Orsay, France*
[2]*Observatoire de la Côte d'Azur, Lab. Cassini CNRS URA1362, 06304 Nice, France*
[3]*Observatoire de l'Université Bordeaux 1, BP 89, 33270 Floirac, France*
[4]*Instituto de Astrofísica de Canarias, 38205 La Laguna, Tenerife, Spain*
[5]*Service d'Astrophysique, DSM/DAPNIA, CE Saclay, 91191 Gif-sur-Yvette, France*
[6]*Astronomy Department, University of California Los Angeles, U.S.A.*
[7]*Teoretisk Astrofysik Center, Danmarks Grundforskningsfond*
[8]*National Solar Observatory, NOAO, PO Box 26732, Tucson AZ-85726, U.S.A.*
[9]*Institut for Fysik og Astronomi, DK 8000 Aarhus C, Denmark*
[10]*Département d'Astrophysique, CNRS URA709, Université de Nice, 06034 Nice, France*
[11]*Centre Nationale de la Recherche, CNCPRST, Rabat, Morocco*

(Received 19 March 1997; accepted 18 June 1997)

Abstract. GOLF in-flight commissioning and calibration was carried out during the first four months, most of which represented the cruise phase of SOHO towards its final L1 orbit. The initial performance of GOLF is shown to be within the design specification, for the entire instrument as well as for the separate sub-systems. Malfunctioning of the polarising mechanisms after 3 to 4 months operation has led to the adoption of an unplanned operating sequence in which these mechanisms are no longer used. This mode, which measures only the blue wing of the solar sodium lines, detracts little from the detection and frequency measurements of global oscillations, but does make more difficult the absolute velocity calibration, which is currently of the order of 20%. Data continuity in the new mode is extremely high and the instrument is producing exceptionally noise-free p-mode spectra. The data set is particularly well suited to the study of effects due to the excitation mechanism of the modes, leading to temporal variations in their amplitudes. The g modes have not yet been detected in this limited data set. In the present mode of operation, there are no indications of any degradation which would limit the use of GOLF for up to 6 years or more.

1. Introduction

GOLF is designed to measure the Global Oscillations of the integrated solar disk, by determining the line-of-sight velocity of the photosphere as a function of time. Although the instrument is sensitive to a wide range of frequencies 10^{-7} to 10^{-2} Hz, the priority objectives concentrate on the frequency range 10^{-5} to 10^{-3} Hz, where are found the lower frequency p modes and the predicted g modes. This is the range in which ground-based measurements are severely penalised by the disturbance of the Earth's atmosphere and the lack of continuity in the observations. By taking full advantage of the unique platform offered by SOHO, GOLF has proved capable

Solar Physics **175:** 207–226, 1997.

of enhancing the sensitivity and decreasing the observational noise, especially at lower frequencies.

For a full description of the GOLF instrument concept and design, the reader is referred to a detailed article published before launch (Gabriel *et al.*, 1995). A pre-launch description of the SOHO platform and mission can also be found (Domingo, Fleck, and Poland, 1995). SOHO maintains its payload in a full-Sun orbit around the Lagrange point L1, from where it remains pointed at the centre of the Sun throughout its normal operating sequences. In this way, the environment has a long-term continuity and stability, in terms of temperature, illumination and pointing, at a level never previously obtained for solar observations.

The GOLF instrument is a development for the space environment of instruments operated for many years on the ground, in particular the BISON (Chaplin *et al.*, 1996) and IRIS (Grec *et al.*, 1991) networks. More closely related to IRIS, GOLF compares the frequency of the solar sodium D Fraunhofer lines with that of an atomic standard sodium vapour cell carried in the instrument. The Doppler shift between these two gives the line-of-sight relative velocity and analysis of its time variation gives the global solar frequencies. Each sodium line is split into two Zeeman components by a strong permanent magnet on board, and these monitor the two wings of the broader solar absorption line. The actual configuration of GOLF differs somewhat from the IRIS concept; in the choice of the plane of polarisation used, in the use of both of the sodium D lines and in the system of polarising mechanisms adopted for switching between the two wings. In addition, GOLF carries small electrical coils, allowing the 5000 G permanent magnetic field to be modulated by a small field of about ± 100 G, to enable also the mean slope of the operating points on the two wings of the solar line to be determined. GOLF carries an additional rotatable quarter-wave plate. This has the dual objective of measuring the intrinsic global solar magnetic field and providing a redundant back-up in the case of failure of the other GOLF polarising mechanism.

Many subsystems of the GOLF instrument carry redundant duplication. These include most of the electronics, some sensitive mechanisms (such as door actuators), local heaters and the photomultiplier detectors. The selection of these is linked globally, so that GOLF provides a choice of either a Channel A or a Channel B.

The pre-launch plan foresaw a 40-s cycle of operations, in which at least one of the two rotating mechanisms and the magnetic field modulation were switched each 5 s. In this way, ratios could be derived giving data points every 40 s, relating to the line-of-sight velocity, the mean slope of the solar absorption line and the global magnetic field.

The SOHO mission was launched with outstanding precision on 2 December 1995. GOLF was switched on some hours later and was subjected to some months of commissioning and calibration during the transfer to L1. In most respects GOLF has been shown to meet fully the scientific specification and is producing data of an outstanding quality, never before obtainable. The malfunctioning of the two rotating mechanisms on GOLF has led us to adopt, after some months, an in-flight sequence which does not use these mechanisms and which was not planned before launch. It is demonstrated that this alternative mode has only a small effect on the sensitivity or noise level of GOLF in detecting solar oscillations, or in measuring their frequencies. It does however considerably increase the task of providing the last 10% to 20% precision on the velocity amplitude of the measured oscillations. In this mode, adopted since April 1997, the secondary objective of measuring the global solar magnetic field cannot be accomplished. However, operating as it does with no mechanical movement, it can be claimed that the reliability for long undisturbed observing sequences is considerable enhanced.

2. Commissioning

2.1. EARLY SWITCH-ON

GOLF was switched on for the first time on 3 December 1995 at 01:53 UT. The primary purpose of this 8 min test was to check that the experiment aperture cover had remained closed through the launch period and was protecting the GOLF entrance filter from contaminating deposits throughout the initial experiment and spacecraft outgassing period. This was also an opportunity to verify that the instrument automatic boot performed nominally and that substitution heaters, directly powered by the spacecraft, provided adequate dissipation when GOLF was not powered. The GOLF collectively controlled units (Power Supply Unit and Data Processing Unit) were respectively at 14.0 and 18.3 °C, close to their eventual operating temperatures. Temperatures ranged satisfactorily between 20 °C for the front end of the sensor unit to 13 °C at the anti-solar end, the filter being at 7.9 °C, its passive radiator at -8.7 °C, the magnet at 16.3 °C and detector cathodes at 14.5 °C. These and all other parameters observable in the telemetry were within their predicted range. The first GOLF command was sent and verified.

2.2. ACTIVITIES WITH DOOR CLOSED

GOLF Channel A was powered ON definitively on 10 December 1995 at 02:14 UT. At switch-on, the sensor mean temperature was close to 20 °C, the filter being at 16 °C, and its passive radiator at −6.6 °C. The Data Processing Unit was at 13.7 °C and the Power Supply Unit at 15.9 °C.

At first, the instrument was set to an almost nominal observation mode except for the resonance cell stem, which was unheated and the aperture cover that was kept closed. On 11 December, the resonance cell stem heater was turned on, completing the successful functional testing of Channel A. Except for a short filter temperature scan, the GOLF Sensor dissipation remained constant up to 16 December to allow the temperatures to stabilise. The instrument was then set into a safe mode while awaiting the end of the first month outgassing. This period has been used to verify the detector background and to study the influence of cosmic rays and any electrical perturbations from the spacecraft or other instruments (see Section 5.2).

2.3. Channel A commissioning

The GOLF entrance door was opened following the SOHO Mid Course Correction 2. The sub-systems full performance characterisation and optimisation were performed between 9 and 24 January 1996. This was continued between 6 and 13 February just before the experiment was set into a safe mode for the duration of injection activities into the Halo Orbit. GOLF was then set into a science observation mode on 19 February, with the polariser mechanism alone providing the polarisation switching necessary to obtain velocity measurements.

Channel A in-flight overall performance, being in very good agreement with the results obtained during ground testing, and these ground measurements having shown that Channels A and B sensitivity were very similar, it was decided to continue gathering science data with Channel A, rather than switch over for commissioning Channel B. This change of policy was in line with the general strategy adopted by most of the SOHO instruments, regarding the testing of redundant systems. It should be recalled that although switching from one channel to another can be done within a few hours, reaching the thermal stability to allow science quality observations takes a few days, in accordance with the predictions of the thermal model, and as verified in flight on 16 February.

However, GOLF did take advantage of a period of disturbed pointing due to spacecraft testing on 21 March 1996, in order to carry out a short functional test of Channel B. All functions were tested and showed very similar performances to those of Channel A. This gave confidence that Channel B is always available, should it ever prove necessary to switch over. After six hours the instrument was returned to the Channel A mode.

3. In-Flight Calibration Activities

Calibration sequences have been carried out for the two most likely environmental disturbances on the instrument performance: temperature and spacecraft pointing. For temperature, the subsystems most sensitive are firstly the cell stem, which determines the sodium vapour pressure, and next the entrance filter.

3.1. Cell temperature

Calibration measurements on the cell temperature were carried out between 12 and 23 January. The stem temperature was progressively increased, while the bulb was maintained at a higher temperature by means of its heater at constant dissipation. The apparent velocity sensitivity reached a maximum at a stem temperature of 170 °C. However, as predicted (Boumier and Damé, 1993), the total PM count rate continued to increase almost linearly with the stem temperature up to the maximum tested of 180.5 °C. At this level, each PM count rate was nearly 8×10^6 per second and the bulb temperature reading was 195.5 °C. It should be emphasised that there is an unknown but constant difference between these thermistor readings and the real cell or vapour temperatures due to contact conductivity effects. At the end of these tests, on 23 January, cell heater currents were adjusted to values which have since remained unchanged. For these settings the stem temperature was close to 170 °C and the bulb temperature reading was 188 °C. The stem temperature increase per heating current commanding step was 0.012 deg (which is close to the temperature reading digitalisation step amplitude). A change in the heating of the bulb modifies the temperature of the stem by about 19% of that obtained at the bulb. The counting rates were close to 5.5×10^6 per second for each detector, with a sensitivity of 126 800 c s^{-1} per one degree change of stem temperature, in accordance with pre-launch calibration. This implies that a stem temperature change of a 0.01 °C digitalisation step corresponds to a counting rate change equal to the statistical error in the counting rate over 1 s.

3.2. Filter Temperature

The filter bandpass is directly influenced by its temperature, as observed and studied before launch. In addition, since the filter area is directly imaged in the vapour cell, any variations across its surface can introduce disturbances in the symmetry and uniformity of the scattering system. As the two photomultipliers cannot view precisely the same scattering volume, this can be tested by comparing their outputs. The filter temperature was scanned very slowly and automatically on 23 and 24 January, to be completed on 6 and 7 February. The range scanned was from 17 to 20 °C. The operating temperature of 17.5 °C was chosen as a result of these scans and has been adopted since 8 February for subsequent observations. At this temperature, the velocity signals from the two PMs are similar and do not vary rapidly with temperature. The filter heater current is maintained constant and has only once been adjusted since, in order to compensate for some very slow temperature drift due to variation in the performance of the various thermal coatings. Monitoring of filter temperatures is maintained at the level of 0.01 °C, and to this precision no variations are observed at frequencies relevant to the GOLF measurements.

3.3. Offset Pointing

Measurements of the sensitivity of GOLF to spacecraft pointing errors was considered to be of great importance, since it had been judged too difficult to carry out meaningful tests on the ground before launch. Moreover, such measurements as had been quoted from tests on IRIS instruments reported a high sensitivity to offset pointing, at levels that could have produced serious problems for the spacecraft platform mounting of GOLF.

GOLF Sensitivity to offset pointing was measured for Yaw and Pitch excursions on 21 and 22 February, respectively. Calibration for Roll had to be postponed to March 15, after SOHO Star Sensor software patching. During 16 continuous hours on each of these days, the spacecraft was oscillated around one of its three reference axis, with a period of 800 s and a square-law amplitude of 30 arc sec in Pitch and Yaw and 90 arc sec in Roll. This is more than an order of magnitude larger than the worse case short-term variations of SOHO observed pointing errors. Should the experiment have been sensitive to such movements, a peak would have appeared at 1.250 mHz in the Fourier spectrum of the detector signals. This frequency being in the area where the solar background 'noise' reaches a minimum provides the best conditions to observe the peak created with the largest signal-to-noise ratio. For all three axes, no peak was detectable above the normal experiment noise level. This confirmed the efficiency of the instrument optical system which has been designed specifically to minimise the sensitivity to spacecraft pointing offsets.

4. Operating Sequence

The nominal pre-planned operating sequence of GOLF is fully described in the earlier instrument publication (Gabriel *et al.*, 1995). In this, switching the orientation of the two mechanisms (quarter-wave plate (QW) and the polariser (POL)) as well as switching the direction of the magnetic field modulation are carried out in a sequence of 40 s total, with at least one component switched each 5 s. By regrouping the data, it is possible to construct signals characteristic of solar velocity, solar magnetic field and slope of the solar line wings each 40 s, or even more frequently if desired.

This nominal sequence was put in operation from the start of science data collection on 18 January 1996. However, it soon became evident that there were occasional glitches in the performance of the QW mechanism. At first not too frequently, these lost only a small number (1 to 5) 40-s sequences per day. Later on, these became more frequent, and more difficult to control. It was decided on February 11 to stop the QW mechanism in its optimum position. With the POL mechanism alone it was then possible to continue the full solar velocity studies, losing only the possibility to extract global magnetic field data.

Towards the end of March 1996, the POL mechanism developed progressively the same symptoms. Early in April, the POL mechanism was also halted in its optimum position. This left the possibility of measuring the solar velocities using only the blue wing of the solar line. The limitations of this mode, which uses no mechanical motions were found to be less serious than anticipated, and will be more fully discussed later. The advantage of freeing the observation of all further risk from mechanical failures persuaded us to leave GOLF in this mode for a substantial period of time and not to risk compromising a long coherent data set by using the flight instrument to attempt to diagnose the fault.

We therefore have today three modes of accumulated GOLF data:

Mode A, the full 40-s pre-planned sequences from 18 January to 11 February. This is the only set usable for global magnetic field studies.

Mode B, the modified sequence using only the Polariser mechanism, from 19 February to 1 April. This enables in principle the full analysis of velocities, using both wings.

Mode C, from 11 April 1996 up to the present. This is the most stable and reliable data with a maximum of continuity.

It should be noted that the period covered by the Mode A and Mode B data includes the later part of the spacecraft and instrument commissioning. Conditions of temperature and pointing are not always stable and this Commissioning Phase data must be treated with caution. However, the data from Mode B is nevertheless uniquely valuable in enabling GOLF to evaluate the limitations and difficulties that might exist in interpreting the subsequent Mode C data. This point will be treated later. Mode C provided five months of data from 11 April to 11 September, with 100% coverage and continuity. This is unique in the history of helioseismological observations. Starting 11 September there was a three-day interruption, for reasons that will be explained later, followed by a resumption in the continuity up to the present.

5. Sub-System Performance

Many aspects of sub-system performance have already been treated in Section 3 dealing with in-flight calibration. In the present section we consider the remaining elements.

5.1. The Vapour Cell

The stray light level in the detection system was measured on 10 December 1995, shortly after switching on the instrument. This measurement was made with the cell bulb heated to 166 °C, but with the stem cold. In effect, conduction within the cell brought the stem temperature up to 48 °C, in accordance with prediction. The measured detection level of 270 000 c s^{-1} on each detector, in excellent agreement with pre-flight tests, is largely due to diffusion of light within the walls of the cell. When the experiment was switched off briefly during tests on 21 March 1996 and during the interruption of 11 September 1996, this measurement was repeated. No significant change in the stray light level was detectable.

5.2. The Detection Sub-System

With the instrument door closed, the detector dark-current could be measured during the early switch-on sequence. For instrument temperatures identical to ground-test conditions of 22.4 °C, the dark-current counting levels were very close to pre-flight levels of 190 c s^{-1} and 130 c s^{-1} for detectors 1 and 2. An important difference in flight was the presence of occasional larger spikes on the counting rate, which can be attributed to luminescence in the photomultiplier systems due to high-energy cosmic particles. These occurred a few tens of times per hour with peak rates up to 500 c s^{-1}. Randomly distributed in time and amplitude, they contribute a mean rate which is only about one tenth of the one sigma statistical uncertainty in the science data counting rates.

From the start of observations in January 1996, the detector 2 counting rate exceeded detector 1 by about 4.1%. In April 1996, this had increased to 4.3%, remaining stable up to mid-July.

Each photomultiplier is equipped with two counting channels. The first, or 'science' channel has its threshold set at 40 mV and is used to provide the primary science data. The second, or 'spectral' channel is normally set with a threshold much higher at 450 mV and serves to emphasise any perturbations on the high voltage or the detector itself. This second channel can also be used occasionally to perform threshold variation scans, giving a pulse-height distribution for the photomultiplier output. These pulse-height distributions, or 'spectra' can be used to monitor the changes in gain in the photomultipliers; changes which can be corrected for by increasing the high voltage.

During the first year of operation, three spectra of the two detectors have been performed, which give a measurement of the rate of change of gain. As a result, a first readjustment of the high voltage was made in mid-March 1996. Measurements of the evolution of gain with time, confirm the prediction that the rate of loss of gain decreases with time. Thus, the pre-launch figure of 0.2% loss per day can be seen to be decreasing and is down to 0.05% to 0.08% per day after 10 months operation in flight. These reducing rates of gain decrease can be extrapolated to predict many years of useful life from Channel A of GOLF, without needing to consider a switch to Channel B. A close monitoring of the count rates from the two active photomultipliers shows on several occasions small jumps in their ratio. These are of the order of 0.05%. Their origin is not yet understood, but they have no detectable effect on the scientific data obtained.

5.3. THE THERMAL SUB-SYSTEM

The GOLF sensor unit is strongly decoupled from the spacecraft both for radiation and conduction. It receives heat from the Sun and radiates heat to deep space. The entrance filter system has its own thermal balance being also somewhat decoupled from the sensor unit and with its own dedicated radiator to space. Heaters control separately the cell stem and bulb and the filter unit. Closed-loop stabilisation of temperature has been excluded in order to avoid thermal oscillations influencing the science data. GOLF relies on the high stability of the SOHO environment and the decoupled sensor thermal system in order to avoid the influence of thermal oscillations in the prime frequency range of study. If either the cell or filter temperatures drift outside their acceptable ranges, the input power from their heaters can be varied manually to bring them back.

Spacecraft and GOLF sensor temperatures drift slowly throughout the observations. This is due in part to variations in the spacecraft/Sun distance and in part to drift in the properties of the thermal coatings. It is difficult to separate these two effects without more than one year's observations. In general, the overall sensor drift is seen directly on the cell temperatures. However, the filter sub-system is more directly influenced by the solar flux and can show larger variations. Since the last thermal adjustment control on 23 January 1996, a very slow positive drift has been observed and should reach about 1.2 °C over the first year. This is due to the ageing of the thermal finishes on the GOLF sensor and on the spacecraft. Seasonal variation of solar flux adds to this drift an annual modulation of about 1 °C amplitude. Thus in early January 1997, when both effects add, the mean sensor temperature is 2.2 °C higher than one year earlier when SOHO was still cruising towards the L1 point.

5.4. The Data System

More than 95% of GOLF data is available in near real time or in playback modes, and thus within a few hours of the observations. This provides a very efficient monitoring of the instrument behaviour, as well as an evaluation of the scientific performance. Final data collection efficiency follows the delivery of the CD-ROMS, typically some 4 weeks later. Over the first 11 months of the mission, this brought the total collection rate up to 99.04%, a mean value, that conceals the fact that on some poor days the figure falls to 90%.

As described in the earlier instrument paper, the GOLF instrument includes its own data buffer, which re-transmits the GOLF data after delays of both 10 hours and 16 hours. These re-transmissions have complemented the data collection by 0.95% and 0.01% respectively, bringing the total rate to 99.998%. This loss of only 6 blocks out of 352995 over the first 11 months represents a unique record of continuity and one of the major strengths of the GOLF project.

On 11 September 1996, the instrument was found to be switched into an unexpected mode, with the cell stem heaters off. This has been attributed to a rare occurrence producing a two-bit error, which the automatic recovery response is not able to correct in all circumstances. Although the normal operations were restored by command intervention within a few hours, the resultant thermal excursion produced a disturbance in the science data quality lasting 2 to 3 days.

6. Scientific Observations

6.1. Sensitivity and Calibration

The most obvious and immediate method of analysing the data consists in first combining the signals from the two active PMs, and then, according to the sequence adopted, in choosing the appropriate algorithm to combine different intensities in order to derive a function which depends on line-of-sight velocity. For the pre-planned sequences of Mode A or B (see Section 4) we showed in the earlier paper (Gabriel *et al.*, 1995) that the velocity is given by

$$v = v_a \frac{I_b - I_r}{I_b + I_r + 2s}, \quad (1)$$

where s is the signal due to stray solar light reaching the detectors without resonant scattering. v_a is the calibration factor, which can be taken approximately as 4 km s^{-1}, but is ultimately calibrated by using either the known slow variation in SOHO/Sun distance, or the calculated shift due to the magnetic modulation of the instrument magnetic field. The function in Equation (1) is insensitive to first order changes in the overall sensitivity of GOLF to the solar intensity. Such changes can arise from changes in the solar intensity or solar distance, or through changes in the instrument, such as the temperature of the cell stem.

For the sequence Mode C, using only the blue wing, the equation becomes

$$v = \alpha_b I_b. \quad (2)$$

The quantity α_b must also be calibrated and again the magnetic modulation provides one method of approach. However it is obvious that, without the denominator, Equation (2) is also sensitive to first order changes in solar intensity or instrument sensitivity. With the exception of the intensity variations due to solar oscillations themselves, most of the variations are slow compared to the 5-min oscillations. Such variations will not contaminate the spectrum of p modes, but they will lead to uncertainty and variation in the velocity calibration to be applied to the p modes. However, for low frequency oscillations, as when searching for g modes, it becomes important to calibrate or to otherwise compensate for the variation of α_b, due to aging or instrumental drifts.

Three different techniques have been used separately in order to correct for these effects; both the first order variations in α_b and the higher order variations in v_a. These can be summarised as :

(a) The use of filtering functions, in order to reject frequencies below those corresponding to periods of the order of 1 day. This corrects for a wide range of causes of calibration variations. It is then necessary to apply an assumed known velocity scale.

(b) Use of the differences in signal due to the magnetic modulation. In principle this corrects for the slope of the solar line wing and includes implicitly a correction for slow intensity variations. However, this small difference between two large quantities has the effect of amplifying the photon noise. It cannot be used directly for individual 40-s data points, but must itself be averaged over a reasonable time period of hours, before being applied. In this way it introduces some of the filter cut-off problems associated with the first method.

(c) The third method consists in trying to model correctly all of the factors influencing the variation in α_b, including orbital motions, solar line profile and measured instrumental temperature changes.

All three methods have given some success and it remains to be seen which will offer long-term advantages.

Whilst the Operating Modes A or B offer in principle the best protection against velocity calibration uncertainties, it must be remembered that these are available only for a short period and that this period is still somewhat disturbed by the final spacecraft commissioning manoeuvres. The precision currently obtained in Mode C for the velocity calibration is of the order 20% or better.

6.2. Background and Noise Level

An attempt has been made to evaluate the early performance of GOLF, by comparing a Fourier spectrum over a wide frequency range with that obtained from one of the best ground-based solar oscillation instruments. The comparison was made with the Mark I instrument at Tenerife (Brookes, Isaak, and van der Raay, 1978). This is a potassium vapour instrument and a component of the BISON network.

In order to improve the validity of the comparison, the spectra were computed using 11-hour periods or 'days' for each. Figure 1 shows the average of 280 days of Mark-I, compared with 70 "days" of GOLF spectra obtained in Mode B. Frequency resolution is limited in this plot by the 11-hour periods used for the Fourier transforms. This approach has been used in order to demonstrate the level of the 'continuous' spectrum, rather than the p modes, which are not well resolved here.

Examination of Figure 1 shows several interesting features. The photon noise flat background instrumental spectrum can be seen in both cases at the high-frequency end. At the low-frequency end the general (frequency)$^{-1}$ variation of the background signal is well demonstrated. This signal has normally been thought of as due to the random velocity fields of solar convection. Differences between the two signals can be attributed to a combination of effects: the lower instrument (photon) noise from GOLF, the absence of terrestrial atmosphere disturbances, and the different height in the solar atmosphere sampled by the sodium and potassium lines. Detailed studies now in hand aim to separate these contributions. The much lower photon noise level of GOLF is in accordance with its design specification and the high PM counting rates adopted.

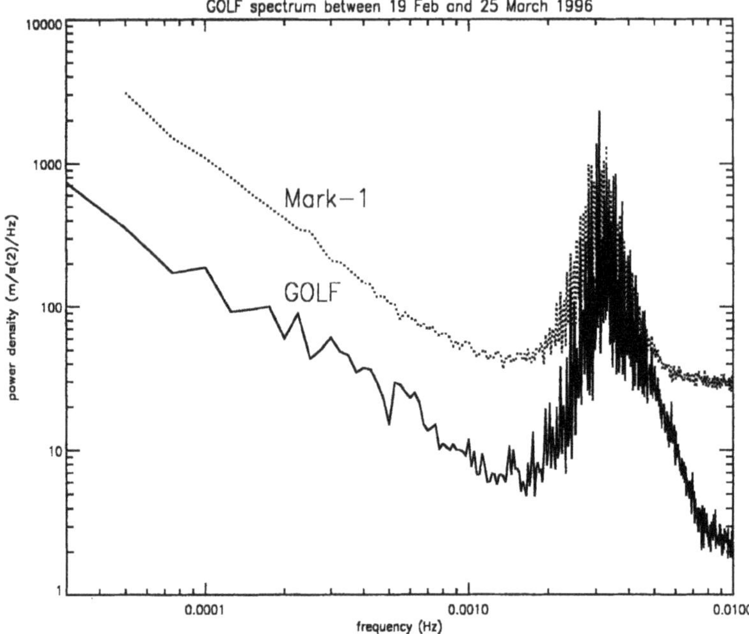

Figure 1. Low-resolution spectrum from GOLF compared with that from a good ground-based instrument Mark 1, showing the lower background level obtained from space.

6.3. VALIDATION OF BLUE-WING ONLY MEASUREMENTS

Following the adoption of Mode C for the routine operation of GOLF, it is relevant to pose the question of what, if any, is the effect of Mode C on the signal-to-noise ratio for the g modes. Although we can put reasonable limits on the loss, it is difficult to determine a precise figure. This is due to uncertainties in the absolute velocity calibration for Mode C, as indicated in Section 6.1 above. The usual technique is to compare the blue-wing-only signal with the blue plus red wings, for the same epoch during which GOLF was operating in Mode B. These two different types of data pose different problems for the calibration. Furthermore, the absolute sensitivity to the oscillation amplitudes is not the same for the two wings, a difference that may vary between p modes and g modes. The problem of comparison reduces to one of normalisation between the two sets. One approach is to normalise so that the p-mode power is the same. This is the technique adopted for the comparison shown in Figure 2. Here we see that the increase in background noise level in the blue-wing-only case is something between 1.4 and 2 in the g-mode region. This would be a valid figure for the loss of signal to noise for the g modes, if we assume that their relative sensitivity on the two wings is the same as for the p modes. Taking account of uncertainties in these assumptions, we cannot say more than that the

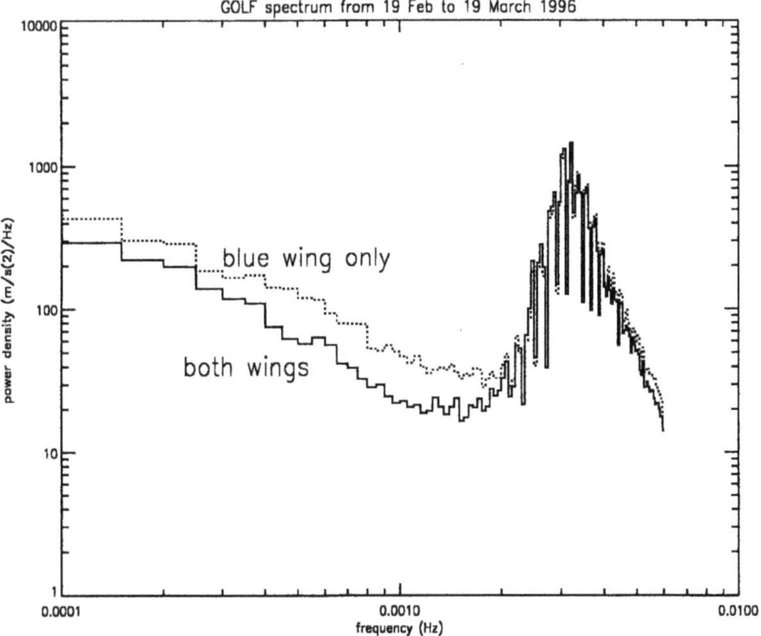

Figure 2. Low-resolution spectrum from blue wing only (dotted), compared with that from blue and red wings (full line).The curves are normalised for the same energy in the 2 to 4 mHz range.

factor loss of detectability for g modes, due to the Mode C operation, is between 1.4 and 2 in units of power.

This small factor justifies the decision to continue operations in Mode C rather than endeavour to re-activate the mechanisms. Effort is instead being invested in improving the resultant difficulties in obtaining a high precision in the oscillation amplitude calibration (see Section 6.1).

6.4. p MODE MEASUREMENTS

The low background level in the region of the g modes is clearly evident in Figure 1, in spite of the low spectral resolution of this plot. Figure 3 shows a section of the p-mode spectrum obtained by Fourier analysis of a 7-month period of GOLF data. Here again one can see the exceptionally clean background, allowing the identification of some very low amplitude modes. A first identification of GOLF p modes has been made, using a variety of techniques, but for the most part by the fitting of Lorenzian profiles by the maximum likelihood technique to a Fourier spectrum. This is the subject of separate more detailed presentations (Grec *et al.*, 1997; Lazrek *et al.*, 1997a, b).

At the time of writing, p-mode identification and frequency measurements have been made for 100 modes in the range 1.1 to 4.6 mHz, covering $n = 7$ to 33 and

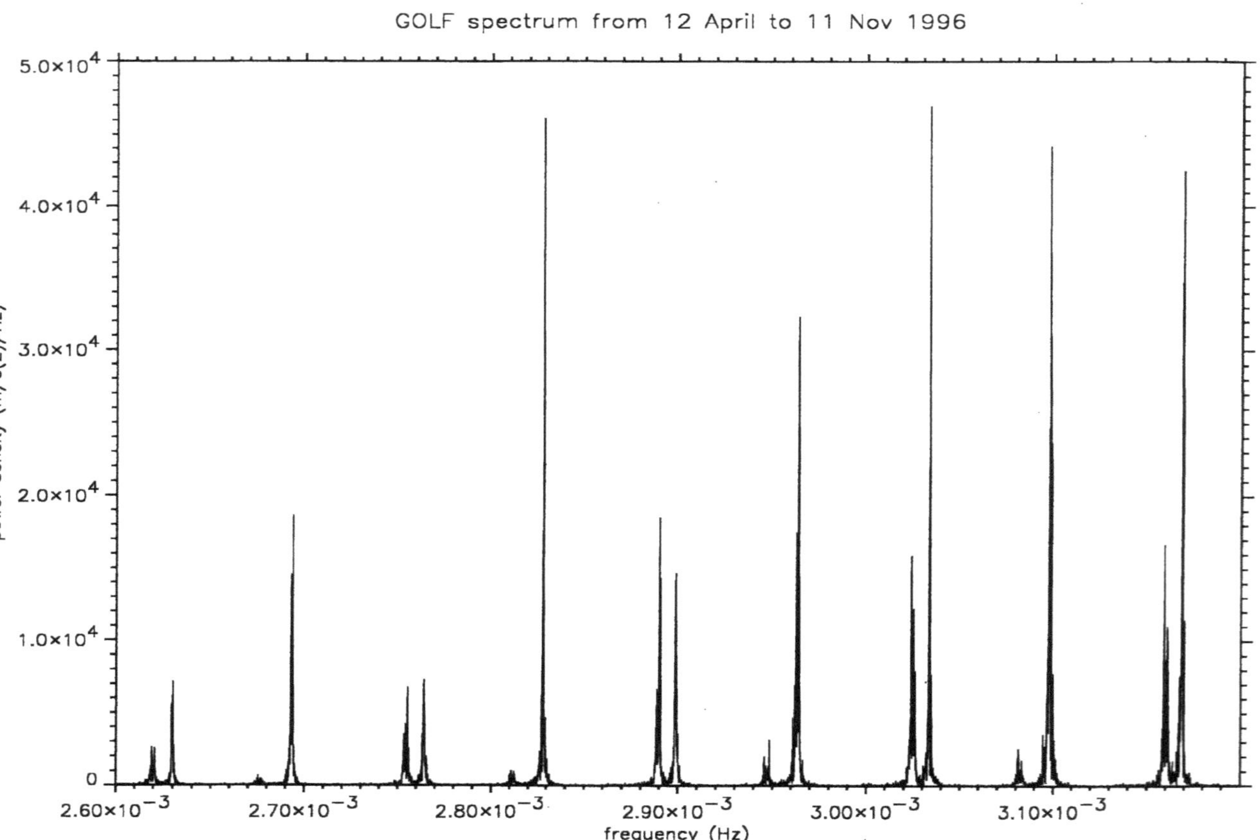

Figure 3. Section of the *p*-mode spectrum obtained from 7 months of data. Note the extremely low background level.

$\ell = 0$ to 4. At this point, with only 10 months of data, the frequency precision is not in general significantly better than that of existing ground-based measurements. The precision quoted is not yet the ultimate, even for the existing data set, since decisions need to be taken concerning which of the present methods of determination should be given the most weight. However, the lower noise levels of the GOLF data have enabled us to identify several new modes at low frequencies. It is clear that with a longer data set from GOLF, significant improvements will be made on both the range and the precision of the measured p-mode frequencies.

In considering the implication of these data for the analysis of the internal structure of the sun, some preliminary inversions have been carried out to determine the sound speed as a function of radial distance. For this work it is necessary to start with some other existing data for the higher degree modes, not available from GOLF. The data used is from the LOWL instrument (Tomczyk *et al.*, 1996) and is for the period February 1995 to February 1996, not fully coincident with the GOLF data. However, the two sets are in remarkable agreement, and the precision obtained can be seen in the plot of sound speed in Figure 4 (Turck-Chièze *et al.*, 1997a). A more complete discussion on the investigation of the core structure is reported in Turck-Chièze *et al.* (1997b).

6.5. Excitation Effects

A time-frequency analysis has been carried out on many of the observed p modes from GOLF (Baudin *et al.*, 1997a). This uses the same techniques developed for analysis of the space observations in solar intensity oscillations from PHOBOS (Baudin *et al.*, 1994, 1996). An example of the time variation observed for an $\ell = 0$ mode (without fine structure or rotational splitting) is shown in Figure 5. In the present case of the low-noise GOLF data, it is much easier to demonstrate that noise from either solar velocity fields or photon statistics does not contribute significantly to the observed effects. We can deduce from the observed variations that there is an important fluctuation in both the amplitude of the mode and in its observed frequency, on time scales as short as a few days. This limit is imposed by the need for a trade-off between time and frequency resolution in the analysis and the importance of avoiding beating between adjacent modes. It is this time variation that produces the characteristic spiky line profile when analysed by Fourier techniques. A number of studies have aimed at interpreting these variations as due to the stochastic nature of the excitation process for the modes (Chang, 1995; Chang and Gough, 1995). García *et al.* (1997), developing the theory of Woodard (1984) and Duvall (1990), have modelled the excitation of a damped oscillator by stochastic impacts, and are able to reproduce Fourier profiles with similar appearance to those of Figure 5. According to García et al, the observed frequency shifts can be explained by excitation effects, without requiring shifts in the frequency of the solar oscillator, although this remains an open question. What is clear is that the time – frequency analysis, as well as the complex profile of the Fourier spectrum, contain important information on the basic excitation mechanism of the modes.

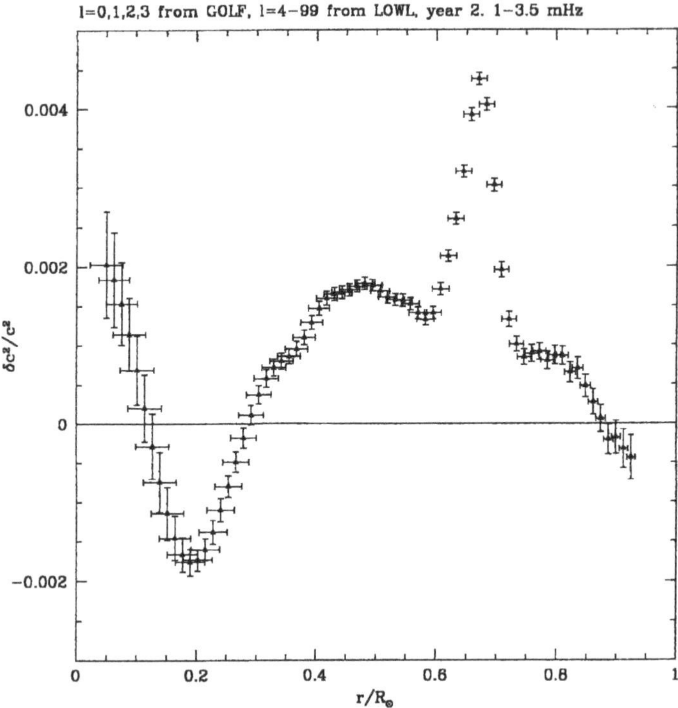

Figure 4. Difference in the square of the sound speed in the Sun and in the Model S of Christensen-Dalsgaard (1996), in the sense (Sun − model), obtained from 4 months of GOLF data, combined with LOWL.

These manifestations of stochastic excitation impose important constraints on the precision obtainable in the frequency measurements of the modes. It is generally accepted that with a long enough time-series, a mean profile of a modified Lorenzian form can be accurately applied. However, it is not clear how rapidly we can converge to this limit, or whether real frequency shifts due, for example, to solar cycle effects will intervene before we reach a high precision. With the present data set of some 10 months, there is clearly a limitation in precision imposed by the stochastic excitation.

This difficulty is particularly troublesome when measuring the difference between two close resonances, as is the case in determining rotational splitting. Reported rotational splittings in the literature cover a very large range of values, much wider than the precision quoted in fitting the Lorenzian profiles. Values are now beginning to emerge from the GOLF data, but it is obvious that with less than one year of data, it is too soon to make an important contribution to the debate concerning solar internal rotation rates.

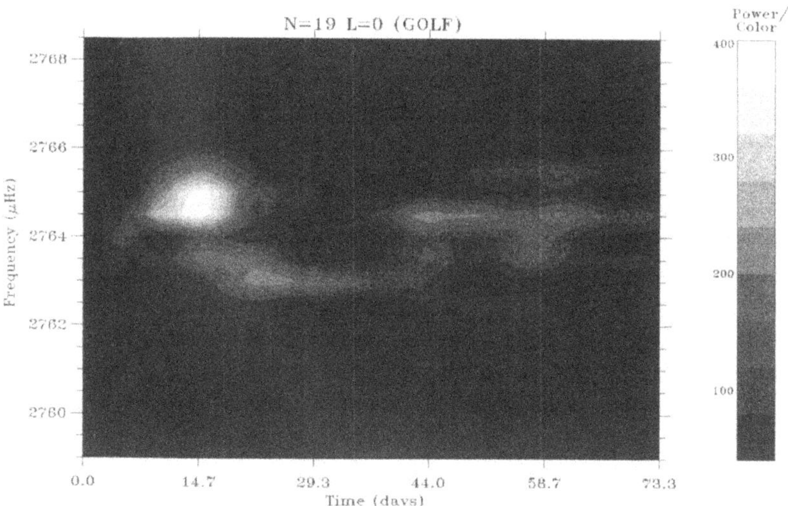

Figure 5. Time/frequency analysis of a single GOLF $\ell = 0$ mode, showing variations in amplitude and apparent changes in frequency.

Both the Fourier techniques and to some extent the time-frequency analysis are based on the assumption of a continuous or smoothly varying oscillation amplitude. If the solar response is really the result of repeated impulse excitation of a damped resonator, then it is possible to derive analysis techniques more closely adapted to the solar situation. This is the basis of a method termed 'homomorphic deconvolution' (HD), which had been developed originally by geophysicists for application to terrestrial seismology. In the HD analysis, the solar response is assumed to be the convolution of an excitation function (consisting of a delta function, randomly distributed in amplitude and in time) and a damped sine wave solar response function. The HD method serves to isolate the damped sine wave component, enabling its Fourier transform, which is a smooth Lorentzian, to be derived, unencumbered by the familiar complex structure due to the stochastic excitation. These techniques have been developed for solar seismology, and applied initially to IPHIR data (Baudin, Gabriel, and Gibert, 1993). Now they are being applied to GOLF data (Baudin *et al.*, 1997b; Lazrek *et al.*, 1997). For the determination of frequency, HD gives much cleaner profiles with more confident Lorenzian fits. For the measurement of rotational splitting, the apparent advantage is sometimes dramatic, yielding two clean separated lines, where otherwise it was difficult to recognise the doublet structure. Work is continuing in order to investigate whether the adoption of this technique in all cases can be justified. At present it is one of a number of techniques being used for the analysis of GOLF data.

6.6. Search for *g* Modes

All of the *p*-mode frequencies are influenced predominantly by the outer layers of the Sun. Inversion of *p*-mode frequencies to obtain the core properties is therefore an ill-conditioned problem. In practice, this means that a much higher degree of precision is required in the *p*-mode frequencies, in order to extract a reasonable precision in the core parameters. GOLF, with its low-noise data, is capable of this high precision, provided a long enough data set can be accumulated. We therefore anticipate important advances in this area, but this will need many more months of data. On the other hand, *g* modes are all concentrated towards the core. This means that knowledge of *g* modes, even with rather approximate frequency values would give quite precise core parameters. However *g* modes have not so far been identified.

The discovery of *g*-mode oscillations from GOLF could never be confidently anticipated, since there are no reliable predictions of their expected amplitudes. However, it was always clear that their detection would provide an inestimable advantage for the interpretation of conditions in the solar core. The search for *g* modes is therefore an important priority for GOLF, whose low-noise, long-term stability offers the best chance for their detection. The solar spectrum in the expected *g*-mode range is dominated by the more random velocity fields produced by the various scales of convection. The approximate form of this continuous spectrum has been predicted by Harvey (1985). When we examine the Fourier transform of GOLF data in this region, we see the background shape predicted by Harvey, with a number of superposed peaks, well above the background level. Although, this is somewhat what one might expect from *g* modes, this interpretation has not yet withstood critical tests concerning long-term invariance of the features. Since such tests depend strongly on the length of the available data base, it is with more months of data that we can expect to arbitrate with more confidence. At the time of writing, with some 10 months of data, it can be stated that no definitive identification of solar *g* modes has been made. However, as the time passes, it will be possible to decrease progressively the upper limit in amplitude for the existence of solar *g*-mode oscillations. For the moment, this is around 1 cm s^{-1} in velocity.

7. Conclusions

The GOLF instrument has been successfully launched and is operating for the most part very well. The exception concerns the rotating polariser mechanisms. After persistent malfunctioning, these have been stopped in optimum rotation orientations and an unforeseen mode of instrument operation has been adopted. This mode, which now involves no moving mechanisms, allows the full determination of frequencies to be performed, with little or no loss of precision. It does however pose difficulties for the absolute velocity calibration, which is currently of the order of 20% or better. In all other respects, GOLF is maintaining its full design specification and is showing no degradation which would impede its continued operation for up to 6 years or more.

The p-mode oscillations from GOLF have a considerably improved signal to noise ratio, which has already allowed the identification of some new resonances. The precision today for the determination of p-mode frequencies must be regarded as preliminary. Significant advances are anticipated, both from improved analysis techniques on the existing data base as well as of course a substantial prolongation of this base. Continuity of the data set is so far quite exceptional, offering ideal opportunities for the study of excitation processes and solar-cycle effects. The search for g-mode oscillations is continuing, but they have not so far been identified. All indications at this point are that GOLF is capable of continuing operations into the next solar maximum.

Acknowledgements

The GOLF programme is based upon a consortium of institutes involving a large number of scientists and engineers, as enumerated in the earlier instrument publication (Gabriel *et al.*, 1995). The analysis work described here is also a team activity, involving many workers outside of the original instrument teams. The work of S. Basu and J. Christensen-Dalsgaard was supported in part by the Danish National Research Foundation through its establishment of the Theoretical Astrophysics Center.

The exceptional performance of the GOLF instrument is in large part due to the unique facilities offered by the SOHO platform, in terms of long-term stability of temperature and pointing and an extremely reliable and accurate clock. The continuous and dedicated attention of the ground system, including the Flight Operations Team at GSFC and the Deep Space Network have enabled an exceptional continuity in the data reception, essential for this project. Our thanks are due to these vital contributions from ESA, NASA and Matra Marconi.

We gratefully acknowledge the continued support (financial and otherwise) of the Centre National d'Études Spatiales (CNES), the Centre National de la Recherche Scientifique (CNRS), the Commissariat à l'Énergie Atomique (CEA) and the DGICYT under grant 95–0028-C.

References

Baudin, F., Gabriel, A. H., and Gibert, D.: 1993, *Astron. Astrophys.* **276**, L1.
Baudin, F., Gabriel, A. H., and Gibert, D.: 1994, *Astron. Astrophys.* **285**, L29.
Baudin, F., Gabriel, A. H., Gibert, D., Pallé, P. L., and Régulo, C.: 1996, *Astron. Astrophys.* **311**, 1024.

Baudin, F., Gabriel, A. H., García, R. A., Foglizzo, T., Gavryusev, V. G., Gavryuseva, E. A., Gough, D., Ulrich, R., and the GOLF Team: 1997a, *Posters from the IAU Symposium No. 181, Nice*, Kluwer Academic Publishers, Dordrecht, Holland, in press.
Baudin, F., Régulo, C., Gabriel, A. H., Roca Cortés, T., and the GOLF Team: 1997b, *Posters from the IAU Symposium No. 181, Nice*, Kluwer Academic Publishers, Dordrecht, Holland, in press.
Boumier, P. and Damé, L.: 1993, *Exp. Astron.* **4**, 87.
Brookes, J. R., Isaak, G. R., and van der Raay, H. B.: 1978, *Monthly Notices Roy. Astron. Soc.* **185**, 19.
Chaplin, W. J., Elsworth, Y., Howe, R., Isaak, G. R., McLeod, C. P., Miller, B. A., van der Raay, H. B., Wheeler, S. J., and New, R.: 1996, *Solar Phys.* **168**, 1.
Chang, H.-Y. and Gough, D.: 1995, in R. K. Ulrich, E. J. Rhodes, and W. Däppen (eds), *GONG '94 Helio- and Astro-Seismology*, p. 512.
Chang H.-Y.: 1995, Ph.D. Thesis, University of Cambridge, Cambridge.
Christensen-Dalsgaard, J., Däppen, W., Ajukov, S. V., Anderson, E. R., Antia, H. M., Basu, S., Baturin, V. A., Berthomieu, G., Chaboyer, B., Chitre, S. M., Cox, A. N., Demarque, P., Donatowicz, J., Dziembowski, W. A., Gabriel, M., Gough, D. O., Guenther, D. B., Guzik, J. A., Harvey, J. W., Hill, F., Houdek, G., Iglesias, C. A., Kosovichev, A. G., Leibacher, J. W., Morel, P., Proffitt, C. R., Provost, J., Reiter, J., Rhodes, E. J., Jr., Rogers, F. J., Roxburgh, I. W., Thompson, M. J., and Ulrich, R. K.: 1996, *Science* **272**, 1286.
Domingo, V., Fleck, B. and Poland, A.: 1995, *Solar Phys.* **162**, 1.
Duvall, T. L.: 1990, in B. Berthomieu and M. Cribier (eds), *Inside the Sun*, Kluwer Academic Publishers, Dordrecht, Holland, p. 253.
Gabriel, A. H., Grec, G., Charra, J., Robillot, J. M., Roca Cortés, T., Turck-Chièze, S., Bocchia, R., Boumier, P., Cantin, M., Cespédes, E., Cougrand, B., Crétolle, J., Damé, L., Decaudin, M., Delache, P., Denis, N., Duc, R., Dzitko, H., Fossat, E., Fourmond, J. J., García, R. A., Gough, D., Grivel, C., Herreros, J. M., Lagardère, H., Moalic, J. P., Pallé, P. L., Pétrou, N., Sanchez, M., Ulrich, R., and Van der Raay, H. B.: 1995, *Solar Phys.* **162**, 61.
García, R. A., Foglizzo, S., Turck-Chièze, S., Baudin, F., Boumier, P. and the GOLF Team: 1997, *Posters from the IAU Symposium No. 181, Nice*, Kluwer Academic Publishers, Dordrecht, Holland, in press.
Grec, G., Fossat, E., Gelly, B., and Schmider, F. X.: 1991, *Solar Phys.* **133**, 13.
Grec, G., Turck-Chièze, S., Lazrek, M., Roca Cortés, T., Bertello, L., Baudin, F., Boumier, P., Charra, J., Fierry-Fraillon, D., Fossat, E., Gabriel, A. H., García, R. A., Gouiffes, C., Régulo, C., Renaud, C., Robillot, J. M., and Ulrich, R.: 1997, *Proceedings of the IAU Symposium No. 181, Nice*, Kluwer Academic Publishes, Dordrecht Holland, in press.
Harvey J.: 1985, in Rolfe and Battrick (eds), *Proceedings of the ESA Workshop on Future Missions in Solar Heliospheric and Space Plasma Physics*, ESA SP 235, p. 199.
Lazrek, M., Régulo, C., Baudin, F., Bertello, L., García, R. A., Gouiffes, C., Grec, G., Roca Cortés, T., Turck-Chièze, S., Ulrich, R., Robillot, J. M., Gabriel, A. H., Boumier, P., Charra, J. and the GOLF Team : 1997a, *Posters from the IAU Symposium No. 181, Nice*, Kluwer Academic Publishers, Dordrecht, Holland, in press.
Lazrek, M., Roca Cortés, T., Régulo, C., Baudin, F., Bertello, L., García, R. A., Gouiffes, C., Grec, G., Boumier, P., Charra, J., Gabriel, A. H., Pérez Hernández, F., Renaud, C., Robillot, J. M., Turck-Chièze, S., and Ulrich, R.: 1997b, *Solar Phys.* **175**, 227 (this issue).
Tomczyk, S. *et al.*: 1996, *Solar Phys.* **159**, 1.
Turck-Chièze, S., Basu, S., Brun, S., Christensen-Dalsgaard, J., Eff-Darwich, A., Gabriel, M., Henney, C. J., Kosovichev, A. G., Lopes, I., Paternò, L., Provost, J., Ulrich, R. and the GOLF Team : 1997a, *Posters from the IAU Symposium No. 181, Nice*, Kluwer Academic Publishers, Dordrecht, Holland, in press.
Turck-Chièze, S., Basu, S., Brun, S., Christensen-Dalsgaard, J., Eff-Darwich, A., Lopes, I., Pérez Hernandez, F., Berthomieu, G., Provost, J., Ulrich, R. K., Baudin, F., Boumier, P., Charra, J., Gabriel, A. H., Garcia, R., Grec, G., Renaud, C., Robillot, J. M., Roca Cortés, T.: 1997b, *Solar Phys.* **175**, 247 (this issue).
Woodard, M.: 1984, Ph.D. Thesis, University of California, San Diego.

FIRST RESULTS ON p MODES FROM GOLF EXPERIMENT

M. LAZREK[2,7], F. BAUDIN[1,8], L. BERTELLO[3], P. BOUMIER[5], J. CHARRA[5], D. FIERRY-FRAILLON[9], E. FOSSAT[9], A. H. GABRIEL[5], R. A. GARCÍA[4], B. GELLY[9], C. GOUIFFES[4], G. GREC[2], P. L. PALLÉ[1], F. PÉREZ HERNÁNDEZ[1], C. RÉGULO[1], C. RENAUD[2], J.-M. ROBILLOT[6], T. ROCA CORTÉS[1], S. TURCK-CHIÈZE[4] and R. K. ULRICH[3]

[1]*Instituto de Astrofísica de Canarias, 38205 La Laguna, Tenerife, Spain*
[2]*Observatoire de la Côte d'Azur, Lab. Cassini CNRS URA1362, 06304 Nice, France*
[3]*Astronomy Department, University of California Los Angeles, U.S.A.*
[4]*Service d'Astrophysique, DSM/DAPNIA, CE Saclay, 91191 Gif-sur-Yvette, France*
[5]*Institut d'Astrophysique Spatiale, CNRS/Université Paris XI, 91405 Orsay, France*
[6]*Observatoire de l'Université Bordeaux 1, BP 89, 33270 Floirac, France*
[7]*Centre National de la Recherche, CNCPRST, Rabat, Morocco*
[8]*National Solar Observatory, NOAO, PO Box 26732, Tucson AZ-85726, U.S.A.*
[9]*U.M.R. 6525, C.N.R.S. et Université de Nice-Sophia Antipolis, F-06108 Nice Cedex 2, France*

(Received 19 March 1997; accepted 17 June 1997)

Abstract. The GOLF experiment on the SOHO mission aims to study the internal structure of the Sun by measuring the spectrum of global oscillations in the frequency range 10^{-7} to 10^{-2} Hz. Here we present the results of the analysis of the first 8 months of data. Special emphasis is put into the frequency determination of the p modes, as well as the splitting in the multiplets due to rotation. For both, we show that the improvement in S/N level with respect to the ground-based networks and other experiments is essential in achieving a very low-degree frequency table with small errors (~ 2 parts in 10^5). On the other hand, the splitting found seems to favour a solar core which does not rotate slower than its surface. The line widths do agree with theoretical expectations and other observations.

1. Introduction

Helioseismology uses the temporal and spatial properties of solar oscillations to study the solar interior. The quality of analysis as measured by the precision of the deductions and the ability to probe all parts of the Sun's interior is dependent on the acquisition of observations of the widest possible range of solar oscillation modes and with the greatest precision possible. The SOHO satellite includes three helioseismology instruments which aim to cover in a complementary manner a wide range of oscillation measurements. GOLF is one of these helioseismology instruments and emphasizes the lower frequencies, utilising global oscillations of the line-of-sight velocity vector of the solar photosphere as measured by the full-disk integrated light.

At present only higher radial orders of the pressure or p modes have been well identified in the solar oscillation spectrum, whether the observation is spatially integrated (Claverie *et al.*, 1979; Grec, Fossat, and Pomerantz, 1980) or spatially resolved (Duvall and Harvey, 1983). The mode amplitudes favour the observations of periods around 5 min, these resulting in some of the best data reported to date

Solar Physics **175:** 227–246, 1997.

(Anguera Gubau *et al.*, 1992; Elsworth *et al.*, 1994; Gelly *et al.*, 1997; Chaplin *et al.*, 1996c). Spatially integrated data favours the low-degree modes, which are limited to $\ell = 0$, 1, 2, and 3. The possibility of detecting $\ell = 4$ in GOLF data was predicted before launch (Régulo, 1995) and nowadays the highest modes are barely detectable in ground-based networks (Chaplin *et al.*, 1996b) and better in our preliminar analysis (Lazrek *et al.*, 1997).

Frequencies below 2 mHz are very difficult to observe from the ground due to a combination of decreasing amplitude and increasing terrestrial and solar noise. The lowest-frequency p mode with $\ell = 0$ and $n = 1$ has a period of the order of 60 min, and has never been identified. An important objective of GOLF is to measure the p modes below 1.5 mHz and to study their entire predicted range up to the cut-off frequency.

A number of factors determine the shape and width of the observed p modes. These include the limited coherence time of the oscillations and variations in amplitude and phase with time, due to the physical process of excitation of the modes. A different and time-invariant reason for fine structure in certain resonances lies in the splittings due to solar rotation. Such splittings have been well-observed for intermediate and high-ℓ p modes and found to correspond approximately to the solar rotation period (Duvall and Harvey, 1983). Although for low-ℓ modes the splitting have been measured by several groups (Claverie *et al.*, 1979; Toutain and Fröhlich, 1992; Jiménez *et al.*, 1994; Lazrek *et al.*, 1996; Chaplin *et al.*, 1996a) the precision is not good enough yet, showing a high scatter which does not help much in constraining the solar core rotation (Fossat, 1995).

Moreover, the above parameters are susceptible to systematic variation with the 11-year solar activity cycle. Frequencies of very low ℓ-value p modes (between 2 and 5 mHz) have been found to vary with time, on scales of the solar activity cycle (Woodard and Noyes, 1985; Gelly *et al.*, 1988). This correlates very well with several solar activity indices, such as sunspot number, magnetic field measurements, UV radiation and 10.7 cm radio flux (Pallé, Régulo, and Roca Cortés, 1989, 1990a, b; Elsworth *et al.*, 1990b; Régulo *et al.*, 1994). The experimental data converge to an average value, over radial orders n, for the frequency shift of 0.45 ± 0.04 μHz (peak to peak), the frequencies being higher at solar activity maximum. This limits the length of the data series to be jointly analysed as it would throw more noise in the line shapes of the modes. Notice that the data used in this work has been taken at the minimum of solar activity, therefore forming a coherent data set.

2. Data, Time Series, and Power Spectra

The technical principle employed by GOLF is to measure the Doppler wavelength shift of the solar Na D Fraunhofer lines by comparison with an absolute standard given by a sodium vapour cell within the GOLF instrument. It uses the optical resonance technique to isolate a narrow region of the solar sodium resonance line. The solar absorption traverses a sodium vapour cell which has an intrinsic (thermal) absorption linewidth of $\sim$ 25 mÅ. Thus a 25 mÅ slice of the solar line is absorbed and re-emitted in all directions. Part of the signal re-emitted at 90 deg is recorded by suitably positioned detectors. By placing the cell in a longitudinal magnetic field of 5000 G, the absorption line is Zeeman split into two components displaced from the original by $\pm \sim$ 108 mÅ (D1 case). If the incoming solar flux is now analysed circularly using a linear polariser followed by a quarter-wave plate, it is possible to select first one, then the other absorption component, respectively left and right circularly polarised, and thus measure the amplitude of a point on each of the two wings of the solar profile at: I_b and I_r. If the cell magnetic field is now increased/decreased sequentially, by a small amount, we can measure two more points, further separated in wavelength: I_b^-, I_b^+ and similarly in the red wing of the line. This technique is similar, in principle, to that carried out in ground-based measurements of solar oscillations, as for example in the IRIS or BiSON networks (Grec *et al.*, 1991; Brookes, Isaak, and van der Raay, 1978); however, some innovations have been introduced which are fully described in Gabriel *et al.* (1995) as well as the whole GOLF instrument.

However, due to occasional malfunction of the rotating mechanisms which allow the polarising elements to rotate, it was decided to stop them on 10 April 1996 at positions which only select the blue wings of the lines (Na doublet). In this way we ensure the continuity in the acquisition of measurements having a duty cycle of nearly 100% but sacrifice observations on the red side of the lines (Gabriel *et al.*, 1997). In this blue-wing-only operating mode, the working cycle of the instrument is now basically I_b^-, I_b^+ measured sequentially at 5-s intervals by two different PM tubes. With these measurements several signals can be defined, of which we underline two:

$$Y^B = (I_b^+ + I_b^-) - \langle I_b^+ + I_b^- \rangle \,, \tag{1}$$

$$X^B = (I_b^+ + I_b^-)/\langle I_b^+ - I_b^- \rangle \,, \tag{2}$$

where $\langle\rangle$ stands for a low-frequency filter with a cut-off at $\sim$22 μHz, that is over a time much longer than the signals to be measured (i.e., longer than $\sim$10 hours) that has been applied to the time series in its Fourier space (see García, 1996; Régulo *et al.*, 1997). Notice that both signals can be calibrated in velocity units in a procedure that is being improved and it will be described shortly elsewhere; in fact, in this paper the calibrations are judged to be accurate to within 20%. Data analysed here starts on 14 April 1996 and lasts until mid December for a total of around 8 months, with a duty cycle greater than 99%. The data have been analysed independently by at least four groups using any of the signals described above and different data lengths, being all from 8 to 10 months. Yet preliminary results based on four months of data have been shown in Grec *et al.* (1997).

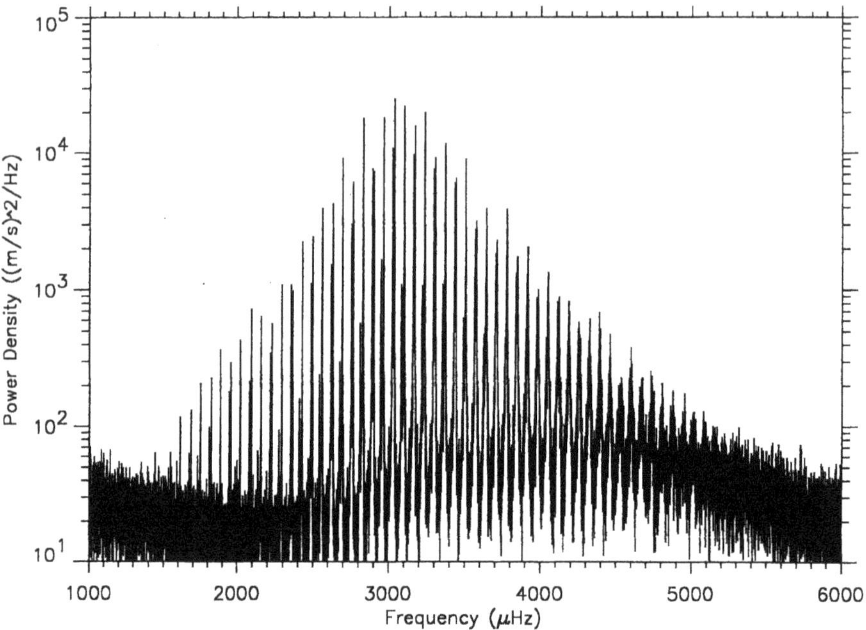

Figure 1. The acoustic p-mode spectrum of the Sun, as measured using the first eight months of GOLF data. At 3 mHz the ratio S/N is ~3000.

To calculate the power density spectrum, each group has used the methods available and with which it has more experience. The standard method consists in taking an FFT of the time series, calculating the power density spectrum and analysing it by using a maximum likelihood technique to fit Lorentzians to the peaks present (Lazrek *et al.*, 1996; Gelly *et al.*, 1997; Anderson, Duvall, and Jefferies, 1990). Also the cross-spectrum between the signals obtained by each PM tube separately has been calculated and used to show the low noise level achieved and to help in the search for modes at low S/N levels. No oversampling of the data has been used by any of the groups participating in the analysis.

The power spectra look very similar independently of the signal and the method used. One example can be seen in Figure 1, where the overall spectrum of the p modes can be seen. The regularity in the spacing of the peaks is clearly shown as predicted by the asymptotic theory of global oscillations, as well as the increasing linewidth of the peaks as frequency increases denoting lower damping times as n increases. A slice of this spectrum at the region of highest power is shown in Figure 2, where discrete modes can be seen; these modes are readily identified as having different ℓ values, which from left to right are 3, 1, 2, and 0. Notice also that no sidebands are visible due to the high continuity in the data. Also plotted is the same spectrum multiplied by 100; this is done for two reasons: the first one is to show the very low noise level between these modes, and the second one to better

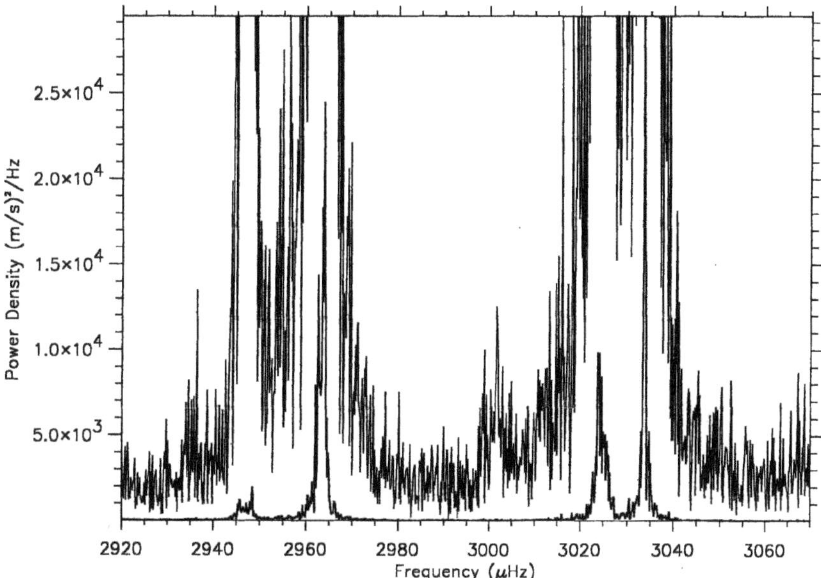

Figure 2. A slice of the acoustic p-mode spectrum of the Sun. The bold line corresponds to the spectrum in the units shown on the vertical axis, while the thin one represents the same spectrum multiplied by 100. Notice the clear presence of modes of different degree: $\ell = 3, 1, 2$, and 0 (*from left to right*) in the bold line and the $\ell = 4$ and 5 in the thin one. Also hints of the $\ell = 6$ and 7 are visible (see text).

look at the 'noise' between peaks and actually discover the broad peaks $\ell = 4$, at $\sim$3000 μHz, clearly present, and also traces of $\ell = 5$, at $\sim$3050 μHz; less clear however, but already some hints of $\ell = 6$ and $\ell = 7$ situated at $\sim$2961 μHz and $\sim$3005 μHz on the wings of much higher peaks.

Acoustic modes with frequencies less than 2 mHz are very difficult to observe from the ground due to a combination of decreasing amplitude of the modes and increasing terrestrial noise and also 'solar noise' produced mainly in the convection zone. From space, free from terrestrial noise, the so-called 'solar noise' also prevents the identification of p modes below 1.5 mHz up to the present (see Figure 3). Notice however, that one is tempted to suggest that below 800 μHz there is enough signal so that some of the peaks might be identified as real modes. It is expected that with more data, the solar non-coherent signals that make up the 'solar noise' will be washed out and the modes will stand up because of their increasingly high coherence time as frequency decreases (Elsworth *et al.*, 1990a).

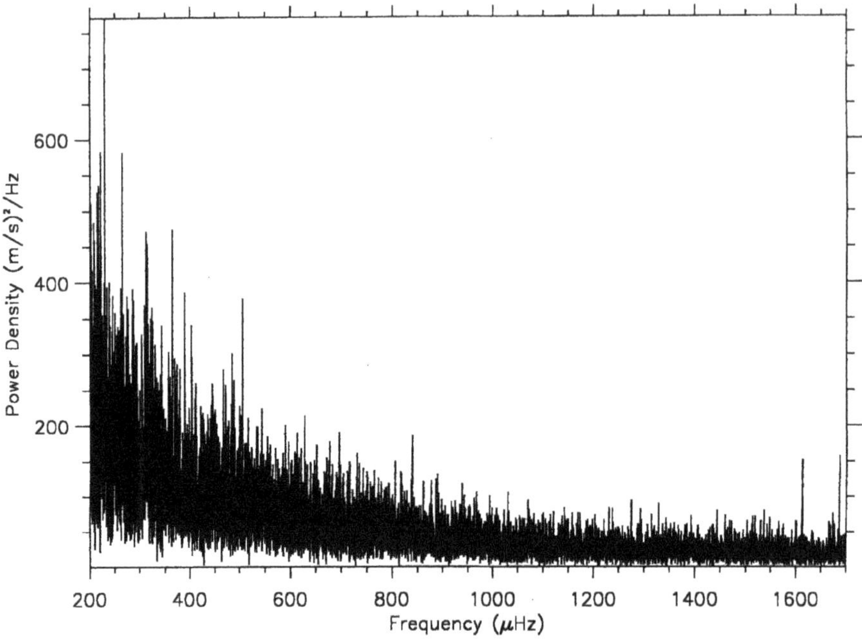

Figure 3. The p-mode spectrum at lower frequencies where the decrease in amplitude of the modes merge with the increase in the so-called 'solar noise', non-coherent solar velocity signals, resulting in a $S/N \leq 1$.

3. Frequencies

The shape of each line in the multiplets is a spiky Lorentzian profile as a consequence of the successive random excitations of the mode, which limits the coherence time of the oscillations. Therefore their mean power density spectrum is of the form

$$P_{nl}(\nu) = \sum_{m=-l}^{m=l} \frac{S_{nlm}(\Gamma_{nlm}/2)^2}{(\nu - \nu_{nlm})^2 + (\Gamma_{nlm}/2)^2} + N_{nl} \, . \tag{3}$$

Consequently, by fitting the spectrum to this model, one can find the frequencies ν_{nlm}, line widths Γ_{nlm}, power amplitudes S_{nlm} and background noise N_{nl}.

Although some $\ell = 5$ have been identified, only modes of degree up to $\ell = 4$ have been systematically searched for and independently fitted. The result of the different fits ended with four frequency tables which are essentially identical, as would be expected because they come from the same observations and therefore they are not completely independent; however, the noise can be quite different as different signals have been processed and even part of the signal can be independent too, since up to 20% independent points might have been used between one series and another. The errors on each frequency determination are the statistical errors

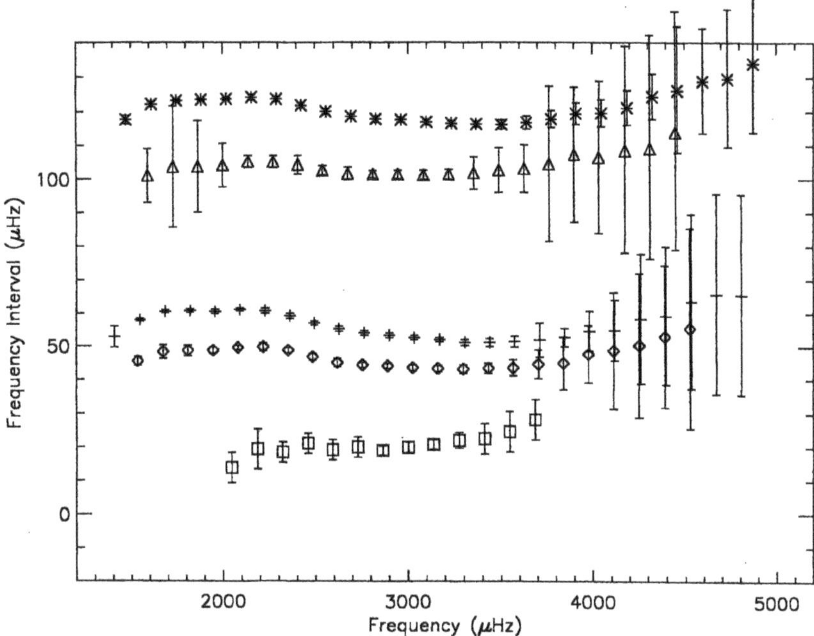

Figure 4. 'Echelle' diagram (with a fixed frequency interval of 135.5 μHz) of the frequencies found (see text and Table I). Error bars are plotted as $\pm 10\sigma$ for clarity. Crosses, asterisks, diamonds, triangles and squares stand for $\ell = 0, 1, 2, 3$, and 4, respectively

that come out from the fitting programs by using appropriately the covariance matrix of the fits. Therefore, it is not surprising that more than 90% of the frequencies of each table coincide with the others within 1σ (combined errors). To end up with one table, they have been merged in the following way: the frequencies found for each (n, ℓ) multiplet have been averaged and the final error for each frequency has been set as the average value of the errors.

The final frequency results are shown in Table I. Notice that towards lower frequencies the identification become increasingly difficult due a poor signal-to-noise level (see Figure 3); here the dispersion found among each group's identifications is larger than the standard errors quoted by each of them, showing clearly that different identifications were made (specially in some $\ell = 3$). In these cases, we quoted the findings inside brackets, avoiding to the mention of other equally possible identifications down to $n = 1$. Probably, new data will help in undisputedly identifying such modes whose errors will be much lower than those shown because their linewidths are smaller than higher overtones of the same degree.

However, at frequencies higher than 3.4 mHz, the linewidths become too broad; the fitting of only one multiplet at a time is no longer a good choice since their wings largely affect each other. Consequently, they have been fitted in pairs: $\ell = 0, 2$ and separately $\ell = 1, 3$. In Figure 4 an 'echelle' diagram (with a fixed 135.5 μHz frequency interval) is made, showing the regularity in the spacings between modes of like ℓ as predicted by the asymptotic theory; the error bars are plotted as $\pm 10\sigma$ for clarity, specially for modes up to 3800 μHz.

On the other hand, it should be stressed that during the time when the data analysed here were obtained, the solar activity on the Sun remained at a very low level (the average of the monthly averaged sunspot number is ~ 8) and therefore its contribution to the variation of frequencies is negligible.

4. Splittings

The identification of the degree of the modes fitted can be done by comparison with the theoretical predictions coming from the eigenvalues of the linearised partial differential equations which describe the oscillations of the Sun. However, there is still another observational way of determining the degree of the modes due to the splitting of the lines caused by the loss of spherical symmetry, already assumed by the theory mentioned above. This loss of symmetry can be produced by the rotation of the Sun and/or the presence of intense global magnetic fields in its radiative interior.

Taking the solar rotation as an axial perturbation, each acoustic mode is split into $(2\ell + 1)$ modes, each with a different m (azimuthal degree) value which goes from ℓ to $-\ell$, each separated by an amount proportional to the rotational velocity value of the solar layers in the cavities where the modes propagate. This is

$$\delta\nu_{nlm} = \nu_{nlm} - \langle \nu_{nl} \rangle \simeq m \frac{\bar{\Omega}}{2\pi} , \tag{4}$$

where $\langle \nu_{nl} \rangle$ is the frequency averaged over m (or ν_{nl0}) and $\bar{\Omega}$ is an appropriate average of $\Omega(r, \theta)$ over the cavity sampled by the modes.

However, due to the angular symmetry of the spherical harmonics, which are the angular part of the eigenfunction solutions of the oscillations problem, only the modes with $(\ell + m)$ even, will be observable. Therefore, modes with degree 1, 2, 3, and 4 will be observable as multiplets with 2, 3, 4, and 5 components, respectively. The relative height of each component can be calculated and predicted (Christensen-Dalsgaard, 1989) and, although they will vary during the year due to the apparent precession of the solar rotation axis, such variation is very small; the one due to the relative movement of the satellite and the Sun could be a bit higher, which will slightly modulate the height of each yearly (n, ℓ, m). This is being studied quantitatively for the particular case of the GOLF blue-wing operating mode.

Table 1

Frequencies and errors, in μHz, as measured by GOLF instrument. Bracketted frequencies are possible identifications with S/N < 1 (see text).

n	$\ell = 0$	$\ell = 1$	$\ell = 2$	$\ell = 3$	$\ell = 4$	$\ell = 5$
8		(1329.49±0.64)	(1393.51±0.49)			
9	(1407.77±0.32)	1472.71±0.13	(1535.95±0.13)	(1591.57±0.80)		
10	1548.40±0.04	1612.74±0.08	1674.31±0.21	(1729.68±1.80)		
11	1686.54±0.04	1749.29±0.04	1810.17±0.16	(1865.26±1.37)		
12	1822.20±0.05	1885.08±0.04	1945.73±0.09	(2001.21±0.65)	(2046.30±1.50)	
13	1957.45±0.06	2020.81±0.04	2082.06±0.07	2137.75±0.19	(2187.40±2.00)	
14	2093.51±0.04	2156.80±0.09	2217.80±0.12	2273.23±0.19	2321.94±1.00	
15	2228.69±0.07	2291.80±0.08	2352.32±0.08	2407.85±0.28	2460.10±1.00	
16	2362.69±0.08	2425.48±0.08	2485.89±0.10	2541.65±0.14	2593.70±1.00	
17	2496.04±0.06	2559.21±0.08	2619.64±0.12	2676.19±0.19	2730.00±1.00	(2778.10±0.50)
18	2629.82±0.08	2693.27±0.07	2754.43±0.08	2811.46±0.11	2864.52±0.47	2913.70±0.30
19	2764.05±0.07	2828.01±0.10	2889.64±0.08	2946.95±0.12	3000.94±0.53	3050.20±0.30
20	2898.83±0.07	2963.33±0.08	3024.65±0.09	3082.28±0.14	3137.24±0.49	(3185.77±0.30)
21	3033.68±0.06	3098.10±0.08	3159.91±0.09	3217.91±0.15	3274.01±0.77	
22	3168.53±0.06	3233.22±0.08	3295.22±0.13	3353.88±0.48	3410.00±1.50	
23	3303.22±0.09	3368.55±0.09	3430.97±0.15	3490.24±0.67	3547.60±2.00	
24	3438.72±0.12	3503.86±0.14	3566.62±0.24	3626.21±0.71	3686.70±2.00	
25	3574.50±0.16	3640.02±0.20	3703.23±0.43	3763.05±2.30		
26	3710.54±0.51	3776.46±0.25	3838.99±0.79	3901.29±2.00		
27	3846.77±0.27	3913.53±0.32	3977.19±0.86	4036.00±2.25		
28	3984.09±0.60	4049.20±0.40	4113.82±1.75	4173.55±3.05		
29	4119.95±0.91	4186.25±0.51	4250.88±2.16	4309.66±3.30		
30	4258.79±1.94	4325.00±0.66	4389.01±2.13	4450.00±3.50		
31	4395.22±2.07	4462.28±1.84	4527.00±3.00			
32	4535.00±2.62	4600.42±1.52				
33	4672.70±3.00	4736.60±2.00				
34	4808.00±3.00	4876.50±2.00				

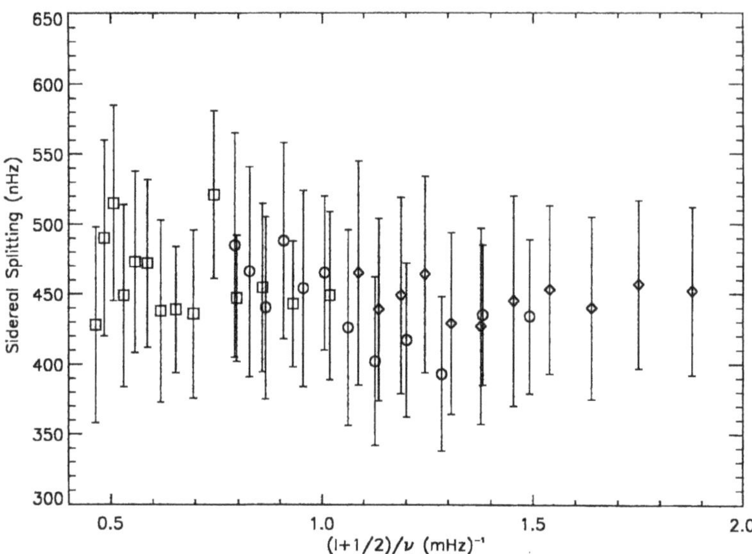

Figure 5. The sidereal splitting found for each mode plotted against $(\ell + \frac{1}{2})/\nu$ which is directly related to the lower turning point of the modes. Open squares, open circles and diamonds are for $\ell = 1, 2$, and 3, respectively.

4.1. INDIVIDUAL DETERMINATIONS

For this reason this multiplet structure (n, ℓ, m) is what is fitted in the power spectra. There is a variety of methods to solve the problem of the nonlinear fitting, and the use of several of them will certainly increase the confidence in the results and reduce the amount of uncertainty that the numerical methods could throw into the result. However, the statistical uncertainty which will only depend on the number of realisations of the excitation of the mode will still remain unchanged. Another problem resides in the fact that the fit is nonlinear, and therefore the result might depend on the starting values needed for the parameters to be fitted. This has been checked carefully and we arrive at the conclusion that the variations in the results are smaller than the errors quoted (obtained from the fit).

In these determinations we use the outputs of the analysis done to find the frequencies of the multiplet lines, in which we define the splitting as half the difference between two consecutive components of a given multiplet, this difference being set equal for all components. Notice that only for $\ell \geq 3$ can we have information on the splitting of different m components of the multiplet other than those with $\ell = m$; these will give information on the latitudinal dependence of the rotation rate. The results can be found in Figure 5 where they are plotted as a function of $(\ell + 0.5)/\nu$, which is directly related to the inner turning point of the modes $2\pi(r/c)_t$. Also weighted averages over all splitting values of the modes

Table II
Sidereal splittings and errors, in nHz, as obtained using the methods described in the text

	$\ell = 1$	$\ell = 2$	$\ell = 3$
Order	9–22	10–21	11–21
Collective	452±14	438±20	456±20
Individual	458±15	437±18	447±20

with the same ℓ have been computed and they are presented in Table II. The error bars shown here are the internal errors of the weighted average (Topping, 1972).

4.2. Collective Methods

On the other hand, to increase the statistical significance and decrease the uncertainties at the minimum, a single measurement for all multiplets of different n and equal ℓ can be calculated. We call these methods collective measurements; they are based on the assumption that the frequency variation of the rotational splitting of modes with the same degree (due to their slightly higher penetration into the solar core) is very small compared to the precision with which it can be measured; therefore these variations will be statistical fluctuations of the energy distribution within each multiplet. As a consequence, the end product here will be one splitting value for all multiplets with the same degree.

Several methods have been applied here, which are improved versions already extensively described in Lazrek *et al.* (1996), also applied to IRIS ground-based data. The methods are:

(1) Iterative method. This consists in fitting all the multiplets with enough S/N and whose linewidth is smaller or of the same order as the splitting to be measured. It is made in two steps: the first one consists in assuming a given value for the splitting of all multiplets; then they are individually fitted, and the central frequency, linewidth, power and noise are determined. In a second step, the values fitted are injected as starting values for the fit and, on a single fit for all multiplets, a single value for the splitting is finally determined. This value is then used to restart again the process with step one. The method converges quite rapidly within a few iterations to a unique value; the process is then repeated for each ℓ value multiplets.

(2) Average spectrum method. In this case, the central frequency of each multiplet is determined from a standard fit to the spectra. Then each multiplet is normalised in order that all multiplets have the same energy, and they are all averaged around their own central frequencies; an averaged multiplet line profile is obtained. Then the splitting is obtained after a least squares fit of this averaged profile. Notice that this method relies heavily on a good determination of the central frequencies of each multiplet. See Figure 6 for a plot of the averaged spectrum for the quadrupole multiplets.

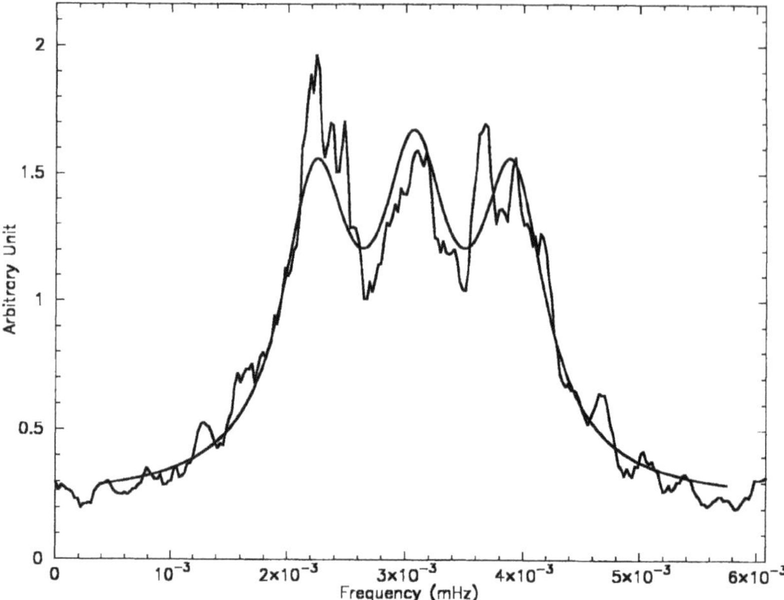

Figure 6. A folded line shape for $\ell = 2$ multiplet showing the triplet structure from which the splitting can be easily calculated.

(3) Average of auto-correlations. Here the normalised auto-correlation function of each multiplet is calculated along an interval of a few μHz. Then the average of these autocorrelation functions is made and the splitting is found by least-squares fitting the appropriate number of Gaussians or Lorentzians to this average. Notice that here there is no need of having previously fitted any function to the multiplet, so no *a priori* knowledge is needed; however the fit is less precise because the function is much smoother than before.

(4) Average of cross-correlations. Here the method is a mixture of the two mentioned above. The time series is divided into two slices and the spectrum is calculated for each one. Afterwards, the crosscorrelation between two frequency intervals, equally large and centred on the same central frequency of a given multiplet, is made. The average of the crosscorrelation functions is taken for all multiplets of a given degree and the splitting is calculated with the method described above (3).

The result of applying all these methods to the spectra is presented in Table II where the weighted average of the four outputs of the methods is given. The estimated error bar is calculated starting from the statistical error of one frequency measurement and taking into account the number of components, the S/N ratio and the number of modes. All the methods gave values within one standard deviation, which is quoted as the error bar.

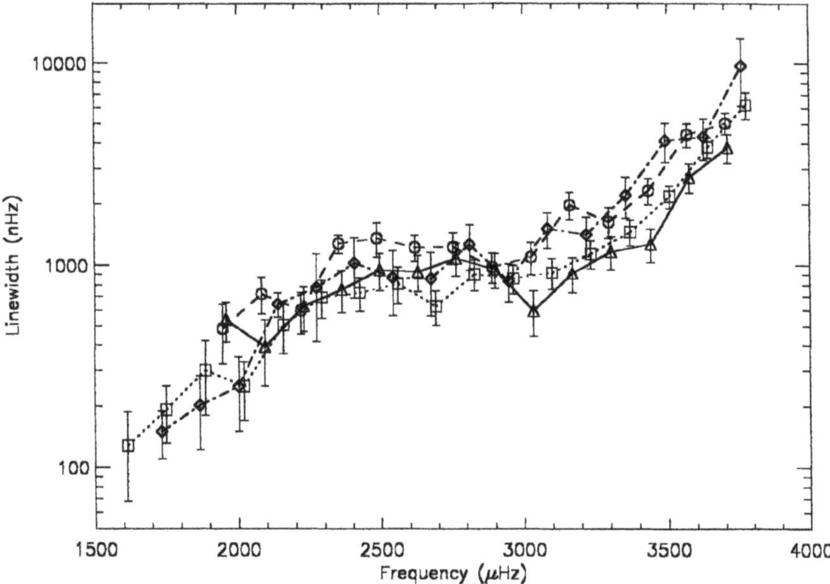

Figure 7. Line widths of the p modes from GOLF experiment. Triangles, open squares, open circles and diamonds are for $\ell = 0, 1, 2$, and 3, respectively.

Notice the excellent agreement between individual and collective methods. Moreover, it is convenient to note that two other groups that have also calculated the splittings of some modes, using methods similar to the individual method described above, gave values whose averages coincide with those of Table II, within errors, but had a higher dispersion than the one shown in Figure 4.

The referee has drawn our attention to the work of Appourchaux *et al.* (1997) in which, using Monte-Carlo simulated data they show the possibility of the existence of bias in the splitting measurements, specially when the profiles of the components of a given multiplet are significantly overlapping or when using some fitting procedures (i.e., considering the amplitudes of the Lorentzians as free parameters or having their relative height fixed at the theoretically predicted value). In our opinion the art of simulating the p modes is far from being easy because the detailed physics of the excitation and damping mechanisms of the modes is not well understood, not only theoretically but also observationally. It is probably one of the most difficult problems and this is not adressed here.

However, great care has been taken in all methods of analysis used here to calculate the splitting of the modes, in order to ensure that if a bias does exist it is certainly small, much below the quoted error bars. For instance, all analyses have been redone separately in the limited low frequency range (below 2.5 mHz) where the line profiles are sharper and thus do not overlap. No single one of all these separate analyses has provided a result beyond 1σ of the average (see Table II) in any of the two directions (plus or minus). The same has also been made by using only the higher part of the frequency range, where the mode profiles do overlap, and the results also remain within 1σ. Therefore the bias should be very small, if any. Moreover, in each one of individual or collective methods, many different fits have been tested. The way to fit the amplitudes has been varying depending on the group making it; in fact, relative amplitudes in a given multiplet have been kept equal while others have let them all vary as free parameters and the results coming from one or the other way have not shown any systematic bias agreeing well within errors. The fits are made using the maximum likelihood estimator in the cases of individual splittings and of iterative global method, while the fits are made using least-squares in other collective cases (averages of many examples). It seems unlikely that with all these different procedures a bias, common in all of them, could persist.

5. Line widths

The estimation of the Γ_{nlm} (see Equation (3)) of the modes can be obtained from the fit. However an *ad hoc* fit using also the maximum likelihood technique on the power spectrum has been used. Each multiplet (n, ℓ) was fitted with $(\ell + 1)$ components as one should expect for full-disk observations. In such fits, neither the central frequency nor the separation of the multiplet components were free parameters. The central frequency was imposed on a tabulated value given by a predetermined frequency table in full agreement with Table I. The separation between the multiplet components was derived from the $\ell = 1$ averaged splitting value in the 'collective' method and was identical for all modes. This partial fitting allows one to distinguish better the uncertainties arising from the frequencies and splitting measurements from the actual uncertainties in the widths.

A value for the width is given only when we think that it conveys a physical meaning: above $n = 22$ the width becomes greater than the separation due to the splitting; but still a correct estimation can be performed. Above $n = 26$ the width becomes greater than the separation between $\ell = 0$ and $\ell = 2$, the multiplets loose their identity as such and this fit does not work, so we do not calculate the widths any further. At low frequencies the low S/N ratio for two of the even groups did not permit to give a reliable estimate. The error bars have been computed using a model of the spectrum and a Monte-Carlo simulation of the statistical dispersion of the results. This dispersion is then fitted by a modified Poisson law giving the σ of the distribution (Fierry-Fraillon *et al.*, 1997). The theoretical formulae found by Toutain and Appourchaux (1994) to calculate the error bars, are not suitable here because half the parameters of the fit are fixed in our case and Monte-Carlo method takes better account of the possible destructive interference both with noise and between multiplet components.

Our final values are shown in Figure 7. The values found here agree very well with the ones measured by Elsworth *et al.* (1990a) using a similar experiment. Also, within the 1-σ error bars, the absolute values are perfectly comparable with the measured IPHIR widths (Toutain and Fröhlich, 1992) and more recently with the VIRGO measurements (Fröhlich *et al.* 1997). The exponential behaviour of the widths is clearly visible for all degrees, and supports well the hypothesis of stochastically excited p mode through the turbulent convection (Goldreich and Murray, 1994). In this context, the departure from this trend, or 'plateau', between 2.4 to 3.2 mHz first described by Libbrecht (1988) is fairly visible. The dip in the 'plateau' at about 3.1 mHz reported by Fröhlich *et al.* (1997) is mostly visible on the $\ell = 0$ mode and suggests a decrease in the damping of the modes in the region where they have their highest amplitude. Longer data sets will be necessary to measure widths below 1.5 mHz, as we are getting near the resolution limit for the smaller widths. The longest apparent lifetime, or phase coherence, measured so far is about 28 days.

6. Discussion

The frequency table presented here is the most complete one done to date, having identified over 100 multiplets. We can compare this list with the one obtained with the BiSON ground-based network (Elsworth *et al.*, 1994) completed with the low-n and $\ell = 4$ values recently reported (Chaplin *et al.*, 1996b, c). We obviously take the list measured at the minimum of solar activity. Out of the 75 frequency determinations in common with ours, 99% coincide within 3σ (combined errors), 87% within 2σ and 52% within 1σ, most of the differences being concentrated in modes of even degree. Taking into account that the two determinations do not come from a coeval set of data, we believe the agreement is quite good and reflects that in periods of low solar activity the Sun seems to have the same structure. The small differences, which appear to concentrate on even degree modes, can be due to the presence of some sideband structure, even having duty cycles of $\sim$ 70% such as those obtained by the integrated Sun ground-based networks, which increases the background noise.

A comparison with the other ground-based network IRIS is not directly possible as the frequency tables published (Gelly *et al.*, 1995) correspond to 1991, 1992, and 1993 – all years with high solar activity which would shift the frequencies towards higher values than the ones found here by an amount several tens of times the errors quoted. However, it can also be compared to a relatively older frequency list obtained by Anguera Gubau *et al.* (1992), coming from only one observing site. Out of 88 frequency determinations in common with ours, 84% coincide within 3σ, 64% within 2σ and only 36% within 1σ showing the great difficulty of measuring frequencies in a spectrum full of sidebands coming from the lack of continuity in the data taken from one observatory. The comparison of frequencies with those

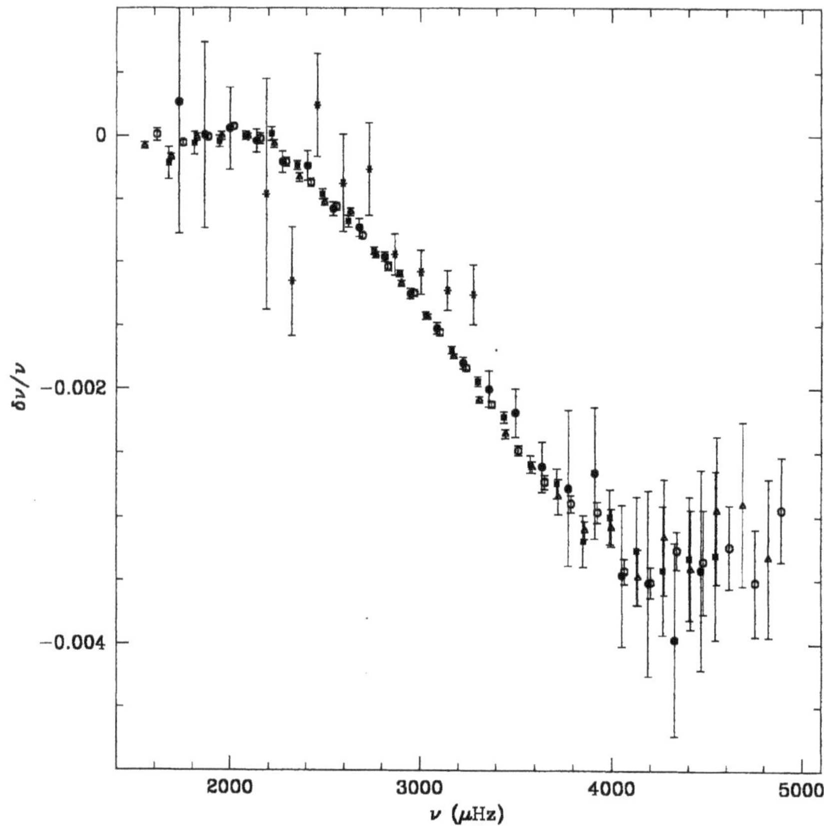

Figure 8. Relative frequency differences between the observed frequencies ($\pm 1\sigma$ error bars) and the predicted ones by a solar model including diffusion (Christensen-Dalsgaard *et al.*, 1996). Triangles, open squares, open circles, full circles and stars are for $\ell = 0, 1, 2, 3$, and 4, respectively. The systematic difference shown is mainly due to a surface effect, because the solar models do not explain correctly the effect of the upper boundary layers.

obtained from VIRGO and MDI experiments on SOHO is not done here because it is done in another paper (Toutain *et al.*, 1997) in this volume.

Finally, we could compare the frequency table found here with the one predicted by a standard solar model, such as the model S calculated by Christensen-Dalsgaard *et al.* (1996), which includes the best parameters to that date and also considers the diffusion of the He and heavy elements in the interior of the Sun. The relative frequency differences (in the sense observed minus predicted) are shown in Figure 8 where it can be seen that up to frequencies of 2.2 mHz, the differences are within observational errors (around 20 nHz), however from here on the differences increase up to about $\sim$3.5 parts in 10^3, and it stabilises (see Turck-Chièze *et al.*, 1997, for a more complete analysis).

The splittings found here (see Table II and Figure 5) can be compared with those found by both ground-based networks using the same technique: IRIS (Lazrek *et al.*, 1996) and BiSON (Chaplin *et al.*, 1996d). With the first one, the agreement is excellent for all degrees, which agree within the errors. However, with the second, even considering the same multiplets, the agreement between the average of the three splitting measurements is within 3σ (combined errors), but their average value, at the closest epoch to us (year 1995), is lower than ours by ~20–50 nHz, depending on which of their averages is chosen. However, their value obtained near the latest solar activity minimum (1985–1987) seems to be closer to the value obtained here. In another study Chaplin *et al.* (1996a) give independent averages for each ℓ up to 4; of these, the $\ell = 2$ is within 1σ while for $\ell = 1$ and 3 their averages are ~40 nHz smaller than ours. On the other hand, they can also be compared with the results found by Toutain and Kosovichev (1994) using data collected by the IPHIR Sun-photometer on board the PHOBOS satellite, back in 1988. In fact, their results agree well with ours, being only some 10 nHz higher, much lower than the combined error bars.

Assuming that there is no significant bias in our splitting measurements, these results can tentatively be interpreted in terms of a non-slowly rotating solar core. They are consistent with a rigid or slighty faster-than-surface core rotation. It is well known that a small error in these measurements will have a devastating effect on the deduced core rotation. So, no conclusion can yet be definitive. However, any possible bias in our measurements seems to be small, and it will not take the results down to a slower-than-surface core rotation. If confirmed by the longer integrations to come soon, this has implications on the quadrupole moment of the Sun and the angular momentum transport in its interior. These splittings should be injected in a rotation inversion programme, with data from higher-ℓ modes, which could allow a complete set to infer the solar rotation profile with depth and latitude; this exercise will be done in the near future using the excellent data that the three helioseismology experiments on-board SOHO are obtaining (Toutain *et al.*, 1997).

7. Conclusions

The first analysis of ~8 months of observations taken by GOLF on board SOHO reveals the quality of the data being stored. The S/N level of the signal at the 5-min oscillations band (where the p modes lie) is the best ever achieved, reaching several thousand, which is readily checked by the clear presence amongst the noise of modes of $\ell \geq 5$, never seen before in integrated sunlight experiments. This supports the idea of having such an instrument to detect solar oscillation modes in stars other than the Sun.

The frequency list presented is the most complete to date, even being a result where the full information of the spectrum has not been extracted yet. At low frequencies, a precise identification is still not possible due to the 'solar noise' throwing peaks into the spectrum which do not allow an unambiguous identification; however it is expected that with more data this will be possible and the full spectrum of p modes will be reachable by GOLF. The first results on splittings have been extracted and the values found seem to favour a rigid, or a slightly faster-than-surface, rotating core; the precise rotation of the solar core remains to be calculated through inversion of data that requires information on many more modes than the ones found here (those of higher ℓ values), which are being observed by the other two helioseismology instruments on board SOHO.

Other parameters that require more attention are being studied at the time of writting this report, such as the energy of the modes, the relative amplitude between the multiplet components and the noise. New methods of analysing the data are also being used which seem to yield more accurate fits to the power spectra.

Acknowledgements

The development of GOLF over a period of several years owes a great deal to the skills and dedication of engineering and technical personnel at each of the participating institutes. The list of these numerous colleagues was published in Gabriel *et al.* (1995). We gratefully acknowledge the continued support (financial and otherwise) of the Centre National d'Études Spatiales (CNES), the Centre National de la Recherche Scientifique (CNRS), the Commissariat à l'Énergie Atomique (CEA), the DGICYT under grant 95–0028-C. M. Lazrek did most of this work at the Observatoire de la Côte d'Azur at Nice, thanks to a 'Henri Poincaré' grant, and thanks to financial support from the Institut d'Astrophysique Spatiale d'Orsay, and from the Université de Nice. He is very grateful to these institutes.

References

Anderson, E. R., Duvall, T. R., Jr., and Jefferies, S. M.: 1990, *Astrophys. J.* **364**, 699.

Anguera Gubau, M., Pallé, P. L., Pérez Hernández, F., Régulo, C., and Roca Cortés, T.: 1992, *Astron. Astrophys.* **255**, 363.

Appourchaux, T., Gough, D. O., Sekii, T., and Toutain T.: 1997, in F. X. Schmider and J. Provost (eds), 'Sounding Solar and Stellar Interiors', *IAU Symp.* **181**, in press.

Brookes, J. R., Isaak, G. R., and van der Raay, H. B.: 1978, *Monthly Notices Roy. Astron. Soc.* **185**, 19.

Christensen-Dalsgaard, J.: 1989, *Monthly Notices Roy. Astron. Soc.* **239**, 977.

Christensen-Dalsgaard, J., Däppen, W., Ajukov, S. V., Anderson, E. R., Antia, H. M., Basu, S., Baturin, V. A., Berthomieu, G., Chaboyer, B., Chitre, S. M., Cox, A. N., Demarque, P., Donatowicz, J., Dziembowski, W. A., Gabriel, M., Gough, D. O., Guenther, D. B., Guzik, J. A., Harvey, J. W., Hill, F., Houdek, G., Iglesias, C. A., Kosovichev, A. G., Leibacher, J. W., Morel, P., Proffitt, C. R., Provost, J., Reiter, J., Rhodes, E. J., Jr., Rogers, F. J., Roxburgh, I. W., Thompson, M. J., and Ulrich, R. K.: 1996, *Science* **272**, 1286.

Chaplin, W. J., Elsworth, Y., Howe, R., Isaak, G. R., McLeod, C. P., Miller, B. A., and New, R.: 1996a, *Monthly Notices Roy. Astron. Soc.* **280**, 849.

Chaplin, W. J., Elsworth, Y., Howe, R., Isaak, G. R., McLeod, C. P., Miller, B. A., and New, R.: 1996b, *Monthly Notices Roy. Astron. Soc.* **280**, 1162.

Chaplin, W. J., Elsworth, Y., Isaak, G. R., Lines, R., McLeod, C.P., Miller, B. A., and New, R.: 1996c, *Monthly Notices Roy. Astron. Soc.* **282**, L15.
Chaplin, W. J., Elsworth, Y., Isaak, G. R., McLeod, C. P., Miller, B. A., and New, R.: 1996d, *Monthly Notices Roy. Astron. Soc.* **283**, L31.
Claverie, A., Isaak, G. R., McLeod, C. P., van der Raay, H. B., and Roca Cortés, T.: 1979, *Nature* **282**, 591.
Duvall, T. and Harvey, J.: 1983, *Nature* **302**, 24.
Elsworth, Y., Isaak, G. R., Jefferies, S. M., McLeod, C. P., New, R., Pallé, P. L., Régulo, C., and Roca Cortés, T.: 1990a, *Monthly Notices Roy. Astron. Soc.* **242**, 135.
Elsworth, Y., Howe, R., Isaak, G. R., McLeod, C. P., and New, R.: 1990b, *Nature* **345**, 322.
Elsworth, Y., Howe, R., Isaak, G. R., McLeod, C. P., Miller, B. A., New, R., Speake, C. C., and Wheeler, S. J.: 1994, *Astrophys. J.* **434**, 801.
Fierry-Fraillon, D., Gelly, B., Schmider, F. X., Fossat, E., and Plantel, A.: 1997, in F. X. Schmider and J. Provost (eds), 'Sounding Solar and Stellar Interiors', *IAU Symp.* **181**, in press.
Fossat E.: 1995, in J. T. Hoeksema, V. Domingo, B. Fleck, and B. Battrick (eds), *Helioseismology*, 4th SOHO workshop, ESA/SP 376-1, Asilomar, Ca., p. 229.
Fröhlich, C., Andersen, B., Appourchaux, T., Berthomieu, G., Crommelynck, D. A., Domingo, V., Fichot, A., Finsterle, W., Gómez, M. F., Gough, D., Jiménez, A., Leifsen, T., Lombaerts, M., Pap, J. M., Provost, J., Roca Cortés, T., Romero, J., Roth, H.-J., Sekii, T., Telljohann, U., Toutain, T., and Wehrli, C.: 1997, in F. X. Schmider and J. Provost (eds), 'Sounding Solar and Stellar Interiors', *IAU Symp.* **181**, in press.
Gabriel, A. H., Grec, G., Charra, J., Robillot, J.-M., Roca Cortés, T., Turck-Chièze, S., Bocchia, R., Boumier, P., Cantin, M., Céspedes, E., Cougrand, B., Crétolle, J., Damé, L., Decaudin, M., Delache, P., Denis, N., Duc, R., Dzitko, H., Fossat, E., Fourmond, J.-J., García, R. A., Gough, D. O., Grivel, C., Herreros, J. M., Lagardére, H., Moalic, J.-P., Pallé, P. L., Petrou, N., Sánchez, M., Ulrich, R. K., and van der Raay, H. B.: 1995, *Solar Phys.* **162**, 61.
Gabriel, A. H., Charra, J., Grec, G., Robillot, J.-M., Roca Cortés, T., Turck-Chièze, S., Ulrich, R. K., Basu, S., Baudin, F., Bertello, L., Boumier, P., Charra, M., Christensen-Dalsgaard, J., Decaudin, M., Dzitko, H., Foglizzo, E., Fossat, E., García, R. A., Herreros, J. M., Pallé, P. L., Pétrou, N., Renaud, C., Régulo, C.: 1997, *Solar Phys.* **175**, 207 (this issue).
García, R. A.: 1996, Ph.D. Thesis, Univ. La Laguna, Tenerife, Spain.
Gelly, B., Fossat, E., Grec, G., and Pomerantz, M.: 1988, in S. Frandsen and J. Christensen-Dalsgaard (eds), 'Advances in Helio- and Asteroseismology', *IAU Symp.* **123**, 21.
Gelly, B., Fossat, E., Pallé, P. L., Appourchaux, T., Ehgamberdiev, S., Fierry-Fraillon, T., Grec, G., Hoeksema, J. T., Kalikov, I., Lazrek, M., Loudagh, S., Pantel, A., Régulo, C., Sánchez, E. L., and Schmider, F. X.: 1955, in J. T. Hoeksema, V. Domingo, B. Fleck, and B. Battrick (eds), '4th SOHO Workshop, ESA SP 376, *Helioseismology*, Asilomar, Ca., p. 373.
Gelly, B., Fierry-Fraillon, D., Fossat, E., Pallé, P. L., Cacciani, A., Ehgamberdiev, S., Grec, G., Hoeksema, J. T., Khalikov, S., Lazrek, M., Loudagh, S., Pantel, A., Régulo, C., and Schmider, F. X.: 1997, *Astron. Astrophys.*, in press.
Goldreich, P. and Murray N.: 1994, *Astrophys. J.* **424**, 480.
Grec, G, Fossat, E., and Pomerantz, M.: 1980, *Nature* **288**, 541.
Grec, G., Fossat, E., Gelly, B., and Schmider F. X.: 1991, *Solar Phys.* **133**, 13.
Grec, G., Turck-Chièze, S., Lazrek, M., Roca Cortés, T., Bertello, L., Baudin, F., Boumier, P., Charra, J., Fierry-Fraillon, D., Fossat, E., Gabriel, A. H., García, R. A., Gouiffes, C., Régulo, C., Renaud, C., Robillot, J. M., and Ulrich, R. K.: 1997, in F. X. Schmider and J. Provost J. (eds), 'Sounding Solar and Stellar Interiors' *IAU Symp.* **181**, in press.
Jiménez, A., Pérez Hernández, F., Claret, A., Pallé, P. L., Régulo, C., and Roca Cortés, T.: 1994, *Astrophys. J.* **434**, 384.
Lazrek, M., Pantel, A., Fossat, E., Gelly, B., Schmider, F. X., Fierry-Fraillon, D., Grec, G., Loudagh, S., Ehgamberdiev, S., Khamitov, I., Hoeksema, J. T., Pallé, P. L., Régulo, C.: 1996, *Solar Phys.* **166**, 1.
Lazrek, M., Régulo, C., Baudin, F., Bertello, L., García, R. A., Gouiffes, C., Grec, G., Roca Cortés, T., Turck-Chièze, S., Ulrich, R. K., Robillot, J. M., Gabriel, A. H., Boumier, P., Charra, J. and the

GOLF Team: 1997, in F. X. Schmider and J. Provost J. (eds), 'Sounding Solar and Stellar Interiors' *IAU Symp.* **181**, in press.
Libbrecht K. G.: 1988, in E. Rolfe (ed.), *Seismology of the Sun and Sun-like Stars*, ESA SP-286, p. 3.
Pallé, P. L., Régulo, C., and Roca Cortés, T.: 1989, *Astron. Astrophys.* **224**, 253.
Pallé, P. L., Régulo, C., and Roca Cortés, T.: 1990a, in G.Berthomieu G. and M. Cribier (eds), 'Inside the Sun', *IAU Colloq.* **122**, 349.
Pallé, P. L., Régulo, C., and Roca Cortés, T.: 1990b, in H. Shibahashi (ed.), *Progress on Seismology of the Sun and Stars, Lecture Notes in Physics*, p. 129.
Régulo, C. 1995 Annual GOLF Meeting,Tenerife.
Régulo, C., Jiménez, A., Pallé, P. L., Pérez Hernández, F., and Roca Cortés, T.: 1994, *Astrophys. J.* **434**, 384.
Régulo, C., Roca Cortés, T., Boumier, P., García, R. A., Robillot, J. M., Turck-Chièze, S., Ulrich, R. K. and the GOLF Team:1997 in F. X. Schmider and J. Provost J. (eds), 'Sounding Solar and Stellar Interiors' *IAU Symp.* **181**, in press.
Topping, J.: 1972, *Errors of observation and Their Treatment*, Chapman & Hall Ltd., London.
Toutain, T. and Appourchaux, T.: 1994, *Astron. Astrophys.* **289**, 649.
Toutain, T. and Fröhlich, C.: 1992, *Astron. Astrophys.* **257**, 287.
Toutain, T. and Kosovichev, A.: 1994, *Astron. Astrophys.* **284**, 265.
Toutain, T., Appourchaux, T., Baudin, T., Fröhlich, C., Gabriel, A. H., Scherrer, P., Andersen, B. N., Bogart, R., Bush, R., Finisterle, W., García, R. A., Grec, G., Henney, C. J., Hoeksema, J. T., Jiménez, A., Kosovichev, A., Roca Cortés, T., Turck-Chièze, S., Ulrich, R. K., and Wehrli, C.: 1997, *Solar Phys.* **175**, 311 (this issue).
Turck-Chièze, S., Basu, S., Brun, S., Christensen-Dalsgaard, J., Eff-Darwich, A., Lopes, I., Perez Hernandez, F., Berthomieu, G., Provost, J., Ulrich, R. K., Baudin, F., Boumier, P., Charra, J., Gabriel, A. H., García, R. A., Grec, G., Renaud, C., Robillot, J.-M., and Roca Cortés, T.: 1997, *Solar Phys.* **175**, 247 (this issue).
Woodard, M. F. and Noyes, R. W.: 1985, *Nature* **318**, 449.

FIRST VIEW OF THE SOLAR CORE FROM GOLF ACOUSTIC MODES

S. TURCK-CHIÈZE[1], S. BASU[2], A. S. BRUN[1], J. CHRISTENSEN-DALSGAARD[2,3], A. EFF-DARWICH[4], I. LOPES[5], F. PÉREZ HERNÁNDEZ[6], G. BERTHOMIEU[7], J. PROVOST[7], R. K. ULRICH,[8], F. BAUDIN[6,11], P. BOUMIER[9], J. CHARRA[9], A. H. GABRIEL[9], R. A. GARCIA[1], G. GREC[7], C. RENAUD[7], J. M. ROBILLOT[10] and T. ROCA CORTÉS[6]

[1]*Service d'Astrophysique, DSM/DAPNIA, CE Saclay, 91191 Gif-sur-Yvette, France*
[2]*Teoretisk Astrofysik Center, Danmarks Grundforskningsfond, Danmark*
[3]*Institut for Fysik og Astronomi, Aarhus Universitet, DK 8000 Aarhus C, Denmark*
[4]*Harvard-Smithsonian Center for Astrophysics, 60 Garden St., Cambridge, MA 02138, MS16, U.S.A.*
[5]*Institute of Astronomy, Madingley Road, Cambridge CB3 OHA, U.K.*
[6]*Instituto de Astrofisica de Canarias, E-38205, La Laguna, Tenerife, Spain*
[7]*Departement Cassini, OCA, BP 229, 06304 Nice Cedex 4, France*
[8]*UCLA, Department of Physics and Astronomy, Los Angeles, U.S.A.*
[9]*Institut d'Astrophysique Spatiale, Orsay, France*
[10]*Observatoire de l'Université Bordeaux 1, BP 89, 33270 Floirac, France*
[11]*National Solar Observatory, National Optical Astronomy Observatories, PO Box 26732, Tucson AZ 85726 U.S.A.*

(Received 19 March 1997; accepted 16 July 1997)

Abstract. After 8 months of nearly continuous measurements the GOLF instrument, aboard SOHO, has detected acoustic mode frequencies of more than 100 modes, extending from 1.4 mHz to 4.9 mHz. In this paper, we compare these results with the best available predictions coming from solar models. To verify the quality of the data, we examine the asymptotic seismic parameters; this confirms the improvements achieved in solar models during the last decade. Using the GOLF set of frequencies for $l = 0, 1, 2, 3$ combined with the LOWL second year data set for $l > 3$, we then carry out inversions to infer properties of the solar core. This largely confirms the previous results down to around 0.1 $R_{\odot}$, while there remain differences, even closer to the centre, where the present study shows an extreme sensitivity of the inversion results to the values of the frequencies. We finally consider physical processes which may influence directly or indirectly the solar core structure.

1. Introduction

Helioseismology has emerged as an impressive tool to investigate the internal properties of the Sun. By placing stringent constraints on solar structure, it has enabled the use of the Sun as a physics laboratory, forcing substantial progress on the equation of state, the opacities and the gravitational settling of elements (for reviews see, for example, Turck-Chièze *et al.*, 1993; Berthomieu, 1996; Christensen-Dalsgaard, 1996). Nevertheless, constraints on the solar core are elusive, due to the small number of modes penetrating this region and the limited time they spend there (typically less than 10% of the travelling time for the radial modes). There have been important recent advances due to the global oscillations networks BiSON (Chaplin *et al.*, 1996a, b) and IRIS (Gelly *et al.*, 1997) observing modes of low

Solar Physics **175:** 247–265, 1997.

degree, the impressive LOWL instrument (Tomczyk *et al.*, 1995) which covers the range of degrees l from 0 to 99 and the recent GONG network which is providing frequencies of modes from $l = 0$ to 250 (Hill *et al.*, 1996). With space observations, one can expect substantially improved constraints, by enlarging the range of modes accessible in a shorter period of observation and by improving the accuracy of the frequency determination. Initial results have been already obtained from VIRGO (Fröhlich *et al.*, 1997) and MDI (Kosovichev *et al.*, 1997).

The stability of the SOHO spacecraft and orbit, together with the very high counting rate of GOLF (typically $8-12 \times 10^6$ counts s^{-1}) leads to a very low statistical noise at high frequencies. This can be further reduced by exploiting the incoherent signal between the detectors located on the two sides of the sodium cell or between the signals obtained every 10 s at two close positions in the blue wing of the sodium lines. The final result is a gain by a factor 5 to 10 in the level of noise at high frequencies, compared with existing ground-based networks. At lower frequencies, the p-mode spectrum benefits from the absence of atmospheric perturbations and the duty cycle of nearly 100% (Gabriel *et al.*, 1997). We note also that the present set of GOLF data represents the first time that such a complete list of frequencies of low-degree modes has been obtained in such a short period of time (8 months compared with 3 years for the low frequencies). This is a very important aspect when determining the variation of the frequencies with the solar cycle (Libbrecht and Woodard, 1990). Even though this is small at low frequencies, it will be important to try to measure it.

A preliminary analysis of early GOLF data (Lazrek *et al.*, 1997a; Grec *et al.*, 1997) was presented by Turck-Chièze *et al.* (1997). Here we carry out a more detailed analysis of GOLF frequencies obtained over an 8-months period, to test the quality of the data and provide initial results on the structure of the solar interior. We first compare the GOLF frequencies (Lazrek *et al.*, 1997b) with model frequencies as well as with the classical helioseismic variables deduced from asymptotic approximations. In the second part we carry out an inversion of the GOLF frequencies together with complementary LOWL data, obtained during a reasonably close period of time, to derive the sound-speed profile in the solar core. In the final part, we discuss the physical processes which may be constrained by the currently available observed frequencies.

2. Direct Frequency Comparisons

The GOLF experiment has identified more than one hundred acoustic modes which penetrate the region of solar energy production ($r < 0.3\ R_{\odot}$). A list of the resulting frequencies is given by Lazrek *et al.* (1997b). It contains results for modes of degree $l = 0$ (radial orders $n = 9$–35), $l = 1$ ($n = 9$–34), $l = 2$ ($n = 9$–33), $l = 3$ ($n = 9$–33), $l = 4$ ($n = 13$–21), and $l = 5$ ($n = 17$–20) for the period from 11 April to the end of December 1996.

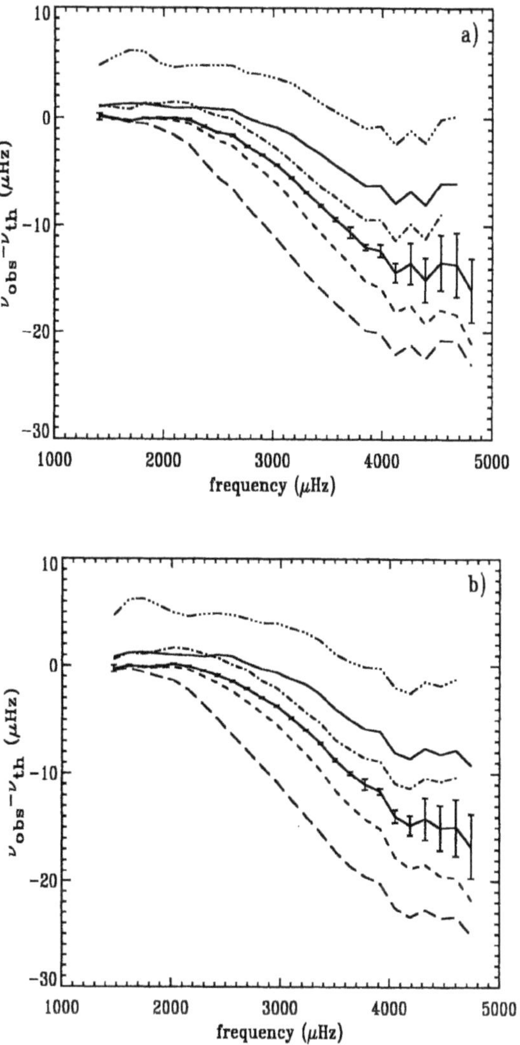

Figure 1. Frequency differences between GOLF observations and different solar models for radial modes (a) and modes with $l = 1$ (b), in the sense (Sun−theory): — · · · —: Turck-Chièze and Lopes (1993); full line: Gabriel and Carlier (1997); — · —: Guzik *et al.* (1996); full and experimental error bars: Christensen-Dalsgaard *et al.* (1996); — — —: Morel, Provost and Berthomieu (1997); long — — —: Brun *et al.* (1997).

We first compare the acoustic mode frequencies, for $l = 0$ and 1, obtained with GOLF with those deduced from different solar models. Figure 1 shows the resulting frequency differences; for clarity, GOLF error bars are superimposed for one model only. Three clear general trends appear:

(1) the agreement is good at low orders ($n < 15$) (frequency below 2.2 mHz) for the models that include updated physics (Livermore equation of state and opacities) and gravitational settling of elements (Christensen-Dalsgaard *et al.*, 1996; Guzik, Cox, and Swenson, 1996; Gabriel and Carlier, 1997; Brun *et al.*, 1997; Morel, Provost, and Berthomieu, 1997), in comparison with models where these effects where not yet introduced (Turck-Chièze and Lopes, 1993). In this last case, the difference is approximatively of 6 μHz from the present solar frequencies.

We note that the low-order modes can be computed with some confidence, because their external reflection is located sufficiently deep below the photosphere, in a region where the physical processes are reasonably well understood.

(2) At higher frequencies (the order $n = 15$–30 or ν between 2.2 and 4.2 mHz), the models show deviations from the experimental values which depend on the details of the treatment of the external layers. Effectively, the frequency dependence of those modes is mainly correlated to the variation of the outer turning point as determined by the cut-off frequency. Present studies show that a number of phenomena must be correctly treated in this photospheric region, about 200 km deep: the effect of opacity, the equation of state, turbulent pressure, improved treatment of convection and non-adiabatic correction (Kosovichev, 1995; Henney and Ulrich, 1995; Guzik, Cox, and Swenson, 1996; Rosenthal *et al.*, 1995) in order to derive precise reflection of the modes in these outer layers. One may notice that the models for which the slope is reduced are those where the structure equations are solved above the solar effective radius without introducing a specific atmosphere (Gabriel and Carlier (1997); Turck-Chièze and Lopes (1993); see also Brun *et al.* (1997) for a discussion on this point).

(3) Finally there is a tendency for the differences to flatten off at the highest frequencies (above 4 mHz). This has to be confirmed, however, owing to the difficulty in measuring such modes as a result of their small lifetime (several hours). Consequently one observes very broad profiles, partly mixed with neighbouring ones, with a shape complicated by the effects of the stochastic excitation, although the very low instrumental noise of GOLF allows the modes to be seen clearly. This difficulty has been discussed in recent analysis of the IRIS data (Gelly *et al.*, 1997).

3. Seismic Indicators of the Internal Solar Structure

The direct comparison with models of the mode frequencies gives little information on the solar internal structure, and mainly encourages improvements of solar modelling for the outer layers. To obtain further information about solar structure, and to test the consistency of the data, asymptotic descriptions have been found very useful (e.g., Deubner and Gough, 1984; Gough, 1993; Lopes and Turck-Chièze,

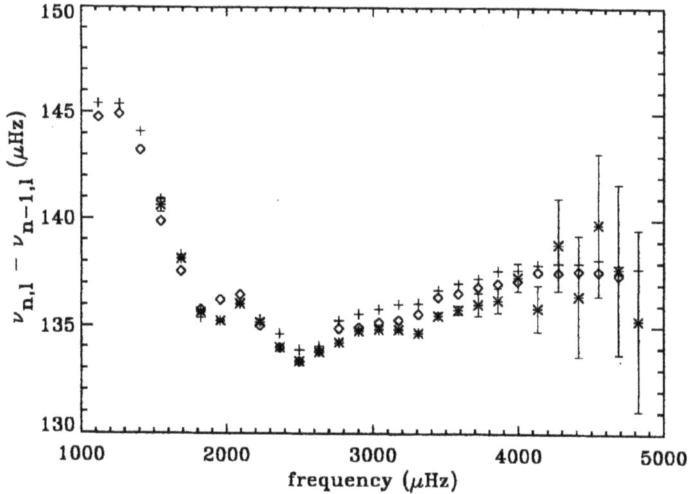

Figure 2. Large separation for $l = 0$. GOLF observations are represented by asterisks with error bars and solar models by crosses for Christensen-Dalsgaard *et al.* (1996) and by diamonds for Turck-Chièze and Lopes (1993).

1994). For low-degree modes the following simple expression for the frequencies may be obtained (Tassoul, 1980; see also Gough, 1986):

$$\nu_{n,l} = \nu_0(n + l/2 + \alpha) - A\frac{l(l+1)+\delta}{n + l/2 + \alpha}, \tag{1}$$

with

$$\nu_0 = \left(2\int_0^{R_\odot} \frac{dr}{c}\right)^{-1}, \quad A = \frac{1}{4\pi^2}\left(\frac{c(R_\odot)}{R_\odot} - \int_0^{R_\odot} \frac{dc}{dr}\frac{dr}{r}\right), \tag{2}$$

α depending on the reflecting properties of the surface and δ on the internal properties.

This simple formulation has the advantage of emphasizing the importance of the sound speed and is extremely useful for the identification of modes. It predicts that the frequencies can be characterized by two variables, the large separation

$$\Delta\nu_{n,l} = \nu_{n,l} - \nu_{n-1,l}, \tag{3}$$

according to which modes with the same degree and consecutive radial order are almost equally spaced, with a separation close to 135 μHz, and the small separation

$$\delta\nu_{n,l} = \nu_{n+1,l} - \nu_{n,l+2}, \tag{4}$$

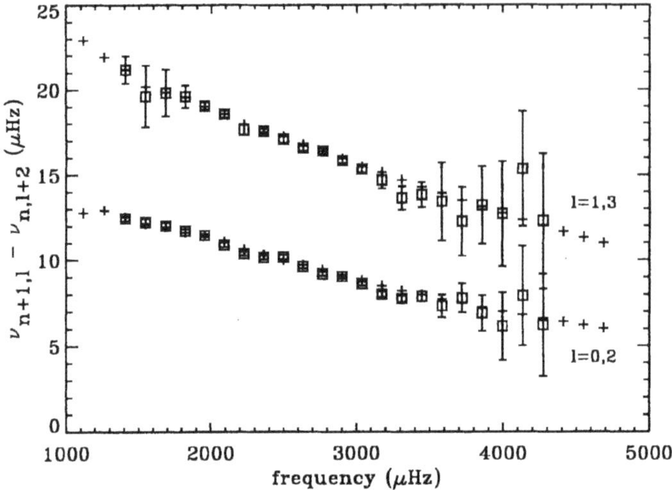

Figure 3. Small separation between frequencies of modes with degree $l = 0, 2$ and $l = 1, 3$. GOLF observations are represented by squares with error bars and the theoretical predictions from the solar model of Christensen-Dalsgaard *et al.* (1996) by crosses.

which depends on the second term of the expression (1) and consequently constrains the derivative of the sound speed near the core. Figures 2 and 3 compare these variables with values obtained from frequencies of solar models.

Unlike earlier models neglecting these effects, the present solar models including microscopic settling and diffusion are clearly in very good agreement with the observed values of the small separation. Indeed, with the present level of accuracy it is necessary to move beyond the small separation to obtain further information on the structure of the solar core; as discussed in Section 4 below, such subtle details require the use of inversion techniques.

The general trend of the large separation is the same for the models and for the Sun, and clearly indicates that Equation (1) is too simple. However, the behaviour can be understood from a more complete asymptotic description where the effects of the outer layers of the Sun are explicitly taken into account (Lopes and Turck-Chièze, 1994). In particular, the effect of the partial ionization of helium and hydrogen appears at relatively low frequencies; these effects are responsible for the oscillations in $\Delta\nu$ clearly visible in Figure 2. In this region, one may notice the good agreement between GOLF data and models (such as the model of Christensen-Dalsgaard *et al.*, 1996) which include the Livermore equation of state (Rogers, Swenson, and Iglesias, 1996). At higher frequency, above 2.5 mHz, one may distinguish the influence of the very external layers; not surprisingly, the agreement with the data is better for models for which the direct frequency differences in Figure 1 are smaller (such as for the model of Turck-Chièze and Lopes, 1993).

The region of partial ionization of the light elements can be investigated also in terms of the derivative of the phase shift of the global acoustic oscillations (Vorontsov, 1991). The resulting variable β is generally used to describe the outer layers from high-degree modes but a generalized formalism has been developed which allows the extraction of similar information from low-degree modes (Lopes and Turck-Chièze, 1994). In practice, one can obtain a simple estimate of this variable from a list of frequencies, by determining the partial derivatives of the frequency with respect to the degree and order of the mode (see Brodsky and Vorontsov, 1988; Pamyatnykh, Vorontsov, and Däppen, 1991):

$$\beta(\nu, l) = \frac{\omega - n(\mathrm{d}\omega/\mathrm{d}n) - L(\mathrm{d}\omega/\mathrm{d}L)}{\mathrm{d}\omega/\mathrm{d}n}\,, \tag{5}$$

with $\omega = 2\pi\nu$ and $L = l + \frac{1}{2}$. The computation of these derivatives is delicate in the case of acoustic modes of low degree, due to the small number of frequencies. Therefore, an algorithm has been developed that allows the derivatives to be estimated with the largest number of frequencies available, typically 5 points.

As shown in Figure 4 the frequencies obtained with GOLF yield a fairly precise estimate of $\beta(\nu, l)$, which is in good agreement with the solar model. One sees, even more clearly than in Figure 2, the oscillation related to the variation of the adiabatic exponent, which is nicely reproduced by the Livermore equation of state in the model of Christensen-Dalsgaard *et al.* (1996). Once more, the region above 3 mHz, sensitive to the description of the atmospheric layers and the treatment of the superadiabatic region, needs to be improved, but is better reproduced by the model of Turck-Chièze and Lopes (1993).

We note that $\beta(\nu, l)$ would be particularly useful for asteroseismological investigations of the external properties of stars, for which only the low-degree modes can be observed (Lopes *et al.*, 1997).

The analysis in this section of the asymptotic properties of the observed frequencies demonstrates the quality of the measured data. Comparisons with corresponding quantities computed from model frequencies show that the progress in solar modelling has been such that little further information about solar structure can be extracted from the simple asymptotic quantities; the main exception is the description of the external layers which is not yet totally reliable. Hence, to fulfil the principal goal of GOLF, i.e., the study of the detailed structure of the solar core, we must move beyond the simple description and carry out inverse analyses of the frequencies.

4. Determination of the Internal Sound Speed

The first inversion of the GOLF data was carried out by combining the preliminary GOLF data set corresponding to a period of about 4 months beginning on April 11th 1996 (Lazrek *et al.*, 1997a; Grec *et al.*, 1997) with the LOWL data for the

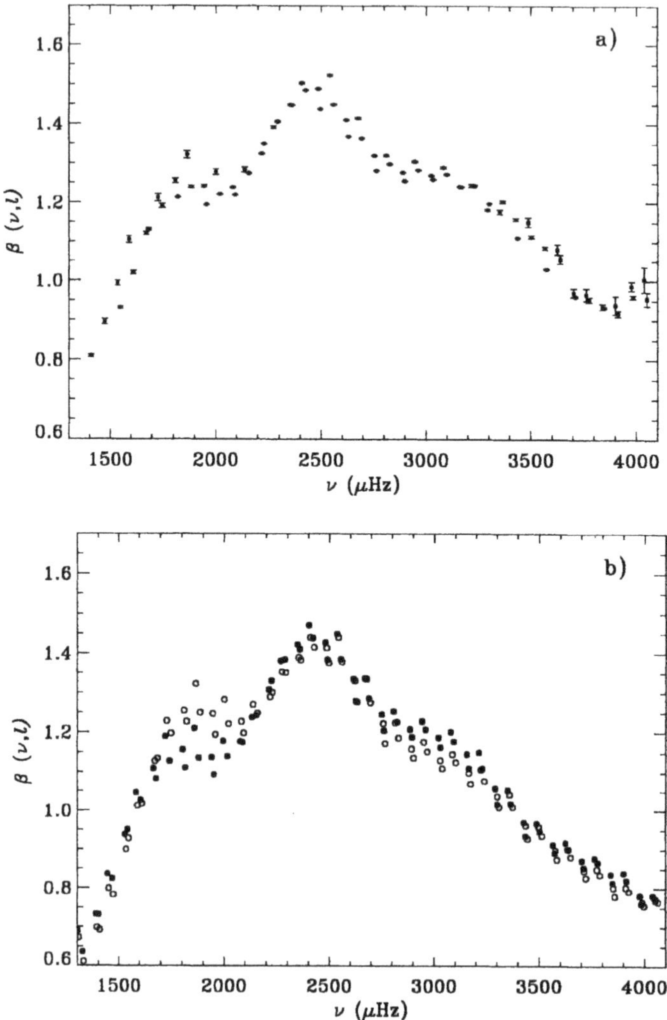

Figure 4. Frequency dependence of $\beta(\nu, l)$, calculated using modes with degrees smaller than 4. (a) Results based on GOLF data. (b) Results for the model of Christensen-Dalsgaard *et al.* (1996; white dots), and for the model of Turck-Chièze and Lopes (1993; black dots).

period of February 1995 to February 1996 (Tomczyk *et al.*, 1995). The results of this inversion have been published by Turck-Chièze *et al.* (1997) and Gabriel *et al.* (1997), and are in good agreement with the results obtained by the combination of BiSON and LOWL (Basu *et al.*, 1997).

Although our main goal is to probe the structure of the core, the inversion requires that modes of higher degree be included, in order to account correctly for contributions from the outer parts of the Sun to the observed frequencies. Thus

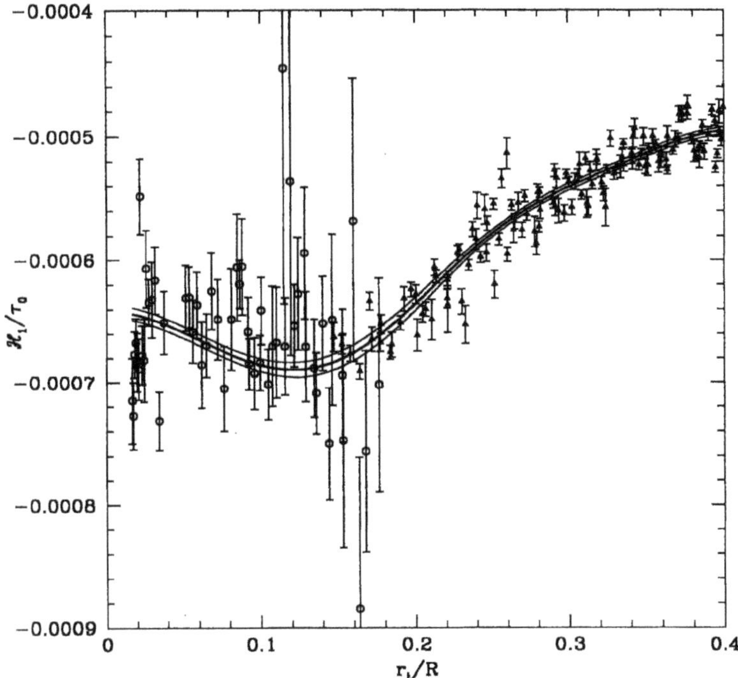

Figure 5. The variable $\mathcal{H}_l$ is obtained from a fit to the asymptotic expression (6) (solid line, with 1-σ errors indicated by the thin lines). GOLF data correspond to open circles, LOWL to triangles. The results have been normalized by the acoustical radius τ_0 of the Sun, such as to correspond approximately to relative frequency differences.

we use the GOLF frequencies for $l = 0$–3 (Lazrek *et al.*, 1997b) and the LOWL frequencies for $l = 4$–99; in the latter case the data were restricted to modes with frequency $\nu \leq 3500\mu$Hz. Any inversion for solar structure must correct for the contributions to the frequency differences from the near-surface region (see also Figure 1); thus modes in the required frequency range must be available for a substantial range in degree. For this reason the inversion had to be restricted to modes below 3500 μHz, where LOWL frequencies were available, even though the GOLF data extend to higher frequencies.

Here we first compare the GOLF data with the LOWL set to check the consistency and the scatter of the data. Afterwards we consider the impact of the GOLF data on the inversion for the solar sound speed.

The overlap between modes in the GOLF and LOWL sets is relatively modest; thus it is difficult to ascertain the consistency between the two sets from a direct comparison of their frequencies. Instead, we base the test on the asymptotic properties of the differences between the solar and model frequencies. From the general asymptotic description of solar acoustic modes (e.g., Gough, 1993) it follows that

p-mode frequency differences between a model and the Sun can be expressed as the sum of three functions:

$$S_{n,l}\frac{\delta\omega_{n,l}}{\omega_{n,l}} \simeq \mathcal{H}_1(\omega_{n,l}/L) + \mathcal{H}_2(\omega_{n,l}) + \frac{\mathcal{H}_3(\omega_{n,l}/L)}{\omega_{n,l}^2} \,. \tag{6}$$

Here the scale factor $S_{n,l}$ is related to the mode inertia. The functions $\mathcal{H}_1$, $\mathcal{H}_2$, and $\mathcal{H}_3$ can be obtained from a weighted least-squares fit to the data. Christensen-Dalsgaard, Gough, and Pérez Hernández (1988) and Christensen-Dalsgaard, Gough, and Thompson (1989) considered only $\mathcal{H}_1$ and $\mathcal{H}_2$. However, the inclusion of the function $\mathcal{H}_3$ for low-degree modes improves significantly the asymptotic fit and the corresponding inversion for the sound-speed differences in the core.

In Figure 5, the solid line shows $\mathcal{H}_1$ obtained by fitting the combined LOWL and GOLF frequencies to Equation (6), as a function of the turning point r_t, defined by $r_t/c(r_t) = L/\omega$. Model S of Christensen-Dalsgaard *et al.* (1996) has been used as the reference. The thin lines correspond to the errors in this quantity derived from a Monte Carlo simulation. Also shown are the differences

$$S_{nl}\delta\omega_{nl}/\omega_{nl} - \mathcal{H}_2(\omega) - \mathcal{H}_3(\omega/L)/\omega^2 \,, \tag{7}$$

with the corresponding observational errors; according to Equation (6) these should correspond asymptotically to $\mathcal{H}_1$.

It is evident from Figure 5 that there is indeed a smooth transition between the overall trends of the GOLF and LOWL frequencies, at the observational precision. In particular, we note that replacing the $l = 3$ modes by those from the LOWL set does not introduce any significant change. The data points in Figure 5 show scatter around the smoothly fitted $\mathcal{H}_1$ of a magnitude exceeding the estimated observational error. When plotted as a function of frequency the residuals for both the GOLF and LOWL data show an oscillatory pattern which probably arise from sharp features in the sound speed at the base of the convection zone (e.g., Monteiro, Christensen-Dalsgaard, and Thompson 1994; Basu and Antia, 1994; Basu, 1997). This behaviour appears also to be consistent between the two datasets.

This analysis indicates that the two datasets are largely consistent and hence justifies performing the inversion with the combined set. This favourable situation is probably due to two factors: the two instruments measure Doppler velocities, thus avoiding the apparent frequency differences between velocity and intensity measurements (Toutain *et al.*, 1997); and the two measurements are very close in time around solar minimum, where the outer layers are reasonably stable.

After this preliminary analysis, a complete inversion has been performed. The inversion was carried out by means of the Subtractive Optimally Localized Averages Technique of Pijpers and Thompson (1992). The frequency differences were expressed in terms of the differences $\delta c^2/c^2$ and $\delta\rho/\rho$ in squared sound speed and density, and the inversion was carried out to infer $\delta c^2/c^2$. Details of the implementation were provided by Basu *et al.* (1996a). Model S of Christensen-Dalsgaard *et al.* (1996) was also used as the reference model.

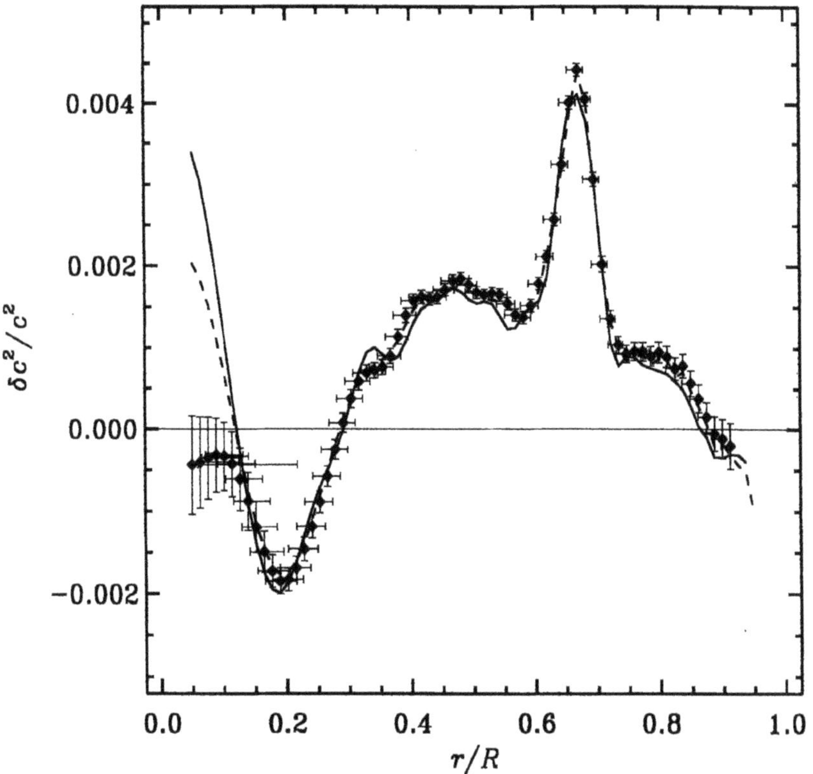

Figure 6. Difference between the square of the sound speed in the Sun and in Model S of Christensen-Dalsgaard (1996), in the sense (Sun−model). The vertical bars indicate $1-\sigma$ errors in the inferred differences, while horizontal bars provide a measure of the resolution of the inversion. For comparison, the solid line shows the results obtained by Basu *et al.* (1997) from analysis of a combination of BiSON and LOWL data, while the dashed line shows the results obtained with just 4 months of GOLF data, combined with LOWL data (Turck-Chièze *et al.*, 1997).

Figure 6 shows the result of the inversion, using the table of Lazrek *et al.* (1997b). In the figure are superimposed the previous determination of the sound-speed difference obtained with 4 months of data (Turck-Chièze *et al.*, 1997), as well as results obtained by using BiSON data in combination with LOWL (Basu *et al.*, 1997). The results are very nearly the same except below 0.1 $R_{\odot}$ where only modes with $l = 0$ and 1 may penetrate.

A critical analysis of the data confirms that the solution below 0.1 $R_{\odot}$ is extremely sensitive, even to changes involving just a few modes.

To investigate the origin of the substantial difference in the present inversion results and those of Basu *et al.* (1997), Figure 7 shows differences between the GOLF frequencies and those used by Basu *et al.* One observes that for a few modes with $l = 1$ the GOLF frequencies are systematically lower than those from BiSON;

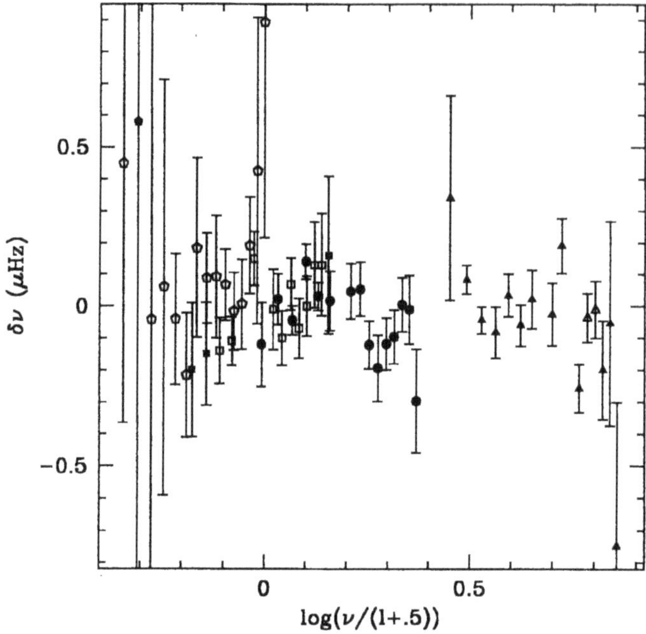

Figure 7. Differences between the present GOLF frequencies and the so-called 'Best Set' of Basu *et al.* (1997) resulting from a combination of BiSON and LOWL data, in the sense (GOLF−other). Filled symbols show GOLF − LOWL and open symbols are GOLF − BiSON. Triangles show modes with $l = 0$, circles are for $l = 1$, squares for $l = 2$ and pentagons for $l = 3$.

if these modes are removed, the inversions of the two sets are essentially consistent. In contrast, a similar comparison of the previous set of GOLF frequencies (Lazrek *et al.*. 1997a) and the BiSON frequencies shows little overall difference for these modes. We note that as the differences between the different datasets are small (0.1 −0.2 μHz) and concern few modes, definite conclusions about the structure of the core evidently require more data and further investigations of the accuracy of the frequency determinations.

We note also that to confirm the insignificance of the potential inconsistency caused by the GOLF and LOWL data being obtained at different times, we have performed an inversion combining the GOLF frequencies with $l > 3$ MDI values, taken during the same period of time (Kosovichev *et al.*, 1997); the results of this inversion are essentially the same as those shown here for the GOLF and LOWL combination.

It is evident from the figure that the sound-speed difference in the very central part is still somewhat uncertain and that possible sources of systematic error in the inversion must be investigated. It seems that the differences between the different solutions may well reflect the difficulty in determining properly the frequencies to the level of accuracy that we need. It is clear that with a longer series and the use of the total range of frequencies, one may improve the situation. Even so, it is remarkable that we are able to obtain measurements of conditions in the inner core of the Sun with a precision of a few parts in a thousand.

5. Discussion of Physical Processes

With only 8 months of data, we are able to confirm using GOLF the profile of the sound speed obtained with the best ground-based systems, at least down to 0.1 $R_\odot$. It appears clear that the GOLF data will contribute to the determination of this profile in considerable detail. Through helioseismology, we are now on the threshold of new investigations of the physics of the solar core. This is extremely important in order to test for possible astrophysical solutions to the neutrino puzzle. In this section, we compare the solar sound speed with theoretical values for different solar models, in order to investigate the sensitivity to some of the ingredients of solar modelling. In all these models, the microscopic diffusion is included.

Some aspects of solar modelling are illustrated in Figure 8. The comparison of the Sun with three 'standard' solar models (Christensen-Dalsgaard *et al.*, 1996; Morel, Provost, and Berthomieu, 1997; Brun *et al.*, 1997) shows that slight differences in the physics used in the calculations (nuclear reaction rates or opacity coefficients or the effect of microscopic diffusion), not surprisingly, influence the sound-speed profile at the level of accuracy reached with present data.

However, all the differences presented in Figure 8, corresponding to comparisons with standard solar models, share a pronounced peak in $\delta c^2/c^2$ at about 0.67 $R_\odot$. This peak is observed using the LOWL data (Basu *et al.*, 1996b) and is also present when GONG data (Gough *et al.*, 1996) or MDI data (Kosovishev *et al.*, 1997) are used. So, it is clearly of interest to investigate what might cause that and, in particular, whether eliminating it may have an effect on the structure of the solar core.

We have checked that the peak is not the signature of a variation of the adiabatic exponent in the region of partial ionization of the CNO elements, since their abundances are too small. Figure 9 illustrates two modifications applied to the model of Brun *et al.* (1997) to test the consequence for the sound-speed profile. The peak appears to be too broad to be suppressed by just the introduction of convective penetration according to Zahn (1991). It is possible that it could be a signature that the microscopic diffusion is not handled entirely correctly, in particular in the transition zone between the transport by radiation and convection (Turck-Chièze and Brun, 1997), with significant effects on the opacity. At the base of the convective transition, oxygen plays a crucial role in determining the opacity; according to Figure 32 of Turck-Chièze *et al.* (1993) the peak begins at the position where the opacity increases due to bound-bound processes involving this element. Hence the peak could indicate that the composition of CNO in the model (including the effects of diffusion) is not correct in this region of the Sun.

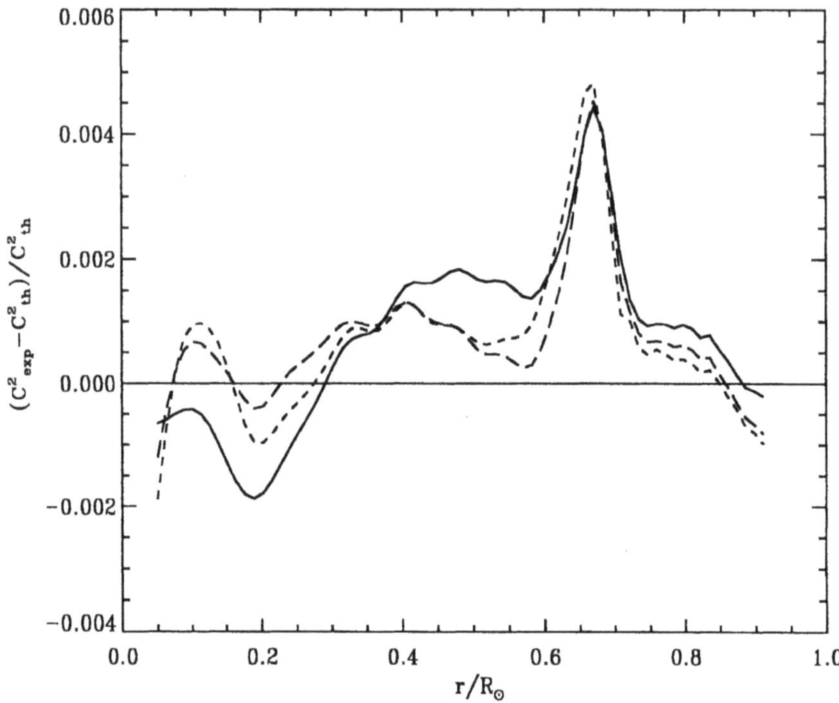

Figure 8. Difference between the square of the solar sound speed and several solar models: continuous line for the model of Christensen-Dalsgaard *et al.* (1996); — — — for Brun *et al.* (1997); long —— for Morel, Provost and Berthomieu (1997).

Of course the opacity contributions from heavy elements (iron, neon, oxygen) play a crucial role in the whole intermediate region (Courtaud *et al.*, 1990). The figure shows that an opacity modification of the order of 5% may be sufficient to remove this peak. We cannot exclude that differences in the sound-speed profile in other parts of the radiative interior might be caused by errors in the opacity: indeed, Tripathy, Basu and Christensen-Dalsgaard (1997) showed that the sound-speed difference can largely be accounted for by a fairly modest modification of the opacity. This illustrates how the present profile of the sound speed may constrain the physics of the solar interior. It is interesting to note that the crude correction to opacity illustrated in Figure 9 does not affect at all the structure of the core. Of course, effects of rotation, particularly mixing induced by instabilities related to the solar spin-down (e.g., Chaboyer, Demarque, and Pinsonneault, 1995) may also contribute to the observed sound-speed differences.

One important ingredient of the neutrino predictions is of course the description of the nuclear reaction rates which produce these neutrinos. It has been shown that the predicted neutrino fluxes still have a substantial sensitivity to this part of the calculation, due to the difficulty in determining cross sections with better than 10%

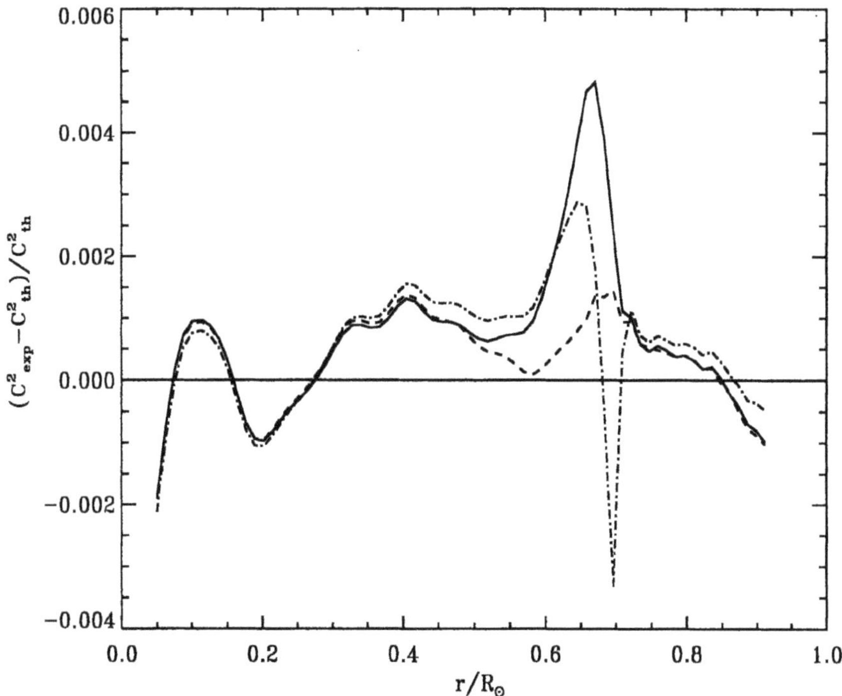

Figure 9. Difference between the square of the solar sound speed and several solar models: continuous line for the model of Brun *et al.* (1997); — · — · — for a model of Brun *et al.* including a large effect of penetrative convection; — — — for a model of Brun *et al.* including a variation of opacity of +5% for temperature between 2.4 and 2.8 × 10^6 K.

accuracy (Dzitko *et al.*, 1995). We compare in Figure 10 the solar sound-speed difference inferred from the GOLF data (Figure 6) with the sound-speed changes obtained in models where we have modified the pp reaction rate by +5% or the (^{3}He,^{4}He) by −30% in order to simulate what we have previously called 'the minimum nuclear solar model': a model, compatible with the present knowledge of the nuclear reaction rates, which minimizes the neutrino fluxes (Dzitko *et al.*, 1995; Turck-Chièze, 1996). In the case of the (^{3}He,^{4}He) reaction rate, we have used a change of three times the uncertainty to amplify the effects on the sound speed, which are very small except very near to the centre.

The effect on the sound speed is clearly visible for a 5% variation of the pp reaction rate, even though such modification of the nuclear core affects the value of the frequencies by no more than 0.5 μHz in absolute values (Brun *et al.*, 1997). However, any reasonable modification of the other reaction rates of the pp chain will be difficult to detect, due to their small effect on the total energy generation and hence the luminosity.

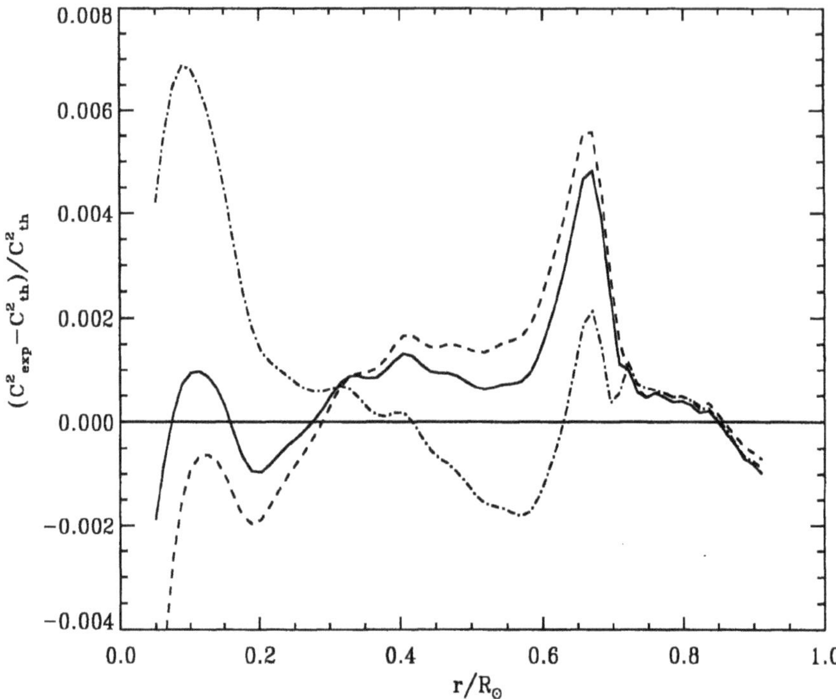

Figure 10. Sound-speed difference between the Sun (GOLF + LOWL acoustic mode frequencies) and different solar models: continuous line for the reference model of Brun *et al.* (1997), — · — · — idem with pp reaction rate modified by +5%; — — — idem with the reaction rate (^{3}He, ^{4}He) reduced by 30%.

It is visible here that this kind of modification affects not only the nuclear core but also the region between 0.2 and 0.7 $R_\odot$.

6. Conclusions

First results of the GOLF space observations give today a detailed profile of the sound speed in the inner core of the Sun. We have tried to demonstrate in detail the direct impact of the present accuracy of the acoustic frequencies measurement on what we learn on the solar structure. It seems that the main limitation today on this precision is no longer the instrumental noise, and not really the solar 'noise', as far as the acoustic modes are concerned, but is due to the stochastic character of the excitation, which the Fourier transform analysis techniques are unable to treat completely.

The sound-speed profile obtained here is not in contradiction with the previous results from ground-based instruments. The available range of frequencies is enlarged, even though in this first study the combination of GOLF data with the ground-based LOWL data has not been able to utilize this improvement. We can hope in the near future to examine the solar core with a far greater precision, when the two complementary instruments GOLF and MDI are ready to carry out a combined very precise inversion using the whole range of frequencies available, obtained over exactly the same period of time.

We are now carrying out extensive studies to investigate the role in the analysis of the errors in the frequencies, the inversion procedures and the interpretation of the resultant sound-speed profile in terms of physical processes. Such investigations will surely extend the classical framework of solar modelling.

Acknowledgements

The performance of the GOLF instrument is due to a team of scientists, engineers and technicians from a consortium of institutes, who worked for many years to achieve this end (see Gabriel *et al.*, 1995). The present authors would like to express their warm acknowledgements to their colleagues, engineers and technicians for their dedication and motivation in the preparation and construction of the instrument. We would like to thank especially N. Pétrou who was responsible for the ground software used for real time commanding and operational data evaluation since the beginning of data acquisition. We are also grateful to S. Tomczyk and his collaborators for giving access and use of the LOWL data. We thank P. Morel for providing and updating the CESAM stellar evolution code and for very useful discussions. We are finally grateful to the anonymous referee for extensive comments on an earlier version of the paper which, we hope, have helped to improve substantially the presentation. The work of S. Basu and J. Christensen-Dalsgaard was supported in part by the Danish National Research Foundation through its establishment of the Theoretical Astrophysics Center. Ilidio Lopes has been supported by a grant from the PPARC of the U.K. to D. O. Gough.

References

Basu, S.: 1997, *Monthly Notices Roy. Astron. Soc.*, in press.

Basu, S. and Antia, H. M.: 1994, *Monthly Notices Roy. Astron. Soc.* **269**, 1137.

Basu, S., Christensen-Dalsgaard, J., Pérez Hernández, F., and Thompson, M. J.: 1996a, *Monthly Notices Roy. Astron. Soc.* **280**, 651.

Basu, S., Christensen-Dalsgaard, J., Schou, J., Thompson, M. J., and Tomczyk, S.: 1996b, *Bull. Astron. Soc. India* **24**, 147.

Basu, S., Chaplin, W. J., Christensen-Dalsgaard, J., Elsworth, Y., Isaak, G. R., New, R., Schou, J., Thompson, M. J., and Tomczyk, S.: 1997, *Monthly Notices Roy. Astron. Soc.*, submitted.

Berthomieu, G.: 1996, in A. Noels, M. Gabriel, N. Grevesse, and P. Demarque (eds), *Stellar Evolution: What Should be Done?*, Liège, p. 263.

Brodsky, M. A. and Vorontsov, S. V.: 1988, in J. Christensen-Dalsgaard and S. Frandsen (eds), 'Advances in Helio- and Asteroseismology', *Proc. IAU Symp.* **123**, 137.

Brun, A. S., Lopes, I., Morel, P., and Turck-Chièze, S.: 1997, in F.-X. Schmider and J. Provost (eds), 'Sounding Solar and Stellar Interiors', *Poster Volume, IAU Symp.* **181**, in press.
Chaboyer, B., Demarque, P., and Pinsonneault, M. H.: 1995, *Astrophys. J.* **441**, 865.
Chaplin, W. J., Elsworth, Y., Howe, R., Isaak, G. R., McLeod, C. P., Miller, B. A., and New, R.: 1996a, *Solar Phys.* **168**, 1.
Chaplin, W. J., Elsworth, Y., Howe, R., Isaak, G. R., McLeod, C. P., Miller, B. A., and New, R.: 1996b, *Monthly. Notices Roy. Astron. Soc.* **282**, L15.
Christensen-Dalsgaard, J.: 1996, in T. Roca Cortés and F. Sanchez (eds), *The Structure of the Sun*, Cambridge University Press, Cambridge, p. 49.
Christensen-Dalsgaard, J., Gough, D. O., and Pérez Hernández, F.: 1988, *Monthly Notices Roy. Astron. Soc.* **235**, 875.
Christensen-Dalsgaard, J., Gough, D. O., and Thompson, M. J.: 1989, *Monthly Notices Roy. Astron. Soc.* **238**, 481.
Christensen-Dalsgaard, J., Däppen, W., Ajukov, S. V., Anderson, E. R., Antia, H. M., Basu, S., Baturin, V. A., Berthomieu, G., Chaboyer, B., Chitre, S. M., Cox, A. N., Demarque, P., Donatowicz, J., Dziembowski, W. A., Gabriel, M., Gough, D. O., Guenther, D. B., Guzik, J. A., Harvey, J. W., Hill, F., Houdek, G., Iglesias, C. A., Kosovichev, A. G., Leibacher, J. W., Morel, P., Proffitt, C. R., Provost, J., Reiter, J., Rhodes, Jr., E. J., Rogers, F. J., Roxburgh, I. W., Thompson, M. J., and Ulrich, R. K.: 1996, *Science* **272**, 1286.
Courtaud, D., Damamme, G., Genot, E., Vuillemin, M., and Turck-Chièze, S.: 1990, *Solar Phys.* **128**, 49.
Deubner, F.-L. and Gough, D. O.: 1984, *Ann. Rev. Astron. Astrophys.* **22**, 593.
Dzitko, H., Turck-Chièze, S., Delbourgo-Salvador, P., and Lagrange, C.: 1995, *Astrophys. J.* **447**, 428.
Fröhlich, C., Andersen, B. N., Appourchaux, T., Berthomieu, G., Crommelynck, D. A., Domingo, V., Fichot, A., Finsterle, W., Gomez, M. F., Gough, D., Jiménez, A., Leifsen, T., Lombaerts, M., Pap, J. M., Provost, J., Roca Cortés, T., Romero, J., Roth, H., Sekii, T., Telljohann, U., Toutain, T., and Wehrli, C.: 1997, *Solar Phys.* **170**, 1.
Gabriel, M. and Carlier, F.: 1997, *Astron. Astrophys.* **317**, 580.
Gabriel, A. H., Grec, G., Charra, J., Robillot, J.-M., Roca Cortés, T., Turck-Chièze, S., Bocchia, R., Boumier, P., Cantin, M., Céspedes, E., Cougrand, B., Crétolle, J., Damé, L., Decaudin, M., Delache, P., Denis, N., Duc, R., Dzitko, H., Fossat, E., Fourmond, J.-J., García, R. A., Gough, D. O., Grivel, C., Herreros, J. M., Lagardére, H., Moalic, J.-P., Pallé, P. L., Petrou, N., Sánchez, M., Ulrich, R. K., and van der Raay, H. B.: 1995, *Solar Phys.* **162**, 61.
Gabriel, A. H., Charra, J., Grec, G., Robillot, J.-M., Roca Cortés, T., Turck-Chièze, S., Ulrich, R., Basu, S., Baudin, F., Bertello, L., Boumier, P., Charra, M., Christensen-Dalsgaard, J., Decaudin, M., Dzitko, H., Foglizzo, T., Fossat, E., Garcia, R. A., Herreros, J. M., Lazrek, M., Pallé, P. L., Pétrou, N., Renaud, C., and Régulo, C.: 1997, *Solar Phys.* **175**, 207 (this issue).
Gelly, B., Fierry-Fraillon, D., Fossat, E., Pallé, P. L., Cacciani, A., Ehgamberdiev, S., Grec, G., Hoeksema, J. T., Khalikov, S., Lazrek, M., Loudagh, S., Pantel, A., Régulo, C., and Schmider, F. X.: 1997, *Astron. Astrophys.* **323**, 235.
Gough, D. O.: 1986, in Y. Osaki (ed.), *Hydrodynamic and Magnetohydrodynamic Problems in the Sun and Stars*, Department of Astronomy, University of Tokyo, p. 117.
Gough, D. O.: 1993, in J.-P. Zahn and J. Zinn-Justin (eds), *Astrophysical Fluid Dynamics, Les Houches Session XLVII*, Elsevier, Amsterdam, p. 399.
Gough, D. O., Kosovichev, A. G., Toomre, J., Anderson, E. R., Antia, H. M., Basu, S., Chaboyer, B., Chitre, S. M., Christensen-Dalsgaard, J., Dziembowski, W. A., Eff-Darwich, A., Elliott, J. R., Giles, P. M., Goode, P. R., Guzik, J. A., Harvey, J. W., Hill, F., Leibacher, J. W., Monteiro, M. J. P. F. G., Richard, O., Sekii, T., Shibahashi, H., Takata, M., Thompson, M. J., Vauclair, S., and Vorontsov, S. V.: 1996, *Science* **272**, 1296.
Grec, G., Turck-Chièze, S., Lazrek, M., Roca Cortés, T., Bertello, L., Baudin, F., Boumier, P., Charra, J., Fierry-Fraillon, D., Fossat, E., Gabriel, A., Garcia, R. A., Gouiffes, C., Régulo, C., Renaud, C., Robillot, J. M., and Ulrich, R. K.: 1997, in F.-X. Schmider and J. Provost (eds), 'Sounding Solar and Stellar Interiors', *Proc. IAU Symp.* **181**, in press.
Guzik, J., Cox, A. N., and Swenson, F. J.: 1996, *Bull. Astron. Soc. India* **24**, 161.

Henney, C. J. and Ulrich, R. K.: 1995, in J. T. Hoeksema, V. Domingo, B. Fleck, and B. Battrick (eds), *Proc. Fourth SOHO Workshop: Helioseismology*, Vol. 2, ESA SP-376, ESTEC, Noordwijk, p. 3.
Hill, F., Stark, B., Stebbins, R. T., Anderson, E. R., Antia, H. M., Brown, T. M., Duvall, T. L., Haber, D. A., Harvey, J. W., Hathaway, D. H., Howe, R., Hubbard, R. P., Jones, H. P., Kennedy, J. R., Korzennik, S. G., Kosovichev, A. G., Leibacher, J. W., Libbrecht, K. G., Pinter, J. A., Rhodes, E. J., Schou, J., Thompson, M. J., Tomczyk, S., Toner, C. G., Toussaint, R., and Williams, W. E.: 1996, *Science* **272**, 1292.
Kosovichev, A. G.: 1995, in J. T. Hoeksema, V. Domingo, B. Fleck, and B. Battrick (eds), *Proc. Fourth SOHO Workshop: Helioseismology*, Vol. 1, ESA SP-376, ESTEC, Noordwijk, p. 165.
Kosovichev, A. G., Schou, J., Scherrer, P. H., Bogart, R. S., Bush, R. I., Hoeksema, J. T., Aloise, J., Bacon, L., Burnette, A., De Forest, C., Giles, P. M., Basu, S., Christensen-Dalsgaard, J., Däppen, W., Rhodes, E. J., Duvall, T. L., Howe, R., Thompson, M. J., Gough, D. O., Sekii, T., Toomre, J., Tarbell, T. D., Title, A. M., Mathur, D., Morrison, M., Saba, J. L. R., Wolfson, C. J., Zayer, I., and Milford, P. N.: 1997, *Sol. Phys.* **170**, 43.
Lazrek, M. and the GOLF Team: 1997a, in F.-X. Schmider and J. Provost (eds), 'Sounding Solar and Stellar Interiors', *Poster Volume, IAU Symp.* **181**, in press.
Lazrek, M., Baudin, F., Bertello, L., Boumier, P., Charra, J., Fierry-Fraillon, D., Fossat, E., Gabriel, A. H., Garcia, R. A., Gelly, B., Gouiffes, C., Grec, G., Pérez Hernández, F., Régulo, C., Renaud, C., Robillot, J. M., Roca Cortés, T., Turck-Chièze, S., and Ulrich, R. K.: 1997b, *Solar Phys.* **175**, 311 (this issue).
Libbrecht, K. G. and Woodard, M. F.: 1990, *Nature* **345**, 779.
Lopes, I. and Turck-Chièze, S.: 1994, *Astron. Astrophys.* **290**, 845.
Lopes, I., Turck-Chièze, S., Michel, E., and Goupil, M. J.: 1997, *Astrophys. J.* **480**, 794.
Morel, P., Provost, J., and Berthomieu, G.: 1997, *Astron. Astrophys.*, in press.
Monteiro, M. J. P. F. G., Christensen-Dalsgaard, J., and Thompson, M. J.: 1994, *Astron. Astrophys.* **283**, 247.
Pamyatnykh, A. A., Vorontsov, S. V., and Däppen, W.: 1991, *Astron. Astrophys.* **248**, 263.
Pijpers, F. P. and Thompson, M. J.: 1992, *Astron. Astrophys.* **262**, L33.
Rogers, F. J., Swenson, F. J., and Iglesias, C. A.: 1996, *Astrophys. J.* **456**, 902.
Rosenthal, C. S., Christensen-Dalsgaard, J., Houdek, G, Monteiro, M. J. P. F. G., Nordlund, Å., and Trampedach, R.: 1995, in J. T. Hoeksema, V. Domingo, B. Fleck, and B. Battrick (eds), *Proc. Fourth SOHO Workshop: Helioseismology*, ESA SP-376, Vol. 2, ESTEC, Noordwijk, p. 459.
Tassoul, M.: 1980, *Astrophys. J. Suppl.* **43**, 469.
Tomczyk, S. Streander, K., Card, G., Elmore, D., Hull, H., and Cacciani, A.: 1995, *Solar Phys.* **159**, 1.
Toutain, T., Appourchaux, T., Baudin, F., Frölich, C., Gabriel, A., Scherrer, P., Andersen, B. N., Bogart, R., Bush, R., Finsterle, W., Garcia, R. A., Grec, G., Henney, C. J., Hoeksema, J. T., Jimenez, A., Kosovichev, A., Roca Cortés, T., Turck-Chièze, S., Ulrich, R. K., and Wehrli, C.: 1997, *Solar Phys.* **175**, 311 (this issue).
Tripathy, S. C., Basu, S., and Christensen-Dalsgaard, J.: 1997, in F.-X. Schmider and J. Provost (eds), 'Sounding Solar and Stellar Interiors', *Poster Volume, IAU Symp.* **181**, in press.
Turck-Chièze, S.: 1996, in *Neutrinos, Dark matter and the Universe, Conf. Blois*, ed. Frontières, in press.
Turck-Chièze, S. and Brun, A. S.: 1997, in W. Hampel (ed.), *Fourth International Neutrino Conference*, Heidelberg, in press.
Turck-Chièze, S. and Lopes, I.: 1993, *Astrophys. J.* **408**, 347.
Turck-Chièze, S., Däppen, W., Fossat, E., Provost, J., Schatzman, E., and Vignaud, D.: 1993, *Physics Rep.* **230** (2–4), 57.
Turck-Chièze, S., Basu, S., Brun, A. S., Christensen-Dalsgaard, J., Eff-Darwich, A., Gabriel, M., Henney, C. J., Kosovichev, A. G., Lopes, I., Paterno, L., Provost, J., Ulrich, R. K., and the GOLF team: 1997, in F.-X. Schmider and J. Provost (eds), 'Sounding Solar and Stellar Interiors', *Poster Volume, IAU Symp.* **181**, in press.
Vorontsov, S. V.: 1991, *Astron. Zh.* **68**, 808 (English translation: *Soviet Astron.* **35**, 400).
Zahn, J.-P.: 1991, *Astron. Astrophys.* **252**, 179.

IN-FLIGHT PERFORMANCE OF THE VIRGO SOLAR IRRADIANCE INSTRUMENTS ON SOHO

CLAUS FRÖHLICH[1], DOMINIQUE A. CROMMELYNCK[2], CHRISTOPH WEHRLI[1], MARTIN ANKLIN[1], STEVEN DEWITTE[2], ALAIN FICHOT[2], WOLFGANG FINSTERLE[1], ANTONIO JIMÉNEZ[3], ANDRÉ CHEVALIER[2] and HANSJÖRG ROTH[1]

[1]*Physikalisch-Meteorologisches Observatorium Davos, World Radiation Center, CH-7260 Davos Dorf, Switzerland*
[2]*Institut Royal Météorologique de Belgique, B-1180 Bruxelles, Belgium*
[3]*Instituto de Astrofísica de Canarias, Universidad de La Laguna, E-38071 La Laguna, Tenerife, Spain*

(Received 19 March 1997; accepted 12 June 1997)

Abstract. The in-flight performance of the total and spectral irradiance instruments within VIRGO (Variability of solar IRradiance and Gravity Oscillations) on the ESA/NASA Mission SOHO (SOlar and Heliospheric Observatory) is in most aspects better than expected. The behaviour during the first year of operation of the two types of radiometers and the sunphotometers together with a description of their data evaluation procedures is presented.

1. Introduction

The aim of the VIRGO investigation (Variability of solar IRradiance and Gravity Oscillations) on SOHO (SOlar and Heliospheric Observatory) is to determine the characteristics of acoustic and internal gravity oscillations by observing irradiance and radiance variations, to measure the solar total and spectral irradiance and to quantify their variability over periods of days to the duration of the mission. The irradiance part of VIRGO contains two different active-cavity radiometers – DIARAD (DIfferential Absolute RADiometer) and PMO6-V (V indicating the VIRGO version of the PMO6-type radiometers) – for monitoring the solar 'constant', and two three-channel sunphotometers (SPM) for the measurement of the spectral irradiance at 402, 500, and 862 nm with a bandwidth of 5 nm, each.

SOHO was successfully launched on 2 December 1995 and VIRGO was switched on a few days later. The covers were released, but kept closed in order to outgas the experiment thoroughly during the following 7 weeks without having the optical surfaces exposed to the Sun. On Christmas Eve all covers were opened for a short period of time in order to test the performance of the instruments in a 'first light' experiment. Observations started mid-January for the radiometers and end of January 1996 for the sunphotometers.

The instrumentation of VIRGO has been described in detail by Fröhlich *et al.* (1995) and first results have been presented in Fröhlich *et al.* (1997); the observed in-flight performance and operational aspects of the irradiance observations are the subject of this paper.

Solar Physics **175:** 267–286, 1997.

2. Total Solar Irradiance Measurements

Absolute radiometers are based on the measurement of a heat flux by using an electrically calibrated heat flux transducer. The radiation is absorbed in a cavity which ensures a high absorptivity over the spectral range of the Sun. During practical operation of the instrument, an electronic circuit maintains the heat flux constant by controlling the power fed to the cavity heater. This is called the active mode of operation; hence also the name 'active cavity radiometer'. As a first approximation the irradiance can be calculated from the shaded and irradiated electrical powers P_s and P_i according to: $S = C_{rad} \times (P_s - P_i)$, with C_{rad} being the reciprocal of the aperture area times a correction factor for the deviations from ideal behaviour.

Although the designs of both instruments – DIARAD and PMO6-V – are based on the same principle, the physical realization is different (for a general description of the VIRGO radiometers see Fröhlich *et al.* (1995), for details of DIARAD, Crommelynck and Domingo (1984), and for the PMO6, Brusa and Fröhlich (1986)). Major differences between them are the arrangements of the compensating cavities and the forms and coatings of the cavities. Moreover, in the DIARAD exists the possibility of assessing the quality of the electrical measurements in flight.

2.1. Operation

The left channel of DIARAD (DIARAD-L) has operated continuously since the beginning of the measurements on 18 January 1996, whereas the right channel (DIARAD-R) is used only from time to time (once every 60 days) to check for degradation of the left channel. DIARAD is operated with an alternately shaded and irradiated cycle of 90 s, providing a solar irradiance measurement every 3 min. The electrical signals are measured twice during 10 s at the end of each cycle. The two PMO6-V radiometers (PMO6-VA and PMO6-VB) were designed to operate with a shaded and irradiated phase of 60 s providing a solar irradiance measurement every 2 min. This normal operation of the PMO6-V radiometers, however, was used only during a short period of time, owing to the failure of the shutters. The shutter of PMO6-VB failed in mid January when normal operation was supposed to start. PMO6-VA was operated normally for about 2 weeks before its shutter failed also. The reason for the failure does not seem to be a mechanical problem of the actuator. The shutters of both PMO6-V radiometers can be opened, but when they are closed again the electronic fuse is triggered and automatic operation is inhibited. So a new operation mode had to be devised which uses the protective cover instead. Such covers are located in front of each instrument within VIRGO, and can be opened and closed by telecommand. In order to avoid an excessive number of cover operations by time tagged commands, we decided to close it for 21 min three times a day. The time of the closure is placed randomly within ± 2 hours around 4, 12, 20 hours of the day. The randomization softens the periodic

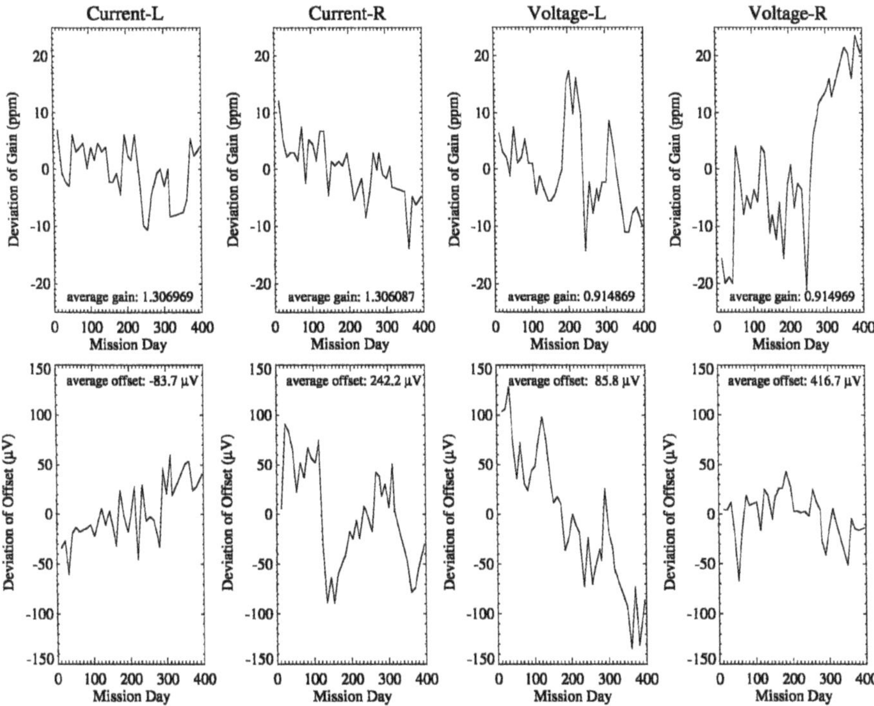

Figure 1. Gain and offset of the four electrical DIARAD channels during the first 400 days of the mission.

modulation of the time series. This mode of operation has the advantage of a 1-min sampling rate with a measurement during 20 s (duty cycle of 30%). It permits the measurements of p modes, although somewhat influenced by the low duty cycle. The rather infrequent reference phases, however, result in a somewhat increased noise at low frequency. This operation mode started on 23 February 1996 with a closed phase of PMO6-VA of 21 min, during which PMO6-VB was opened to fill the gaps. After some time it was realized that the gap filling was not really needed if the closed period of PMO6-VA was shortened. Thus, since 6 July 1996 the closure time of PMO6-VA has been reduced to 6 min and PMO6-VB takes only one measurement per week, which helps to keep its degradation low.

2.2. Electrical measurements

The electrical power (current and voltage) from the left and right channels of DIARAD are amplified with four differential amplifiers before being fed to the VIRGO data acquisition system (DAS). Knowledge of the gains and offsets is a crucial element in determining the accuracy of the observed solar irradiance. For this reason, the inputs of the four amplifiers are calibrated every few days

Table I

Gains and offsets for voltage (V) and current (C) channels of DIARAD and PMO6-V. For DIARAD the inflight measurements are listed with their standard deviations from Figure 1, and for PMO6-V the values determined before launch.

Channel	Gain (V/V)	Offset (μV)
DIARAD V left	0.914869 ± 7.1 ppm	85.8 ± 66.6
DIARAD V right	0.914969 ± 11.8 ppm	416.7 ± 22.6
DIARAD C left	1.306969 ± 6.1 ppm	− 83.7 ± 66.6
DIARAD C right	1.306087 ± 6.5 ppm	242.2 ± 51.5
PMO6-VA V	2.002530 ± 13.0 ppm	1319.5 ± 12.2
PMO6-VA C	3.027608 ± 12.6 ppm	−439.8 ± 7.3
PMO6-VB V	1.996693 ± 12.9 ppm	838.8 ± 13.2
PMO6-VB C	3.026542 ± 16.5 ppm	−380.8 ± 9.2

with an accurate voltage reference built into DIARAD (at the beginning this was performed every second day, since 3 August 1996 every 5 days). The measured gains and offsets are plotted in Figure 1 for the first 400 days of the mission, and the mean values with their standard deviations are shown in Table I. As the gain and offsets are quite stable, mean values are used for the evaluation of the DIARAD electrical signals, and any further development will be taken into account in the level-2 data. In summary, the uncertainty of the electrical measurements within DIARAD is about ±50 ppm, which is an excellent achievement in space.

The PMO6-V uses pre-launch calibrations for the gain and offsets as listed in Table I. During the data analysis it was realized that a problem with the offsets influences the calculated irradiance. The value of the heater resistance as a function of the heater power can be used to determine possible changes of the offsets. The observed $R_{\mathrm{o,c}} = u_{\mathrm{o,c}}/i_{\mathrm{o,c}}$ can be 'adjusted' by adding the same offsets $\Delta_{\mathrm{u,i}}$ to the measured open and closed signals $u_{\mathrm{o,c}}$ in order to achieve $\overline{R}_{\mathrm{o}} = \overline{R}_{\mathrm{c}}$:

$$\overline{R}_{\mathrm{o}} = \frac{u_{\mathrm{o}} + \Delta_{\mathrm{u}}}{i_{\mathrm{o}} + \Delta_{\mathrm{i}}} = \overline{R}_{\mathrm{c}} = \frac{u_{\mathrm{c}} + \Delta_{\mathrm{u}}}{i_{\mathrm{c}} + \Delta_{\mathrm{i}}} . \tag{1}$$

Here, the subscripts o and c denote open and closed, and u and i mean voltage and current. This assumes that only the offsets are subject to change, and not the gains. We assume also that the value of the heater resistor does not change between the open and closed readings. From the observed $R_{\mathrm{o,c}}$ one can determine one of the currently prevailing offsets Δu or Δi, assuming the other remains constant. Based on the analysis of PMO6-V radiometers, it was shown that the contribution of the U-channel to the offset is about an order of magnitude larger than the contribution of the I-channel. Accordingly we set $\Delta_{\mathrm{i}} = 0$, and Equation 1 is solved for Δ_{u}:

$$\Delta_{\mathrm{u}} = \Delta R \left(\frac{1}{i_{\mathrm{c}}} - \frac{1}{i_{\mathrm{o}}} \right)^{-1} \quad \text{with } \Delta R = R_{\mathrm{o}} - R_{\mathrm{c}}. \tag{2}$$

With these assumptions one can also calculate the difference in irradiance from the difference in the observed heater resistance

$$\begin{aligned}\Delta S &= S_{\text{corr}} - S = C_{\text{rad}}\left((u_{\text{c}} + \Delta_{\text{u}})i_{\text{c}} - (u_{\text{o}} + \Delta_{\text{u}})i_{\text{o}} - u_{\text{c}}i_{\text{c}} + u_{\text{o}}i_{\text{o}}\right) = \\ &= C_{\text{rad}}\Delta_{\text{u}}\,(i_{\text{c}} - i_{\text{o}}) = C_{\text{rad}}\Delta R\,(i_{\text{c}} - i_{\text{o}})\left(\frac{1}{i_{\text{c}}} - \frac{1}{i_{\text{o}}}\right)^{-1},\end{aligned} \tag{3}$$

which is plotted for the PMO6-VA in Figure 2. Note the small offset of about 0.02 W m^{-2}, corresponding to 15 ppm, at the beginning of the measurements indicating that the transfer from ground to space is indeed smooth. After the large excursion observed symmetrically around the greatest distance from the Sun around DoY96 200 (Day of the Year 1996) the curve is again smooth, but with a slightly different offset of 0.08 Wm^{-2} or about 60 ppm. The rms noise in the quiet periods is about ± 0.03 W m^{-2}, or 25 ppm, which is similar to the measured variance of the electrical channels of DIARAD. The reason for the large excursion is not understood, but it seems to be correlated with the increasing open electrical power in the cavity to compensate for the lower solar irradiance. It is, however, difficult to imagine how such a small change can influence the offset of the voltage channel so strongly, and in such a nonlinear fashion: increasing first up to DoY96 173 and then suddenly decreasing until the maximum of the electrical power is reached around DoY96 198; the whole is then repeated in the opposite direction. In order to cope with this problem, the electrical power in the cavity is calculated from the current, and the resistance value measured during the closed phase as $P_{\text{o,c}} = R_{\text{c}} i_{\text{o,c}}^2$, where P denotes power. The omission of the voltage signal in the open phase removes the strong excursion in the solar irradiance, as shown in Figure 3. Only some small remnants can be seen, e.g. between DoY96 220 and 230. The uncertainty of the electrical measurements in PMO6-V with this method is about ± 80 ppm.

2.3. Determination of the Irradiance

The difference of the electrical power during the open and the mean of the two adjacent closed phases is converted into irradiance by the use of the individual radiometric constant C_{rad}, which will be discussed in Section 2.4.

As the cavity does not see the Sun alone, the measured irradiance has to be corrected for the changes in the infrared radiation of the thermal environment within the radiometer. The main advantage of the active operation is that the pace of the open/closed phases are short enough not to change the thermal environment of the radiometer. In the case of the DIARAD the influence of changing thermal conditions on the irradiance is mainly compensated by the symmetrical arrangement of the cavities, both seeing the same thermal environment. Thus only a correction for the shutter emission in the closed phase needs to be applied.

In the new operation mode of PMO6-V the stability of the thermal environment is degraded. Since the open phases last long enough for the instrument to reach its thermal equilibrium, the measurements, which are carried out shortly after the

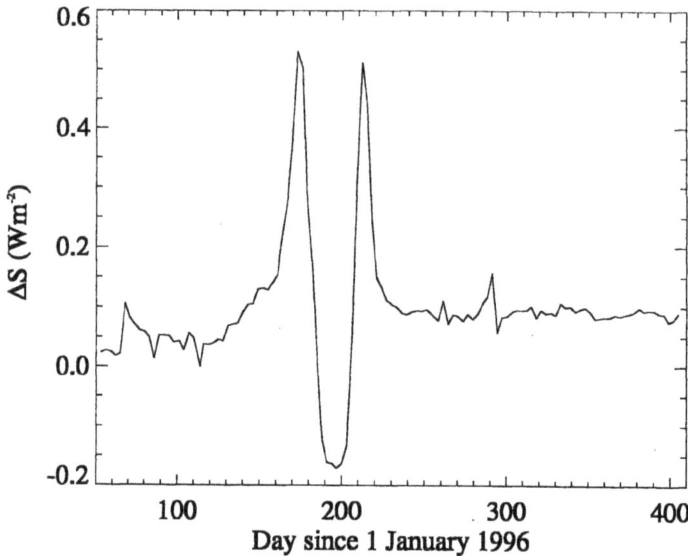

Figure 2. Change of solar irradiance caused by changes of the offset of the voltage channel of PMO6-VA during one year of the new operation.

opening of the cover, see a changing thermal environment which is different from that seen later. Consequently there are substantial relaxation tails in the irradiance time series whenever the cover changes its state. These changes can be accounted for by modelling the influence of the thermal environment seen by the cavity. The model uses five surfaces, taking into account their temperatures, their radiative properties and the view factors to estimate the contributions to the irradiance. The temperature within the radiometer is deduced from the two sensors in the heat sink and the front part; the temperature of the cover is calculated from an energy-balance model. The irradiance during the 8 hours of the open phase is calculated using a 'cover closed' value as reference, which is determined at each measurement in two steps. First, the closed values of the operational radiometer before and after the 8-hour period are linearly interpolated to the time of the observation. Secondly, small deviations of the power from the straight line are deduced from the closed (backup) radiometer, then smoothed and applied to the measurements of the open one.

The irradiance of the Sun as measured by the radiometer has to be normalized to a reference distance (1 AU) and zero velocity in order to reveal the proper variability of the Sun. The normalization to 1 AU is straightforward and is achieved by dividing the irradiance by the square of the heliocentric distance in AU. A Lorentz transformation reduces the irradiance to that in a frame at rest with respect to the Sun. The dependence of irradiance on orbital velocity was first illustrated by Hudson (1988) using SMM/ACRIM irradiance data (Willson, 1984). The correction factor given there as $(1+\beta^2)$ is incorrect; thus we give a brief derivation here. For a Planck radiating source $B_\nu(T)$ ($\sim \nu^3$) the irradiance is

$$S = B_\nu(T)\,\mathrm{d}\nu\,\mathrm{d}\Omega\,. \tag{4}$$

The Lorentz transformation changes ν and $\mathrm{d}\Omega$ by a factor f and f^{-2}, respectively, where

$$f = \frac{1-\beta\cos\theta}{\sqrt{(1-\beta^2)}} \tag{5}$$

with $\beta = v/c$ and θ denoting the angle between the velocity and the line of sight. With $\beta_r = c^{-1}\,\mathrm{d}r/\mathrm{d}t$ and the fact that S is proportional to ν^4 and $\beta \ll 1$ the reduced irradiance S_0 becomes

$$S_0 = S_{\mathrm{obs}}\frac{1}{(1-\beta_r)^2} = S_{\mathrm{obs}}(1+2\beta_r). \tag{6}$$

For observations from SOHO this effect is rather small ($< \pm 0.2$ ppm), but for Earth-bound satellites with radial velocities of up to ± 9 km s^{-1} it is important (± 60 ppm). The heliocentric distance and velocities of SOHO provided by the spacecraft operations are used to perform this normalization.

2.4. Absolute Accuracy of the Solar Irradiance

The radiometric constant C_{rad} is the reciprocal of the aperture area times a correction factor for the deviations from ideal behaviour. The correction factor accounts for different effects such as the reflectivity of the cavity, the losses due to diffraction at the apertures, stray light in the view-limiting muffler, heating of leads, and the non-equivalence between electrical and radiative heating for PMO6-V and the air and vacuum cavity efficiency for DIARAD. The procedure for determining these factors is called characterization of the radiometer, and it provides both, the correction factors and their uncertainty, the sum of the latter determining the absolute accuracy of the radiometer. For the DIARAD and PMO6-V the accuracy is $\pm 0.15\%$ (Crommelynck, 1988) and $\pm 0.17\%$ (Brusa and Fröhlich, 1986). For measurements in space (vacuum) the uncertainty is reduced, as the correction factor for, e.g., the non-equivalence between electrical and radiative heating is much smaller or the cavity efficiency is better.

The DIARAD has been characterized in the radiometer characterization laboratory of IRMB, the PMO6-V at PMOD. For the evaluation of the DIARAD data the radiometric characterization constants as determined at IRMB are used. The evaluation of PMO6-V can be based on radiometric constants determined either by characterization (PMO6-VA: 50697.5 m^{-2}; PMO6-VB: 50687.4 m^{-2}) or by indirect comparison with cryogenic radiometers (Romero, Fox, and Fröhlich, 1996) on the ground (PMO6-VA: 50599.8 m^{-2}; PMO6-VB: 50681.1 m^{-2}). One way of

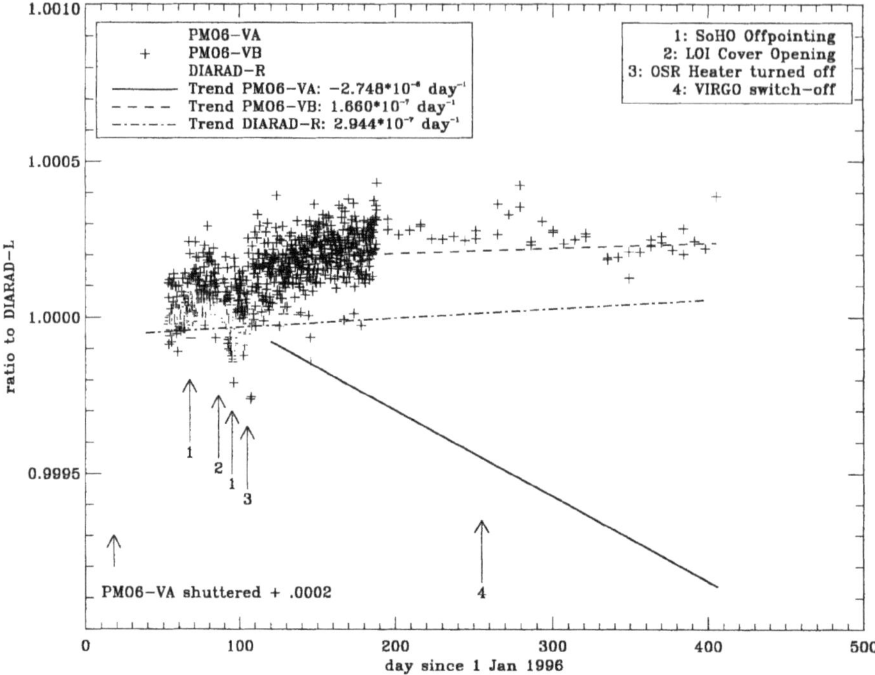

Figure 3. Comparison of the ratio of the solar irradiance measured by the four VIRGO radiometers to that measured by DIARAD-L. The labelled arrows show 3 major spacecraft events and No. 4 was the experiment switch-off due possibly to a hit by an energetic particle.

assessing the uncertainty of the transfer of the instruments to space is by comparing the relative instrument readings during ground tests and in space. The ratio DIARAD-L/DIARAD-R was 0.994645 on the ground and 0.995790 in space; there a difference of +1145 ppm is observed, which is within the uncertainty of the ground comparison. The ratio PMO6-VA/PMO6-VB changed from 1.000000 on the ground (both adjusted to the cryogenic scale) to 0.999330 in space, differing by −670 ppm, which is also within the uncertainty. Thus, the radiometers have not changed significantly during the integration, test and launch.

The relations between the four radiometers during DoY96 52–61 are presented in Table II. Adjusted time series are plotted as ratios to DIARAD-L in Figure 3. The most obvious deviation is in the early phase of the PMO6-V data. These data were taken in the shuttered mode. Relative to DIARAD the PMO6-V radiometer shows an increase, which seems to be exponentially damped with a time constant of about 40 days. There is no obvious reason for such behaviour, but it could be related to the non-equivalence of the PMO6-V radiometers which is slowly decreasing, due, e.g., to a change in the IR-emissivity of the outside wall of the cavity. The values of the shuttered operation differ from the later ones by about 200 ppm, with

Table II

Ratio of the four VIRGO radiometers to the mean value of DIARAD-L and PMO6-V determined during DoY96 52–61

Radiometer	Factor
DIARAD-L	1.000339
DIARAD-R	1.004569
PMO6-VA	0.999995
PMO6-VB	0.999325

the former being lower. In Figure 3 the shuttered data have been moved up by this amount. In view of the substantial change in operation, this is a rather small difference, and confirms the fact that the thermal model adopted for the cover mode is satisfactory. The observed difference of 0.423% between the two channels of DIARAD is unusual and certainly outside the uncertainty. The higher reading on the right channel may be due to an unidentified internal reflection on the right side of DIARAD. To maintain the independent character of the left and right channels, DIARAD-R is used only to measure the aging of DIARAD-L.

Besides the degradation discussed in Section 2.5, there are several other differences between the radiometers. These are mostly related to operations, such as the change of DIARAD after the switch-off in September 1996 (DoY96 255–261, see also Figure 5) which relaxes after about a month. It is not clear what caused this effect; though, some changes in gains, offsets and/or DAS references could be responsible. The variation in the PMO6-V value around DoY96 210–220 may be still some remnant of the offset problem discussed in Section 2.2.

The absolute value of the irradiance will be compared with measurements of similar radiometers on a planned stratospheric balloon flight in spring/summer 1998 and in August 1997 with the experiment SOLCON on board STS 85 with the NASA Hitchhiker Program (Crommelynck *et al.*, 1996). These comparisons will be complemented by a detailed study of the time series of simultaneous measurements of DIARAD-L and PM06-VA in VIRGO and ACRIM-II on the Upper Atmosphere Research Satellite (UARS). These comparisons will also permit us to refer the VIRGO observations to the Space Absolute Radiometer Reference (SARR) as defined by Crommelynck *et al.* (1995).

In summary the evaluation of DIARAD and PMO6-V signals is a straightforward conversion from measured pulses to voltages, taking into account the electrical calibration of the data acquisition system, which are then transformed into electrical power using the electrical calibration for each radiometer. The power difference between the open and the mean of the two adjacent closed measurements is used together with VIRGO housekeeping data in the DIARAD and PM06-V algorithms to obtain total irradiance values in W m^{-2}. These raw irradiances are then reduced

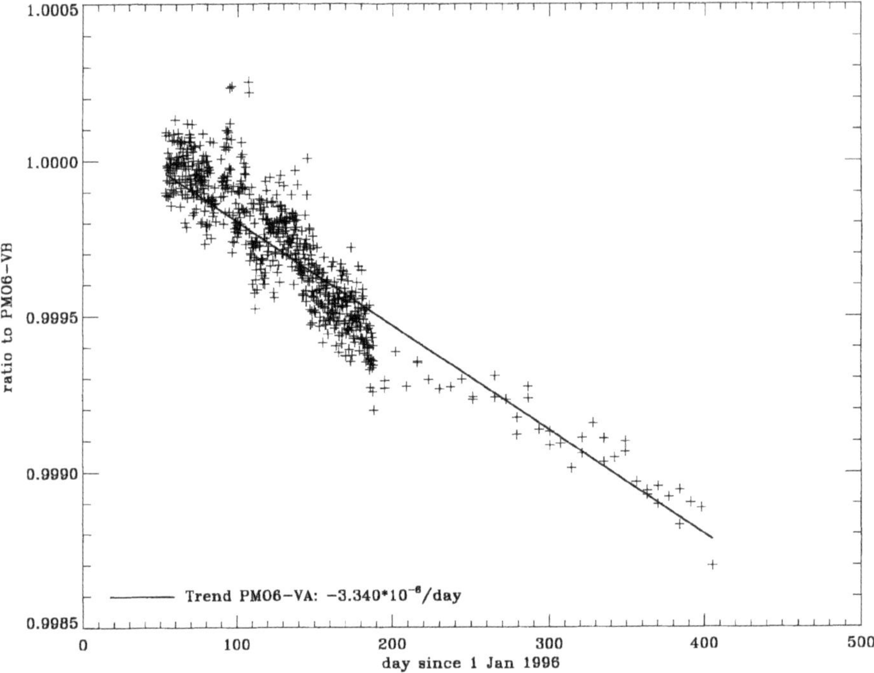

Figure 4. Determination of the degradation from the ratio of PMO6-VA/PMO6-VB irradiance for the first year of operation. Note that PMO6-VB is also corrected for its own time-dependent exposure degradation.

to 1 AU and corrected for the radial velocity as described in Section 2.3, using SOHO ancillary data.

2.5. Determination of the Degradation

The continuously exposed DIARAD left channel shows a very low degradation which was not measurable up to about the end of 1996. Data comparison between DIARAD-L and DIARAD-R over the full year shows a trend of -0.29 ppm day^{-1} deduced from linear regression of the ratio. For the PMO6-VA the rate of change is -3.34 ppm day^{-1} from the comparison of PMO6-VA and PMO6-VB shown in Figure 4. It is interesting to note that radiometers of both types (DIARAD and PMO6-V) on EURECA showed a degradation similar to that of the VIRGO PMO6-V, if referred to the same exposure times, whereas on SOHO the DIARAD behaves quite differently.

The irradiance measures of the radiometers DIARAD-L and PMO6-V, individually corrected for degradation, are shown in Figure 5. The agreement between the two time series is very good, especially after DoY96 270, when the standard deviation between the two time series is less than 40 ppm. As expected from the new operation mode, the PMO6-V has more low frequency noise, and for the period before DoY96 150 its behaviour is still dominated by the slowly levelling off of the increase observed since the beginning of the measurements. It should also be noted that the amplitude of the activity-related modulation of the solar irradiance measured by PMO6-V is larger than that observed by DIARAD by up to 20%. This may suggest that DIARAD is less sensitive to UV-EUV irradiance. The mean ratio DIARAD-L/PMO6-V for the whole period shown in Figure 5 amounts to 1.000573, which is very close to the original ratio of 1.000674 determined between DoY96 53 and 62. This difference of 100 ppm reflects the fact that the PMO6-V radiometer still increased slightly during the first 100 days. The correlation between the two radiometers at the 3-min sampling level is 0.8421 for the whole period and 0.8949 for the last 5 months (after switch-off/on in September).

The observations of DIARAD-L and PMO6-V are an excellent basis for studies of the solar variability on time scales longer than a few hours. The DIARAD-L, with its low degradation is best suited for keeping the long time record, and the PMO6-V may yield more complete information on the active-region related variability, if it turns out that the higher amplitude is indeed due to UV-EUV radiation. This is an excellent basis for producing an unbiased estimate of the solar irradiance time series for the study of the medium and long-term variability, as it was set forth as VIRGO science objective by having two different radiometers. The production of such a time series is based on the two series individually corrected for degradation as described in this Section. The VIRGO irradiance is composed of the PMO6-VA values 'adjusted' to the long-term behaviour of the DIARAD-L. This is performed by fitting a smooth curve to the ratios PMO6-VA/DIARAD-L during periods of quiet Sun. This time dependent ratio is then used to adjust each value of PMO6-VA for its long-term changes. These are then normalized to VIRGO values by using the corresponding factor in Table II. The use of only quiet-Sun periods prevents the correction from being influenced by the different sensitivity of the two radiometers to medium and short-term solar variability (see above). This is certainly feasible during periods of low activity, as prevailing now. With increasing activity, however, this method may need some modification. The level-2 VIRGO irradiance data contain the three time series (hourly or daily values): VIRGO, PMO6-V and DIARAD-L.

For studies of the high-frequency behaviour of the total irradiance (above about 25 μHz) the level-1 PMO6-VA values should be used. In the 5-min range, e.g., the noise level of the PMO6-VA data is only slightly higher than the one of the SPM data (Fröhlich *et al.*, 1997). This means that the PMO6-VA noise in this range is only limited by the low sampling duty cycle (30%) of the originally actively operated radiometer. Towards lower frequencies this influence decreases and the observed noise becomes predominately solar.

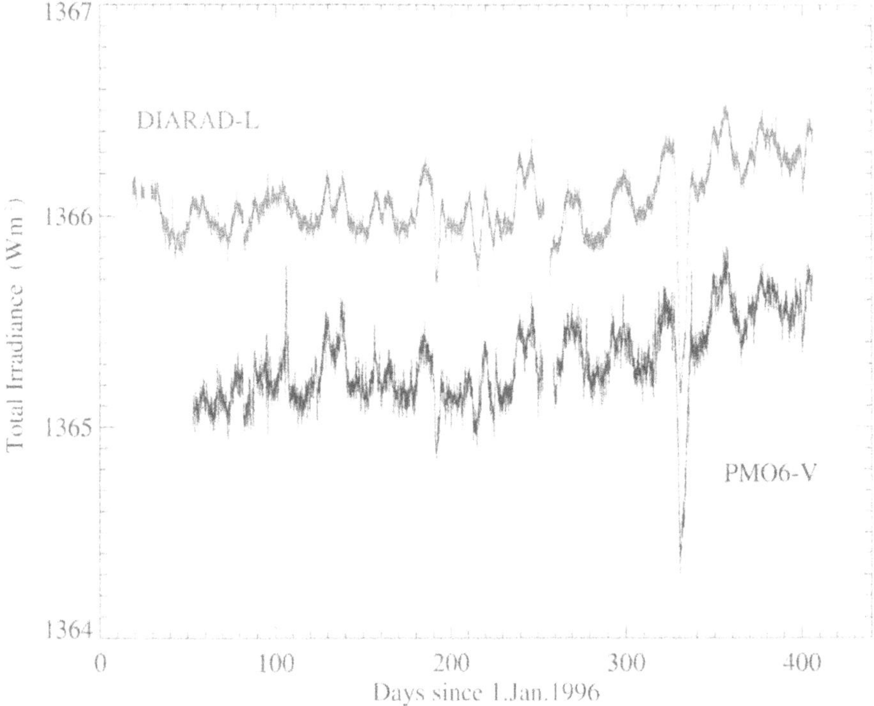

Figure 5. Time series of DIARAD-L and PMO6-V after removing their individually determined rate of degradation.

3. Spectral Irradiance Measurements

The spectral irradiance measurements within VIRGO are performed with two sun-photometers (SPM), SPMA and SPMB. These are spectral radiometers with three channels with 5nm bandwidth centered at 402, 500, and 862 nm, using interference filters and silicon photodiode detectors. Two apertures of 3.0 and 6.9 mm diameter on each channel define the radiometric area and the view geometry (4° full view and 1° slope angle). The filters and detectors are heated to a few degrees above the temperature of the heat sink to reduce condensation of gaseous contaminants on the optical surfaces. The SPM is not intrinsically an absolute radiometer, as it is not self-calibrating, but calibrated by two FEL irradiance standard lamps traceable to the National Institute of Standards and Technology (NIST) to an accuracy of 2–3%. A detailed description of the SPM instrument and its calibration is given by Wehrli and Fröhlich (1991) and their implementation in VIRGO by Fröhlich *et al.* (1995).

Table III
Temperature coefficients (TC) of the SPM as determined in space

	862 nm	500 nm	402 nm
TC in 0.1% K^{-1}	1.774	1.054	1.056

3.1. Corrections for Temperature and Pointing

SPMA is operated continuously with a 1-min cadence, while SPMB is exposed briefly every few weeks to check the degradation of the primary instrument. In addition, SPMB may be used as a backup in the event of a failure of SPMA. The actual detector temperature is monitored by two thermometers and used to correct the sensitivity during evaluation of the data. After launch, the SPM heaters were fully on during the outgassing period, and the sensor temperature was 6.5 K above ambient. When SPMA's cover was opened on 17 January 1996, its temperature rose by another 3 K due to solar irradiance absorbed by the radiometric aperture. On 2 February 1996 the heater power for SPMA was reduced to $\frac{1}{3}$, resulting in a temperature drop of 7.5 K. This temperature step was then used to determine the temperature dependence of the SPM *in situ*. The results are listed in Table III. They are quite different from the ones which were determined on the ground. Partly, this is due to the fact that they were measured in a seperate setup and not in the VIRGO sensor package with its data aquisition and its specific thermal environment. In the data evaluation the coefficients from the space determination (Table III) are used. Because the degradation rate showed a temperature dependence, at least for the 862 nm channel, the heater for SPMB was also switched off completely on DoY96 92. Thanks to solar heating, SPMA still remains 3 K above the heat sink temperature. SPMB is only 0.7 K warmer than the heat sink, but its entrance is protected by the cover. Outgassing should be low after 4 months in space. Figure 6 shows the course of the temperature of the SPM and of the VIRGO sensor package to illustrate the quiet thermal environment on SOHO.

During a joint observing program the spacecraft was depointed by 3.3 arc min south and a slight increase was detected in the red channel and not in the other channels. After the original pointing direction was restored, the red signal returned back to its original value, indicating that this channel shows a non-negligible sensitivity to pointing. From 3–5 April 1996 the spacecraft performed an off-pointing exercise sweeping from 10′ north to 10′ south and then from 10′ east to 10′ west braced by rapid excursions to 20′ in each direction. Relative variations of the red channel, determined by subtracting a linear fit and dividing by the mean value, were fitted for both angular directions by cubic polynomials in the depointing angle. These coefficients are listed in Table IV, and can be used to correct the red channel's signal. Incidentally, the N/S sensitivity is almost exactly 1 per radian,

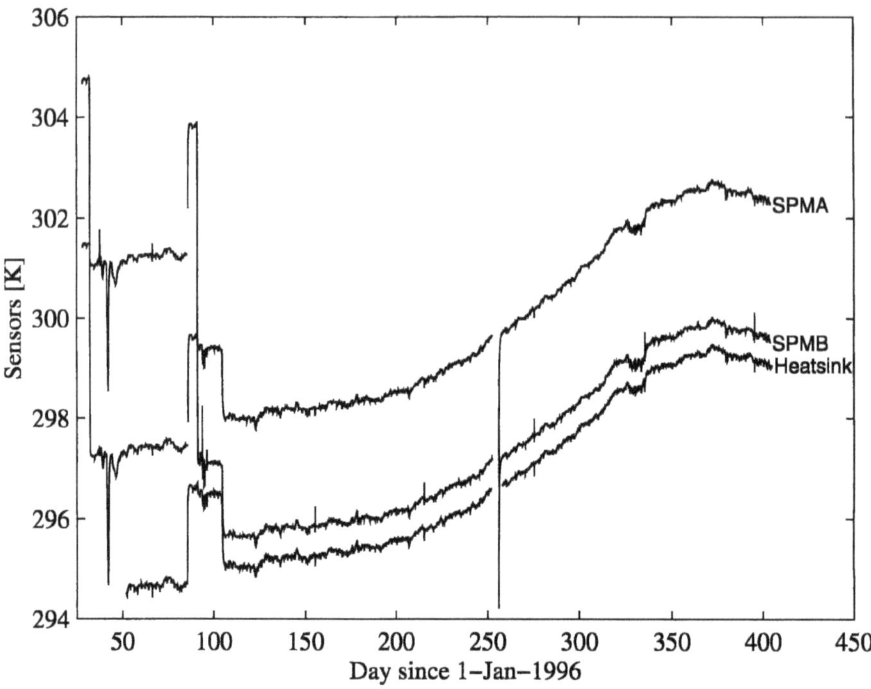

Figure 6. Temperature of the SPM sensors and the VIRGO sensor package during the first year of operation.

Table IV

Coefficients of the third-order polynomial $a_{0...3}$ used to correct the pointing sensitivity of the red channel of SPMA with the angular deviation in radians (positive towards N and E, respectively)

Direction	a_0	a_1	a_2	a_3
N/S	-6.4861×10^{-5}	$+1.1022$	$+46.915$	-3.4962×10^4
E/W	-1.0482×10^{-4}	$+0.2712$	$+22.249$	-1.4660×10^4

which is easy to remember. The absolute pointing offsets for SPMA amount to $-13''$ in pitch and $-22''$ in yaw angle. On 17 April 1996 the SOHO pitch offset adjustment of 9.6 mrad caused a signal change of 10^{-3}, which could be corrected to an accuracy of a few ppm. Both other channels are much less sensitive and need no correction.

3.2. Determination of the Spectral Irradiance

During the 'first light' experiment, all three spectral irradiances were about 2.9% below the values of the spectrum of Neckel and Labs (1984) as listed in Table V.

Table V

Calibration and 'first light' results from SPMA and SPMB compared with Neckel and Labs (1984) values. Calibration values are in A m^{-2} nm^{-1}, and irradiances in W m^{-2} nm^{-1}.

	SPMA			SPMB		
	862 nm	500 nm	402 nm	862 nm	500 nm	402 nm
Calibration	0.280	0.515	0.507	0.270	0.507	0.511
N&L irradiance	0.995	1.900	1.705	0.995	1.900	0.705
'1st light' irradiance	0.967	1.844	1.661	0.964	1.836	1.665
Difference	−2.8%	−3.0%	−2.6%	−3.1%	−3.4%	−2.3%

While this is still within the combined errors of the NIST spectral irradiance scale (NBS-1973: 1.5–1.7%), lamp scatter (<0.3%) and the reference spectrum (r.m.s. <1%), the general lower values hint at a systematic error during calibration. A possible reason is that the calibrations were performed in the clean room which has bright walls, rather than in the optics laboratory with dark curtains and more baffles. Any additional stray light measured would have decreased the calibrated response, and would therefore have led to a systematically lower value of solar irradiance measurements. The ratios SPMA/SPMB during the 'first light' experiment were 1.0031, 1.0043, and 0.9976 for the red, green and blue channels. They agree within less than 0.5%, which indicates that their calibrations have been transferred to space with very good accuracy.

The absolute values will be checked during the stratospheric balloon flight planned for 1997 by comparing with a similar SPM calibrated before and after the flight. These calibrations will be based on the recently developed, very accurate radiometric method (e.g., Friedrich, Fischer, and Stock, 1995) rather than on sources such as standard lamps.

In summary, the evaluation of SPM signals is a straightforward conversion from measured pulses to voltages, taking into account the electrical calibration of the data acquisition system, which are then multiplied by a calibration factor (Table V) to obtain spectral irradiance values in W m^{-2} nm^{-1}. These raw irradiances are corrected for temperature, normalized to 1 AU and corrected for the radial velocity as for the radiometers (Section 2.3) using VIRGO housekeeping and SOHO ancillary data. Additionally the red channel is corrected for its pointing sensitivity if needed.

3.3. Determination of the Degradation

The sensitivity of the main instrument has slowly degraded with mean rates of about −64 ppm day^{-1} for the 862 nm, −313 ppm day^{-1} for the 500 nm and −620 ppm day^{-1} for the 402 nm channel as shown in Figure 7. These rates are determined by fitting a straight line to the data of Figure 7, which is the same as assuming a constant solar irradiance over the period of evaluation. The total decrease of sensitivity during the first year of operation amounts to 3, 12, and 23%

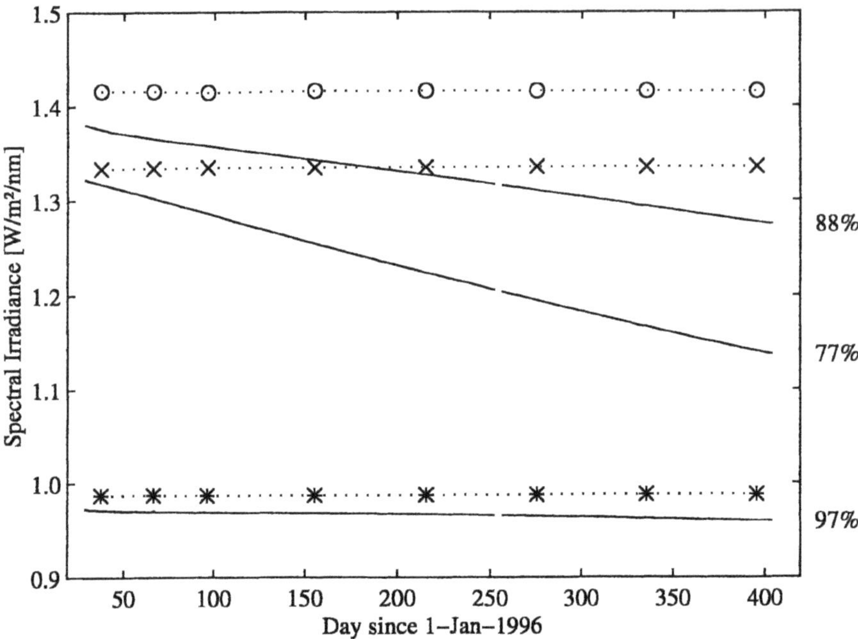

Figure 7. Summary of the measurements of the SPMA and SPMB for the first year of operation. The full lines are for the operational instrument and the dotted lines connect the measurements of the back-up channels. The lines starting below the symbols and the symbols correspond to the following channels: o green, × blue, and ⋆ red (from top to bottom).

for the red, green, and blue channels, respectively. The degradation is one order of magnitude lower than what was experienced by similar SPM instruments on EURECA or PHOBOS; more stable filters and strict cleanliness control during construction of VIRGO and the spacecraft have indeed paid off. It is interesting to note that the three channels have very different initial degradation. It was already noted during the EURECA flight that the first few hours of exposure can decrease the sensitivity substantially more severely than later on. This effect is strongest in the green channel with 4.8% change between 1st light and first comparison with SPMB. It is 3.8% for the red and only 1.6% for the blue.

During this first year the spare instrument (SPMB) has been exposed only 8 times for 21-min intervals each plus 406 min during the off-pointing tests. The measurements are plotted in Figure 8. The red and blue channels show increases similar to the one observed in total irradiance (Figure 5); in contrast, the green channel decreased slightly. These differences in behaviour may be explained with an initial rapid decrease in sensitivity as observed in SPMA for the green channel. These data are obviously not accurate enough to be used to correct the SPMA for degradation to a level of 100 ppm needed to assess the long-term solar changes.

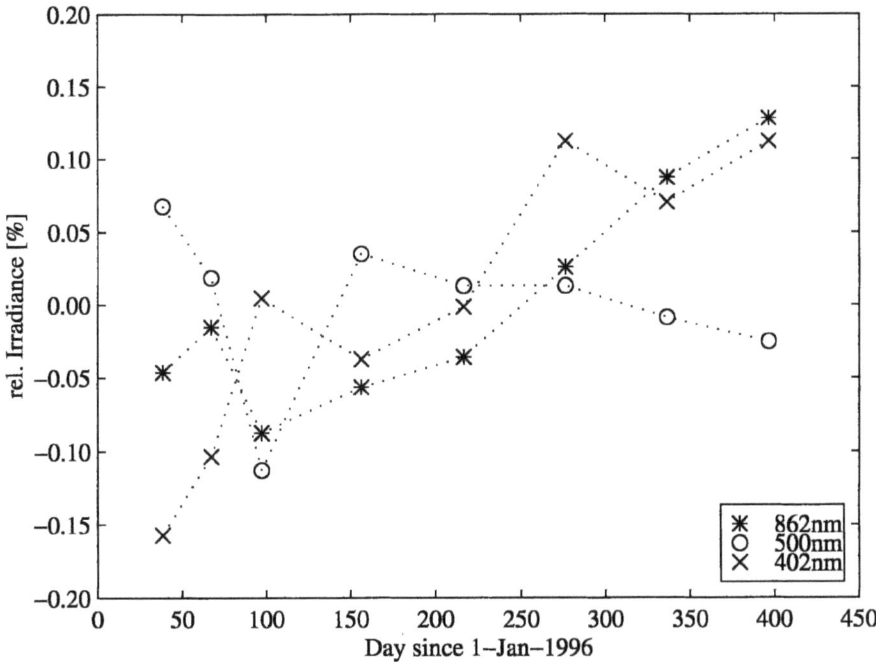

Figure 8. Measurements of SPMB for the first year of operation.

More detailed studies and possibly a longer time series are needed. For the analysis of solar oscillations and the variability on time scales of up to a few months, detrending by polynomials or cubic splines is adequate as shown in Figure 9. Such time series are the basis for level-2 SPM data. As the mission goes on and more data are gathered, the whole time series may have to be updated, resulting in a new version level-2 data.

4. Data Products

The VIRGO Data Centre (VDC) at IAC in Tenerife receives the data from the SOHO Operation Center (SOC) at Goddard Space Flight Center (GSFC) near Washington D.C. and produces and archives level-0, level-1, and level-2 data. The data can be accessed by the VIRGO team according to the rules of the Data Rights (Fröhlich *et al.*, 1995). The level-0 data are files containing the original counts, but sorted by instrument together with the VIRGO HK data needed for the evaluation of this instrument. The level-1 data are time series which are corrected for all *a priori* known effects such as the influence of temperature and pointing, instrument-Sun distance and relative velocity, and are calculated according to the algorithms developed by the instrument co-investigators and described in Sections 2 and 3.

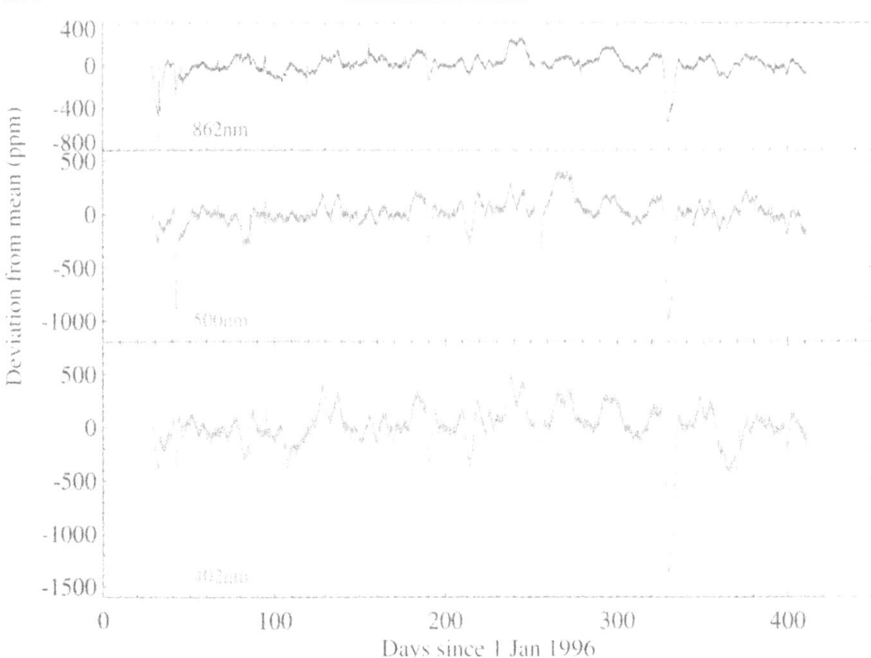

Figure 9. Spectral solar irradiance for the first year of operation from the 3 SPMA channels. The degradation is taken out by fitting cubic splines.

They are stored in daily FITS files with the 3-min data of DIARAD and the 1-min data of PMO6-V and SPM. For medium-term time series analysis to up to a month these data can be used directly. For analysis of longer periods, level-2 data should be used. The production of these data depends strongly on the reliability of the knowledge of the degradation. As stated in Section 2.5, this is well established for the total irradiance. The problems with spectral irradiance are discussed in Section 3.3 and only preliminary level-2 data are produced. The level-2 data as hourly means are stored in monthly FITS files for the total irradiance (VIRGO, DIARAD-L and PMO6-VA values) and the spectral irradiance (RED, GREEN and BLUE) separately. According to data release rules (Fröhlich *et al.*, 1995) these data are only available to the VIRGO team (including guest-investigators) until the end of the second year, when the first year of data will be released. A time series of daily means of the total solar irradiance will be made available to the public soon.

5. Conclusions

The irradiance instruments in the VIRGO package work very well with performances in general exceeding expectation. The total irradiance data show solar variability on all time scales with a precision of a few ppm down to periods of hours, and continuing well below in the 5-min range. Owing to the low degradation, the present high precision can possibly be maintained into the solar maximum. The values of DIARAD-L with their excellent precision together with the PMO6-V data with the higher sensitivity to the overall spectrum are the basis for the VIRGO total irradiance data base as set forward in the science objective by having two different radiometers. The very good signal-to-noise ratio of the PMO6-V at frequencies above about 25 μHz can be used for studies of solar oscillations, supergranulation, mesogranulation and granulation. The spectral irradiance observations with the SPM have excellent signal-to-noise ratio in the 5-min range, where the instrumental noise is at least one order of magnitude below the solar noise (Fröhlich *et al.*, 1997). The stability for periods over a month, however, is not yet established, as the detailed understanding of the degradation mechanisms is still lacking. In summary, the excellent performance, the very quiet environment of SOHO and the continuity of the observations enables us to obtain time series of a quality never previously achieved. The first results presented by Fröhlich *et al.* (1997) demonstrate the potential of these data.

Acknowledgements

VIRGO is an investigation on the Solar and Heliospheric Observatory, SOHO, which is a mission of international cooperation between ESA and NASA. The experiment is a cooperative effort of many individual scientists and engineers at several institutes in Europe and the USA as listed by Fröhlich *et al.* (1995). Without the continuous and concerted efforts of the team this experiment would never have reached the success it demonstrates today. Thanks are extended to various members of the VIRGO science and technical team for many helpful discussions during the compilation of the material for this paper. The irradiance part of the VIRGO team has been supported by several national and international funding agencies which are gratefully acknowledged: The PMOD/WRC by the Swiss National Science Foundation under grants 2.860–0.88, 20–28779.90, 20–33941.92, 20–40589.94, 20–47039.96 and PRODEX, the IRMB by the Fonds de la Recherche Fondamentale Collective d'Initiative Ministérielle and PRODEX, the IAC by the CICYT through PNIE under grants ESP88–0354 and ESP90–0969.

References

Brusa, R. W. and Fröhlich, C.: 1986, *Appl. Optics* **25**, 4173.

Crommelynck, D.: 1988, in N. Fox (ed.), *New Developments and Applications in Optical Radiometry*, Inst. Phys. Conf. Ser. 92, London, p. 19.

Crommelynck, D. and Domingo, V.: 1984, *Science* **225**, 180.

Crommelynck, D., Fichot, A., Lee III, R. B., Romero, J.: 1995, *Adv. Space Res.* **16**, (8), 17.
Crommelynck, D., Fichot, A., Domingo, V., Lee III, R. B.: 1996, *Geoph. Res. Letters* **23**, 2293.
Friedrich, R., Fischer, J., and Stock, M.: 1996, *Metrologia* **32**, 509.
Fröhlich, C., Romero, J., Roth, H., Wehrli, C., Andersen, B. N., Appourchaux, T., Domingo, V., Telljohann, U., Berthomieu, B., Delache, P., Provost, J., Toutain, T., Crommelynck, D., Chevalier, A., Fichot, A., Däppen, W., Gough, D. O., Hoeksema, T., Jiménez, A., Gómez, M., Herreros, J., Roca-Cortés, T., Jones, A. R., Pap, J., and Willson, R. C.: 1995, *Solar Phys.* **162**, 101.
Fröhlich, C., Andersen, B., Appourchaux, T., Berthomieu, G., Crommelynck, D. A., Domingo, V., Fichot, A., Finsterle, W., Gómez, M. F., Gough, D. O., Jiménez, A., Leifsen, T., Lombaerts, M., Pap, J. M., Provost, J., Roca Cortés, T., Romero, J., Roth, H., Sekii, T., Telljohann, U., Toutain, T., Wehrli, C.: 1997, *Solar Phys.* **170**, 1.
Hudson, H. S.: 1988, *Ann. Rev. Astron. Astrophys.* **26**, 473.
Neckel, H. and Labs, D.: 1984, *Solar Phys.* **90**, 205.
Romero, J., Fox, N., and Fröhlich, C.: 1996, *Metrologia* **32**, 523.
Wehrli, Ch. and Fröhlich, C.: 1991, *Metrologia* **28**, 285.
Willson, R. C.: 1984, *Space Sci. Rev.* **38**, 203.

MEASUREMENTS OF FREQUENCIES OF SOLAR OSCILLATIONS FROM THE MDI MEDIUM-l PROGRAM

E. J. RHODES, JR.[1], A.G. KOSOVICHEV[2], J. SCHOU[2], P.H. SCHERRER[2] and J. REITER[3]

[1] *Department of Physics and Astronomy, University of Southern California, Los Angeles, CA 90089, and Space Physics Research Element, Jet Propulsion Laboratory, Pasadena, CA 91101, U.S.A.*
[2] *W.W. Hansen Experimental Physics Laboratory, Stanford University, Stanford, CA 94305, U.S.A.*
[3] *Mathematisches Institut, Technische Universität München, D-80333 Münich, Germany*

(Received 21 March 1997; accepted 18 June 1997)

Abstract. Inversions of solar internal structure employ both the frequencies and the associated uncertainties of the solar oscillation modes as input parameters. In this paper we investigate how systematic errors in these input parameters may affect the resulting inferences of the sun's internal structure. Such systematic errors are likely to arise from inaccuracies in the theoretical models which are used to represent the spectral lines in the observational power spectra, from line blending, from asymmetries in the profiles of these lines, and from other factors. In order to study such systematic effects we have employed two different duration observing runs (one of 60 days and the second of 144 days) obtained with the Medium-l Program of the Michelson Doppler Imager experiment onboard the SOHO spacecraft. This observing program provides continuous observations of solar oscillation modes having angular degrees, l, ranging from 0 to $\sim$ 300. For this study intermediate- and high-degree p-mode oscillations having degrees less than 251 were employed.

In the first of our tests we employed two different methods of estimating the modal frequencies and their associated uncertainties from the 144-day observational power spectra. In our second test we also repeated both methods of frequency estimation on the 60-day time series in order to assess the influence of the duration of the observed time series on the computed frequencies and uncertainties. In a third test we investigated the sensitivity of the computed frequencies to the choice of initial-guess, or 'seed' frequencies that are used in the frequency estimation codes. In a fourth test we attempted to investigate the possible systematic frequency errors which are introduced when the observational asymmetry in the p-mode peaks is ignored. We carried out this particular test by fitting simple models of asymmetric line profiles to the peaks in the observational power spectra. We were then able to compute the differences between those frequencies and our previous frequencies which had been obtained using the assumption that all of the observational peaks were symmetric in shape.

In order to study the possible influence of the two different frequency estimation methods upon the radial profile of the internal sound speed, we carried out four parallel structural inversions using the different sets and subsets of frequency estimates and uncertainties as computed from the 144-day observing run as inputs. The results of these four inversions confirm the previous finding by the GONG project (Gough *et al.*, 1996) and by the MDI Medium-l Program (Kosovichev *et al.*, 1997) that, in a thin layer just beneath the convection zone, helium appears to be less abundant than predicted by theory. However, differences in our four inverted radial sound speed profiles demonstrate that the currently-available techniques for determining the frequencies of the Medium-l oscillation peaks introduce systematic errors which are large enough to affect the results of the structural inversions. Moreover, based upon the differences in these four inverted sound speed profiles, it appears that the choice of which subset of modes is included in a particular inversion and which modes are not included may also be introducing systematic errors into our current understanding of solar internal structure. Hence, it appears to be very important that consistent sets of modal selection criteria be employed.

Finally, at least one of the two frequency estimation codes which we used was not sensitive to changes in the input 'seed' frequencies which were employed as initial guesses for that code. This result allays fears that the difference in the helium abundance between the sun and the reference solar model in the thin layer beneath the convection zone which was mentioned above might have been due to the particular seed frequencies which were employed in the earlier inversions. Since this thin layer

Solar Physics **175:** 287–310, 1997.

may likely be the place where the solar dynamo operates, it will be extremely important to observe any possible evolution of this transition layer throughout the upcoming 11-year activity cycle.

1. Introduction

The two principal goals of helioseismology are: (1) the inference of the thermodynamic structure and (2) the inference of the dynamical motions of the solar interior as functions of both position within the Sun and of time. The principal means by which both of these goals are now being sought is the inversion of the observed properties of the solar normal modes of oscillation. In the case of the thermodynamic structure it is the oscillation frequencies themselves (weighted by their uncertainties) which are the input data for the inversions, while in the case of the dynamical motions it is the frequency splittings (again properly weighted) which are inverted. As is described elsewhere (Kosovichev *et al.*, 1997), it is only the frequencies and splittings of the solar f- and p-mode oscillations have been inverted until now. Because of the stochastic nature of solar oscillations (illustrated in the solar sounds files on the CD-ROM), one of the most important and difficult problems in the frequency measurements is how to account for the stochastic component in the oscillation spectra. In this paper, we discuss two different spectral fitting procedures: the so-called mean-multiplet and averaged-spectra techniques.

Our main goal in this paper is to demonstrate how these two different fitting techniques, which we have employed to estimate the frequencies of the normal modes from the MDI Medium-l data, may influence the inferred radial profile of the solar internal sound speed. Our additional goals are: (1) to learn how the frequencies and uncertainties depend upon the duration of the observing run from which they are computed, (2) to learn how the asymmetry which is present in the observational power spectra will alter the frequencies which are obtained under the assumption of symmetric shapes for those peaks, (3) to learn whether or not the frequencies which have been computed might be sensitive to changes in the initial guesses which are employed in the frequency estimation programs, and (4) to study the effects of modal selection critera upon the inverted results.

For all of these tests we have employed observations which were obtained by the Solar Oscillation Investigation/ Michelson Doppler Imager (SOI/MDI) experiment on board the Solar and Heliospheric Observatory (SOHO) spacecraft. In particular, we have employed time series of oscillation observations which were made with the MDI Medium-l Program. This program has been described in detail by Scherrer *et al.* (1996).

In Section 2 we will describe the different MDI observing runs which we employed in our study. In Section 3 we will describe the two different methods which we employed in the estimation of the intermediate-l frequencies from these observing runs. We will also describe in this section how we evaluated the effects of the observational asymmetry in the observed power spectral peaks upon the frequency estimates. Next we will describe how two different sets of 'seed' frequencies were employed as input parameters for one of the two different frequency estimation programs.

The inferred radial profiles of the solar internal sound speed which resulted from the different inversions we carried out will be presented in Section 4, while in Section 5 we will discuss the relative importance of all of these different effects upon the resulting sound speed profiles.

2. The MDI Medium-l Observations

As described by Scherrer *et al.* (1996) the Medium-l Program of the SOI/MDI experiment is dedicated to the nearly-continuous monitoring of the Doppler velocity field of the visible solar hemisphere. The Medium-l data are transmitted though the low-rate (5 kbps) telemetry channel of SOHO. It is this relatively low bit rate which limits the spatial resolution of the Medium-l observations. This low data rate also requires that some initial processing of the original 1024×1024-pixel CCD full-disk Dopplergrams be carried out on board the MDI instrument. Specifically, for both Medium-l observing runs which we employed in this study, the Dopplergrams were binned with Gaussian weights on a square 5×5-pixel grid. The binned images which resulted from this procedure were then generated on board the spacecraft once per minute. With the exception of rare instrumental problems and occasional telemetry drop outs these binned Doppler maps were then transmitted to the Earth on a minute-by-minute basis.

The first of the two Medium-l observing runs which we employed was obtained on 60 consecutive days running from 25 May 1996, through 24 July 1996. The overall duty cycle of this run was 0.9801. The second of the two runs we used covered a span of 144 days which ran from 9 May 1996 through 29 September 1996. The overall duty cycle of this run was 0.9547 (see a 5-hour movie of the time series on the CD-ROM). An autoregressive gap-filling technique, Brown (1996), was employed to fill-in relatively short gaps in the time series generated from the second run and its overall duty cycle was raised to 0.9639. As can be noted from the dates given above these two data sets were not independent of each other. Rather, the shorter-duration run is a sub-set of the longer run. This choice of time intervals was made in order to minimize the possible influence of any solar-cycle dependent frequency shifts on our study.

3. Analysis of the Medium-l Data

3.1. Power Spectra

The procedures which we employed to convert the time series of Medium-l observations into power spectra are described in Kosovichev *et al.* (1997) in this volume.

For both the 60-day and 144-day observing runs we computed power spectra for all $2l+1$ individual m-values at every degree ranging from $l=0$ through $l=299$. Hence, for every non-zero degree greater than two, each set of power spectra consisted of one zonal (i.e., $l=0$), two sectoral (i.e., $m=-l$ and $m=+l$), and $2l-2$ tesseral (i.e., l not equal to m for non-zero m) power spectra. The spectra which were computed from the 60-day observing run contained 43 201 points each. The spectra computed from the 144-day run contained 103 680 points each. The formal frequency resolution of the 43 201-point power spectra was 0.192901= μHz, while the resolution of the 103 680-point spectra was 0.0803755 μHz.

In addition to these two complete sets of power spectra, we also computed two sets of m-averaged power spectra. In the procedure, which collapsed each set of spectra into a single averaged spectrum, we approximated the frequency splitting which is introduced into each multiplet of modes by the solar internal rotation and asphericity with a polynomial expansion similar to that of Duvall, Harvey, and Pomerantz (1986):

$$\nu_{nlm} = \nu_{nl} + \sum_{k=1}^{36} a_{nl}^{(k)} \mathcal{P}_k^{(l)}(m)\,, \tag{1}$$

where ν_{nl} is the mean frequency of a mode multiplet, and $\mathcal{P}_k^{(l)}(m)$ are orthogonal polynomials of degree k defined by

$$\mathcal{P}_k^{(l)}(l) = l \quad \text{and} \quad \sum_{m=-l}^{l} \mathcal{P}_i^{(l)}(m)\mathcal{P}_j^{(l)}(m) = 0 \quad \text{for} \quad i \neq j\,. \tag{2}$$

The polynomials, $\mathcal{P}_k^{(l)}(m)$, can be expressed in terms of the Clebsch–Gordan coefficients (Ritzwoller and Lavely, 1991). At $l \gg k$, $\mathcal{P}_i^{(l)}(m) \rightarrow LP_k(m/L)$, where $L = l + 1/2$, and P_k are the Legendre polynomials. In this limit, coefficients $a_{nl}^{(k)}$ are equivalent to the coefficients introduced by Duvall, Harvey, and Pomerantz (1986). Since

$$\sum_{m=-l}^{l} \mathcal{P}_k^{(l)}(m) = 0 \quad \text{for} \quad k > 0\,, \tag{3}$$

the mean frequencies, ν_{nl}, depend only on the spherically symmetric component of the solar structure. Frequency splitting coefficients, $a_{nl}^{(k)}$, for even k depend on the aspherical component of the structure, while the coefficients for odd k measure the rotation rate (e.g., Gough, 1993). Therefore, in the method in which we collapsed the individual power spectra into m-averaged spectra we wanted to remove the effects of solar rotation from the individual spectra and to accomplish this goal we used non-zero values for the three odd-k splitting coefficients (i.e., $a_{nl}^{(1)}$, $a_{nl}^{(3)}$, and

$a_{nl}^{(5)}$). In particular, the three even-k splitting coefficients (i.e., $a_{nl}^{(0)}$, $a_{nl}^{(2)}$, and $a_{nl}^{(4)}$) were each set equal to zero because they were each extremely small in comparison with the three odd-k coefficients. Also, no higher-k coefficients were included in the expansion because they were all very small in magnitude in comparison with the three coefficients which were employed.. Finally, in the procedure which averaged together the shifted tesseral power spectra, we used a single frequency shift for each separate tesseral or sectoral spectrum. We did not apply a separate frequency shift for each multiplet within that spectrum. The $2l$ shifted spectra were then averaged together with the unshifted zonal spectrum. In this procedure each of the separate spectra was included with equal weight. No attempt was made to weight each separate spectrum according to its m-value. The 301 m-averaged power spectra are shown as a two-dimensional $l - \nu$ diagram on the CD-ROM (Figure 1(b)).

3.2. Frequency Estimation Techniques

3.2.1. *Mean-Multiplet Technique*

The first of our two different frequency estimation techniques was the so-called 'mean-multiplet' technique of Schou (1992). In this technique the power spectral peaks are assumed to have a symmetric Lorentzian shape and a maximum likelihood method is employed to determine the parameters of Lorentzian profiles. Furthermore, in this method the peaks are fit simultaneously in all of the $2l + 1$ individual power spectra for each multiplet so that the effects of overlapping peaks can be included in the fits. These $2l + 1$ frequencies are then averaged to yield a single frequency, ν_{nl}, for that multiplet. In addition, an associated frequency uncertainty, σ_{nl}, and a set of frequency splitting coefficients, $a_{nl}^{(k)}$, for which k runs from 1 to 36, are obtained for the same multiplet. The mean frequencies which resulted from the use of this technique on the power spectra from the 144-day MDI Medium-l observing run were illustrated in Figure 11 of Kosovichev *et al.* (1997).

3.2.2. *Averaged-Spectrum Technique*

The second frequency estimation technique which we employed has recently been described by Reiter and Rhodes (1997). This technique is an elaboration and refinement of the least-squares fitting technique employed earlier by Korzennik (1990) in the analysis of data obtained with the Mt. Wilson Observatory's 60-Foot Solar Tower. This technique is becoming known as the 'averaged-spectrum' method because it employs the m-averaged power spectra rather than the $2l + 1$ individual power spectra which the mean-multiplet method employs.

The application of this second technique results in a single frequency for each multiplet rather than a set of $2l + 1$ frequencies which must later be averaged together to yield an estimate of the average multiplet frequency. (We note here that the mean-multiplet technique includes includes fits to several of the spatial sidelobes which are located on both sides of each true peak within a given multiplet. Hence,

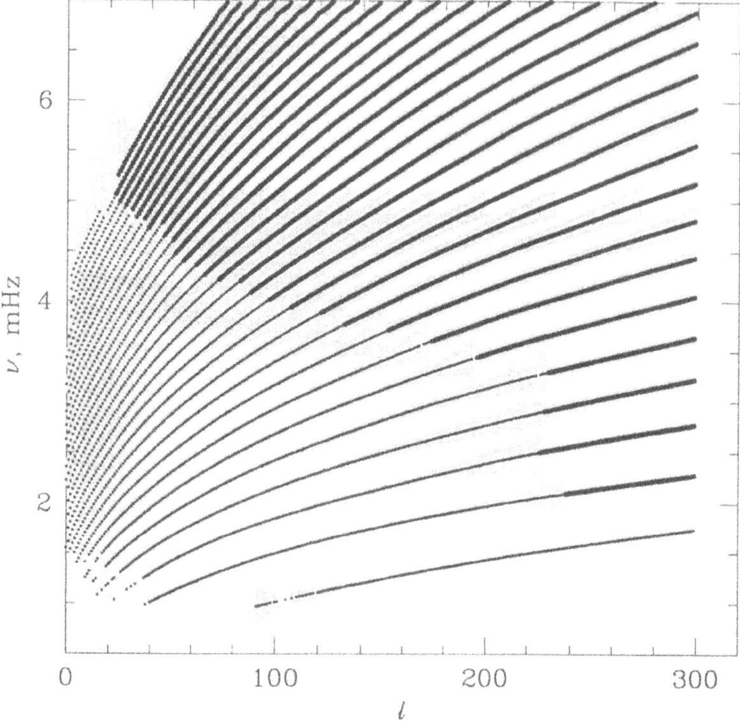

Figure 1. Multiplet frequencies determined from m-averaged power spectra obtained from the MDI 144-Day Observing Run using the averaged-spectrum method are shown as the small dots. Ridge-fit frequencies determined with the averaged-spectrum method as applied to entire p-mode ridges rather than individual modal peaks are shown as the larger dots along the extensions of the ridges toward higher degrees and frequencies. The formal frequency uncertainties obtained from the averaged-spectrum method are superimposed upon both the modal and ridge-fit frequencies in gray. For frequencies below 5000 μHz the frequency uncertainties of the modal fits were multiplied by a factor of 1000 before being plotted. For frequencies greater than 5000 μHz the uncertainties of the ridge fits were multiplied by 100 instead. The uncertainties of the ridge fits below 5000 μHz were multiplied by 1000. A color version of this figure (Figure 1(a)) is on the CD-ROM. In Figure 1(a) the modal-fit frequencies are shown in blue, the ridge-fits are shown in red, and both sets of uncertainties are shown in yellow. Figure 1(a) may be compared with the m-averaged spectrum (Figure 1(b)) on the CD-ROM. Samples of the MDI full-disk (Figure 1(c)) and Medium-l Dopplergrams (Figure 1(d)) are also shown on the CD-ROM.

this technique fits many more peaks in its estimation of multiplet frequencies than does the averaged-spectrum technique.)

3.3. AVERAGED-SPECTRUM RESULTS

The coverage in the $l-\nu$ plane which resulted from the application of the averaged-spectrum fitting technique to the set of m-averaged power spectra computed from the 144-day observing run is shown here in Figure 1. We note that there were only

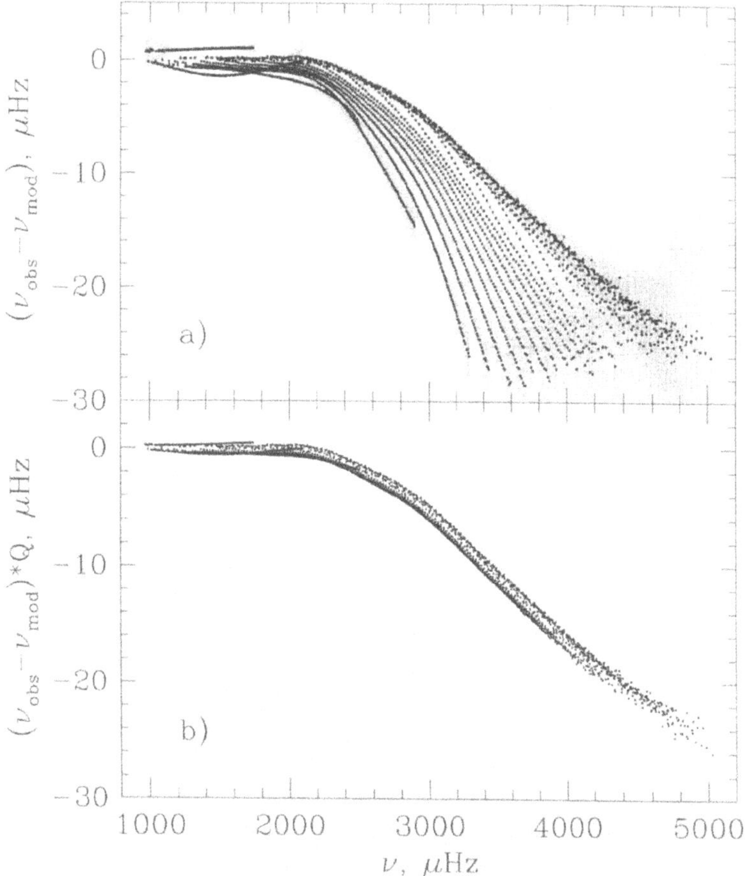

Figure 2. (a) Frequency dependence of differences between averaged-spectrum frequency estimates and reference solar model frequencies (the frequency uncertainties were multiplied by a factor of 10 before being plotted in gray). (b) The same frequency differences scaled with mode inertia Q.

a small number of spectral peaks for which we were able to obtain a converged fit in the spectrum computed from the 144-day run but for which we were unable to obtain a corresponding converged fit in the 60-day spectrum.

In addition to the 2794 modal frequencies which are shown in Figure 1 as the small dots (and which are shown in blue in the color version of this figure on the CD-ROM), we have also included in this figure an additional 4555 so-called 'ridge-fit' frequency estimates which are shown as the larger dots (and which are shown in red on the CD-ROM). These ridge-fit frequency estimates were obtained from the application of the averaged-spectrum method to the entire f- and p-mode ridges instead of to individual modes as was done for the modal fits. Such ridges arise when the spatial sidelobes which occur for several successive degrees overlap so closely in frequency that individual modes can no longer be individually identified. The analysis of such ridge-fit frequencies is now in progress. The results of that analysis will be presented in a future paper.

3.4. Comparison of Frequency Datasets

3.4.1. *Comparison with Reference Solar Model*

The tables of both sets of modal frequencies and their associated frequency uncertainties are presented on the CD-ROM. Once we had generated both frequency estimate datasets, we then compared both of them with a set of theoretical frequencies computed from a reference solar model. The solar model which we chose to use was Model S of Christensen-Dalsgaard *et al.* (1996). This model was a standard evolutionary model computed using the most recent information on nuclear reaction rates (Bahcall and Pinsonneault, 1995) and radiative opacity and the equation of state (Rogers and Iglesias, 1996). The gravitational settling and diffusion of helium and heavier elements were taken into account in the computation of this model following the theory by Michaud and Proffitt (1993).

The comparison of the observed frequency estimates obtained from the averaged-spectrum method with the theoretical model frequencies is shown here in Figure 2. The raw frequency differences (in the sense of the observed minus the theoretical frequency estimates) are shown as a function of frequency in Figure 2(a). We have also included the frequency uncertainties in Figure 2(a); however, in order to make them visible at the scale of the figure, we have multiplied these uncertainties by a factor of 10 prior to plotting them. We show a similar comparison between the averaged-spectrum estimates and the theoretical model frequency estimates in which we have normalized the frequency differences by the inertia of the different p modes in Figure 2(b). Once again the frequency uncertainties have been multiplied by a factor of 10 prior to their inclusion in Figure 2(b). We have not included a separate pair of plots for the mean-multiplet frequency estimates in Figure 2 as the differences between them and the averaged-spectrum estimates were small enough that they were indistinguishable at the scale of the figure.

3.4.2. *Inter-Comparison of 144-Day Observational Frequency Estimates*

In addition to comparing both sets of observed frequency estimates with theoretical frequencies, we also inter-compared the two sets of observational frequency estimates directly. We obtained a total of 1940 different modes for which converged solutions were found for both of our two frequency estimation programs. The differences in the two sets of frequency estimates obtained from the 144-day observing run are shown here in Figure 3. The frequency dependence of the raw frequency differences (in the sense of the averaged-spectrum set minus the mean-multiplet set) are shown in Figure 3(a), where we have added the unscaled one-sigma frequency uncertainties. The degree dependence of the same frequency differences is shown in Figure 3(b) (again with one-sigma errors superimposed).

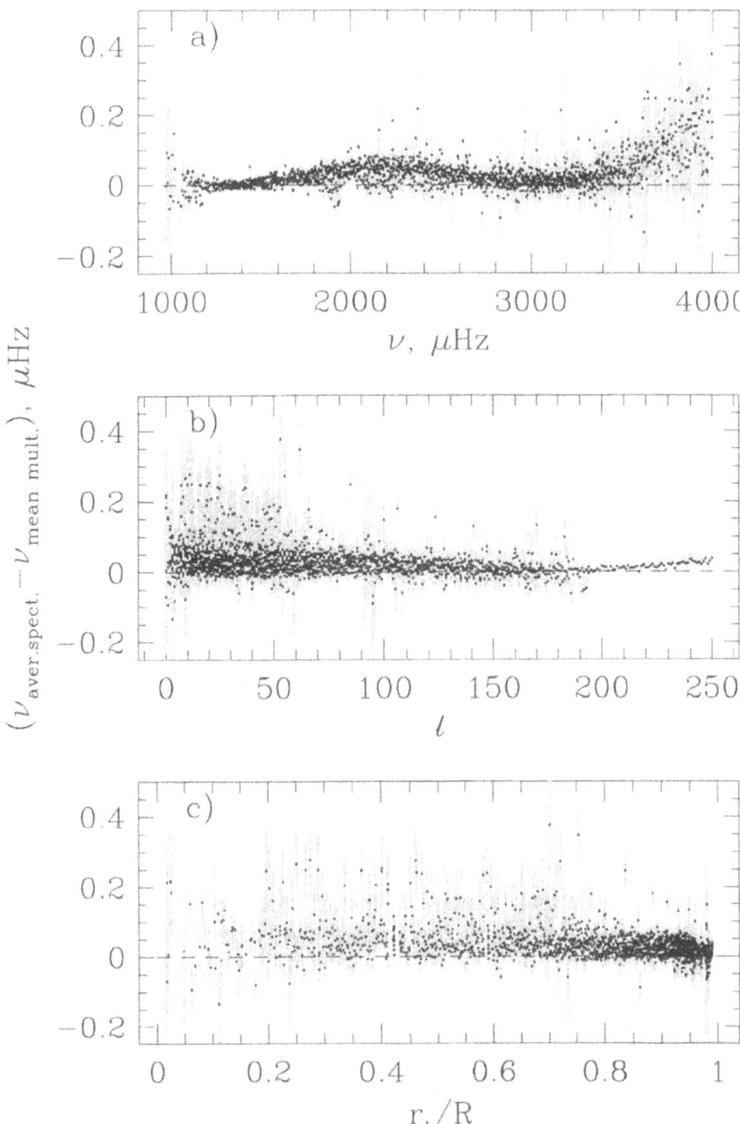

Figure 3. (a) Frequency dependence of differences between averaged-spectrum and mean-multiplet frequency estimates. (b) The degree dependence of the same frequency differences: (c) The dependence of the frequency differences upon the inner turning point radius of the corresponding p modes. The unscaled frequency uncertainties have been added to all three panels in gray.

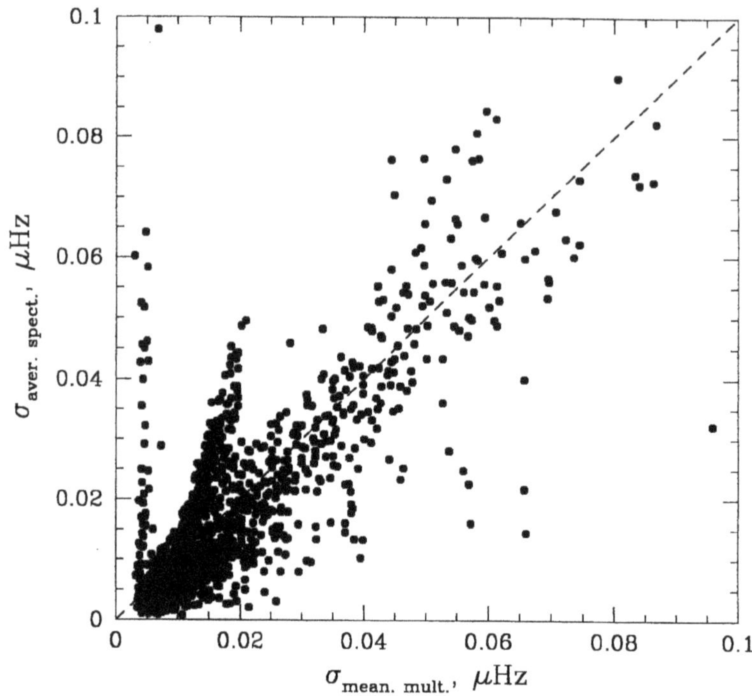

Figure 4. Comparison of 1σ frequency uncertainties inferred from mean-multiplet and averaged-spectrum methods. Most of the points scatter around the diagonal of the figure. This concentration of points shows that in the large majority of cases the two uncertainties are nearly equal.

The inner turning point dependence of these frequency differences is shown in Figure 3(c). It should be noted that the vertical scale of Figure 3 is much smaller than was the scale used in Figure 2. Nevertheless, Figure 3 still shows that there were small, but significant differences in the frequency dependence of the two frequency datasets. The largest differences between the two datasets can be seen at frequencies near 2400 and 3400 μHz. All three panels of Figure 3 illustrate that the averaged-spectrum frequency estimates tended to be larger than those determined from the mean-multiplet method.

Figure 3(c) also illustrates that the two sets of frequency estimates agreed most closely for modes with inner turning points near the solar surface (i.e., where the fractional radii of the inner turning points are close to unity), and for modes having deep inner turning points (e.g., for modes having inner turning point radii near $0.2R$). For modes having inner turning point radii located in between these two extremes the differences in the two sets of frequencies were larger. The largest frequency differences can be seen to have occured for modes having inner turning point radii near $0.5R$.

In addition to comparing the two sets of frequency estimates from our two different methods, we also compared the estimated frequency uncertainties. The comparison that we made of these uncertainties (also referred to as frequency errors) is shown here in Figure 4. In this figure the uncertainties we obtained from the averaged-spectrum method are plotted as a function of the corresponding uncertainties for the same modes as determined by the mean-multiplet method. It is evident in Figure 4 that most of the uncertainties were very similar in size; however, there was one group of points which fell near the vertical axis of the figure. These points corresponded to cases in which the uncertainties were considerably larger in the averaged-spectrum method than they were in the mean-multiplet method. This may have been due to the wider peaks in some of the averaged spectra in comparison with the narrower peaks in the unaveraged tesseral power spectra which were used in the mean-multiplet method. The larger uncertainties of these few averaged-spectrum frequency estimates meant that they were assigned less weight in the inversions than were the corresponding mean-multiplet estimates. Fortunately, there were relatively few such cases.

3.5. Comparison of 60-day and 144-day Frequency Estimates

The effects of changes in the duration of the observing run from which the power spectra are computed are illustrated here in Figures 5 and 6. The frequency dependence of the differences in two sets of averaged-spectrum frequencies which were computed from the 144-day and from the 60-day MDI Medium-l observing runs is shown in Figure 5(a), while the frequency dependence of the differences in the two sets of mean-multiplet frequencies is shown in Figure 5(b). Figure 5(a) indicates that the frequency estimates computed from the averaged-spectrum method do not show any systematic shift between the 144- and 60-day observing runs. In contrast, Figure 5(b) shows that there was a systematic frequency variation in the two sets of frequencies which were computed using the mean-multiplet method. We note that the frequency variation shown in Figure 5(b) is very similar in shape to that shown in Figure 3(a). It is not at all clear why the frequencies computed from one of our two methods would depend upon the duration of the observing run when those computed from the other method show no such variation. We plan to investigate this issue in the future.

An increase in the duration of the observing run from which the modal parameters are estimated would be expected to result in a decrease in the measured widths of the power spectral peaks and also in a corresponding decrease in the magnitudes of the formal frequency uncertainties of the modes. In order to determine whether or not such expectations were realized during the MDI Medium-l observing program, we also computed the full-width at half maximum (FWHM) of the different power spectral peaks in the two sets of averaged power spectra. We then computed the ratios of the FHWM values computed from the 144-day and 60-day observing

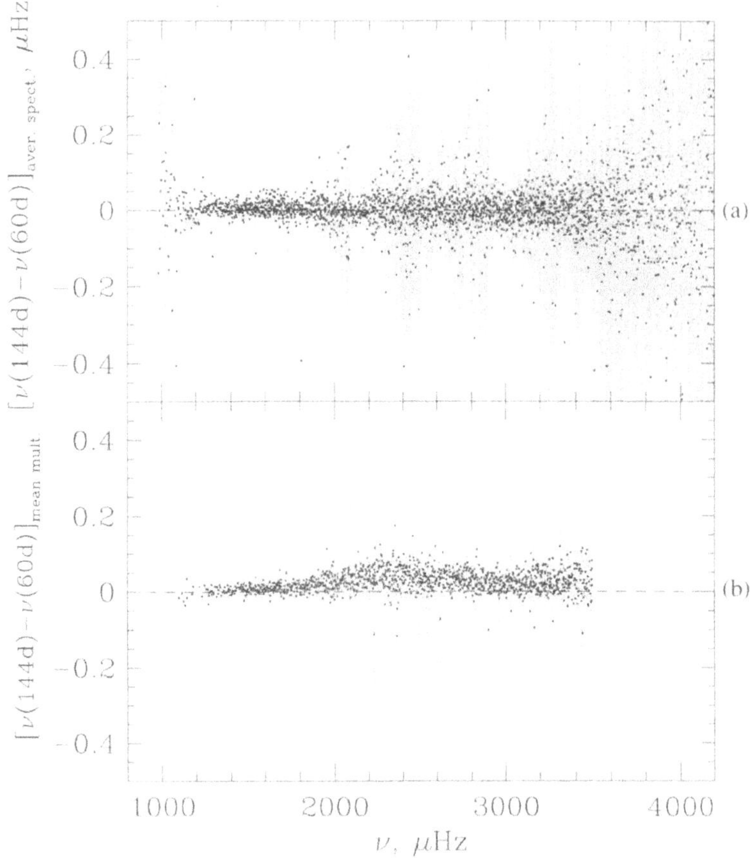

Figure 5. (a) Frequency dependence of frequency differences as determined by the averaged-spectrum method applied to power spectra computed from the 144- and 60-day observing runs. (b) Frequency dependence of the frequency differences obtained by the mean-multiplet method applied to the same two observing runs. The unscaled frequency uncertainties are superimposed upon both sets of frequency differences in gray.

runs. When we examined these FWHM ratios, we found that most of these FWHM ratios were indeed less than unity, as was expected.

In a similar fashion we also computed the ratios of the formal frequency uncertainties which resulted from the fits to our two sets of averaged power spectra. The frequency dependence of the ratios in the frequency uncertainties is shown here in Figure 6. The average ratio of the 144-day and 60-day frequency uncertainties was equal to 0.669. Since the frequency uncertainties are expected to scale with the square root of the duration of the time series from which they are computed, we would have anticipated the ratio of the 144-day and 60-day uncertainties to be equal to 0.646. The close agreement of the above two numbers does indeed

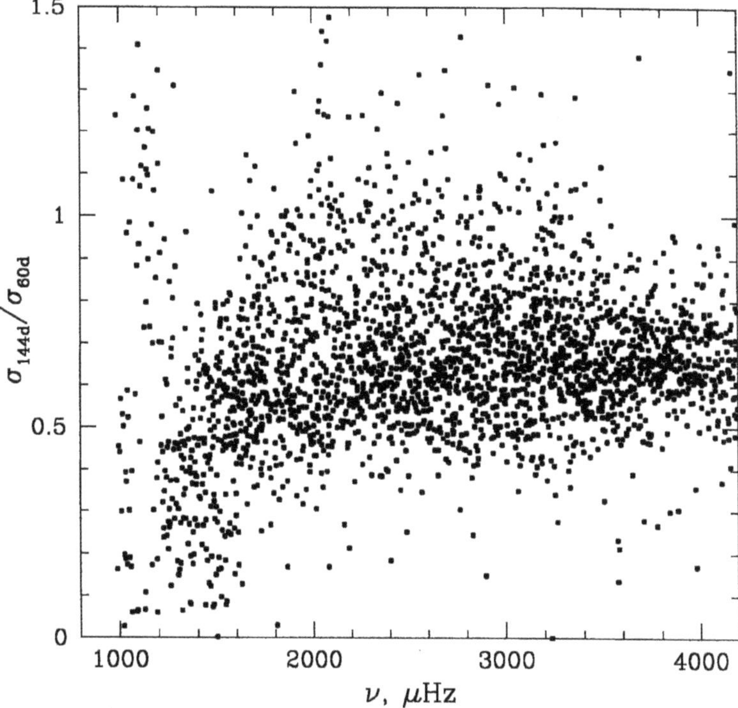

Figure 6. Frequency dependence of the ratios of the frequency uncertainties as computed using the averaged-spectrum method for both the 144- and 60-day observing runs. The ratios of these uncertainties are almost all less than unity and the average ratio was 0.669. Based upon the durations of these two runs the average ratio would be expected to be equal to 0.646.

suggest that the frequency uncertainties scale as predicted as the duration of the observing run is increased. We plan to test this scaling of the uncertainties again in the near future when Medium-l power spectra are available for a much longer 360-day observing run.

3.6. Frequency Effects of Asymmetric Power Peaks

Because of the significant reduction in the noise, the Medium-l data have revealed interesting characteristics of the line profiles of the oscillation power spectra. The most interesting feature is the asymmetry of the line profiles. Though the asymmetry has been noticed in the ground-based data (Duvall *et al.*, 1993), frequencies of solar modes are usually determined by assuming that the line profile is symmetric and can be fitted by a Lorentzian, which would be the case if the solar p modes were damped simple harmonic oscillators excited by a stochastic source. However, this leads to systematic errors in the determination of frequencies (Hill *et al.*, 1996; Abrams and Kumar, 1996). Several authors have studied this problem theoretically and have

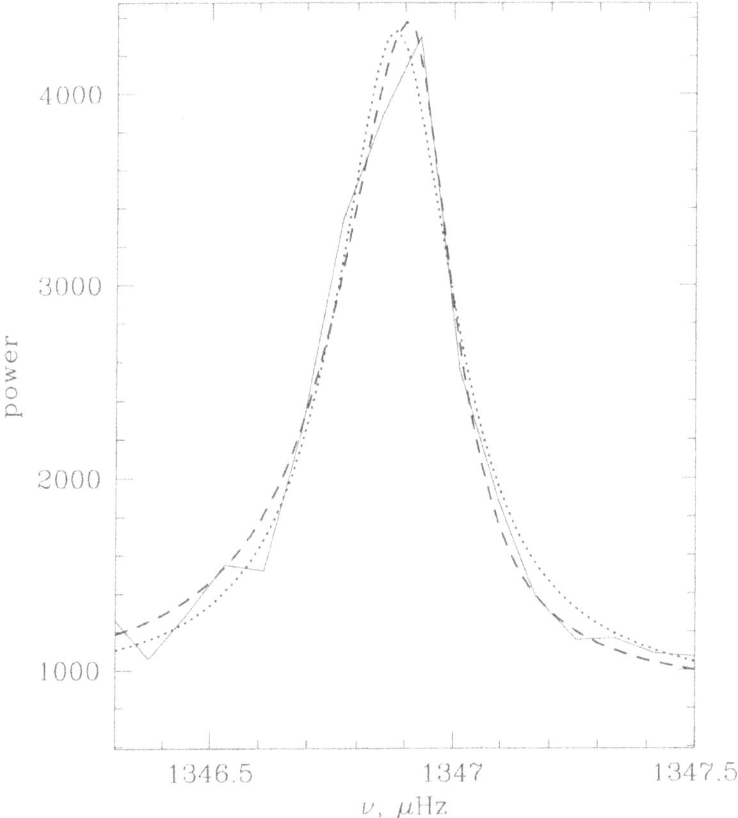

Figure 7. Comparison of observed peak from m-averaged power spectrum obtained from 144-day observing run for l = 178, n = 0 (solid line) and the numerical fits to this peak computed from the averaged-spectrum method assuming a symmetric fit (dotted line) and assuming an asymmetric Lorentzian profile (dashed line).

found that there is an inherent asymmetry whenever the waves are excited by a localized source (Gabriel, 1992, 1993, 1995; Kumar *et al.*, 1994; Roxbourgh and Vorontsov, 1995). Physically, the asymmetry is an effect of interference between an outward direct wave from the source and a corresponding inward wave that passes through the region of wave propagation (Duvall *et al.*, 1993). Figure 10 of Kosovichev *et al.* (1997) shows a theoretical power spectrum of p modes of $l = 200$ obtained by Nigam and Kosovichev (1996). This theoretical model is in good qualitative agreement with the observations. The degree of the asymmetry depends on the relative locations of the acoustic sources and the upper reflection layer of the modes.

An example of an asymmetric peak in the 144-day averaged spectrum is shown as the solid line in Figure 7. The peak shown in this figure is the $l = 178$, $n = 0$

peak. Also shown in this figure are two fitted profiles for this peak. The dotted line is the fit of a symmetric Lorentzian with a linear background term, while the dashed line is the fit of an asymmetric Lorentzian profile in which the two different halves of the profile are both Lorentzian half-profiles each having different half-widths at half-maximum and different amplitudes above the background. The differences between the asymmetric and asymmetric fits are apparent even at the scale of Figure 7. In particular, the peaks of the two different fitted functions are located at different frequencies.

In order to make initial estimates of the errors introduced into the frequency estimates due to the presence of asymmetric peak profiles, we modified the numerical algorithm used in the average-spectrum frequency estimation program to fit four different-shaped profiles to all of the observed peaks in our 144-day spectra. We then ran this modified version of the program and for each observed peak we selected the model fit which minimized the residuals about that fit as the 'best' model for that peak. The first of these four different models was our original symmetric Lorentzian profile with a linearly-varying background. The second model was an asymmetric Lorentzian with linearly-varying background of the type shown in Figure 7. The third model was an asymmetric Lorentzian with a quadratically-varying background and the fourth model was an asymmetric Gaussian with a quadratic background.

The results of running this modified averaged-spectrum program on the 144-day spectra are shown in Figure 8. In Figure 8(a) we show the ratios of the widths of the small-frequency half-profiles (i.e., the widths of the left halves of the profiles given the traditional increase of frequency to the right in most plots) to the widths of the large-frequency half-profiles (i.e., the widths of the right halves of the profiles) plotted as a function of frequency. For the case of symmetric peaks this ratio would be equal to one. Inspection of Figure 8(a) shows that the large majority of the peaks were fit with model profiles which had larger widths on the small-frequency sides of the maxima. In Figure 8(b) we show the frequency dependence of the frequency differences between the asymmetric and symmetric fits for all of the modes for which we obtained converged solutions with both asymmetric and symmetric model fits. These frequency differences show a very similar frequency variation to that shown earlier in Figure 3(a) for the differences between the averaged-spectrum and mean-multiplet frequencies which were obtained using the assumption of symmetric profiles in both cases. The amplitude of the frequency variation shown in Figure 8(b) is about one-half of that shown in Figure 3(a) and so it is possible that some of the systematic differences shown in Figure 3(a) are in fact due to the neglect of asymmetry in both frequency estimation methods. If so, then this would seem to imply that the peaks in the m-averaged power spectra are more asymmetric in shape than are the corresponding peaks in the unaveraged spectra. This is an interesting possibility that we are planning to look into shortly. Also, since there is a theoretical explanation for the observed profile shapes, there is also a mathematical function for the asymmetric shapes of the peaks which may be more suitable to use

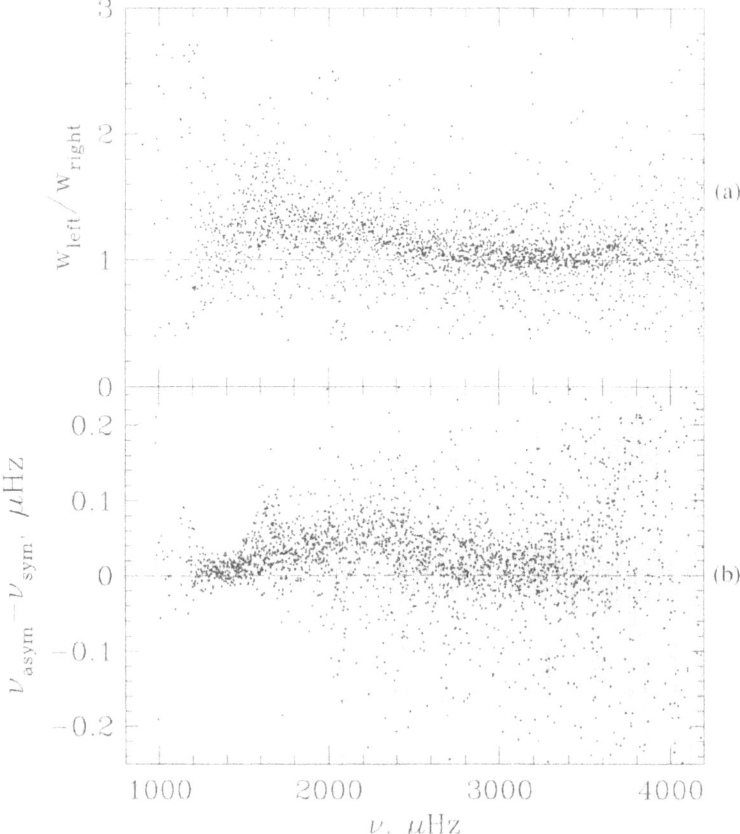

Figure 8. (a) Frequency dependence of the ratios of the widths of the left and right halves of the asymmetric profiles used to estimate the amount of asymmetry in the observed power spectral peaks. Symmetric peaks would have ratios equal to unity. Ratios greater than unity correspond to peaks which are wider at smaller frequencies than they are at higher frequencies. (b) Frequency dependence of the differences in the frequencies as determined by the averaged-spectrum method using the asymmetric fits and using the assumption of symmetric profiles.

than the asymmetric Lorentzians and Gaussians which we employed in generating Figure 8. We plan to investigate the use of such a theoretical function as a more accurate model of asymmetry in the near future.

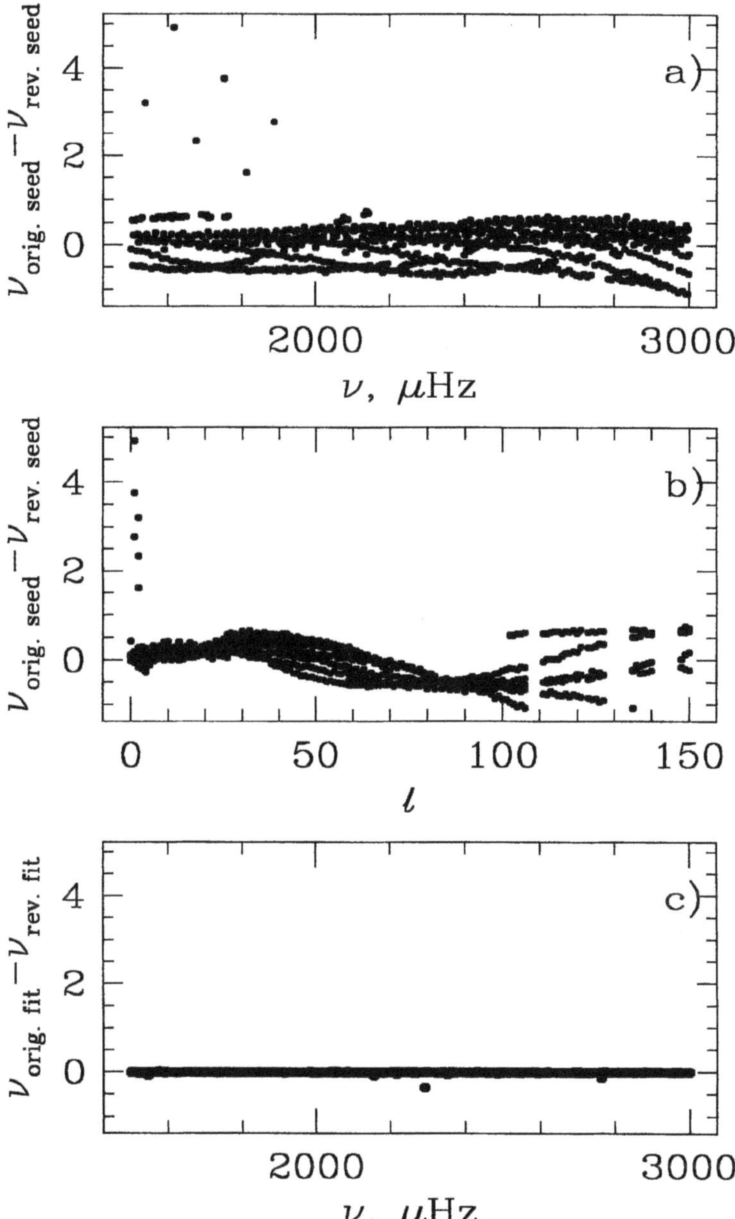

Figure 9. (a) Frequency dependence of the differences between the two different sets of 'seed' frequencies which were used as inputs to the averaged-spectrum method. (b) Degree dependence of the differences in the two sets of seed frequencies. (c) Frequency dependence of the differences in the final computed frequencies which resulted from using the two different sets of seed frequencies as inputs to the averaged-spectrum program.

3.7. Comparison of Frequency Estimates Computed from Different 'Seed' Frequency Sets

Both of the frequency estimation methods which we employed for the tests described in this paper use sets of initial-guess, or so-called 'seed,' frequencies as input parameters. These sets of 'seed' frequencies provide starting points for the two methods which are hopefully close enough to the desired frequencies that the methods will converge to the proper modes rather than to adjacent sidelobes of them, or to noise in the spectra. Specifically, in order for the estimation methods to converge to the proper peaks, these tables of 'seed' frequencies must be within a few microhertz of the final solutions for most modes.

Our next test was a test of the sensitivity of our final frequency estimates to changes in our 'seed' frequency tables. We carried out this test because the radial profile of the squared differences between the inferred sound speed and our reference solar model which is published by Kosovichev *et al.* (1997, Figure 12) shows several interesting departures from the reference solar model and we were interested in learning if any of these interesting features could be caused by systematic errors in the 'seed' frequencies themselves.

In order to test such a possibility we then conducted a 'blind' test in which one of us (AGK) generated a new set of specially-altered 'seed' frequencies, after which two others of us (JR and EJR) used this new set of 'seed' frequencies as inputs to the averaged-spectrum frequency estimation program. Before showing the results of this test, we wish to illustrate how large the differences in the two sets of 'seed' frequencies actually were. The frequency dependence of the differences between the original and the revised 'seed' frequencies is shown here in Figure 9(a), while the degree dependence of these same 'seed' frequency differences is shown in Figure 9(b). Inspection of Figure 9(b) shows that there was a pronounced degree dependence in the differences in the two sets of 'seed' frequencies, with a peak-to-peak amplitude of three microhertz between $l = 30$ and $l = 110$.

Once we had computed the additional set of modal parameter estimates, we compared them with the parameters obtained from the original set of 'seed' frequencies. For the frequency estimates of the modes we simply matched up common modes in our two tables and then subtracted the two estimates. The results of this comparison are shown here in Figure 9(c). Figure 9(c) shows the frequency dependence of the differences in the two sets of frequency estimates. A quick glance at this figure shows that, with the exception of a very small number of modes for which the frequency differences ranged between -0.05 and -0.4 microhertz, all of the remaining frequency differences were less than ± 20 nanohertz. Such tiny frequency differences will have no measurable effects on the sound speed profiles which will be shown in the next section.

4. Inversion Results

4.1. Radial Stratification from Mean-Multiplet Frequencies

We have determined the spherically symmetric structure of the Sun by using the optimally localized averaging techniques (Backus and Gilbert, 1968; 1970) to invert three different sets of frequencies, ν_{nl}, obtained from the mean-multiplet method. Figure 10(a) shows the relative differences between the square of the sound speed in the Sun and model S of Christensen-Dalsgaard *et al.* (1996) for three different sets of mean-multiplet frequencies. The large dots with the superimposed horizontal and vertical error bars are the results of the inversion of the full set of 2047 frequencies obtained from the 144-day observing run. The dashed curve (which is almost indistinguishable from the dots over most of the figure) shows the results of a similar inversion of the subset of those modes which were common to the averaged-spectrum method. This subset contained 1940 modes. (This common mode subset of the mean-multiplet data set is also included on the CD-ROM for comparison with the full mode set.) The curve comprised of small dots with no errors bars gives the result of an inversion of the set of 1473 modes obtained from the 60-day observing run which was previously published by Kosovichev *et al.* (1997).

It is important to note that the inversion results represents localized estimates of the sound speed, which can be interpreted as a result of convolution of the solar sound-speed profile with localized averaging kernels, a sample of which is shown in Figure 11 (a data file with the kernels is available on the CD-ROM). These kernels are obtained as linear combinations of the sensitivity functions of individual modes with no a priori assumption about the solar sound-speed profile. When the solar models are compared with the inversion results, the model sound-speed profiles must be convolved with the averaging kernels.

In Figure 10, the inversion results are characterized by four parameters: the localized average of the squared sound-speed difference between the Sun and the reference model, the central position of the averaging kernels, the formal error estimate obtained from the errors of the frequencies of the individual modes, and the characteristic width of the kernals ('spread'). For the central position and for the width, we have adopted the definitions by Backus and Gilbert (1968, 1970), which fully account for the sidelobes of the localized kernels and their asymmetry that are particularly important near the center and the surface.

Occasionally, the effects of the sidelobes and asymmetry are ignored in presentations of inversion results when some other definitions for the central location and for the width are employed (e.g., Turck-Chieze *et al.*, 1997; Basu *et al.*, 1996) resulting in a misleading impression of improvement of the resolution of the core structure or the structure of subsurface layers. However, the true resolution can be judged only through examining the averaging kernels, also because quite often an attempt to obtain a more narrow central peak leads to higher sidelobes, in particular, near the surface. Since the subsurface structure is not well determined yet, doing so may result in substantial systematic errors in the localized averages representing the core structure.

All three of these inverted profiles in Figure 10(a) show that the maximum difference in the square of the sound speed between the model and the Sun is only 0.4%. Nevertheless, this difference is very important for understanding solar evolution and physical processes inside the Sun. One feature that is common to all three of these sound-speed profiles is particularly notable. This is the narrow peak centered at 0.67 R, just beneath the convection zone. This peak was previously detected in the LOWL (Basu *et al.*, 1996) and GONG data (Gough *et al.*, 1996) and is most likely due to a deficit of helium in the sun in comparison with the reference solar model in this narrow region. The deficit of helium decreases the mean molecular weight and thus increases the sound speed. The deficit of helium could result from additional mixing of the material in the layer with the surrounding plasma if turbulence is generated in this layer because of rotational shear. Indeed, as was shown by Kosovichev *et al.* (1997), there is a strong radial gradient of the rotation rate in this layer.

A second interesting feature that is common to all three profiles is the decrease in the sound speed that occurs between $0.30R$ and $0.25R$, or just outside the energy-generating core. The three curves begin to deviate from one another inward of $0.25R$. In particular, the two inversions from the 144-day dataset both remain low, while the profile from the 60-day run shows a turn-up toward $0.075R$. The absence of this increase in sound speed toward the solar center in the two 144-day inversions is most likely due to the removal of some low-degree modes which were included in the inversion of the 60-day dataset. Hence, the reality of this turn-up in sound speed inward of the minimum near $0.20R$ must await the accumulation of additional observations of the low-degree modal frequencies which were removed from the 144-day mean-multiple dataset. These low-degree frequencies were deleted because they did not pass the reliability criteria for the 144-day run (see, e.g., Schou, 1992).

4.2. Radial Stratification from Averaged-Spectrum Frequencies

We also obtained additional estimates of the spherically-symmetric structure of the solar internal sound speed by using the same inversion techniques, but by applying them to invert the frequency estimates obtained from the averaged-spectrum method. Figure 10(b) shows the relative differences between the square of the sound speed in the Sun and the same model S of Christensen-Dalsgaard *et al.* (1996) as determined by inversions of the full set of 2794 averaged-spectrum modes (the large dots with the horizontal and vertical errors) and of the 1940 modes common to the mean-multiplet data set (the dashed curve). (This common mode subset of the averaged-spectrum frequencies is also included on the CD-ROM for comparions with the full mode set.) These two additional estimates of the radial profile of the

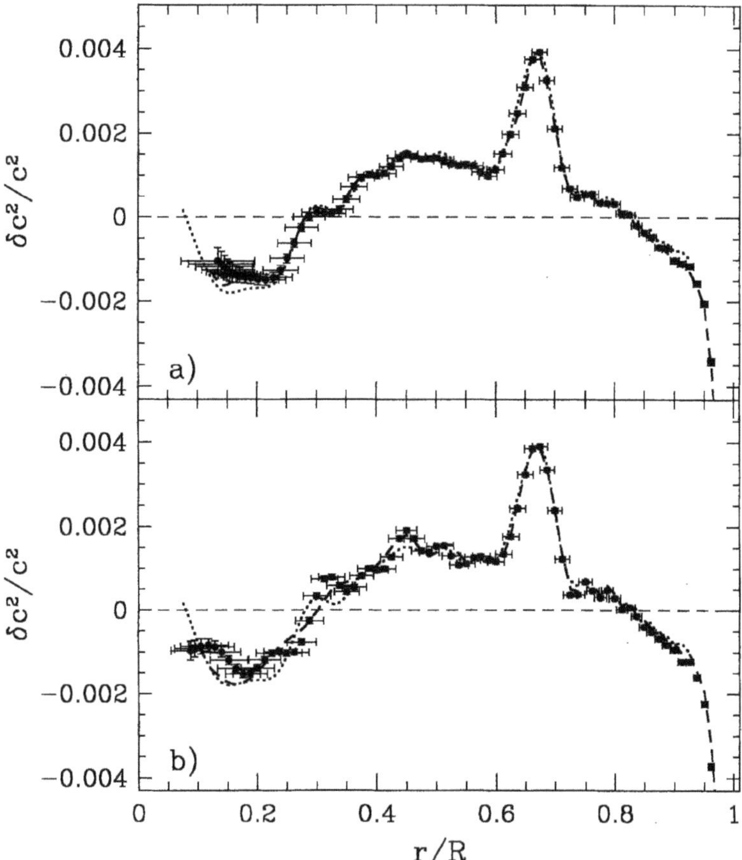

Figure 10. (a) Relative differences between the squared sound speed in the Sun and a standard solar model as inferred from three inversions of the frequency estimates computed with the mean-multiplet method from the MDI Medium-l data. The curve comprised of large dots with superimposed error bars came from the inversion of the entire set of 2047 frequencies obtained from the 144-day observing run. The dashed curve came from an inversion of a subset of 1940 of these modes that was common to the mode set determined from the averaged-spectrum method. The curve of small dots came from the inversion of the 1473 frequencies obtained from the 60-day observing run. (b) Radial profiles of similar inferred squared sound speed deviations from the standard solar model as inferred from the inversion of the averaged-spectrum frequency estimates. The large dots with the error bars were from from the inversion of the entire 144-day set of 2794 modes. The dashed curve was from the inversion of the set of 1940 modes that was common to the mean-multiplet mode set. The small dotted curve is repeated from the top panel for reference. In both panels the horizontal bars show the spatial resolution of the inferred values, while the vertical bars are the formal error estimates. These inversion results are compared in the color version of this figure on the CD-ROM. The four new sound-speed profiles are also included on the CD-ROM in tabular format.

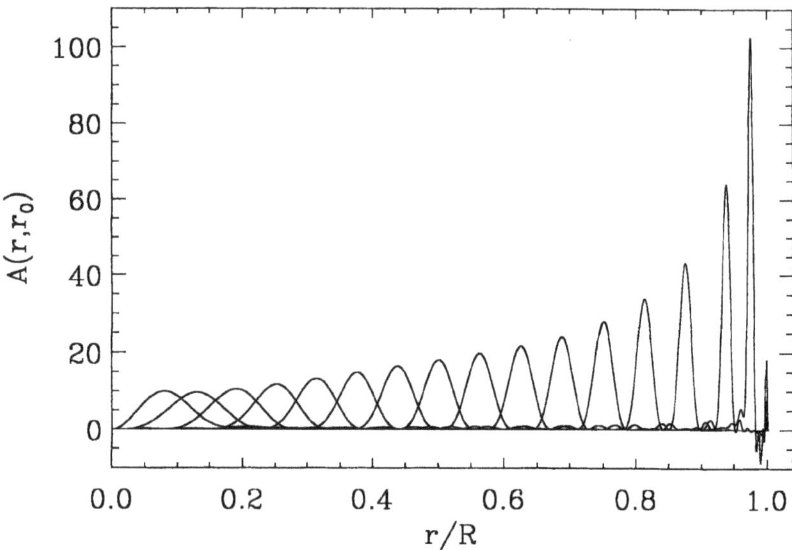

Figure 11. Localized averaging kernels for the squared sound-speed estimates.

internal sound speed show several similarities to the profiles shown in Figure 10(a). Most importantly, both of these profiles also show the same sharp increase in the square of the sound speed at a radius of $0.67R$ which was seen in the radial profiles which resulted from the inversions of the mean-multiplet frequency estimates shown in Figure 10(a).

Furthermore, both of these radial profiles of the solar sound speed show evidence for the decrease in the sound speed outside the solar core that is shown in Figure 10(a). However, the inverted profile obtained from the full mode set shows more radial variation in this region than does the profile obtained from the common mode set. Any verification of this extra variation will have to await the acquisition and analysis of additional observations. Also, both averaged-spectrum profiles show that the sound speed reaches a similar minimum value to that found in the mean-multiplet profiles inside of $0.20R$.

Finally, both of the averaged-spectrum profiles show some evidence for the turn-up toward the center that is seen in the Kosovichev *et al.* (1997) 60-day inversion. However, this turn-up is less prominent in the new inversions. Since the inversion results near the center depend only on a few low-degree modes (e.g., Gough, Kosovichev, and Toutain, 1995) that are particularly difficult to measure, the turn-up must be viewed with caution until we obtain more precise measurements of the low-degree mode frequencies.

5. Conclusions

The results of the different numerical tests which we have reported above suggest that we must continue to work diligently in order to learn whether or not the systematic differences in the frequency estimates which we have found are in fact due to the effects of the asymmetry of the observational peaks or whether they are due to another, as yet undetermined, cause. In particular, we must do so before we will be able to have confidence in the entire radial range of the sound speed profile which we have obtained from our current inversion computations. The potential importance of every deviation of the solar sound speed profile from that in our standard reference solar models is so great that we must continue our attempts at finding and learning about all possible sources of systematic errors in the input datasets which will be used as inputs to future inversion investigations. We must also put great effort into determining the validity of the low-frequency and low-degree modal fits which have a large influence on the sound speed in the solar core.

On the other hand, we have also demonstrated that we can anticipate that the formal frequency uncertainties which are employed as additional inputs to the inversion programs will continue to decrease as additional days of observations are included in the time series from which the observational power spectra are computed. This decrease in frequency uncertainties will increase our confidence in the inverted profiles once we have been able to eliminate the systematic errors which still exist.

We have also demonstrated that one of the more interesting features in the radial profile of the solar internal sound speed which has been found in the past few years from both GONG and MDI data, namely the rather narrow peak in the deviation of the sound speed from that of the current reference solar model beneath the convection zone, is not due to either any peculiarity in the choice of the seed frequencies which were employed in the determination of the frequencies which went into those inversions nor is it due to the particular method of frequency estimation that is employed.

Finally, all five of the inversions we have presented here also show evidence for a similar minimum value in the sound speed relative to the standard solar model in the solar core. The question of whether the sound speed remains low throughout the core or whether it once again increases toward the very center of the sun is a matter that must await the acquisition additional observations to be resolved.

Acknowledgements

The authors acknowledge many years of effort by the engineering and support staff of the MDI development team at the Lockheed Palo Alto Research Laboratory (now Lockheed-Martin) and the SOI development team at Stanford University.

SOHO is a project of international cooperation between ESA and NASA. This research is supported by the SOI-MDI NASA contract NAG5–3077 at Stanford University. The portion of this research which was conducted at USC and at the Technical University of Munich was supported by Stanford University Sub-Contract Number PR 6914 to USC and by NASA Grant NAGW-13 to USC. Portions of the computations for this research were carried out using the facilities of USC's Campus Computing Center. J.R. is grateful to R. Bulirsch for his generous support and hospitality.

References

Abrams, D. and Kumar, P.: 1996, *Astrophys. J.* **472**, 882.
Backus, G. and Gilbert, F. 1968, *Geophys. J. Roy. Astron. Soc.* **16**, 169.
Backus, G. and Gilbert, F. 1970, *Phil. Trans. Roy. Soc.* **A266**, 123.
Bahcall, J. N. and Pinsonneault, M. H.: 1995, *Rev. Mod. Phys.* **67**, 781.
Basu, S., Christensen-Dalsgaard, J., Schou, J., Thompson, M. J., and Tomczyk, S.: 1996, *Bull. Astron. Soc. India* **24**, 147.
Brown, T.: 1996, private communication.
Christensen-Dalsgaard, J., Däppen, W., and the GONG Team: 1996, *Science* **272**, 1286.
Duvall, T. L., Harvey, J. W., and Pomerantz, M. A.: 1986, *Nature* **321**, 500.
Duvall, T. L., Jr., Jefferies, S. M., Harvey, J. W., Osaki, Y., and Pomerantz, M. A.: 1993, *Astrophys. J.* **410**, 829.
Gabriel, M.: 1992, *Astron. Astrophys.* **265**, 771.
Gabriel, M.: 1993, *Astron. Astrophys.* **274**, 935.
Gabriel, M.: 1995, *Astron. Astrophys.* **299**, 245.
Gough, D. O.: 1993, in J. P. Zahn and J. Zinn-Justin (eds), *Astrophysical Fluid Dynamics*, Elsevier, Amsterdam, p. 399.
Gough, D. O., Kosovichev, A. G., and Toutain, T.: 1995, *Solar Phys.* **157**, 1.
Gough, D. O., Kosovichev, A. G., and the GONG Team: 1996, *Science* **272**, 1296.
Hill, F., Stark, P. B., Stebbins, R. T., and the GONG Team: 1996, *Science* **272**, 1292.
Korzennik, S. G.: 1990, Ph.D. Dissertation, University California, Los Angeles.
Kosovichev, A. G., Schou, J., Scherrer, P. H., Bogart, R. S., Bush, R. I., Hoeksema, J. T., Aloise, J., Bacon, L., Burnette, A., DeForest, C., Giles, P. M., Nigam, R., Rubin, R., Basu, S., Christensen-Dalsgaard, J., Däppen, W., Rhodes, E. J., Jr., Duvall, T. L., Jr., Howe, R., Thompson, M. J., Gough, D. O., Sekii, T., Toomre, J., Tarbell, T. D., Title, A. M., Mathur, D., Morrison, M., Wolfson, C. J., and Zayer, I.: 1997, *Solar Phys.* **170**, 43.
Kumar, P., Fardal, M. A., Jefferies, S. M., Duvall, T. L., Jr., Harvey, J. W., and Pomerantz, M. A.: 1994, *Astrophys. J.* **422**, L29.
Michaud, G. and Proffitt, C. R.: 1993, in W. Weiss and A. Baglin (eds), *Inside the Stars*, A.S.P. Conf. Ser., Vol. 40, Astron. Soc. Pacific, S.F., CA, p. 246.
Nigam, R. and Kosovichev, A. G.: 1996, in J. Provost and F. X. Schmieder (eds), 'Sounding Solar and Stellar Interiors', *Proc. IAU Symp.* **181**, in press.
Reiter, J. and Rhodes, E. J., Jr.:1997, SOI Technical Note, in preparation.
Ritzwoller, M. H. and Lavely, E. M.: 1991, *Astrophys. J.* **369**, 557.
Rogers, F. J. and Iglesias, C. A.: 1996, *Astrophys. J.* **456**, 902.
Roxbourgh, I. W. and Vorontsov, S. V.: 1995, *Monthly Notices Roy. Astron. Soc.* **272**, 850.
Scherrer, P. H., Bogart, R. S., Bush, R. I., Hoeksema, J. T., Kosovichev, A. G., Schou, J., Rosenberg, W., Springer, L., Tarbell, T. D., Title, A., Wolfson, C. J., Zayer, I. and the MDI Engineering Team: 1996, *Solar Phys.* **162**, 129.
Schou, J.: 1992, *On the Analysis of Helioseismic Data*, Thesis, Aarhus University.
Turck-Chieze, S., Basu, S., Brun, S., Christensen-Dalsgaard, J., Eff-Darwich, A., Lopes, I., Perez Hernandes, F., Berthomieu, G., Provost, J., Ulrich, R. K., Baudin, F., Boumier, P., Charra, J., Gabrial, A. H., Garcia, R. A., Grec, G., Renaud, C., Robillot, J. M., and Roca Cortes, T.: 1997, *Solar Phys.*, submitted.

TRI-PHONIC HELIOSEISMOLOGY: COMPARISON OF SOLAR *p* MODES OBSERVED BY THE HELIOSEISMOLOGY INSTRUMENTS ABOARD SOHO

T. TOUTAIN[1], T. APPOURCHAUX[2], F. BAUDIN[3], C. FRÖHLICH[4], A. GABRIEL[5], P. SCHERRER[6], B. N. ANDERSEN[7], R. BOGART[6], R. BUSH[6], W. FINSTERLE[4], R. A. GARCÍA[8], G. GREC[1], C. J. HENNEY[9], J. T. HOEKSEMA[6], A. JIMÉNEZ[10], A. KOSOVICHEV[6], T. ROCA CORTÉS[10], S. TURCK-CHIÈZE[8], R. ULRICH[9] and C. WEHRLI[4]

[1]*Observatoire de la Côte d'Azur, 06304 Nice Cedex, France*
[2]*Space Science Department, ESTEC, NL-2200 AG Noordwijk, The Netherlands*
[3]*National Solar Observatory, National Optical Astronomy Observatories, PO Box 26732, Tucson AZ 85726, U.S.A.*
[4]*Physikalisch-Meteorologisches Observatorium Davos World Radiation Center, CH-7260 Davos Dorf, Switzerland*
[5]*Institut d'Astrophysique Spatiale, CNRS/Université Paris XI, 91405 Orsay, France*
[6]*W.W. Hansen Experimental Physics Laboratory, Center for Space Science and Astrophysics, Stanford University, Stanford, CA 94305-4085, U.S.A.*
[7]*Norwegian Space Centre, P.O. Box 113 Skoyen, N-0212 Oslo, Norway*
[8]*Service d'Astrophysique, DSM/DAPNIA, CEA Saclay, 91191 Gif-sur-Yvette, France*
[9]*Astronomy Department, University of California, Los Angeles, U.S.A.*
[10]*Instituto de Astrofísica de Canarias, 38205 La Laguna, Tenerife, Spain*

(Received 10 March 1997; accepted 23 June 1997)

Abstract. The three helioseismology instruments aboard SOHO observe solar *p* modes in velocity (GOLF and MDI) and in intensity (VIRGO and MDI). Time series of two months duration are compared and confirm that the instruments indeed observe the same Sun to a high degree of precision. Power spectra of 108 days are compared showing systematic differences between mode frequencies measured in intensity and in velocity. Data coverage exceeds 97% for all the instruments during this interval. The weighted mean differences $(V - I)$ are -0.1 μHz for $l = 0$, and -0.16 μHz for $l = 1$. The source of this systematic difference may be due to an asymmetry effect that is stronger for modes seen in intensity. Wavelet analysis is also used to compare the shape of the forcing functions. In these data sets nearly all of the variations in mode amplitude are of solar origin. Some implications for structure inversions are discussed.

1. Introduction

On board the SOHO spacecraft are three helioseismology instruments: GOLF (Global Oscillations at Low Frequency), VIRGO (Variability of solar IRradiance and Gravity Oscillations) and MDI (the Solar Oscillations Investigation's Michelson Doppler Imager). MDI observes the Sun in velocity and intensity using a detector with 1024×1024 pixels. Unresolved observations of the Sun-as-a-star are made in velocity by GOLF and in three different wavelength bands and in total irradiance by VIRGO. In addition, VIRGO's LOI instrument images the Sun near 500 nm onto 12 pixels. All three instruments are described in detail in the special

Solar Physics **175:** 311–328, 1997.

issue of *Solar Physics* on SOHO (Scherrer *et al.*, 1995; Gabriel *et al.*, 1995; and Fröhlich *et al.*, 1995).

In this paper, we report on the first comparison of the time series and spectra measured by these instruments. The directly compared time series begin on 24 May 1996 and end on 23 July 1996. The spectra are computed from a longer time series of 108 days that ends on 8 September 1996. Finalization of the low-l intensity component of the MDI structure program determined the start date of the time series. The ending date for the power spectra was dictated by a temporary interruption of VIRGO data acquisition on 9 September 1996, probably caused by high energy particle irradiation. In the next section, we describe how the various data sets were obtained. In the two following sections, we compare the time series, the spectra, and the p-mode frequencies derived from the 108-day power spectra; time-frequency analysis is also used for comparison.

2. The Data Sets

2.1. GOLF VELOCITY

The performance of the GOLF instrument has been reported by Gabriel *et al.* (1997), and the p-mode analysis by Lazrek *et al.* (1997). GOLF measures full-disk integrated solar velocity using a sodium resonance cell (Approximate Observation Height (AOH) above τ_{500} = 1:500 km) to detect intensity at wavelengths near the D1 and D2 spectral lines. Since 11 April 1996, the instrument has been able to measure the intensity in only the blue wing of the sodium lines. This means that the signal obtained by GOLF is actually a combination of the true velocity signal, caused by displacement perturbations, and an intensity signal, caused by temperature and opacity variations. This has to be taken into account when one wants to measure the phase shift between the GOLF 'velocity' signal and the VIRGO or MDI 'intensity' signals. Phase analysis of the GOLF data (when this was possible, prior to April 1996) shows that the actual signal is mostly due to a Doppler shift; the remaining pure 'intensity' perturbation is minute, being responsible for a phase shift of only a few percent (Régulo *et al.*, 1996). Therefore the observed signal has been calibrated into velocity with an absolute precision of about 20%. The basic sampling rate of the instrument is five seconds with a duty cycle of 80%. For this study. the data have been averaged in to 60-s bins. The data coverage for the 108-day time series was better than 99.99%.

2.2. VIRGO INTENSITIES

The overall performance of the VIRGO instrument has been reported by Fröhlich *et al.* (1997). The VIRGO instrument has several components: sunphotometers, radiometers, and LOI. Each is described below.

2.2.1. *Sunphotometers (SPM)*

The VIRGO/SPM comprises two sunphotometers with three channels at 402 nm, 500 nm, and 862 nm that view the Sun as a star (AOH for all three: a few tens of km). The sampling rate of VIRGO/SPM is 60 s with a 94% duty cycle. The VIRGO sampling is offset by 30 s relative to the other two SOHO helioseismology instruments. This is because VIRGO is the only instrument synchronized to the SOHO daily pulse sent every midnight. The daily pulse was synchronized to Temps Atomique International before 27 March 1996, and thereafter to Universal Time. At that time the two standards were 30 s apart. The data coverage for the 108-day time series is 99.17%.

2.2.2. *Total Irradiance*

On VIRGO there are two radiometers that measure the radiative output of the Sun: DIARAD and PMO6 (Fröhlich *et al.*, 1997). The DIARAD instrument has a sampling rate of 120 s and is therefore not suitable for p-mode analysis. On the other hand the PMO6 is sampled every 60 s with a 31% duty cycle with samples taken 20 s later than the PMO6. This modification from the original mode of operation (Fröhlich *et al.*, 1995) is primarily due to the shutter not working as expected. Only the PMO6 is used for the p-mode comparison (AOH: close to red channel of the SPM). The data coverage for the 108-day time series is 97.08%.

2.2.3. *Luminosity Oscillations Imager (LOI)*

The performance of the VIRGO/LOI instrument has been reported by Appourchaux *et al.* (1997). The VIRGO/LOI images the Sun at 500 nm on a 12-pixel photodiode allowing detection of modes with $l \leq 8$ (AOH: same as green channel of the SPM). The sampling rate of VIRGO/LOI is 60 s with a duty cycle of 99% for the scientific pixels, and 5% for the guiding pixels. The LOI is in phase with the other instruments of VIRGO. The LOI data from the 12 scientific pixels can be combined to obtain a measurement of the Sun seen (nearly) as a star. The data coverage for the 108-day time series was better than 99.9%; most of the gaps are due to spacecraft momentum managements that create brief spikes in the data string.

2.3. MDI VELOCITY AND INTENSITY

The performance of the Michelson Doppler Imager instrument has been reported by Scherrer *et al.* (1997). MDI measures velocity in the 676.8 Ni I line using a pair of tunable Michelson interferometers with a narrow fixed passband Lyot. The velocity is computed from filtergrams taken at four wavelengths across the spectral line (AOH: 250–300 km). The continuum intensity and line depth are computed from these and an additional filtergram at another wavelength.

The data used here come from a proxy that sums the million full-disk two-arc sec pixels into 180 bins that subdivide the pixels of the VIRGO/LOI (Appourchaux *et al.*, 1997). The 12 LOI scientific pixels are divided as follows: the 4 north–south pixels are each divided into 17 bins and the four east–west and four central pixels into nine bins each; the four guiding pixels are each divided into 10 bins (see Appourchaux *et al.*, 1997 for the nomenclature of the pixels). For comparison with the full disk integrated signals of VIRGO/SPM and GOLF, we averaged the 140 non-limb LOI-proxy bins to determine a 'full disk' signal. Each LOI scientific pixel was also compared directly with its MDI counterpart, by averaging over the corresponding bins. For this analysis, the sampling rate of MDI is 60 s in phase with GOLF, normally with a 5% duty cycle. The data coverage for the 108-day time series is 97%.

3. Time Series Comparison

These measurements are of interest because they serve as diagnostics of the solar interior, most directly through analysis and interpretation of the p-mode spectra. Another important use of these data is to understand the longer-term variations that are caused by incoherent solar signals (sometimes referred to as solar noise). This will be essential to the search for g modes, yet another potentially powerful tool for sounding the deepest solar interior. To identify g modes, we must find a way to identify and then remove the comparatively large signals generated by the Sun through non-oscillatory, long time-scale processes. Only in this way will we be able to isolate the g modes. For periods longer than about 1 hour, the overlapping frequencies and the expected low amplitude for the g modes makes the search difficult. This is the domain where we most need to develop models of incoherent solar processes.

Each of the helioseismology instruments on SOHO measures a solar signal that samples the solar surface in a different way. Comparison of the time dependence shown in Figure 1 makes it clear that these incoherent processes influence each instrument in distinctly different ways. The VIRGO instruments respond rather distinctly to the passage of sunspots, especially between days 45 and 49. Note the apparent temporal offset between GOLF and VIRGO in responding to the sunspot. This is due to the fact that the sunspot affect the velocity and intensity signals differently. The GOLF signal is also more perturbed by the passage of the sunspots than MDI. This is because MDI is capable of correcting at a very high spatial resolution the effect of the sunspot on the velocity, while GOLF see only the integrated effect. Two ground-based indicators of activity are included: the Magnetic Plage Strength Index and the Mt. Wilson Sunspot Index (Ulrich, 1991).

Figure 2 shows 100 minutes of the full-disk time series observed by several of the instruments on 24 May 1996. The agreement among the instruments in intensity and in velocity is quite striking. It shows that all three instruments do observe the same Sun, and they do so to high precision. The two VIRGO instruments agree very well with each other; the MDI intensity is somewhat noisier due to the taking

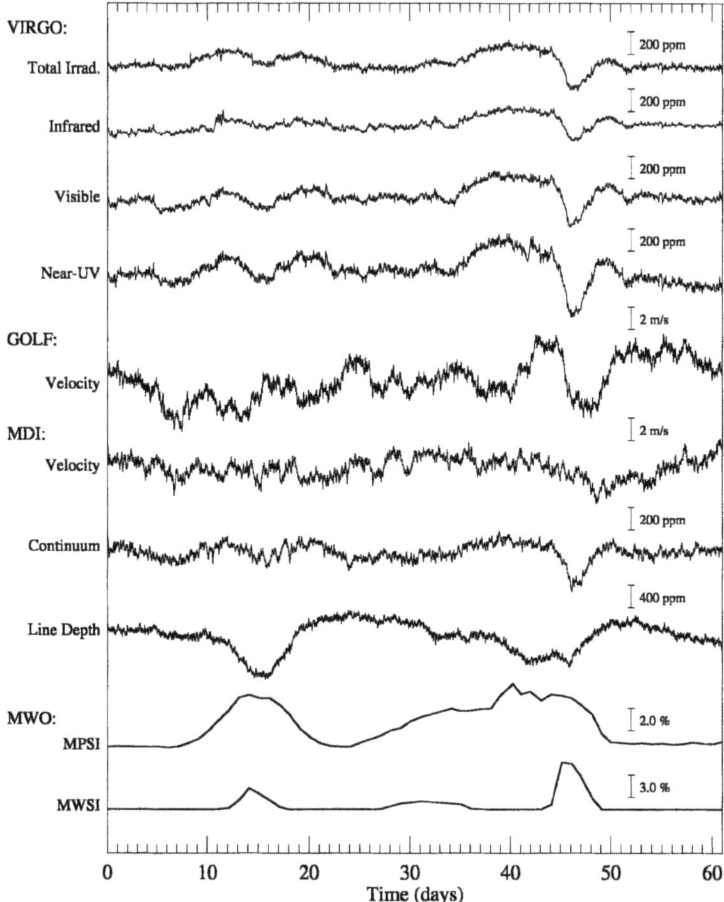

Figure 1. Comparison of the long-time trends for various observable from VIRGO, GOLF, and MDI during the period of 24 May through 23 July 1996. Reductions and calibrations of these quantities are still under study and future modifications of the results shown are possible. Each MDI signal was detrended with a 3rd-order polynomial. The GOLF velocity during this period was extracted from a 283-day long data set which was detrended with an 8th-order polynomial. The VIRGO data were detrended with a 5th-order polynomial. Also included in the plot are the Magnetic Plage Strength Index (MPSI) and the Mount Wilson Spot Index (MWSI) which show a relative measure of plage and spot activity, respectively. The percent change shown for MPSI and MWSI are relative to their typical respective values during solar maximum. All the above signals were smoothed using a 1-hour boxcar window except for the MPSI and the MWSI values which are daily averages.

of magnetograms. The intensity and velocity time series look different due to a higher sensitivity to p modes seen in velocity and to different characteristics of the low-frequency noise.

Figure 3 shows the time series of selected LOI pixels and the equivalent averages of the corresponding MDI LOI-proxy continuum bins. For each of the scientific

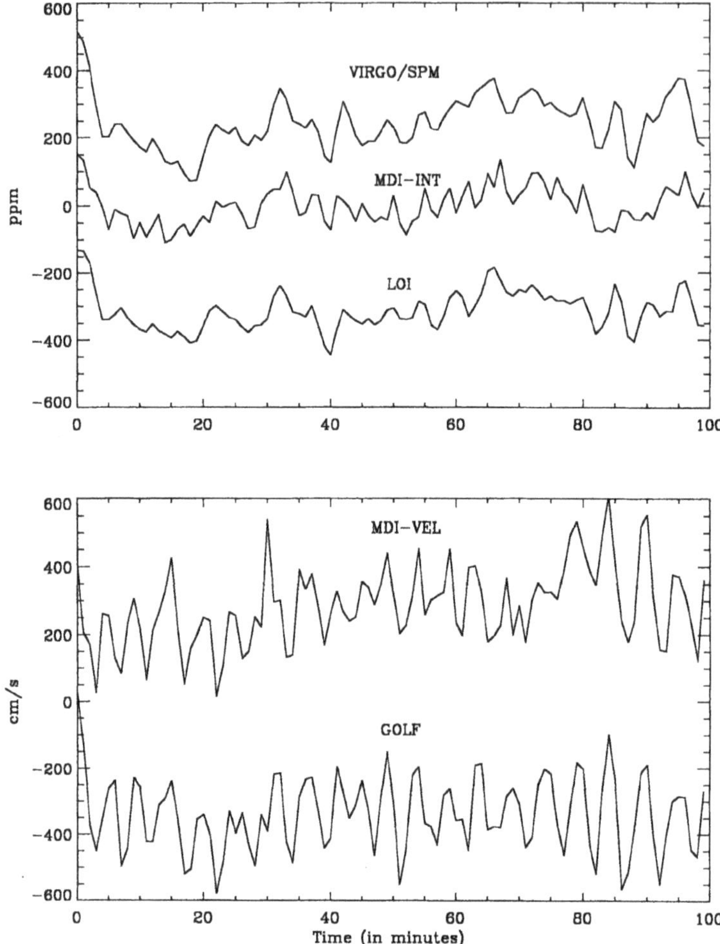

Figure 2. Upper panel: time series of the full disk integrated intensity of the VIRGO/SPM blue channel (*top*), the MDI LOI-proxy continuum (*middle*), and the VIRGO/LOI (*bottom*) for 24 May 1996 starting at 06:40 UT; the curves have been displaced by 300 ppm relative to each other. *Lower panel*: comparison of the GOLF velocity time series (*bottom*) and the averaged MDI LOI-proxy velocity (*top*) for 24 May 1996.

pixels the agreement is similar to Figure 2. For the guiding pixels there is a clear discrepancy. This may be due to the fact that for the guiding pixels the LOI and MDI data are not taken simultaneously. Though the duty cycle for each is 5% at the limb, the MDI sampling takes place throughout the minute.

In Tables I and II we have summarized how similar the Sun is observed by the various SOHO instruments. The VIRGO shifts were corrected for computing the correlations. The p mode correlations between the intensity signals is very high; the

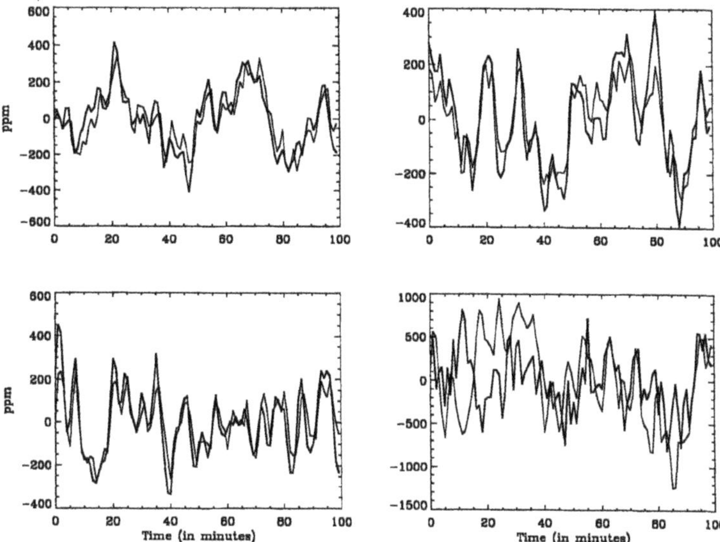

Figure 3. Time series of individual LOI pixels (bold line) and the corresponding MDI LOI-proxy bins (thin line) for 24 May 1996 starting at 06:40 UT. *Top-left*: pixel 1 (a north–south pixel). *Top-right*: pixel 5 (an east–west pixel). *Bottom-left*: pixel 9 (a central pixel). *Bottom-right*: pixel 13 (a guiding pixel). See Appourchaux *et al.* (1997) for the pixel numbering scheme.

lowest correlation is obtained with PMO6 intensity signal which is also the noisier. The high temporal shift between intensity and velocity is a phase effect. For the p modes this phase shift is between $-90°$ and $-110°$; it is in good agreement with the correlation found in Table I. For frequencies below 1 mHz, lower correlations are related to the solar noise and other low frequency noise. For instance, the LOI has a higher noise at very low frequency than the SPM, due to daily perturbations of the shape of the solar image (Appourchaux *et al.*, 1997). It is worth pointing out that the correlations between intensity and velocity have a different signature below 1 mHz than above 1 mHz.

The remarkable agreement among the signals detected by the different instruments, both for the 'noise' and for the p modes, can be used to our advantage to isolate different phenomena. For instance, we could use a reference instrument to correct the low-frequency noise. The goal is to reduce the solar noise and enhance the g modes. This might be less difficult if the g modes had somewhat different signatures in the different instruments.

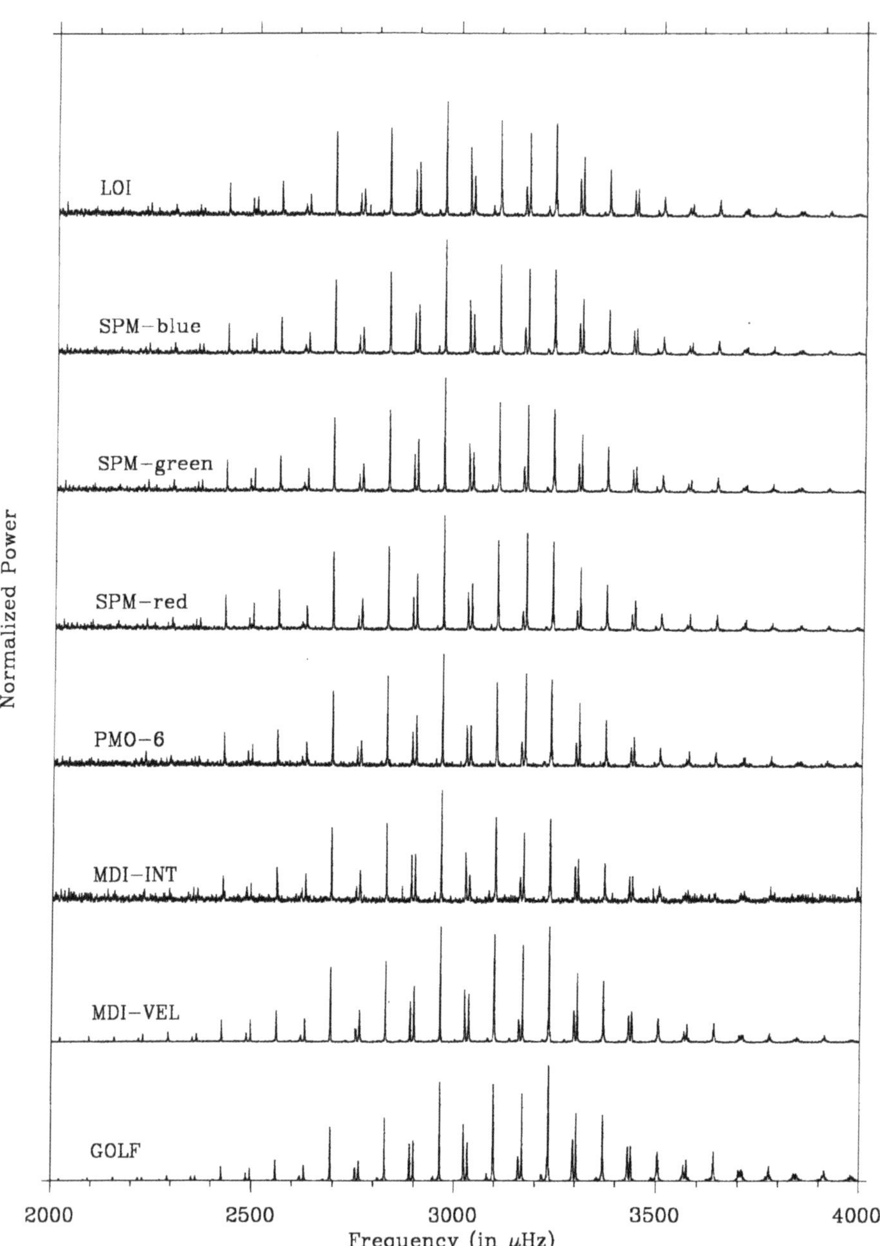

Figure 4. Full-disk power spectra derived from the time series of the various instruments. Each spectrum has been normalized to the maximum power in the frequency range and offset by an equal amount.

Table I

Correlation for frequencies above 1 mHz (p modes) for a 31-day time series starting 24 May 1996 at 00:00 UT. *Top part*: maximum correlation. *Bottom part*: shift in seconds for the maximum correlation. All shifts in seconds have been corrected from the various offsets in the different samples. Please note that the time shifts are given for completeness; they should not be used to deduce a mean phase shift between intensity and velocity.

Ins.	LOI	Blue	Green	Red	PMO6	MDI_I	MDI_V	GOLF
LOI	1.000	0.942	0.939	0.908	0.589	0.594	0.389	0.369
Blue		1.000	0.981	0.948	0.613	0.618	0.400	0.382
Green			1.000	0.953	0.613	0.618	0.400	0.378
Red				1.000	0.605	0.610	0.372	0.344
PMO6					1.000	0.390	0.252	0.239
MDI_I						1.000	0.246	0.237
MDI_V							1.000	0.753
GOLF								1.000
LOI	0.00	−2.43	−0.43	−8.09	−6.30	−3.79	+124.02	+250.15
Blue		0.00	+1.19	−6.56	−2.59	−3.23	+125.71	+251.83
Green			0.00	−8.42	−4.85	−5.40	+122.34	+262.76
Red				0.00	+2.93	+1.19	+148.82	+273.69
PMO6					0.00	−1.00	+122.37	+242.61
MDI_I						0.00	+142.10	+244.69
MDI_V							0.00	+36.74
GOLF								0.00

Table II

Correlation for frequencies below 1 mHz (solar noise) for a 31-day time series starting 24 May 1996 at 00:00 UT

Ins.	LOI	Blue	Green	Red	PMO6	MDI_I	MDI_V	GOLF
LOI	1.000	0.407	0.467	0.386	0.154	0.517	−0.072	0.221
Blue		1.000	0.933	0.760	0.586	0.407	−0.044	0.239
Green			1.000	0.654	0.376	0.391	−0.031	0.235
Red				1.000	0.583	0.443	−0.048	0.241
PMO6					1.000	0.191	−0.029	0.106
MDI_I						1.000	−0.081	0.226
MDI_V							1.000	0.505
GOLF								1.000

4. *p*-Mode Comparison

4.1. FREQUENCIES

Examples of 108-day full-disk and $l = 0$ spectra are shown in Figures 4 and 5 (resolution: 0.107167 μHz). No oversampling or window taper was used for com-

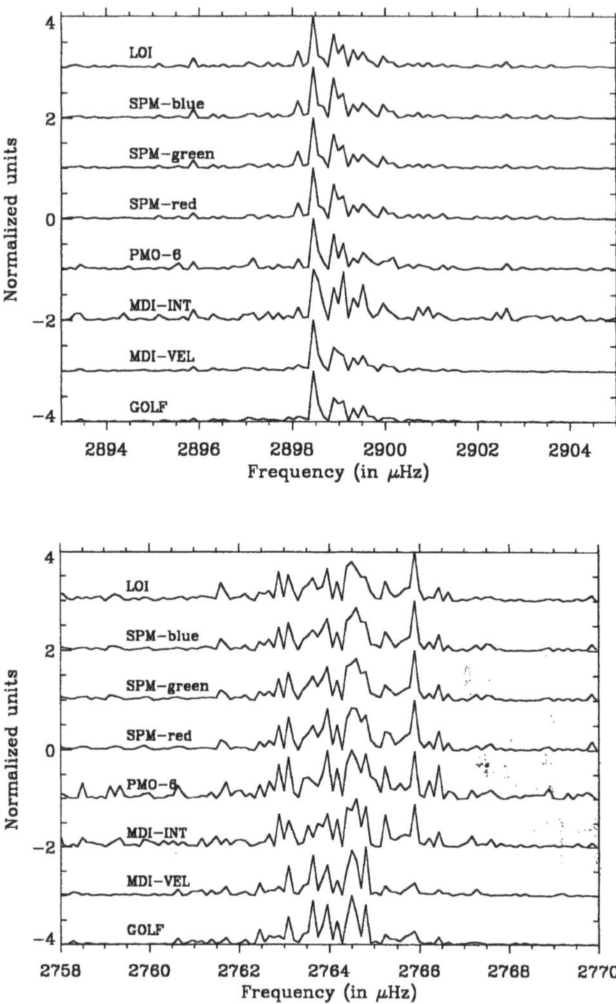

Figure 5. Power spectra of $l = 0$ modes observed by various instruments. *Top*: $n = 20$. *Bottom*: $n = 19$. Each spectrum has been normalized to the maximum power in the frequency range and offset by an equal amount.

puting the spectra; this ensures that the frequency bins are statistically independent (Gabriel, 1994). The p-mode frequencies have been measured using maximum likelihood estimators (Anderson, Duvall, and Jefferies, 1990). The full disk integrated spectra were fit assuming equal amplitudes for all the components in a multiplet; statistically all the components in a multiplet are cited to the same amplitude. All of the fits were performed with the same code. The 'reference' frequencies are those of the VIRGO/LOI 108-day power spectra. Figure 6 shows the frequency

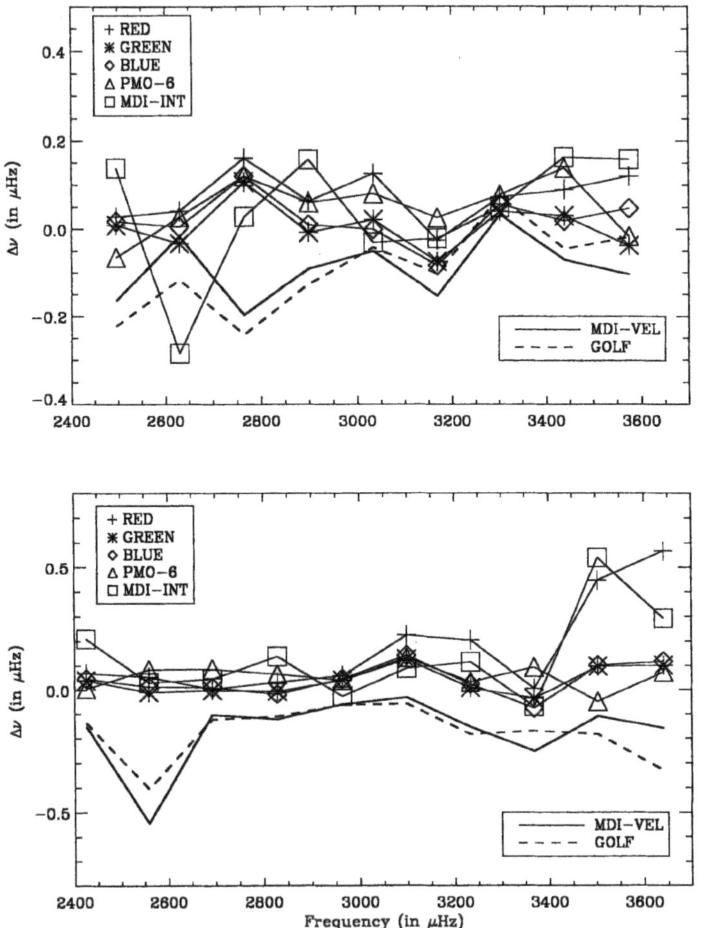

Figure 6. The frequency differences of GOLF, the MDI LOI-proxies, and VIRGO/SPM relative to VIRGO/LOI for $l = 0$ (*top panel*) and $l = 1$ (*bottom panel*). This is for the 108-day time series. A systematic shift is clearly visible between the intensity and velocity measurements. It is premature to determine whether the systematic difference has a frequency dependence.

comparison. The $l = 0, 1$ spectra show systematic differences; the mean frequency differences are given in Table III. The systematic difference manifests itself in Figure 5 as increased power at higher frequencies in intensity relative to velocity.

4.2. Time – Frequency Analysis

Another way to compare helioseismic time series is to look at the temporal behavior of the p modes, which are known to vary strongly in power (Baudin, Gabriel, and Gibert, 1994; Leifsen *et al.*, 1995). The method used here is derived from wavelet

Table III

Frequency differences with respect to the LOI intensity modes. This is a weighted mean over 9 modes ($17 \leq n \leq 25$) for $l = 0$ and over 10 modes for $l = 1$ ($16 \leq n \leq 25$). The error bars are 0.049 μHz and 0.035 μHz for $l = 0$ and $l = 1$, respectively. The error bars are derived from the LOI frequency error bars averaged optimally over the modes.

Instrument	$l = 0$ (in μHz)	$l = 1$ (in μHz)
SPM/Blue	+0.010	−0.001
SPM/Green	+0.003	+0.008
SPM/Red	+0.058	+0.063
PMO6	+0.044	+0.069
MDI intensity	+0.046	+0.032
MDI velocity	−0.094	−0.174
GOLF	−0.101	−0.147

analysis (Baudin *et al.*, 1996). It has been applied to the initial 61-day time series of the various instruments in 10 μHz bands centered on the frequencies of the p modes identified above. An example is given in Figure 7.

The observed power variations are consistent with convective excitation of p modes (Woodard, 1984); as are the apparent frequency variations (Garcia *et al.*, 1996).

In previous data sets, the non-solar noise present in the data was expected to yield spurious variations of power and/or frequency. The comparison of the data from the helioseismic experiments on board SOHO shows their excellent quality. For the modes of high amplitude (in the frequency range 2700 $\mu\text{Hz} \leq \nu \leq 3500\,\mu\text{Hz}$), the distribution of power versus time and frequency is strikingly similar for the different time series, suggesting that the variations are of purely solar origin. It must be noted that this applies for *all* of the time series, including those with intrinsically lower signal-to-noise ratio. The analysis of lower amplitude modes may show discrepancies between velocity and intensity, the latter being noisier. Nevertheless, the power distributions in intensity measurements from different instruments are very similar, showing clearly that the origin of this distribution is solar, not instrumental.

Figure 8 shows also how the amplitude envelopes of an $l = 0$ mode varies with time for the different instruments. The envelopes look qualitatively very similar to those measured by Toutain and Fröhlich (1992). Table IV summarize the correlation between the envelope. The correlation between intensity and velocity is lower than for intensity/intensity or velocity/velocity.

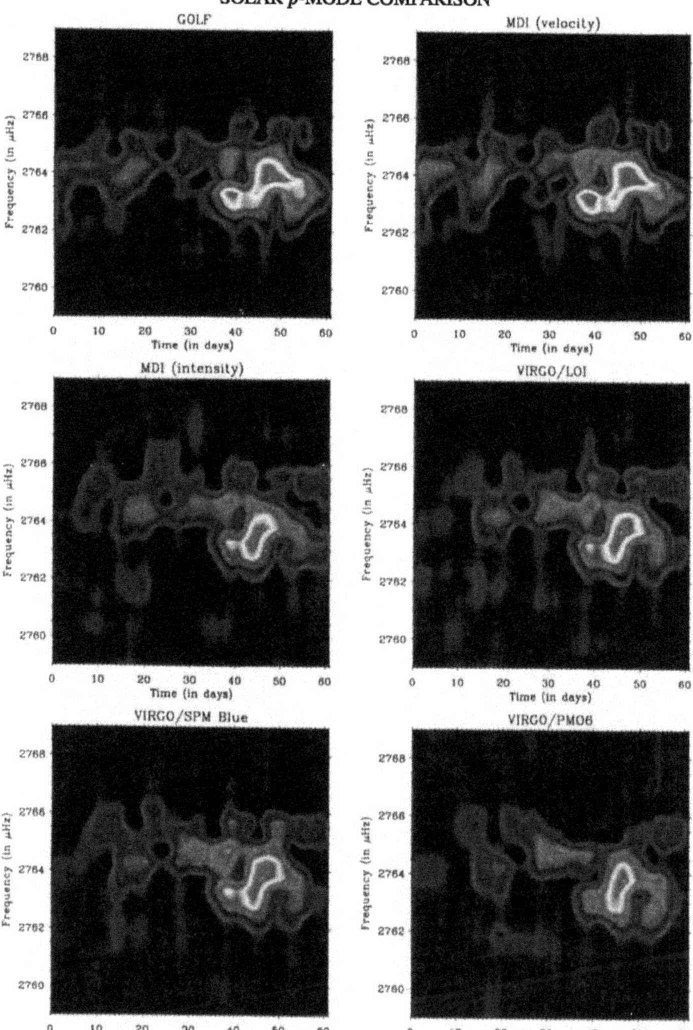

Figure 7. Wavelet analysis of the $l = 0$, $n = 19$ mode for the different SOHO instruments. This analysis was performed with Lorentzian wavelets of full width at half maximum of 1 μHz, corresponding to a 5.1-day resolution in time. The apparent synchronicity of the excitation of this mode on day 48 and the passage of a sunspot (see Figure 1) is probably coincidental, as it does not occur for other modes.

5. Discussion

The systematic difference between the frequencies of the $l = 0, 1$ modes measured in intensity and velocity is a very significant result of this analysis. This difference is also present in the time-frequency analysis: the velocity has slightly more power at lower frequencies than the intensity. The results presented in Table III clearly

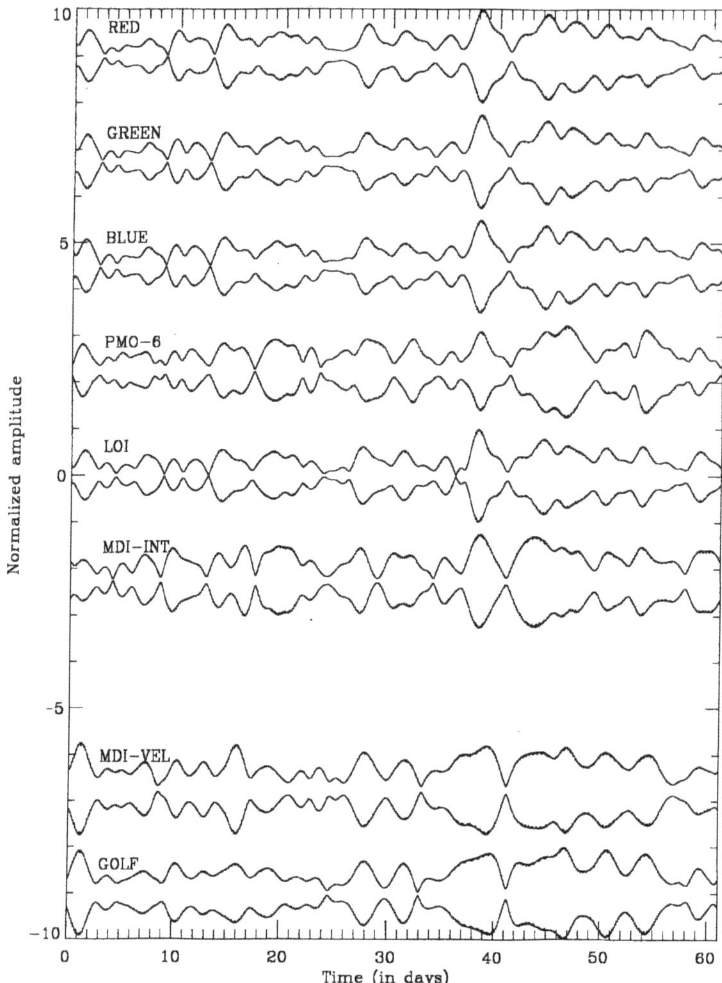

Figure 8. Amplitude envelope of the $l = 0$, $n = 19$ mode as a function of time for the various instruments. The envelope of the amplitude was extracted from the mode time series using a Hilbert transform. The mode time series was derived by computing the inverse Fourier transform of a spectrum filtered around a p mode with a gaussian window of width 10 μHz.

show a shift between the measured frequencies of the velocity and intensity modes. There are also indications of shifts seen in different colors. The channels with the longest effective wavelengths (SPM Red, PMO6, and MDI) give higher frequencies (+0.05 μHz with uncertainties of 0.03 and 0.02 μHz for $l = 0, 1$, respectively) than the green channel of the SPM and the LOI. The comparison between the velocity measurements should properly be made with intensity measurements at these longer effective wavelengths. This results in an average frequency difference

Table IV
Correlation for the envelopes shown in Figure 8

Ins.	LOI	Blue	Green	Red	PMO6	MDI_I	MDI_V	GOLF
LOI	1.00	0.96	0.96	0.96	0.77	0.79	0.63	0.69
Blue		1.00	0.99	0.97	0.81	0.81	0.64	0.70
Green			1.00	0.99	0.79	0.80	0.64	0.70
Red				1.00	0.77	0.79	0.63	0.72
PMO6					1.00	0.66	0.55	0.64
MDI_I						1.00	0.49	0.54
MDI_V							1.00	0.93
GOLF								1.00

between modes seen in intensity and velocity of about -0.15 μHz and -0.2 μHz for $l = 0, 1$ modes, respectively.

Several causes for these differences can be proposed. They all depend on the basic fact that the observations are caused by the response of the solar atmosphere to the standing wave patterns of the modes. The response depends both on the effective formation height in the atmosphere and on whether the observations are made in velocity or intensity. These frequency differences should increase with deteriorating quality of the oscillation cavity.

The changing shift with wavelength may be caused by the difference in formation height of the emergent radiation. At disc center the height difference between the red and green channels of the SPM is a few tens of kilometers. The observed formation height increases toward the limb. The shift caused by a difference in formation height will further increase because the green channel is less weighted towards the limb.

The formation height arguments do not explain the frequency difference between velocity and intensity, since the mode frequencies shift in the *opposite* direction with formation height. The formation heights of the lines where the velocity is measured are a few hundred kilometers above the green channel.

Since we observe the p modes indirectly, through the response of the atmosphere, the difference could be attributed to an effect of the observed atmosphere. One of the main effects of the atmosphere on the p modes is to modify the natural line profiles of the modes. For example mode interferences, like those in a Fabry–Pérot cavity, can cause a line asymmetry, as reported by Duvall *et al.* (1993) and Kosovichev *et al.* (1997). The frequency differences among the SOHO instruments are in rough agreement with the profiles observed by Duvall *et al.* (1993) for high degree modes: frequencies of p modes derived from intensity observations are higher than those observed in velocity. This is mainly due to line asymmetry effects pulling the mode frequencies in opposite directions (Figure 5). On the other hand, it is premature to conclude that asymmetries will always have such systematic effects: Nigam,

Kosovichev, and Scherrer (1996) showed that the line asymmetry varies with mode frequency and degree.

Another 'side effect' of the frequency shifts is related to structure inversions. Appourchaux *et al.* (1997) reported that there seems to be a systematic difference between their results and the LOWL inversion, as well as with inversions using contemporaneous data from MDI (Kosovichev *et al.*, 1997). These inversion differences may be due to line asymmetries, as suggested by Kosovichev (1996, private communication), that lead to systematic errors in the determination of the frequencies. We hope that by understanding the line asymmetry we will be able to reduce systematic errors and thereby better interpret the results of inversions derived from VIRGO intensity measurements.

6. Conclusion

The time series of all three helioseismology instruments agree extremely well with each other. Solar p-mode frequencies show a systematic difference between intensity and velocity. This systematic difference for low l modes is attributed to a line asymmetry that is stronger in intensity and gives higher measured mode frequencies. Fitting p modes with an asymmetrical profile might help reduce the discrepancy, and may produce inversion results from intensity measurements that agree better with those from velocity. The uniformity of the data from the various instruments should be advantageous in identifying g modes.

Acknowledgements

All three instruments benefit from the quiet and well run SOHO platform built by Matra-Marconi Space. SOHO is an international collaboration programme of the European Space Agency and the National Aeronautics Space Administration. F.B. and the SOI/MDI Team at Stanford acknowledge support from NASA under Contract NAS 5–30386.

VIRGO is a cooperative effort of many individual scientists and engineers at several institutes in Europe and U.S.A. The VIRGO team has been supported by several national and international funding agencies which are gratefully acknowledged: the PMOD/WRC by the Swiss National Science Foundation and PRODEX, the IRMB by the Fonds de la Recherche Fondamentale Collection d'initiative ministérielle and PRODEX, the SSD/ESA by their annual funds from the Science Directorate, the IAC by the CICYT through PNIE, and the OCA by the CNES and CNRS.

References

Appourchaux, T., Andersen, B. N., Fröhlich, C., Jiménez, A., Telljohann, U., and Wehrli, C.: 1997, *Solar Phys.* **170**, 27.

Appourchaux, T., Sekii, T., Gough, D.O., Lee, U., Wehrli, C., and the VIRGO team: 1996, in in F.-X. Schmider and J. Provost (eds), 'Structure Inversion with the VIRGO Data', *IAU Colloq.* **181**, in press.

Anderson, E. R., Duvall, T. L., Jr., and Jefferies, S. M.: 1990, *Astrophys. J.* **364**, 699.

Baudin, F., Gabriel, A. H., and Gibert, D.: 1994, *Astron. Astrophys.* **285**, L29.

Baudin, F., Gabriel, A. H., Gibert, D., Pallé, P. L., and Régulo, C.: 1996, *Astron. Astrophys.* **311**, 1024.

Duvall, T. L., Jr., Jefferies, S. M., Harvey, J. W., Osaki, Y., and Pomerantz, M. A.: 1993, *Astrophys. J.* **410**, 829.

Fröhlich, C., Romero, J., Roth, H., Wehrli, C., Andersen, B. N., Appourchaux, T., Domingo, V., Telljohann, U., Berthomieu, G., Delache, P., Provost, J., Toutain, T., Crommelynck, D., Chevalier, A., Fichot, A., Däppen, W., Gough, D. O., Hoeksema, T., Jiménez, A., Gómez, M., Herreros, J., Roca-Cortés, T., Jones, A. R., Pap, J., and Willson, R. C.: 1995, *Solar Phys.* **162**, 101.

Fröhlich, C., Andersen, B. N., Appourchaux, T., Berthomieu, G., Crommelynck, D., Domingo, V., Fichot, A., Finsterle, W., Gómez, M. F., Gough, D. O., Jiménez, A., Leifsen, T., Lombaerts, M., Pap, J., Provost, J., Roca-Cortés, T., Romero, J., Roth, H., Sekii, T., Telljohann, U., Toutain, T., and Wehrli, C.: 1997, *Solar Phys.* **170**, 1.

Gabriel, A. H., Grec, G., Charra, J., Robillot, J.-M., Roca Cortés, T., Turck-Chièze, S., Bocchia, R., Boumier, P., Cantin, M., Cespédes, E., Cougrand, B., Crétolle, J., Damé, L., Decaudin, M., Delache, P., Denis, N., Duc, R., Dzitko, H., Fossat, E., Fourmond, J.-J., García, R. A., Gough, D. O., Grivel, C., Herreros, J. M., Lagardère, H., Moalic, J.-P., Pallé, P. L., Pétrou, N., Sanchez, M., Ulrich, R., and Van der Raay, H. B.: 1995, *Solar Phys.* **162**, 61.

Gabriel, A. H., Charra, J., Grec, G., Robillot, J.-M., Roca Cortés, T., Turck-Chièze, S., Ulrich, R., Basu, S., Baudin, F., Bertello, L., Boumier, P., Charra, M., Christensen-Dalsgaard, J., Decaudin, M., Dzitko, H., Foglizzo, E., Fossat, E., García, R. A., Herreros, J. M., Lazrek, M., Pétrou, N., Renaud, C., and Régulo, C.: 1997, *Solar Phys.* **175**, 207 (this issue).

Gabriel, M.: 1994, *Astron. Astrophys.* **287**, 685.

Garcia, R., Foglizzo, T., Turck-Chièze, S., Baudin, F., Boumier, P., and the GOLF Team: 1996, in F.-X. Schmider and J. Provost (eds), 'Time Evolution of p Modes in the GOLF Data', *IAU Colloq.* **181**, in press.

Kosovichev, A. G., Schou, J., Scherrer, P. H., Bogart, R. S., Bush, R. I., Hoeksema, J. T., Aloise, J., Bacon, L., Burnette, A., DeForest, C., Giles, P. M., Leibrand, K., Nigam, R., Rubin, M., Scott, K., Williams, S. D., Basu, S., Christensen-Dalsgaard, J., Däppen, W., Rhodes, E. J., Jr., Duvall, T. L., Jr., Howe, R., Thompson, M. J., Gough, D. O., Sekii, T., Toomre, J., Tarbell, T. D., Title, A. M., Mathur, D., Morrison, M., Saba, J. L. R., Wolfson, C. J., Zayer, I., and Milford, P. N.: 1997, *Solar Phys.* **170**, in press.

Lazrek, M., Baudin, F., Bertello, L., Boumier, P., Charra, J., Fierry-Fraillon, D., Fossat, E., Gabriel, A. H., García, R. A., Gelly, B., Gouiffes, C., Grec, G., Pallé, P. L., Peréz Hernandez, F., Régulo, C., Renaud, C., Robillot, J.-M., Roca Cortés, T., Turck-Chièze, S., and Ulrich, R.: 1997, *Solar Phys.* **175**, 227 (this issue).

Leifsen T., Hanssen, A., Andersen, B. N., and Toutain, T.: 1995, in R.Ulrich, E. J.Rhodes, Jr., and W. Däppen (eds), 'Wavelet Analysis of IPHIR Data', *Helio- and Astero-Seismology from the Earth and Space*, ASP Conf. Series 76, p. 520.

Nigam, R., Kosovichev, A. G., and Scherrer, P. H.: 1996, in F.-X. Schmider and J. Provost (eds), 'Study of Line Asymmetries of Solar Oscillations Above and Below the Acoustic Cut-Off Frequency', *IAU Colloq.* **181**, in press.

Régulo, C., Roca Cortés, T., Boumier, P., Robillot, J.-M., Garcia, R. A., Turck-Chièze, S., Ulrich, R. K., and the GOLF Team: 1996, in F.-X. Schmider and J. Provost (eds), 'Velocity and/or Intensity Signal in the GOLF One Wing Signal?', *IAU Colloq.* **181**, in press.

Scherrer, P. H., Bogart, R. S., Bush, R. I., Hoeksema, J. T., Kosovichev, A. G., Schou, J., Rosenberg, W., Springer, L., Tarbell, T. D., Title, A., Wolfson, C. J., Zayer, I., and the MDI Engineering Team: 1995, *Solar Phys.* **162**, 129.
Scherrer, P. H. and the MDI Team: 1997, *Solar Phys.*, submitted.
Toutain, T. and Fröhlich, C.: 1992, *Astron. Astrophys.* **257**, 287.
Ulrich, R.: 1991, *Adv. Space Res.* **11**, 217.
Woodard, M.: 1984, *Short-Period Oscillations in the Total Solar Irradiance*, Ph.D. thesis, University of California, San Diego,, California.

MODELING THE DISTRIBUTION OF MAGNETIC FLUXES IN FIELD CONCENTRATIONS IN A SOLAR ACTIVE REGION

CAROLUS J. SCHRIJVER[1], ALAN M. TITLE[1], HERMANCE J. HAGENAAR[2] and RICHARD A. SHINE[1]

[1] *Stanford-Lockheed Institute for Space Research, Dept. H1-12, Bldg. 252, 3251 Hanover Street, Palo Alto CA 94304, U.S.A.*

[2] *Astronomical Institute, University of Utrecht, P.O. Box 80,000, 3508 TA, The Netherlands*

(Received 14 February 1997; accepted 3 June 1997)

Abstract. Much of the magnetic field in solar and stellar photospheres is arranged into clusters of 'flux tubes', i.e., clustered into compact areas in which the intrinsic field strength is approximately a kilogauss. The flux concentrations are constantly evolving as they merge with or annihilate against other concentrations, or fragment into smaller concentrations. These processes result in the formation of concentrations containing widely different fluxes. Schrijver *et al.* (1997, Paper I) developed a statistical model for this distribution of fluxes, and tested it on data for the quiet Sun. In this paper we apply that model to a magnetic plage with an average absolute flux density that is 25 times higher than that of the quiet network studied in Paper I. The model result matches the observed distribution for the plage region quite accurately. The model parameter that determines the functional form of the distribution is the ratio of the fragmentation and collision parameters. We conclude that this ratio is the same in the magnetic plage and in quiet network. We discuss the implications of this for (near-)surface convection, and the applicability of the model to stars other than the Sun and as input to the study of coronal heating.

1. Introduction

The magnetic field in the solar outer atmosphere is rooted in an ensemble of relatively small concentrations of flux in the photosphere. These concentrations are embedded in the granular and supergranular flows, as reviewed by Spruit, Nordlund, and Title (1990). The shaking and braiding of the field lines at the photospheric level and below results in outer atmospheric heating (e.g., Parker, 1994) that is ubiquitous among magnetically active cool stars that, like the Sun, have a convective envelope immediately below their photosphere. The details of the field generation and dynamics can only be studied for the Sun. Understanding the physics of these processes is the key to understanding how outer atmospheres of cool stars are structured and heated by magnetic fields emerging through their surfaces. After all, the distribution of magnetic flux over the stellar surface, on both the small and large scale, determines the total non-radiative heating that occurs in stellar outer atmospheres (e.g., Schrijver and Harvey, 1989; Schrijver, 1993).

The large-scale, low-resolution distribution of the field is described quite well by a random-walk diffusion (e.g., Wang and Sheeley, 1991, and references therein) in which the field is assumed to respond passively to the convective flows, differential rotation, and meridional flow. On the smallest observable scales, the solar

Solar Physics **175:** 329–340, 1997.

photospheric field is bunched into compact regions that are generally considered to be (clusters of) nearly vertical flux tubes. The intrinsic field strength within these flux tubes is 1–2 kG. On scales of at most a few thousand kilometers, these flux tubes appear to positioned randomly at least within magnetic plages (Schrijver *et al.*, 1992), by an interaction with the granular flows that position the field in the downflows of surrounding cells.

The present paper deals with the intermediate scale of a few thousand kilometers up to scales exceeding some 100 000 km. At a resolution of 1 to 2 arc sec (700 to 1500 km), the photospheric magnetic field is again bunched into clusters of flux tubes, forming what we shall refer to as flux concentrations: at the resolution of a magnetograph (here the SOHO Michelson Doppler Imager – MDI, Scherrer *et al.*, 1995) these concentrations appear as coherent patches of photospheric magnetic field. These concentrations are found in areas of strong flow convergence and downflow (see the review by Spruit, Nordlund, and Title, 1990).

Flux concentrations in the photosphere are not long-lived entities moving about with a well-defined identity. They are deformed, broken up, and driven together to merge or cancel in the evolving flow field, which may itself depend on the strength and geometry of the field. The intrinsically dynamic character makes it hard to study the behavior of concentrations of flux, and therefore also hampers the interpretation of the field dynamics. Schrijver *et al.* (1997, hereafter referred to as Paper I) developed a model that quantifies these processes in a statistical sense. Their model incorporates three processes for the evolution of concentrations of magnetic flux: (i) flux concentrations can collide with other concentrations of the same polarity to merge into a larger concentration; (ii) they can collide with concentrations of the opposite polarity, in which case the flux in the smallest concentration is assumed to cancel against the same amount of flux of the larger concentration; (iii) they can fragment into smaller pieces. These processes of merging and fragmentation refer to apparent merging or splitting on the scale of the instrumental resolution. The processes are described in terms of collision frequencies and fragmentation rates. The model predicts a (nearly) exponential decrease in the number density $N(\Phi)$ of concentrations as a function of flux Φ in quiet network, as is observed, when the fragmentation rate increases roughly in proportion to the flux in the concentration, i.e., bigger concentrations tend to fragment more rapidly than smaller ones. Schrijver *et al.* (1997) estimate that the mean time scale for fragmentation of concentrations in the quiet network is about 3 to 6 days for an average concentration containing 3×10^{18} Mx.

If that model is to be applied to stars with levels of magnetic activity that differ from that of the Sun, the model must be tested in solar regions of different activity. Paper I tests the model for very quiet network. In the present paper we use the same description to model the distribution $N(\Phi)\, \mathrm{d}\Phi$ for flux concentrations in a relatively large magnetic plage. The mean magnetic flux density there is 25 times higher than in the quiet network studied in Paper I.

2. Observations

We use a sequence of magnetograms obtained with MDI to determine the number density of concentrations as a function of their flux content. We concentrate on Active Region 7962 and its surroundings, using a sequence of 316 magnetograms that were obtained in the high-resolution mode between 20:08 UT on 12 May 1996 and 01:30 UT on 13 May 1996. The magnetograms were binned onboard the spacecraft into groups of 2 by 2 pixels, where each original pixel is an 0.60 arc sec square. The spatial resolution of MDI is about 1.2 arc sec, i.e., equal to our bin size. The MDI magnetograms are calibrated to magnetic fluxes per pixel by a conversion factor of 5×10^{15} Mx per 'data number' (DN), or flux densities of 2.8 G/DN (T. D. Tarbell, private communication).

At the time of observation, the active region was located near disk center. A typical magnetogram of the region is shown in Figure 1. Only the leading polarity contained sizeable spots. These spots and their surrounding moats were exluded from our analyses.

We identify concentrations of magnetic flux in the MDI magnetograms with an algorithm that uses both a threshold and a curvature. The identification procedure, described in detail by Hagenaar, Schrijver, and Title (1997), can be summarized as follows. We first apply a Gaussian smoothing with a FWHM of two 1.2 arc sec image pixels, or about 1700 km. A pair of logical (bit)maps is then created by determining the sets of image pixels for which the local curvature in all directions (determined from the signal in that pixel and its nearest neighbors) is negative for positive polarity, or positive for negative polarity. We thus locate pixels at peak values in the flux density, but also on slopes adjacent to peaks. All pixels within a cluster defined as a set of adjacent nearest-neighbor pixels in these maps are subsequently given the same unique label. The final map is then created by preserving only those clusters which have at least one pixel exceeding a threshold. This threshold is set to 17 G, or a detection limit of approximately 0.5×10^{17} Mx for a 2 by 2 pixel concentration. Because the segmentation algorithm is based on curvature, only the flux in the convex core of the concentration is included. Hagenaar, Schrijver, and Title (1997) estimate that the total flux in concentrations is underestimated by a factor of about three. We apply this same factor of three as a correction factor in the computation of total fluxes in the present study, because owing to the relatively high number density of concentrations in the plage, an insufficient number of essentially isolated concentrations could be identified to determine this factor independently.

3. Summary of the Model

The model developed in Paper I describes the number density $N(\Phi)\,\mathrm{d}\Phi$ of the concentrations of each polarity containing a total flux $|\Phi|$ as a function of the average

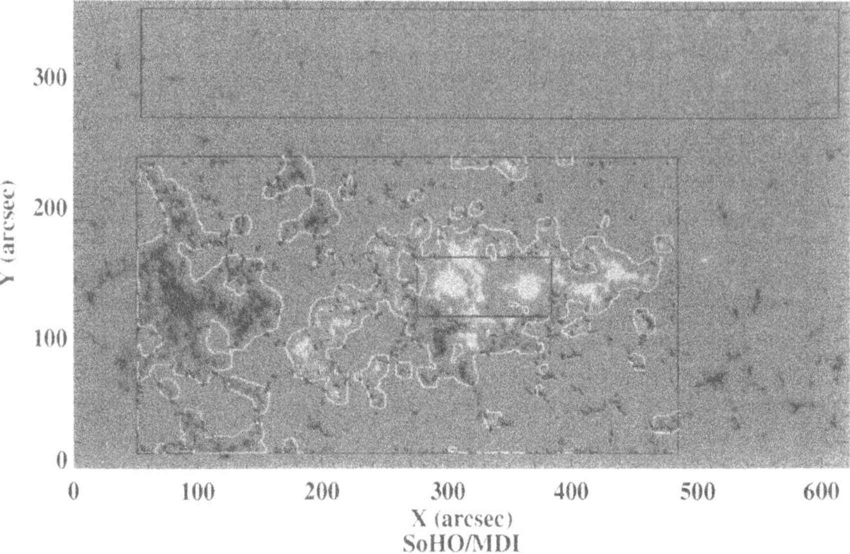

Figure 1. Part of a SOHO/MDI time-averaged magnetogram obtained on 13 May 1996. Five subsequent images, taken one minute apart from 00:53 UT to 00:57 UT, were averaged. The value of the flux density is shown as shades of gray, that saturate into white or black at ±300 gauss. The horizontal and vertical axes are aligned with the east–west and north–south directions, respectively. The reference quiet-Sun region is enclosed by the upper box, and the plage area in the lower box. The white plage contours outline the area within which concentrations of magnetic flux were identified for the magnetic plage (the contours are determined from a 50 gauss threshold on a smoothed abolute-value magnetogram, with a box width of 10 pixels, or 12 arc sec). The two major spots and their surrounding moats (contained in the small rectangle within the plage) have been excluded from our study. Markers show the locations where concentrations with fluxes of less than 10^{19} Mx occur at any time during the 5-hour run on 12–13 May 1996.

flux densities in the two polarities within a sufficiently large area (typically a few supergranules or more) by balancing loss and gain terms for $N(\Phi)\,\mathrm{d}\Phi$ for a flux range from Φ to $\Phi + \mathrm{d}\Phi$. The frequency ν of all collisions between flux concentrations of a given polarity but of any size per unit area of the photosphere is assumed to vary quadratically with the number density of concentrations, i.e., $\nu = \ell N_{\mathrm{t}}^2$. The fragmentation frequency is described by the rate $k(\Phi_1, \Phi_2)$ for concentrations of flux $\Phi_1 + \Phi_2$ to split into two parts Φ_1 and Φ_2. With this definition, the total fragmentation probability per time interval for a concentration with flux Φ is given by the integral over all possible pair formations, $K(\Phi) = \int_0^\Phi k(x, \Phi - x)\,\mathrm{d}x$.

Although the model in Paper I is formulated to study the temporal evolution of the ensemble of flux concentrations, the time scale on which a (near-) equilibrium is reached is so short (likely of the order of a few days in the quiet network), that only (quasi-)stationary solutions needed to be studied. If the environment is strictly unipolar, such as is very nearly the case in each of the polarities of a magnetic plage, a stationary situation can exist without the need for a source of emerging flux, because no flux is being cancelled. Paper I shows that if the total fragmentation rate is proportional to flux, i.e., $K(\Phi) = k\Phi$, the slope β of the resulting exponential distribution $P(\Phi) \sim \exp(-\beta\Phi)$ is given by

$$\beta = \left(2\frac{k}{\ell}\right)^{1/2} |\varphi|^{-1/2} \equiv \beta_0 |\varphi|^{-1/2} , \tag{1}$$

for an average absolute flux density $|\varphi|$ that is determined over an area of a few supergranules or more.

In a mixed-polarity case, the stationary solution is in principle sensitive to the spectrum of the emerging flux concentrations that replace the canceling flux. In Paper I we show that the distribution $N(\Phi)d\Phi$ is relatively insensitive to the spectrum of the emerging flux, provided that it is not a narrow, spiked function. If there is as much positive flux as there is negative, if $K(\Phi) = k \cdot \Phi$, and if collisions between equal and opposite polarities are equally likely, the slope β of the approximately exponential distribution equals

$$\beta = \frac{\pi}{\sqrt{6}} \beta_0 |\varphi_\pm|^{-1/2} . \tag{2}$$

Note that this differs by less than 30% from the unipolar solution in Equation (1).

It follows from the above description that the solutions are controlled by the ratio k/ℓ. It is in principle possible to measure k and ℓ directly from solar observations, each as a function of the fluxes of the two concentrations involved in the pair interaction. This would require large data sets for a statistically significant determination. Instead, we estimate the collision frequency in Paper I by using the argument that concentrations of magnetic flux are constrained to move in the supergranular network formed by lanes where opposing outflows from neighboring upwellings meet and are deflected both downward and sideways: supergranulation is thus described as a tesselation formed by cells with a range of sizes, and flux moves only on cell perimeters. In that paper we assumed that concentrations that are driven towards each other in this flow geometry collide regardless of their horizontal extent, simply because they cannot pass each other. For this particular geometry, the collision parameter ℓ equals $v/2\sqrt{\rho}$, where v is the average displacement velocity and ρ the number density of supergranular upflow centers. This collision parameter takes into account the details of the network geometry. The value of $\ell = v/2\sqrt{\rho}$ for the quiet network lies in the range of $\approx$ 1400–2800 km^2 s^{-1}. If the flow pattern were to be affected by the presence of magnetic fields or would change with effective temperature and gravity (see Section 4), the value of ℓ would be expected to reflect such changes.

From the above estimate of the collision parameter ℓ and the empirically determined value of β_0 for the quiet network, we estimate in Paper I that k should equal approximately (0.4–0.8) $\times 10^{-24}$ Mx s^{-1}.

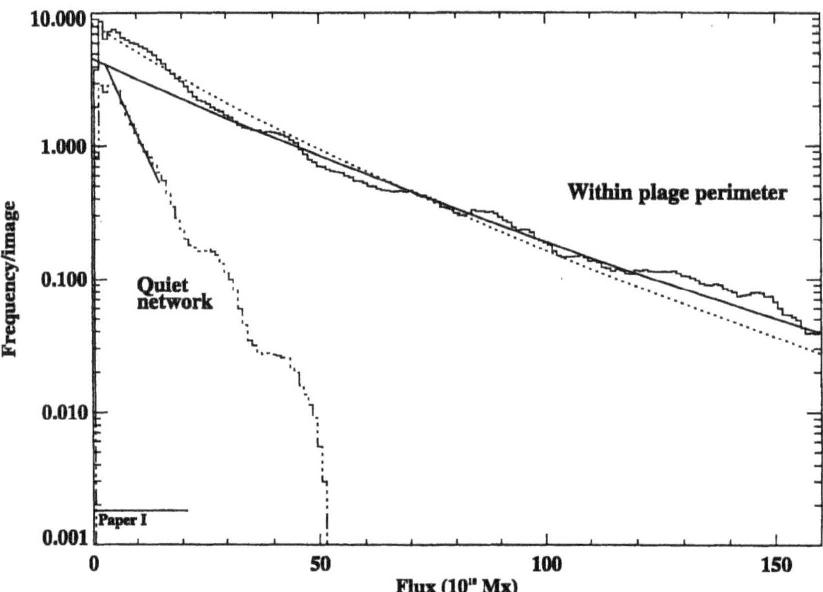

Figure 2. Histograms of the number of concentrations of total magnetic flux Φ observed in a SOHO/MDI magnetograms sequence for two different regions within the same field of view: the plage area within the white plage perimeter in Figure 1, and the quiet-network region enclosed by the upper box in Figure 1. The frequencies are given per image, per interval of 10^{18} Mx. The distributions are averages for the two different polarities, and are smoothed with a box width of 5×10^{18} Mx. The lower solid line is an exponential fit to the network distribution. The upper solid curve is the model prediction based on the time-averaged magnetogram with a spatial smoothing of 1700 km, while the dotted curve is for a spatial smoothing of 8300 km, as discussed in Section 4. The flux range for which the model was initially derived in Paper I is shown in the lower left-hand corner.

4. Comparison of Model and Observations

We identify two distinct regions within the field of view: quiet network and plage interior. We do not study the network in the vicinity of the plage, on the one hand because the model requires a statistically homogeneous environment, and on the other hand because the distribution function for that area depends very sensitively on the criteria used to separate plage and surrounding network.

The histogram of fluxes Φ of quiet-network concentrations is plotted in Figure 2. Within the range of total fluxes that was studied in Paper I, i.e., from 3×10^{18} Mx up to about 20×10^{18} Mx, this distribution can indeed be approximated by an exponential $\sim \exp(-\beta\Phi)$. The slope $\beta = 0.17 \times 10^{-18}$ Mx^{-1} of the exponential fit to the histogram is in fair agreement with the value 0.23×10^{-18} Mx^{-1} that is predicted by Equation (2) for this region with a mean absolute flux density of $|\varphi| \approx 6.3$ G. This agreement shows that the results are not very sensitive to the

2 by 2 pixel onboard binning that was applied to the present data but not to the data set analyzed in Paper I.

Figure 2 shows the quiet-network distribution of fluxes over a range in number density that is substantially larger than in Paper I. This is in part a consequence of the higher mean flux density, which equals 3.1 G per polarity compared to the 1.5 G per polarity in the quiet network region studied in Paper I. The model predicts that the number density of concentrations is proportional to $\sqrt{|\varphi|}$, so that the number density in the currently studied network is approximately 40% higher than in the region studied in Paper I, which improves the statistics for the sparse relatively large objects. The larger set of magnetograms presently studied also improves the statistics.

For relatively large flux concentrations in the quiet network, the high-flux tail of the observed histogram exceeds the exponential fit. This has also been noted in Paper I, and may reflect a modest change in the ratio k/ℓ of fragmentation and collision parameters compared to the constant value for smaller flux values. Further study of this high-flux tail requires large-number statistics, requiring data sets from substantially larger areas than a single MDI field of view.

The flux histograms for the plage (excluding spots and moats) and for its surrounding enhanced network are also shown in Figure 2. We define a plage perimeter based on the mean local activity level. Schrijver (1987) demonstrated that a contour level at 25 G to 75 G agrees well with an eye estimate of plage perimeters, with little difference within that threshold range. We use a plage perimeter threshold of 50 G, applied after a spatial smoothing with a box width of 10 pixels, or 12 arc sec. We chose that length scale somewhat arbitrarily in order to obtain a relatively smooth perimeter.

The comparison of the plage data with the model prediction is simplified by the essentially unipolar character of the plage within each large-scale polarity domain. We point out that small concentrations of opposite polarity may exist within the two polarity regions of the magnetic plage, for instance caused by ephemeral regions that may emerge within the magnetic plage perimeter. This would not affect the model solution significantly, however: Paper I showed that in situations substantially out of flux balance, the distribution function of the majority species is approximated accurately by an exponential with a slope given by Equation (1). Some effects of the presence of an opposite polarity must occur, however, because in other magnetic plages flux concentrations have been observed to cancel across the neutral line that separates the two polarities (e.g., Martin, Livi, and Wang, 1990). We neglect such flux annihilation and any source of flux within the well-developed plage, because our present model does not allow steep spatial gradients in the mean flux density.

The comparison of plage data and model prediction is complicated, on the other hand, by the inhomogeneneity of the plage environment: substantial differences in the local mean flux density exist within the plage on length scales that smooth over a number of flux concentrations. Equation (1) predicts that the detailed functional dependence of the distribution of fluxes depends on the average magnetic flux density on a sufficiently large scale. If these gradients are weak enough and persist for a time long enough to establish a local quasi-equilibrium distribution anywhere within the plage, then instead of a single exponential distribution, a compound distribution is to be expected that is a sum of local distributions.

If we assume that statistical equilibrium holds locally in spite of the density gradients within the plage, the compound distribution for the plage as a whole can be predicted as follows. Let the flux density $|\varphi|$ within the magnetic plage have a distribution given by $\Psi(|\varphi|)\,\mathrm{d}|\varphi|$ when measured at a spatial resolution that is coarse enough to smooth over a number of flux concentrations, yet good enough to preserve the more gradual variations in the flux density (further discussed below). The compound distribution of fluxes is then given by

$$N^{*}_{\pm}(\Phi)\,\mathrm{d}\Phi = \int_{0}^{\infty} \Psi(|\varphi|) 2\frac{k}{\ell} e^{-\beta(|\varphi|)\Phi}\,\mathrm{d}|\varphi|\,\mathrm{d}\Phi\,, \quad (3)$$

with the value of $\beta(|\varphi|)$ given by Equation (1). The upper bound of this integral is, of course, physically limited to a maximum local flux density of 1–2 kG.

The histogram of local flux densities $\Psi(|\varphi|)\,\mathrm{d}|\varphi|$ is to be determined allowing for the specific geometry of the photospheric flows. Using a simple spatially smoothed magnetogram, for instance, would assign cell interiors, where flux cannot go, a finite flux density. This would adversely affect the compound distribution $\Psi(|\varphi|)\,\mathrm{d}|\varphi|$. A more appropriate smoothed plage magnetogram can be made by averaging over time: as time progresses, flux concentrations move about within the plage, and if their motions are, as expected, restrained by the convective flow, then a time-averaged magnetogram provides a smooth representation of the flux density that incorporates where flux can go and where it cannot go. We average over the entire $5\frac{1}{2}$ hr of the data set. The cells of the larger scales of convection will not have evolved very much within this time interval. We apply a subsequent spatial smoothing with a box of 2 and 10 bins (or 1700 and 8300 km) to smooth out residual strong spikes in the smoothed flux density. The effects of this spatial smooting are limited, as shown below. Histograms $\Psi(|\varphi|)\,\mathrm{d}|\varphi|$ are then determined for these two differently smoothed, time-averaged magnetograms, using only the plage interior inside the 50 G contour.

Figure 2 shows that the agreement of the compound distribution predicted by Equation (3) and that extracted from observations is surprisingly good for concentrations exceeding about $\Phi = 2 \times 10^{19}$ Mx, particularly for the lower spatial smoothing. Note that there are no free scaling parameters that were used to fit the model to the observational data, but that the results do depend somewhat on the details of the spatial smoothing. We therefore conclude that the model of random splitting and merging of objects apparently explains the observed distribution of fluxes even within a plage environment.

Below 2×10^{19} Mx, the model based on a spatial smoothing of 2 pixels predicts fewer small concentrations than are actually observed. This may in part be the result of the strong flux density variation across the plage perimeter: a larger smoothing lowers the mean flux density near the plage perimeter, where many of the smaller concentrations are found (Figure 1), thus resulting in a higher prediction for the frequency of small concentrations. The better agreement for the larger smoothing would then imply that the local distribution is sensitive to the flux within a radius of at least about 8000 km. On the other hand, the relatively small concentrations that are found within the relatively small 'plagette' countours around network fragments will themselves enhance the low flux end of the histogram; the smallest of the two spatial smoothings does not properly account for the relative isolation of the network fragments, thus leading to a frequency prediction that is too low at low fluxes. A larger sample of relatively large magnetic plages will be required to resolve this issue.

5. Discussion and Conclusions

We conclude that the observed distribution function of fluxes in magnetic concentrations in the quiet solar photosphere as well as within active regions can be described successfully by a model of random collisions and fragmentations leading to a quasi-steady distribution of fluxes. The quantitative agreement between model and observation for both quiet network and magnetic plage suggests that the basic processes that are involved in establishing the flux distribution are the same in these environments. Specifically, this agreement requires that the controlling parameter, β_0, involving the ratio of the fragmentation and collision parameters (Equation (1)), is the same within the network and plage environments. This does not, however, necessarily mean that the flow scales and patterns are the same.

There is, in fact, observational evidence for a marked difference in the convective flows between plage and network. One such difference is associated with the large-scale convective pattern. Zwaan (1978) points out that cellular openings within plages appear to be significantly smaller than the supergranulation in the quiet Sun. If this were the only change, one would expect that the collision parameter ℓ in plages would be smaller than in the network (Section 3), because if the number density of cells is larger, the total length of the network lanes per unit area and thus the average spacing between concentrations is larger. If the flow pattern in plage and network would differ topologically, a change in the value of the collision parameter ℓ would be even more likely. The collision parameter is, moreover, a function of the mean velocity of the concentrations. Schrijver and Martin (1990) and Schrijver *et al.* (1996) find no significant difference in the instantaneous velocities of concentrations in plage and network, but they do find a difference in the rms displacements between plage and network. This effect remains unexplained, but could reflect a different convection pattern at or below the surface.

Another difference between the flows in plage and network is associated with the granulation (e.g., Spruit, Nordlund, and Title, 1990): within plages, the scale of the granulation is substantially reduced, the pattern of downflows is markedly different with smaller scales and less clearly defined cells, and the intensity contrast between up- and downflows is much lower than in the quiet network.

We are thus faced with explaining why the parameter β_0 is the same despite substantial differences in the flows in which the field is embedded. Two possible solutions come to mind. First, it is possible that the flows that are actually driving the interactions of concentrations of magnetic flux are those at some substantial depth, rather than those at the surface. Such deep flows may not be significantly affected by the presence of magnetic flux underneath network or plage. An alternative explanation that would lead to a cancellation of effects in collision and fragmentation parameters upon taking their ratio, and thus to a parameter β_0 independent of average flux density, rests on the plausible assumption that the flows that drive the concentrations together are identical to those that move the parts apart after fragmentation. After all, the streamlines that converge to bring flux together will also take the fragments on diverging paths some time later. We stress that the fragmentation may well be a superposition of two different physical processes. One of these, operating at a scale below the current instrumental resolution, involves the detailed physics of interaction of flux and flows. Granular buffeting and shuffling, for instance, could be instrumental in rearranging the flux in sub-resolution flux concentrations (see also Ryutova, Kaisig, and Tajima, 1996), putting them on diverging paths in the large-scale flows. In addition to this, the magnetic interaction between the field concentrations themselves may be instrumental in the breakup of concentrations in response to being pushed together (as shown in numerical simulations, M. Ryutova, private communication). The second process that is involved in what is observed as a fragmentation is the larger-scale observable separation of the sub-resolution flux concentrations in diverging flows.

The applicability of the statistical model developed in Paper I to the very different environments of network and magnetic plage suggests first of all that the model holds anywhere on the Sun where gradients in the mean activity level can be neglected. From this it immediately follows that the same applicability can be expected for any other star with effective temperature and surface gravity similar to those of the Sun. But even if the convection were very different in geometry or magnitude, as expected for cool stars well away from the Sun's location in the HR diagram, the model itself is still likely to hold. Whether or not the same controlling parameter β_0, and thus the same functional character of the histogram of fluxes in concentrations should be expected depends on which of the above explanations for the solar case is correct. If the convergence and divergence of stream lines at the surface control the rates of the processes of collision and fragmentation, then one may expect the same value of β_0 even in stars that differ substantially from the Sun. If, however, an unperturbed subsurface convection is the reason why β_0 is the same in network and plage, then this parameter is likely to depend on fundamental stellar parameters.

In Paper I we argued that the quasi-steady state for the distribution function of fluxes in field concentrations is reached within a few days. The large-scale dispersal of flux that escapes from active regions takes place on a time scale of weeks to months. Because of this substantial difference in time scales, these processes occur largely independently of each other, so that a simple combination of the two models should describe the photospheric fields of cool stars in general with sufficient accuracy to predict the radiative losses from their outer atmospheres: the random-walk model developed by Sheeley and colleagues (e.g., Wang and Sheeley, 1994, and references therein) can be used to model the flux dispersal on large scales, and the model developed in Paper I can then be applied subsequently to compute the local distribution of flux concentrations provided that it is expanded to include spatial gradients: regions in which strong gradients occur (near neutral lines internal to magnetic plages, but also large-scale neutral lines in the network) are not accurately handled by the model developed in Paper I. But the model may provide an acceptable first estimate because the relatively small spatial filling factor of regions with strong density gradients makes their impact on the disk-averaged radiative losses likely relatively weak.

Our model for the distribution of magnetic field in photospheric flux concentrations can also be useful in detailed modeling of coronal heating in general, and the properties of X-ray bright points in particular: the model provides the frequencies of collisions between flux of opposite polarity, and the statistical effects of other concentrations in the proximity. Detailed models of field reconnection, such as that developed by Priest, Parnell, and Martin (1994), can then be used to model brightness distributions of X-ray bright points.

Acknowledgements

We are grateful to the SOI-MDI team for all their efforts from building and operating the instrument to the collection and initial processing of the data. We thank R. Ryutova and C. Zwaan for insightful comments. This research was supported in part by NSF grant ATM-953-1778 and by the MDI-SOI NASA contract NAS5-30386 (through Stanford contract PR 9162.)

References

Hagenaar, H. J., Schrijver, C. J., and Title, A. M.: 1997, *Astrophys. J.*, in preparation.
Martin, S. F., Livi, S. H. B., and Wang, J.: 1985, *Australian J. Phys.* **38**, 929.
Parker, E. N.: 1994, *Spontaneous Current Sheets in Magnetic Fields*, Oxford University Press, Oxford.
Priest, E. R., Parnell, C. E., and Martin, S.F.: 1994, *Astrophys. J.* **427**, 459.
Ryutova, M. P., Kaisig, M., and Tajima, T.: 1996, *Astorphys. J.* **459**, 744.

Scherrer, P. H., Bogart, R. S., Bush, R. I., Hoeksema, J. T., Kosovichev, A. G., Schou, J., Rosenberg, W., Springer, L., Tarbell, T. D., Title, A., Wolfson, C. J., Zayer, I., and The MDI Engineering Team: 1995, *Solar Phys.* **162**, 129.
Schrijver, C. J.: 1987, *Astron. Astrophys.* **180**, 241.
Schrijver, C. J.: 1993, *Astron. Astrophys.* **269**, 395.
Schrijver, C. J. and Harvey, K. L.: 1989, *Astrophys. J.* **343**, 481.
Schrijver, C. J. and Martin, S. F.: 1990, *Solar Phys.* **129**, 95.
Schrijver, C. J., Zwaan, C., Balke, A. C., Tarbell, T. D., and Lawrence, J. K.: 1992, *Astron. Astrophys.* **253**, L1.
Schrijver, C. J., Shine, R. A., Hagenaar, H. J., Hurlburt, N. E., Title, A. M., Strous, L. H., Jefferies, S. M., Jones, A. R., Harvey, J. W., and Duvall, Th. L: 1996, *Astrophys. J.* **468**, 921.
Schrijver, C. J., Title, A. M., van Ballegooijen, A. A., Hagenaar, H. J., and Shine, R. A.: 1997, *Astrophys. J.* in press.
Spruit, H. C., Nordlund, Å, and Title, A. M.: 1990, *Ann. Rev. Astron. Astrophys.* **28**, 263.
Wang, Y.-M. and Sheeley N. R., Jr.: 1991, *Astrophys. J.* **375**, 761.
Wang, Y.-M. and Sheeley, N. R., Jr.: 1994, *Astrophys. J.* **430**, 399.
Zwaan, C.: 1978, *Solar Phys.* **60**, 213.

BURSTS OF EXPLOSIVE EVENTS IN THE SOLAR NETWORK

D. E. INNES[1], P. BREKKE[2], D. GERMEROTT[1] and K. WILHELM[1]
[1]*Max-Planck-Institut für Aeronomie, D-37189 Katlenburg-Lindau, Germany*
[2]*Institute of Theoretical Astrophysics, University of Oslo, Blindern, N-0315 Oslo, Norway*

(Received 14 March 1997; accepted 28 May 1997)

Abstract. Observations of the quiet-Sun network in the UV emission line Si IV 1393 Å over a time period of two hours are presented. Bursts of explosive events, highly Doppler-shifted emission, seem to be sporadically emitted from the brighter regions of the network lanes. Individual events have typical lifetimes of $\approx 1-6$ min and come in bursts of up to 30 min. The most spectacular burst in this dataset, shown in the accompanying movie, lasts ≈ 30 min and shows a wide variety of line profiles with both red and blue shifts ≈ 180 km s^{-1}. There appears to be no characteristic form or evolutionary pattern to the line profiles in either the individual events or series of events. There are about twice as many blue shifts as red shifts.

1. Introduction

Explosive events are frequent, small-scale bursts of high-velocity plasma seen throughout the quiet-Sun network (Brueckner and Bartoe, 1983; Dere, Bartoe, and Brueckner, 1984). They show up as highly broadened and/or shifted line profiles from ions with characteristic temperatures in the range $2 \times 10^4 - 5 \times 10^5$ K. Investigations of explosive events with the NRL High-Resolution Telescope and Spectrograph (HRTS) on several rocket flights and *Spacelab 2* have shown that at any one time there are up to 30 000 events on the Sun, distributed mainly in or near the network lanes (Dere, 1994). Typically, explosive events are characterised by Doppler shifts ≈ 100 km s^{-1}, a lifetime ≈ 60 s and size ≈ 2000 km. These observations, however, show a large variety of line profiles (Dere, Bartoe, and Brueckner, 1986, 1989; Cook *et al.*, 1987). There are about equal numbers of profiles that show red, blue and simultaneous red and blue Doppler shifts. When the red and blue shifts appear together they are sometimes offset by a couple of arcseconds along the spectrometer slit. This was one of the features that led Dere *et al.* (1991) to speculate that the red and blue shifts indicate two oppositely directed jet streams. Recent observations with the high-resolution ultraviolet (UV) spectrometer SUMER on board SOHO show that this is indeed the case (Innes *et al.*, 1997). The profile shifts change from red to blue when scanning across a central high intensity source and the spatial extent of the red and blue shifts moves outwards from the source during the lifetime of the event.

There is considerable evidence from coordinated high-resolution UV and magnetic field measurements that explosive events are related to magnetic reconnection in the solar network (Dere, 1994). HRTS observations show that explosive events occur in the network lanes at the boundaries of supergranulation cells and along

Solar Physics **175:** 341–348, 1997.

the neutral line separating regions of opposite magnetic polarity (Dere *et al.*, 1991; Porter and Dere, 1991). The magnetic network boundary contains many intense, low-lying flux tubes. If flux tubes of opposite polarity are forced together by, for example, convective motions at their footpoints, the anti-parallel fields may reconnect. Under conditions in the chromospheric network, where the plasma β is low, theory predicts a bi-directional jet emanating from the reconnection site (Petschek, 1964; Priest, 1982).

The explosive event lifetime is, however, much shorter than the time scale for photospheric flux cancellation which is of the order if a couple of hours (Livi, Wang, and Martin, 1985). Dere (1994) has suggested that explosive (reconnection) events occur in short bursts at discrete locations along the network boundary. The available 4-min rocket flight and *Spacelab 2* data could not provide a long-term study of explosive events. The aim of this paper is to show the network behaviour as seen in the Si IV emission line over a period of two hours. As predicted by Dere, we see that explosive events occur at small discrete locations in sporadic bursts. The accompanying movie gives an impressive display of how the line profiles along a $1'' \times 105''$ strip of the network vary over a 30-min period.

These events may be related to the coronal high-velocity events that have been observed in lines with higher characteristic temperatures, $5 \times 10^5 - 3 \times 10^6$ K, and at larger velocities, 200–1200 km s^{-1}(Kjeldseth-Moe and Cheng, 1990). So far, only a few of these coronal events have been detected and analysed, and it has not yet been possible to establish their relationship to the lower temperature network events.

2. Instrumentation

The observations described here were obtained with SUMER on 28 May 1996 at Sun centre between 8 and 10 UT. A full description of the instrument and its performance is given in Wilhelm *et al.* (1995, 1997) and Lemaire *et al.* (1997). Briefly, SUMER is a UV telescope and spectrometer with a wavelength resolution element of 44 mÅ over the range 800–1610 Å (in first order). Along the N–S directed entrance slit the spatial resolution is $1''$. In the E–W direction the resolution depends on the slit width and step size. For the observations shown here the resolution is $1''$.

The data cover fifty wavelength pixels (2 Å) centred on the Si IV 1393 Å line along the $1'' \times 120''$ entrance slit. During our observations the lower $15''$ of the slit were obscured and therefore appear dark in the images. Consecutive 10 s exposures were made for a full two-hour period. The same position on the Sun was tracked using the standard rotation compensation scheme outlined in Wilhelm *et al.* (1995). In effect, this resulted in a step of $0.76''$ every $\approx$ 4.5 min. We have checked the data and can see no obvious changes in the intensities or line profiles associated

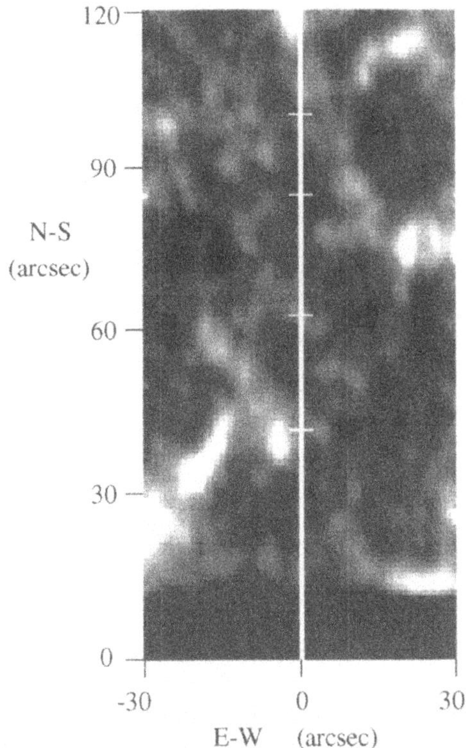

Figure 1. The position of the SUMER slit with respect to the chromospheric network seen in Si II 1309 Å before the sequence. The horizontal tick marks at 42″, 63″, 85″, and 100″ indicate the position of the central pixels used to construct the profiles in Figure 2.

with this movement. The data in the movie covers 190 frames equivalent to 30 min of evolution.

Before and after the two-hour sequence, a map (60″ × 105″) of the area in Si II 1309 Å was made. The position of the slit with respect to the Si II 1309 Å chromospheric network at the start is shown in Figure 1. This shows a typical quiet-Sun region. In Si II the bright emission is clearly concentrated in network lanes that delineate dark cell-like structures. The SUMER slit crosses bright Si II regions at the top and a quarter of the way up from the bottom. Nearer the slit centre the slit seems to cross a fainter, not so well defined, boundary region. The Si II network pattern is very similar at the end of the two hours. The supergranular cell structure and the positions of the bright and dark network lanes moved at the standard rotation velocity.

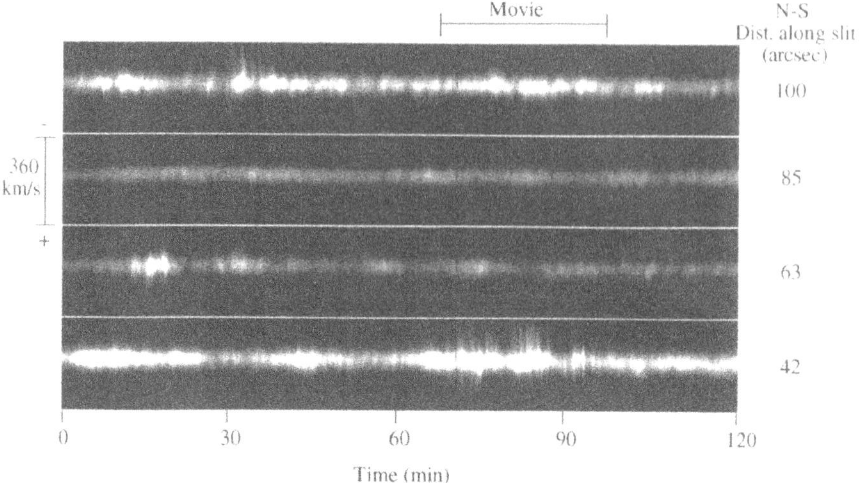

Figure 2. The time evolution of the Si IV line profiles at four positions along the slit. The profiles are produced by binning 10 pixels centred on the position given on the right of the image. Blue shifts are directed upwards.

3. Explosive Events

Previous analyses of the positions of microflares and explosive events observed in the UV have shown that they tend to occur along the boundaries of supergranular cells. The observations presented in this paper show, and it is clearly seen in the accompanying movie, that explosive events are produced at all positions along the slit where it crosses network lanes. There is, however, a tendency for them to be associated with the brighter network lane regions.

The time evolution of the line profiles at four 10″ long sections of the slit are displayed in Figure 2. The four positions are marked on Figure 1. The positions selected, from top to bottom, correspond to (i) a position on the edge of a bright network region (ii) a dark-cell centre region (iii) a faint network lane (iv) a bright network lane.

The bright network lane, in the bottom section, shows the greatest activity. At this position there are three well-defined bursts with velocities greater than 100 km s^{-1} that occur between 60 and 100 min. They are all associated with brightening and show both red (positive) and blue shifts. Blue shifts become dominant after the first 10 min. This is the most dramatic series of bursts seen in this dataset. For this reason they have been chosen for the movie. At other times at this position, there are several additional brightenings associated with lower velocity, $\approx$ 60 km s^{-1}, shifts and also some shifts that last about 2 min and are not associated with brightening. An example of the latter occurs around 35 min.

The series along the top also shows several events accompanied by brightening. The event at $\approx$ 30 min shows an interesting line profile development. The figure shows a very thin streak to positive velocities at the beginning of this event. Inspection of the individual spectra reveals that the event producing this red shift lasts for 2 min and the high-velocity red shifts last for at most 40 s. The intensity then declines. The subsequent brightening and blue shift look like a separate event.

The fainter Si II network lane region is also fainter and less active in Si IV. Along this section, there are several events lasting 5–10 min each. The first of these produces the strongest red shifts in the dataset. All the other events are comparatively faint and not necessarily associated with brightening. The time series along the dark-cell centre shows considerably less variation than at the other positions. The intensity variations are less than a factor 2.5 and all the shifts are less than 60 km s^{-1}.

In general, both widely symmetric and strongly asymmetric profiles are recorded. The profiles and the intensities change significantly between one frame and the next; however over 3 or 4 frames the changes usually show a definite trend. For example, the intensity increases or decreases steadily and shifts seen in one frame can also be seen in neighbouring frames.

4. The Movie

The movie on the CD-ROM shows a half-hour sequence of Si IV data. The times are indicated in Figure 2. One frame from the movie is shown in Figure 3. On the left, the image of the Sun in He II 304 Å taken on 29 May 1995 by EIT (Delaboudinière *et al.*, 1995) shows that the Sun was very quiet at this time. The central section displays the spectrum along the slit after each 10 s exposure. This image is the raw image to which the standard on-board flat-field has been applied. No corrections have been made to rectify the geometric distortion. The velocity scale is measured relative to the centroid of the line profile summed along the slit at a time when no explosive event could be seen along the slit. This scale is only accurate to 20 km s^{-1}. Red shifts (positive velocities) are on the right and blue shifts are on the left. The white tick marks along the slit direction indicate the region of the spectrum that has been summed together to produce the profile on the right. This profile shows the average number of counts in the marked section of the image as a function of wavelength. Wavelength increases to the right. The dashed line is the average profile along the slit at a time when no explosive events could be seen.

The frame displayed in Figure 3 is chosen because it was one of the most spectacular in the movie. It shows a very strong event in the bright network lane, three closely spaced in the faint network lane and a fifth event on the edge of the bright network region at the top of the slit. Most frames show only one or two events.

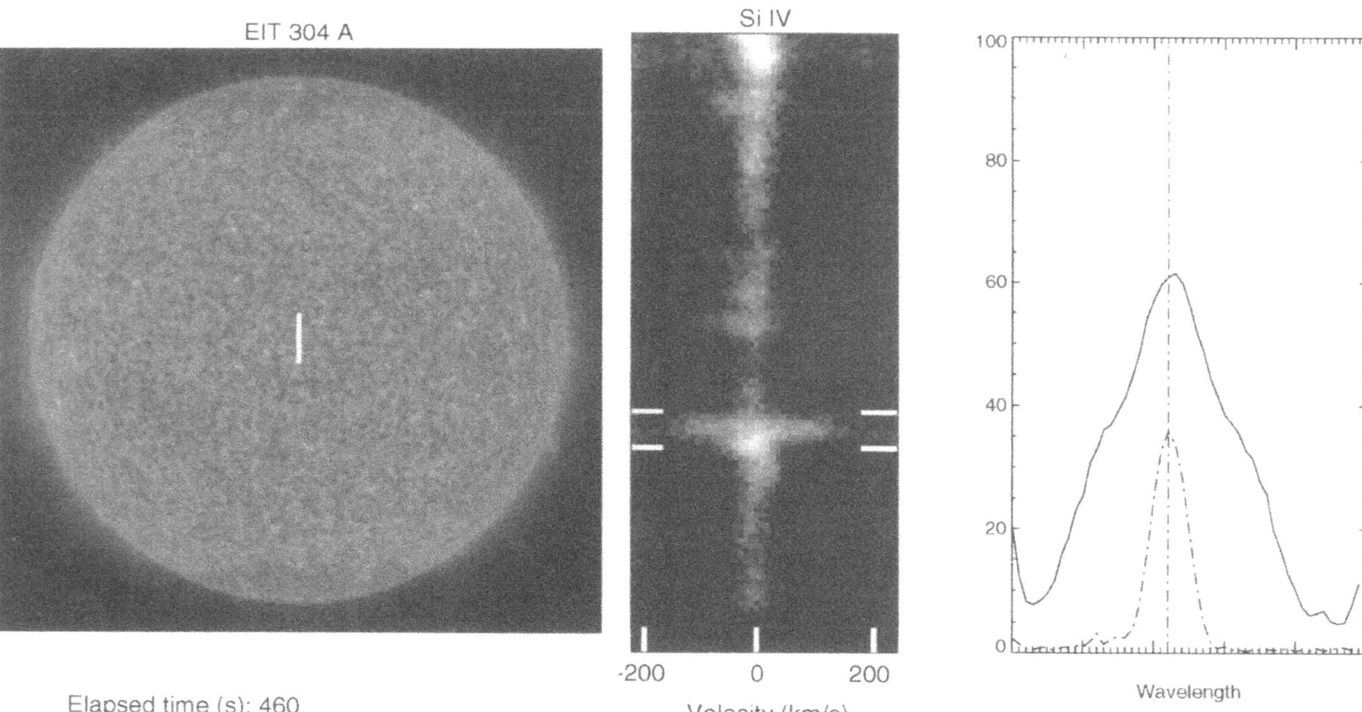

Figure 3. A frame from the movie showing, from left to right, the EIT He II 304 Å image taken 28 May 1996 with the position of the SUMER slit marked, the Si IV spectra along the slit at 460 s from the beginning of the movie, and the average number of counts across the line for the section of the slit indicated in the central section. (EIT image by courtesy of J.B. Gurman and the EIT consortium).

5. Conclusion

It seems as though explosive events occur in bursts as predicted by Dere (1994). The individual events last 1–6 min and can occur several times. The brightest regions show the most activity. The highest shifts come from the brightest parts. Brightenings do not always produce shifts.

If one looks at the profiles in detail it can be seen that each event has a unique evolution. The number of blue shifts outnumber the red shifts by a factor of roughly 2. This predominance for blue shifts was not seen in the *Spacelab* data (Cook *et al.*, 1987) however that was a large-area Sun survey. If the explosive events are bi-directional jets (Innes *et al.*, 1997), it is not surprising that at Sun-centre blue shifts are seen significantly more often than red shifts. If the jet source is about 1000 km above the photosphere (Dere, 1994), any jet stream directed towards the Sun will be stopped by the high-density chromospheric/photospheric material whereas jets directed outwards from the Sun can stream into the corona. The blue stream is therefore expected to be longer and blue shifts will be more plentiful.

These observations lend support to the furnace model, proposed by Axford and McKenzie (1992), in which it is postulated that frequent small-scale reconnection events in the network are the source of wave energy that drives the high speed solar wind.

Acknowledgements

We would like to thank Udo Schühle and Detlef Gigas for their assistance with the SUMER operations at Goddard. The SUMER project is financially supported by DARA, CNES, NASA and the ESA PRODEX programme (Swiss contribution). SUMER is part of SOHO, the Solar and Heliospheric Observatory, of ESA and NASA.

Appendix. CD-ROM Movie Caption

Time evolution of Si IV 1393 Å profiles taken with SUMER at Sun centre on 28 May 1996 around 9 UT. The spectra, taken with an exposure time of 10 s, show a 2 Å wide spectral window centred on 1393 Å, along the $1'' \times 120''$ N–S directed entrance slit. During the observations the Sun was tracked at a rate given by the standard rotation compensation described in Wilhelm *et al.* (1995). On the left is the image of the Sun in He II 304 Å taken with EIT showing the position of the SUMER slit. On the right is the average line profile along the section of the slit marked with white tick marks on the vertical axis of the box around the SUMER spectra.

References

Axford, W. I. and McKenzie, J. F.: 1992, in E. Marsch and R. Schwenn (eds), *Solar Wind Seven*, Pergamon Press, London, p. 1.
Brueckner, G. E. and Bartoe, J.-D. F.: 1983, *Astrophys. J.* **272**, 329.
Cook, J. W. *et al.*: 1987, in J. L. Linsky and R. E. Stencel (eds), *Cool Stars, Stellar Systems and the Sun, Lecture Notes in Physics* **291**, 150.
Delaboudinière, J.-P. *et al.*: 1995, *Solar Phys.* **162**, 291.
Dere, K. P.: 1994, *Adv. Space Res.* **14**, 13.
Dere, K. P., Bartoe, J.-D. F., and Brueckner, G. E.: 1984, *Astrophys. J.* **281**, 870.
Dere, K. P., Bartoe, J.-D. F., and Brueckner, G. E.: 1986, *Astrophys. J.* **310**, 456.
Dere, K. P., Bartoe, J.-D. F., and Brueckner, G. E.: 1989, *Solar Phys.* **123**, 41.
Dere, K. P., Bartoe, J.-D. F., Brueckner, G. E., Ewing, J., and Lund, P.: 1991, *J. Geophys. Res.* **96**, 9399.
Innes, D. E., Inhester, B., Axford, W. I., and Wilhelm, K.: 1997, *Nature* **386**, 811.
Kjeldseth-Moe, O. and Cheng, C. C.: 1990, in P. Maltby and E. Leer (eds), *Proceedings of the Mini-Workshop on Physical Processes in the Solar Transition Region and Corona*, University of Oslo, Oslo, p. 37.
Lemaire, P. *et al.*: 1997, *Solar Phys.* **170**, 105.
Livi, S. H. B., Wang, J., and Martin, S. F.: 1985, *Australian J. Phys.*, **38**, 855.
Petschek, H. E.: 1964, *AAS-NASA Symp. on Solar Flares*, NASA SP-50, 425.
Priest, E. R.: 1982, *Solar Magnetohydrodynamics*, D. Reidel Publ. Co., Dordrecht, Holland.
Porter, J. G. and Dere, K. P.: 1991, *Astrophys. J.* **370**, 775.
Wilhelm, K. *et al.*: 1995, *Solar Phys.* **162**, 189.
Wilhelm, K. *et al.*: 1997, *Solar Phys.* **170**, 75.

DOPPLER SHIFTS IN THE QUIET-SUN TRANSITION REGION AND CORONA OBSERVED WITH SUMER ON SOHO

P. BREKKE[1], D. M. HASSLER[2] and K. WILHELM[3]
[1]*Institute of Theoretical Astrophysics, University of Oslo, Oslo, Norway*
[2]*High Altitude Observatory, National Center for Atmospheric Research, Boulder, Colorado, U.S.A.*
[3]*Max-Planck-Institut für Aeronomie, D-37189 Katlenburg-Lindau, Germany*

(Received 14 March 1997; accepted 14 June 1997)

Abstract. New observations of systematic red shifts of transition region and coronal lines obtained with SUMER (Solar Ultraviolet Measurements of Emitted Radiation) on SOHO (the Solar and Heliospheric Observatory) are presented. With the extensive wavelength coverage of SUMER it is possible to extend the measurements of the red shifts to much higher temperatures compared to previous instruments. We find lines formed in the upper transition region (e.g. O V, S V, and S VI) to be red-shifted similar to lower temperature lines ($T \leq 1.8 \times 10^5$ K). Even hotter lines such as O VI, Ne VIII and Mg X show systematic red shifts on the order of 5 km s^{-1} in the quiet Sun. This is a new and significant result since previous measurements of the red shifts were less well constrained.

The behavior of the red shifts above $T = 10^5$ K has been somewhat controversial. In some earlier investigations the magnitude of the red shift has been found to increase with temperature, reaching a maximum at $T = 10^5$ K and then to decrease toward higher temperatures. Thus, our results will put new constraints on theoretical models. The measured shifts are compared to recent observations of red-shifted emission in stellar spectra obtained with the Hubble Space Telescope.

1. Introduction

One of the most puzzling problems in solar physics is the apparent net red shift of emission lines in the transition region. During the last decades, this phenomenon has been observed with several UV instruments with different spatial resolution (e.g., *Skylab:* Doschek, Feldman, and Bohlin 1976; Feldman, Cohen, and Doschek 1982; OSO 8: Roussel-Dupré *et al.*, 1976; Lites *et al.*, 1976; Roussel-Dupré and Shine, 1982; SMM: Gebbie *et al.*, 1981; Athay *et al.*, 1983a; Athay, Gurman, and Henzen 1983; Klimchuk, 1986, 1987, 1989; Henze and Engvold, 1988, 1992; HRTS: Dere, 1982; Dere, Bartoe, and Brueckner, 1984, 1986; Athay and Dere, 1989; Brekke, 1993; Brekke and Hassler, 1995; Achour *et al.*, 1995; LASP EUV: Rottman *et al.*, 1990; Hassler, Rottman, and Orral, 1991). Systematic red shifts have also been observed in stellar spectra of late type stars, first with the *International Ultraviolet Explorer* (e.g., Ayres *et al.*, 1983; Ayres, Jensen, and Engvold, 1988; Engvold *et al.*, 1988) and recently by the Hubble Space Telescope (Wood *et al.*, 1996, 1997). Both solar and stellar flows in the transition region have been most extensively studied at temperatures around $T = 10^5$ K. For the Sun, the typical value of the average downflow velocity, derived from the C IV lines at 1550 Å, is 5–10 km s^{-1}. Measurements of the variation of the flow with temperature are somewhat ambiguous, however.

Solar Physics **175:** 349–374, 1997.

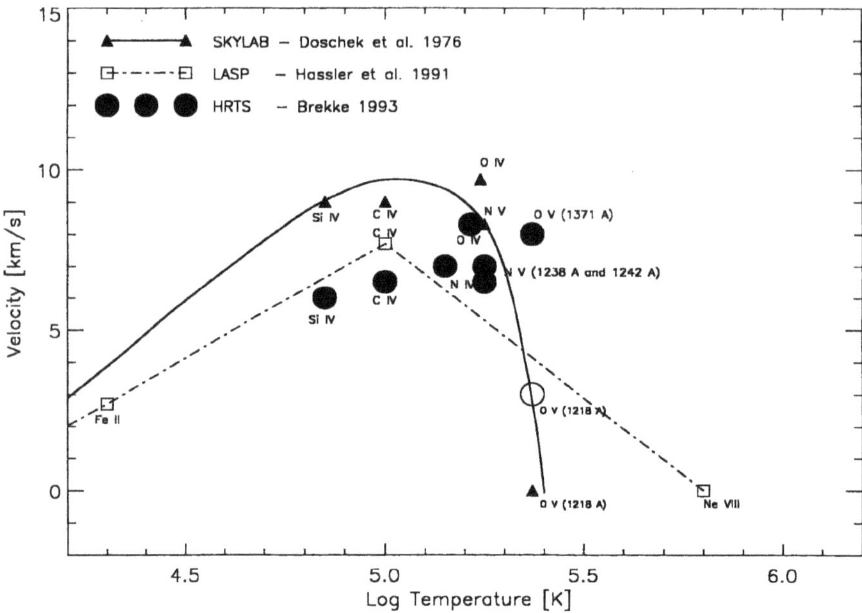

Figure 1. The variation of the red shift with temperature of formation in the quiet Sun as observed with *Skylab*, LASP, and HRTS. This plot summarizes the previous picture of the red shifts as a function of temperature. The solid line illustrates the best fit to the *Skylab* measurements. The O V 128 Å value reported from HRTS (open circle) is most probably affected by uncertainties in the rest wavelength as discussed in the text.

In earlier investigations the magnitude of the red shift has been found to increase with temperature, reaching a maximum at $T = 10^5$ K, and then to decrease toward higher temperatures as shown in Figure 1 which summarizes some of the previous observations. Doschek, Feldman, and Bohlin (1976) found no significant shift in the O V line at 1218 Å at disk center and the commonly quoted average velocity variation with temperature above 10^5 K depends to a large extent on this particular observation of the 1218 Å line. This result has also been extensively used in comparison with numerical simulations of flows in the transition region (e.g., Spadaro, Antiochos, and Mariska, 1991; Hansteen, 1993). Some authors present models where the red-shifted emission is replaced by blue-shifted emission at temperatures around 2.5×10^5 K where O V is formed (e.g., Mariska 1988; Spadaro, Antiochos, and Mariska, 1991). However, results from the observed wavelength of this line, as presented by Doschek, Feldman, and Bohlin (1976) should be regarded with caution since the strong underlying Lα continuum was not subtracted. This will tend to shift the O V 1218 Å line towards shorter wavelengths as demonstrated by Brekke (1994).

The LASP EUV coronal spectrometer recorded high-resolution spectra of the Sun along an entire solar diameter at the equator during a rocket flight on 27 July

1987 (Hassler, Rottman, and Orrall, 1991). The instrument included an on-board platinum hollow cathode calibration lamp, thus the wavelength calibration of the observed spectra are excellent.

As shown in Figure 1, systematic red shifts were found in the Fe II line at 1563 Å and the C IV line at 1548 Å while no substantial shifts were found in Si II at 1533 Å. More importantly, no systematic shift was observed in the Ne VIII line at 770 Å formed at 8×10^5 K (Hassler, Rottmann, and Orrall, 1991).

Also, based on the assumption that systematic horizontal motions cancel statistically so that the line-of-sight velocities approach zero at the limb, Rottman *et al.* (1990) found the rest line positions of the two C IV lines to be 1548.195 ± 0.008 Å and 1550.770 ± 0.008 Å. These values are in agreement with the most recent laboratory determined rest wavelengths made by Bockasten, Hallin, and Huges (1963). In fact, these values are measured with an uncertainty better than those determined in the laboratory by Bockasten, Hallin, and Huges (1963) who report $\pm$ 0.010 Å.

During a later LASP rocket flight on 12 March 1988 the spectrometer was tuned to cover the spectral ranges 1200–1280 Å in the first order and 600–640 Å in the second order. Systematic shifts were found in both N V at 1238 Å and O V 630 Å. (Hassler, Brekke, and Rottman, 1997). The measured velocities in the N V and O V lines were 3.5 ± 2.0 km s^{-1} and 4.0 ± 1.0 km s^{-1}, respectively. This result was derived using the same assumptions as discussed above.

Rottman, Orrall, and Klimchuk (1982) reported blue-shifted emission in Mg X (624 Å) and O V Å above a equatorial coronal hole. The shifts corresponded to an outflow velocity of 12 km s^{-1} and 7 km s^{-1}, respectively. These observations were, however, measured relative to the average quiet-Sun position of the same lines.

The net red shift of transition region lines has also been studied in spectra from the High-Resolution Telescope and Spectrograph – HRTS – during a rocket flight on 21 July 1975 (Brekke, 1993; Achour *et al.*, 1995). During the flight the slit extended from the solar center to the limb covering both quiet and active regions. The results from the HRTS observations are also summarized in Figure 1. The emission lines from S IV, C IV, O IV all fit fairly well to previous observations, but for the higher temperature lines large discrepancies are found, most probably due to errors in the laboratory wavelengths. We also note that significant shifts were observed at and above the limb. One should therefor be very careful to compare shifts on the disk relative to the limb positions using a single perpendicular slit position. A extended raster is required to derive an average shift on the limb.

The O V line at 1371 Å is clearly red-shifted with an average line-of-sight velocity of 8 km s^{-1} in the quiet regions. The O V line at 1218 Å was found to be much less red-shifted when using the laboratory wavelength from Brown (1980) of 1218.344Å. This is approximately 5 km s^{-1} less than the value obtained from the line at 1371 Å. The difference is probably due to an error in the adopted laboratory wavelength of the 1218 Å line, which is an inter-system line and thus, more difficult to measure in the laboratory. The result suggests a laboratory wavelength close to 1218.325 Å if the two lines are to have the same red shift.

Another interesting result is the observed shifts of the two N V lines at 1238 Å and 1242 Å. The measured shifts of these two lines, using the most recent laboratory wavelengths (Hallin, 1966), gives considerable smaller red shifts than observed at both lower and higher temperatures. If we instead use the laboratory wavelength presented by Edlén (1934) we obtain shifts that are comparable to the other transition region lines. This also is consistent with the results of Hassler, Rottman, and Orrall (1991) showing a center-to-limb red shift of N V 1238 Å compared to the in-flight calibration lamp. One should also note that Bockasten, Hallin, and Huges (1963) published laboratory wavelengths of N V, in which the 1238 Å wavelength was similar to those published by Edlen (1934), while the 1242 Å wavelength was close to what Hallin (1966) published. From the behavior of the other transition region lines Brekke and Hassler (1995) suggest that the laboratory wavelength presented by Edlen may be the most accurate. This is supported by the observations of N V by Doschek, Feldman, and Bohlin (1976) since their result was measured relative to the extreme limb position of the same line and is therefore independent on any reference values.

Comparison of line shifts in quiet regions with plage areas shows that all the transition region lines, except the O V lines, show larger Doppler shifts in the plage relative to the average quiet region. In fact the observed shifts in the plage areas are approximately twice the speed found in quiet Sun while the O V lines have nearly the same velocities in both quiet regions and the plage. No reasonable explanations have been suggested yet.

In this paper we present measurements of Doppler shift in emission lines observed with SUMER (the Solar Ultraviolet Measurements of Emitted Radiation instrument) on SOHO (the Solar and Heliospheric Observatory). SUMER is a high-resolution telescope and spectrometer designed to obtain line profiles with high spatial and spectral resolution ($1''$ and 40 mÅ pixel^{-1}) as well as high temporal resolution over the wavelength range 465–1610 Å. Combined with the extensive wavelength coverage, SUMER allows us to observe emissions from neutrals and ions of various elements in the temperature range from $T_e = 1 \times 10^4$ to 2×10^6 K. Thus, it is possible to extend the measurements of the red shifts to much higher temperatures compared to previous instruments.

We find the red shift to extend to much higher temperatures than previously reported. The results are discussed, in particular the difficulties that arise due to discrepancies found in reported laboratory wavelengths. Our results will put new constraints on models of the red shifts in the solar atmosphere. Several hypotheses to explain the red-shifted emission have been proposed so far and they include the return of spicular material, siphon flows in coronal loops, nanoflares, a different temporal duration of upflows and downflows, and asymmetric heating of coronal loops. None of them has been found to be entirely satisfactory. These hypothesis will be discussed in more detail in Section 5.

2. Instrument and Data Acquisition

SUMER is part of the ESA/NASA Solar and Heliospheric Observatory (SOHO) which was launched on 2 December 1995. It was injected into a halo orbit around the first Lagrange point, L1, on 14 February 1996, where, in constant view of the Sun, it accompanies the Earth at a sunward distance of approximately 1.5×10^6 km.

This section describes the data acquisition and instrumental details which are relevant for the interpretation of the data. A more detailed instrument description is given by Wilhelm *et al.* (1995) and first inflight performance characteristics are given by Wilhelm *et al.* (1997) and Lemaire *et al.* (1997).

2.1. Instrument

SUMER is a stigmatic normal-incidence spectrograph operating in the range from 400 to 1610 Å. The off-axis parabola telescope mirror can be moved in two dimensions around the focal point to allow pointing of the instrument. Four slits with angular dimensions of 4×300, 1×300, 1×120, and 0.3×120 arc sec^2 are available. During the observations used for this particular study the 1×120 arc sec slit was used.

Both 1st and 2nd diffraction orders are superimposed on the SUMER detector, and the dispersion is wavelength dependent varying from 45 mÅ pixel^{-1} (first order) and 22.5 mÅ pixel^{-1} (second order) at 800 Å to 41.8 mÅ pixel^{-1} and 20.9 mÅ pixel^{-1} at 1600 Å.

The instrument is equipped with two photon-counting detectors (A and B) operating in Cross Delay Line technique (XDL), for details see Siegmund *et al.* (1994). However, only one detector can be operated at a time. Each detector has 1024 spectral pixels and 360 spatial pixels. The central area of the detector is coated with KBr which increases the quantum efficiency up to a factor of >10 in the range from 900 Å to 1500 Å. Observation of lines on both sections of the photocathode can be used to discriminate second-order lines, since they appear with similar intensity on both parts of the photocathode.

Both detectors show non-uniformity effects typical for MCP intensifiers. These effects stem from the MCP structure, the inhomogeneity of the HV electric field and the analog nature of the XDL electronics. These effects can be accounted for to a large extent by a flat-field correction as discussed in Section 3.1.

2.2. Observations

The results presented in this communication were derived from an observing sequence designed to record a so-called reference spectrum. A reference spectrum covers the entire spectral range of the spectrometer and allows us to study a large number of lines formed at different temperatures. The observing sequence consists of obtaining a series of full detector readouts at different wavelengths. To cover the wavelength range 780–1590 Å a set of 60 spectral sections, each offset by 13 Å were obtained. In this way each wavelength will be recorded two times on different parts of the KBr and one time on each of the bare parts of the detector.

The observing sequence used to obtain a reference spectrum begins with a raster sequence to build up monochromatic images of the area of interest. The spectrometer slit is moved perpendicular to the slit direction to map the region of 120 × 120 arc sec^2. This particular raster contains 159 slit positions and with a step size of 0.76 arc sec. The exposure time at each spatial position is 10 s and the duration of the raster is 1590 s. Three windows centered on selected emission lines are extracted from the detector at each slit position. Monochromatic images can be constructed from the resulting data as shown in Figure 2. From left is shown H I (1025 Å), C II (1037 Å), and O VI (1031 Å). These images can be used to find the exact slit position during the following observing sequence as well as to co-align the SUMER observations with observations obtained with other instruments.

The two vertical lines in Figure 2 mark the area that was covered by the slit during the 1570 s observing sequence. Thus, the first exposure (shortest wavelength) in the reference spectrum was obtained where the right most vertical dashed lines is. The consecutive exposures cover the features toward the left as they drift into the 1 × 120 arc sec^2 field of view due to the solar rotation. The last exposure (longest wavelength) represents the area closest to the left dashed lines.

We note that this type of observing sequences introduces some uncertainties in the following interpretation. First, different solar features are recorded as the Sun rotates throughout the observing sequence. In addition the spectra in different wavelength ranges are recorded at different times. This makes a direct comparison between different spectral features more difficult and care should be taken. In this particular work we assume the solar features covered by the slit are comparable and that the average spectrum along the slit does not vary significantly during the observing sequence.

The quiet-Sun spectrum was obtained on 12 August 1996, 01:13–03:19 UT (SUMER fits files: sum_960812_011303.fits to sum_960812_024958.fits). Using the 1 × 120 arc sec slit the 780–1590 Å spectral range was recorded on the upper part of detector A. The reason for using detector A was to record the Ne VIII 770 Å and 780 Å lines in 2nd order around 1550 Å. This region includes a number of reliable reference lines which is not the case in the wavelength region around 770 Å (1st order). This wavelength range, as observed with detector B, includes a number of strong and unblended transition region lines such as O IV, O V, S V, N IV, and Ne VIII. However, the absence of chromospheric reference lines makes it impossible to accurately measure line shifts relative to the chromosphere.

The integration time of each exposure was 115 s and the entire observing sequence was completed in 115 min. As mentioned above, different wavelength ranges represent slightly different solar features. A section of the detector readout around the S VI 933 Å is shown in Figure 3. It shows a quiet-Sun spectrum between 924–941 Å which, in addition to the S VI line, includes several H I Lyman series

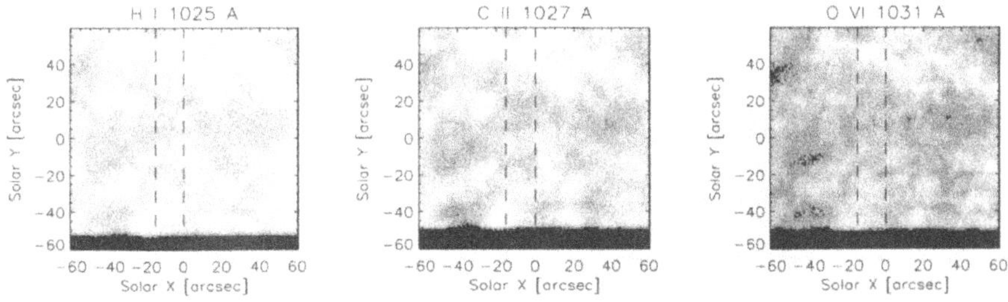

Figure 2. Monochromatic images in H I (1025 Å), C II (1037 Å), and O VI (1031 Å) obtained during the pre-scan at disk center. The vertical lines marks the area that drifted through the 1 × 120 arc sec^2 slit during the subsequent 126 min observing sequence. The right dashed line represents the area recorded during the first exposure (shortest wavelength) while the left dashed line represents the area recorded during the last exposure (longest wavelength).

lines. The reference lines used for deriving the wavelength scale are marked with R. The lower panel shows the average spectrum along the slit.

3. Data Reduction

The raw counts as a function of pixel position have been converted to radiance as a function of wavelength and spatial position through a set of correction and calibration routines. This includes correction for flat field and geometrical distortion as well as wavelength and photometric calibration. The methods used to correct the data are described below.

3.1. CORRECTION FOR FLAT FIELD AND GEOMETRICAL DISTORTION

As mentioned in Section 2.1, the sensitivity of the detector is nonuniform on scales about 20 px or less. A correction for the flat field is important for precise photometry. At regular intervals flat-field raw data are being acquired by a long exposure (approximately 3 hours) in the Lyman continuum at 880 Å while the spectrometer grating is defocused. The properties of the detector vary with time as the detector is being 'scrubbed'. Thus, the flat-field data need to be updated quite frequently.

It turns out that the fixed pattern of the detector drifts with time both in the spectral and spatial direction (Carlsson and Judge, 1996, private communication). Different techniques have been developed to compensate for this drift between flat fields and the observations of interest. The data used in this study were, however, corrected using the last standard flat field exposure obtained prior to our observing sequence. This should be sufficient since the data we finally use represent an average along an extended area along the instrument slit.

The fring fields in the detector MCP-anode gap lead to a geometrical distortion which makes the image of the slit shorter at the center of the detector compared

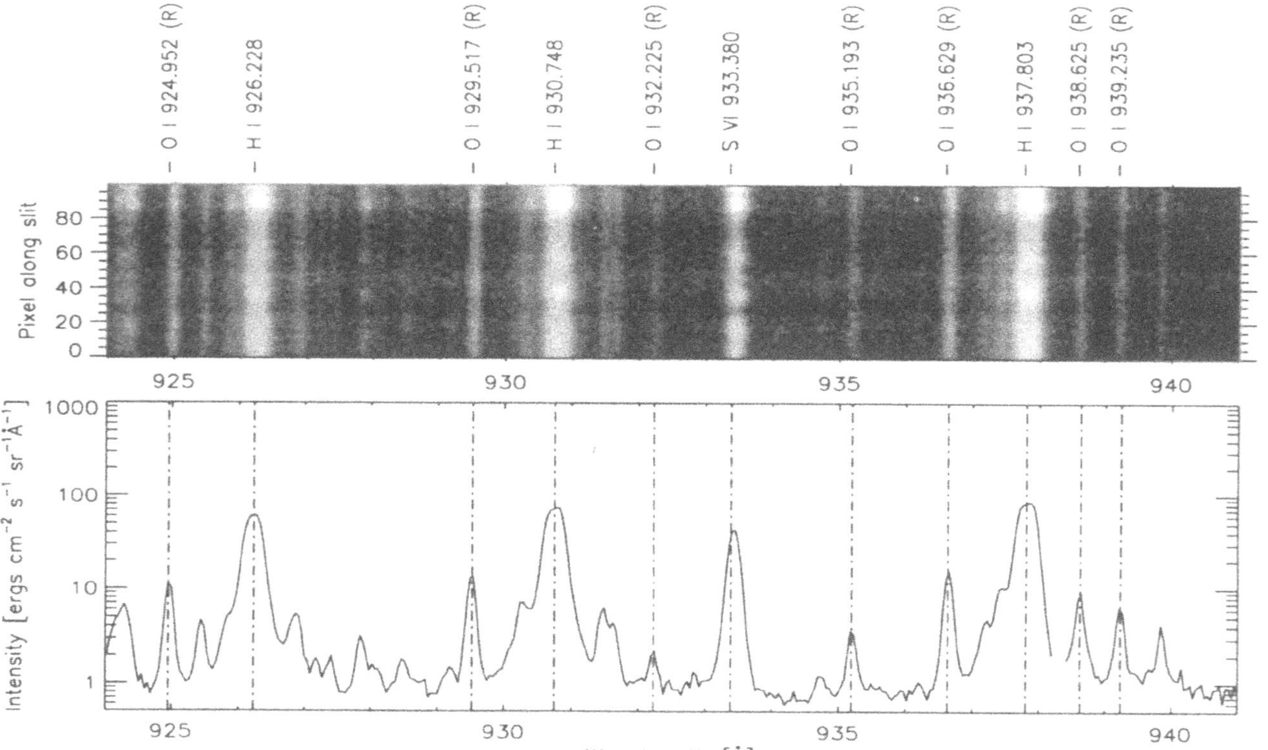

Figure 3. A typical quiet-Sun spectrum between 924 – 941 Å which includes the S VI line at 933 Å and several H I Lyman series lines. The top panel shows the the detector plane with the recorded spectrum. The position along the slit is in the y direction and north is up. The lower panel shows the average spectrum along the slit. The reference lines (marked with R), as well as the other most prominent lines are marked. The dot-dashed lines represent the rest wavelengths.

to areas closer to the edges. Also, the spectral lines are curved. An artificial 'rectangular grid' has been produced using averaged stable solar spectral lines and continuum data, from which the distortion has been parameterized (Moran, 1996, private communication). This correction is then applied to the data to correct for this shortcoming.

3.2. Wavelength Calibration

The determination of flow velocities requires that an accurate wavelength scale be established. Since there is no absolute wavelength reference available on the spectrograph, a wavelength scale was derived using a set of selected solar chromospheric lines as wavelength references. We have mainly chosen reference lines from neutral and singly ionized atoms. The selected lines are fairly strong and unblended in the solar spectrum and laboratory wavelengths are known with fairly high precision. Since the reference lines are formed over a limited range in temperature and other physical conditions, one would expect that they would have near zero line-of-sight Doppler shift relative to each other. This, furthermore, seems a reasonable assumption considering the observed small velocity variations along the slit in these lines, particularly in the quiet solar region. Lines formed in the chromosphere are also known to show relatively small average absolute shifts (e.g., Samain, 1991) and should therefore allow the determination of an absolute wavelength scale. Comparison of the wavelengths of these lines with an accurate in-flight wavelength calibration source suggests this is a reasonable assumption (Hassler, Rottman, and Orrall, 1991).

The wavelength calibration is based on identifying the position of chromospheric lines on the detector. The identification of reference lines is sometimes difficult due to the presence of many blended lines and the fact that both first and second-order lines are prominent in the SUMER spectrum. By comparing the KBr spectrum with spectrum obtained on the bare part of the detector second-order lines can be isolated. Once the identification is made, the pixel position of the line is calculated by fitting a single Gaussian plus a constant background to the line profiles. The positions of the lines were derived from an average over 80 spatial pixels along the slit. A new dispersion relation is derived from a linear polynomial fit of the centroids to the laboratory values. We have also performed quadratic polynomial fits to the line centroids without finding significant differences as long as the calibration is performed over a limited portion of the detector.

Table I lists all the lines used to derive the wavelength scale within each spectral band (given in column 1) which include the high temperature ions discussed in this paper (also given in column 1). The laboratory wavelengths were taken mainly from Kelly (1987) except the O I lines which were taken from Eriksson and Isberg (1968). After the new wavelength scale was derived the positions of the reference wavelengths were re-calculated. The solar positions of the reference lines are also listed in Table I as well as the difference between the laboratory wavelengths and the

Table I
Reference lines used for the local wavelength calibration

Wavelength band Ions	ID	λ_{lab} (Å)	λ_{solar} (Å)	$\Delta\lambda$ (Å)	σ (Å)	σ (km s^{-1})
925–940	O I	925.442	925.447	0.005	0.0043	1.4
S VI	O I	929.517	929.513	−0.004		
	O I	932.225	932.227	0.002		
	O I	935.193	935.198	0.005		
	O I	938.625	938.624	−0.001		
	O I	939.235	939.231	−0.004		
	O I	939.841	939.846	0.005		
1027 – 1042	O I	1027.431	1027.432	0.001	0.0032	0.9
C II, O VI	O I	1028.157	1028.156	−0.001		
	O I	1039.230	1039.230	0.000		
	O I	1040.943	1040.950	0.007		
	O I	1041.688	1041.688	0.000		
1241–1256	C I	1244.535	1244.532	−0.003	0.0045	1.1
N V, Mg X	C I	1245.943	1245.947	0.004		
	C I	1249.004	1249.008	0.004		
	C I	1249.405	1249.398	−0.007		
	C I	1252.208	1252.207	−0.001		
	C I	1254.513	1254.516	0.003		
1244–1271	C I	1244.535	1244.533	−0.002	0.0043	1.0
O V	C I	1252.208	1252.211	0.003		
	C I	1254.513	1254.518	0.005		
	Si I	1258.795	1258.802	0.007		
	S I	1262.860	1262.859	−0.001		
	C I	1267.596	1267.591	−0.005		
	S I	1270.782	1270.781	−0.001		
1392–1413	Fe II	1392.817	1392.817	0.000	0.0072	1.5
O IV, Si IV	S I	1396.112	1396.115	0.003		
	Ni I	1399.026	1399.028	0.002		
	S I	1409.337	1409.332	−0.005		
	Ni II	1411.071	1411.062	−0.009		
	Fe II	1412.842	1412.854	0.012		
1526–1545	Si II	1526.708	1526.708	0.000	0.0038	0.7
N IV, Ne VIII	C I	1542.177	1542.183	0.006		
	Si I	1545.575	1545.574	−0.001		
1545 – 1564	Si I	1545.575	1545.572	−0.003	0.0039	0.8
C IV, Ne VIII	Fe II	1550.274	1550.277	0.003		
	Fe II	1556.527	1556.526	−0.001		
	Fe II	1559.085	1559.092	0.007		
	Fe II	1563.790	1563.790	0.000		
1563–1589	Fe II	1563.790	1563.790	−0.010	0.0091	1.7
O IV, S V	Fe II	1569.674	1569.689	0.015		
	Fe II	1570.244	1570.252	0.008		
	Fe II	1584.952	1584.955	0.003		
	Fe II	1588.290	1588.290	0.000		

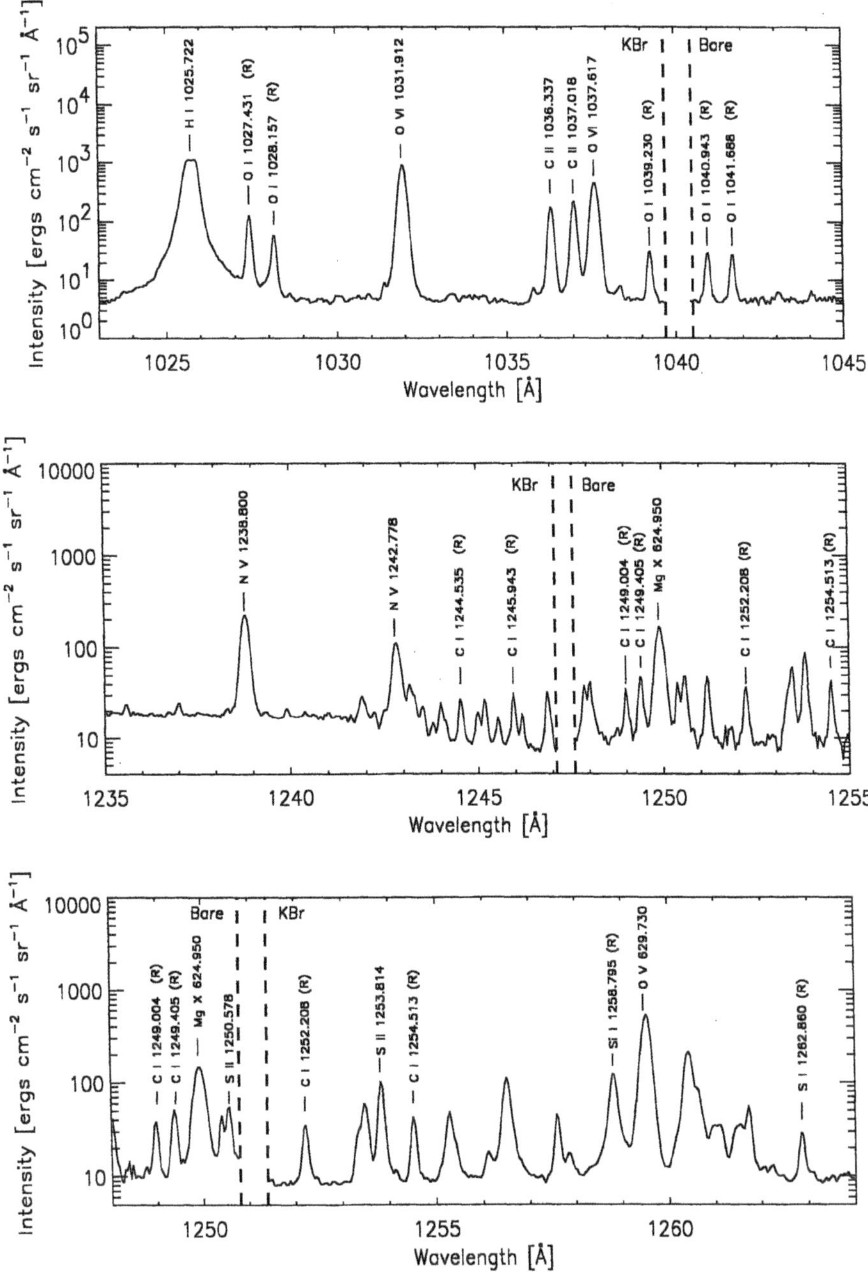

Figure 4. Spectral scans around selected high-temperature lines. The top panel shows the average spectrum around the O VI lines (1031 and 1037 Å), the center panel displays the spectral region around the N V lines (1238 and 1242 Å) and the Mg X 624 Å line (Mg X is observed in second order). The bottom panel shows the Mg X (624 Å) and O V 629 Å, the latter also in second order. The raw counts have been converted to physical units. The spectral ranges recorded on the KBr and the bare part of the detector are marked in the plot. The reference lines used for establishing the wavelength scale are marked with an *R*.

solar wavelengths. This residual gives an indication of the accuracy of the derived wavelength scale. For most of the wavelength bands selected in this investigation the wavelength scale is accurate to approximately ± 1 km s^{-1} on a velocity scale. However, again we emphasize that any shifts measured in the spectra are relative to the velocity in the chromosphere where the neutral and singly ionized reference lines are formed.

Spectral scans around some selected high temperature lines are shown in Figure 4. The top panel shows the average spectrum around the O VI lines (1031 and 1037 Å),. The center panel displays the spectral region around the N V lines (1238 and 1242 Å) and the Mx X 624 Å line (Mg X is observed in second order). The bottom panel shows the Mg X (624 Å) and O V 629 Å, the latter also in 2nd order. The spectral ranges recorded on the KBr and the bare part of the detector are marked in the plot. The reference lines used for establishing the wavelength scale are marked with an R.

During the wavelength calibration we found a non-linearity in the dispersion just to the left of the KBr section of the detector (pixel positions between 150–300 counted from the left of the detector). This asymmetry is being investigated but in the present study we avoided this part of the detector when deriving a calibrated wavelength scale.

3.3. Photometric Calibration

The SUMER instrument was initially calibrated in the laboratory (Hollandt *et al.*, 1996). In-flight calibration updates have thereafter been performed during the operations of SUMER (Wilhelm *et al.*, 1997). The observations presented here have been converted from raw counts to physical units (ergs cm^{-2} s^{-1} sr^{-1} Å^{-1}) as illustrated in Figure 4.

4. Observed Shifts vs Temperature

This section will summarize the observed shifts of lines formed at different temperatures. In particular the lines formed in the upper transition region ($T \geq 1.8 \times 10^5$) will be discussed since they represent new results compared to previous observations. The lower temperature lines also have been discussed numerous times in previous publications. Our results are summarized in Table II and Figure 6 which also includes the commonly quoted variation of the line shifts reported by Doschek, Feldman, and Bohlin (1976)

The reported line shifts in this study represents an average along the spectrograph slit. However, it should be noted that significant variations in line shifts are observed along the slit even in the quiet Sun as illustrated in Figure 5. The shifts were determined after integrating along the slit to obtain one average line profile. It should be pointed out that this yields an intensity-weighted result. The average shift was also derived from the measured line position in each pixel along the slit. No significant difference was found between these two methods in the quiet-Sun data. However, such a difference has been reported by Brynildsen *et al.* (1997) near an active region observation where the emission is much higher compared to the quiet Sun values. Similar correlation in the quiet Sun has also been presented by Brynildsen *et al.* (1995) while conflicting results have been presented (see Dere, Bartoe, and Brueckner, 1984, and references therein). In this paper we were mainly interested in the average shift of the lines in the quiet Sun and have assumed the observations represent typical quiet-Sun conditions. Based on the pre-scan showed in Figure 2, and full disk images from both EIT and ground based observatories we regard this to be a good assumption.

Table II lists the reference wavelengths, observed wavelengths and the corresponding Doppler shifts converted to line-of-sight velocities. The use of velocity to describe the line shifts should not be taken to mean we restrict possible interpretations to gas flows only. Other possible interpretations will be discussed in Section 5. The error in the measured Doppler velocities derived from Gaussian fits to the lines listed in Table II are on the order of ± 1.5 km s^{-1}. In some cases more than one reference wavelength can be found in the literature. Such lines are marked with a dagger in Table II. During the observing sequence each wavelength is recorded several times on different parts of the detector. It therefore was possible to confirm independently the observed line shifts. As mentioned in Section 3.2 we found a nonlinearity in the dispersion just to the left of the KBr section of the detector and that portion was avoided.

The lines are sorted according to temperatures of formation taken from Arnaud and Rothenflug (1985). The references for each rest wavelength are listed. Some of the measured lines are discussed below.

4.1. O IV

The O IV lines at 787 Å and 1401 Å show average shifts of $+13$ and $+12$ km s^{-1}, respectively. These values are comparable with other lines formed in the same temperature regime. The O IV 790 Å line however, shows for some reason a considerably smaller red shift compared to the other two O IV lines with an average value of $+7$ km s^{-1}. We have no reason to mistrust the wavelength calibration in this region. Thus, we could question the reference value of the O IV 787 Å line as listed by Kelly (1987). This line is marked with an open circle in Figure 6.

4.2. O V

Previous measurements of O V were discussed in Section 1. In our study we have measured the position of the O V 629 Å line. The O V 1371 Å line is too weak for reliable measurements, and the wavelength calibration around the O V 1218 Å line is difficult due partly to lack of reference lines and partly to the sloping Lα wing.

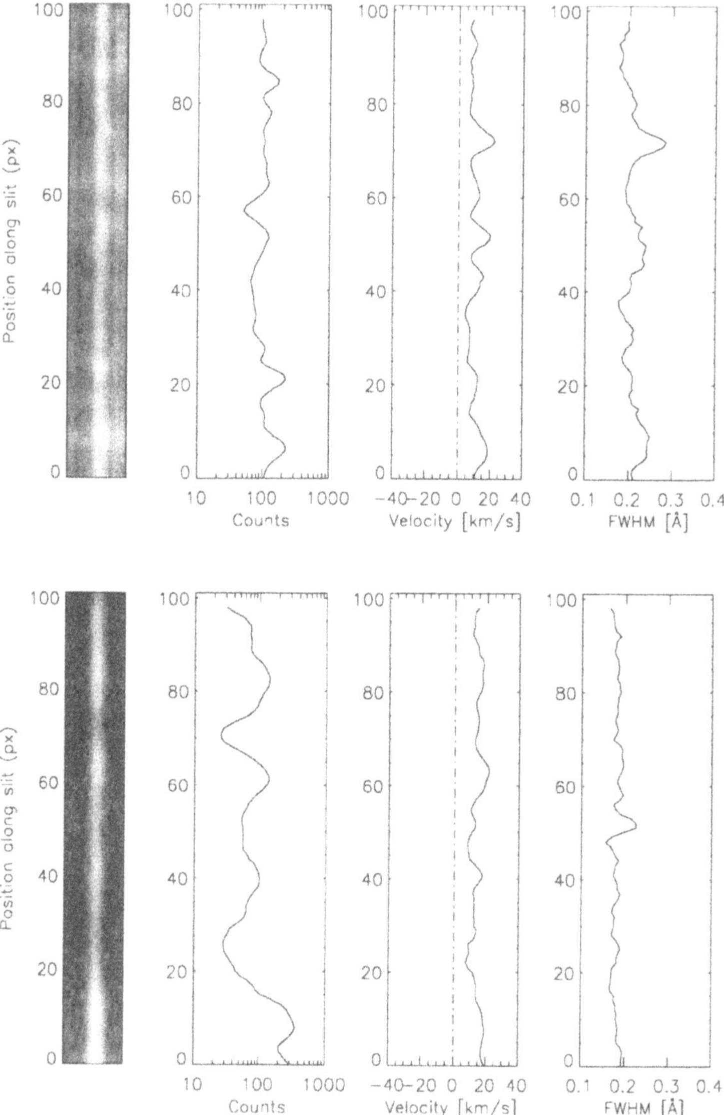

Figure 5. Variation in the line parameters along the SUMER slit. The left panels shows the spectrum of S VI 933 Å and N V 1242 Å. The plots shows from left to right: the corresponding peak intensity, line-of-sight velocity, and the line width. The line parameters were derived from a Gaussian fit to the line profiles. A running mean of 2 pixels was used to suppress some of the noise. North is down on this figure.

Table II
Observed shifts as a function of temperature of maximum ionic abundance

Element	λ_{lab} (Å)	λ_{solar} (Å)	Log T (K)	Velocity (km s^{-1})	Ref.	Remarks
C II	1036.3367	1036.347^{1st}	4.36	3 ± 1.5	1	
Si IV	1402.769	1402.802^{1st}	4.80	7 ± 2.0	8	
C IV	1548.195	1548.225^{1st}	5.01	6 ± 1.5	7	
C IV	1550.770	1550.811^{2nd}	5.01	8 ± 1.5	7	
N IV	765.148	765.171^{2nd}	5.15	9 ± 1.5	5	
O IV	787.711	787.746^{2nd}	5.24	13 ± 2.0	5	
O IV	1401.156	1401.211^{1st}	5.24	12 ± 2.0	5	
O IV	790.199	790.218^{2nd}	5.24	7 ± 2.0	5	
S V	786.470	786.508^{2nd}	5.24	14 ± 2.0	2	†
N V	1242.778	1242.824^{1st}	5.25	11 ± 1.5	4	†
S VI	933.380	933.424^{1st}	5.31	14 ± 2.0	2	
O V	629.730	629.765^{2nd}	5.37	16 ± 3.0	6	
O VI	1031.926	1031.941^{1st}	5.46	5 ± 1.5	3	†
O VI	1037.617	1037.636^{1st}	5.46	6 ± 1.5	3	†
Ne VIII	770.409	770.423^{2nd}	5.80	5 ± 1.5	5	
Ne VIII	780.324	780.341^{2nd}	5.80	6 ± 3.0	5	
Mg X	624.950	624.962^{2nd}	6.05	6 ± 1.5	5	†

1 – Kaufman and Edlén, 1974.
2 – Kaufman and Martin, 1993.
3 – Kaufman and Martin, 1989.
4 – Edlén, 1934.
5 – Kelly, 1987.
6 – Bockasten and Johansson, 1968.
7 – Rottman *et al.*, 1990.
8 – Moore, 1965.
9 – More and Gallagher, 1993.
† – Conflicting reference wavelengths have been reported (see text).

The O V 629 Å line is observed in second order in the SUMER spectrum. We find the average position to be 629.765 Å, corresponding to a line-of-sight Doppler shift of +16 km s^{-1}. This is higher than any of the other lines and might be influenced by a relatively strong blend of S II at 1259.530 Å, positioned 0.07 Å from the rest wavelength of O V (as observed in first order). Thus, the measured position of the O V line should be considered with care. Based on the intensities of two nearby S II lines (1250.578 Å, 1253.814 Å), comparable in magnitude to S II 1259.530 Å in the quiet Sun (Sandlin *et al.*, 1986), we estimate that OV is approximately six times stronger than S II. Thus, only a small fraction of the measured shift of the O V line probably is attributable to S II. The total uncertainty in the measured shift is about 3 km s^{-1}.

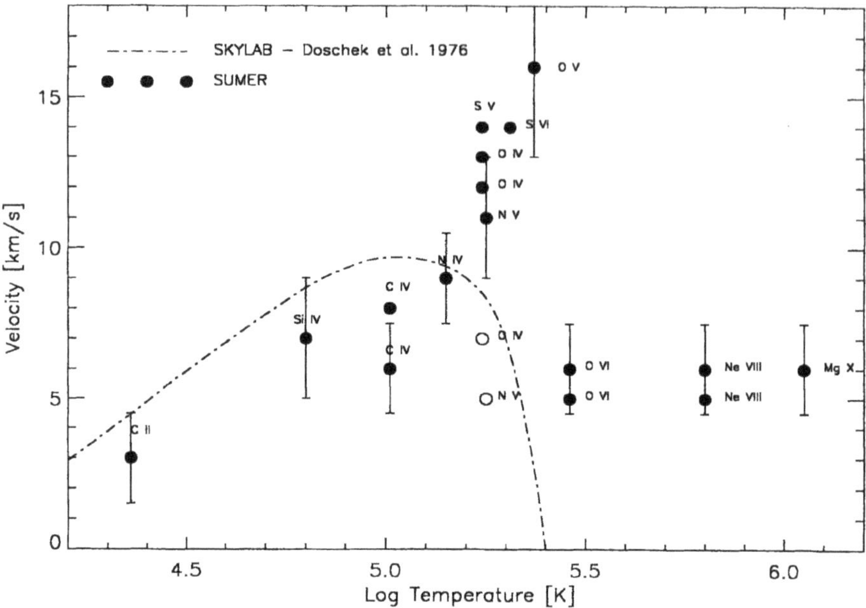

Figure 6. The variation of the red shift with temperature of formation in the quiet Sun as observed with SUMER on 12 August 1996. The measured shift of the O IV 790 Å line deviates from the other two O IV lines and is also marked with an open circle. The uncertainties of selected lines are shown. The earlier commonly quoted average velocity variation, as given in Figure 1, is shown for comparison.

4.3. N V

The wavelength calibration of the region around the N V lines at 1238 and 1242 Å is affected by lack of reference lines, in particular on the shortward side of the N V lines (see Figure 4). Thus, in this paper we report measurements only of the 1242 line even though it is surrounded by several weak lines including Fe XII 1240.03 Å. These lines are fitted to separate Gaussians and minimally influence our measurements of the N V line. We adopted the rest wavelength 1242.778 Å as reported by Edlen (1934) and obtain a red shift of $+11$ km s^{-1}. This velocity is in agreement with other lines formed at similar temperatures. However, it should be noted that the measured shift of the 1242 Å line based on the most recent laboratory wavelength (Hallin 1966), gives a red shift of only $+5$ km s^{-1}. A similar discrepancy, as mentioned before, has been reported for N V 1238 Å as discussed in detail by Brekke and Hassler (1995) and Achour *et al.* (1995). Our results point to an error in the most recent laboratory wavelength of N V.

4.4. S V

The S V 786 Å line is found to be shifted by $+14$ km s^{-1}. This value was derived using a rest wavelength of 786.470 Å as given by Kaufman and Martin (1993). However, Kelly (1987) reports a rest value of 786.480 Å which would yield a shift of $+10$ km s^{-1}.

4.5. S VI

The average position of the S VI 933.380 Å was found to be 933.424 Å corresponding to a shift of $+14$ km s^{-1}, the same as found for S V. The reference wavelength was taken from Kaufman and Martin (1993). The wavelength range around S VI is shown in Figure 3 with the nearby reference lines marked.

4.6. O VI

The O VI lines at 1032 and 1037 Å are formed at 2.9×10^5 K according to Arnaud and Rothenflug (1985). The spectral region around the O VI lines is displayed in Figure 4. In addition to O VI and C II, there are Lβ (1025.72 Å) and five O I lines which are used to derive an accurate wavelength scale.

We find the average observed wavelengths of the two O VI lines to be 1031.941 Å and 1037.636 Å with a separation between the two lines of 5.695 Å. The observed separation is in agreement with the difference reported by Kaufman and Martin (1989). Adopting the reference wavelengths of 1031.9261 ± 0.003 Å and 1037.6167 ± 0.003 Å suggested by Kaufman and Martin (1989) we find the O VI lines to be shifted by $+5$ and $+6$ km s^{-1}, respectively.

A discussion of the different reference wavelength available for the O VI lines has been given by Warren *et al.* (1997). They find the 1037 Å line to be shifted by approximately $+5$ km s^{-1} while the 1031 line yields either red shifts or blue shifts depending on which rest wavelength adopted. This discrepancy is most probably caused by the nonlinearity of the dispersion on the left side of the detector as discussed in Section 3.2.

4.7. NE VIII

The Ne VIII lines at 770.409 and 780.324 Å are both red-shifted, according to the rest wavelength reported by Kelly (1987). The Ne VIII 780 Å line is blended with a fairly strong C I line at 1560.683 Å and a very weak Si I line at 1560.929 Å. The Ne VIII and C I lines are well separated in the quiet-Sun spectrum and we accounted for the C I emission when deriving the shift of Ne VIII. The Si I line does not contribute significantly. Ne VIII 770 Å is blended with a weak Si I line at 1540.963 Å, but it should not influence our result.

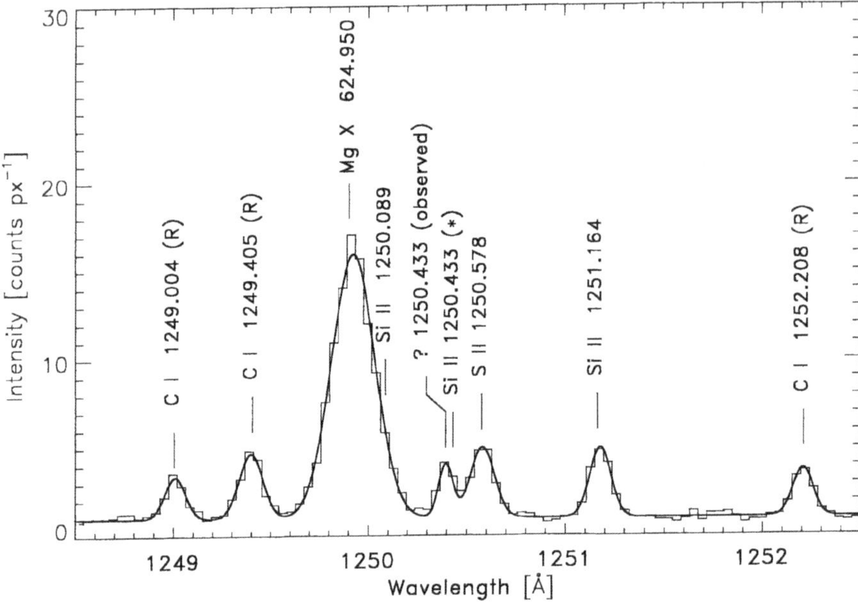

Figure 7. Spectral scan around the Mg X 624 Å line. The data were obtained on the bare part of the detector and the Mg X line (as observed in second order) is very strong compared to the first-order lines. The thin solid line shows the raw counts while the thick solid line represents the sum of the individual Gaussian profiles fitted to the data.

4.8. Mg IX

The average position of the Mg IX 706 Å line ($T \sim 9.5 \times 10^5$ K) is 706.022 Å. It is difficult to estimate possible shifts of Mg IX because there is no accurate laboratory wavelength. Kaufman and Martin (1991) report 706.06 Å, based on a solar observation reported by Ridgeley and Burton (1972). The wavelength is cited to only two decimals places, thus it is not suitable as a reference. Also, since it is based on a solar observation any systematic red shift will bias the reference value. Possible contributions from O III (705.762, 706.224 Å) and S VI (706.48 Å) further complicate accurate measurements of Mg IX. Thus, we did not include Mg IX in our study. We should also mention that Kelly (1987) reports a reference wavelength of 705.72 Å for this Mg IX line.

4.9. Mg X

The wavelength interval around the Mg X 624 Å line is shown in Figure 4 and in more detail in Figure 7. The data in Figure 7 were obtained on the bare part of the detector and the Mg X line (as observed in second order) is very strong compared to the first order lines. The thin histogram depicts the raw counts while the thick solid line shows the spectrum derived from Gaussian fits to the profiles. The reference lines are marked with R. The observed position of the Mg X line is 624.962 Å corresponding to a line-of-sight velocity of +6 km s^{-1} using a rest wavelength of 624.950 Å (Kelly 1987). Kaufman and Martin (1991) report a rest wavelength of 624.943 Å which implies a red shift of +9 km s^{-1}. The latest rest wavelength is, however, based on a solar observation published by Behring *et al.* (1976).

The Mg X line is blended with a weak Si II line at 1250.089 Å as illustrated in Figure 7. However, we have checked the wavelength of Mg X measured on the KBr portion of the detector, which has significant first-order contribution, with the measured wavelength on the bare portion with negligible contribution from first-order lines. We did not find any significant difference in the position of the Mg X line. We conclude that the Si II line does not influence our result.

4.10. Fe XII

It is also tempting to measure the position of the Fe XII line at 1242.03 Å. This line is fairly strong in active regions and on the extreme limb as well as above the limb. However, the line is very weak in the quiet Sun and affected by a blend with S I 1241.905 Å. The latter line contributes significantly to the emission of the Fe XII profile tending to shift it toward shorter wavelengths. The result is a blue-shifted profile, based on the rest wavelength 1242.00 Å reported by Sandlin *et al.* (1986) if the blend is not compensated. The reference value is based, however, on previous solar observations (Burton, Ridgeley, and Wilson, 1974; Sandlin, Brueckner, and Tousey, 1977). Kelly (1987) reports a rest wavelength of 1242.03 Å, again a value based on previous solar observations (Feldman and Doschek, 1977). The Fe XII line at 1349.40 Å is more suitable for determining velocities since it is not blended. However, the line is much weaker than 1242 Å and is difficult to observe in the quiet Sun. Brekke (1993) and Achour *et al.* (1995) found the Fe XII 1349 Å line to be red-shifted in a plage area observed with HRTS.

5. Discussion

Based on the present data used we find the average red shift to persist to higher temperatures compared to previous investigations. Shifts in the range +10–16 km s^{-1} are observed in lines formed at T = 1.3–2.5 × 10^5 K. Even coronal lines (O VI, Ne VIII, and Mg X) show systematic red shifts corresponding to velocities around +5 km s^{-1}. Most previous observations show a peak in the red shift around C IV (100 000 K) while in the present data it peaks around 250 000 K. This, and the fact that we observe systematic red shifts also in the coronal lines are new results. Our measured shift of O VI is consistent with the values reported by Warren *et al.* (1997).

One should note that our reported shifts are based only on a single observing sequence representing a limited area of the quiet Sun on that particular day.

However, the region appeared to represents a typical quiet-Sun area based on the pre-raster sequence (Figure 2) and full-disk images from EIT and different ground-based observations. Comparison of the systematic line shifts between different solar features, such as equatorial coronal holes and active regions will be presented in a later paper. The coronal hole observations will be used to test if the coronal holes reveal, on the average, more blue-shifted emission in the hotter lines compared to the average quiet-Sun values. The comparison between active regions and the quiet Sun should be interesting based on the results previously presented by Brekke (1993) and Achour *et al.* (1995). They find that the differential red shifts between the plage areas and the surrounding quiet Sun increase smoothly with temperature, reaching a maximum around 1.35×10^5 K then falling abruptly to zero at 2.3×10^5 K (O V). This observation is independent of the laboratory wavelengths. The issue is an important one, because the origin of the transition-region line red shift is poorly understood, and the dependence of the red shifts on line formation temperature provides a potentially useful test of theoretical models.

The average red-shifted emissions appear to be a persistent phenomenon in both quiet and active regions. The estimated downward mass flux around $T = 10^5$ K is sufficient to empty the corona in a few minutes. It therefore seems likely that upflows also exist which balance the observed downflow unless the Doppler shifts are due to waves propagating through the transition region. The red shifts often are interpreted as a net downflow of gas, although that conclusion does not directly follow. Both downflows and upflows may occur where, for some reason, the downflowing plasma is slightly brighter than the upflowing plasma. Several explanations of the net red shift has been put forward:

– The red-shifted EUV emission may be attributed to the return of spicular material (e.g., Pneuman and Kopp, 1978; Athay and Holzer, 1982; Athay, 1984). The rising spicules are cool and not observable in the hot transition-region lines. More recent work by Cheng (1992) gives support to this idea. Working with a rebound shock model for the spicule, he showed that the rebound shock train could give rise to an average downward directed velocity in the transition region. However, taking into account that the line emission varies with time and position, Hansteen and Wikstøl (1994) showed that the resulting hydrodynamic evolution leads to preceived upflow and blueshift in the transition-region lines, even though the average velocity in the line-forming region is directed downward.

– The net red shift may be caused by siphon flows in coronal loops. Flows can be driven by an asymmetry in the loop such as an imbalance in heating or a pressure difference between the two legs of the loop (Boris and Mariska, 1982) or a difference in footpoint area (Mariska and Boris 1983). Boris and Mariska (1982) find a transition-region velocity ($T \approx 10^5$ K) of less than 1 km s^{-1} while Mariska and Boris (1983) give a peak velocity of ~ 4 km s^{-1}.

– Asymmetric heating of coronal loops may drive the flow along the magnetic field lines where emission from the downflowing leg dominates that of the upflowing leg (McClymont and Craig, 1987; Mariska, 1988; McClymont, 1989). In these models the energy input must have an extreme spatial dependence. McClymont and Craig modeled steady flows in lower temperature loops, including thermal conduction but neglecting gravity. They claim that under extremely asymmetric heating in very short loops the flows reproduce many of the observed features, and that the steady flow hypothesis can explain the lack of Doppler shifts at coronal temperatures as suggested by some of the previous observations.

– Mariska (1988) computed UV emission line profiles and Doppler shifts from flows in coronal loops. He claims that because of the highly asymmetric localization of the heating there will be a cutoff temperature above which the red-shifted emission will be replaced by only blue-shifted emission. He obtains downflow velocities of 8 km s^{-1} and 5 km s^{-1} for C IV and O IV, respectively, but predict an upflow of -18 km s^{-1} in O V. That contradicts the red shift observed in O V in the present material.

– Spadaro *et al.* (1991) adopt the same loop model as Mariska (1988), but include the effects of non-equilibrium ionization. Motion of plasma through a steep temperature gradient causes the ions to undergo variations of temperature on time scales much shorter than the relevant ionization and recombination times. This can cause significant departures from ionization equilibrium in the flowing plasma (e.g., Borrini and Noci, 1982; Noci *et al.*, 1989). The model produces only a small red shift at the point of maximum specific intensity of C IV but a pronounced asymmetry of the profile gives a net blue shift. The hotter lines like O V show somewhat less upflow compared to Mariska (1988).

– A hypotheses that downflows driven by radiatively-cooling condensations in the solar transition region are able to produce significant red shifts has been proposed by Reale, Peres, and Serio (1996, 1997). Their approach employed a 2-D hydrodynamic code to model the evolution of an otherwise steady coronal loop (typical either of active or of quiet regions), perturbed with an isobaric condensation placed in the transition region, and they synthesize from the model results (i.e., density, temperature and velocity) the emission and Doppler shifts of UV lines. They find significant red-shifted emission (5–10 km s^{-1}) in some representative UV lines (Si IV)., C IV, O IV, O V) both in quiet and active regions with somewhat larger shifts in active region plasma. The latter is consistent with observations presented by Achour *et al.* (1995). However, it is not clear from this model how the initial condensation is produced.

– A new theory to explain the net red shift in the transition region has been suggested by Hansteen (1993). He considers nanoflares near the apex of coronal loops, generating wave disturbances that propagate downward along the magnetic fields toward and through the transition region. He has calculated the effect of the disturbances on transition-region lines, and obtains red shifts. This is caused partly by particle motions in the wave coupled with the variation in line emission, and partly by wave pressure introducing a temporary displacement of the transition region. The model suggests velocities that are too low compared to the observed ones, however.

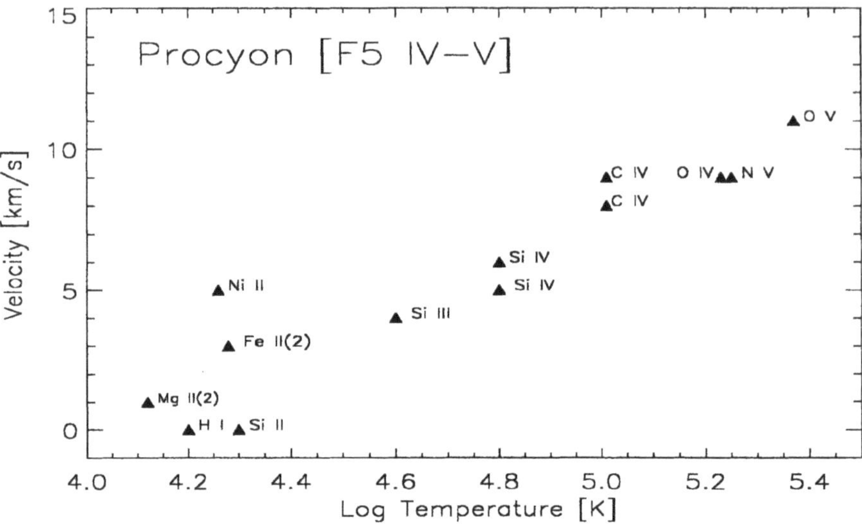

Figure 8. Red shifts of transition region lines observed in spectra from Procyon with the HST/GHRS spectrometer (Wood *et al.*, 1997). Procyon is a hotter and somewhat more active star compared to the Sun.

– The previous idea was extended by introducing a magnetic field by Hansteen, Maltby and Malagoli (1996). The interpretation is that the observed line shifts are caused by MHD waves propagating along magnetic field lines from the corona downward towards the chromosphere. A nanoflare 'pulse' strikes the transition region where a large red shift is produced while the intensity increases rapidly. After the pulse has passed the transition region bounces back and a blue shift ensues. The intensity during this phase is reduced due to lower density as the transition region decompresses. The correlation between compression and displacement from several such pulses produce an average red shift of the transition region lines. The average red shift of C III and C IV found from this calculation is 15 and 7.5 km s^{-1} respectively. However, they find the coronal line (Mg IX) to show a net blue shift of -15 km s^{-1}, inconsistent with our measured shift of the Mg X line if we assume that lines of similar exitation should show approximately the same shift.

Reproducing the observed net red shift by flow through coronal loops seems difficult based on the predicted velocities at higher temperatures. The predicted large blue shift at higher temperatures does not agree with the observed red shifts. The most promising of the theories mentioned are the models by Hansteen, Maltby and Malagoli (1996) and Reale, Peres, and Serio (1997). However, the work by Hansteen, Maltby and Malagoli (1996) produces upflows in Mg IX which is not consistent with the results presented here for Mg X assuming the two lines to show approximately the same shifts. Thus, other mechanism should still be considered to explain the observed red shifts at higher temperatures.

We assume the emitting plasma to be confined to magnetic structures with size below the resolving power of the instrument. Observations from HRTS point to the magnetic field as the key parameter in determining the magnitude of the red shifts (Brynildsen *et al.*, 1996). Observations also suggests that only a small fraction of the observed volume contains emitting gas (e.g., Kjeldseth Moe and Nicolas, 1977; Feldman, 1983; Kjeldseth-Moe *et al.*, 1984; Dere 1987; and Brekke *et al.*, 1992). If the emitting area, or area filling factor, is larger for the downflowing plasma compared to the upflowing plasma this can result in asymmetric red-shifted profiles.

Detailed studies of the red-shifted emission observed on the Sun also are important for the understanding of similar observations of stellar spectra. Recent observations of red-shifted transition region lines in spectra obtained with the GHRS on Hubble Space Telescope were presented by Wood *et al.* (1996, 1997). Below temperatures of about 1.6×10^5 K, the line red shifts of the Sun, α Cen A, α Cen B, and Procyon are all very similar. However, at higher temperatures both α Cen A and α Cen B suggest dropoffs in red shifts similar to that suggested by the solar data presented by Achour *et al.* (1995). In comparison, no dropoff in the red shifts was found in the spectra from Procyon, which is a hotter and somewhat more active star. The observed shift of O v 1218 Å was derived using the rest wavelength of 1218.344 Å (Brown, 1980). Adopting the rest wavelength 1218.325 Å, as suggested by Brekke (1993), gives an observed shift of $+16$ km s^{-1}. It should be pointed out that the measurements in Figure 8 were in some cases based on slightly different rest wavelengths compared with Table II. For more details about the stellar observations we refer to Wood *et al.* (1997). Comparison of the red shifts observed in active regions and the quiet Sun, as well as in coronal holes are being investigated. Such studies could give new information that can explain the recent stellar observations.

Observations of the solar UV spectrum with high-resolution spectrometers have demonstrated that there is a need for improved measurements of laboratory wavelengths of a number of spectral elements. The uncertainties in some of the reference wavelengths complicate the interpretation of the current observations with SUMER as shown in this paper. Such uncertainties have also been discussed earlier by Brekke (1993) and Brekke and Hassler (1995). A general problem when calculating shifts of high-temperature ions is the uncertainties in some of the reference wavelengths as discussed earlier by Brekke (1993), Brekke and Hassler (1995) and recently by Warren *et al.* (1997). To fully understand and interpret the current and future measurements of line shifts, we suggest that the accuracies of the laboratory wavelengths and energy levels of the most important lines in the SUMER/CDS wavelength ranges (100–1600 Å) be evaluated, and if necessary, re-measured in the laboratory.

Acknowledgements

We would like to thank the many members of the SUMER hardware and operation teams. We acknowledge useful discussion with Harry Warren, U. Schühle, Werner Curdt, Nils Brynildsen, Tom Ayres, and Jonchul Chae during the preparation of this paper. This work was supported by the The Research Council of Norway. A portion of the work of D.M.H. has been supported by NASA Grant NAG5–2903 to the National Center of Atmospheric Research. SOHO is a project of international cooperation between ESA and NASA.

References

Achour, H., Brekke, P., Kjeldseth-Moe, O., and Maltby, P.: 1995, *Astrophys. J.* **453**, 945.
Arnaud, M. and Rothenflug, R.: 1985, *Astron. Astrophys.* **60**, 425.
Athay, R. G.: 1984, *Astrophys. J.* **287**, 412
Athay, R. G. and Dere, K. P.: 1989, *Astrophys. J.* **346**, 514
Athay, R. G. and Holzer, T. E.: 1982, *Astrophys. J.* **255**, 743
Athay, R. G., Gurman, J. B., and Henze, W.: 1983, *Astrophys. J.* **269**, 706.
Ayres, T. R., Jensen, E., and Engvold, O.: 1988, *Astrophys. J. Suppl.* **66**, 51.
Ayres, T. R., Stencel, R. E., Linsky, J. L., Simon, T. E., Jordan, C., Brown, A., and Engvold, O.: 1983, *Astrophys. J.* **274**, 801.
Behring, W. E., Cohen, L., Feldman, U., and Doschek, G. A.: 1976, *Astrophys. J.* **203**, 521.
Bockasten, K. and Johansson, K. B.: 1968, *Arkiv Fys.* **38**, No. 31, 563.
Bockasten, K., Hallin, R., and Hughes, T. P.: 1963, *Proc. Phys. Soc. London* **81**, 522.
Boris, J. P. and Mariska, J. T.: 1982, *Astrophys. J.* **258**, L49.
Borrini, G. and Noci, G.: 1982, *Solar Phys.* **77**, 153.
Brekke, P.: 1993, *Astrophys. J.* **408**, 735.
Brekke, P.: 1994, *Space Sci. Rev.* **70**, 97.
Brekke, P. and Hassler, D. M.: 1995, in A. J. Sauval, R. Blomme, and N. Grevesse (eds), *Laboratory and Astronomical High Resolution Spectra, ASP Conf. Ser.* **81**, 589.
Brekke, P., Brynildsen, N., Kjeldseth-Moe, O., Maltby, P., Brueckner, G. E. 1992, *Proc. 26th ESLAB Symposium on Study of the Solar-Terrestrial System*, ESA SP-346, p. 211.
Brown, C. M.: 1980, *Astron. Astrophys.* **88**, 273.
Brynildsen, N., Kjeldseth-Moe, O., and Maltby, P.: 1996, *Astrophys. J.* **462**, 534.
Brynildsen, N., Kjeldseth-Moe, O., and Maltby, P.: 1995, *Astrophys. J.* **455**, L81.
Burton, W. M., Ridgeley, A., and Wilson, R.: 1967, *Monthly Notices Roy. Astron. Soc.* **135**, 207.
Cheng, Q. Q.: 1992, *Astron. Astrophys.* **262**, 581.
Dere, K. P.: 1982, *Solar Phys.* **77**, 77.
Dere, K. P.: 1987 ,*Solar Phys.* **144**, 367.
Dere, K. P.: 1982, *Solar Phys.*, **75**, 189.
Dere, K. P.: 1987, *Solar Phys.*, **114**, 367.
Dere, K. P., Bartoe, J.-D. F., and Brueckner, G. E.: 1984, *Astrophys. J.* **281**, 870.
Doschek, G. A., Feldman, U., and Bohlin, J. D.: 1976, *Astrophys. J.* **205**, L177.
Edleń, B.: 1934, *Nova Acta R. Soc. Sci. Uppsala (IV)* **9** (6).
Engvold, O., Ayres, T. R., Elgarøy, Ø., Jensen, E., Jorås, P. B., Kjeldseth-Moe, O., Linsky, J. L., Schnopper, H. W., and Westergaard, N. J.: 1988, *Astron. Astrophys.* **192**, 234.
Eriksson, K. B. S. and Isberg, H. B. S.: 1968, *Arkiv Fys.* **37**, 221.
Feldman, U.: 1983, *Astrophys. J.* **275**, 367.
Feldman, U. and Doschek, G. A.: 1977, *J. Opt. Soc. Am.* **67**, 726.
Feldman, U., Cohen, L., and Doschek, G. A.: 1982, *Astrophys. J.* **255**, 325.
Gebbie, K. G. *et al.*: 1981, *Astrophys. J.* **251**, L115.
Hallin, R.: 1966, *Ark. Fys.* **31**, 511.

Hansteen, V. H.: 1993, *Astrophys. J.* **402**, 741.
Hansteen, V. H. and Wikstøl, Ø.: 1994, *Astron. Astrophys.* **290**, 995.
Hansteen V. H., Maltby, P., and Malagoli, A.: 1996, in R. D. Bentley and J. T. Mariska (eds), *Magnetic Reconnection in the Solar Atmosphere*, PASP Conference Series 111, in press.
Hassler, D. M., Brekke, P., Rottman, G. J.: 1997, in preparation.
Hassler, D. M., Rottman, G. J., and Orrall, F. Q.: 1991, *Astrophys. J.* **372**, 710.
Henze, W. and Engvold. O.: 1988, *Bull. Am. Astron. Soc.* **20**, 703.
Henze, W. and Engvold. O.: 1992, *Solar Phys.*, **141**, 51.
Hollandt, J., Schühle, U., Paustian, W., Curdt, W., Kühne, M., Wende, B., and Wilhelm, K.: 1996, *Appl. Opt.* **35** (25), 5125.
Kaufman, V. and Edleń, B.: 1974, *J. Phys. Chem. Ref. Data* **3**, 825.
Kaufman, V. and Martin, W. C.: 1989, *J. Opt. Soc. Am.* **6**, 1769.
Kaufman, V. and Martin, W. C.: 1991, *J. Phys. Chem. Ref. Data* **20**, No. 1, 83
Kaufman, V. and Martin, W. C.: 1993, *J. Phys. Chem. Ref. Data* **22**, Ni. 2, 279
Kelly, R. L.: 1987, *J. Phys. Chem. Ref. Data* **16**, Suppl. No. 1.
Kjeldseth Moe, O. and Nicolas, K. R.: 1977, *Astrophys. J.* **211**, 579.
Kjeldseth-Moe, O., Andreassen, Ø., Maltby, P., Bartoe, J.-D. F., Brueckner, G. E., and Nicolas, K. R.: 1984, *Adv. Space Res.* **4** (8), 63.
Klimchuk, J. A.: 1986, Ph.D. Thesis, University of Colorado and NCAR, NCAR/sla CT-96.
Klimchuk, J. A.: 1987, *Astrophys. J.* **323**, 368.
Klimchuk, J. A.: 1989, *Solar Phys.* **119**, 19.
Lites, B. W., Bruner, E. C., Chipman, E. G., Shine, R. A., Rottman, G. J., White, O. R., and Athay, R. G.: 1976, *Astrophys. J.* **210**, L111.
Lemaire, P., Wilhelm, K., Curdt, W., Schühle, U., Marsch, E., Poland, A. I., Jordan, S. D., Thomas, R. J., Hassler, D. M., Vial, J.-C., Kühne, M., Huber, M. C. E., Siegmund, O. H. W., Gabriel, A., Timothy, J. G., Grewing, M.: 1997, *Solar Phys.* **170**, 105.
Mariska, J. T.: 1988, *Astrophys. J.* **334**, 489.
Mariska, J. T. and Boris, J. P.: 1983, *Astrophys. J.* **267**, 409.
McClymont, A. N.: 1989, *Astrophys. J.* **347**, L47.
McClymont, A. N. and Craig, I. J. D.: 1987, *Astrophys. J.* **312**, 402.
Moore, C. E.: 1965, *NSRDS-NBS* **3**, Section 1.
Moore, C. E. and Galagher, J. W.: 1993, *Tables of Spectra of Hydrogen, Carbon, Nitrogen and Oxygen Atoms and Ions*, CRC Series in Evaluated Data in Atomic Physics.
Noci, G., Spadaro, D., Zappalá, R. A., and Antiochos, S. K.: 1989, *Astrophys. J.* **338**, 1131.
Pneuman, G. W. and Kopp, R. A.: 1978, *Solar Phys.* **57**, 49.
Reale, F., Peres, G., and Serio, S.: 1996, *Astron. Astrophys.* **316**, 215.
Reale, F., Peres, G., and Serio, S.: 1997, *Astron. Astrophys.*, in press.
Ridgeley, A. and Burton, W. M.: 1972, *Solar Phys.* **27**, 280.
Rottman, G. J., Hassler, D. M., Jones, M. D., and Orrall, F. Q.: 1990, *Astrophys. J.* **358**, 693.
Rottman, G. J., Orral, F. Q., and Klimchuk, J. A.: 1982, *Astrophys. J.* **260**, 326.
Roussel-Dupré, D., Shine, R. A., Chipman, E.,G., Bruner, E. C., Jr., Lites, B. W., Rottman, G. J., Orral, F. C., Athay, R. G., and White, O. R.: 1976, *Bull. Am. Astron. Soc.* **8**, 312.
Roussel-Dupré, D. and Shine, R. A.: 1982, *Solar Phys.* **77**, 329.
Samain, D.: 1991, *Astron. Astrophys.* **244**, 217.
Sandlin, G. D, Brueckner, G. E., and Tousey, R.: 1977, *Astrophys. J.* **214**, 898.
Sandlin, G. D, Bartoe, J.-D. F., Brueckner, G. E., and VanHoosier, M. E.: 1986, *Astrophys. J. Suppl.* **61**, 801.
Siegmund, O. H. W., Gummin, M. A., Stock, J. M., Marsh, D., Raffanti, T., Sasseen, T., Tom, J., Welsh, B., Gaines, G. A., Jelinsky, P., and Hull, J.: 1994, *Proc. SPIE* **2280**, 89.
Spadaro, D., Antiochos, S. K., and Mariska, J. T.: 1991, *Astrophys. J.* **382**, 338.
Warren, H. P., Mariska, J. T., Wilhelm, K., and Lemaire, P.: 1997, in preparation.
Wilhelm, K., Curdt, W., Marsch, E., Schühle, U., Lemaire, P., Gabriel, A., Vial, J.-C., Grewing, M., Huber, M. C. E., Jordan, S. D., Poland, A. I., Thomas, R. J., Kühne, M., Timothy, J. G., Hassler, D. M., Siegmund, O. H. W.: 1995, *Solar Phys.* **162**, 189.

Wilhelm, K., Lemaire, P., Curdt, W., Schühle, U., Marsch, E., Poland, A.= I., Jordan, S. D., Thomas, R. J., Hassler, D. M., Huber, M. C. E., Vial, J.-C., Kühne, M., Siegmund, O. H. W., Gabriel, A., Timothy, J. G., Grewing, M., Feldman, U., Hollandt, J., and Brekke, P.: 1997, *Solar Phys.* **170**, 75.

Wood, B. E., Harper, G. M., Linsky, J. L., and Dempsey, R. C.: 1996, *Astrophys. J.* **458**, 761.

Wood, B. E., Linsky, J. L., and Ayres, T. R.: 1997, *Astrophys. J.*, in press.

OBSERVATIONS OF POLAR PLUMES WITH THE SUMER INSTRUMENT ON SOHO

D. M. HASSLER[1,*], K. WILHELM[2], P. LEMAIRE[3] and U. SCHÜHLE[2]
[1]*High Altitude Observatory/NCAR, P.O. Box 3000, Boulder, CO 80307, U.S.A.*
[2]*Max-Planck-Institut für Aeronomie, D-37189 Katlenburg-Lindau, Germany*
[3]*Institut d'Astrophysique Spatiale, Unité Mixte CNRS – Université, Paris XI, 91405 Orsay, France*

(Received 13 March 1997; accepted 16 June 1997)

Abstract. We present new observations of O VI 1032 Å line profiles in polar plumes, and inter-plume regions, on the disk and above the limb in the north coronal hole obtained with the SUMER (Solar Ultraviolet Measurements of Emitted Radiation) instrument on the SOHO (Solar and Heliospheric Observatory) spacecraft. On 22 May 1996, a 5 × 5 arc min spectroheliogram was scanned above the north polar coronal hole with the entrance slit extending from 1.03 to 1.33 solar radii with 1.5 arc sec spatial resolution and ≈ 0.044 Å per pixel spectral resolution in the wavelength range 1020–1040 Å. Detailed plume structure in O VI 1032 Å can be seen extending beyond 1.3 solar radii, with intensities in the plume regions 10–50% brighter, but line widths 10–15% narrower, than the inter-plume regions. Possible explanations for this observed anti-correlation between line width and intensity in the plume and inter-plume regions are discussed. We conclude that the source of the high-speed solar wind may not be polar plumes, but the inter-plume lanes associated with open magnetic field regions of the chromospheric network.

1. Introduction

Knowledge of the physical conditions in coronal holes, and the observed structures such as polar plumes seen in polar coronal holes, are essential to understanding their relationship with the high-speed solar wind and its source regions, as well as understanding the fundamental mechanisms responsible for the acceleration of these high-speed streams. The detailed structure in polar coronal holes has been observed and analyzed in white light using both eclipse observations (cf., van de Hulst, 1950; Saito, 1958, 1965; Koutchmy, 1977) and coronagraph observations from Spartan 201, SOHO/LASCO, and MLSO/Mk III (e.g., Fisher and Guhathakurta, 1994, 1995; DeForest *et al.*, 1997; Hassler *et al.*, 1997), as well as in the extreme ultraviolet (EUV) from *Skylab*, sounding rockets and SOHO/EIT (e.g., Bohlin, Sheeley, and Tousey, 1975; Ahmad and Withbroe, 1977; Ahmad and Webb, 1978; Walker *et al.*, 1993; DeForest, 1995; Delaboudinière *et al.*, 1995). The relationship between structures such as polar plumes, observed in white light and in the EUV, have been well established (Bohlin, Sheeley, and Tousey, 1975; DeForest *et al.*, 1997), and temperatures and densities have been inferred from white-light and EUV intensity scale heights (Ahmad and Withbroe, 1977; Walker *et al.*, 1993; DeForest, 1995), as well as from the presence or absence of plume

*Now at Southwest Research Institute, Boulder, CO, U.S.A.

structures observed in images of EUV emission lines and line ratios from sounding rockets and SOHO/EIT (Walker *et al.*, 1993; DeForest, 1995; DeForest *et al.*, 1997; Delaboudinière *et al.*, 1995). Until now, however, there have been no detailed spectroscopic observations in coronal holes with sufficient spatial and spectral resolution to resolve structures such as polar plumes, and provide essential diagnostic information necessary for constraining the thermodynamic models which describe the outflow and acceleration of the solar wind.

We present here new observations from the SUMER (Solar Ultraviolet Measurements of Emitted Radiation) telescope and spectrometer on the Solar and Heliospheric Observatory (SOHO) spacecraft to provide constraints for current models. As with other EUV imagers, SUMER reveals detailed physical structure in coronal holes, but with the added advantage of providing detailed spectroscopic diagnostic information for each structure at the same time, with spatially-resolved information on such physical quantities as thermal and nonthermal plasma velocities, and upper limits on wave velocity amplitudes, which can be directly compared with in-situ observations. The unique capability of SUMER to produce spectroheliograms with the combination of both high spatial resolution (1 arc sec pixels) and high spectral resolution (40 mÅ per pixel) on the solar disk and above the limb out to 1.5 solar radii make it ideally suited for studying the physical processes in coronal holes and the mechanisms for solar wind acceleration. In this context, we discuss observations of line profiles and line widths of O VI 1032 Å in polar plumes and inter-plume regions, both on the disk and above the limb in the north polar coronal hole, and attempt to relate them to observations from other instruments on SOHO and elsewhere (both remote and in-situ) to obtain a better understanding of these ubiquitous structures, and their relationship to the fast solar wind.

2. Observations

SUMER is one of twelve instruments on the ESA/NASA Solar and Heliospheric Observatory (SOHO) spacecraft (Fleck, Domingo, and Poland, 1995), which was launched on 2 December 1995 by an Atlas IIA Centaur rocket from Cape Canaveral. SOHO is in a halo orbit around the Lagrange point L1, roughly 1.5 000 000 km sunward of the Earth with a constant view of the Sun. The SUMER instrument is comprised of an off-axis silicon carbide (SiC) telescope mirror feeding a stigmatic Wadsworth configuration grating spectrograph operating in the wavelength range 500–1610 Å. The telescope mirror can be scanned along two axes out to ±32 arc min from Sun center to provide pointing and raster images using one of several possible entrance slits, with angular dimensions of 4×300, 1×300, 1×120, and 0.3×120 arc sec^2. Overlapping first- and second-order stigmatic spectra are imaged onto one of two 1024×360 pixel Cross Delay Line microchannel plate detectors (Siegmund *et al.*, 1994) with a dispersion of 42–45 mÅ pixel^{-1} in first order and 21–22.5 mÅ pixel^{-1} in second order. A grating scan mechanism permits

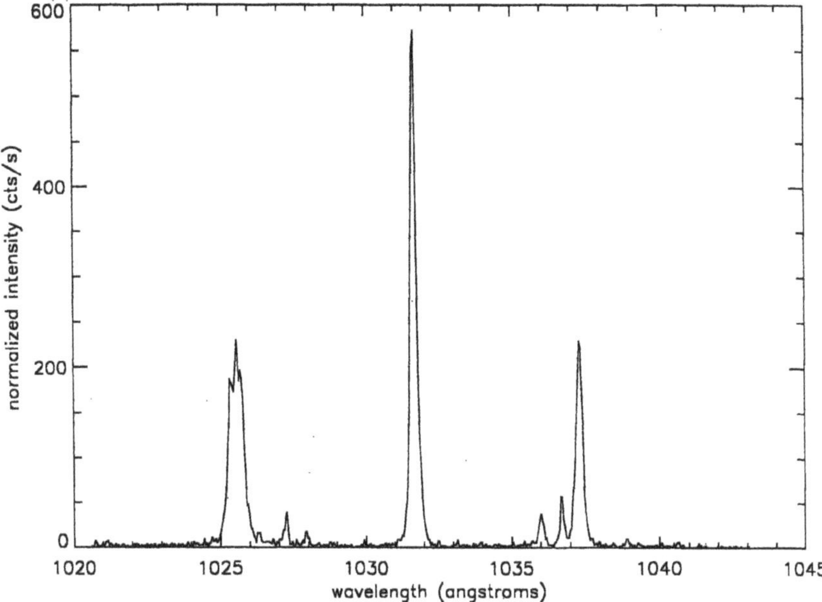

Figure 1. The spectral bandpass selected for this study, shown in this figure, extends from 1021 to 1043 Å (utilizing half of the detector), which includes the strong lines of O VI 1032/1037 Å and Lβ 1025 Å, as well as several weaker lines such as Fe X 1028 Å and C II 1036 Å (to assess the contribution of disk stray light scattered by the telescope mirror).

selection of the individual bandpass (40 Å in first order, 20 Å in second order) which is imaged onto the detector at a given time. A more complete description of the SUMER instrument can be found in the paper by Wilhelm *et al.* (1995) and the inflight performance during the first six months of operation is discussed by Wilhelm *et al.* (1997a) and Lemaire *et al.* (1997).

The majority of the observations discussed here were obtained between 19:20 UT 21 May and 14:15 UT 22 May 1996, although many other observations of polar plumes were also obtained during the first year of SUMER observations. Preliminary observations taken March 13, 1996 in the south coronal hole and presented by Wilhelm *et al.* (1997b) will be mentioned here briefly. The observational sequence performed by SUMER on 21–22 May 1996 consisted of two parts, the first of which was intended to trace the plume structures onto the disk and attempt to relate them to disk features such as X-ray bright points or macrospicules. The second part of the sequence was intended to obtain high signal-to-noise line profiles in the plume and inter-plume regions above the limb out to 1.3 solar radii.

The spectral bandpass selected for this study, shown in Figure 1, extends from 1021–1043 Å (utilizing half of the detector), which includes the strong lines of O VI 1032/1037 Å and Ly-β 1025 Å, as well as several weaker lines such as Fe X 1028 Å and C II 1036 Å (to assess the contribution of disk stray light scattered by the telescope mirror). Although SUMER has the capability to select 25 or 50 pixel wide spectral bandpasses to optimize the allocated telemetry rate (for studies where high temporal resolution is required), we chose to select a bandpass extending over 20 Å or 512 pixels ($\frac{1}{2}$ detector) in order to improve the flat fielding registration and calibration, and permit an assessment of the telescope stray light using the weak chromospheric lines interspersed between the strong coronal lines.

The first part of the sequence used the 0.3×120 arc sec slit to make two raster images or spectroheliograms (16×2 arc min^2) with 1.5 arc sec steps extending ± 8 arc min from central meridian, shown in the bottom half of Figure 2. (The scale in the figure is 1.5 arcsec pixel^{-1} along the x-axis and 1 arc sec pixel^{-1} along the y-axis). Each raster image took roughly 20 min to obtain. The purpose of taking two raster images 20 min apart was to provide a crude indication of the lifetimes of the observed small scale structures, such as bright points and macrospicules, with respect to the long-lived polar plumes seen above the limb. In these raster images, one can see narrow (3–5 arc sec wide) jets or macrospicules originating from bright points near the limb and extending 30–60 arc sec above the limb on both the east and west side of the central meridian. The second part of the sequence, using the 1×300 arc sec slit, extending from 30 arc sec above the limb out to 1.33 solar radii, obtained a single raster image or spectroheliogram extending ± 5 arc min from central meridian, shown in the top half of Figure 2. One can see in this image, along with the larger-scale plume structures, two small jets or macrospicules near central meridian which appear to originate at the limb in the lower images and extend 15–20 arc sec into this upper image. A raster step size of 1.5 arc sec along with an integration time of 150 s per step position provided both adequate spatial resolution to resolve fine scale structures and excellent signal to noise and counting statistics to perform high-resolution spectroscopy.

The relationship of polar plumes above the limb and the physical structures observed on the disk (bright points, macrospicules, network cell boundaries) has attracted much attention, with the most recent discussion given by DeForest *et al.* (1997). Additional insight can be obtained with the observations discussed here. It is clear from these observations that there is no obvious relationship between macrospicules or the small-scale jets observed at the limb and polar plumes higher up, other than they both benefit from the open magnetic field configuration in the coronal hole. As can be seen in the upper portion of Figure 2, these macrospicules appear, but do not seem to be related in any spatial sense, to the larger scale polar plumes. DeForest *et al.* (1997) suggest (from EIT and MDI image overlays) that polar plumes do seem to be related to, and originate from, small EUV bright points corresponding to quiescent, unipolar magnetic flux concentrations on network cell boundaries. However, the relationship is not one-to-one, not all magnetic flux concentrations are plume footpoints, and visa versa. Others (i.e., Golub *et al.*, 1974; Wilhelm *et al.*, 1997c) suggest that polar plumes may have a closed loop structure, similar to very small-scale streamers. In either case, the long lifetimes of polar plumes compared with the shorter lifetimes of X-ray and EUV bright points,

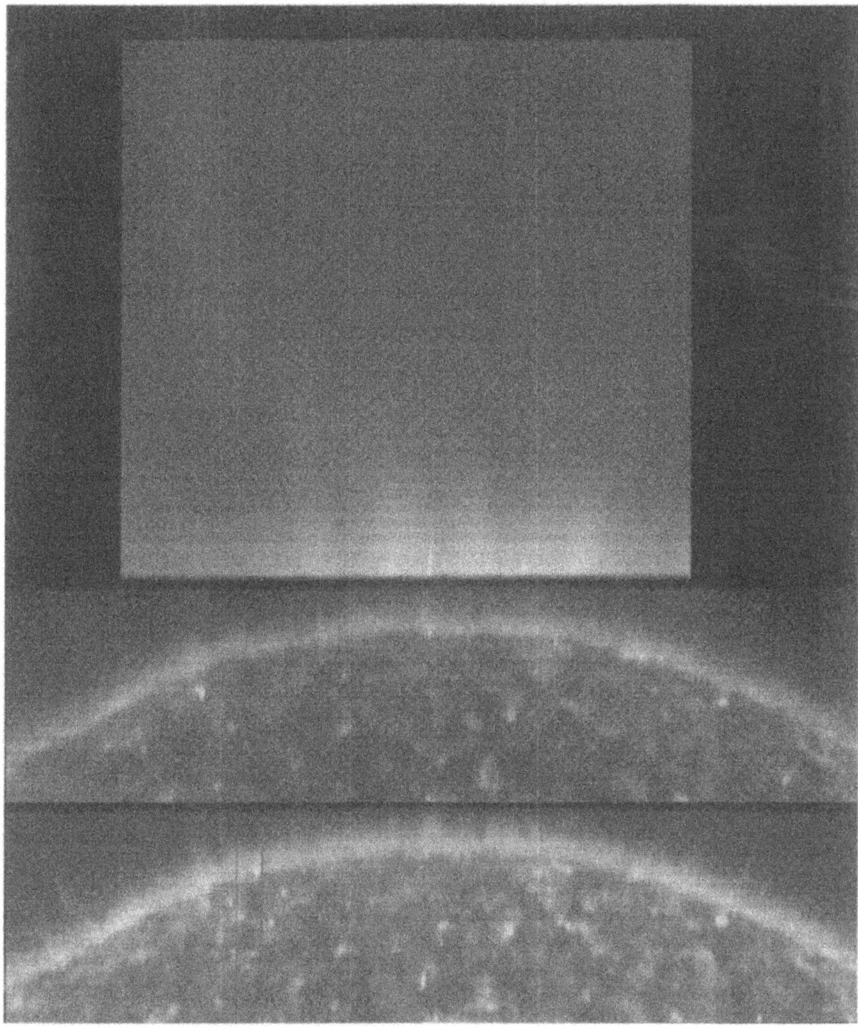

Figure 2. The first part of the sequence used the 0.3 × 120 arc sec slit to make two raster images or spectroheliograms (16 × 2 arc min^2) with 1.5 arc sec steps extending ±8 arc min from central meridian, shown in the bottom half of the figure. (The scale is 1.5 arc sec pixel^{-1} along the x-axis and 1 arc sec pixel^{-1} along the y-axis). In these raster images, one can see narrow jets or macrospicules originating from bright points near the limb and extending 30–60 arc sec above the limb on both the east and west side of central meridian. The second part of the sequence, using the 1 × 300 arc sec slit, extending from 30 arc sec above the limb out to 1.33 solar radii, obtained a single raster image or spectroheliogram extending ±5 arc min from central meridian, shown in the top half of the figure. One can see in this image, along with the larger-scale plume structures, two small jets or macrospicules near central meridian which appear to originate at the limb in the lower images and extend 15–20 arc sec into this upper image. A color version of this figure is included on the CD-ROM.

Figure 3. The corresponding SOHO/EIT image in Fe XII at 195 Å taken at 02:06 UT 22 May 1996 shown for reference. As can be seen in the figure, the polar coronal holes in both the north and south extend far enough onto the disk (toward low latitudes) to permit unambiguous line-of-sight observations above the limb without risk of contamination from foreground or background structures. (Courtesy of the EIT Consortium.) The color version of this figure is included on the CD-ROM.

as well as the size and distribution of plumes observed above the limb (with a size or spatial modulation of 20–30 arc sec, similar to the size of chromospheric network cells), suggest that polar plumes are extensions of the chromospheric network into the open magnetic field regions of the coronal hole, fixed or rooted at the network cell boundaries. These results reinforce the relationship between plumes and network cell boundaries and X-ray/EUV bright points which occur at these network cell boundaries.

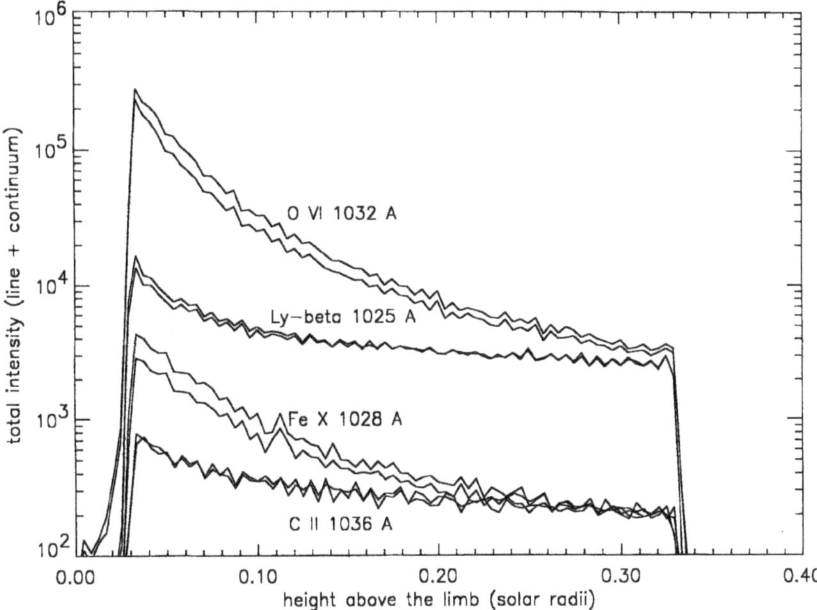

Figure 4. The relative line intensity of four emission lines (O VI 1032, Ly-β 1025, Fe X 1028, and C II 1036, respectively from top to bottom) as a function of height above the limb near the central meridian in both a plume and inter-plume region. The contrast between the plume and inter-plume regions in OVI 1032 Å changes with height from 50% to 20%, whereas the contrast in Ly-β disappears above 1.1 solar radii, and no contrast whatsoever can be seen in the chromospheric line of C II 1036 Å, since its contribution is due entirely to stray light.

The corresponding SOHO/EIT image in Fe XII at 195 Å taken at 02:06 UT 22 May 1996 is shown in Figure 3 for reference. As can be seen in the figure, the polar coronal holes in both the north and south extend far enough onto the disk (toward low latitudes) to permit unambiguous line-of-sight observations above the limb without risk of contamination from foreground or background structures.

Figure 4 shows the relative line intensity of four emission lines (O VI 1032, Ly-β 1025, Fe X 1028, and C II 1036) as a function of height above the limb near the central meridian in both a plume and inter-plume region. The contrast between the plume and inter-plume regions in O VI 1032 Å changes with height from 50% to 20%, whereas the contrast in Ly-β disappears above 1.1 solar radii, and no contrast whatsoever can be seen in the chromospheric line of C II 1036 Å, since its contribution is due entirely to stray light. In fact, the shape of the intensity curve in C II and Ly-β above 1.15 solar radii is a good illustration of the shape of the instrumental stray light curve as a function of height above the limb, and is useful in understanding the contribution of stray light to the coronal profiles. Figure 5 shows the total line intensity of the O VI 1032 Å line in both a plume and inter-plume region, along with an estimate of the stray light contribution to the total line

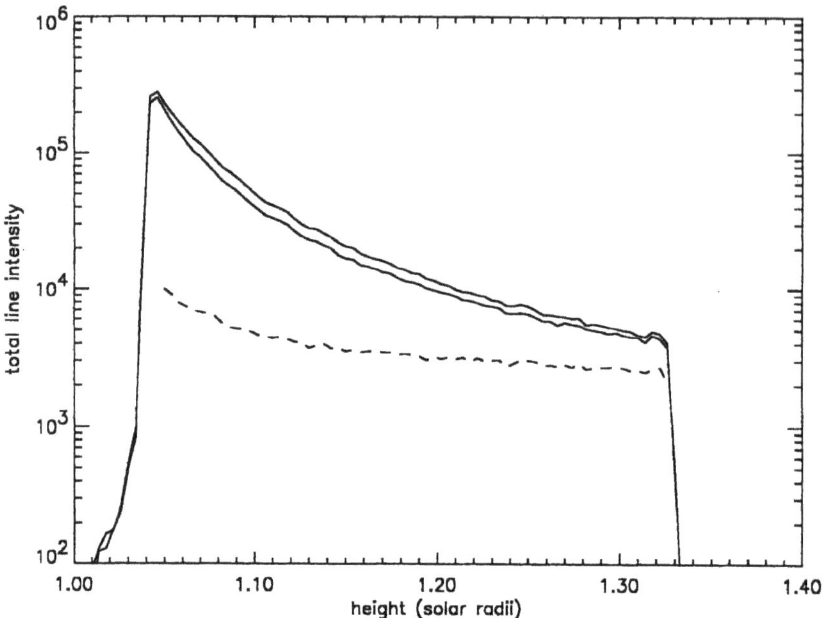

Figure 5. The total line intensity of the O VI 1032 Å line in both a plume and inter-plume region, along with an estimate of the stray-light contribution to the total line intensity (dashed curve). If we make the assumption that the contrast between the plume and inter-plume regions disappears in the O VI 1032 Å line when the coronal signal becomes dominated by instrumental stray light, then the plume/inter-plume contrast will go to zero when these two curves (solid and dashed) converge.

intensity (dashed curve). If we make the assumption that the contrast between the plume and inter-plume regions disappears in the O VI 1032 Å line when the coronal signal becomes dominated by instrumental stray light, then the plume/inter-plume contrast will go to zero when these two curves (solid and dashed) converge.

Figure 6 shows three O VI 1032 Å spectral line profiles obtained in a polar plume at 1.05, 1.10, and 1.15 solar radii along with the estimated stray light profile contribution (dashed line). As can be seen in the figure, the contribution of stray light to these coronal O VI 1032 Å profiles approaches 20–30% at heights of 1.2 to 1.3 solar radii. Figure 7 shows a sample of a single Gaussian fit to the O VI 1032 Å line profiles at 1.05 solar radii for both a plume and inter-plume region. As can be seen in the figure, the center of the O VI plume profiles appear to be very nearly Gaussian, with small deviations from Gaussian in the wings of the profile. (In theory, these emission line profiles should be Voigt profiles, represented by a Gaussian in the center and a Lorentzian in the wings.) In an effort to better characterize these broad wings, Wilhelm *et al.* (1997b) have fit a double Gaussian to these profiles, with the total intensity of the broad line about 10% of the narrow one. The most probable velocities reported by Wilhelm and co-workers for a double Gaussian fit were 48 km s^{-1} and 104 km s^{-1} for the narrow and broad components,

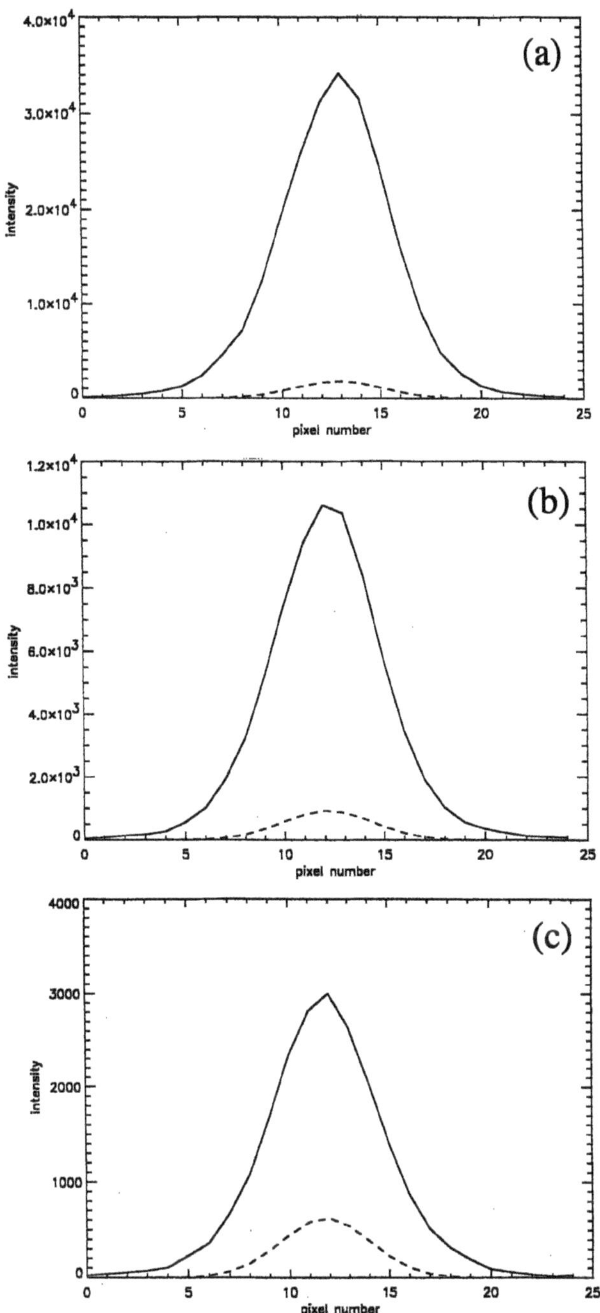

Figure 6. Three O VI 1032 Å spectral line profiles obtained in a polar plume at 1.05 (a), 1.10 (b), and 1.15 (c) solar radii along with the estimated stray light profile contribution (dashed line). As can be seen in the figure, the center of the O VI plume profiles appear to be very nearly Gaussian, with small deviations from Gaussian in the wings of the profile.

respectively. Understanding further the significance of this deviation from Gaussian is in progress and will be discussed in a future work. The estimated stray light profile contributions shown here were obtained by fitting a Gaussian to stray light profiles above 1.6 solar radii, assuming no coronal contribution at this height, similar to the "Sun as a Star" observations (Lemaire, 1997), and then normalizing the total line intensity for each specific height to that calculated in the previous figure. Again, the assumption of a Gaussian profile shape for both the O VI profiles and stray light profiles is an idealization which is not necessarily valid and will be discussed further in a future work.

Figure 8 shows an image of the O VI 1032 Å line intensity and line width (assuming a single Gaussian profile) as a function of height and position above the limb obtained during the raster scan on 22 May 1996. The most notable feature is the striking anti-correlation between line intensity and line width near central meridian. Figure 9 shows a plot of line intensity and line width along two strips tangent to the limb at 1.05 and 1.15 solar radii, to illustrate more clearly this anti-correlation. The modulation in intensity can be seen to be on the order of 50% whereas the modulation in line width is on the order of 10%. This observed anti-correlation between intensity and line width in polar plumes has also been reported by Antonucci (1997) and Poletto (1996) and Noci *et al.* (1997) with the SOHO/UVCS (Ultraviolet Coronagraph Spectrometer) instrument, and analysis of those data are in progress.

The line widths shown in Figure 9, in a polar plume at 1.05 solar radii near central meridian, after correcting for instrumental width, correspond to Doppler (e^{-1}) widths ($\Delta\lambda$) of approximately 152 mÅ, or a most probable velocity of 44 km s^{-1}. The e^{-1} widths in the inter-plume region to the east of central meridian at 1.05 solar radii are $\Delta\lambda = 165$ mÅ, corresponding to a most probable velocity of 48 km s^{-1}. The statistical variation or relative uncertainty in an individual line width measurement is on the order of 2–4% (one sigma), which is significantly smaller than the observed variation of line widths between the plume and inter-plume regions (10–15%). Other observations with SUMER, obtained on 13 March 1996 in the south coronal hole, reported by Wilhelm (1997) show Doppler widths of $\Delta\lambda = 165$ mÅ corresponding to a most probable line-of-sight velocity of 48 km s^{-1} in a polar plume, and $\Delta\lambda = 192$ mÅ or 56 km s^{-1} in an inter-plume region at one arcminute above the limb (1.05 solar radii). For reference, the thermal velocity for O VI corresponding to an ion temperature of 3×10^5 K, which is the temperature of maximum abundance for O VI assuming ionization equilibrium, is 17 km s^{-1}. The thermal (most probable) velocity for O VI, assuming a temperature of 1×10^6 K, is 32 km s^{-1}.

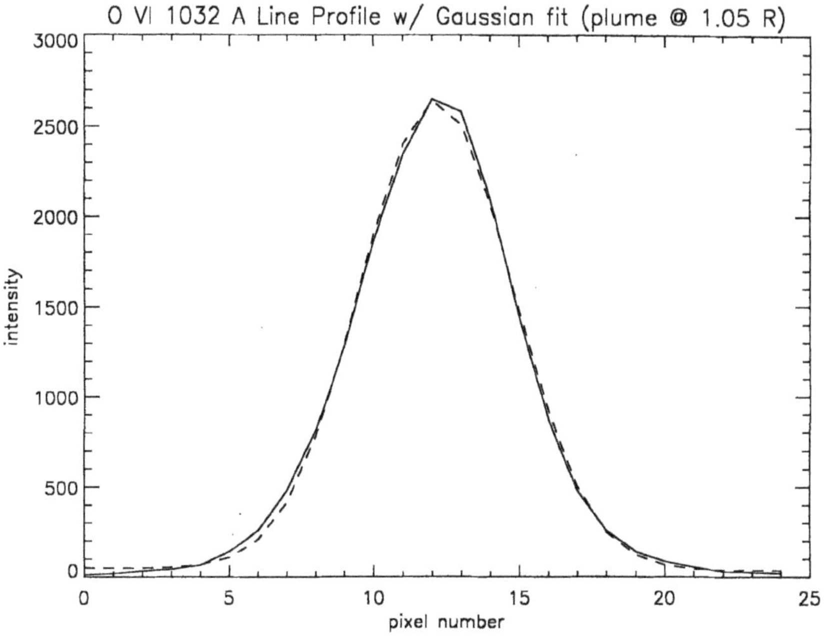

O VI 1032 A Line Profile w/ Gaussian fit (plume @ 1.05 R)

intensity

3000
2500
2000
1500
1000
500
0

0 5 10 15 20 25

pixel number

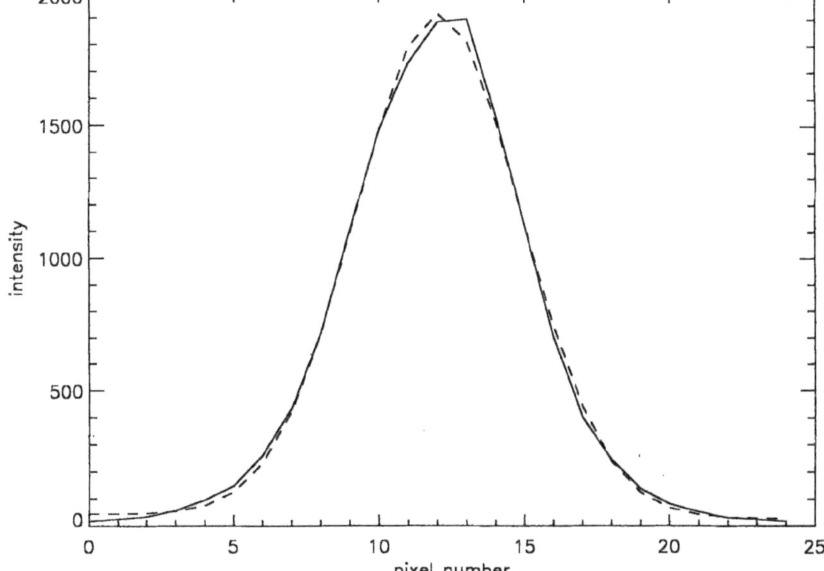

Figure 7. Sample of a single Gaussian fit (dashed curve) to the O VI 1032 Å line profiles at 1.05 solar radii in both a plume and inter-plume region. As can be seen in the figure, the center of the O VI profiles appear to be very nearly Gaussian, with small deviations from Gaussian in the wings of the profile.

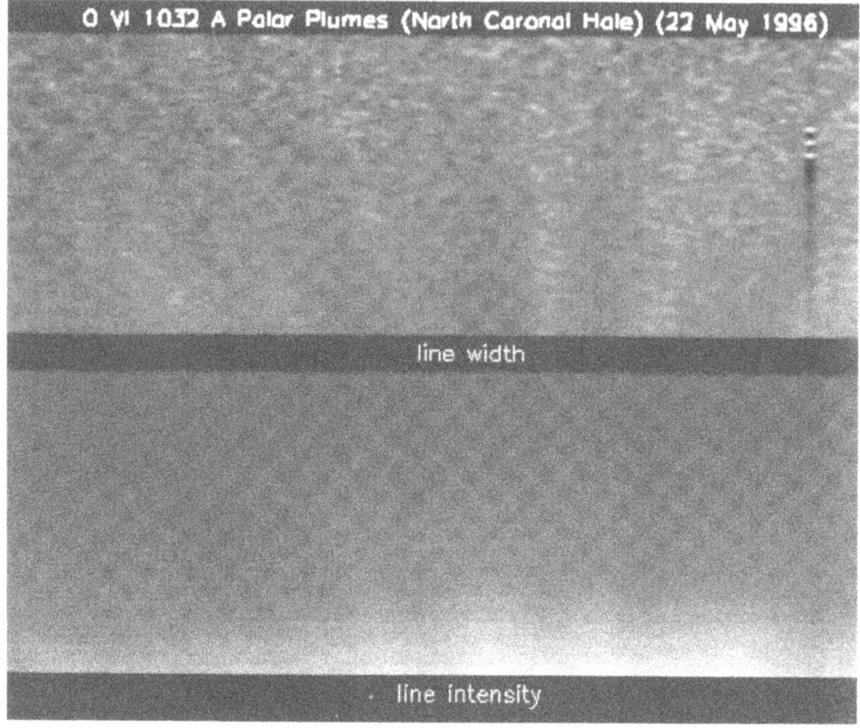

Figure 8. An image of the O VI 1032 Å line intensity and line width (assuming a single Gaussian profile) as a function of height and position above the limb obtained during the raster scan on 22 May 1996. The most notable feature is the striking anti-correlation between line intensity and line width near central meridian. A color version of this figure is included on the CD-ROM.

3. Discussion

The new result from these observations is the observed anti-correlation between line width and intensity in the plume/inter-plume region line profiles at heights below 1.3 solar radii. We will discuss here several possible explanations for this observed anti-correlation, and their consequences. The first possible explanation is related to the geometry, structure, and physical characteristics of polar plumes and their effect on the integration of the line profile along the line of sight. Other explanations involve the two line broadening components (thermal and non-thermal) which contribute to a spectral line profile. If we assume a Gaussian distribution, the Doppler width is equal to the Gaussian width of the distribution, described by the equation

$$I_\lambda = \frac{I}{\pi^{1/2}\Delta} \exp\left(\frac{-(\lambda - \lambda_0)^2}{\Delta^2}\right) , \tag{1}$$

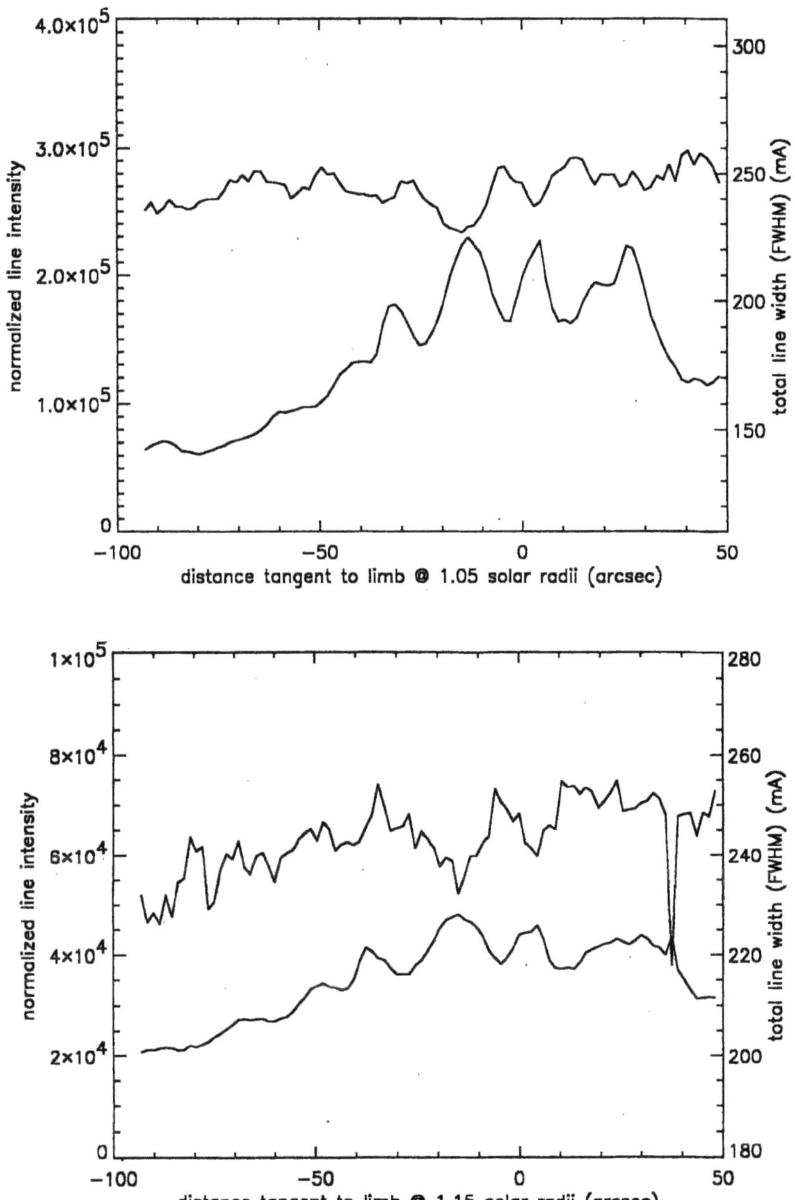

Figure 9. A plot of line intensity (bottom curve) and line width (top curve) along two strips tangent to the limb at 1.05 and 1.15 solar radii, to illustrate more clearly the anti-correlation between line width and intensity. The modulation in intensity can be seen to be on the order of 50% whereas the modulation in line width is on the order of 10%.

where Δ for thermal and nonthermal broadening is given by

$$\Delta = \frac{\lambda}{c}\left(\frac{2kT}{M} + v_{\rm NT}^2\right)^{1/2} , \tag{2}$$

where M and T are the ion mass and ion temperature, respectively, and $v_{\rm NT}$ is the nonthermal velocity of the ion.

The first possible explanation for the observed anti-correlation between line width and intensity is related to the effects of integration along the line of sight. If we assume that plumes are cylindrical in shape, then the majority of the observed plume emission will come from a more localized region than the inter-plume regions, which appear much less structured and therefore have a more uniform contribution along the line of sight. However, this broadening mechanism only works if there are significant systematic radial or super-radial flows along the line of sight. In this case, the line-of-sight component of the flow velocity varies, so that emission from different points along the line of sight produces a Gaussian line shape with the line center shifted in accordance with the local line-of-sight component of the flow velocity. The intensity of each Gaussian depends on the local contribution function, and the observed line width is determined by the superposition of various lines along the line of sight. If this effect is significant, then the line broadening should increase with height due to greater and greater outflow velocities, as can be seen in Figure 8, which does not occur to any large extent (<10%). (However, the relative contribution of the stray light profile, whose line width remains roughly constant with height, to the total observed profile also increases with height, which can have the tendency to mask any observable coronal line broadening.) In addition to increased broadening with height, significant sytematic flows in the more localized plume regions should be observable as Doppler shifts due to the line of sight velocity component of the individual plumes either into or out of the plane of the sky. However, only small (<3 km s^{-1}) Doppler shifts are observed in the plumes, suggesting very modest (<10 km s^{-1}) radial flow velocities out to 1.2 solar radii.

A second explanation for the observed anti-correlation between intensity and line width is that plumes are cooler in temperature than the inter-plume region. Mg IX (706/750 Å) and Si VIII (1440/1445 Å) line ratio observations from SUMER (discussed by Wilhelm *et al.*, 1997c), suggest electron temperatures which are lower and decrease with height in the relatively higher density plume structures compared to the lower density inter-plume regions. Fe XII (195 Å) to Fe X (171 Å) line ratio images from SOHO/EIT (Thompson *et al.*, 1997) also show electron temperatures which appear to be cooler in polar plumes than inter-plume regions, and DeForest *et al.* (1997) have found the ionization temperature in plumes to be on the order of 30% cooler than inter-plume regions.

Another explanation is that the thermal temperature is roughly the same, but that the non-thermal broadening (due to waves or turbulence) is greater in the inter-plume regions than in the plume regions. Other techniques, such as scale height temperatures derived from white-light polarized brightness (pB) images from the

MkIII coronagraph on Mauna Loa, for example, show higher density but roughly constant temperature in the plume regions compared to the inter-plume regions. In this case, the observed excess line broadening in the inter-plume regions would be due to nonthermal processes, such as Alfvén waves. Although more difficult to distinguish between these broadening mechanisms, the implications are significant, because polar plumes have been suggested to be the source region for the high-speed solar wind (Ahmad and Withbroe, 1977; Ahmad and Webb, 1978; Mullan and Ahmad, 1982).

To fully understand the physical conditions in the solar corona and the mechanisms responsible for accelerating the solar wind, we must discuss these observations in the context of other observations (both remote and *in-situ*) from other instruments on SOHO, other satellites and spacecraft, and ground-based observatories.

For example, it is interesting to compare the elemental abundances of low FIP (first ionization potential) ions observed in polar plumes from *Skylab* (Widing and Feldman, 1992a, b; Feldman, 1992) with those measured in-situ in the fast solar wind from *Ulysses* (Geiss *et al.*, 1992). The elemental abundances of low FIP ions in the fast solar wind observed with the SWICS instrument on Ulysses (Geiss *et al.*, 1992; Gloeckler, 1997) do not appear to be enriched relative to photospheric values, with a FIP enhancement close to 1 (i.e., no 'FIP effect' in fast flows). However, Widing and Feldman (1992a, b) and Feldman (1992) report that the enrichment by low FIP ions in polar plumes is large (a strong 'FIP effect'), suggesting that perhaps it is not polar plumes which are responsible for accelerating the fast solar wind. Feldman and Doschek (1997) have recently made similar observations with SUMER to study abundances of the low FIP ions in polar plumes, and analyses of those data are in progress.

This relationship between measured abundances of low FIP ions in polar plumes and in situ measurements in high-speed solar wind streams, together with both lower electron and effective ion temperatures (line widths), and the relatively small Doppler velocities observed in polar plumes, suggest that the origin of the high-speed solar wind may indeed not be the polar plume structures, but more likely the inter-plume regions. This conclusion is also consistent with recent solar wind model calculations (Habbal *et al.*, 1995). Based on electron density profiles obtained with the White-Light Coronagraph on the Spartan 201 spacecraft (Fisher and Guhathakurta, 1995), they derive slower solar wind flow speeds from the 'denser' plume structures with respect to the 'ambient' coronal hole, due to lower effective 'scale height' temperatures. Although these results are self-consistent, further analysis of these and other data may yield more clues. More insight may be gained, for example, by comparison of observations of line profiles from ions of different mass, such as Ne VIII 770 Å, Mg X 625 Å, Fe XII 1242 Å. However, further analysis of these data is ongoing and will be presented in a future work.

Acknowledgements

The authors would like to thank and acknowledge the essential and untiring help of Nicolas Morisset for solving all of the operations and software problems and system management bugs with enthusiasm and good humor. The authors would also like to extend sincere thanks to Dermott Mullan for many useful and insightful conversations. The SUMER project is financially supported by DARA, CNES, NASA and the ESA PRODEX program (Swiss Contribution). A portion of the work of D.M.H. has been supported by NASA under grant NAG 5–2903 to the National Center for Atmospheric Research.

References

Ahmad, I. A. and Withbroe, G. L.: 1977 *Solar Phys.* **53**, 397.
Ahmad, I. A. and Webb, D. F.: 1978, *Solar Phys.* **58**, 323.
Antonucci, E.: 1997. private communication.
Bohlin, J. D., Sheeley, N. R., and Tousey, R.: 1975, in M. J. Rycroft (ed.), *Space Research XV*, Akademie-Verlag, Berlin, p. 651.
Brueckner, G. E. *et al.*: 1995, *Solar Phys.* **162**, 357.
Delaboudinière, J.-P. *et al.*: 1995, *Solar Phys.* **162**, 291.
DeForest, C. E.: 1995, Doctoral thesis, Stanford University, submitted.
DeForest, C. E., Hoeksema, J. T., Gurman, J. B., Thompson, B., Howard, R, Plunkett, S. P., Harrison, R., and Hassler, D. M.: 1997, *Solar Phys.*, this issue.
Dupree, A. K., Penn, M. J., and Jones, H. P.: 1996, *Astrophys. J.* **467**, L121.
Feldman, U.: 1992, *Physica Scripta* **46**, 202.
Fisher, R. R. and Guhathakurta, M.: 1994, *Space Sci. Rev.* **70**, 267.
Fisher, R. R. and Guhathakurta, M.: 1995, *Astrophys. J.* **447**, L139.
Fleck, B., Domingo, V., and Poland, A. I. (eds): 1995, *Solar Phys.* **162**.
Geiss, J., Ogilvie, K. W., von Steiger, R., Mall, U., Gloeckler, G., Galvin, A. B., Ipavich, F., Wilken, B., and Gliem, F.: 1992, in E. Marsch and R. Schwenn (eds), *Solar Wind Seven*, Goslar, Germany, September 16–20, 1991, Pergamon Press, Oxford, p. 341.
Geiss, J., Gloeckler, G., von Steiger, R., Balsiger, R., Fisk, L. A., Galvin, A. B., Ipavich, F., Livi, S., McKenzie, F. J., Ogilvie, K. W., and Wilken, B.: 1995, *Science* **268**, 1033.
Golub, L., Krieger, A. S., Silk, J. K., Timothy, A. F., and Vaiana, G. S.: 1974, *Astrophys. J.*, **189**, L93.
Habbal, S. R.: 1992, *Ann. Geophys.* **10**, 34.
Habbal, S. R., Esser, R., and Arndt, M. B.: 1993, *Astrophys. J.* **413**, 435.
Habbal, S. R., Esser, R., Guhathakurta, M., and Fisher, R. R.: 1995, *Geophys. Res. Letters* **22**, 1465.
Hassler, D. M.: 1997, in preparation.
Hassler, D. M. and Moran, T. G.: 1994, *Space Sci. Rev.* **70**, 373.
Hassler, D. M., Rottman, G. J., Shoub, E. C., and Holzer, T. E.: 1990, *Astrophys. J.* **348**, L77.
Hassler, D. M., Wilhelm, K., Lemaire, P., and Schühle, U.: 1997, *Astrophys. J.*, in preparation.
Holzer, T. E., Shoub, E. C., Hassler, D. M., and Rottman, G. J.: 1990, *EOS* **70**, 1252.
Isenberg, P. A. and Hollweg, J. V.: 1983. *J. Geophys. Res.* **88**, 3923.
Khabibrakhmanov, I. K. and Mullan, D. J.: 1994, *Astrophys. J.* **430**, 814.
Koutchmy, S.: 1977, *Solar Phys.* **51**, 399.
Lemaire, P.: 1997b, private communication.
Lemaire, P., Wilhelm, K., Curdt, W., Schühle, U., Marsch, E., Poland, A. I., Jordan, S. D., Thomas, R. J., Hassler, D. M., Huber, M. C. E., Vial, J.-C., Kühne, M., Siegmund, O. H. W., Gabriel, A., Timothy, J. G., and Grewing, M.: 1997, *Solar Phys.* **170**, 105.
Mullan, D.: 1997, private communication.
Mullan, D. J. and Ahmad, I. A.: 1982, *Solar Phys.* **75**, 347.

Noci, G. *et al.*: 1997, *Adv. Space Res.*, in press.
Osterbrock, D. E.: 1961, *Astrophys. J.* **134**, 347.
Poletto, G.: 1996, private communication.
Saito, K.: 1958, *Publ. Astron. Soc. Japan* **10**, 49.
Saito, K.: 1965, *Publ. Astron. Soc. Japan* **17**, 1.
Siegmund, O. H. W., Gummin, M. A., Sasseen, T., Jelinsky, P., Gaines, G. A., Hull, J., Stock, J. M., Edgar, M., Welsh, B., Jelinsky, S., and Vallerga, J.: 1995, *Proc. SPIE* **2518**, 344.
Thieme, K. M., Schwenn, R., and Marsch, E.: 1989, *Adv. Space Res.* **9**, (4)127.
Thieme, K. M., Marsch, E., and Schwenn, R.: 1990, *Ann. Geophys.* **8**, 713.
Thompson, B. *et al.*: 1997, in preparation.
van de Hulst, H. C.: 1950, *Bull. Astron. Inst. Neth.* **11**, 150.
Walker, A. B. C., DeForest, C. E., Hoover, R. B., and Barbee, T. W.: 1993. *Solar Phys.* **148**, 239.
Widing, K. G. and Feldman, U.: 1992, *Astrophys. J.* **392**, 715.
Widing, K. G. and Feldman, U.: 1992. in E. Marsch and R. Schwenn (eds), *Solar Wind Seven*, Goslar, Germany, September 16–20, 1991, Pergamon Press, Oxford, p. 405.
Wilhelm, K.: 1997, *Chinese J. Space Sci.*, in press.
Wilhelm, K., Curdt, W., Marsch, E., Schühle, U., Gabriel, A., Lemaire, P., Vial, J.-C., Grewing, M., Hassler, D. M., Huber, M. C. E., Jordan, S. D., Poland, A. I., Thomas, R. J., Kühne, M., Timothy, J. G., and Siegmund, O. H. W.: 1995, *Solar Phys.* **162**, 189.
Wilhelm, K., Lemaire, P., Curdt, W., Schühle, U., Marsch, E., Poland, A. I., Jordan, SD. D., Thomas, R. J., Hassler, D. M., Huber, M. C. E., Vial, J.-C., Kühne, M., Siegmund, O. H. W., Gabriel, A., Timothy, J. G., Grewing, M., Feldman, U., Hollandt, J., and Brekke, P.: 1997a, *Solar Phys.* **170**, 75.
Wilhelm, K., Curdt, W., Gabriel, A., Grewing, M., Hassler, D. M., Huber, M. C. E., Jordan, S. D., Kühne, M., Lemaire, P., Marsch, E., Poland, A. I., Schühle, U., Siegmund, O. H. W., Thomas, R. J., Timothy, J. G., and Vial, J.-G.: 1997b, *Astron. Soc. Pacific Conf. Series* **118**.
Wilhelm, K., Marsch, E., Hassler, D. M., Gabriel, A., Lemaire, P., and Huber, M. C. E.: 1997c, in preparation.
Williams, L. L. and Mullan, D. J.: 1996, *Astrophys. J.* **457**, L95.

POLAR PLUME ANATOMY: RESULTS OF A COORDINATED OBSERVATION

C. E. DeFOREST[1], J. T. HOEKSEMA[1], J. B. GURMAN[2], B. J. THOMPSON[3], S. P. PLUNKETT[4,*], R. HOWARD[4], R. C. HARRISON[5] and D. M. HASSLER[6,†]

[1] *W.H. Hansen Experimental Physics Laboratory, Center for Space Science and Astrophysics, Stanford University, Stanford, CA 94305–4085, U.S.A.*

[2] *NASA/Goddard Space Flight Center, Greenbelt, MD 20771, U.S.A.*

[3] *Applied Research Corporation, U.S.A.*

[4] *E.O. Hurlburt Center for Space Research, Naval Research Laboratory Washington, DC 20375, U.S.A.*

[5] *Rutherford Appleton Laboratory, U.S.A.*

[6] *High Altitude Observatory, Boulder, CO, U.S.A.*

(Received 18 March 1997; accepted 18 June 1997)

Abstract. On 7 and 8 March 1996, the SOHO spacecraft and several other space- and ground-based observatories cooperated in the most comprehensive observation to date of solar polar plumes. Based on simultaneous data from five instruments, we describe the morphology of the plumes observed over the south pole of the Sun during the SOHO observing campaign. Individual plumes have been characterized from the photosphere to approximately 15 $R_\odot$, yielding a coherent portrait of the features for more quantitative future studies. The observed plumes arise from small (∼2–5 arc sec diameter) quiescent, unipolar magnetic flux concentrations, on chromospheric network cell boundaries. They are denser and cooler than the surrounding coronal hole through which they extend, and are seen clearly in both Fe IX and Fe XII emission lines, indicating an ionization temperature between 1.0–1.5 × 10^6 K. The plumes initially expand rapidly with altitude, to a diameter of 20–30 Mm about 30 Mm off the surface. Above 1.2 $R_\odot$, plumes are observed in white light (as 'coronal rays') and extend to above 12 $R_\odot$. They grow superradially throughout their observed height, increasing their subtended solid angle (relative to disk center) by a factor of ∼10 between 1.05 $R_\odot$ and 4–5 $R_\odot$, and by a total factor of 20–40 between 1.05 $R_\odot$ and 12 $R_\odot$. On spatial scales larger than 10 arc sec, plume structure in the lower corona ($R < 1.3\ R_\odot$) is observed to be steady-state for periods of at least 24 hours; however, on spatial scales smaller than 10 arc sec, plume XUV intensities vary by 10–20% (after background subtraction) on a time scale of a few minutes.

1. Introduction

Polar plumes are linear structures that are apparent over the solar poles in visible light (e.g., Saito, 1965; Koutchmy, 1977), in extreme ultraviolet (e.g., Bohlin, Sheeley, and Tousey, 1975; Walker *et al.*, 1988, 1993), and in soft X-rays (e.g., Ahmad and Webb, 1978). They are thought to be denser than the surrounding media, and have been shown to diverge superradially with altitude (Ahmad and Withbroe, 1977; Fisher and Guhathakurta, 1995). While some observations (Harvey, 1965; Newkirk and Harvey, 1968; Lindblom, 1990; Allen, 1994) have suggested that polar plumes land on unipolar magnetic structures, the prevailing model has been

* Dr Plunkett is employed by the Universities Space Research Association.

† Dr Hassler is now employed by Southwest Research Institute, Boulder, CO.

Solar Physics **175:** 393–410, 1997.

one with small dipoles at the footpoints (e.g., Ahmad and Withbroe, 1977; Habbal, 1992; Saito, 1965). The relation of the polar plumes to the solar wind has been in dispute: some authors (e.g., Walker, 1990) believe that the plumes are a likely source for the high speed solar wind; while others (e.g., Habbal, 1992) believe otherwise.

The SOHO observing campaign of 7 and 8 March 1996 was the most comprehensive observation of polar plumes to date. The observation was designed to obtain a complete picture of the morphology of the plumes, from photospheric footpoint to outer corona; to provide plasma diagnostics to constrain semi-empirical plume and solar wind models; and to identify possible heating mechanisms – in particular, small-scale magnetic reconnection and MHD waves in the 2–10-min frequency band. For descriptions of the participating SOHO instruments, refer to Fleck, Domingo, and Poland (1995).

Figure 1 diagrams the observation plan that was used for all participating SOHO instruments and one of the ground-based observatories. The Michelson Doppler Imager ('MDI') recorded magnetic and Doppler information with 0.6 arc sec resolution and 1-min cadence; the Coronal Diagnostic Spectrometer ('CDS') observed the chromospheric roots of the plumes in He I, O V, and Mg IX spectral lines (intensity only) with a 1-min cadence; the EUV Imaging Telescope ('EIT') generated 3-min-cadence XUV images of the coronal hole in Fe IX/X and He II line emissions; the Solar Ultraviolet Measurement of Emitted Radiation spectrometer ('SUMER') rapidly scanned across the edge of a single observed plume to record differences in the O VI and Mg IX line profiles within and outside the plume; the UltraViolet Coronal Spectrometer ('UVCS') recorded Doppler dimming information and high cadence Lyα and O VI spectra across the hole at a single altitude (1.8 $R_{\odot}$) to obtain outflow rates and detect possible wave motions; and the Large Angle Spectrometric COronagraph ('LASCO') recorded still images of the middle and outer corona with its C2 and C3 coronagraphs before and after the bulk of the observations. Daily average images from the White-Light Coronagraph Mk 3 ('Mk 3') at HAO's Mauna Loa Solar Observatory were also incorporated into the SOHO data set.

In this paper, we concern ourselves primarily with a complete description of the directly observable parameters of the observed plumes. We use EIT, Mk 3, and LASCO to trace the plumes' morphology; CDS to identify the location of individual plumes' footpoints relative to the chromopheric network; and MDI for information on the magnetic structure of the plumes' lower reaches. As instrument calibrations become available, the EIT and LASCO data will be used in future analyses to extract density as a function of height in the plumes, while SUMER and UVCS data will be used to search for signatures of wave heating and plasma acceleration at moderate and high altitudes.

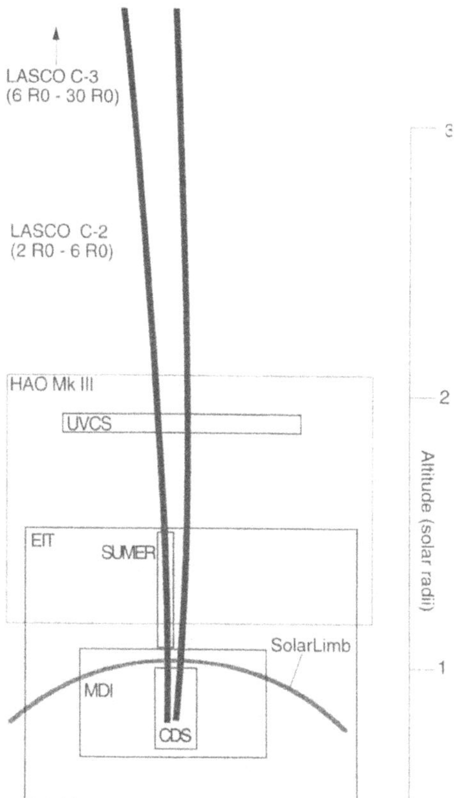

Figure 1. Schematic diagram of the SOHO March 1996 plume observation. Each instrument's field of view is marked, roughly (but not exactly) to scale. Because the LASCO fields of view are large compared to the disk, they are not drawn (only notated). MDI, CDS, EIT, SUMER, and UVCS each generated data at cadences of 1–4 min; the outer coronagraphs recorded single images before and after the bulk of the campaign observations.

2. Observations

2.1. EIT: POLAR PLUMES IN EUV/XUV

EIT (Delaboudinière *et al.*, 1995) recorded images of the south pole in the EUV/XUV portion of the spectrum for most of the observation, affording views of plume morphology and evolution in the lowest reaches of the corona. Because the off-point maneuver of the plumes campaign required LASCO to close its front aperture doors, all of the telemetry bandwidth normally allocated to LASCO was available to EIT. This in turn allowed EIT to record solar images with a much higher cadence, and a longer duration, than is normally possible. EIT's principal observations were a five-hour-long sequence of exposures at 171 Å (Fe IX/X) with

3-min time resolution from ~16:30 UT to ~21:30 UT on 7 March; a slightly shorter (2.5 hour) sequence of images at 304 Å (He II) starting at 7 March 07:22 UT; and several sets of full-disk exposures that cycle through the four available passbands. The plumes are readily visible in the Fe IX/X (171 Å) movie and the Fe XII (193 Å) images; they are less visible (but still apparent) in the Fe XV (284 Å) exposures, and do not appear in the He II (304 Å) images.

Figure 2 shows two images of the coronal hole taken 5 min apart: one in the 171 Å (Fe IX/X) passband, which is sensitive to plasma near 1.0×10^6 K; the other in the 195 Å (Fe XII) passband, which 'sees' plasma at temperatures of about 1.6×10^6 K. The plumes visible in Fe IX/X can be described as roughly linear features that subtend about 2.5° relative to Sun center as they cross the limb. The observed plumes appear to expand radially from a radiant that is between Sun center and the limb of the Sun, following the observations of Fisher and Guhathakurta (1995). However, the base of the plume can best be described as a 'blunt pencil' shape, suggestive of Newkirk and Harvey's (1968) model of magnetic flux near plume footpoints. Each plume arises from a small (2–4 Mm; ~4 arc sec) footpoint and expands rapidly with height before assuming its characteristic slow expansion at about 20–30 Mm off the surface (with a diameter of about 20–30 Mm at that altitude).

While calculation of ionization temperatures from this dataset must wait for the final intensity calibration for EIT, one may draw some conclusions from the uncalibrated data. With the assumption that each plume is approximately isothermal across its width, one may conclude that the plume temperatures are between about $1.0-1.5 \times 10^6$ K: otherwise, they would be clearly visible in one or the other, but not both, of the images in Figure 2. This is because the Fe XII telescope is not sensitive to plasmas under about 1.0×10^6 K and the Fe IX/X telescope is not sensitive to plasmas above about 1.5×10^6 K. Furthermore, one may determine qualitatively which regions are hotter than others by comparing the relative brightnesses of features in the Fe IX/X and Fe XII passbands: hotter structures emit proportionally more light from the higher ionization state.

Figure 3 is a composite image of the two panels in Figure 2: its pixel values are calculated by dividing the Fe XII intensity by the Fe IX/X intensity, to show a rough temperature map. Most of the plumes appear darker than the surrounding media, indicating proportionally less 195 Å emission than 171 Å emission, or a lower ionization temperature than their surrounding media. Because the surrounding media are also visible in both passbands, the temperatures of both the interplume hole and the plumes themselves must be in the $1.0-1.5 \times 10^6$ K range; hence, the plumes are cooler than the background by up to (but not more than) 30%.

The result that plumes are relatively cool is consistent with scale height measurements made by Walker *et al.* (1993) from a 1987 XUV rocket observation, which showed a slightly higher scale height in the interplume regions than in plumes themselves in Fe IX/X emission; with recent observations of line width in O VI plumes by SUMER (Hassler *et al.*, 1997), wherein the O VI spectral line was

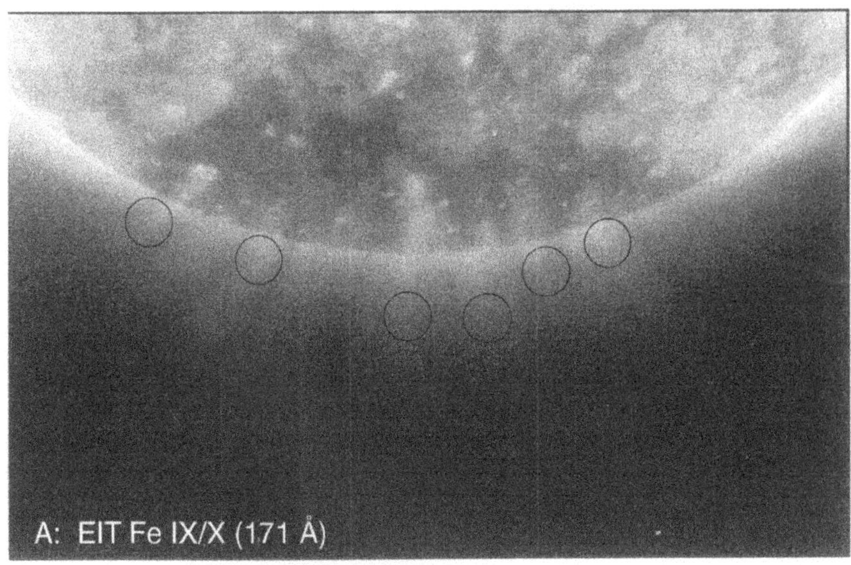

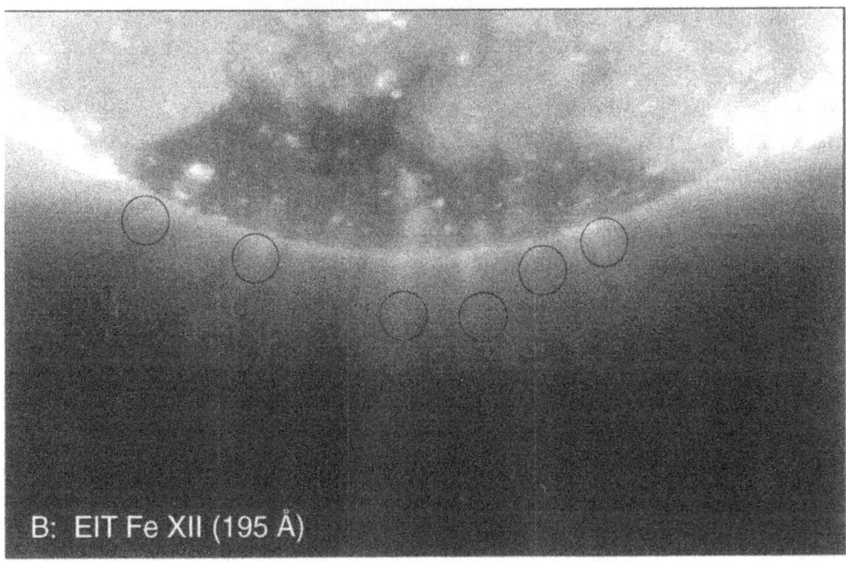

Figure 2. (A) The EIT Fe IX/X (171 Å) image from 16:45 UT, 7 Marrch 1996, showing polar plumes over the south polar hole at 1.0×10^6 K. (B) The EIT Fe XII (195 Å) image from 16:34 UT, 7 March 1996. This passband is sensitive to plasmas at 1.6×10^6 K. Some key locations have been circled for intercomparison between the two images.

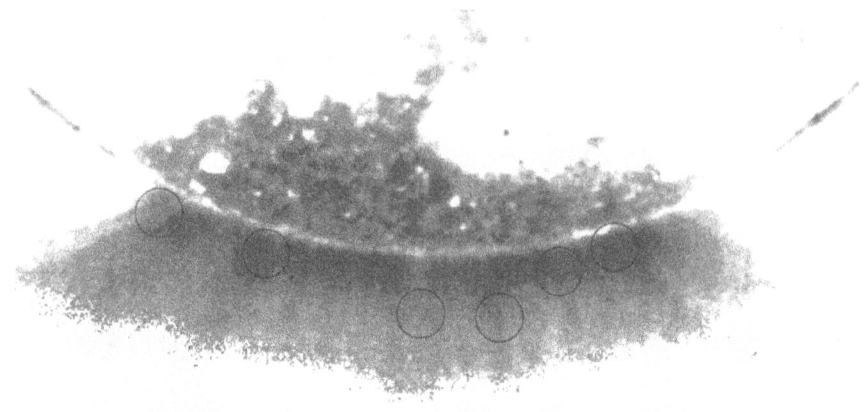

Figure 3. EIT ratio image: this image is a ratio of the two images in Figure 2, showing plasma temperature variations. Briqht features are proportionally brighter in Fe XII than in Fe IX, indicating higher temperature; dark features are proportionally dimmer in Fe XII than in Fe IX, indicating lower temperature. This ratio is sensitive over a range of about $0.9-1.8 \times 10^6$ K. All but two of the marked plumes appear dark in this image, indicating that they are somewhat cooler than the surrounding interplume background.

found to be narrower in plumes than in the interplume region; and with predictions made by Wang (1994). Walker *et al.* (1993) found plumes to have densities of $(10^9 - 10^{10})$ cm^3, and local thermalization times on the order of a few seconds, near the limb of the Sun; thus, it is reasonable to speak of a single temperature of the plume plasma at the low altitudes imaged by EIT.

The most remarkable temporal characteristics of the plumes are their constant change on the small scale and their static nature on larger scales. Figure 4 shows several images and a light curve of the two plumes near the center of Figure 2. The shape of each plume's core changes dramatically on a time scale of less than 10 min, with small filamentary structures ($\sim$5 arc sec across) brightening and fading apparently randomly. These intensity variations amount to $\sim$10% of the plume's background-subtracted brightness, and propagate coherently outward at speeds of 300–500 km s^{-1}. The propagation speed is near the estimated Alfvén speed, suggesting some type of compressional waves possibly driven by small, impulsive events at the base of the plume. A more detailed time-domain analysis is being performed by DeForest and Gurman (1997).

On time scales of hours to days, and spatial scales of >30 arc sec, the observed plumes' overall shapes appear fixed: we observed little or no change in the plumes' large-scale shapes over the course of the 16-hour observation, and the brightest plumes maintained the same overall structure in the EIT images as early as one day before and as late as one day after the main high cadence sequence. However, plumes may change their overall intensity rather suddenly: several 'plumewise

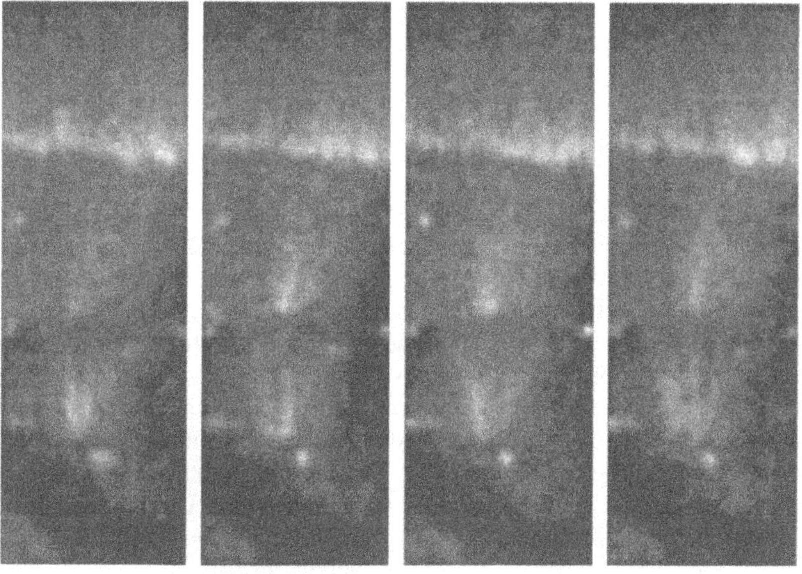

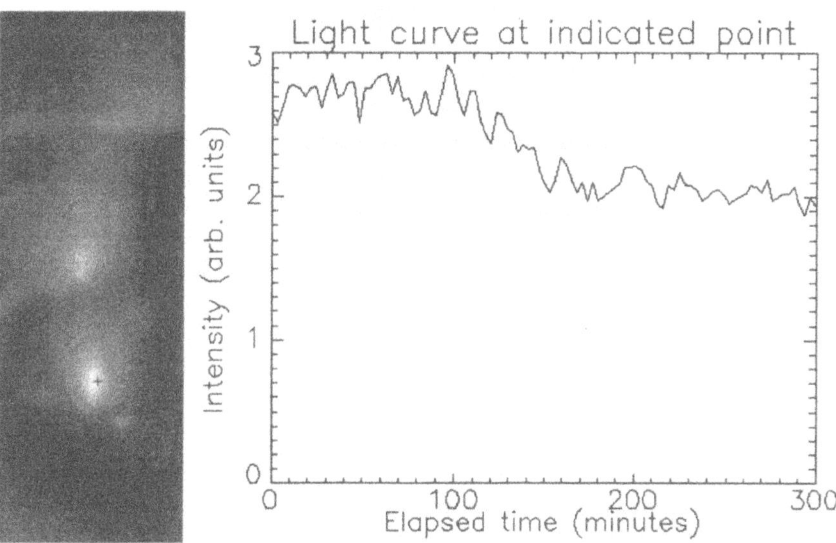

Figure 4. Time variability of EIT images. *Top*: a sequence of four frames, taken 9 min apart, of the two bright plumes at the center of Figure 2 in the EIT Fe IX/X (171 Å) band. A radially-varying model background corona has been subtracted from the images. ***Bottom***: a time-averaged image of the above two plumes for all five hours (104 frames) of the high-cadence period. The frames have been translated and rotated in the image plane to keep the lower plume stationary (causing some registration error, hence blurring, in the upper plume). The graph at right shows intensity variation at the marked '+'. The fluctuation amplitude is ~20%; this is ~10 times the noise level in the individual images.

global' events were observed, in which a whole plume permanently changed its Fe IX/X brightness by up to a factor of 2 over a span of ~2 hours, without significantly changing shape.

2.2. MDI: MAGNETIC FOOTPOINTS OF PLUMES

MDI (Scherrer *et al.*, 1995) recorded high-resolution mode (0.6 arc sec pixel) magnetograms, Dopplergrams, and line-depth images of the photosphere in the Ni I 6768 Å absorption line. Figure 5 shows two views of a single south pole magnetogram made by averaging 5 min worth of data. The top is scaled so that black and white correspond to ±200 G along the line of sight; the second is scaled to show the details of the weaker field structure. The line-of-sight field has been divided by the cosine of the surface projection angle, giving an estimate of the radial field at each point; this is why the region near the limb of the Sun appears 'noisier' than the regions farther away. The most apparent feature in the upper image is the magnetic flux concentrations that dot the surface of the Sun. The chief difference between the coronal hole and 'quiet-Sun' regions in this image is the presence (outside the coronal hole) or lack (within the coronal hole) of balance in the strength of the flux concentrations of opposing sign.

Figure 6(A) shows the distribution of field strength by percentile pixel value in the boxed region in Figure 5. The strongest flux concentrations in the coronal hole, which account for ~2% of the covered area, have a peak radial field strength of ~300 G (the average field inside the flux concentrations depends on how the boundaries of the concentrations are treated). The net average field into the indicated region is ~5 G. The r.m.s. line-of-sight weak field (outside the flux concentrations) is ~8 G (with the instrument's inherent 0.6 arc sec pixel averaging); after binning the image into 1.2 arc sec pixels, the r.m.s. line-of-sight weak field is 3.5 G.

The flux concentrations account for about half of the magnetic flux that escapes the coronal hole. Figure 6(B) shows the average flux through the boxed region in Figure 5 when the strongest pixels (white and black) are removed. Neglecting the strongest 5% of the image area removes half of the total net magnetic flux.

Figure 7 is an example of an MDI image that is co-aligned with, and overlain upon, a simultaneous EIT image in 171 Å light. The two images have been scaled to the same resolution, and regions of the EIT image where the measured line-of-sight magnetic field was stronger than 20 G were replaced with the MDI signal, in blue (negative; white in Figure 5) or red (positive; black in Figure 5). Several plumes are visible. Every plume that lands on a place that is visible in the magnetogram (i.e., not closer than ~1 arc min to the limb) lands on a unipolar magnetic flux concentration. The relationship is not one-to-one: not all of the flux concentrations are plume footpoints.

The finding that plume footpoints occur on unipolar magnetic flux concentrations permits one to understand the geometry of the lowest portion of the plume in terms of the magnetic field configuration (as in Newkirk and Harvey, 1968):

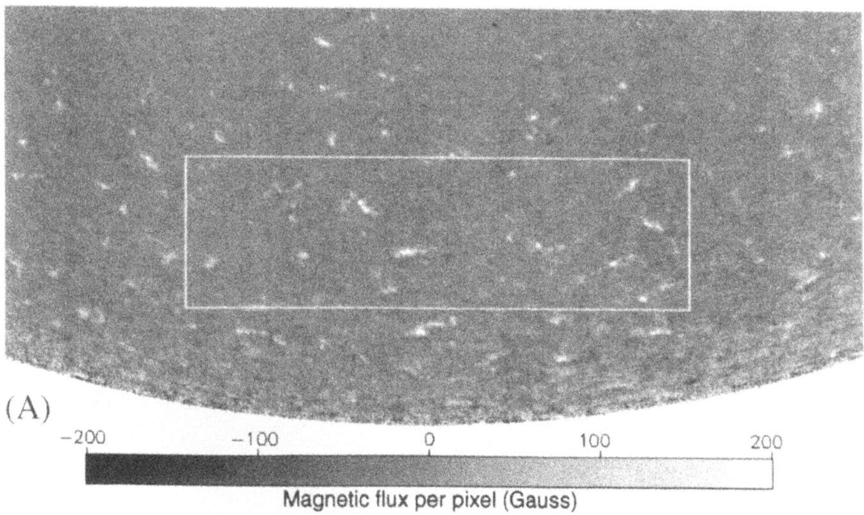

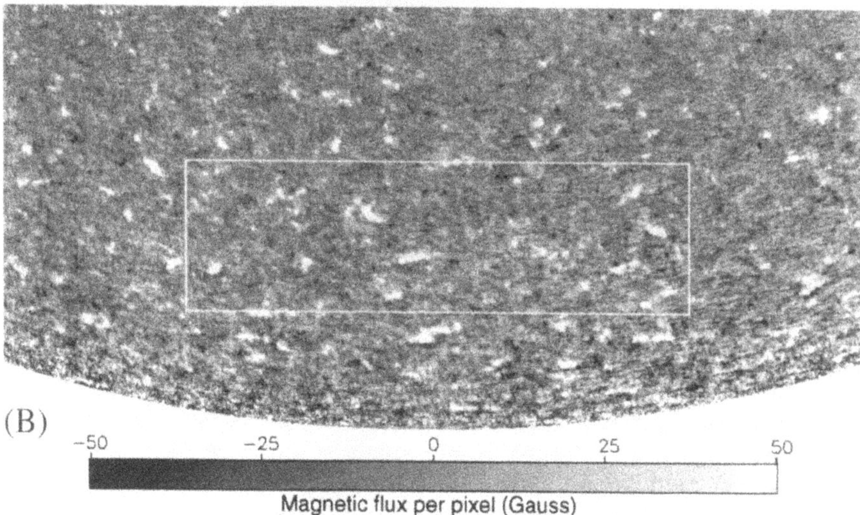

Figure 5. MDI high-resolution magnetogram showing the flux concentrations in the coronal hole. (A) This image has been scaled so that the brightest spots (±200 G along the line of sight) are black/white, to illustrate the details of the strong field. (B) This image has been scaled to ±50 G to illustrate the weak background field outside the stronger flux concentrations. The white rectangle marks the region used to generate the plots in Figure 5. Both images have been adjusted for the projection angle of local vertical into the line of sight.

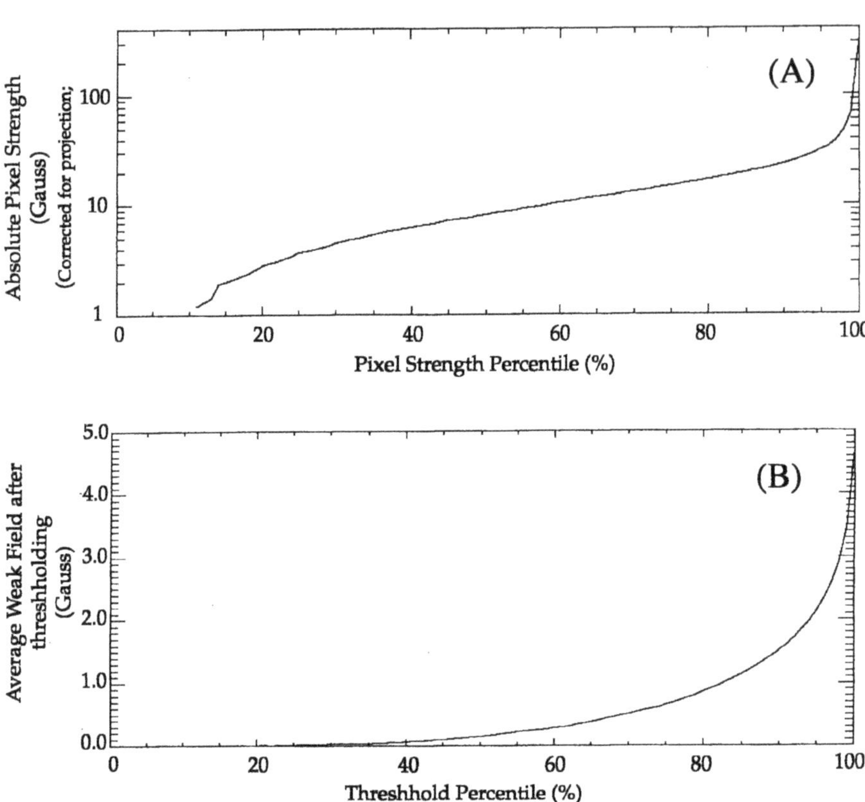

Figure 6. Statistical breakdown of measured line-of-sight magnetic flux within the white rectangle in Figure 4. The top plot (A) shows the distribution of absolute angle-corrected field strength vs percentile. Only a small percentage of the pixels have values >30 G. The lower plot (B) shows the calculated average field over the whole region, when the strongest pixels are removed from the average. The strongest 5% of the pixels account for over half of the average field.

assuming a density of $\sim 10^{10}$ cm^{-3} (Walker *et al.*, 1993) and electron temperature of 10^6 K inside the base of a plume gives (in a 100 G magnetic field) a plasma β parameter of 4×10^{-3}, so that plume geometry should be close to the shape of a force-free field. Suess *et al.* (1997) have extensively modelled polar plume expansion in the potential-field geometry; the principal limitation of the initial superradial expansion is the presence of other open magnetic field lines (either from a smooth background field or other magnetic flux concentrations).

The footpoints that give rise to plumes appear to be more complex than the ones that do not give rise to plumes. For example, the two central plumes in Figure 7 both have complicated footpoints with several local maxima of field strength. Several equally strong but morphologically simpler concentrations are visible in Figure 7; and these flux concentrations do not appear to give rise to plumes.

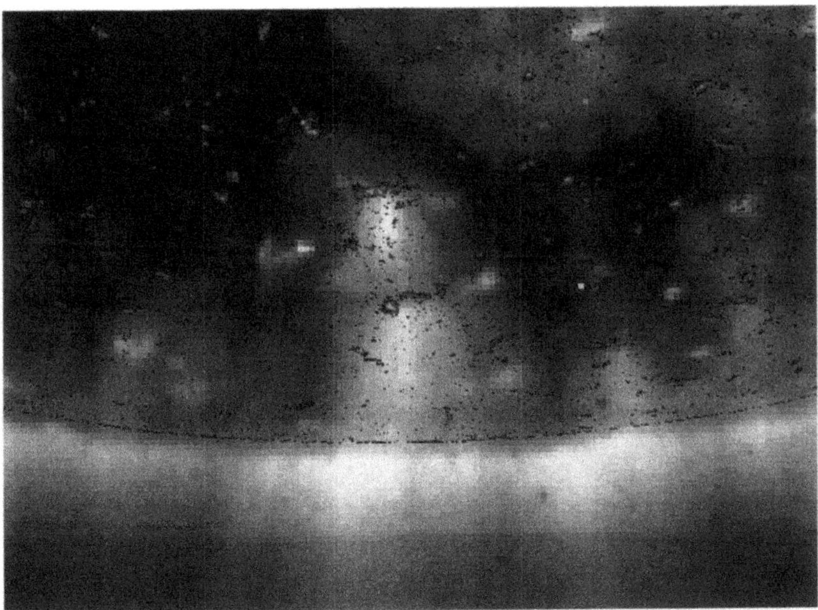

Figure 7. Overlay of an MDI magnetogram with a simultaneous image from EIT in Fe IX (171 Å). Regions where the photospheric field is strong have been replaced with magnetic information in a blue/red color scheme. Fields between −20 and +20 G are not visible; the colored spots are graded in intensity from 20 to 160 G (uncorrected) along the line of sight in each direction. Each polar plume whose footpoint is visible lands on a unipolar blue flux concentration. The brightest plumes appear to arise from flux concentrations with more complex geometry; however, in no case does a plume footpoint correspond to a balanced dipole magnetic structure.

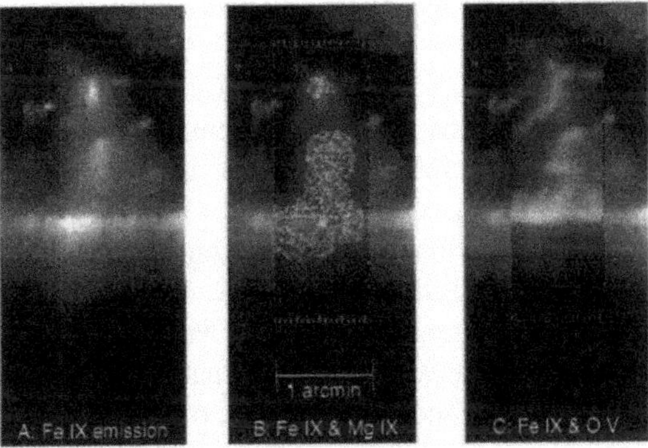

Figure 8. Plume footpoints as seen by EIT and CDS. (A) The central region of Figure 3(A), showing the footpoints of the two central plumes in Fe IX emission (171 Å), as seen by EIT. (B) Overlay of a simultaneous CDS image in Mg IX light (368 Å; contours) with (A). The plumes and the limb of the Sun are congruent between the two images. (C) Overlay of a simultaneous CDS image in O V (629 Å; orange color table) with (A). The plumes are visible but appear orange as they 'shine through' the orange color table of the O V. The plume footpoints are on the chromospheric network (on cell boundaries); however, there is nothing in the O V image to distinguish the plumes' footpoints from other parts of the network.

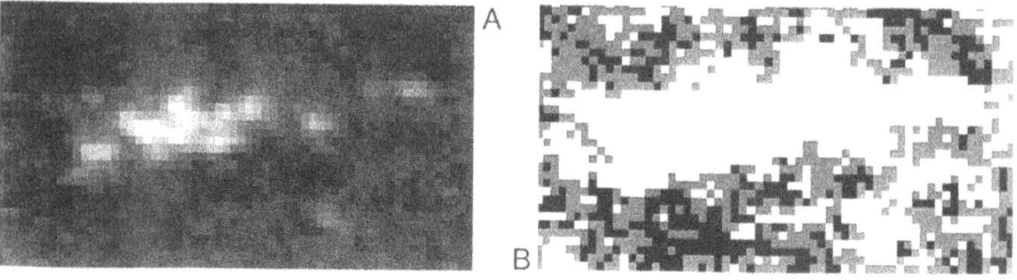

Figure 9. Magnetogram close-up of a plume footpoint as seen by MDI. (A) The footpoint of the brightest plume in Figure 7 (near the center of the image). (B) Mask indicating where the field is within one sigma of zero (grey); into the page (white); or out of the page (black).

Figure 9 is a close-up of the flux concentration under one of the two central plumes in Figure 7, illustrating the mainly unipolar nature of the footpoint and the detailed structure of the field. Figure 9(A), at left, shows the several local maxima that comprise the main footpoint of the plume; Figure 9(B), at right, is a three-level mask indicating whether the field is positive, within 8 G (i.e., one standard deviation) of neutral, or negative. For scale, the squares in Figure 9(B) are individual MDI high-resolution pixels, subtending 0.6 arc sec each. Note that, while there are weak opposing-flux regions near the base of the plume, there is no overall dipole nature to the footpoint. The total measured, radial-projection-corrected flux into regions that are white in Figure 9(B) is 8.6×10^{19} Maxwells; the flux into the black regions is -9.5×10^{18} Maxwells, or about an order of magnitude less than the forward flux.

2.3. CDS: PLUME FOOTPOINT LOCATION IN THE CHROMOSPHERIC NETWORK

The two central plumes in Figure 2 were within the CDS field of view, allowing the plumes' footpoints to be located relative to chromospheric features as seen in the He I (584 Å) and O V (629 Å) emission lines (Harrison *et al.*, 1995). CDS acquired intensity data in its three brightest lines, the He I, O V, and Mg IX (368 Å) lines for a period of 1 hour near and following 17:00 UT on 7 March 1996.

Figure 8(A) shows the central portion of the EIT field of view; Figure 8(B) shows the same image, with simultaneous CDS Mg IX intensity contours overlaid upon it. Mg IX and Fe IX are present at approximately the same temperature. The two central plumes are visible in the CDS Mg IX data, and their shape agrees well with the EIT image.

Figure 8(C) is a similar overlay, with Fe IX in blue-white and O V in orange–white. The plumes' footpoints clearly land within the chromospheric network, on the boundaries of network cells. This is in agreement with the observation that plumes arise from magnetic flux concentrations: magnetic flux has been shown by Berger and Title (1996) to be carried rapidly to the edges of network cells by normal supergranular convection.

While the plumes seem to land on bright portions of the chromospheric network, there is no distinguishing characteristic of those particular brightenings in the CDS images: there are many small brightenings visible in the images, and without the Fe IX overlay, it is impossible to determine where the footpoints lie.

2.4. MK-3 AND LASCO: OUTER STRUCTURE OF PLUMES

Coronal rays, the linear structures visible in the polar coronal holes with white-light coronagraphs, have been studied extensively (see, for example, Fisher and Guhathakurta, 1995; Newkirk and Harvey, 1968). The rays are density structures that are visible in Thomson-scattered white light in the low to mid corona. To compare near-simultaneous coronal rays and plumes, and to trace the plumes out as far as possible, the LASCO coronagraph (Brueckner *et al.*, 1995) acquired white-light brightness images of the corona in its C2 ($2\ R_\odot - 6\ R_\odot$) and C3 ($5\ R_\odot - 30\ R_\odot$) telescopes. Because the spacecraft off-point during the campaign required LASCO to shut its doors (due to concerns about direct solar exposure past the occulters), LASCO images were not taken during the high cadence portion of the campaign. A complete set of images was recorded before the spacecraft pointing was initially adjusted, and again after it had returned to the nominal direction.

The LASCO data are unpolarized white-light images, and thus include Thomson scattered light from the K-corona and specularly scattered light from the F-corona. The F-corona is a slowly-varying signal added onto the K-corona signal; it was removed from the raw C2 and C3 images by subtraction of a radially varying, smooth function that was fit to the radial dropoff in background intensity. The resulting images were subjected to more traditional radial filtering (in which they were multiplied by a slowly varying function of radius) to raise the contrast near the outer edges of the images.

Coronal rays were clearly visible over the South pole in the processed C2 and C3 white-light images. The Mk-3 daily average images from 7 and 8 March were co-aligned with the SOHO instruments' images to bridge the gap between the LASCO C2 field of view ($>2.0\ R_\odot$) and the EIT field of view ($<\sim 1.3\ R_\odot$); and to allow a detailed comparison of the white-light rays and the XUV plumes.

Unfortunately, bad seeing conditions at MLSO on 7 March raised the noise level of that day's images above the intensity level of the rays, so simultaneous image comparisons are not possible between the SOHO and the Mk-3 data. Instead, we use the Mk-3 image from 8 March (mean time 22:00 UT 8 March 1996), whose collection time is some 16 hours later than the last SOHO images from the campaign.

Figure 10 is a montage image made by co-aligning images from EIT (171 Å), Mk-3 (white-light polarization brightness, daily average), and LASCO's C2 camera, showing fair to good correspondence between the plumes (in EIT) and the rays (in Mk-3 and LASCO). We identify the XUV polar plumes seen in EIT with the coronal rays seen by Mk-3 and LASCO. The slight positional mismatches are

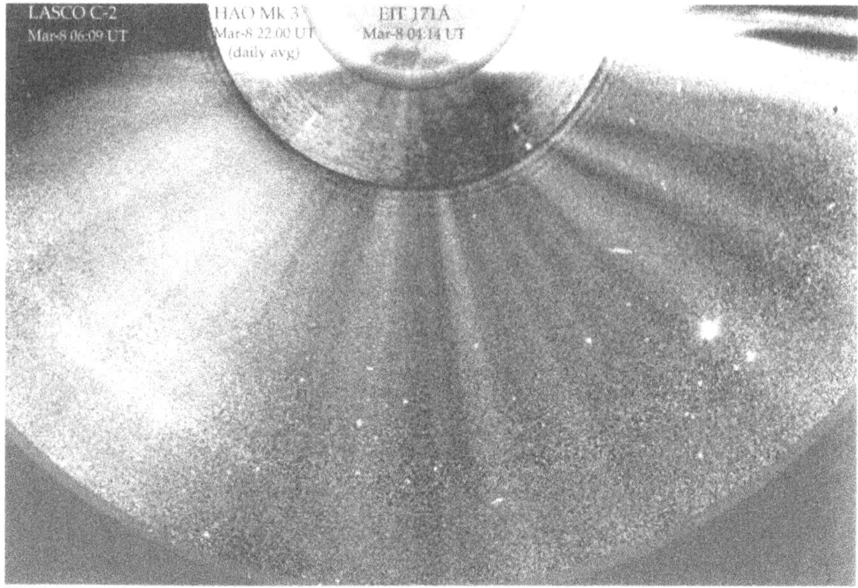

Figure 10. Embedded images of the corona as seen by EIT (Fe IX/X line emission), the MLSO Mk-3 white-light coronameter (white-light polarized brightness), and LASCO's C2 camera (unpolarized white light). The two outer images have been subjected to radial filtering. Despite the 16-hour difference between the Mk-3 and SOHO data collection times, there is good correspondence between plumes observed in the three fields of view.

attributed to solar rotation during the 16-hour time difference between the Mk-3 and the other two images: the plumes whose projection angle out of the image (as determined by footpoint location) is small match the Mk-3 better than the ones whose projection angle is large; and the apparent mismatches go in opposite directions between EIT–Mk-3 and Mk-3–C-2.

We find that the plumes/rays appear to diverge radially from a point that is between the solar center and the South pole of the Sun; this agrees with the observations of Fisher and Guhathakurta (1995) with SPARTAN data and DeForest (1995) with the Stanford/Marshall Space Flight Center Rocket Spectroheliogram.

To clarify the expansion behavior of the observed plumes, and to fit the huge LASCO C3 field of view into the same figure with EIT, it was necessary to transform the plume images into radial coordinates. Figure 11 presents the same data as Figure 10, but extended into the LASCO C3 field and mapped into image-plane radial coordinates. The transformation preserves the shape of small features (such as plume footpoints), and directly represents a feature's subtended linear angle relative to disk center in the image plane. For example, the coronal hole subtends approximately 50° at the limb of the Sun (near the bottom of the image); but expands to subtend closer to 150° by 3 $R_{\odot}$. Straightness of lines is not preserved: Figure 10, which is in Cartesian image coordinates, shows that the rays appear

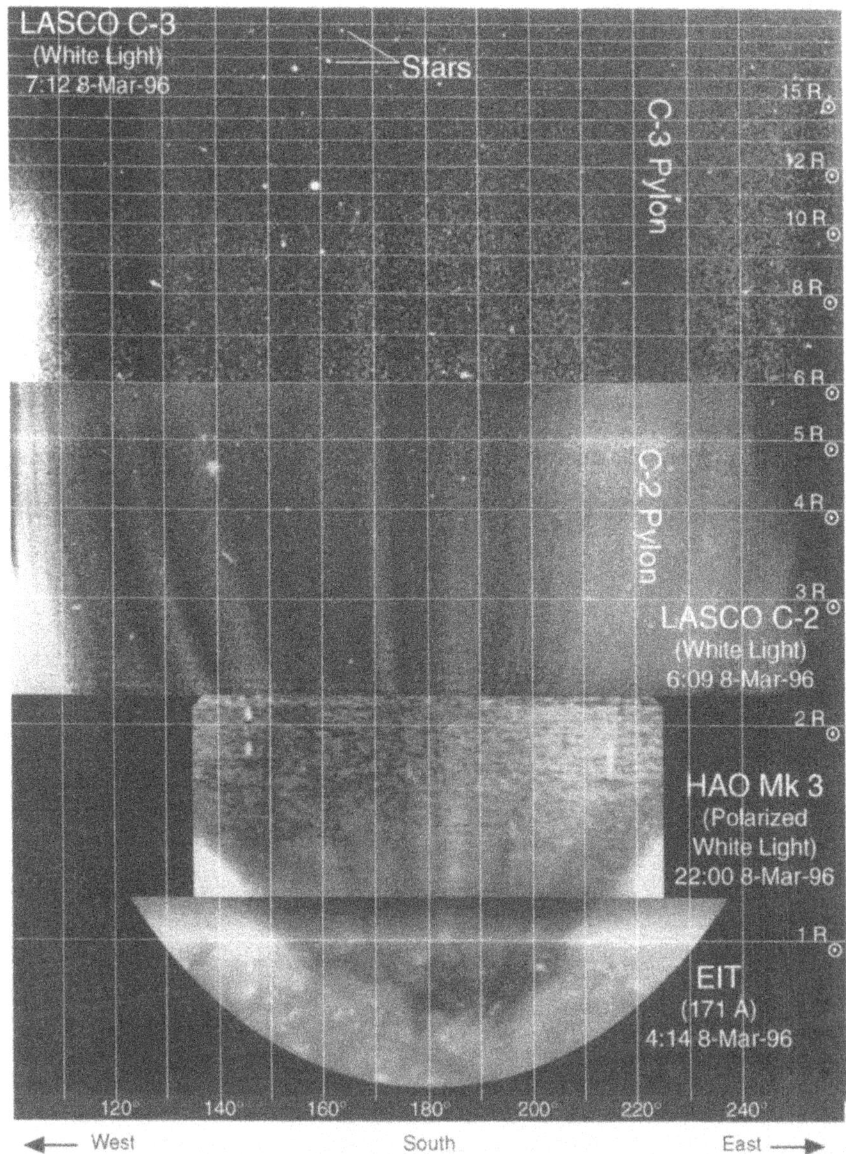

Figure 11. Overlain images of the data in Figure 10 and an image from the LASCO C3 camera, in an image-plane radial coordinate system. Circles concentric with the solar disk have been mapped to horizontal lines; lines radial to the Sun have been mapped to vertical lines. (Note that lines that are radial in 3-space are also radial in 2-D projection.) The superradial expansion of the coronal hole, and of individual plumes, can be seen directly in this image. The logarithmic scaling of radius renders the coordinate transformation 'conformal', meaning that the shape of small features is preserved though scale changes with altitude.

straight; in the transformed image, they appear curved. All three of the visible light coronagraph images have been subjected to radial filtering and/or background subtraction; however, none of the images has been treated with generic unsharp masking. In particular, great care has been taken only to introduce radial (and not angular) gradients, so as to preserve the apparent size and position of the plumes at all altitudes. The C3 image has been divided by a modelled vignetting function that varies only near the marked pylon.

Fisher and Guhathakurta (1995) found that rays observed by SPARTAN appear to expand radially, subtending the same 2.5-deg angle relative to Sun center, regardless of altitude; the LASCO data are not consistent with that observation. We find that, above the rapid-phase expansion at the footpoint, plumes expand superradially in two broad phases outside the image-plane solar limb. The difference in the two phases lies in the comparison between the plumes' expansion ratio and the global expansion ratio of the coronal hole. In the lower phase, below 5 $R_\odot$, the plumes and the coronal hole expand at roughly the same rate. Above about 5 $R_\odot$, the coronal hole has expanded to fill nearly the entire hemisphere (as the streamer belts pinch off), and begins to expand radially. However, the brightest plumes in Figure 2 continue to expand superradially between 5 $R_\odot$ and 15 $R_\odot$, where they fade into the background noise of the C3 image. The FWHM of a typical bright plume is 2.5° at 1.05 $R_\odot$; ~7° at 5 $R_\odot$; and ~15° at 15 $R_\odot$, for a linear expansion ratio of 1, ~3, and ~6 at the three altitudes. Over the same altitude range, the coronal hole expands from 50° on the limb to 150° at 5 $R_\odot$ and 160° at 15 $R_\odot$, for a linear expansion ratio of 1, 3, and 3 at the three altitudes.

3. Conclusions

Combining the data from several SOHO instruments has resulted in a coherent picture of polar plumes by allowing a complete 'snapshot' of a particular set of the plumes to be recorded. Plumes originate in unipolar magnetic flux concentrations that lie on cell boundaries in the chromospheric network. They expand rapidly superradially (with a half-cone angle of ~45°) in their lowest 20–30 Mm, and more slowly above that, for a linear superradial expansion factor of 3 at 4–5 $R_\odot$ and of ~6 at 15 $R_\odot$. Plumes (as seen in XUV) are density structures that are also seen as coronal rays (in Thomson-scattered white light) in the mid to upper corona. They are denser than the surrounding media (as observed by Walker *et al.*, 1993) and are discernable to an altitude of at least 10–15 $R_\odot$. In their lower reaches, plumes are slightly cooler than the surrounding interplume coronal hole, by a factor not exceeding ~30%. On distance scales of >15 arc sec, plumes are remarkably quiescent structures, maintaining their form essentially unchanged for days at a time; at smaller distance scales, they exhibit outward flowing fluctuations in density and/or temperature on a time scale of a few minutes and a length scale of ~5–15 arc sec, up to an altitude of at least 1.2 $R_\odot$.

4. Direction of Future Analyses

Several of the results presented here lend themselves to continued study. Because this observing campaign was run during SOHO's commisioning phase, several of the instruments in the SOHO plumes campaign had not yet been absolutely calibrated; detailed analysis has had to wait for the final calibration. In particular, derived densities and temperatures (e.g., DeForest, 1995) from the EIT data in the lower corona will be important for comparison with past results. Likewise, a detailed intercomparison of line width temperatures from SUMER and ionization state temperatures from EIT should be made. That work is in progress, and is currently awaiting absolute intensity calibrations from EIT.

Suess *et al.* (1997) have developed detailed models of the magnetic structure inside coronal holes, which should be compared with the actual morphology at the footpoints of plumes seen in EIT. Suess's model includes rapid flux tube expansion with altitude above flux concentrations; the flux tube expansion is limited by the total background field. Comparing plume geometry and flux concentration statistics to the model will indicate the proportions of closed field lines emanating both from flux concentrations and the 'quiet coronal hole' in between them.

The observed rapid small-scale brightness variation in plumes, coupled with their large-scale stability, is puzzling. A careful comparison of the variations with bulk photospheric motions and/or magnetic field changes at the footpoints may help explain plume heating and acceleration. Ofman and Davila (1997) have proposed a heating model in coronal holes that invokes nonlinear Alfvén-compressional wave coupling in the 5-min frequency band; the correspondence of plumes with the more complicated flux concentrations suggests that plumes may be heated by small magnetic reconnection events at their footpoints. Time domain analysis of the existing data is needed to test these ideas.

Acknowledgements

Participation in this observation has been broad-based, and many individuals not named on the author list have contributed to the observation and analysis described by this article. Special thanks are due to the principal investigators of the SOHO instruments, for allowing this science campaign to be run during SOHO's commissioning period; and to the flight operations teams of the spacecraft and of the individual instruments. Stanford's Rick Bogart initially pointed out the possibility of an MDI polar high-resolution campaign during the SOHO commissioning phase. Giannina Poletto and Steve Suess provided insightful commentary and conversation on plume and magnetic field structure; Madhulika Guhathakurta and Dick Fisher helped understand the data from the HAO instrument.

SOHO is a project of international co-operation between NASA and ESA.

References

Ahmad, I.A. and Webb, D. F.: 1978, *Solar Phys.* **58**, 323.
Ahmad, I. A. and Withbroe, G. L.: 1977, *Solar Phys.* **53**, 397.
Allen, M. J.: 1994, 'The First Flight of the MSSTA', doctoral dissertation, Stanford University.
Berger, T. E. and Title, A. M.: 1996, *Astrophys. J.* **463**, 365.
Bohlin, J. D., Sheeley N. R., Jr., and Tousey, R.: 1975, in M. J. Rycroft (ed.), *Space Research XV*, Akademie-Verlag, Berlin, p. 651.
Brueckner, G. E. *et al.*: 1995, *Solar Phys.* **162**, 357.
Delaboudinière, J.-P. *et al.*: 1995, *Solar Phys.* **162**, 291.
DeForest, C. E.: 1995, 'High Resolution Multi-Spectral Observations of Solar Coronal Open Structures: Polar and Equatorial Plumes and Rays', doctoral dissertation, Stanford University.
DeForest, C. E. and Gurman, J. B.: 1997, *Astrophys. J.*, in press.
DeForest, C. E. *et al.*: 1990, *Opt. Eng.* **30**, 1126.
Fisher, R. and Guhathakurta, M.: 1995, *Astrophys. J.* **447**, L139.
Fleck, B., Domingo, V., and Poland, A. I. (eds): 1995, *Solar Phys.* **162**.
Habbal, S. R.: 1992, *Ann. Geophys.* **10**, 34.
Harrison, R. A. *et al.*: 1995, *Solar Phys.* **162**, 233.
Harvey, J. W.: 1965, *Astrophys. J.* **141**, 832.
Hassler, D. M., Wilhelm, K., Lemaire, P., and Schuehle, U.: 1997: *Solar Phys.*, in press.
Koutchmy, S.: 1977, *Solar Phys.* **51**, 399.
Lindblom, J. F.: 1990, 'Soft X-Ray/Extreme Ultraviolet Image of Solar Atmosphere with Normal Incidence Multilayer Optics', doctoral dissertation, Stanford University.
Newkirk, G., Jr., and Harvey, J.: 1968, *Solar Phys.* **3**, 321.
Ofman, L. and Davila, J. M.: 1997, *Astrophys. J.* **476**, 357.
Saito, K.: 1965, *Publ. Astron. Soc. Japan* **17**, 1.
Scherrer, P. H.: 1995, *Solar Phys.* **162**, 129.
Suess, S. T.: 1982, *Solar Phys.* **75**, 145.
Suess, S. T., Poletto, G., Wang, A.-H., Wu, S. T., and Cuseri, I.: 1997, *J. Geophys. Res.*, submitted.
Walker, A. B. C. *et al.*: 1988, *Science* **241**, 1781.
Walker, A. B. C., DeForest, C. E., Hoover, R. B., and Barbee, T. D. W.: 1993, *Solar Phys.* **148**, 239.
Wang, Y.-M.: 1994, *Astrophys. J.* **435**, L153.

ERUPTIVE PROMINENCE AND ASSOCIATED CME OBSERVED WITH SUMER, CDS AND LASCO (SOHO)

J. E. WIIK[1,2], B. SCHMIEDER[1,3], T. KUCERA[4,5], A. POLAND[5], P. BREKKE [1] and G. SIMNETT [6]

[1]*ITA, P.O. Box 1029, Blindern, N-0315 Oslo, Norway*
[2]*Observatoire de la Côte d'Azur, BP 229, F-06304 Nice Cedex 04, France*
[3]*Observatoire de Meudon, DASOP, F-92195 Meudon Cedex, France*
[4]*Applied Research Corporation, 8201 Corp. Dr., Landover, MD 20785, U.S.A.*
[5]*NASA/GSFC, Code 682, Greenbelt, MD 20771, U.S.A.*
[6] *University of Birmingham, Birmingham B15 2TT, U.K.*

(Received 19 March 1997; accepted 12 June 1997)

Abstract. Observations of an eruptive prominence were obtained on 1 May 1996, with the SUMER and CDS instruments aboard SOHO during the preparatory phase of the Joint Observing Programme JOP12. A coronal mass ejection observed with LASCO is associated temporally and spatially with this prominence. The main objective of JOP12 is to study the dynamics of prominences and the prominence–corona interface. By analysing the spectra of O IV and Si IV lines observed with SUMER and the spectra of 15 lines with CDS, Doppler shifts, temperatures and electron densities (ratio of O IV 1401 to 1399 Å) were derived in different structures of the prominence. The eruptive part of the prominence consists of a bubble (plasmoid) of material already at transition region temperatures with red shifts up to 100 km s^{-1} and an electron density of the order of 10^{10} cm^{-3}. The whole prominence was very active. It developed both a large helical loop and several smaller loops consisting of twisted threads or multiple ropes. These may be studied in the SUMER movie (movie 2). The profiles of the SUMER lines show a large dispersion of velocities ($\pm$50 km s^{-1}) and the ratio of the O IV lines indicates a large dispersion in electron density (3×10^9 cm^{-3} to 3×10^{11} cm^{-3}). The CME observed by LASCO left the corona some tens of minutes before the prominence erupted. This is evidence that the prominence eruptions are probably the result of the removal of the restraining coronal magnetic fields which are in part responsible for the original stability of the prominence.

1. Introduction

Prominences are objects discussed extensively during the first part of the century by D'Azambuja (1948) and partially understood by magnetohydrodynamic models (Tandberg-Hanssen, 1974, 1995; Priest, 1989). They are sustained and confined by magnetic field lines above the chromosphere, and exhibit a strong coupling between magnetic forces and thermodynamics.

However, many problems remain unresolved. It is unknown whether a stable prominence consists of many threads of different temperatures (Poland and Tandberg-Hanssen, 1983; Wiik, Dere, and Schmieder, 1993) or if there is a significant shell at the transition zone temperature wrapping the cool core of the prominence (Schmieder, 1992). Also, no mechanism has yet been presented which is sufficient to sustain the material in the leg of the prominence – those parts of the structure which tie it to the photosphere. Also, the scale height of the prominence plasma, not larger than 500 km, is too small to explain the prominence height at the altitudes at which they are observed. Understanding the dynamical processes in prominences is a key to resolving these issues.

Once activated, the prominence becomes denser (signaled by the broadening of the Hα line), rises, and erupts. Such events are referred to as 'disparitions brusques' (Schmieder, 1989; Démoulin and Vial, 1992). There has been much debate concerning the mechanisms involved. Some prominences disappear in chromospheric lines but are still visible in hotter lines, as was shown with *Skylab* observations (Hiei and Widing, 1979; Mouradian *et al.* 1987). Others lose their magnetohydrostatic equilibrium and erupt at high velocities (Rompolt, 1990). The latter are frequently associated with coronal mass ejections (CME) in Solar Maximum Mission (SMM) observations (Illing and Hundhausen, 1986; Mein *et al.* 1982) although the relationship between them is not completely understood. The relationship between prominence eruption and CMEs is not clear. Harrison (1995) argues that both are consequences of the same magnetic 'disease' and that one is not the trigger of the other. However, the close proximity of the two events in both space and time makes this assertion unlikely. With the LASCO CME observations we have a much better constraint on the timing of the CME, and for the event on 1 May 1996, it is clear that the CME onset preceded the associated prominence eruption. It has been observed that CMEs are often offset equatorward from associated prominences (Webb *et al.* 1994; Simnett and Harrison, 1985). This is then consistent with a model in which the main part of the CME is within the equatorial streamer belt, and that it removes, or weakens, the coronal magnetic field at higher latitudes, thereby affecting any underlying structures at these latitudes which have been relying on the coronal field for their stability.

In this paper we present a good example of the destabilization of a prominence and its associated CME. We have taken advantage of observations in a large range of wavelengths and over a large range of temperatures. We discuss the observed phenomena in terms of the transport of mass and energy. We have observed high Doppler shifts (100 km s^{-1}) in the prominence which indicate a real transport of mass along the field lines. We do not see any cool structures becoming progressively visible in hotter and hotter lines, demonstrating that no heating of the prominence plasma is taking place.

Two movies as well as the colour figures 1, 2, 3, 4, 7, 8, and 9, and descriptions (see index.htm) are included on the CD-ROM provided with this volume of *Solar Physics*. One of the movies (movie 1) shows the CME observed with the LASCO (Large Angle Spectroscopic Coronagraph) C3 coronagraph; the other (movie 2) illustrates the variation of mass motion in the prominence (SUMER (Solar Ultraviolet Measurements of Emitted Radiation) raster).

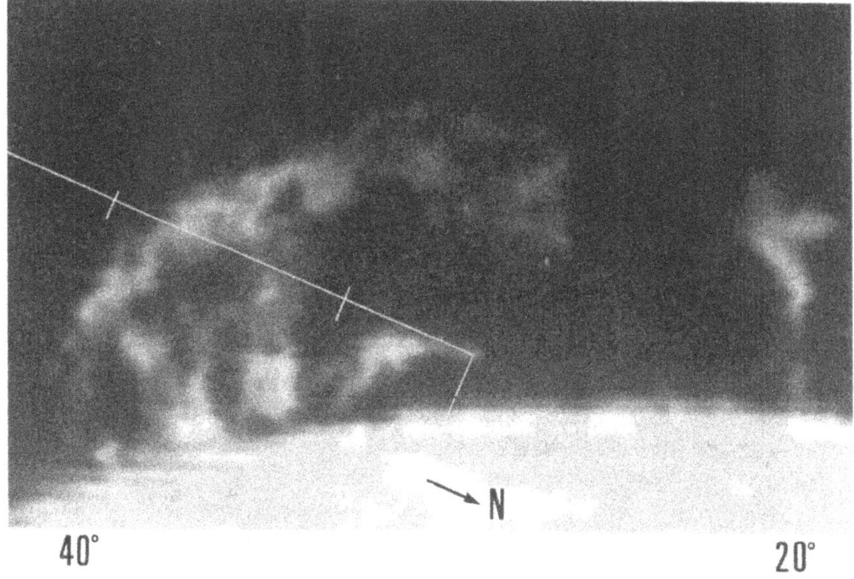

Figure 1. The huge prominence observed on 1 May 1996 (SE 20–40) as seen in the Ca II K3 line at 07:25 UT (Meudon spectroheliogram, courtesy of Z. Mouradian). CDS (see the box) and SUMER (indicated by crosses in the CDS box) cover mostly the southernmost part of the prominence (to the left).

2. Observations

The prime observations we have used are those from SUMER, CDS (Coronal Diagnostic Spectrometer), EIT (Extreme-Ultraviolet Imaging Telescope) and LASCO on the Solar and Heliospheric Observatory (SOHO). The 1 May 1996, observations of the prominence and its environment were obtained during a test run of the SOHO Joint Observation Program 12 (JOP 12). There were no coordinated observations from the ground, but the normal synoptic data from ground-based observatories are available. For example, Meudon synoptic data was obtained with the spectroheliograph and a large prominence is clearly visible in Ca II (Figure 1 and Wiik *et al.* 1997). We also use data from the *Yohkoh*/Soft X-ray Telescope (SXT).

2.1. SUMER

The SUMER instrument is described by Wilhelm *et al.* (1996), Wilhelm, Lemaire, and Curdt (1997). In JOP 12 a staggering mode was used for rastering the region of interest. The prominence was rastered with a slit of 1″ and a large step size (12″). The position of the origin of each raster was shifted by 3″ compared to the origin of the previous raster, for four successive rasters. Four such rasters are necessary to obtain a complete raster covering the region with a spatial resolution of 3″. This mode of observation was selected in order to detect high velocity events occurring in a large field-of-view in a reasonable time ($\sim$ 8 min). For each position of the slit we obtained a spectra on the detector A. A total of six lines (in six windows of 50×120 pixels each, defined to reduce the telemetry requirement) were studied from the ions O IV and Si IV.

2.1.1. *Data reduction*

High signal to noise flat-field exposures contribute significantly to the degradation of the detector and are therefore not taken more often than about once per month. The flat-field exposures closest to our observing run on May 1 were taken on 19 April and 15 May. Unfortunately it was found that the flat-field pattern of sensitivity-variations ($\pm$50%) was slightly displaced in the observations compared with the flat-field exposures. Thus we construct a special flat-field image using the 1 May data. The following procedure was employed to obtain an optimum removal of pixel-to-pixel sensitivity variations.

In each of the first four rasters there were six exposures taken with the slit located entirely on the solar disk. We summed the data from these exposures to create arrays containing the average response of the detector to radiation as a function of position on the detector. Each observed was treated separately and each of the resulting arrays was divided by the column mean (the average along the slit as a function of wavelength) to remove the spectral lines. Each row was thereafter divided by the row mean, removing intensity variations along the slit. The remaining images contain the small scale sensitivity variation pattern with its large scale and small scale grid. These images were cross-correlated with the flat-field image using Fourier transforms to determine an optimum shift of the flat-field. Such a shift was determined individually for each line and the shifted flat-field was used subsequently. Comparing the 1 May flat-field image to the regular flat-field images we found that the 15 May flat-field image was better adapted to our observations than the 19 April flat-field image, provided a shift of one pixel in the north-south direction (slit direction) was introduced.

The curvature of the lines depends on their position on the detector. It is necessary to correct for detector pin-cushion distortion. We 'destretched' the spectra by using the SUMER analysis software procedure destretch.pro provided by T. Moran. The destretching was performed based on corrections calculated by in flight observations of O I lines and the Lyman continuum emission positioned at many locations on the detector.

Wavelength calibration used the S I lines at 1401.5136 and 1412.8726 Å in the full spectra observed during the first part of the observing run (item 1). From the position of these two lines we derive the mean dispersion between them to be 0.0426 Å pixel^{-1}. We find that the measured wavelengths are shifted relative the official wavelengths (SUMER Scientific Report, or 'red-book') by 0.022 Å for Si IV 1393.755, 0.027 Å for Si IV 1402.770, and 0.024 Å for O IV 1399.774. These differences correspond to nearly half a pixel ($\sim$10 km s^{-1}). By using the mean images observed entirely on the disk we obtained average profiles which could be compared to the SUMER atlas (Brekke *et al.* 1997). We find shifts around $-3\,\mathrm{km\,s^{-1}}$. The offsets were $-1.4\,\mathrm{km\,s^{-1}}$ for O IV 1401 Å, $11.2\,\mathrm{km\,s^{-1}}$ for O IV 1399 Å, and $-3.6\,\mathrm{km\,s^{-1}}$ for Si IV 1393 Å. These shifts are consistent with the shifts found in the full spectra. The accuracy is better than one pixel so we expect to derive Doppler shifts with an accuracy of $\pm 5\,\mathrm{km\,s^{-1}}$.

2.1.2. *Intensity and Velocity Images*

Applying the corrections outlined in Section 2.1.1 we constructed images of integrated intensity over the line profile of Si IV for the five complete rasters, which are shown in Figure 2. The pointing was shifted by 48.44″ eastward after each complete raster) The time series, from 07:13 UT–09:55 UT, allows us to see the evolution of the prominence within the field of view of SUMER shown in Figure 1. The Si IV 1393 Å images show considerable evolution of the loops closest to the limb, which are shown in the two upper panels of Figure 2. A low compact loop is visible close to the disk in (b) and (c) with a huge loop above visible in (c), (d), and (e). Below the compact loop, a low density structure, referred to as 'bubble A', is visible in (a). Point B is in the leg of the compact loop. It took 32 min to scan this field-of-view with a slit of 1″ × 120″ and a step size of 3″; thus the images are undersampled. However, because of an optical misalignment, only 107″ of the slit length appeared on the active part of the detector. The dimension of the panels is 135″ × 107″. The CDS image of the prominence is presented in the O V line obtained between 07:07 and 07:58 UT and shows the compact loop.

Doppler images were computed from the displacement of the centroid of the gaussian profiles fitted to the observed ones compared to a reference value (Figure 3(a–c)). This reference value was determined by the average line shifts measured over the disk.

2.1.3. *Spectra*

The six lines observed with SUMER during JOP 12 on 1 May are Si IV 1393.755 and 1402.770Å, and O IV 1399.774, 1401.156, 1404.812 and 1407.386 Å. The O IV lines have a temperature of formation around $1.7\text{–}2.0 \times 10^5$ K, while the Si IV lines are formed around $6\text{–}8 \times 10^4$ K. Six windows were extracted from the detector. In the first window we have the Si IV 1393.760 Å line. The next five windows contain adjacent wavelengths starting at 1398.9 Å ending at 1409.2 Å.

In Figure 4 we present the spectral evolution for 3 successive rasters of the prominence observed in the Si IV 1393 Å line. The staggering mode mixes the time of adjacent slit positions used to make an image. At the top of Figure 4(a) eight slit positions observed between 07:12 and 07:44 UT are displayed. This spectrum corresponds to the lower part of the prominence close to the disk. In the middle, respective bottom, 24 slit positions, respective 40 are displayed showing the temporal evolution of the prominence (Figures 4(b) and 4(c)). On the same vertical row the spectra correspond to the same spatial position in the prominence.

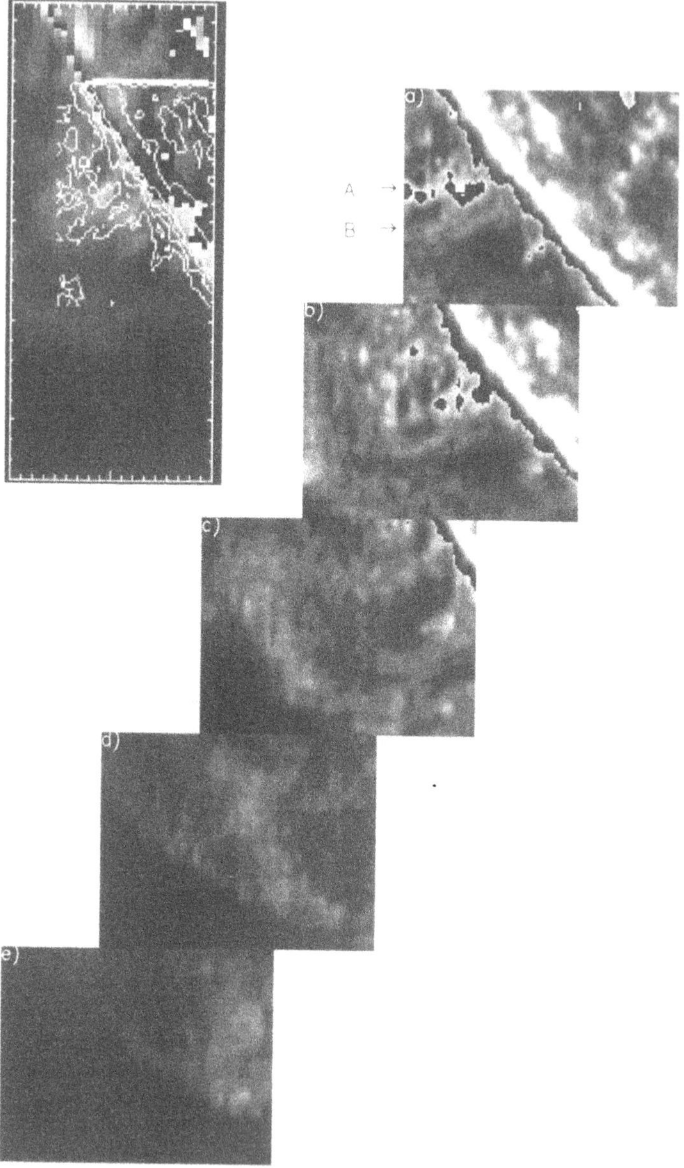

Figure 2. SUMER observations of the prominence on 1 May 1996 in the Si IV 1393 Å line (*right panels*). The prominence is observed between (a) 07:13 and 07:44 UT; (b) 07:45 and 08:16 UT; (c) 08:16 and 08:50 UT, (d) 08:50 and 09:21 UT, (e) 09:22 and 09:55 UT. North is up, and the right corner in the upper panels shows the disk. The dimension of the panels is 135″ × 107″. The CDS image of the prominence is presented in the O V line obtained between 07:07 and 07:58 UT in the *upper left* panel with the SUMER contour image.

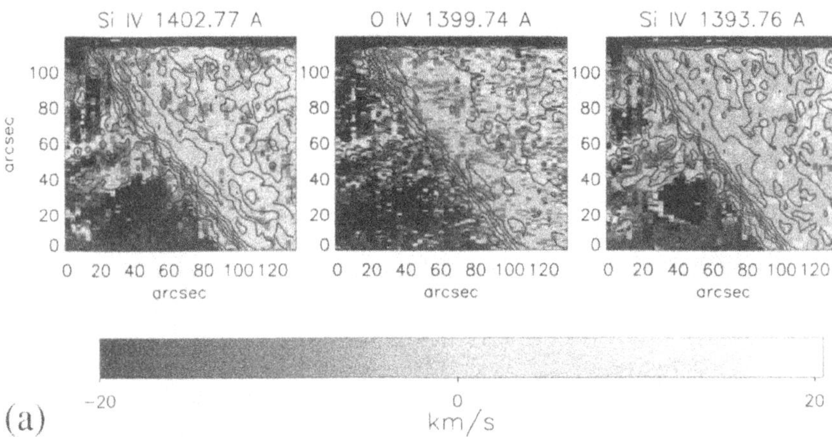

Figure 3a.

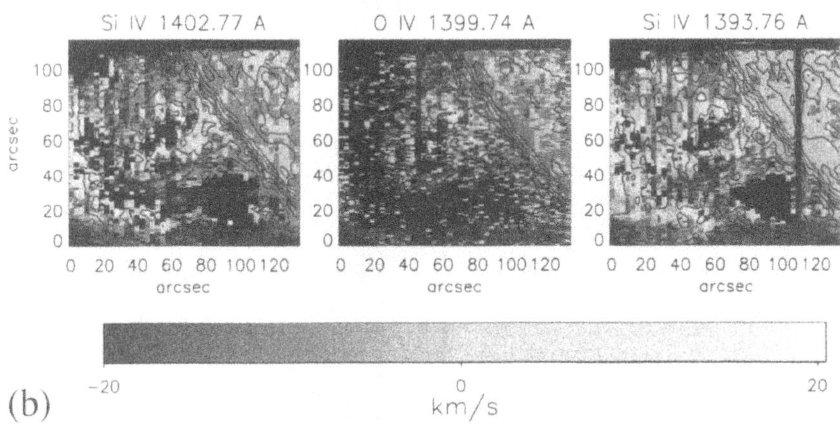

Figure 3b.

The east–west axis is from left to right in Figure 4. The easternmost spectra in the top panel (obtained at 07:28:45, 07:13:09, 07:27:20 UT) show dense red-shifted material visible in point A (position 58). We named this particular feature 'bubble A'. Looking at the same three spatial positions in the middle panel we no longer see the bubble with high red shift at position A. Instead, there is blue-shifted material in the leg of a loop at point B (position 36), about 20″ to the south. A third point, point C, in Figure 4(b) exhibits a double structure with a high red-shifted component.

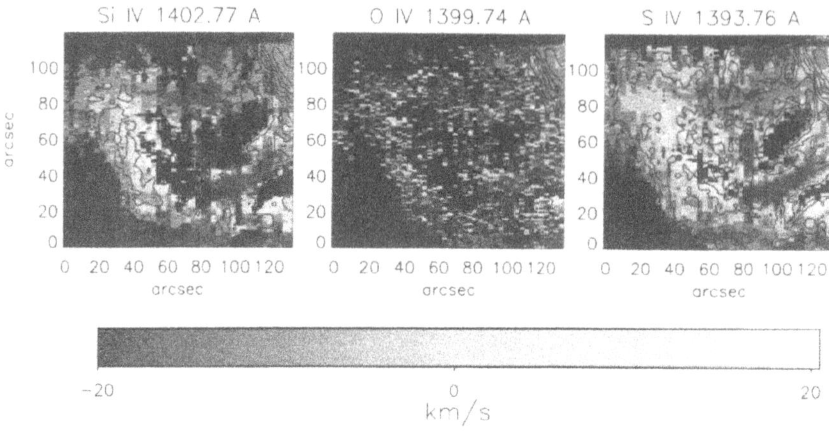

Figure 3c.

Figure 3a–c. Doppler shifts observed in the prominence. The shifts were obtained by using the SUMER spectra in Si IV and O IV lines for the 3 first rasters. We notice that the low loop has mixed velocities during the two first times, and later consists of two blue legs. The huge loop is red and blue along its axis, suggesting twisting motion.

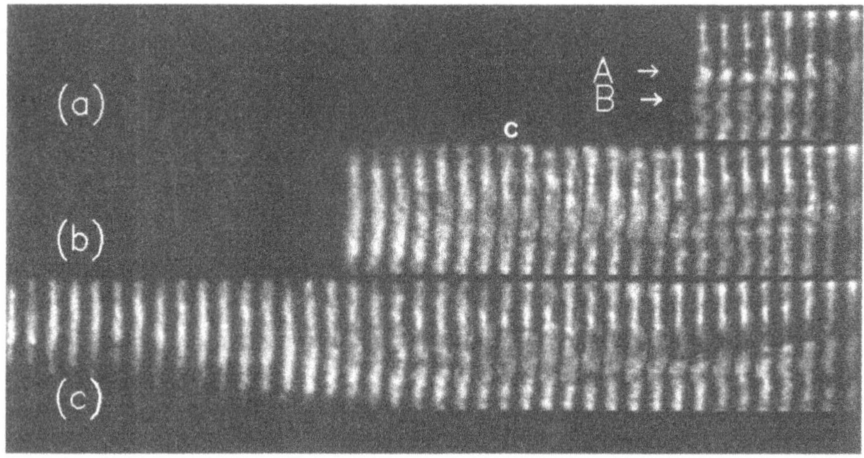

Figure 4. Example of SUMER spectra of Si IV line through the prominence (horizontal direction) displayed with their individual intensity scale for three raster periods: (a) between 07:12 and 07:44 UT, (b) between 07:48 UT and 08:16 UT, (c) between 08:16 UT and 08:50 UT. The data has been corrected for detector pin-cushion distortion. The arrows indicate the horizontal position where points A, B, C can be observed.

Table I

Doppler shifts (v_1, v_2) and line widths (W_1, W_2) of two different components identified in 'bubble A' at 07:28:45 UT, in the leg of a loop (point B) at 08:03:58 UT and at the top of a low loop (point C). Negative values are blue shifts and positive values red shifts.

Profile	v_1 km s^{-1}	W_1 Å	v_2 km s^{-1}	W_2 Å
Point A	9.4	0.22	50.8	0.23
Point B	−40	0.17	8	0.25
point C	1.2	0.22	95	0.21

2.1.4. *Doppler Shifts*

The line profiles show multiple structures (Figure 4), illustrating mass motions, twisted threads or multiple ropes. They are presented on the CR-Rom provided with this volume. Individual profiles are relatively noisy, particularly those in O IV, because of the low count number. We derived quantitative Doppler shifts of the different components of some observed Si IV profiles after averaging 5 neighbouring profiles. We used the procedure provided by S.V. Haughan called 'xcfit'. We fit the profiles by using one, two or three gaussians with a line width each of 0.17 to 0.25 Å (4–6 pixels). The Doppler shifts can be determined with an accuracy of at least 15 km s^{-1} because the solution is not unique. Figure 5(a–c) shows some examples of the profiles whose characteristics are presented in Table I. In some structures we find Doppler shifts as high as 100 km s^{-1}. More commonly Doppler shifts between ± 50 km s^{-1} are derived.

2.1.5. *Electron Density*

The O IV lines were chosen for density diagnostics in the prominence–corona interface at 1.7×10^5 K (Wiik, Dere, and Schmieder, 1993). The nearby Si IV lines are formed at a lower temperature of 6.3×10^4 K and are therefore closer to the cooler plasma observed in Hα as it is confirmed by the images (see Section 3.1). The O IV 1404 and 1407 Å lines are contaminated by second order O III lines (Brage, Judge, and Brekke, 1996). Therefore we are left with only one line pair for density diagnostics; O IV: 1401/1399. In the lower right corner of Figure 6 we show the theoretical ratio of these two lines (CHIANTI, Dere *et al.* 1997).

Using line-ratio diagnostics we have estimated the electron densities in the red-shifted bubble and in the blue-shifted leg (point A and B in Figure 6) along one scan crossing the structures. At 07:28:45 UT we find values of the electron density in the red-shifted bubble (point A) in the range $1-3 \times 10^{10}$ cm^{-3}. The dispersion is much larger for the densities found in the blue-shifted leg (point B). Here, the values range from 2×10^9 to 6×10^{10} cm^{-3}. At 08:03:58 UT (upper right corner)

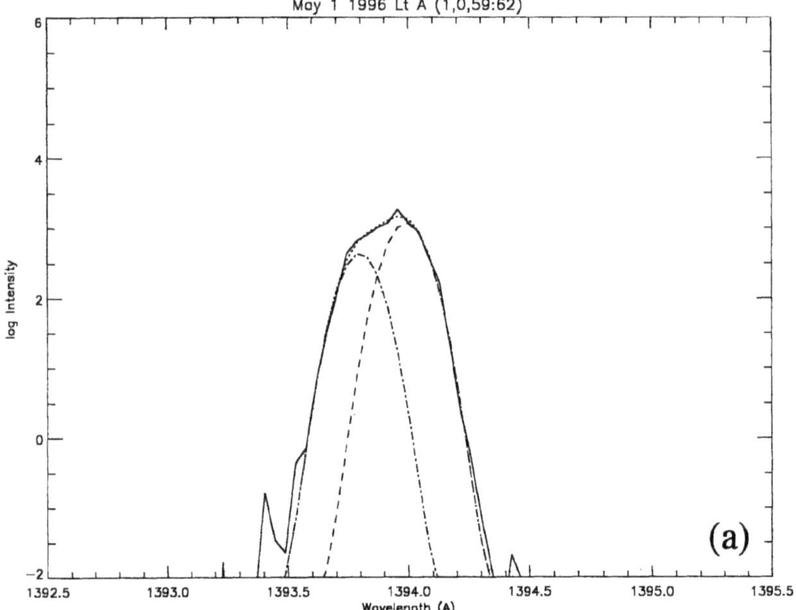

Figure 5a.

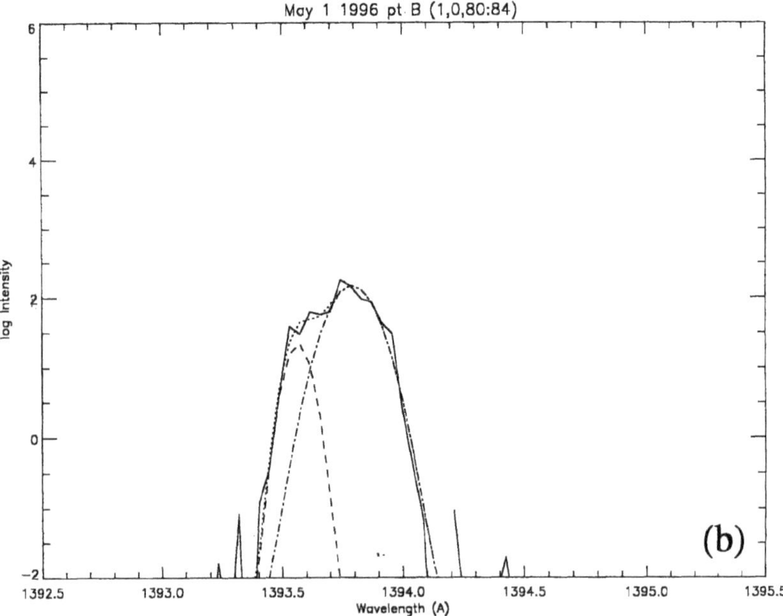

Figure 5b.

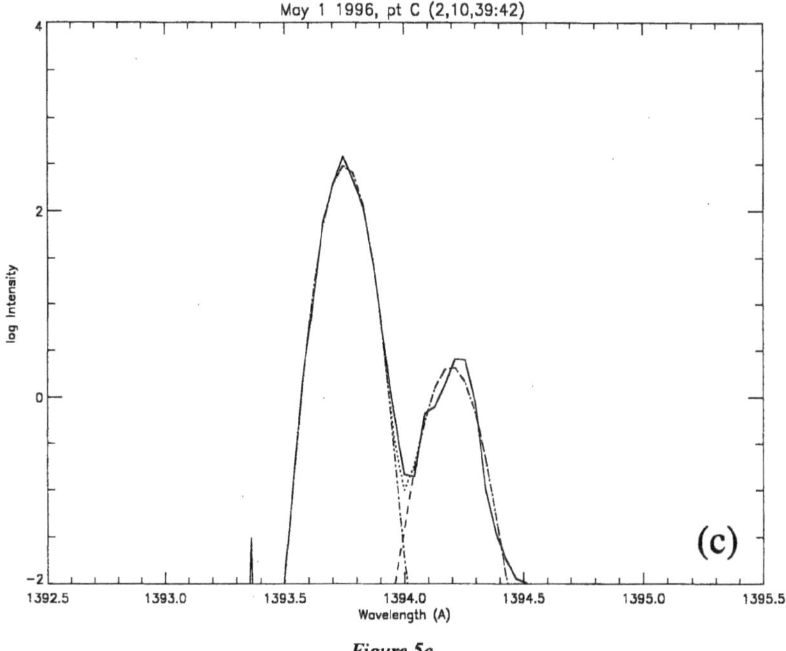

Figure 5c.

Figure 5a–c. Examples of multi-component profiles of Si IV 1393 Å line corresponding to points A, B, C (see Figures 2 and 4), The solid line is the observed profile, the dashed lines the elementary gaussians, the dotted line represent the result of the fit with 2 gaussians. (see movie 2 and Figure 14 on the CD-ROM).

the red bubble has disappeared, and at this moment we find a slight increase in electron density in the blue leg. Values range from 3×10^{10} to 3×10^{11} cm^{-3}. At 08:39:12 UT (lower left corner) we find again a large dispersion of densities in the blue leg ranging from 4×10^9 to 3×10^{11} cm^{-3}. It is interesting to note that this large dispersion takes place over not more than 10″. Such a large dispersion reflects the separation of the plasma into very fine-scale structures. The profiles show multiple components at different velocities and apparently these components have very different electron density.

2.2. CDS

The CDS with the normal incident spectrometer (NIS) observed 15 lines in JOP 12. The lines covered a large range of temperatures which are indicated in the third column of Table II. The exposure time was 100 s. Large rasters (240″ × 240″) were first obtained, then smaller rasters (120″ × 240″). A step size of 2″ was used to obtain these rasters. The pixel size along the slit is 1″ and the width of the slit is 2″. The spatial resolution of the instrument is 1.2″–1.5″ (Harrison *et al.* 1995). The

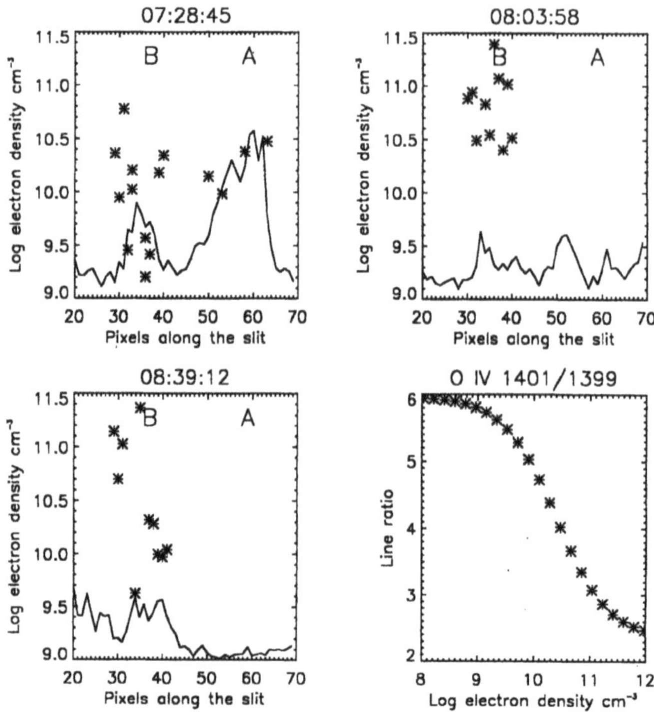

Figure 6. Electron density (stars) in the bubble (point A) and in the loop leg (point B) versus time computed from O IV line ratio. The solid lines represent curves of integrated intensity over the Si IV 1393 line profile in one scan along the slit crossing the structures A and B (arbitrary units). In the right bottom corner is displayed the theoretical curve of the ratio of O IV lines versus electron density. See Section 2.1.5 for further comments.

widths of the windows are 1.334 Å (310–380 Å) and 2.34 Å (520–630 Å) with 21 points in the line profiles. The spectral resolution is respectively 0.0635 Å pixel^{-1} and 0.111 Å pixel^{-1} (60 km s^{-1}). If lines are well fit, motions corresponding to 15 km s^{-1} or even lower can be detected.

A previous paper by Wiik *et al.* (1997) contains a figure showing the prominence in all the observed lines. In Figure 7 we present the prominences as seen in 6 of the lines (He I, O III, O IV, O V, Ne VI, and Mg X). The prominence is in emission for lines formed at $T < 5 \times 10^5$ K (until Ne VI). At chromospheric and transition zone temperatures we see two feet of a huge loop, the southern one is projected onto the disk and appears to be part of a filament extending westward. In coronal lines there is a lack of emission, due to the absorption of the Lyman continuum of hydrogen (around 629 Å) and helium (around 330 Å). On the disk the filament is clearly visible, particularly in O IV and O V.

CDS/NIS spectral data sit on top of a scattered light component due to small irregularities in the grating. The scattered light component is particularly visible

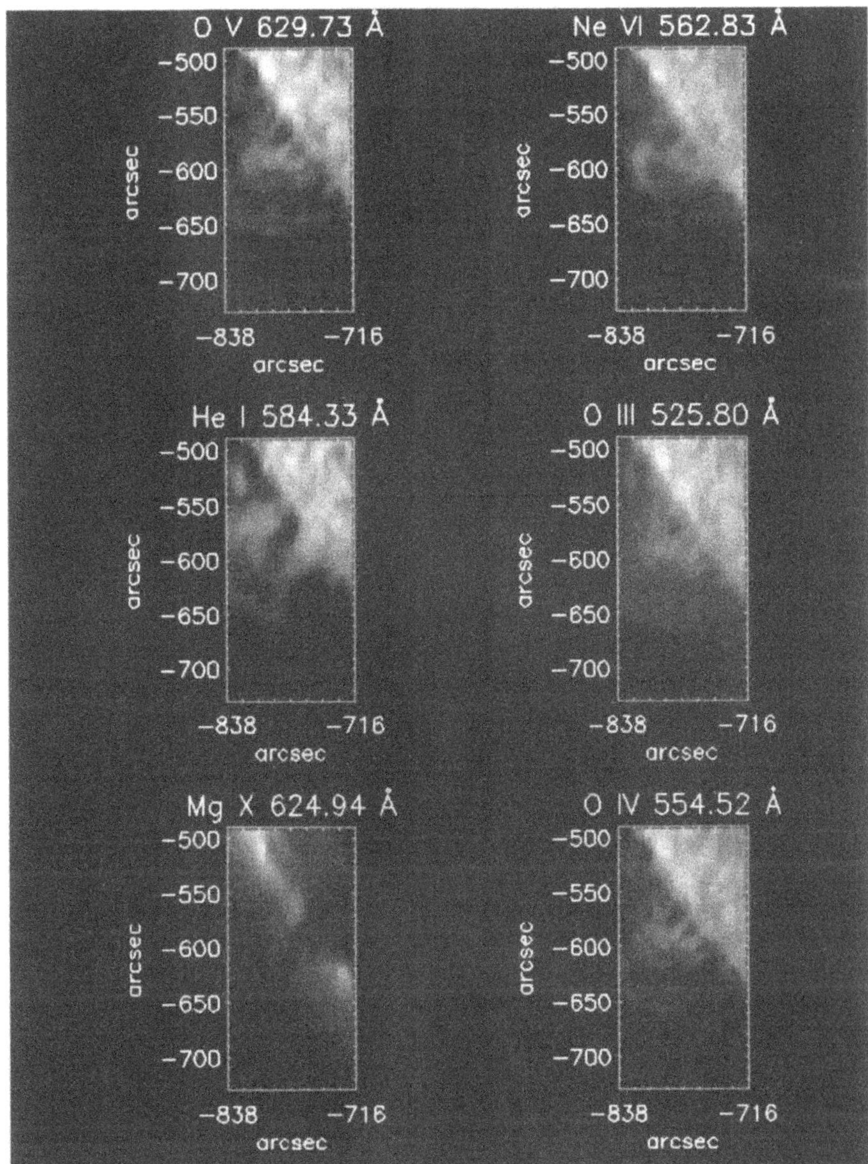

Figure 7. CDS observations of the prominences as seen in 6 lines (He I, O III, O IV, O V, Ne VI, and Mg X). The time indicate the mean of the rasters. The prominence is in emission for lines formed at $T < 5 \times 10^5$ K (until Ne VI). In Mg X we observe a lack of emission due to absorption of chromospheric continua.

Table II

The CDS lines used for the current observation. The second column lists the wavelengths observed, for which the first and third columns indicate the identified spectral line and its typical temperature of formation, respectively.

Ion	Wavelength Å	Temperature K
He I	522.21, 537.03, 584.33	2×10^4
O III	525.80, 599.59	8.5×10^4
O IV	554.52	2×10^5
O V	629.73	2.5×10^5
Ne VI	562.83	4×10^5
Mg IX	368.06	10^6
Mg X	624.94	2×10^6
Fe XIV	334.17	2×10^6

when the line is weak. It can been removed by subtracting the values situated in the extreme ends of the wavelength window from the line profiles. We used the following method.

An algorithm was used to identify pixels with excessive counts due to charged particles passing through the CCD, and replace the values in the affected pixels by interpolation between adjacent ones. A background level was determined for each spectral line by fitting the background count rates on one or, when possible, both sides of the line. These levels were determined for each spatial pixel and subtracted. Images were then made by summing over the wavelengths of the spectral lines.

Line ratios were calculated by dividing the images made with the O III (525.8 Å), O IV (554.52 Å), O V (629.73 Å), and Ne VI (562.83 Å) lines by the He I (584.33 Å) image. The values were then normalized by dividing by a time and space averaged value taken from the top 84 rows of the image (out of 143) for each ratio. Only the top portion was used because it is on the disk and the intensity looks relatively stable.

In Figure 8 the variation with time in two lines is presented: He I 584.33 Å (top panels), and O V 629.73 Å (middle panels). In the bottom panels are the ratios of the two lines. We observe that the small loop corresponding to the low loop observed with SUMER is opening with time in both lines. At the inner edge of the loop, O V material is present, indicated by the crossed fiducial lines. No cool material (He I) is present. This phenomenon is emphasized in the ratio maps.

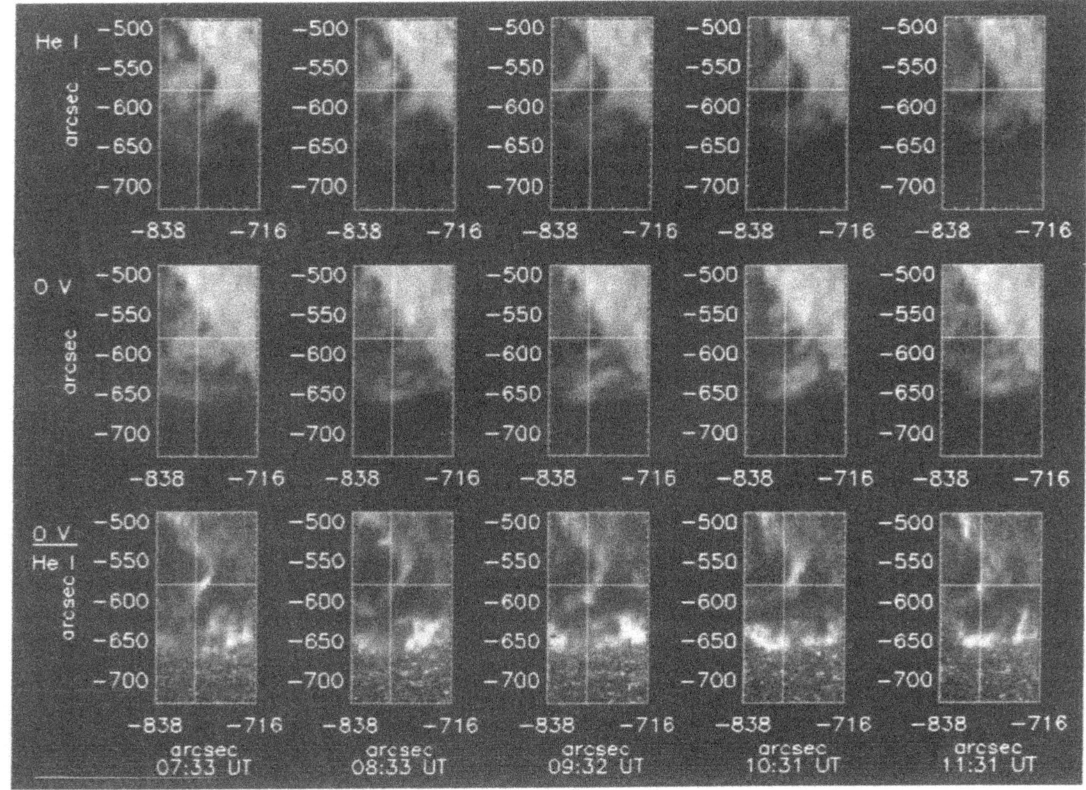

Figure 8. Variation with time of the CDS observations of the prominence in two lines: He I 584.33 A (*top panels*), and O V 629.73 A (*middle panels*). In the bottom panels are the ratios of the two lines. Notice the opening loop. At the inner edge of the loop, O V material is present, indicated by the cross, at 07:33 UT and disappears later. No cool material (He I) is present. The north is up.

3. Evolution of the Prominence Observed on 1 May

3.1. GENERAL BEHAVIOUR

The prominence was located in a long-lived polar-crown filament channel which is still visible during the following solar rotation. The prominence was situated on the east limb, between 20 and 40 deg south. It was huge on 1 May, as can be seen on the Ca II K3 Meudon spectroheliogram taken at 07:25 UT (Figure 1). We estimate its projected altitude to have been ~100 000 km which is around the altitude of marginal stability for a twisted prominence (Vršnak, Ruždjak, and Rompolt, 1992). In Ca II the prominence consists of a huge loop marking the upper boundary of the prominence. We can identify three legs connecting this loop to the chromosphere. Two of them are situated in the southmost part of the prominence at S33 and S36. The northernmost leg is weaker and situated at S20. A low loop is visible between the legs at S33 and S36. In addition another low loop, or maybe a surge, is visible close to the leg at S33, between this one and the leg at S20. The SUMER and CDS observations covered approximately 10 degrees between 30 and 40 deg south, including the two southernmost legs and the two low loops.

We can compare the prominence observations at cool temperatures in the lines of Ca II and He I (CDS) with observations at transition-region temperatures shown in the lines of Si IV (SUMER) and O V (CDS) (Figures 1, 2, 8). The fields of view of CDS and SUMER (before 08:16 UT) were centered on the two southern legs close to the limb. Thus the dynamics and temperature behaviour of this region can be studied particularly well.

The huge loop described above in Ca II is also visible in the Si IV 1393 line. The loop itself is clearly visible in Figures 2(c–e), while its two narrow legs to the south (S33 and S36) are visible in Figures 2(a–c). The smaller loop between the two southernmost legs is identified in Figure 2(b). It is a closed loop which later on opens up, as may be seen in Figure 2(c). No big differences between Ca II and Si IV are observed.

The CDS observations (Figure 8) show different aspects of the prominence at different temperatures. At the beginning of our observations (07:12 UT) the low loop (around 37 000 km high) is visible in He I between the two legs of the prominence. An inner part of this loop, around 20 000 km high, is observed only in transition zone lines (O V and Si IV) and is identified as bubble A. The southern leg at S36 continues as a filament on the disk. This leg is thus an extension of the filament crossing the limb. The filament is not so dense at 09:33 UT and the limb appears brighter in O V. The general behaviour of the legs is globally stable but the inner part of the loop (bubble A) escapes and is no longer visible at 08:33 UT (CDS). During that time the low loop is progressively opening.

3.2. Dynamical Aspect

The dynamical aspect of the prominence can be studied by looking at displacements in the plane of the image or by analyzing the spectra and computing the Doppler shifts. The first method can lead to misinterpretations if we have observations only in one temperature. One must be careful not to interpret a progressive heating of the plasma as motions. In our case we have several images obtained in a large temperature range. When we compare images of the huge loops in Ca II at 07:25 UT to those observed between 08:16 and 09:21 UT in Si IV, we note no apparent changes. We conclude that the global structure of the prominence during these 2 hours is relatively stable for such a very high altitude prominence. In the Doppler-shift images (Figure 3(a–c)), this structure shows blue and red shifts symmetrically to its axis indicating a very large twist which is a condition of eruption (van Ballegooijen and Martens, 1989). Looking at the evolution versus time of smaller features, i.e., the low loop between the two southernmost feet of the huge loop, we note that the loop top (well visible in O V) is rising, expanding and opening (Figure 8). The fact that we see some material in the last panel of Figure 8 at the same place suggests that another part of the filament on the disk escapes and/or appears as a prominence.

The SUMER spectra are very impressive and the reader is urged to look at the movie (movie 2 on the CD-ROM) in order to get the full effect. The spectra show convincingly the fast motions of the prominence, illustrating the existence of multiple and twisted structures. Using the spectra of CDS, we have computed the Doppler shifts and the line widths of the stronger lines (He I and O V, see Figure 9(a–b)), using the moments of the lines. The line width images are not very significant because of the integration of many structures of different velocities along the line of sight.

The Doppler-shift images obtained during the first raster (07:07 to 07:58 UT) show clearly red shifts in the northern part and blue shifts in the southern part. Velocities are around 40 km s^{-1}. These are consistent with the observations of the red bubble A and the blue leg observed with SUMER and discussed in Section 2.1.4. Due to the high spectral and spatial resolution of SUMER we resolve structures with higher velocities, up to 100 km s^{-1} (Table I). These velocities are observed parallel to the solar surface. To observe such high horizontal velocities in a prominence is the signature of activity. Also, it is interesting to note that such high horizontal velocities are usually not observed in the transition region and the corona (Schmieder, 1989).

In Figure 5(a–c) it can be seen that there are several structures in the line of sight with a large velocity dispersion. It cannot be excluded that this is due to the very active aspect of this particular prominence. The behavior of bubble A looks like that of the erupting plasmoid in the model proposed by Raadu *et al.* (1987) to explain filament disappearances. This plasmoid could be also an ejection of material such as surge or jet (Shibata, Nozawa, and Matsumoto, 1992) due to reconnection of

HE_1_584_00 moment
s623r00.fits
Intensity
0
2
4
6
8
Velocity
-40
-20
0
20
40
Line width
0.20
0.30
0.40
0.50
0.60

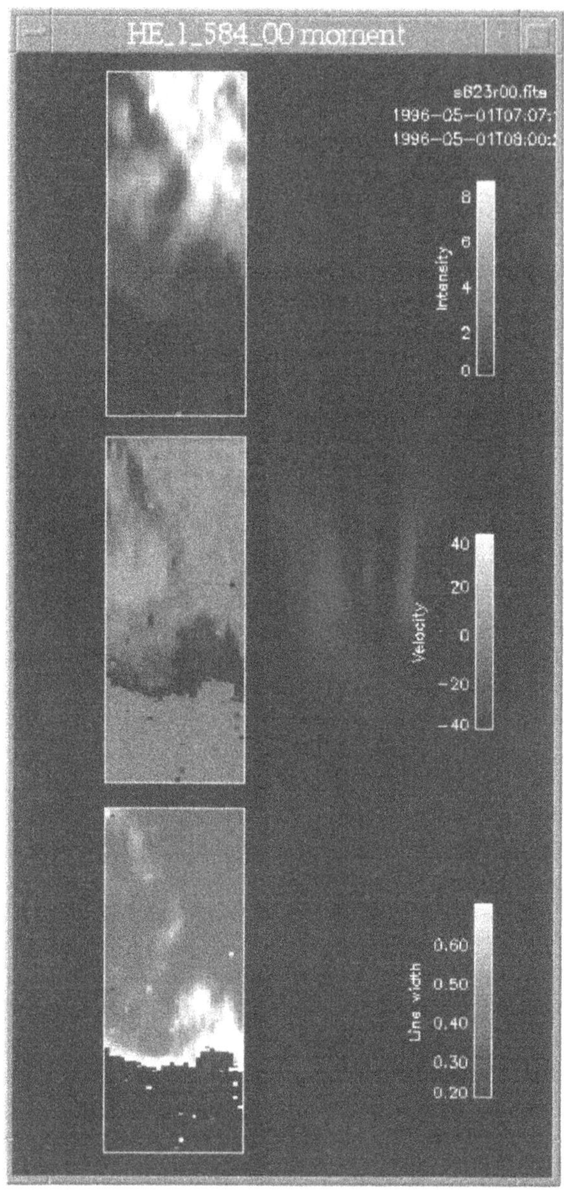

Figure 9a.

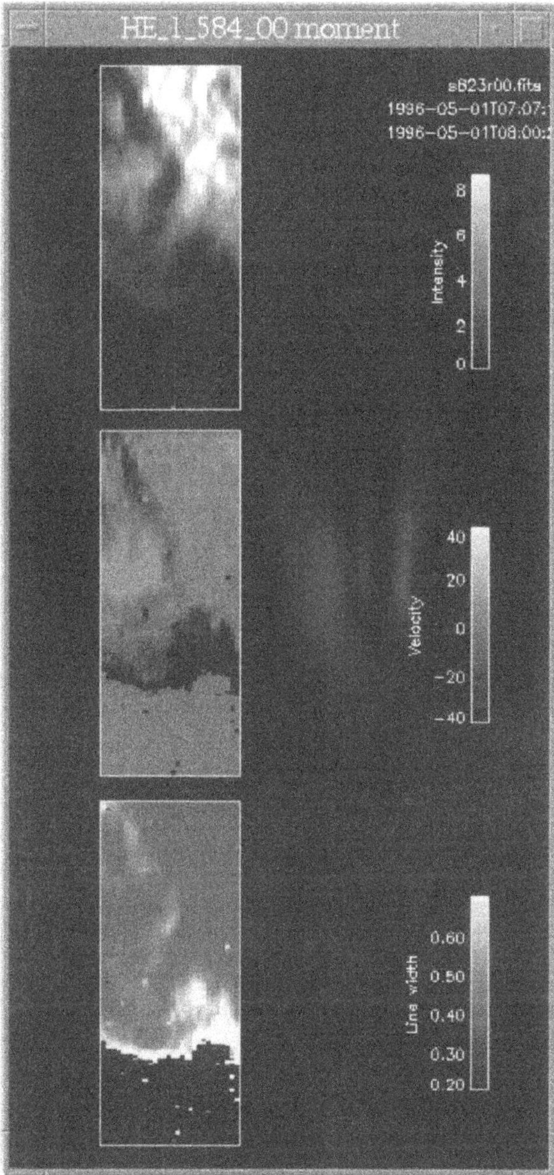

Figure 9b.

Figure 9a–b. CDS observations between 07:07 and 07:58 UT for He I (a) and O V (b) lines. *Top panels*, intensity; *middle panels*, Doppler shift (white/black correspond to red/blue shifts); *low panels*, line width. The red-shifted part corresponds to the SUMER bubble A, the blue-shifted one to the SUMER structure B.

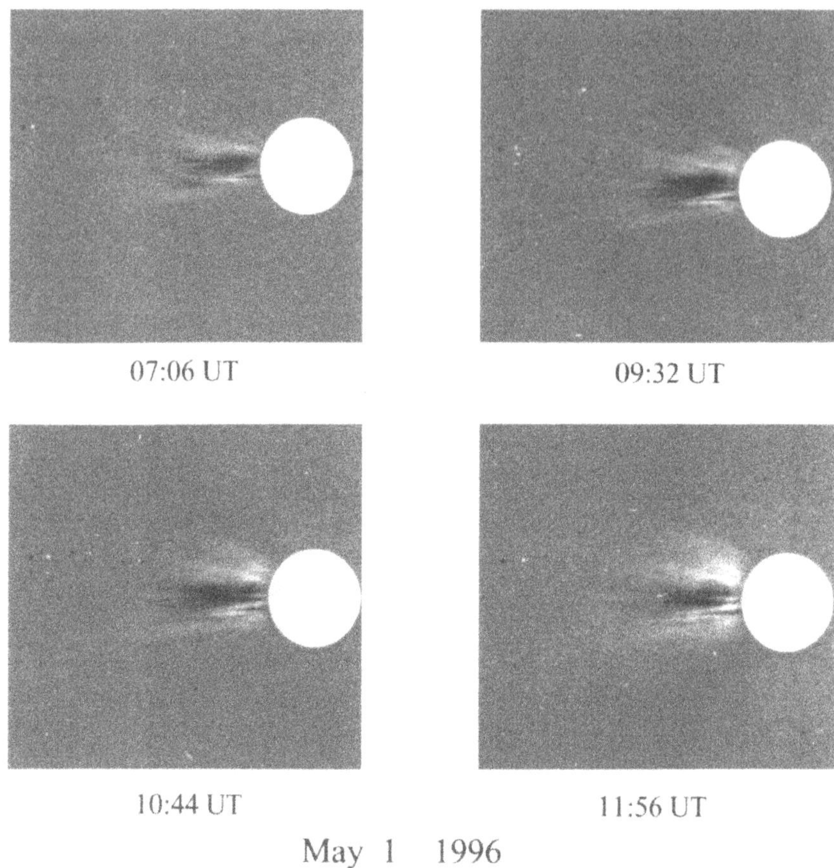

Figure 10. CME observed by LASCO C3 coronagraph. There is a bright feature in the CME images at a position angle of 110 degrees. A sequence of images is shown in Figure 13 (see on the CD-ROM) which illustrates this. The ejection at this position angle first appears above the 3.4 solar radii occulting disc at 09:32 UT. It brightens and expands outwards, such that by 16:48 UT there was a bright region linking back to the Sun. These images contain the pylon for the occulting disc at a position angle of 135 deg, which appears as a streak of bright points extending to the edge of the image. The frame at 18:10 UT shows a rarefaction which is moving out ahead of the bright feature.

magnetic field lines. The high Doppler shifts measured in one leg of the huge loop (point B) could represent twisted threads along the axis of the loop.

The eruption of the prominence was accompanied by a large coronal mass ejection which was seen by LASCO off the east limb (see Figure 10) the outer C3 coronagraph, which has a 3.4 solar radii occulting disc (Brueckner *et al.*, 1995). (LASCO was configured to make observations of comet Hayakutake on 1 May 1996, and thus only C3 images were available to this study.) It extended in the south to a position angle of around 125 deg or S35 (corresponding to the feet of the

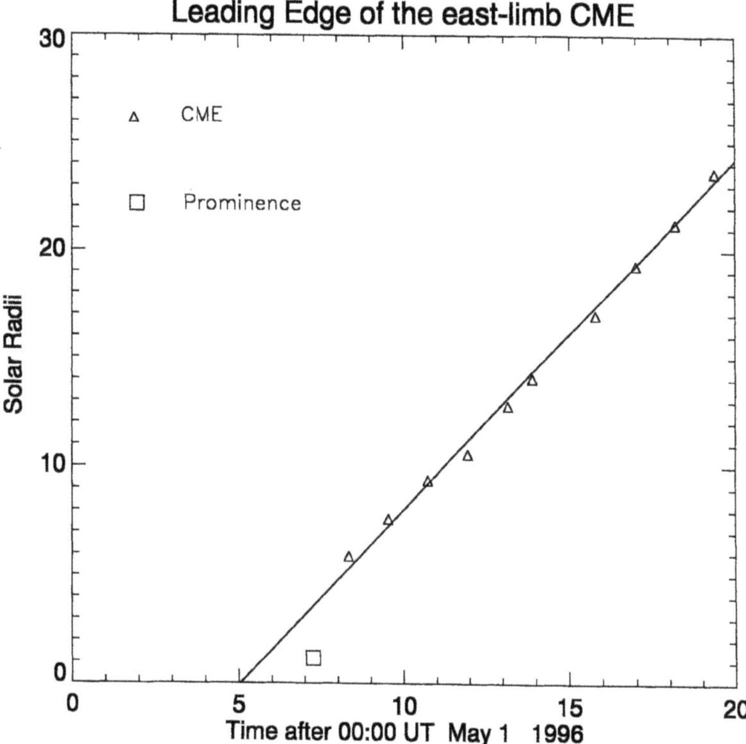

Figure 11. The CME observed by the LASCO C3 coronagraph had significant structure, as seen in this altitude versus time diagram for the leading edge. The top of the prominence at 07:25 UT on 1 May is indicated. The square corresponds to the bubble observed with SUMER in Si IV with large Doppler shift.

prominence: S33, S36) and in the north to about 40 deg or N50. The position angle is measured from the north pole, such that the position angle of the equator at the east limb is 90 deg. Thus the CME had a projected angular extent of 85 deg, with the prominence clearly at the southernmost edge. LASCO did take some images in the Fe XIV line in the C1 coronagraph approximately every four hours on 1 May. Although a number of coronal loops were visible off the south-east limb, including one above the prominence, it was not possible to detect any significant motions in the C1 images, although the loops did appear brighter at the end of the day.

The CME had significant structure, and we have plotted the projected altitude versus time in Figure 11 for the leading edge. For this analysis we used ratios of the images to an average coronal image made from the images taken during the first six hours of 1 May. This technique is chosen because it is relatively free of artifacts. We have also plotted in Figure 11 the top of the prominence at 07:25 UT on 1 May. The CME appears to have started prior to the prominence eruption, but the SUMER/CDS observations begin only at 07:13 UT. However, the CME had a number of separate, identifiable parts, one of which later was visible as an ejection at position angle of 105°. It is probable that the main CME off the east limb, which had a projected onset time of <06:00 UT. (This onset is based on a linear extrapolation of the main event in C3 back to one solar radius. In fact, the altitude-time profile appears to show some slight acceleration, which would therefore extrapolate back to an earlier onset.) This event suggests that the main CME in fact destabilized both the prominence seen by SUMER/CDS and the other discrete structure seen at a position angle of 105°.

3.3. Thermal Aspect

The ratios of line intensities are indicators of the relative abundance of material at a particular temperature within a defined area. These indicators have been used previously by Fontenla and Poland (1989) using data from the Ultraviolet Spectrometer and Polarimeter (UVSP) aboard SMM. The lines were not observed simultaneously and thus motions as well as time evolution of temperature and density affect the ratios. A similar study has been done with this data set of simultaneously observed lines. We have calculated the ratios of O III, O IV, O V, and Ne VI to He I in three different regions in the prominence, which are defined by boxes in the upper right panel of Figure 12. The other panels give the normalised ratios in the three boxes. The variation versus time of the ratios is less than a factor 2. There are no drastic changes and no exceptional heating or cooling. The eruptive part of the prominence, which we observe in lines around 10^5 (O III, O IV, O V) (box 1 and 2 in Figure 12) is brighter than the surrounding prominence in the early image at 07:33 UT. This indicates that the eruptive part is hotter than the stable part. After the ejection of the material the ratio (O V/He I) decreases and increases again when a new low loop is formed or a new part of the filament is crossing the limb. The ratio in box 2 decreases because the box becomes empty, due to the opening of the middle loop. By comparison we have shown box 3 which has a different behaviour. Box 3 corresponds to a small loop or surge visible in Ca II (Section 3.1). In this box the ratio increases until 09:32 UT and then decreases.

The eruptive prominence has a much larger fraction of material around 100000 K than the close by prominence loop; it even had some evidence of higher temperatures, around 4×10^5 K. If bubble A, or the central part of the middle loop, were disappearing due to a thermal process, we would have seen at the same place some material in hotter lines like N IV and/or Mg X (see Table II). In fact no changes were observed in these lines. As a result, we conclude that there is no transfer of heat during the time of our observations.

3.4. Environment of the Prominence and CME

Two full-disk EIT images in He II 304 Å were taken on 1 May, the first at 02:13:56 UT and the second at 11:06:22 UT. In these images very fine twisted structures are

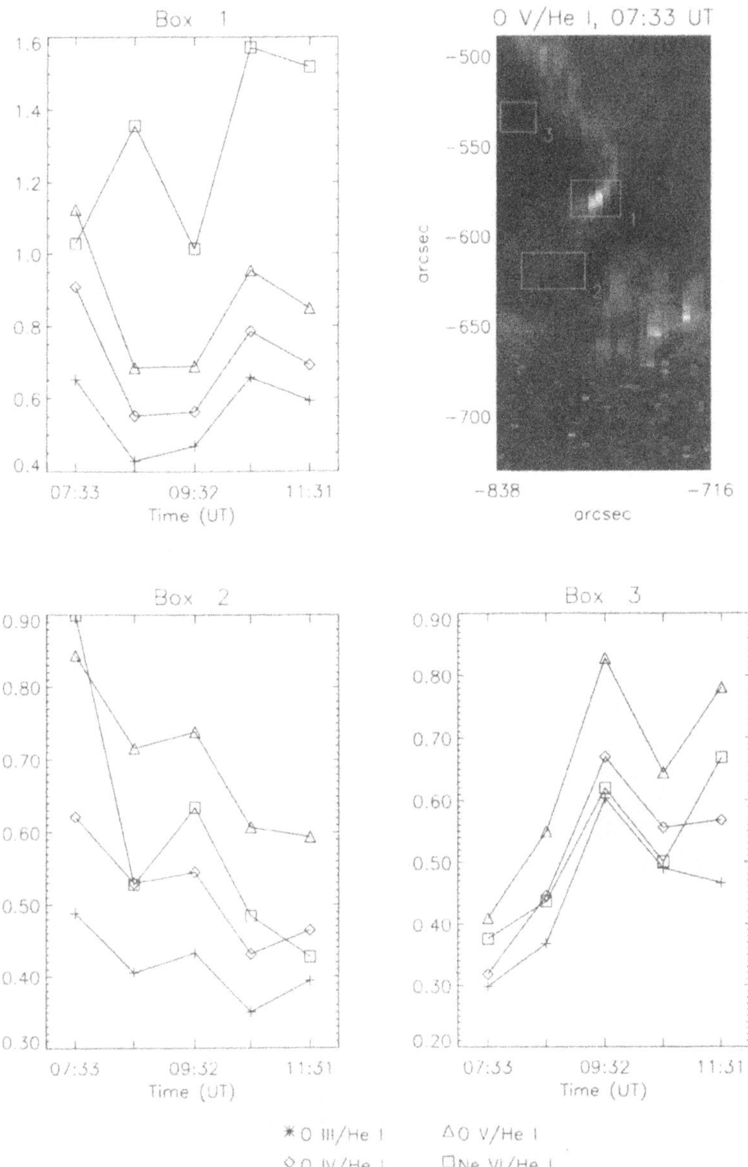

Figure 12. The ratios of O III, O IV, O V, and Ne VI by He I in three different regions in the prominence (see Section 3.3).

seen. The global structure of the prominence is similar at these two times, but the fine structure has changed. We conclude that the eruption discussed here was not observed by EIT because it occurred during a data gap. From CDS we know that by 11:00 UT either the prominence loops are significantly reformed or a new part of the filament appears as a prominence.

In *Yohkoh*/SXT images the region filled by the prominence is very dark, and appears as a lack of emission (dimming). Only small X bright points are visible at the limb, which suggests that there was some emerging flux near the footpoints of the prominence. The nearest active region on the Sun during this time period was on the hidden part of the Sun. It does not appear on the limb until 6 May. Thus, it is unlikely that this active region could be involved in the CMEs of 1 May. Again, we do not see any thermal events such as those commonly observed during the formation of bright cusp structures in X rays (Mc Allister *et al.*, 1996), so we conclude that the process of eruption of this prominence is purely dynamic.

4. Conclusion

We have shown one example of a partial eruption of a prominence associated with a CME, observed in several wavelengths with instruments on SOHO (SUMER, CDS, and LASCO). We have measured Doppler shifts in the prominence corresponding to mass motions as large as 100 km s^{-1}. Conversely, displacement of matter measured in the plane of the sky is more difficult to interpret as transport of mass because it could be due to a change in pressure or density or to the propagation of waves.

We have observations in a large temperature range, and the part of the prominence which disappears is not visible throughout the entire temperature range but only at transition-zone temperatures $0.8-2 \times 10^5$ K). We observe transport of mass and no real heating of the prominence structure. The eruption of the prominence and the CME, which are spatially and temporally related, are interpreted in terms of global destabilization of the magnetic field. One event is not triggering the other, but they are caused by the same magnetic instability. The observed partial eruption is a purely dynamical phenomenon. Several interpretations may be possible. The eruption looks like the ejection of a plasmoid, and the prominence loops are twisted. The prominence eruption that we have observed might be explained by the plasmoid model of a *disparition brusque* of a filament proposed by Raadu *et al.* (1987). In this model the acceleration is sustained by magnetic energy stored in the field of a moving plasmoid. Another possible interpretation could be magnetic reconnection leading to the ejection of material, such as a surge or a jet. Such reconnection could be possible without any signature of very hot plasma (Schmieder, Golub, and Antiochos, 1994).

The filament eruption and the CME could be consequences of some destabilization of the global magnetic structure. The exact time-line of the phenomena is difficult to estimate due to the low cadence of the SOHO SUMER/CDS observations. The large helical structure of this prominence could thus be the consequence of an important instability as suggested in the model of van Ballegooijen and Martens (1989) for the formation and eruption of solar prominences based on cancelling flux. In this case the magnetic flux is transferred from the arcade field which supports the prominence to a helical field.

In our case, the CME appears to have started prior to the prominence eruption. This is consistent with the results of Simnett and Harrison (1985) who studied chromospheric/low coronal phenomena associated with CMEs with data from the Solar Maximum Mission. Not only did the CMEs appear to be the cause of the subsequent activity, but also in the cases they studied the activity was predominantly from one side of the the CME. Thus for the May 1 event, the release of the CME could have weakened the magnetic field which was previously causing the prominence to be stable.

Acknowledgements

We want to thank all the SOHO teams and in particular the PIs of SUMER and CDS, Klaus Wilhelm and Richard Harrison, who permitted the successful run of the prominence programme, and the planners of the day and of each instrument. We also thank the Oslo group for providing extensive help in the data reduction, in particular, Olav Kjeldseth-Moe, Nils Brynhildsen, S. H. Haughan and Mats Carlsson. Jun Elin Wiik acknowledges support from the Research Council of Norway. The SUMER project is financially supported by DARA,CNES,NASA and the ESA PRODEX programme (Swiss contribution). SUMER is part of SOHO, the Solar and Heliospheric Observatory of ESA and NASA.

References

Brage, T., Judge, P. G., and Brekke, P.: 1996, *Astrophys. J.* **464**, 1030.

Brekke, P., Schühle, U., Curdt, W., Wilhelm, K., Kucera, T., and Moran, T.: 1997, *Astrophys. J. Suppl. Series*, submitted.

Brueckner, G. E. *et al.*: 1995, *Solar Phys.* **162**, 357.

Démoulin, P. and Vial, J. C.: 1992, *Solar Phys.* **141**, 289.

D'Azambuja, L. and D'Azambuja, M.: 1948, *Ann.Obs. Paris-Meudon* **6**, 7.

Dere, K. P., Landi, E., Mason, H. E., Monsignori Fossi, B. C., and Young, P.: 1997, *Astron. Astrophys. Suppl. Series*, in press.

Fontenla, J. M. and Poland, A. I.: 1989, *Solar Phys.* **123**, 143.

Harrison, R. A.: 1995, *Astron. Astrophys.* **304**, 585.

Harrison, R. A., Sawyer, E. C., Carter, M. K. *et al.*: 1995, *Solar Phys.* **162**, 233.

Hiei, E. and Widing, K. G.: 1979, *Solar Phys.* **61**, 407.

Illing, R. M. E. and Hundhausen, A. J.: 1986, *J. Geophys. Res.* **91**, 10951.

McAllister, A., Kurokawa, H., Shibata, K., and Nitta, N.: 1996, *Solar Phys.* **169**, 123.

Mein, N., Schmieder, B., Simon, G., Tandberg-Hanssen, E., and Wu, S. C.: 1982, *Astron. Astrophys.* **114**, 192.

Mouradian, Z., Martres, J. M., Soru-Escaut, I., and Gesztelyi, L.: 1987, *Astron. Astrophys.* **183**, 129.

Poland, A. I. and Tandberg-Hanssen, E.: 1983, *Solar Phys.* **84**, 63.

Priest, E. R.: 1989, *Dynamics and Structure of Quiescent Prominences*, Kluwer Academic Publishers, Dordrecht, Holland.

Raadu, M. A., Malherbe, J. M., Schmieder, B., and Mein, P.: 1987, *Solar Phys.* **109**, 59.

Rompolt, B.: 1990, *Hvar Obs. Bull.* **14**, 37.
Schmieder, B.: 1989, in E. R. Priest (ed.), *Dynamics and Structure of Quiescent Prominences*, Kluwer Academic Publishers, Dordrecht, Holland, p. 15.
Schmieder, B.: 1992, *Solar Phys.* **141**275
Schmieder, B., Golub, L., and Antiochos, S.K.: 1994, *Astrophys. J.* **425**, 326.
Shibata, K., Nozawa, S., and Matsumoto, R.: 1992, *Publ. Astron. Soc. Japan* **44**, 265.
Simnett, G. M. and Harrison, R. A.: 1985, *Solar Phys.* **99**291
Tandberg-Hanssen, E.: 1974, *Solar Prominences*, D. Reidel Publ. Co., Dordrecht, Holland.
Tandberg-Hanssen, E.: 1995, *The Nature of Solar Prominences*, Kluwer Academic Publishers, Dordrecht, Holland.
van Ballegooijen, A. A. and Martens, P. C. H.: 1989, *Astrophys. J.* **343**, 971.
Vršnak, B., Ruv̌djak, V., and Rompolt, B.: 1992, *Solar Phys.* **136**, 151.
Webb, D. F., Forbes, T. G., Arass, H., Chen, J., Martens, P., Rompolt, B., Rusin, V., Martin, S. F., and Gaizauskas, V.: 1994, *Solar Phys.* **153**, 73.
Wiik, J. E., Dere, K. P., and Schmieder, B.: 1993, *Astron. Astrophys.* **273**, 267.
Wiik, J. E., Schmieder, B., Kucera, T., and Poland, A.: 1997, in B. Schmieder, J. C. del Toro Iniesta, and M. Vázquez (eds), *Advances in the Physics of Sunspots*, ASP96 Conferences Series **118**, 278.
Wilhelm, K., Lemaire, P., and Curdt, W: 1997, *Solar Phys.* **170**, 75.
Wilhelm, K., Curdt, W., Marsh, E. *et al.*: 1996, *Solar Phys.* **162**, 189.

FIRST OBSERVATIONS OF CORONAL HOLE STRUCTURE AND EVOLUTION USING SOHO-CDS

J. E. INSLEY[1], V. MOORE[1] and R. A. HARRISON[2]
[1]*Space and Atmospheric Physics Group, Blackett Laboratory, Imperial College, London SW7 2BZ, U.K.*
[2]*Rutherford Appleton Laboratory, Didcot, Oxon OX11 0QX, U.K.*

(Received: 15 October 1996; accepted: 26 March 1997)

Abstract. We report on initial observations of coronal hole structure and evolution by the Coronal Diagnostic Spectrometer (CDS) instrument on board the Solar and Heliospheric Observatory (SOHO). The data show that there is coronal activity on time scales of tens of minutes, manifested as brightenings at chromospheric network cell junctions in extreme ultraviolet (EUV) wavelengths. There are also significant differences in structure seen in different wavelengths in coronal holes. Finally, we show coronal hole EUV spectra and compare them to quiet-Sun spectra, also taken by CDS.

1. Introduction

High-resolution (time, space, and spectral) observations of the structure and evolution of coronal holes are important in order to further the understanding of the evolution of the large-scale corona, the solar wind, and the structure of the heliosphere in three dimensions on all time scales. For example, an understanding of the activities at coronal hole boundaries would allow an explanation of how coronal holes grow and decay and, thus, how the open field lines projected into space evolve. This is the case because coronal holes are basically open magnetic field regions (Bohlin and Sheeley, 1978; Zirker, 1977), i.e., the boundary of the heliospheric magnetic and particle environment. They are the sources of high-speed solar wind streams (Krieger, Timothy, and Roelof, 1973; Nolte *et al.*, 1976; Bohlin, 1977; Leer and Holzer, 1985).

Extreme ultraviolet (EUV) observations of the Sun enable the probing of the transition region and the low corona in a way which cannot be achieved at other wavelengths. The EUV spectrum consists of emission lines and line continua, mainly emitted by trace elements in the Sun's atmosphere, superimposed on the background solar black-body spectrum, and observations of these emission lines allow the temperature and density structure of the corona to be investigated. Observations of the corona and coronal holes in the EUV wavelength range have been few and far between as they require space based observations, due to the Earth's atmosphere absorbing these wavelengths. The most notable example of previous EUV observations of the Sun comes from the *Skylab* mission, which ran from May 1973 to February 1974 (see Zirker, 1977, and references therein; Vernazza and Reeves, 1978).

Solar Physics **175:** 437–456, 1997.

EUV observations have been made also by the OSO satellite series, by the CHASE experiment on Spacelab 2 and on some rocket flights (see, e.g., Lang, Mason, and McWhirter (CHASE), 1990; Malinovsky and Héroux (rocket), 1973; Maran and Thomas (OSO 7), 1973). However, time resolutions of a few seconds or less, combined with spatial resolving elements of a few arc sec and spectral resolutions (for emission line separation) of less than 1 Å, have never been achieved.

The Solar and Heliospheric Observatory (SOHO) mission, launched in December 1995, provides an up-to-date platform for studying the UV Sun, with its complement of ultraviolet (UV) and EUV spectrograph and coronagraph instruments. Owing to the lack of previous high time resolution and imaging EUV observations of the corona this is the first opportunity for coronal holes to be studied at a high spatial and temporal resolution in this particular wavelength band. With this new opportunity for observations of the corona comes the ability to study some previously undetermined properties of coronal hole structure and evolution, such as the spatial temperature and density structure of a coronal hole and how a coronal hole boundary evolves on short time scales, e.g., a few hours. The need for images of the coronal hole region in order to determine the structure and evolution of both the coronal hole itself and the coronal hole boundary means that the Coronal Diagnostic Spectrometer (CDS) instrument on board SOHO is most ideally suited for these purposes.

In this paper we present the first results from observations made of coronal hole structure and boundary evolution using CDS. We make use of the CDS Normal Incidence Spectrometer (NIS) which can produce spectra and images in the range 310–630 Å. A NIS spectral atlas showing the emission lines present within a coronal hole is shown, along with images of a coronal hole in some of the identified emission lines, showing the structure within a coronal hole and how some of this structure evolves on a time scale of tens of minutes.

2. Observing Sequences

In order to correctly identify the emission lines which we have chosen to use to study the structure and evolution of coronal holes and their boundaries we need to have a full spectrum from the NIS. From this we can then determine which lines are present in the spectrum and their relative intensities. The coronal hole observations presented here were taken before thorough spectral identification analyses were completed for CDS and therefore the lines chosen are not necessarily ideal for the purpose. CDS observing sequences, or Studies, are given 5-letter labels, which we will use in this paper for ease of identification. For this work the Studies used are called NISAT, BOUND, and CHSTR. The CHSTR and BOUND observations described here were taken on the 1 April, 1996, whereas the NISAT observations were taken on the 24 June.

2.1. SPECTRAL ATLAS – NISAT

The Normal Incidence Spectrometer spectral atlas takes the full NIS spectrum from the two wavelength bands, 308–381 Å and 513–633 Å, over an area of 20 by 240 arc sec, using a 2 × 240 arc sec slit, with an exposure time of 50 s. The target for these particular observations was the north polar coronal hole. The spectra presented in Figures 1 and 2 are averaged spectra over all the points deemed to be within the coronal hole region. Figures 1(a) and 1(b) show the shorter wavelength band at two intensity scales to highlight the brighter (1(a)) and weaker (1(b)) lines. Similarly, Figures 2(a) and 2(b) show the longer wavelength range.

There are a number of lines clearly visible in the shorter wavelength band. The most notable of the emission lines is the Mg IX 368 Å (the peak at 324 Å is caused by an instrument artefact and is not an emission line). The Mg IX emission line is one which has been identified by the authors, from past *Skylab* observations (Feldman, Purcell, and Dohne, 1992), as a good identifier of coronal hole boundaries and is an integral part of the CHSTR and BOUND observing sequences. Previous observations from *Skylab* have also identified the Mg 368 Å line as being almost totally absent within coronal holes (Bohlin, 1977). There are a number of other lines which are used in CHSTR and BOUND for diagnostic purposes which lie within this wavelength region, such as the density sensitive Si IX 349/341 line pair (Mason and Bhatia, 1978). In addition, many lines from highly ionised magnesium, silicon and iron are present.

In general, the lines present in the longer wavelength band, shown in Figure 2(a), are brighter than those in the first wavelength band. A number of lines are clearly visible in this spectrum, from oxygen (O III 599 Å, O IV 554 Å, O V 629 Å), helium (He I 584 Å, He II 304 Å – second order) and magnesium (Mg X 610 Å, 625 Å). A list of well-identified lines from both wavelength ranges is given in Table I. This includes a comparison of coronal hole to quiet-Sun intensities.

The longer wavelength range shows a significant decrease in the helium and magnesium emission relative to the oxygen and neon lines, when compared to quiet-Sun spectra from CDS described by Harrison *et al.* (1997). Indeed, for quiet-Sun conditions the He I line dominates, but in these coronal hole regions the O V line at 629 Å dominates, the O IV 554 Å lines become almost as bright as the He I 584 Å line, and the O III 559 Å line is as bright as the He II 304 Å second-order line. The depletion of hotter lines such as iron and magnesium would be anticipated, as would the depletion of the helium lines due to the influence of coronal electrons and coronal radiation (Wahlstrøm and Carlsson, 1994). A comparison of the shorter wavelength range is more complex. The Mg IX line still dominates but across the wavelength range the emission lines are much weaker since they are mainly from highly-ionised iron, silicon, and magnesium. Some of the higher temperature lines such as from Fe XVI (360 Å) are not evident at all. A list of the identifiable lines is shown in Table I.

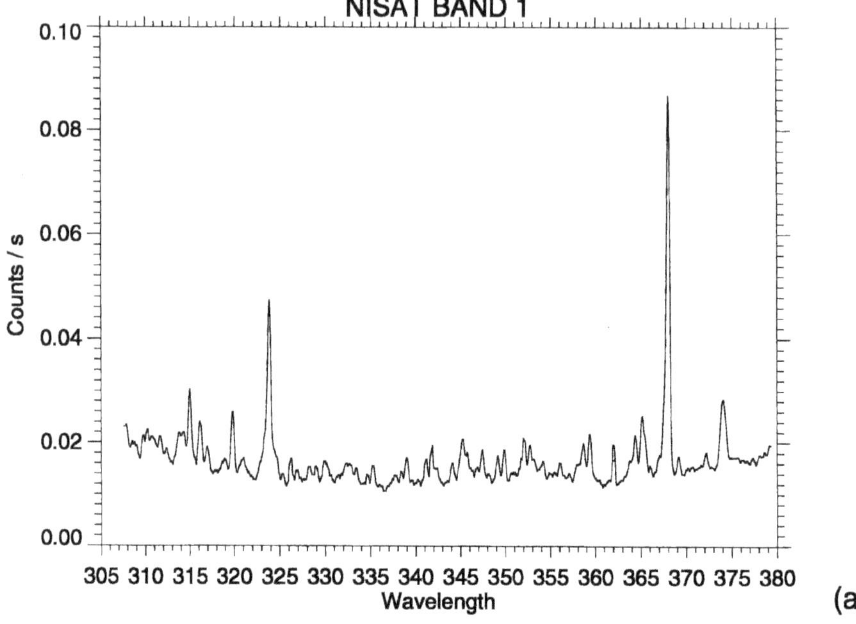

Figure 1a. NISAT observations of NIS spectral band 1: 308–381 Å.

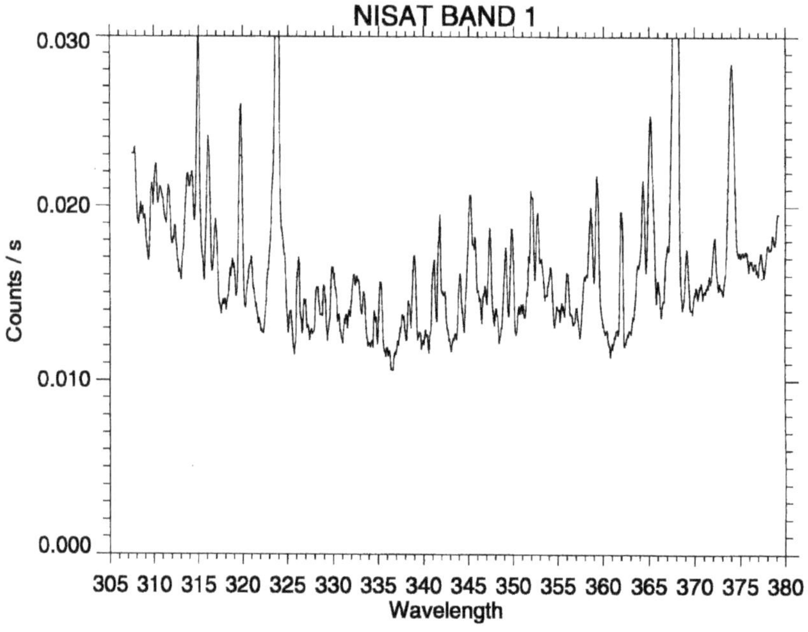

Figure 1b. NISAT observations of NIS spectral band 1, scaled to show weaker lines.

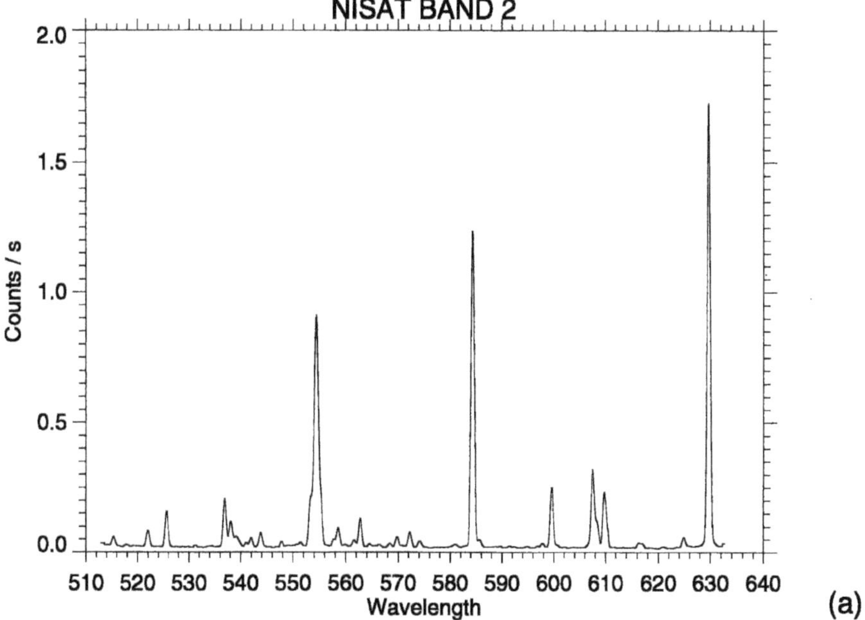

Figure 2a. NISAT observations of NIS spectral band 2: 513–630 Å.

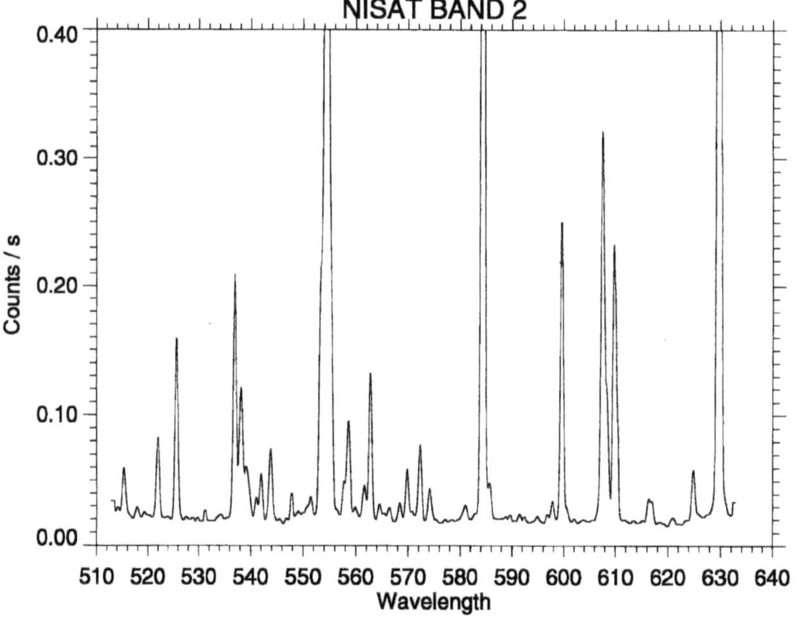

Figure 2b. NISAT observations of NIS spectral band 2, scaled to reveal weaker emission lines.

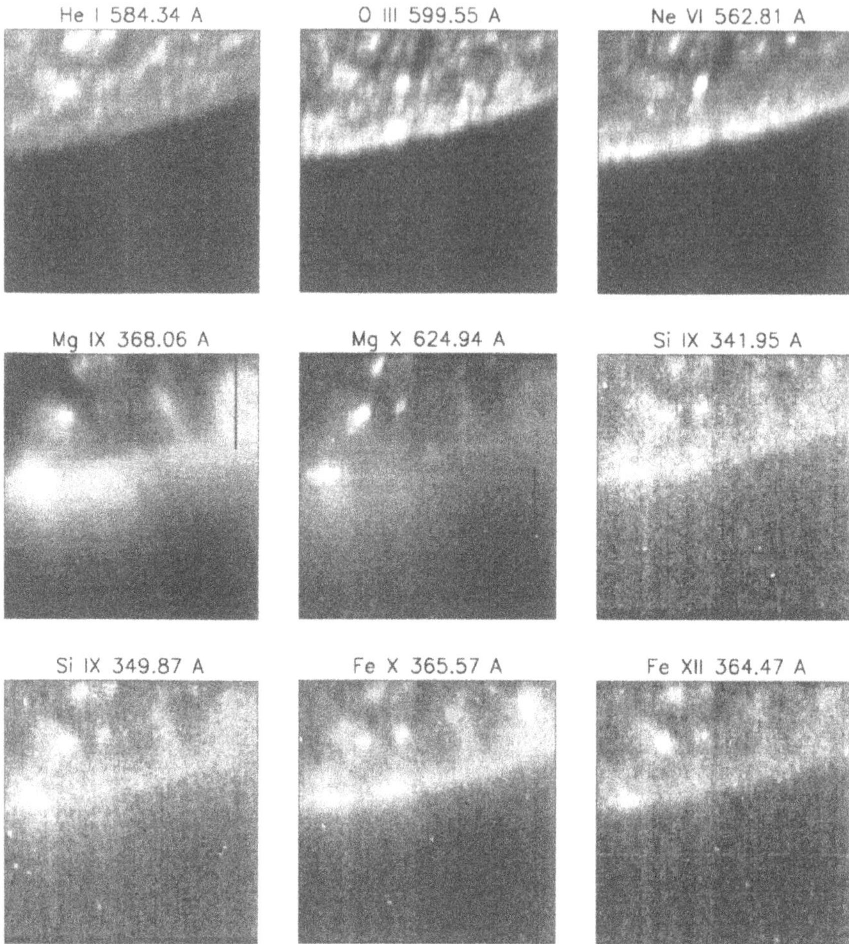

Figure 3. CHSTR observations taken on 1 April, 1996. These show the limb of the Sun and a portion of the southern polar coronal hole, with some radial structures, possibly polar plumes, visible in the Mg IX 368.06 Å image in particular.

2.2. Coronal Hole Structure – CHSTR

High spatial resolution images of coronal holes in a number of emission lines enables us to probe the density and temperature structure of a coronal hole at a level which has previously been impossible. Thus the CDS CHSTR observing sequence was designed for basic high-resolution observation of coronal features, through production of large rasters. The sequences is based on three main requirements:

(i) a good temperature range of ions showing structure within a coronal hole,

(ii) a range of iron ions to provide temperature diagnostics, and

Table I

Well-identified lines in the 308–381 Å and 513–630 Å wavelength bands. The intensity ratios are approximate only.

Ion	Wavelength (Å)	Log T	CH/QS intensity ratio
Mg VIII	315.02	5.9	0.4
Si VIII	316.21	5.9	0.4
Mg VIII	317.01	5.9	0.3
Si VIII	319.83	5.9	0.3
Fe XIV	335.40	6.3	0.3
Mg VIII	339.00	5.9	0.3
Si IX	341.87	6.0	0.4
Si IX	345.13	6.0	0.2
Si X	347.40	6.1	0.2
Si IX	349.87	6.0	0.2
Fe XI	352.67	6.1	0.3
Fe XIII	359.64	6.2	0.3
Fe XII	364.47	6.2	0.2
Fe XI	365.57	6.1	0.3
Mg IX	368.06	6.0	0.3
Unidentified	374.00	?	0.4
He I	516.60	4.3	0.4
He I	522.20	4.3	0.3
O III	525.80	5.0	0.8
He I	537.03	4.3	0.3
Ne IV	543.80	5.2	0.6
O IV	554.52	5.2	0.9
Ne VI	558.59	5.6	0.6
Ne VI	562.83	5.6	0.6
Ne V	569.20	5.5	0.5
Ne V	572.20	5.5	0.6
Ca X	574.00	5.8	0.4
He I	584.33	4.3	0.3
O III	599.59	5.0	0.8
O II	616.60	4.5	0.4
He II	303.78 (2nd order)	4.7	0.3
Mg X/O IV	609.79	6.0/5.2	0.4
Mg X	624.94	6.0	0.2
O V	629.73	5.4	0.8

(iii) a density sensitive line ratio, using Si IX.

CHSTR detects emission from eight Fe ions, Fe VIII 370.43 Å, Fe X 365.57 Å, Fe XI 356.54 Å, Fe XII 364.47 Å, Fe XIII 348.18 Å, Fe XIV 334.176 Å, Fe XVI 335.40 Å, and Fe XII 338.17 Å, to provide temperature diagnostics and a good temperature range. The Si IX line ratio of 349.87/341.95 Å was used to provide density diagnostics. Also included in the observations were Mg IX 368.06 Å, Mg X 624.94 Å, O III 599.55 Å, Ne VI 562.81 Å, and He I 584.34 Å emission lines, to provide coronal hole boundary locations and a view of the cooler low corona and transition region layers below the coronal hole. It is clear from the previous section that some of these early line selections were not ideal. Several of the iron lines were identified in the NISAT observation, but some were not; the silicon pair was seen, as were the magnesium, oxygen, helium, and neon lines.

The images were produced by rastering the 2 × 240 arc sec slit in the solar east–west direction, producing an image 240 × 240 arc sec in area, as shown in Figure 3. Each individual exposure was 60 s long, and the total time taken to build up the images was approximately 2 hr and 8 min. These particular observations were made of the southern polar coronal hole and the off limb area below. It is possible to see bright regions in these images, which can be identified as bright points and the base of polar plumes, and in the Mg IX 368.06 Å image it is possible to see the radial structures normally referred to as polar plumes (see Section 3.1). The magnesium emission in particular shows several bright patches, some of which extend into the corona as plumes. Other than these, there is very little emission. The bright patches are seen in varying degrees in other wavelengths, and this is discussed below.

2.3. Coronal Hole Evolution – BOUND

One of the major unanswered questions relating to coronal hole physics is how coronal holes evolve. In order to address this question an observing sequence was developed for CDS which looks at the coronal hole boundary and how it evolves on time scales of the order of 10 min. The observing sequence was split into three phases:

(i) Phase I – determine the density and temperature structure of the coronal hole boundary, using a similar line selection to that used in the CHSTR observing sequence. The lines used were He I 584.33 Å, O V 629.73 Å, Mg IX 368.06 Å, Si IX 341.95 Å, Si IX 349.87 Å, Si XII 520.67 Å, and Fe XVI 360.76 Å. This phase covers an area of 60 × 240 arc sec, using a 2 × 240 arc sec slit, with a 20 s exposure time. It is centred on the coronal hole boundary, running east–west, as shown in Figure 4. The He I, O V, Mg IX, and Si XII emission is formed at temperatures of approximately 20 000 K, 250 000 K, 1 MK, and 2 MK, respectively. Thus, these give an excellent temperature range in the atmosphere. The Si IX lines are a density sensitive pair.

(ii) Phase 2 – a raster of 60 × 60 arc sec, using 2 × 240 arc sec slit, and an exposure time of 20 s. It is centred in the field of view of phase 1 and repeated twenty times, to study the activity seen at the coronal hole boundary. Six lines were selected to give a wide temperature coverage – He I 584.33 Å (20 000 K), O III 599.59 Å (100 000 K), Ne VI 562.83 Å (400 000 K), Mg IX 368.06 Å (1 MK),

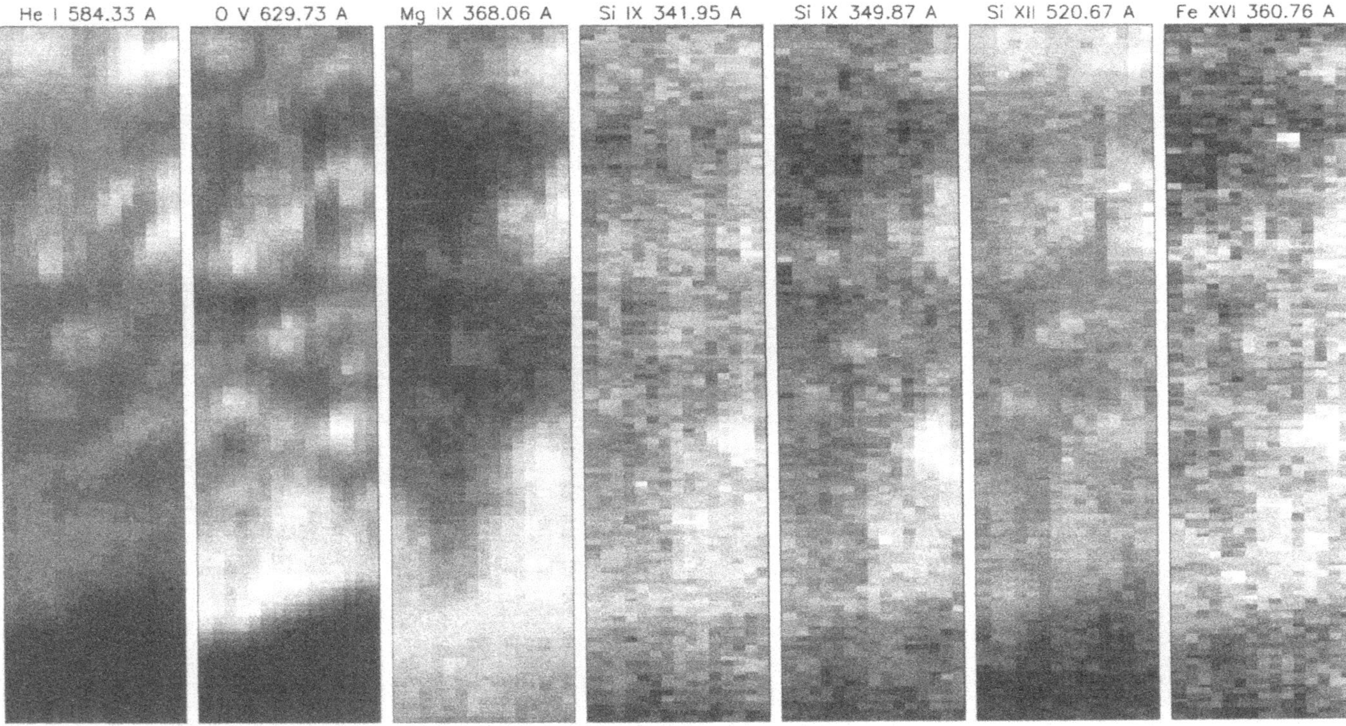

Figure 4. BOUND observing sequence, phase 1. This is used to determine the density and temperature structure of the coronal hole boundary region. The observations were made on 1 April, 1996.

Fe XIII 320.80 Å (1.6 MK), and Fe XVI 335.40 Å (2 MK), as shown in Figure 5. Both of the Fe ions show almost no emission as there are no high temperature regions within the coronal hole.

A time sequence of the four brighter lines are shown in Figures 6(a–d). These show how the intensity of emission can vary on time scales of the order of the time resolution, i.e., ≈10 min, on a spatial scale of a few arc sec, as demonstrated by the two bright regions in these images. These images are the first observations of the evolution of EUV structure within a coronal hole at a high time resolution (≈10 min) and in a variety of emission lines, allowing the probing of different temperatures and heights within the coronal hole region.

(iii) Phase 3 is a repeat of phase 1.

3. Analysis and Results

3.1. Coronal Hole Structure

In Figure 3, the first CHSTR observations on 1 April, 1996, it is possible to see coronal structure in all of the images, although the count rates for all of the Fe and Si emission lines are much lower than for the Mg, O, Ne, and He lines. Therefore, in these images, the background appears to be brighter, as the images have been scaled to show any structure within them. The images are shown in roughly ascending temperature from left to right.

In the images shown in Figures 6(a–d) there are a number of bright regions, ≈10 arc sec across, which are visible in most of the images. Figure 7 shows the Mg IX image again, with the solar *X* and *Y* shown, so that there is a reference point for solar position in these images (Figures 6(a–d)). The bright points are labelled as A, B, C, and D. In the Fe images there are two bright regions on the disc which appear approximately at the same intensity, with a further bright region appearing at the limb in the south–east corner – regions A, B, and C. It is the difference in intensities of these three bright regions in the different wavelengths that provides most interest. In the Mg IX image only one of the disc bright points, point A, is visible, with a large brightening at the limb, point C, occurring where the bright region on the limb is located in the other image. In the Mg X image on the other hand, all three bright regions are visible (points A, B, and C). Also, in the O III image the right-hand most of the two disc bright points is visible, with a new, fourth bright point, point D, showing up directly beneath it on the limb. In the Ne VI image, only one bright region is visible, point B, the right-hand most of the two disk regions. In contrast, the only bright region visible in the He I image is the left-hand most one, point A. Such variations with temperature give an intriguing view of these features.

These structures are probably related to X-ray bright points and/or the base of polar plumes. They may also relate to the so-called *Yohkoh* jets (e.g., Shibata

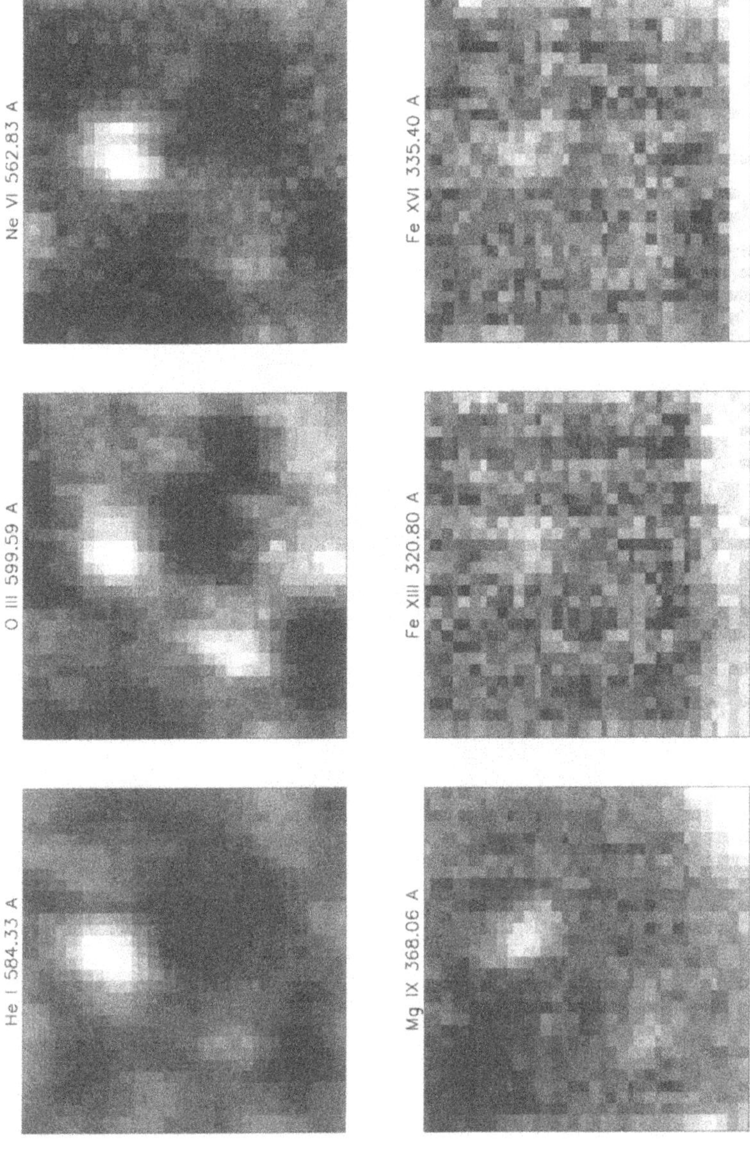

Figure 5. BOUND observing sequence, phase 2. This shows a single raster from the second phase of the BOUND observing sequence. The observations were made on 1 April, 1996. It is possible to see two bright regions and the chromospheric network cells in these images.

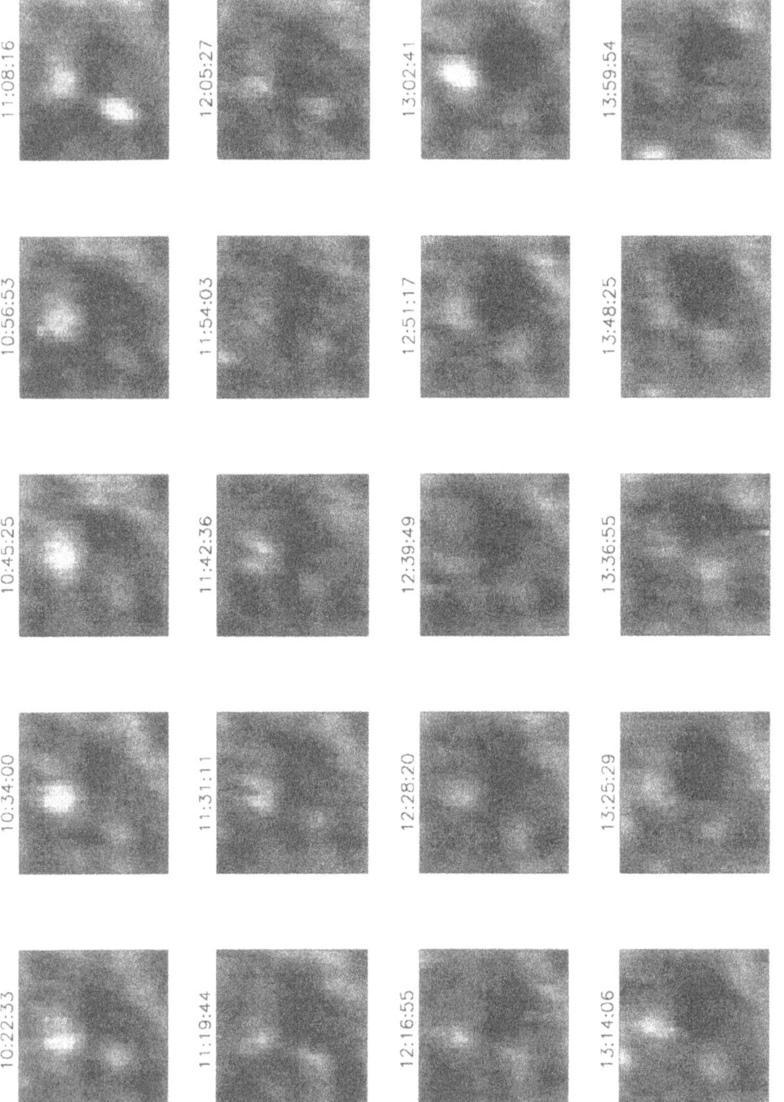

Figure 6a. Time series of He I 584.33 Å rasters taken on 1 April, 1996 by CDS, using BOUND observing sequence.

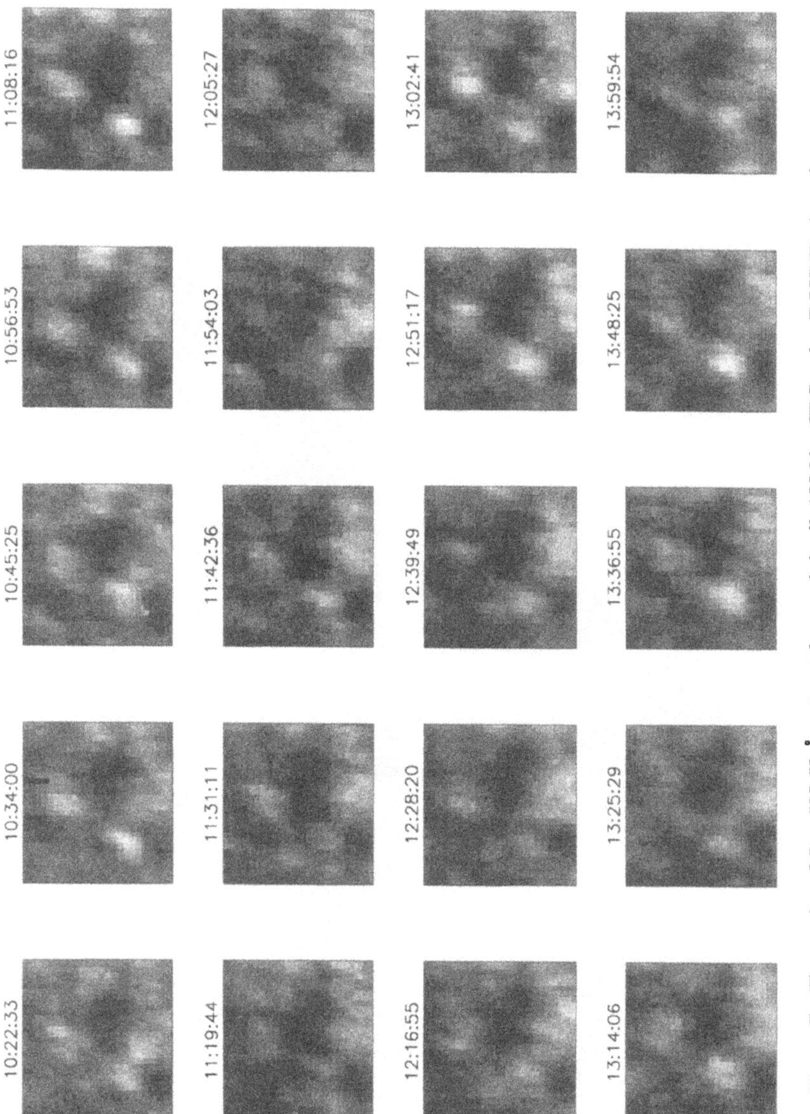

Figure 6b. Time series of O III 599.59 Å rasters taken on 1 April, 1996 by CDS, using BOUND observing sequence.

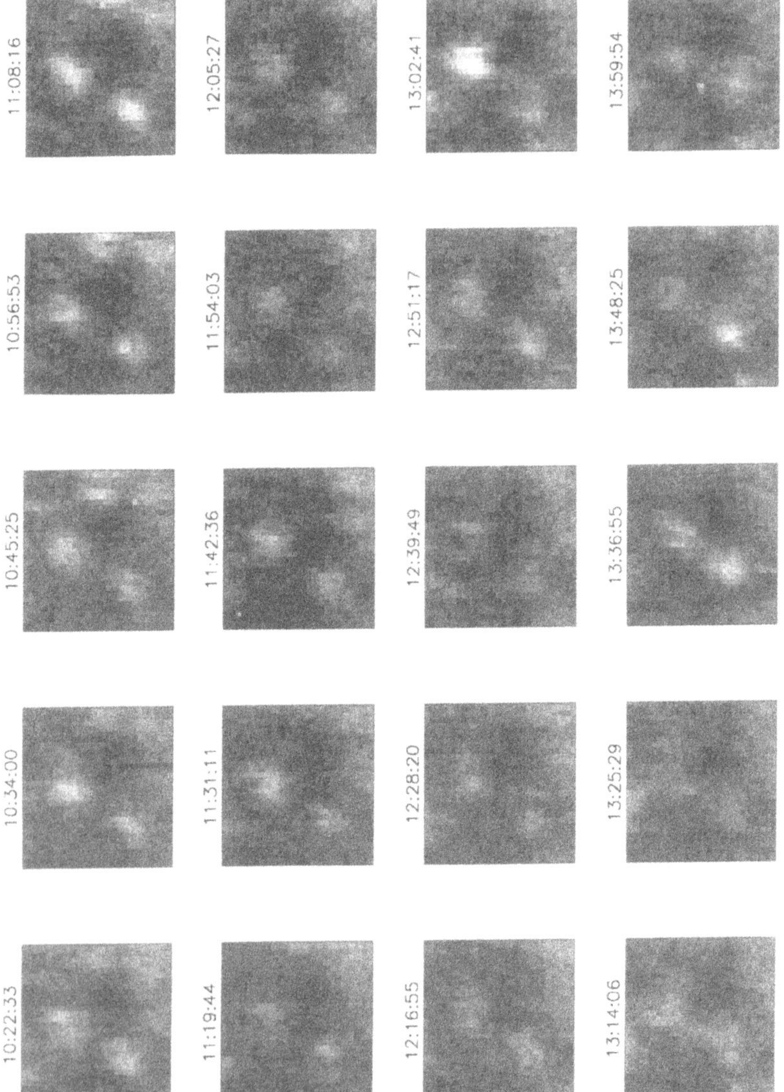

Figure 6c. Time series of Ne VI 562.83 Å rasters taken on 1 April, 1996 by CDS, using BOUND observing sequence.

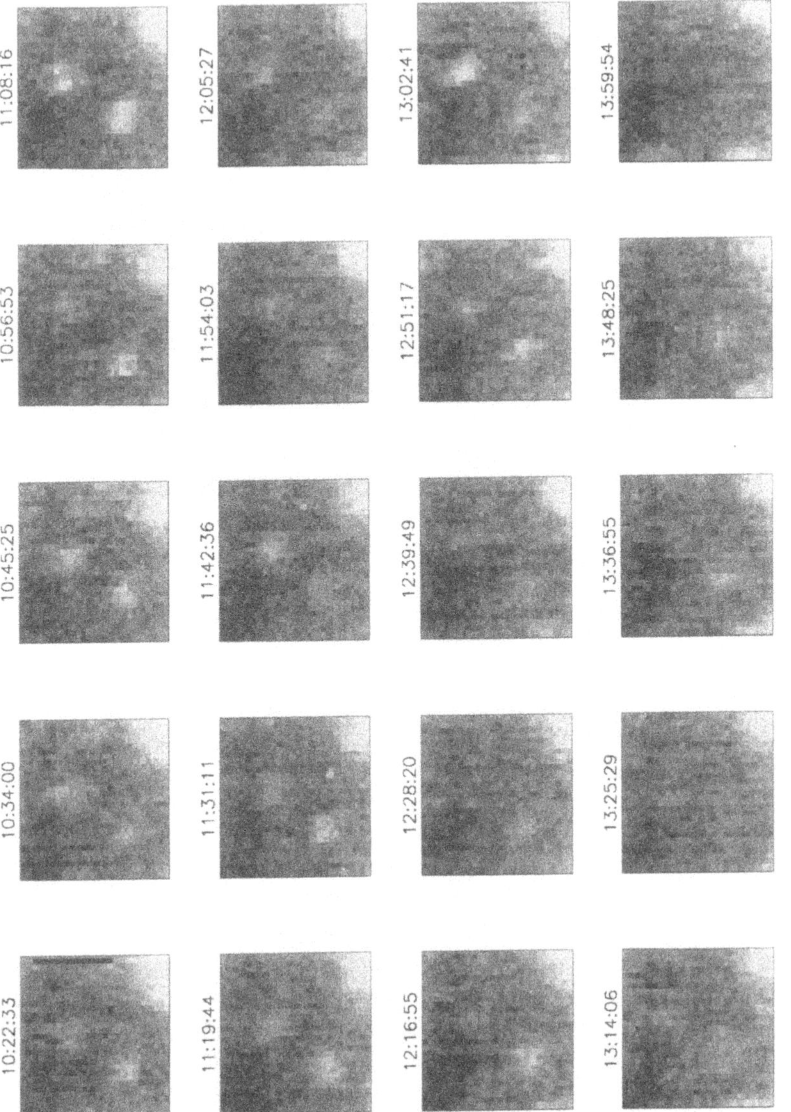

Figure 6d. Time series of Mg IX 368.06 Å rasters taken on 1 April, 1996 by CDS, using BOUND observing sequence.

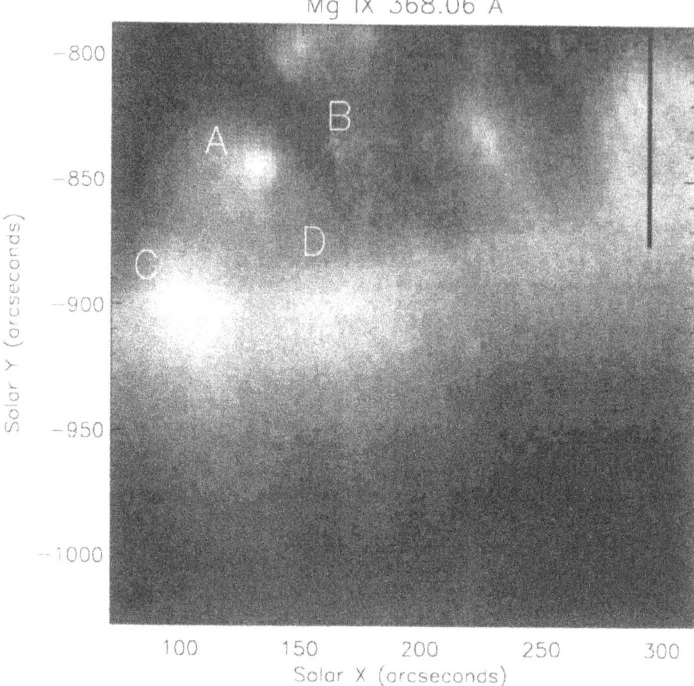

Figure 7. CHSTR observation in Mg IX 368.06 Å, 1 April, 1996. Four regions of interest are marked as A, B, C, and D.

et al., 1994). It is difficult to see how such features can be bright in one line and invisible in another but it may be due to an extremely localized plasma heating, with a well-defined, single temperature. Clearly, further analysis of such structures is required.

In Figure 8 we show a north–south cut through the CHSTR data at 201.1 arc sec west of the central meridian. This figure clearly shows that the emission in the hot Fe ions is much lower than for the other ions and is close to the background level. The ratio of actual emission to background is about 1.5 to 2 in the case of the Fe ions, but greater than 10 for the O III, NeVI, and He I ions. The profiles shown also differ, with the O III and NeVI showing a strong degree of limb-brightening and the Mg IX/X profiles showing a radial decrease in intensity above the limb, which is located at approximately pixel 150 in the profiles. The He I profile shows the strongest emission, with nearly 1500 counts pixel^{-1}. The Fe and Si profiles are all very similar, with a definite step in intensity at the limb, and a constant intensity on the disc.

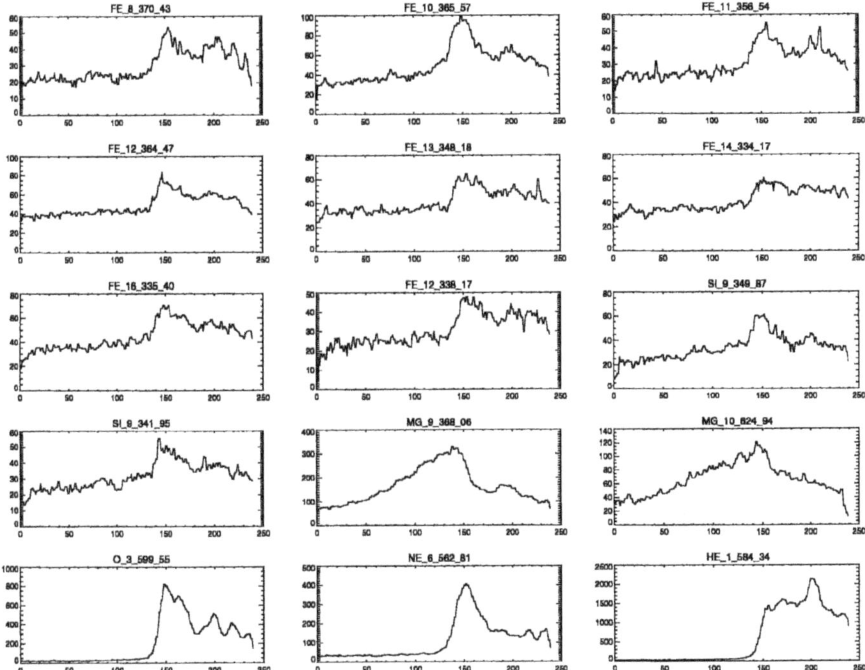

Figure 8. Intensity plot of CHSTR observations, from a north–south slice at 201.1 arc sec east–west.

3.2. Coronal Hole Evolution

The evolution of structures within a coronal hole and the evolution of the coronal hole itself is not well understood. These CDS observations provide the opportunity to study the evolution of coronal hole structure on short time scales and at high spatial and spectral resolutions, with the observations being made at multiple wavelengths simultaneously. Figures 6(a–d) show the evolution of bright regions within a coronal hole, over approximately three hours and in four different wavelengths. These two bright regions are marked as 1 and 2 in Figure 9. The relationship between how the intensity varies in each of the wavelengths for each of these bright regions (region 1 and region 2) is shown in Figures 10(a) and 10(b).

In Figures 10(a) and 10(b) it is possible to see that the intensity variations in each of the four wavelengths are roughly in phase. In particular, in Figure 10(a), the brightening that occurs at approximately 13.02, and in Figure 10(b), the brightening that occurs at 11.08, seem to be in phase. The intensity profiles vary for the brightening in each of the wavelengths, especially in Figure 10(b), and this must be in some related to how and where the energy is being dissipated. It is possible to see, however, that there is an increase of approximately 50 to 100% in the intensity

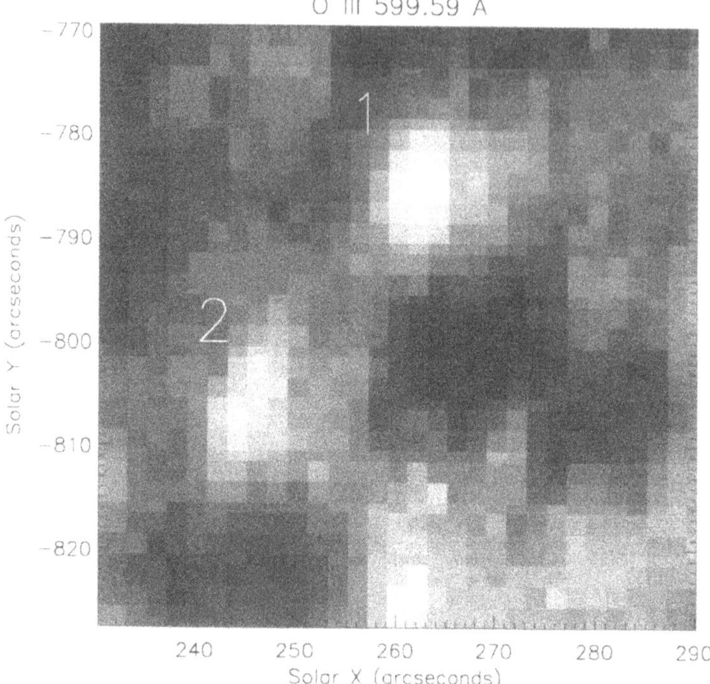

Figure 9. Single raster from BOUND phase 2, showing the two bright regions (1 and 2) and the chromospheric network cells.

during the brightening of these two regions. It is also interesting to note that these brightenings seem to occur at the junctions of the chromospheric network cells, and therefore and probably related to energy dissipation in the chromospheric network. The brightenings in regions 1 and 2 do not seem to be related to one another.

These are the first observations of rapid variations in a coronal hole in the EUV wavelength range. The fact that these bright point which vary in intensity seem to lie at the junctions of chromospheric network cells implies that they are the signature of some periodic energy dissipation, such as magnetic reconnection, caused by fluctuations in the chromospheric network cells themselves.

4. Summary

Initial observations of coronal hole structure and evolution by SOHO-CDS have revealed several new, interesting results:

– We have identified rapid intensity variations in features of scale size 6–10 arc sec within coronal holes, associated with the junctions of chromospheric network boundaries.

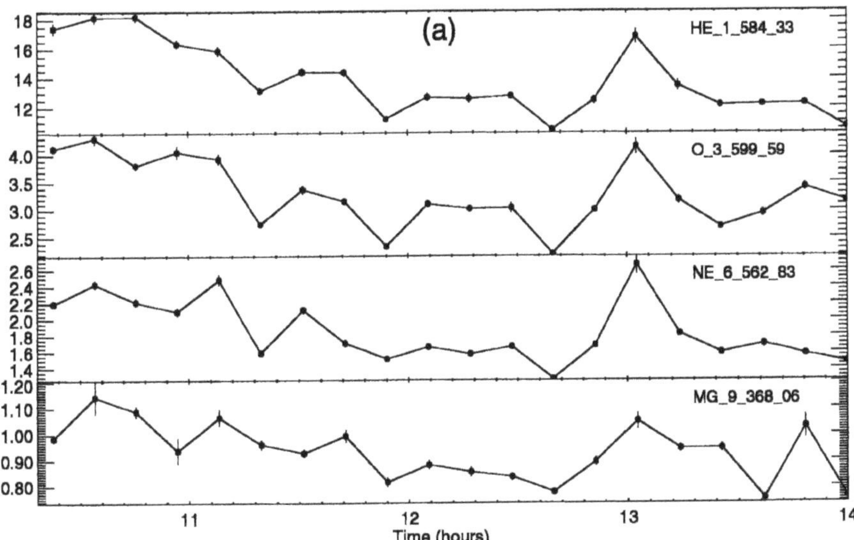

Figure 10a. Intensity plot of region 1, showing how the intensity varies over a five by five arcsecond area around the bright region during the BOUND phase 2 observations, for the four brighter emission lines. The points are the average intensity over the area and the error bars are the standard error.

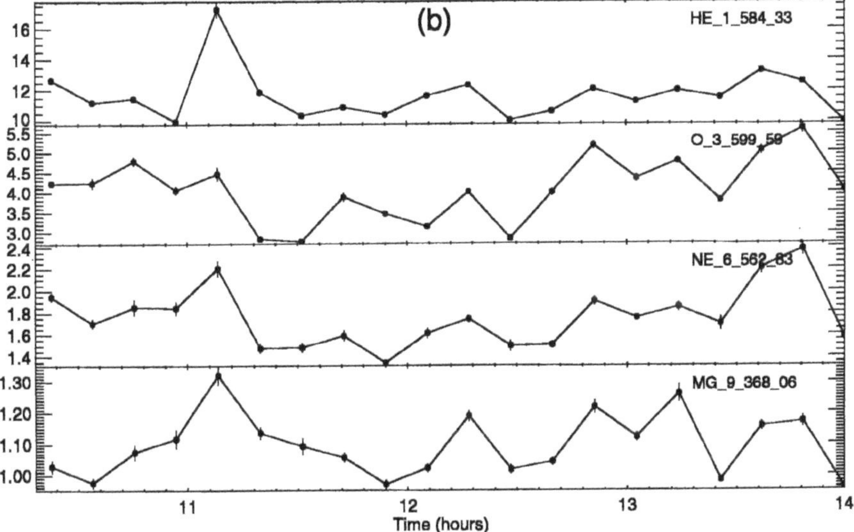

Figure 10b. Intensity plot of region 2, showing how the intensity varies over a five by 5 arc sec area around the bright region during the BOUND phase 2 observations, for the four brighter emission lines.

– We have shown that small bright patches in coronal holes show widely varying emission characteristics with wavelength.

– We have identified bright patches at the base of polar plumes, which appear as bright points.

– We have identified basic differences between the quiet Sun and coronal hole EUV spectra as identified by CDS.

These preliminary observations are the forerunner of further coronal hole studies. The results here have already been used to refine the CDS coronal hole observing sequences ion order to build on this work and produce a better analysis of the features discussed here. The data have served to indicate the potential for such work using CDS observations.

Acknowledgements

JEI acknowledges the financial support of a PPARC postdoctoral research assistantship. RAH thanks Prof. D. J. Southwood for a visiting lectureship at Imperial College.

References

Bohlin, J. D.: 1977, in J. B. Zirker (ed.), *Coronal Holes and High-Speed Wind Streams*, Ch. 2, Colorado Associated Press, Boulder, p. 27.

Bohlin, J. D. and Sheeley, N. R.: 1978, *Solar Phys.* **56**, 125.

Feldman, Purcell, and Dohne: 1992, *An Atlas of Extreme Ultraviolet Spectroheliograms from 170–625 Ångstroms*, NRL.

Harrison, R. A., Fludra, A., Pike, C. D., Payne, J., Thompson, W. T., Poland, A. I., Breeveld, A. A., Culhane, J. L., Kjeldseth-Moe, O., Huber, M. C. E., and Aschenbach, B.: 1997, *Solar Phys.* **170**, 123.

Krieger, A. S., Timothy, A. F., and Roelef, E. C.: 1973, *Solar Phys.* **29**, 505.

Lang, J., Mason, H. E., and McWhirter, R. W. P.: 1990, *Solar Phys.* **129**, 31.

Leer, E. and Holzer, T. E.: 1985, *Future Missions in Solar, Heliospheric Space Plasma Physics*, Session 1, p. 3.

Malinovsky, M. and Héroux, L.: 1973, *Astrophys. J.* **181**, 1009.

Maran, S. P. and Thomas, R. J.: 1973, *Sky Telesc.* **45**.

Mason, H. E. and Bhatia, A. K.: 1978, *Monthly Notices Roy. Astron. Soc.* **184**, 423.

Nolte, J. T., Krieger, A. S., Timothy, A. F., Gold, R. E., Roelef, E. C., Vaiana, G., Lazarus, A. J., Sullivan. J. D., and McIntosh, P. S.: 1976, *Solar Phys.* **46**, 303.

Shibata, K., Nitta, N, Strong, K. T., Matsumoto, R., Yokoyama, T., Hirayama, T., Hudson, H., and Ogawara, Y.: 1994, *Astrophys. J.* **431**, L51.

Vernazza, J. E. and Reeves, E. M.: 1978, *Astrophys. J. Suppl.* **37**, 485.

Wahlstrøm, C. and Carlsson, M.: 1994, *Astrophys. J.* **443**, 417.

Zirker, J. B. (ed.): 1977, *Coronal Holes and High-Speed Wind Streams*, Ch. 1, Colorado Associated Press, Boulder, p. 1.

EUV OBSERVATIONS OF A MACROSPICULE: EVIDENCE FOR SOLAR WIND ACCELERATION?

C. D. PIKE and R. A. HARRISON
Space Science Department, Rutherford Appleton Laboratory, Chilton, Didcot Oxfordshire OX11 0QX, U.K.

(Received 27 February 1997; accepted 28 May 1997)

Abstract. We present a unique observation of a macrospicule, recorded in extreme ultraviolet light on 11 April 1996, using the Coronal Diagnostic Spectrometer (CDS) on board the Solar and Heliospheric Observatory (SOHO). The observation was made by chance as part of a daily, large-area mapping sequence. Although the feature has some characteristics of the class of events which have become known as X-ray jets, we argue that the feature observed here is a macrospicule. This being the case, the observation demonstrates several new features of macrospicule observation. Emission is detected from the macrospicule to temperatures of 1 million degrees. In addition, some footpoint structure is detected at the root of the macrospicule, and the edges or sides of the macrospicule appear brighter than the central regions. A velocity analysis shows high speed flows within the macrospicule. Velocities are seen to increase with altitude until a plateau is achieved. Coincident with this, there is evidence for emission line narrowing. The significance of these observations for solar wind acceleration processes is discussed.

1. Introduction

Bohlin *et al.* (1975) announced the discovery of the macrospicule two decades ago. This phenomenon was identified from the *Skylab* slitless spectrograph data in the He II 304 Å line, and was manifested by a column of cool material extending into the corona, apparently rooted only in coronal holes. Bohlin *et al.* specified macrospicule lengths of between 5–60 arc sec, widths of between 5–30 arc sec, and lifetimes of 5–40 min. Whilst they found significant numbers of macrospicules in He II 304 Å radiation, they found only a few which had coincident radiation in Ne VII. Thus they suggested that whilst the plasma was predominantly at about 50 000 K (temperature of formation of He II line), temperatures generally are less than 300 000 K (temperature of formation of Ne VII line).

Habbal and Gonzalez (1991) reported on the first macrospicule observation in radio wavelengths – specifically at 4.8 GHz. Their interpretation of the data suggested that a macrospicule had a cool core, at 8000 K, surrounded by a thin sheath at 1–200 000 K. They reported also on ballooning at the top of the macrospicule and the possibility of pinching off, resulting in possible outward motion of plasmoids from the upper portions of a macrospicule.

Karovska and Habbal (1994) reanalysed the *Skylab* data using enhanced image processing. Specifically, they looked at the C III 977 Å line from the Harvard Spectroheliometer. This observation had a 1.5-min time resolution with a spatial element size of 2.5 × 5 arc sec. The temperature of formation of C III is 74 000 K.

Solar Physics **175:** 457–465, 1997.

The analysis revealed that the macrospicules contained column-like structures with low-lying arcades at the base.

Many questions remain with regard to macrospicule formation and evolution. How frequent are they? Are they limited to coronal holes? We know little about plasma characteristics in macrospicules – what are the density, temperature and flow properties? Until we can explore such questions we will not know the significance of macrospicules to coronal hole structure and evolution, and to the solar wind flow.

2. The Observations

We report here on a single observation of a macrospicule made by chance during a large-scale scan of the Sun. We note that a coordinated Solar and Heliospheric Observatory (SOHO) Joint Observing Programme led by Shadia Habbal will most likely result in a thorough multiwavelength analysis of this phenomenon, but it is quite possible that the observation reported here may be not only supplementary to the Habbal work, but may be more complete in its extreme ultraviolet (EUV) coverage.

The observation was made on 11 April 1996 using the Coronal Diagnostic Spectrometer (CDS) on board SOHO. This is a double EUV spectrometer operating in the range 151–785 Å. CDS is described in detail by Harrison *et al.* (1995) and first results have been reported which illustrate the operation of the instrument and the nature of the observations (Harrison *et al.*, 1997; Mason *et al.*, 1997; Brekke *et al.*, 1997). The EUV region of the solar spectrum is rich in emission lines from trace elements in the Sun's atmosphere, which can be used to determine plasma properties through analyses of line intensities and intensity ratios, and line profiles, for plasmas in the range 10 000 K to over 2 million K. Thus, we have a very powerful tool for probing the nature of the solar atmosphere.

For the observation in question, the instrument was being used to scan the solar central meridian from north to south pole using a mosaic of 4×4 arc min 'rastered' images with a 2×240 arc sec slit. Using the normal incidence component of the CDS instrument one is able to image along the slit, thus, by scanning the solar image across the slit in 2 arc sec steps one can build up images rapidly, simultaneously in any wavelength ranges which have been selected for transmission to the ground. For this particular operation only four wavelength 'windows' were selected, corresponding to carefully selected emission lines from ions of helium, oxygen, magnesium, and iron.

The macrospicule was seen in the last 4×4 arc min image of the scan, i.e., that taken on the southern polar region. The images, for He I 584.33 Å, O V 629.73 Å, and Mg IX 368.06 Å are shown in Figure 1. Remember that these images were taken simultaneously on the same region of the Sun. The three represent temperatures of

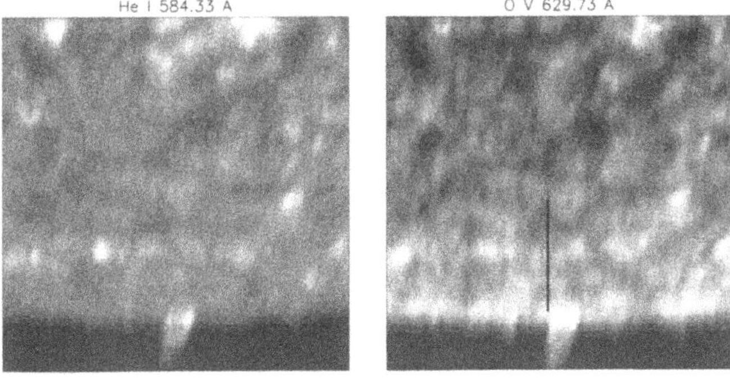

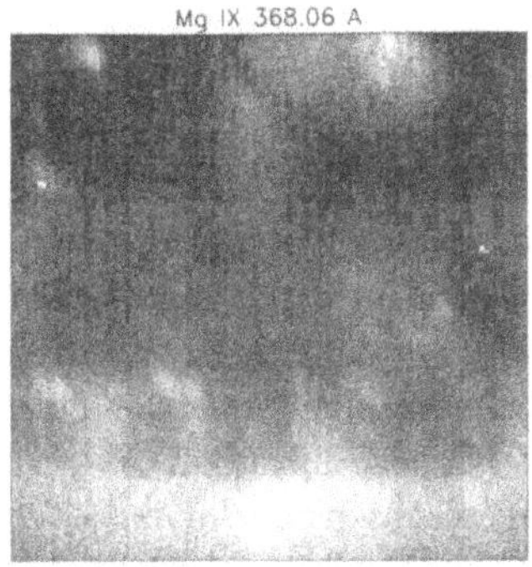

Figure 1. A 4 × 4 arc min CDS rastered image of the south polar region made on 11 April 1996, in light from He I (20 000 K), O V (250 000 K) and Mg IX (1 million K). The macrospicule can be seen in the southern extreme of the image.

20 000 K, 250 000 K, and 1 000 000 K – the formation temperatures of He I, O V, and Mg IX.

The macrospicule is clearly visible off the limb at the bottom of the image. It appears to be rooted just inside the limb and extends to the edge of the image at least. Thus, we can put a minimum altitude of 42 arc sec or 31 000 km. The maximum width appears to be 18 arc sec or 13 300 km. These values are well within the ranges found by Bohlin *et al.* (1975). However, if Bohlin *et al.* were correct about the durations of these events (5–40 min), we were very lucky to detect it at all! It should be noted that since the images shown in Figure 1 are built up by rastering in the E–W direction, the E–W dimension of the macrospicule was

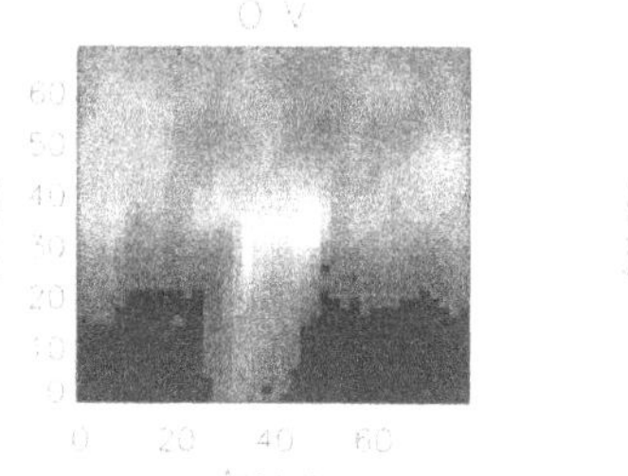

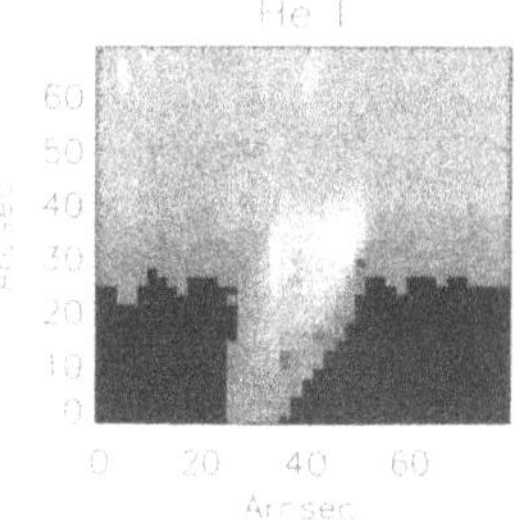

Figure 2. An expansion of the area covering the macrospicule, for the O V and He I images.

sampled progressively in time. The total time taken to raster across the macrospicule from west to east was approximately four minutes.

In the He I image (20 000 K) we see what appears to be a non-radial 'ministreamer'. At the root of the feature we see two 'footpoints' which appear bright. The western footpoint is brighter. Are these really footpoints, perhaps of low-lying loop structures like those suggested by Karovska and Habbal (1994)? There is some evidence that the sides of the macrospicule are brighter than the central regions, even into the corona. Whilst recognising that we do not have a time series, we do not see any sign of fragmentation or ballooning along the length of the macrospicule, which may have been evidence of the pinching off or ballooning at higher altitudes in the macrospicule, as reported for the radio data (Habbal and Gonzalez, 1991).

The O V image (250 000 K) shows the macrospicule to be the same basic form, but the most striking difference is the bright root. The macrospicule shows extremely bright O V emission at its base. In addition, as with the He I data, one can see that there is structure in the macrospicule in that the core is weaker than the sides.

The Mg IX image (1 million K) shows a significant new result. Although the image is noisy due to low counting statistics, one can clearly see a bright region at these temperatures above the limb encompassing the area of the macrospicule. Thus, despite the limit of 300 000 K discussed by Bohlin *et al.* (1975), we detect million degree emission.

Given the He I, O V, and Mg IX images it is hard to support or deny the view that the macrospicule is a cold sheath surrounded by hot layers. However, it clearly is a 'cool' feature reaching into the 'hot' corona, so such an interpretation must be at least partly true.

Figure 2 shows an expansion of the macrospicule region in He I and O V with the colour table set to best view the footpoint region. One can see clearly that both show evidence for two bright patches on either side of the root of the macrospicule.

CDS provides spectral information. The selected wavelength 'windows' in this case provide 1.44 Å across the Mg IX 368 Å line, and 2.52 Å across the O V and He I lines. These are sufficient to examine the line profiles and to determine suitable line fits.

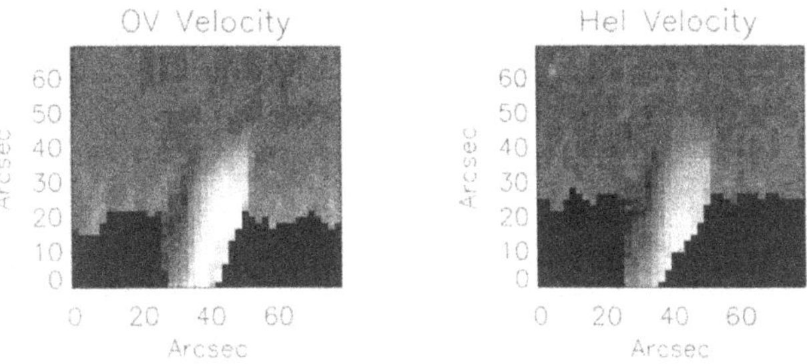

Figure 3. A velocity map of the macrospicule in O V and He I.

Figure 3 shows velocity maps for the O V and He I emission based on the centroids of fitted Gaussians at each pixel. Areas with insufficient count rates to produce a good fit are shown black. Since CDS does not have an intrinsic wavelength calibration, relative velocities have been calculated using those seen on the disk as the zero point. The velocities are shown from black to white. The brightest regions of the O V and He I maps are 150 and 200 km s^{-1}, respectively. These are red-shifted velocities, i.e., they are directed away from the observer. Indeed, the emission from the body of the macrospicule shows predominantly red-shifted emission, though the eastern side shows much smaller velocities. Some small blue-shifts are detected from the eastern edges of the feature.

Since these are from Doppler shifts, we are examining line-of-sight velocities and this must be borne in mind when one considers these velocities which are within a feature which is near the plane of the sky. Interpretation of what we are seeing depends critically on what geometry we accept.

Since the macrospicule appears to be rooted this side of the limb, and is clearly non-radial, we believe that we are seeing the red-shifted component of a fast plasma flow along the macrospicule, which is apparently concentrated to one side.

In order to explore the existence of any velocity changes with height above the footpoints, we have averaged the velocities from the western boundary pixels and plotted them as a function of distance from the footpoint. These data are shown in Figure 4. There is clear evidence for an acceleration of the plasma to a distance of 20 to 25 arc sec. Above that altitude both the He I and O V velocities tend to a constant apparent velocity, namely 200 km s^{-1} for the He I emission and 150 km s^{-1} for the O V emission. The increase in speed is about a factor of two over the 20–25 arc sec, or a gradient of 6.667×10^{-3} km s^{-1} per km in altitude. The error bar is less than ± 10 km s^{-1}.

Also plotted in Figure 4 are the line widths, for the same points, derived from the line profiles. Although not so pronounced, these data suggest that the profiles narrow by some 10% over the region of the acceleration, and shows signs of becoming constant when the velocity curve flattens.

Our belief is that we are witnessing a systematic outward velocity increase with altitude, consistent with plasma acceleration. Since the macrospicule is undoubtedly

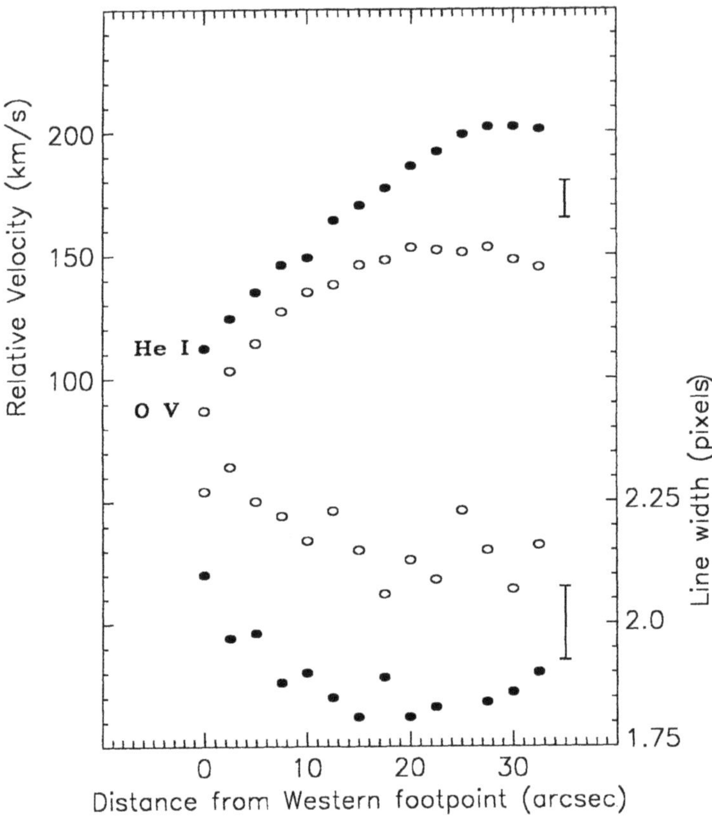

Figure 4. A plot of the He I and O V Doppler velocities against altitude. Also plotted are the line widths for the same data.

non-radial in nature, we are viewing the component of the outward velocity which is in the line of sight.

If the macrospicule was in the plane of the sky we would not see an outward flow; the increase in apparent velocity would be hard to explain. If we were looking at a loop structure, rooted at the two footpoints identified in Figure 2, we might expect to see an apparent velocity increase due to the curvature of the loop, with a minimum velocity at the footpoints, and a maximum at the top of the loop. In our opinion this geometry is not appropriate. For one thing, the acceleration is seen to altitudes of 20–25 arc sec, and the highest velocity is traced to the edge of the image, which would suggest a closed loop structure emitting at up to 1 million degrees, extending up to 40 arc sec above a coronal hole region. In addition, one would expect a smooth, almost symmetrical velocity map as one moves from east to west, with the footpoint velocities similar, and the two sides of the loop (macrospicule) showing similar velocities with altitude. This is very different from what we do see.

So, our interpretation is that we see a non-radial open structure, perhaps overlying a low-altitude loop system identified by the footpoints, and that we witness acceleration of the plasma to an altitude of 20–25 arc sec.

Clearly, a simultaneous observation of clear plasma acceleration and emission line narrowing has exciting consequences for our investigations of solar wind acceleration processes.

3. A Macrospicule or an X-ray Jet?

Before summarising the new results, we would like to discuss the nature of this event. We have discussed the compatibility of the observed features with the macrospicule events seen using *Skylab* data. This is primarily on the basis that we have detected an event which is a bright coronal feature emitting at temperatures characterised by He I , from a coronal hole region. However, Shibata *et al.* (1992) reported on the discovery of 'X-ray jets' from time-series observations using the Soft X-ray Telescope on board *Yohkoh* and we must establish whether the 11 April event could be such an event.

Shibata *et al.*'s study was extended to a statistical study of 100 events by Shimojo *et al.* (1996). They showed that the X-ray jets were typically associated with small flare events (sub-flares or microflares) in active regions or bright points. The bright footpoints of the 11 April event are clearly not associated with an active region. The existence of a 'pre-existing' bright point is difficult to establish and any transient activity (i.e., subflare brightening) at the base of the event cannot be checked with available data.

The upside-down 'Y' shape seen in the event under discussion here is consistent with one class of jet events listed by Shimojo *et al.*, which encompassed 13% of all jets.

Shimojo *et al.* noted (their Table 1) that about 10% of their jets occurred in coronal holes, but all but two were low latitude hole regions (less than 44 deg latitude). The two exceptions were at N75 and N64. They showed also that the jet lengths averaged at 15×10^4 km but could be several tens of 10^4 km, and that the widths averaged at 18×10^3 km. The 11 April event has an altitude of over 3.1×10^4 km and a width of 13.3×10^3 km. This would be a small X-ray jet and, as discussed above, is typical of macrospicule scales.

The Shimojo and Shibata studies list jet velocities, but these are apparent velocities determined by the travel time of intensity 'fronts'. The *Yohkoh* data do not supply spectral information for Doppler analyses.

Shimojo *et al.* discuss the comparison to Hα observation and note that Hα surges are found in association with many jets, though this work is in its infancy. They also note that there are other jet-like events which could be related to the

X-ray jets, and even note that macrospicules may be closely associated. Again, it is too early to confirm any relationship.

Thus, we note that the characteristics of our EUV observations are consistent with aspects of the X-ray jet observations. However, on the basis of detecting He I radiation from an extended structure in a polar coronal hole region, with dimensions consistent with those reported by Bohlin *et al.* (1975), we feel that at this stage we must class our event as a macrospicule and await further multi-wavelength analyses to identify any relationships between the two classes of event.

4. Discussion

A serendipitous observation of a macrospicule in EUV has allowed us to extend our understanding of macrospicules – at least, as displayed by one event.

Our major new findings can be summarized as below:

(1) Emission from the macrospicule is not only detected in radiation from He I and O V (20 000 and 250 000 K) but also in Mg IX which is characteristic of 1 million K.

(2) The He I and O V emissions display two bright patches at either side of the base of the structure, suggestive of magnetic footpoints, and display a 'ministreamer'-like structure with the outer edges or sides of the macrospicule being brighter than the interior regions, as it extends into the corona.

(3) The flow patterns in the macrospicule have been examined. Relative apparent velocities of up to 200 km s^{-1} are detected in He I, which we interpret as outward flows from the macrospicule. These flows are seen from the foot of the macrospicule and are concentrated on the western side of the macrospicule structure. If these are line of sight velocities due to plasma flow within the macrospicule then the geometrical considerations suggest that extremely high wind speeds may be generated in such features.

(4) We see strong evidence for plasma acceleration to a height of 20–25 arc sec, where a plateau is reached. Evidence for coincidental line narrowing is detected. This is consistent with the plasma being driven by the dissipation of waves.

Point number 3 has important implications for the high speed solar wind. We do not know the non-radial nature of the macrospicule out of the plane of the image, and so cannot calculate the actual velocities involved. However, if we assume that the macrospicule is rooted this side of the limb (which it appears to be) and that the macrospicule is tilted from the plane of the sky, and that the combined effect is a 20 deg tilt, then the 200 km s^{-1} velocity translates to an actual wind speed of 580 km s^{-1}.

It is clear that we should not put too much emphasis on one event. However, this observation does suggest that there are some intriguing questions and stresses the importance of examining these events in greater detail, in particular to assess their role in solar wind acceleration.

References

Bohlin, J. D., Vogel, S. N., Purcell, J. D., Sheeley, N. R., Tousey, R., and Van Hoosier, M. E.: 1975, *Astrophys. J.* **197**, L133.

Brekke, P., Kjeldseth-Moe, O., Brynildsen, N., Maltby, P., Haugan, S. V. H., Harrison, R. A., Thompson, W. T., and Pike, C. D.: 1997, *Solar Phys.* **170**, 163.

Habbal, S. R. and Gonzalez, R. D.: 1991, *Astrophys. J.* **376**, L25.

Harrison, R. A., Sawyer, E. C. and 37 co-authors: 1995, *Solar Phys.* **162**, 233.

Harrison, R. A., Fludra, A., Pike, C. D., Payne, J., Thompson, W. T., Poland, A. I., Breeveld, E. R., Breeveld, A. A., Culhane, J. L., Kjeldseth-Moe, O., Huber, M. C. E., and Aschenbach, B.: 1997, *Solar Phys.* **170**, 123.

Karovska, M. and Habbal, S. R.: 1994, *Astrophys. J.* **431**, L59.

Mason, H. E., Young, P. R., Pike, C. D., Harrison, R. A., Fludra, A., Bromage, B. J. I., and Del Zanna, G.: 1997, *Solar Phys.* **170**, 143.

Shimojo, M., Hashimoto, S., Shibata, K., Hirayama, T., Hudson, H. S., **44**, L173.

Shibata, K. and 14 co-authors: 1992, *Publ. Astron. Soc. Japan* and Acton, L. W.: 1996, *Publ. Astron. Soc. Japan* **48**, 123.

EUV BLINKERS: THE SIGNIFICANCE OF VARIATIONS IN THE EXTREME ULTRAVIOLET QUIET SUN

RICHARD A. HARRISON
Space Science Department, Rutherford Appleton Laboratory, Chilton, Didcot, Oxfordshire OX11 0QX, U.K.

(Received 8 January 1997; accepted 12 June 1997)

Abstract. A search for microflare activity in the extreme ultraviolet (EUV) quiet Sun using the Coronal Diagnostic Spectrometer (CDS) aboard the Solar and Heliospheric Observatory (SOHO) spacecraft has not resulted in the identification of microflare activity, but has resulted in the identification of a hitherto unknown phenomenon: enhancements of a factor of 2–3 in the flux of transition region lines at network junctions. A total of some 6 hours of observation of 5 different target areas showed this 'blinker' activity at each area, with durations ranging from 1 to 30 min and averaging 13 min, and thermal energy content of order 10^{-6} that of a 'standard' flare. Assuming that the observations are of typical quiet Sun, and projecting these data to predict a distribution of these events over the entire Sun, the total thermal energy content of these 'blinkers' is insignificant when compared to the energy required to heat the corona. The nature of these events and their significance are discussed in this paper.

1. Introduction

Among the principal goals of the Solar and Heliospheric Observatory (SOHO) mission are an understanding of the heating of the corona and the acceleration of the solar wind. To this end much observational effort is being placed on the observation of small scale, globally distributed activity such as explosive events, jets and micro/nano flares.

The Coronal Diagnostic Spectrometer (CDS) is a double spectrometer operating in the 150–800 Å region, allowing the production of images and spectra from emissions from trace elements in the Sun's atmosphere that can be used to probe the different temperature regimes of the corona and transition region. CDS is described in detail by Harrison *et al.* (1995) and first results papers allow one to view the basic operation of the instrument and the nature of the observations being made (Harrison *et al.*, 1997; Brekke *et al.*, 1997; Mason *et al.*, 1997).

Here, we describe an observation sequence which was run a number of times in November of 1996. Initially designed to detect microflares, this sequence has detected transient brightenings in lines from ions at transition region temperatures in the quiet Sun.

Solar Physics **175:** 467–485, 1997.

2. The Observing Sequence

The observations reported in this paper make use of the CDS sequence called 'FLARE'. All CDS scientific Studies are given a 5 to 8 letter acronym for identification purposes. The FLARE sequence was designed primarily to identify and examine microflare activity and rapid intensity variations. Thus, the sequence consists of relatively small, fast rastered images in a small selection of emission lines which represent significantly different temperature regimes. Telemetry limitations dictate that we cannot return large spectral and spatial ranges when wishing to study rapid variations.

The normal incidence system of CDS is used and the narrow slit, 2×240 arc sec, was chosen in order to ensure the best spectral and spatial resolution. Only 60 arc sec of data along the slit is returned (35 pixels of 1.68 arc sec each) in order to keep a fast cadence time. Using a scan mirror, the image of the Sun is rastered across the slit, giving 5 locations of exposure time 2 s each. Thus, we build up an image of 10 arc sec by 60 arc sec.

The total duration of a single raster is 43 s. This includes any instrumental and telemetry overheads and thus represents the cadence time of the rastered images shown in this work.

The FLARE sequence was run 6 times on 5 and 6 November 1996. Five runs were for 100 rasters over 1 hr 12 min. A sixth run was truncated due to operational schedules.

Data from four emission lines are returned. These are for the lines of He I 537.03 Å, O IV 554.52 Å, Mg IX 368.06 Å, and Fe XIV 334.17 Å. Data from 11 pixels across the line profiles are returned in order to examine the line shapes. For the Mg IX and Fe XIV lines, 1 pixel represents 0.08 Å; for the other two lines 1 pixel represents 0.14 Å (Harrison *et al.*, 1995). For the quiet Sun observations reported here, the Fe XIV observations, which represent plasmas at 1.6 million K, show very weak emission. For this reason we continue the analysis considering only the He I, O V, and Mg IX data.

Ionization balance distributions suggest (Arnaud and Rothenflug, 1985) that the Mg IX emission comes predominantly from plasmas at 1 million K, that the O IV emission is due to plasmas at about 160 000 K ($\log T = 5.2$), and the He I emission is from plasmas at temperatures of about 20 000 K ($\log T = 4.3$) and below. The distributions are fairly sharply defined, so, these three lines are good indicators of structure and evolution in the corona (Mg), transition region (O), and chromosphere (He).

3. An Initial Inspection

To put the observations in context, we show first a typical quiet Sun region in radiation from He I 584.34 Å, O V 629.73 Å, and Mg IX 368.06 Å. This is shown

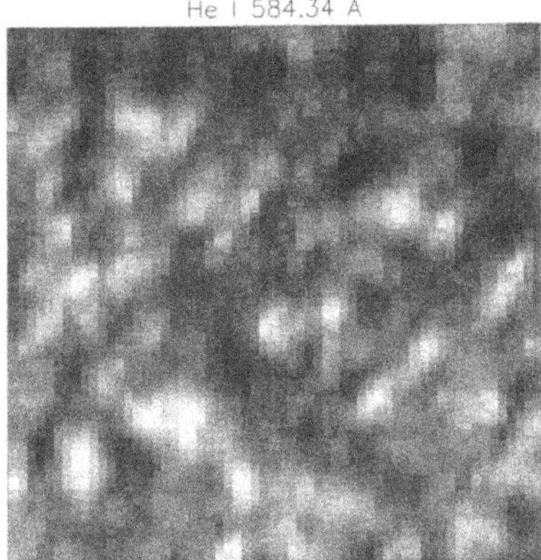

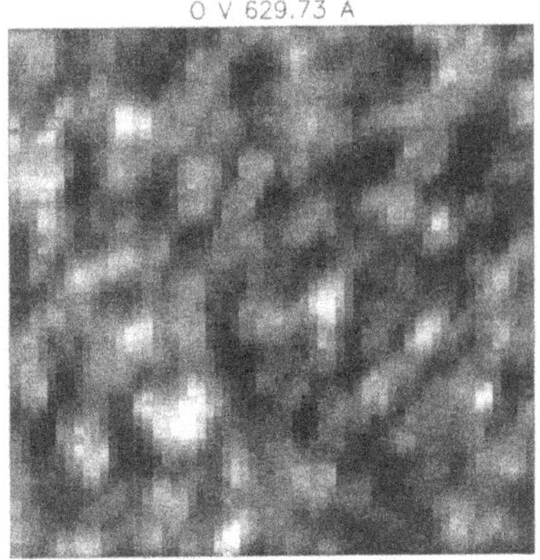

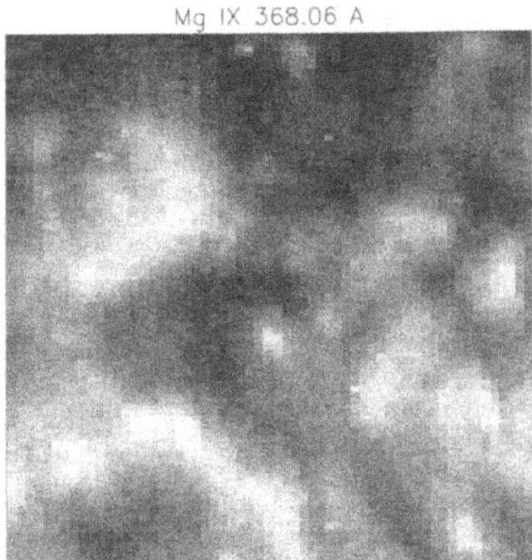

Figure 1. The supergranular network as seen using CDS observations in He I, O V, and Mg IX on 25 August 1996. The images are 4 × 4 arc min in size and are shown with north upwards and west to the right.

in Figure 1. The particular observation shown was made on 25 August 1996 and it is a raster over a 240 × 240 arc sec portion of the Sun at a quiet-Sun location.

In many ways, this is a typical CDS observation, taking 36 min to raster a narrow slit over a relatively large area whilst returning data from a carefully selected group of emission lines (we only show three). Such rasters are ideal for analysing the basic structure of a portion of the Sun, as well as evolution over time scales of more than several tens of minutes. However, any variations of time scales of order 1 min would be lost – and that is the subject of this paper.

Figure 1 shows clearly the supergranular network in the He I image, with 'typical' cell sizes of order 30–40 arc sec. Remember that this is from radiation predominantly less than 20 000 K. The same basic structure is seen in the O V image, which is from a transition region temperature of 250 000 K. Although the network structure is identical, the brightest He I and O V locations do not necessarily correspond. For example, the brightest patch in the top portion of the O V image (top left) does not correspond to a particularly bright He I patch, and the bright He I patch to the left of the brightest He I patch is colocated with a rather weak O V enhancement. Many other variations of this type can be seen. Indeed, the conclusions of this paper suggest that such differences reflect the variability of the oxygen, or transition region, emission in particular.

The million degree Mg IX emission does not 'see' the network pattern. Some of the brighter patches in He I and O V lie within bright regions of the Mg IX emission, but, in general, the structure is of a much larger scale size, and is in such images due to hot plasma trapped in loops and the structure of coronal holes.

We start the analysis with an inspection of one observation, that of 5 November 1996, CDS sequence number 5642. We will examine time varying emissions from the same temperature regimes as shown in Figure 1, but over a much reduced spatial area. This area reduction is necessary in order to make the rapid rastering possible.

Figure 2 shows a summed image of the first 40 images of run 5642 of 5 November, in order to see clearly the structure within which we see variations with time. The images are 10×60 arc sec, i.e., just one 96th of the area of the rasters of Figure 1 ($\frac{1}{4}$ the height and $\frac{1}{24}$ the width). Thus, the bright patches we see in Figure 2 are consistent with the network structure.

Again, we stress that these observations are for quiet Sun. Indeed, they are from Sun centre in this case.

One feature we note immediately is the difference between the images. The He I emission is dominated by a patch in the southern extreme of the image. O IV radiation is seen from the western side of the same patch, as well as from, a patch to the north. The Mg IX radiation bears little resemblance to the He I and O IV features and shows a bright region to the northern extreme.

Having established the basic layout of the features, we now examine the temporal variations.

The top panel of Figure 3 shows the He I 10×60 arc sec rasters for the first 40 images of run 5642. For each display in this format we show 40 images set out in two rows of 20 with the first image at the top left and the last image of the sequence at the bottom right. Thus, Figure 3 consists of three panels, each showing two rows of 10×60 arc sec images. The first raster was taken at 16:10:05 UT on 5 November, and the subsequent images were taken at 43-s intervals.

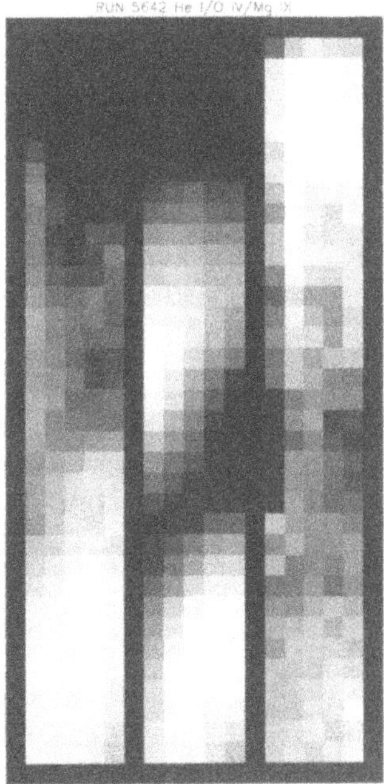

Figure 2. A sum of the first 40 rastered images of run 5642, on November 5, 1996, showing the principal bright regions in the He I 537 Å, O IV 554 Å, and Mg IX 368 Å lines. Each image is 5×35 pixels, i.e., 10×60 arc sec, in size.

Little significant variation is seen in the He I emission. Clearly the lower portion of the rastered area is brighter throughout the sequence (accounting for the bright patch in Figure 2), and the patch is brighter in the first half of the sequence. However, there is little variation and the emission is relatively weak; the statistical 'noise' can be seen throughout the images.

The middle panel of Figure 3 (middle two rows) shows a very different story for the O IV emission. Remember that these rasters were taken simultaneously with the He and Mg observations, so the alignment and timing is identical. Here we see a very significant bright patch to the south at the start of the sequence. It decays over two rasters and brightens significantly some 11–12 rasters (473–516 s) later. It decays again, slightly, and then displays a major brightening in the second half of the sequence, even with some extension to the southern extreme of the image. It decays as we head into the last 7–8 images. Another patch, just above the centre of the image shows a transient brightening about the 10th image in the sequence, and it brightens again between images 31–36 (903–1118 s later). These two 'active' patches show some enhanced background emission throughout the sequence.

The bottom panel of Figure 3 shows the Mg IX emission for the same sequence. We show this for completeness; there is very little emission of significance. One can just determine that the northern regions of the rastered images are brighter, and

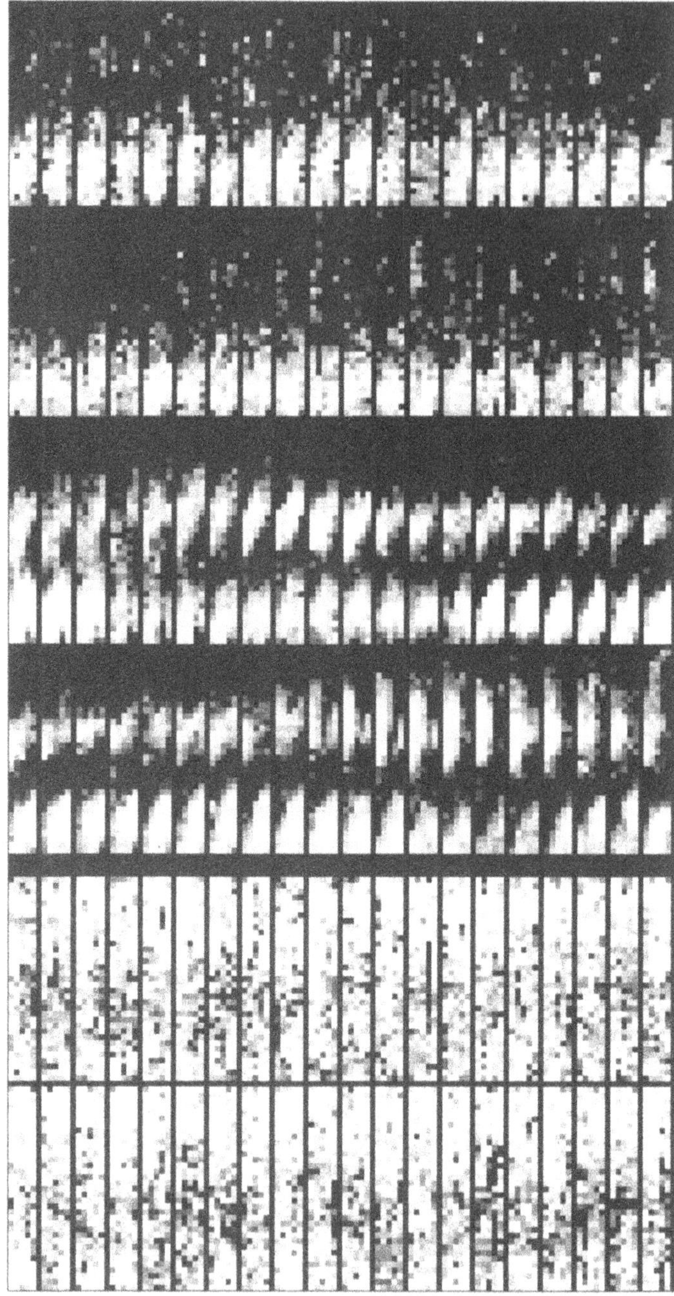

Figure 3. A mosaic of the first 40 rastered images of run 5642, showing (*top*) the He I 537 Å rasters, (*middle*) the O V 629 Å rasters, and (*bottom*) the Mg IX 368 Å rasters.

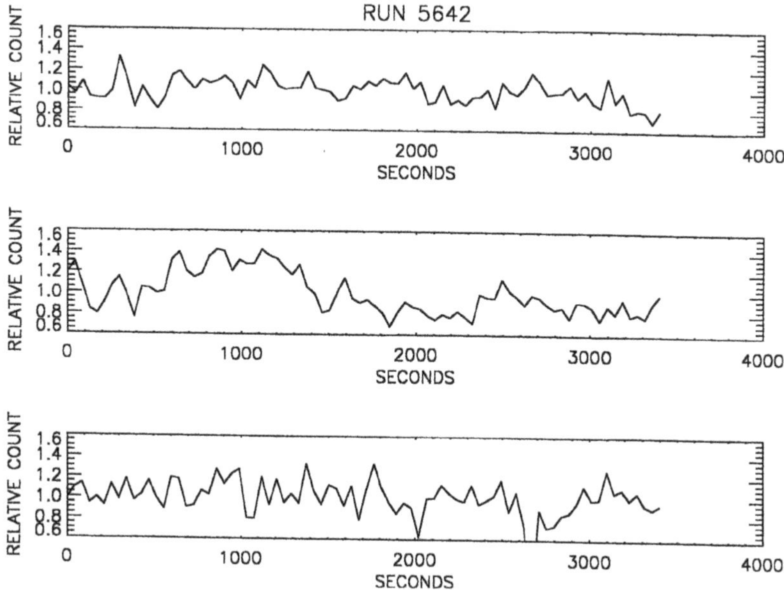

Figure 4. Three traces showing the intensity time history for the southern bright region (pixels (2,4), (3,4), (3,5), (4,4,), (4,5), and (4,6) where (0,0) is the bottom left pixel). Curves are shown for He I 537 Å (*top*), O IV 554 Å (*middle*), and Mg IX 368 Å (*bottom*). The abscissa is time, in seconds, starting at 16:10:05 UT on 5 November 1996. The ordinate is given in relative counts from the mean intensity for each emission line. For the He, O, and Mg data, the mean is 570, 1200, and 1450, respectively.

this is consistent with Figure 2, but there is little evidence for variations or for any emission linked with the He I and O IV features.

To examine the variations in greater detail we plot the time intensity curve for the southern patch. Figure 4 represents a time sequence of the summed intensity from pixels (2,4), (3,4), (3,5), (4,4), (4,5), and (4,6), where we define pixel (0,0) as being the lower left pixel.

Three curves are shown for He I (upper curve), O IV (middle), and Mg IX (lower). The intensity is given in 'relative counts', that is the ratio of the count to the mean count of the sequence for each emission line. The mean counts are 570, 1200, and 1450 counts, for the He I, O IV and Mg IX, respectivey and, to a good approximation, the error can be considered to be of order the root of the intensity count. All three curves show a scatter which is apparently greater than the anticipated 1σ statistical scatter (approximately 0.96–1.04) and this may be due to solar variation. However, the O IV curve shows by far the greatest intensity variation, displaying a range of a factor of two (0.7 to 1.4), and greater coherence in intensity between images. Thus, whereas the helium and magnesium curves are showing statistical 'noise' and, no doubt, some solar fluctuation, the oxygen curve is showing evidence for very significant intensity brightenings.

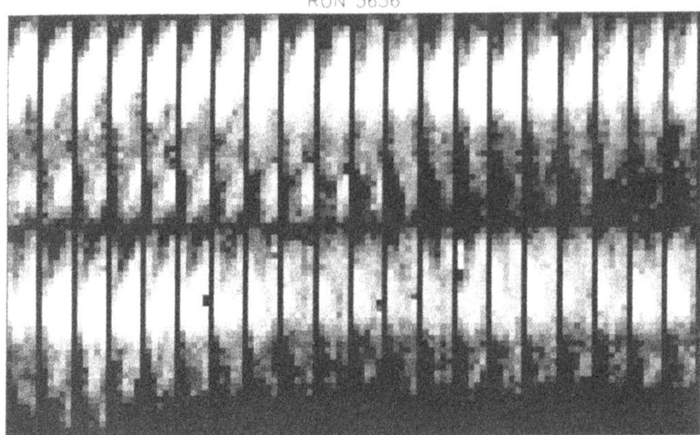

Figure 5. A mosaic of the first 40 rastered images of run 5656, showing the O IV 554 Å rasters.

The plots cover some 80 rasters, the first 40 of which were shown as images in Figure 3. The statistical noise can be seen in the helium and magnesium images as a mottled intensity distribution, and some variation in the brighter regions can be determined from image to image. However, the oxygen data clearly support the images of Figure 3, with the initial brightening followed by the longer bright event (between rasters 10–30).

4. Further Examples

Having 'set the scene' by showing typical quiet-Sun images and then identifying a significant variation in O IV emission with time, the object of this section is to show other examples and assess how common they are.

Figures 5–12 show O IV FLARE rasters and time-intensity profiles for four other observations, all on quiet Sun, shown in the same format as Figures 3 and 4. The CDS run numbers, dates and times are shown in Table I. In addition, columns 3–5 show some basic parameters of the blinker events seen in the traces shown in Figures 6, 8, 10, and 12. Note that this is not the full set of blinkers seen in the images of Figures 5, 7, 9, and 11; the parameters are listed for those events where the time-intensity curves are shown.

Column 3 in Table I is the time in seconds into the sequence at which the peak of the blinker intensity is achieved. Column 4 gives the ratio of the peak intensity to the non-event 'pedestal' intensity, and column 5 gives the duration of the blinker in seconds, measured from the point of the initial sustained rise in intensity to the point at which the intensity falls to previous levels. Clearly these parameters are approximate and subjective, but they do indicate some basic characteristics of the events in question.

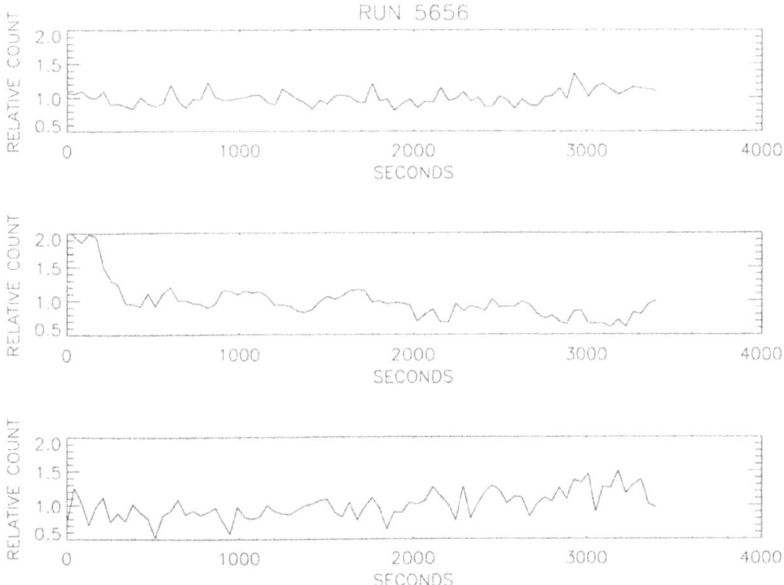

Figure 6. He I (*top*), O IV (*middle*), and Mg IX (*bottom*) traces for run 5656. The ordinate is given in relative counts from the mean intensity for each emission line. For the He, O and Mg data, the mean is 500, 1500, and 1400, respectively.

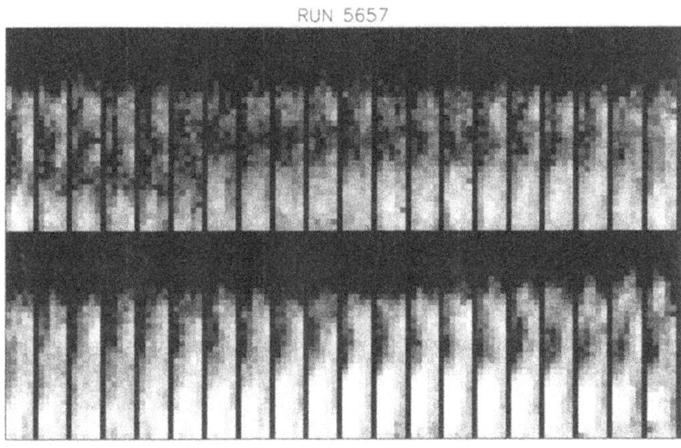

Figure 7. A mosaic of the first 40 rastered images of run 5657, showing the O IV 554 Å rasters.

Run 5656, illustrated in Figures 5 and 6, shows significant intensity variations in an area to the north of the centre of the rastered image, showing variations which include peaks in intensity at the start of the sequence, a third of the way through the sequence, and just into the second half of the sequence. The intensity profile is plotted for the 'core' pixels of the bright patch, namely (0,21), (0,22), (0,23), (1,22),

Figure 8. He I (*top*), O IV (*middle*), and Mg IX (*bottom*) traces for run 5657. The ordinate is given in relative counts from the mean intensity for each emission line. For the He, O, and Mg data, the mean is 450, 1500, and 2100, respectively.

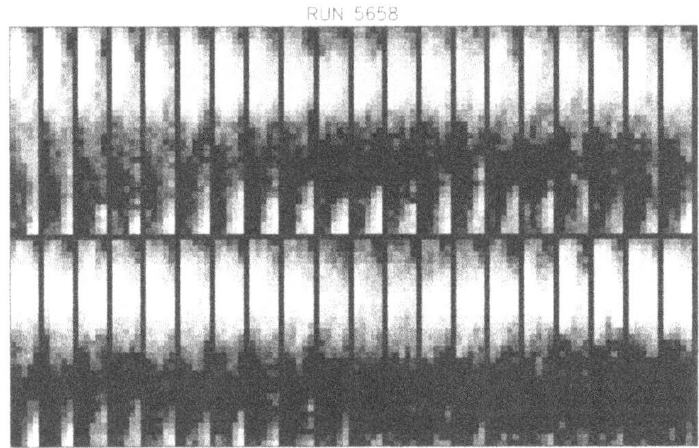

Figure 9. A mosaic of the first 40 rastered images of run 5658, showing the O IV 554 Å rasters.

(1,23), (1,24) using the same convention as above. We show this plot in order to compare the He I, O IV, and Mg IX intensity variations to see if the characteristics are similar to the first event studied. Indeed, the activity is similar. The He I (top panel) and Mg IX curves (bottom panel) show variations of about 20–25% yet the O IV intensity varies by a factor of almost three. This pixel group was the site of the

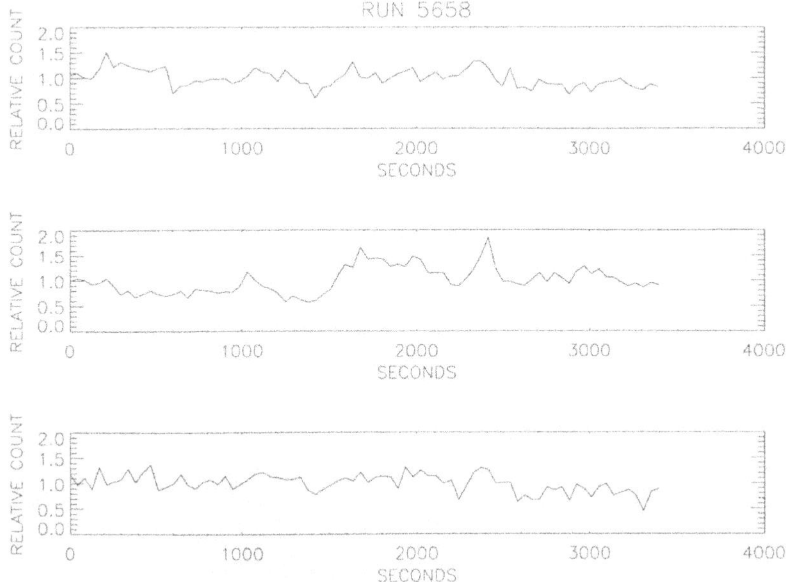

Figure 10. He I (*top*), O IV (*middle*), and Mg IX (*bottom*) traces for run 5658. The ordinate is given in relative counts from the mean intensity for each emission line. For the He, O, and Mg data, the mean is 500, 2400, and 1700, respectively.

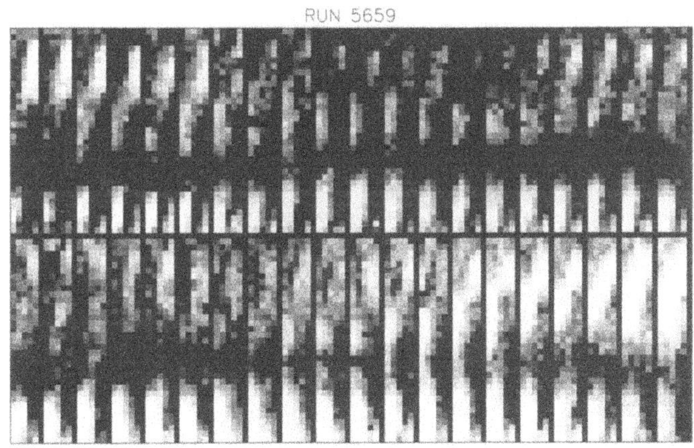

Figure 11. A mosaic of the first 40 rastered images of run 5659, showing the O IV 554 Å rasters.

very significant brightening at the start of the sequence, which decays over some 7–8 images.

Run 5657, illustrated in Figures 7 and 8, shows a similar story, but in this case the brightening takes place over 10–12 rasters in the second half of the sequence. Figure 8 shows this event well, and demonstrates how it relates to the lower and

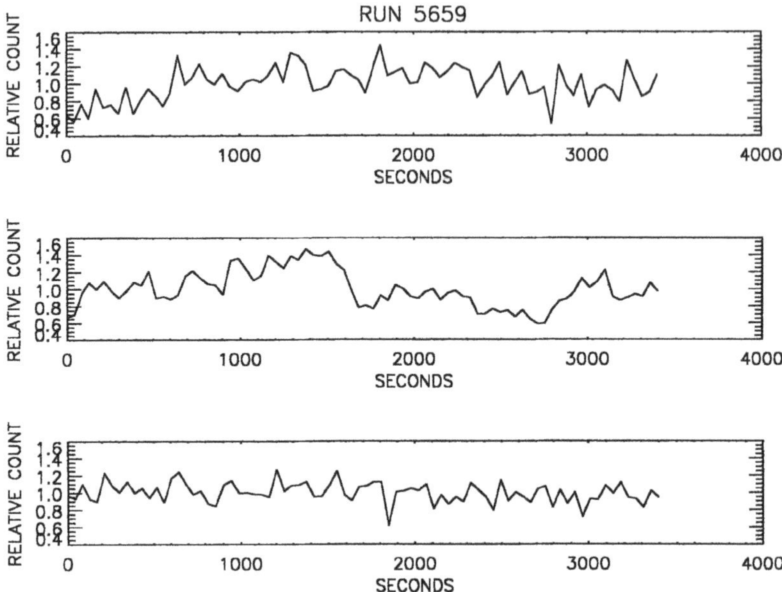

Figure 12. He I (*top*), O IV (*middle*), and Mg IX (*bottom*) traces for run 5659. The ordinate is given in relative counts from the mean intensity for each emission line. For the He, O, and Mg data, the mean is 350, 1700, and 2200, respectively.

Table I

A summary of the CDS 'FLARE' sequence observations (see text for explanation)

CDS run number	Date/time	Peak time (s)	Intensity	Duration (s)
5642	5 Nov./16:10 UT	43	1.6	86+
		1000	1.8	1200
5656	6 Nov./19:07 UT	0	2.3	344+
5657	6 Nov./20:19 UT	1300	2.8	1283
5658	6 Nov./21:31 UT	1677	2.4	774
		2400	2.6	320
5659	6 Nov./22:44 UT	1350	2.1	1677
		3100	1.8	430

higher temperature emission. The plots are for pixel addresses (0,4), (0,5), (0,6), (1,4), (1,5), (1,6). This is, in fact, the most intense of the transient brightenings reported in this work, with a factor of 3.5 increase in intensity in O IV. We see a double peak lasting some 30 rasters (about 20 min). The He I emission does show a more significant variation this time, with a peak almost coinciding (but a little later than) the O IV peak and a smaller peak later in the sequence. The He I intensity varies by a factor of about 1.5. However, the Mg IX intensity remains as in the previous events, with fluctuations at the 20–25% level.

Run 5658 is shown in Figures 9 and 10. Yet again, we see the O IV brightening events. In this case a region to the north of the image centre is bright at the start of the sequence, dies away and, after a couple of minor brightenings, displays a very significant event at the end of the sequence. Figure 10 shows the intensities of pixels (0,25), (0,26), (0,27), (1,25), (1,26), and (1,27). They show a low level of emission in O IV in the first half of the sequence and then a rise by a factor of up to 2. Although the He I emission appears to be bright at the start of the sequence, the variations in intensity for the He I and Mg IX plots are, again, at a 20–25% level.

Finally, run 5659 is shown in Figures 11 and 12. It shows a bright patch in the southern part of the image, which is varying in intensity across the sequence, as well as a brightening just at the end of the sequence in the northern part of the image. Figure 12 is a plot of the intensity from pixels in the southern patch, namely (0,2), (0,3), (0,4), (1,2), (1,3), and (1,4). Whereas the O IV curve shows variations of a factor of 2.5, with the brightest peak just before the half way stage of the sequence, the He I and Mg IX intensities show similar variations to before, at about the 25% level.

5. Discussion

It should be stressed that the data shown are from the five observations made using the CDS FLARE sequence which were run fully (one other sequence was run but was truncated due to instrumental schedule clashes). They were all on quiet Sun, and on differing portions of the Sun. Therefore, they can be assumed to be representative of quiet-Sun conditions.

All show evidence for significant O IV intensity enhancements, of factors between 1.6 and 2.8, usually evident over an area of 10–20 pixels – an area of about 2.2–4.4×10^7 km^2. If we assume that such an event is primarily contained in a 4×4 pixel group, the diameter of 8 arc sec is consistent with the bright patches at the network junctions seen in, for example, Figure 1.

One concern for such small rasters is that solar rotation may create the effect of an intensity 'event' due to a 'long-lived' bright patch moving into the raster field of view. The raster cadence is 43 s and in that time, at Sun centre on the equator, the rotation rate is about 0.1 arc sec per 43 s. The pixel size in the east-west direction is 2 arc sec. Thus with a pixel size of 20 times the rotation distance, we may conclude that such effects are negligible. In any case, one can see by inspection of the images that the bright patches brighten and decay from locations which are well observed.

So, with five sequences and 11 blinker events we appear to detect about 2 of these blinkers in a 3440 s sequence of quiet-Sun observation, in an area 10×60 arc sec. Each one appears to involve an intensity increase of about 2.5 over a period of, on average, 13 min. One of these rasters is 5.33×10^{-5} the surface area of the Sun. So, in any 3440 s, we may expect 37 500 of these events to be starting around the globe. This means that about 3300 should be in progress at any point in time. Of course, this assumption of a global distribution reflected in the data shown here must be tested by further observations in the near future.

It is difficult to estimate a thermal energy content for the blinker events. The majority of them have signatures in only the O IV emission line, of the lines selected. Thus, we cannot give a density estimate and can only assign a temperature of order of the temperature of formation of O IV, i.e., 160 000 K.

We know the typical size of the brightenings, in two dimensions, and can assume a similar depth. If we assume a typical transition region density we can estimate the number density. Thus, to calculate a rough order of magnitude energy content of these events, we can simply assume that each event can be thought of as a blob of plasma at 160 000 K, 2×10^{11} km^3 in size ($4 \times 4 \times 4$ pixels) and that the density is 10^{10} cm^{-3}, i.e., 1.67×10^{-11} kg m^{-3}. This translates to a thermal energy content of approximately 4.4×10^{25} erg per event, i.e., 4.8×10^{26} erg s^{-1} over the entire globe. This does assume that each event is due to the thermalisation of one pulse. If the energy input period is more extended, then the total energy of these blinker events could be much larger.

We note that the thermal energy content from this very rough calculation is of order 10^{-6} that of a 'standard' flare energy. Thus, in the true sense of the word, the blinker events are microflares. However, their durations and frequency are not what is traditionally thought of for microflare events. On the other hand it is very important to note that within the confines of the observations reported on, to time scales of tens of seconds, we do not observe any other class of event which could be viewed as being microflares.

Given the above assumptions, we have an average energy of $\sim 10^4$ erg cm^{-2} s^{-1} when the total thermal energy contained in the blinker events is divided by the surface area of the Sun. This is almost two orders of magnitude lower that the total radiation loss of the upper chromosphere and corona (6×10^5 erg cm^{-2} s^{-1}; see, e.g., Athay and White, 1978). If we demand that there is an extended period of energy input for each blinker event, the total energy content may be significant when considered in the light of coronal heating. However, it is important to note that there is no reason to suppose that a transient event in the Sun's atmosphere that drives brightenings in the transition region cannot dump energy in other forms, whose magnitude has no relationship to that seen in the thermal emission from the transition region plasma. The blinkers may be the most visible products of a process which is energetically significant for mass acceleration or heating.

To investigate this, we need to know more about the nature of the blinker events. Our knowledge is limited because of the need for fast rastering or image taking. However, in the next section, we examine clues which are evident in the first observations and discuss future observations.

We have used the term 'microflare' to describe a globally distributed (not just active region) phenomena, with each event having the energy content of about 10^{-6} that of a flare. This is in keeping with many discussions over the last decade, generated by ideas mainly due to Parker (1983). These are thought to be the fundamental processes leading to global phenomena such as coronal heating and solar wind acceleration. Such events have not been detected to date in quiet Sun observations. Some similar small-scale 'flaring' activities have been detected in active regions (e.g., Porter, Toomre, and Gebbie, 1984; Porter *et al.*, 1986; Lin *et al.*, 1984; Harrison, 1987; Hayes and Shine, 1987), and some of these have been linked to microflares. Indeed, Hayes and Shine (1987) reported on frequent O IV and Si IV intensity variations in active regions. Significant variations were seen at temperatures similar to the quiet-Sun blinkers discussed here.

The blinkers are seen because plasmas along the line of sight in the transition region become more dense, or because plasmas are heated or cooled to transition region temperatures. The lack of response in the He I and Mg IX emissions suggests that the former is the case. Such density increases could be due to magnetic field-line compression in supergranular cell boundaries or junctions. For some reason, particular locations have the conditions for such activity; we do not see such brightening at all of the cell boundaries. Is this related to the age of the cells involved or to the interaction with underlying or emerging cells?

Such questions cannot be answered with this report alone. The suggested multi-instrument studies incorporating magnetic data and high spectral resolution ultraviolet data are required to take this work to the next logical step.

6. Further Clues?

One immediate thought is that the O IV brightenings may be related to the jets and flow events, known as explosive events, seen in transition region lines by Brueckner and co-workers on the flights of the HRTS instrument (e.g. Brueckner and Bartoe, 1983) and, more recently, from the SUMER observations on SOHO (see, e.g., Innes *et al.*, 1997). Typically, such events are short lived (around 60 s), small scale (about 2 arc sec) and occur at a rate of about 600 per second over the Sun's surface (Dere, 1994). Typical line-of-sight velocities of 150 km s^{-1} are seen.

The HRTS/SUMER observations are made at much longer wavelengths, upwards of 1000 Å. However, they both detect emissions lines in similar temperature regimes to the O IV line.

The HRTS/SUMER events appear to be shorter lived by an order of magnitude than the blinker events being reported on here. In addition, the events in this report are much larger in size - by a factor of about 10. Finally, whereas our estimates suggest that there are 3300 blinker events on the Sun at any point in time, the HRTS data suggest that there are over 2 million explosive events on the Sun at any point in time.

Either we are examining different phenomena, or the blinker events represent the larger, longer lived explosive events. One method for checking this is to examine the O IV line shapes from the CDS blinker data.

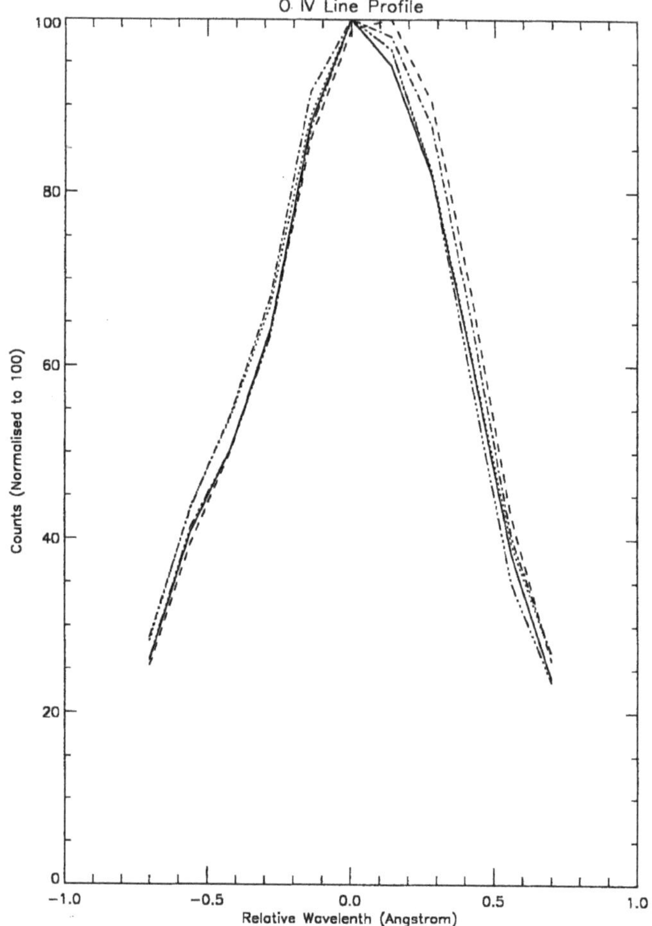

Figure 13. Five O IV 554 Å line profiles taken across the brightest blinker event, that of run 5657. The counts are normalised to 100, for ease of comparison and the abscissa is centred on 554.52 Å. The curves are for ten image accumulations: rasters 10–19 (solid curve), 20–29 (dotted), 30–39 (dashed), 40–49 (dash dot), 50–59 (dash dot dot dot).

However, Figure 13 illustrates a problem with this. The FLARE rasters were designed to be as fast as possible whilst retaining some spatial information and spectral information. The raster area was limited to 10 × 60 arc sec. In the same way, the spectral information was restricted to 11 bins across each line - more to ensure that a good intensity from the line was detected rather than for the study of line shapes. We do not take sufficient data on either side of the emission line to allow good line fitting. However, Figure 13 shows the shape for five periods across the event of run 5657 (the brightest event). The plot shows the profile from a selection of four pixels in the brightest region, with the line peaks normalised to 100 counts, for comparison. The data are summed across rasters 10–19, 20–29, 30–39, 40–49, 50–59, giving five line shapes with good statistics. Figures 7 and 8 show that these should give a measure of the line shapes prior to the blinker event (10–19) and throughout the event.

It is difficult to assign any significant line broadening or shifts to these data. Close inspection suggests that we may see a shift to longer wavelengths (red shift) in the dashed and dash-dot curves by a maximum of 0.07 Å, which translates to about 35–40 km s^{-1}. These two represent the maximum intensity of the blinker event. However, this is a 'subpixel' shift which is statistically questionable and cannot be supported (with these data) with a good fitting procedure. We note that the SUMER and HRTS data suggest that explosive events show Doppler shifts typically of 150 km s^{-1}. A velocity of 100 km s^{-1} would represent a shift of 0.55 Å. Shifts of this order are not seen with these blinker events.

This inspection has been performed with all of the blinker events described above and no clear shifts can be determined for these events.

If we cannot find evidence for line of sight velocities, what about lateral velocities? We may be able to see shifts in the centroids of the emission regions. Of course, it is recognised that motion of brightness may not be due to motion of material.

For the 5642 run, one of the 2–3 events detected shows some migration of the bright emission southwards (Figure 3, southern patch). This occurs as the event decays. This kind of activity is a feature of a few of the events, but cannot be said to be general.

It is clear that we have many unknowns which can be explored by further observations whose characteristics are geared to the temporal and spatial scales of these events.

The first suggestion is that whilst the use of He I, O IV, and Mg IX data in the FLARE runs showed that the He and Mg emissions varied little with respect to the O, and that the brightenings were not necessarily colocated, new runs could be made with 2–3 transition region lines only. Good candidates would be O III 599 Å and O V 629 Å.

It is important to retain the cadence of these rasters, but it may be crucial to obtain more spectral information, to fit line profiles properly (i.e., take more pixels across the lines). Perhaps this could be done by only recording two transition region lines but taking more spectral pixels.

Thirdly, the use of coincident observations with other instruments on SOHO will be essential to tracking down the real nature of these events and their relationship to other features, such as the HRTS/SUMER explosive events. So, coincident, colocated observations with SUMER should be regarded as essential and the SUMER explosive events and CDS blinkers should be compared to one another. Another essential collaboration would be with the MDI instrument on SOHO which can provide high resolution (1.2 arc sec) magnetogram data.

7. Conclusions and Interpretations

We conclude the following:

(1) We have identified a class of subflare brightenings of small patches in transition region temperatures. We have called these brightening events 'blinkers'.

(2) These blinker events typically involve an increase in intensity of a factor of 2.5 and have durations ranging from 86 to 1677 s, with an average duration of 13 min. The brightenings are contained within areas of about 6000×6000 km (4×4 pixels) which appear to be at network cell junctions.

(3) Emission lines from higher and lower temperatures observed at the same time do not display the same variations in brightenings; the blinker activity appears to be pretty much confined to transition region temperatures. In addition, the blinker events do not always appear at the locations of the brightest intensities at higher or lower temperatures.

(4) All five observing sequences taken in November 1996, from five different quiet-Sun regions, display blinker events. This fact suggests that the blinker events are common and that they might be globally distributed. Simple calculations have shown that there may be 37 500 of these events starting on the solar globe in any 3440-s period, if the 'global' assumption is correct. The total thermal energy content appears to be of order 4.4×10^{25} erg per event, or 4.8×10^{26} erg s^{-1} over the entire globe.

(5) Comparisons to the characteristics of explosive events seen at longer wavelengths, but similar temperatures, suggest that the blinker events and explosive events are not related. Within the limitations of the instrumentation, we cannot identify clear Doppler shifts or line broadenings in association with these brightenings.

(6) Blinker events appear to contain about 10^{-6} the thermal energy of a flare – i.e., they could be candidates for microflare activity. However, their durations and frequency may not quite match expectations for such activity. However, if other microflare activity exists, then on time scales of tens of seconds and spatial scales of over 2 arc sec it is not detected in the FLARE sequences which were used for these observations.

The main conclusion is that we are seeing a global distribution of rather small (spatially and energetically) transient events seen in the extreme ultraviolet. Their accumulated thermal energy content is not sufficient to power the solar wind acceleration or coronal heating processes, but they may be a visible signature of globally distributed activities which are responsible for such phenomena.

Acknowledgements

CDS was built and is operated by a consortium led by the Rutherford Appleton Laboartory and including the Mullard Space Science Laboratory, the NASA Goddard Space Flight Center, Oslo University and the Max Planck Institute for Extraterrestrial Physics, Garching. SOHO is a mission of international cooperation between ESA and NASA.

References

Arnaud, M. and Rothenflug, R.: 1985, *Astron. Astrophys. Suppl.* **60**, 425.
Athay, G. and White, O.: 1978, *Astrophys. J.* **226**, 1135.
Brekke, P., Kjeldseth-Moe, O., Brynildsen, N., Maltby, P., Haugan, S. V. H., Harrison, R. A., Thompson, W. T., and Pike, C. D.: 1997, *Solar Phys.* **170**, 163.
Brueckner, G. E. and Bartoe, J.-D. F.: 1983, *Astrophys. J.* **272**, 329.
Dere, K. P.: 1984, *Adv. Space Res.* **14**, 13.
Harrison, R. A.: 1987, *Astron. Astrophys.* **182**, 337.
Harrison, R. A., Sawyer, E. C. and 37 co-authors: 1995, *Solar Phys.* **162**, 233.
Harrison, R. A., Fludra, A., Pike, C. D., Payne, J., Thompson, W. T., Poland, A. I., Breeveld, E. R., Breeveld, A. A., Culhane, J. L., Kjeldseth-Moe, O., Huber, M. C. E., and Aschenbach, B.: 1997, *Solar Phys.* **170**, 123.
Hayes, M. and Shine, R. A.: 1987, *Astrophys. J.* **312**, 943.
Innes, D. E., Inhester, B., Axford, W. I., and Wilhelm, K.: 1997, *Nature* **386**, 811.
Lin, R. P., Schwartz, R. A., Kane, S. R., Pelling, R. M., and Hurley, K. C.: 1994, *Astrophys. J.* **283**, 421.
Mason, H. E., Young, P. R., Pike, C. D., Harrison, R. A., Fludra, A., Bromage, B. J. I., and Del Zanna, G.: 1997, *Solar Phys.* **170**, 143.
Parker, E.: 1983, *Astrophys. J.* **264**, 642.
Porter, J. G., Toomre, J., and Gebbie, K. B.: 1984, *Astrophys. J.* **283**, 879.
Porter, J. G., Reichmann, E. J., Moore, R. L., and Harvey, K. L.: 1986, *Coronal and Prominence Plasmas*, NASA Conf. Publ. 2442, p. 383.

ACTIVE REGIONS OBSERVED IN EXTREME ULTRAVIOLET LIGHT BY THE CORONAL DIAGNOSTIC SPECTROMETER ON SOHO

A. FLUDRA[1], P. BREKKE[2], R. A. HARRISON[1], H. E. MASON[3], C. D. PIKE[1], W. T. THOMPSON[4] and P. R. YOUNG[3]

[1]*Space Science Dept., Rutherford Appleton Laboratory, Chilton, Didcot, Oxfordshire OX11 0QX, U.K.*

[2]*Institute of Theoretical Astrophysics, University of Oslo, Oslo, Norway*

[3]*Department of Applied Mathematics and Theoretical Physics, Silver Street, Cambridge CB3 9EW, U.K.*

[4]*NASA Goddard Space Flight Center, Greenbelt, MD 20771, U.S.A.*

(Received 11 March 1997; accepted 12 June 1997)

Abstract. We present observations of five active regions made by the Coronal Diagnostic Spectrometer (CDS) on the Solar and Heliospheric Observatory (SOHO). CDS observes the Sun in the extreme ultraviolet range 150–780 Å. Examples of active region loops seen in spectral lines emitted at various temperatures are shown. Several classes of loops are identified: those that are seen in all temperatures up to 2×10^6 K; loops seen at 10^6 K but not reaching 1.6×10^6 K; those at temperatures $2–4 \times 10^5$ K and occasionally at 6×10^5 K but not reaching 10^6 K. An increasing loop size with temperature and the relationship between the cool and hot structures is discussed. CDS observations reveal the existence of loops and other unresolved structures in active regions, at temperatures between $1.5–4 \times 10^5$ K, which do not have counterparts in lines emitted above 8×10^5 K. Bright compact sources only seen in the transition region lines are investigated. These sources can have lifetimes of up to several days and are located in the vicinity of sunspots. We study the variability of active region sources on time scales from 30 sec to several days. We find oscillatory behaviour of He I and O V line intensities in an active region on time scales of 5–10 min.

1. Introduction

Coronal signatures of solar activity encompass phenomena with very different spatial and time scales, from small bright points a few arc sec in diameter and with a lifetime of minutes, to active regions which can be as large as 5 arc min across and persist for several solar rotations. Both bright points and active regions can be regarded as a manifestation of the magnetic activity in the corona and transition region.

Active regions result from the emergence of the magnetic flux which, viewed at the photospheric level, rapidly organizes itself into magnetic pores, sunspots and plages. This vast aspect of active region structure and evolution is reviewed in Thomas and Weiss (1992) and Balasubramaniam and Simon (1994).

As the region grows, the magnetic field spreads out into large magnetic loop structures rising to coronal heights, many of them visible as a more intense emission in X-ray and UV wavelengths, apparently confined in closed magnetic loops. The increased coronal emission results from increase in electron temperature and density over the quiet solar atmosphere, suggesting that additional heating is taking place in active regions over and above that responsible for the quiet solar corona.

Solar Physics **175:** 487–509, 1997.

Temperature structures of coronal loops in active regions were exhaustively studied with the X-ray and EUV observations from *Skylab* (cf., 'Solar Active Regions', 1981, F. Q. Orrall (ed.)). It was found (cf., Raymond and Foukal, 1982; Doyle, Mason, and Vernazza, 1985) that the deduced physical parameters of active region loops, based on the assumption of constant pressure, cannot be reconciled with static loop models. In particular, the substantial amounts of cool material observed high in the corona could not be explained. In a further debate concerning whether or not the cool and hot emission was co-spatial, Foukal and colleagues associated the brightest cool loops with sunspots and proposed that some hot loops had *cool cores*. The weight of evidence from the NRL S082A observations, which had somewhat better spatial resolution, indicated that this was not the case. Other studies (e.g., Dere, 1982; Habbal, Ronan, and Withbroe, 1985) concluded that active region loops do not have cool cores and that loops at different temperatures are not cospatial.

The traditional meaning of the transition region, defined as the interface between the plasmas at chromospheric and coronal temperatures, has been expanded to simply represent a temperature regime. Feldman (1987) pointed out the existence of structures at transition region temperatures that are not related to coronal structures, i.e., that the 3×10^4–5×10^5 K radiation originates from plasma structures not associated with the adjacent corona at $T_e \geq 10^6$ K. Feldman and Laming (1994) further discussed the relationship between the transition region and coronal structures observed from *Skylab* data for quiet Sun, coronal holes and active regions, and concluded that the properties of coronal plasmas ($T_e \geq 10^6$ K) cannot be inferred from the properties of the $T_e \leq 7 \times 10^5$ K plasmas.

More recently, the Solar EUV Rocket Telescope and Spectrograph (SERTS) (Neupert *et al.*, 1992) has been successfully flown on several occasions. A spectral atlas (170–450 Å) for an active region was published by Thomas and Neupert (1994) and SERTS images of an active region in He I and Mg IX were published by Brosius *et al.* (1996), together with one dimensional scans across an active region in other EUV lines.

Extensive studies of X-ray emission from active regions have been carried out with *Yohkoh* (cf., Enome and Hirayama, 1994; Fludra *et al.*, 1995; Uchida, Kosugi, and Hudson, 1996). Dynamic events have been observed from small scale jet-like features up to large-scale flares. However, only the high temperature coronal emission is observed by the *Yohkoh* instruments, so the relationship of this to low temperature emission requires additional ground based or EUV observations.

The past observations of loop structure based upon EUV, X-ray, and centimeter wavelength data have recently been summarized by Arndt, Habbal, and Karovska (1994). These results suggest that active region loops may exist in different temperatures, however, they conclude that the reports from different studies are sometimes conflicting.

The Coronal Diagnostic Spectrometer (CDS) on SOHO observes the Sun in the EUV wavelength range 150–780 Å (Harrison *et al.*, 1995). The observations from

CDS offer the unique opportunity to unambiguously relate the structures observed in low temperatures (transition region), with those observed in higher temperatures. CDS has observed a large number of active regions, both off limb and on the disk in a multitude of spectral lines between 3×10^4 K and 3×10^6 K. Some of these active regions had associated sunspots. With CDS observations, we are able to distinguish differences between the loops on a much finer temperature scale than has previously been possible from *Skylab* or broad-band X-ray images. This allows us to identify individual magnetic loops at different temperatures that constitute the active region and to study their time variability on time scales from minutes to many days.

In this paper we survey observations of active regions made by the CDS between March and August 1996. Our aim is to explore the relationship between the hot active region loops, seen in coronal lines, and the cooler loops, seen in the transition region lines. In addition, we use ground based observations to spatially relate the cool emission to sunspots. We select a few examples of active regions both with and without sunspots, to illustrate the different loop structures, the relationship between the transition region and coronal loops and the presence of bright transition region sources near sunspots. We also study variability of transition region and coronal lines on time scales from half a minute to several days and present examples of other transient heating episodes resulting in brightening of EUV lines.

The paper is organized as follows. Section 2 gives a brief summary of instrumental aspects of the CDS relevant to the analysed observations. The relationship between cool and hot loops is illustrated in Section 3. Bright sources at transition region temperatures, their relationship to sunspots, and variability of the EUV emission are addressed in Section 4. Conclusions are given in Section 5.

2. The Coronal Diagnostic Spectrometer

The wavelength range of CDS covers lines and continua emitted at temperatures from the chromosphere at $\approx 2 \times 10^4$ K to the hottest parts of active coronal loops at several million degrees. A full description of the CDS was given by Harrison *et al.* (1995). Here we summarize the information relevant to the observations discussed later in the paper.

The data discussed in this paper were recorded by the Normal Incidence Spectrometer using either a 2×240 arc sec^2 slit or a 4×240 arc sec^2 slit. At each slit position, an exposure of a required duration (typically, 5 to 50 s) was taken and several wavelength 'windows', centered on selected spectral lines were extracted and transmitted. The width of the spectral windows was sufficient to include the full line profile and adjacent background for each selected line. While the CDS is capable of taking the full spectrum in its wavelength band and is sometimes used that way, telemetry limitations make it more efficient to reduce the amount of spectral information and select up to 20 spectral lines of interest. For these active region observations, various CDS *studies* (test6-1, largebp, largebp2, tede22) were used and the line list usually included a selection of lines to cover the temperatures from the chromosphere, transition region to the corona, and to provide also a density diagnostics. The following lines are frequently included: He I 584.3, O III 599.6, O IV 554 (blend), O V 629.7, Ne VI 562.8, Ca X 557.8, Mg IX 368.1, Mg X 624.9, Fe XIV 334.2, Fe XIV 353.8, Fe XVI 360.8, Si X 347.4, Si X 356.0, Si XII 520.7 Å (Harrison *et al.* 1997). After each exposure, the scanning mirror is rotated by one step, equivalent to the width of the slit, so that the adjacent area of the Sun is presented to the slit, another exposure is taken, the mirror is rotated again, and so on, until the whole required area of the Sun is covered. A maximum area that can be covered by the CDS raster without repointing the instrument is 4×4 arc min^2. It requires 120 mirror steps when the 2×240 arc sec^2 slit is used, or 60 mirror steps with the 4×240 arc sec^2 slit. For small targets, it is possible to reduce the image size in the E–W direction by taking fewer mirror steps, and in the N–S direction by transmitting data from less than the full length of the slit. Taking fewer mirror steps reduces the duration of the raster, while reducing the length of the slit reduces the telemetry time required to transmit each line, which in turn either allows to transmit more spectral lines per exposure or may also shorten the raster duration. The spatial dimensions of the raster elements are either 2 or 4 arc sec in the E–W direction (depending whether a 2×240 arc sec^2 or 4×240 arc sec^2 slit was used), and 1.68 arc sec in the N–S direction. The data analysed in this paper were taken using several of the above options, and the relevant information will be given for each data set.

3. Relationship of Hot and Cool Active Region Loops

3.1. Disk Observations

3.1.1. *22 March 1996, AR 7953*

Images of a complex active region recorded by CDS on 22 March 1996 are displayed in Figure 1. The two panels across the top of the image show the active region as observed in Mg IX and Mg X, emitting at temperatures only $\approx$150 000 K apart around 10^6 K. The detailed loop structures are different. CDS is able to distinguish these differences on a much finer temperature scale than possible from broad-band X-ray images, for example, from the Soft X-ray Telescope on *Yohkoh*. The lower two panels show images from the same ions approximately one hour later. One can see that some evolution of the structures has taken place between the two times. The largest loop in Figure 1 appears to be in the process of heating up or being filled up with hotter plasma – the top images only show the left-hand side leg and the top of the loop, while the bottom panels show also the right-hand side leg.

Images are also available in a range of cooler and hotter lines. They give further evidence that the large loop seen in the top frame undergoes an asymmetric heating: only the the left-hand leg and the top of the loop can be seen in Mg IX, while the

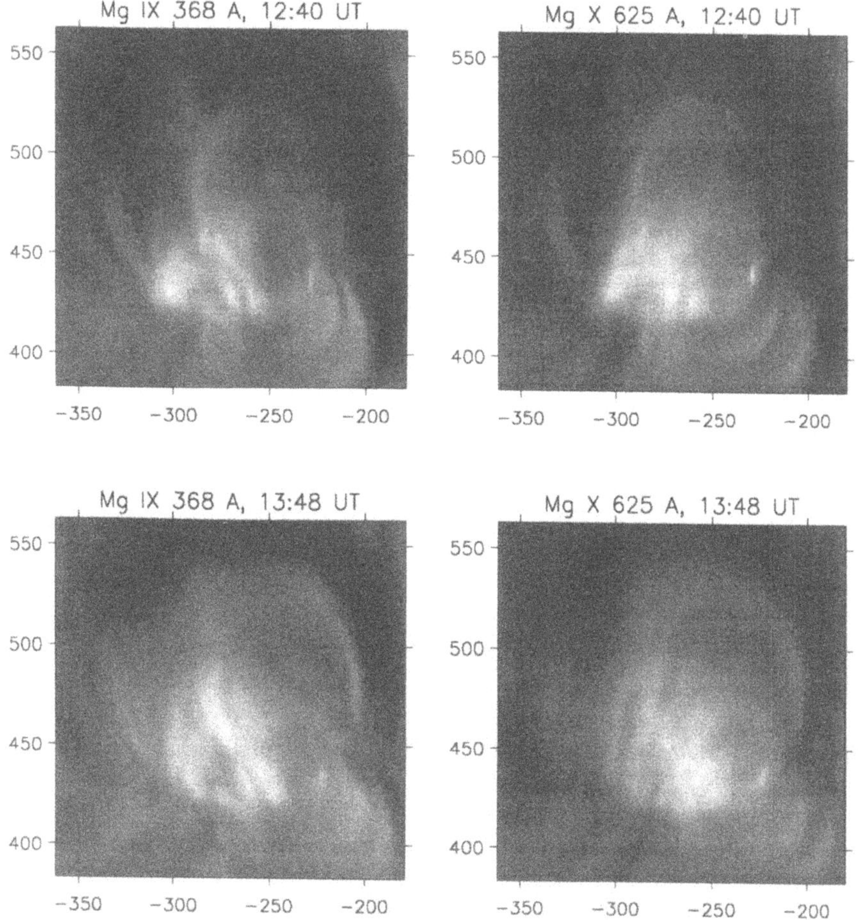

Figure 1. Images of an active region observed on 22 March 1996 at 12:40 UT (*top*) and 13:48 UT (*bottom*). *Left panel*: Mg IX 368 Å. *Right panel*: Mg X 625 Å. The axes are solar X and Y position in arc sec.

right hand leg can be seen in Fe XIV 334.2 Å line. In Mg VI 349.2 Å line, the lower part of the left-hand leg can be seen and there does not appear to be any emission from the Mg VI line in the right-hand part of the loop. The top of the loop appears to be thicker in all lines than the loop legs, and the top cross-section in Fe XIV is largest and most fuzzy. Moreover, if one looks more closely at the bottom part of the left-hand leg, the Mg VI emission does not appear to coincide with the Mg IX emission. This could be interpreted as meaning there are two separate loops.

In the bottom panel of Figure 1, Mg IX emission is now seen in the right-hand leg also. The most striking feature of the bottom frame is that the right-hand-side leg of the large loop is very sharply defined in Mg IX. Figure 2 shows images of this leg in the following ions: Mg VI, Mg VII, Mg IX, Mg X, Fe XII, Fe XIV, with the Mg IX emission denoted in all panels by small white squares. The emission of lower temperature lines (Mg VI) is confined to the lower portion of the loop and it moves higher up in the loop with increasing temperature of line formation. It is surprising, however, that the Fe XIV leg does not coincide with the Mg IX leg.

The overall impression one gets from this dataset is of a bundle of magnetic flux converging on two footpoint regions, with the coronal emission being confined to fairly thick 'strands' within this bundle. The lack of co-spatiality of the Mg VI, Mg IX, and Fe XIV emissions could then be explained by different strands of the bundle having somewhat different temperatures.

3.1.2. *11 May 1996, AR 7962*

The bright coronal lines available in the Normal Incidence range of CDS cover two characteristic temperatures: from $0.9-1.1 \times 10^6$ K in Mg IX and Mg X lines, and from $1.6-2.5 \times 10^6$ K from Fe XVI and Si XII lines. We have found that 10^6 K plasma is often not co-spatial with the 2×10^6 K plasma. In some compact, hot sources the Mg IX/X structures may seem to occupy the same area as the Si XII structures. However, many bright structures in the Si XII and Fe XVI images are often spatially different from Mg IX and Mg X structures. The example below illustrates this in case of an active region that was observed by CDS near the disk centre on 11 May 1996, 17:09 UT.

Figure 3 shows a comparison of Si XII and Mg IX images for this active region. There are three bright loops seen in the Mg IX image. The Si XII image has two circularly-shaped areas with voids in emission exactly corresponding to the position of the two left-hand side Mg IX loops. In addition, a thin gap in Mg IX emission between those two sources corresponds to a thin Si XII loop, and the wide gap between and slightly above the two right-hand-side Mg IX loops corresponds to an area of enhanced Si XII emission. Then, the two elongated areas of intense Si XII emission, one on the left-hand side, and the other at the bottom right-hand side of Figure 3, have almost no Mg IX emission. This illustrates that we observe two classes of loops, one at $\approx 1 \times 10^6$ K and another one at $\approx 2 \times 10^6$ K, which are separated spatially.

The same dataset shows how the size of the emitting structures increases with temperature, from the transition region at 2.5×10^5 K to the corona at 10^6 K. Figure 4 shows a comparison of the three bright structures in O V, Ne VI, Mg VII (from the blue wing of Mg IX), and Mg IX lines. These loops are viewed mostly from the top and it appears as if the common part where all the loops are bright was co-spatial at different temperatures. However, it is possible to envisage a nest of loops which not only expand the cross-section but also increase their height with temperature. In this case, the cooler, smaller loops would lie underneath the hotter, larger loops.

We have estimated the area of each source seen in each of the lines. In agreement with what can be seen in Figure 4, there is a clear dependence of the area on

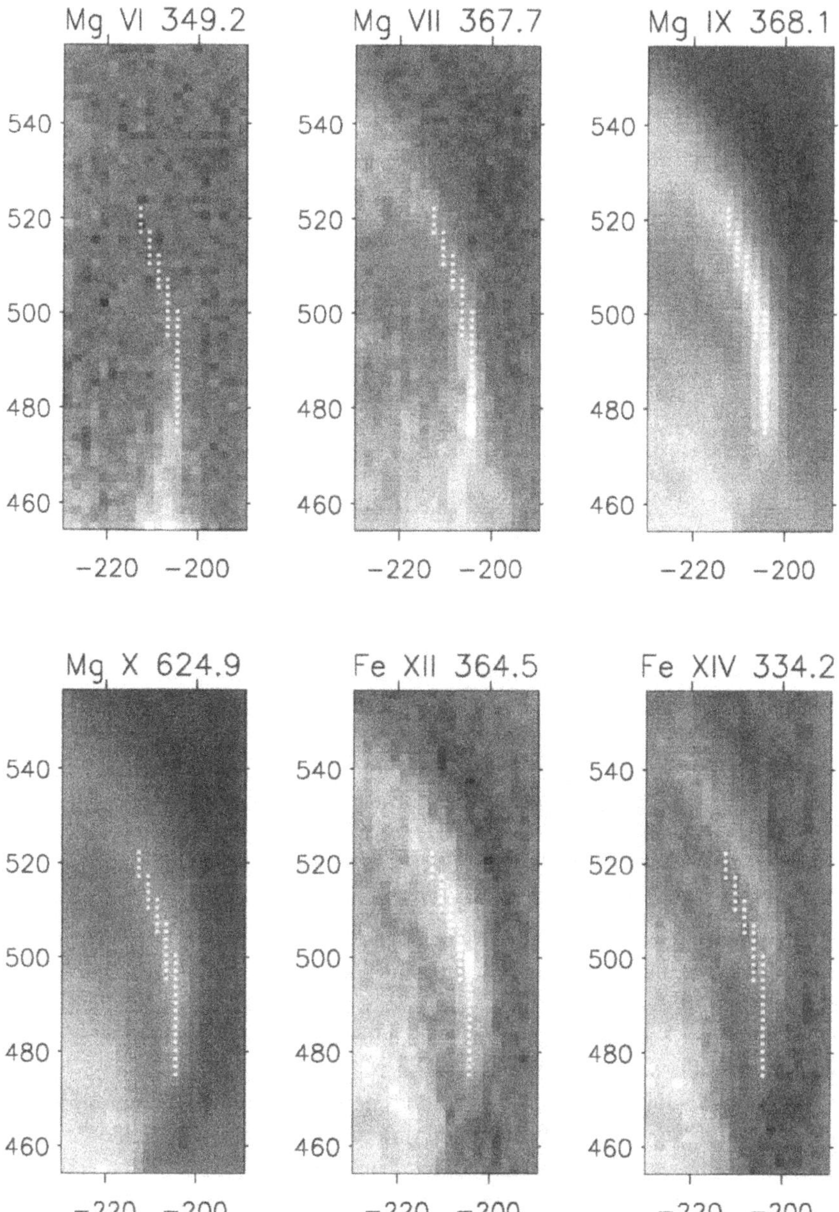

Figure 2. Images of the right-hand-side 'leg' of the loop shown in Figure 1 at 13:48 UT, in the lines from the following ions (*from left to right*): Mg VI, Mg VII, Mg IX (*top panels*), Mg X, Fe XII, and Fe XIV (*bottom panels*). The emission at higher temperatures moves higher up along the loop. The axes are solar X and Y position in arc sec.

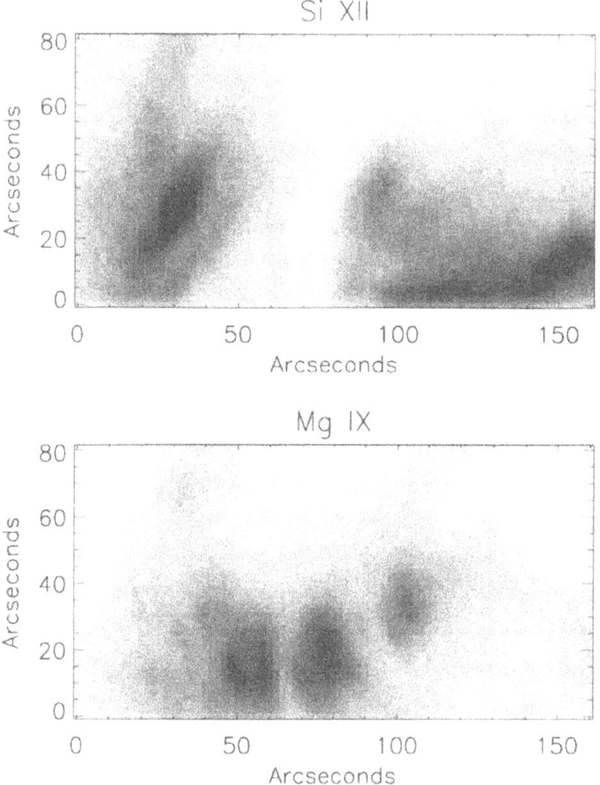

Figure 3. Comparison of Si XII 520 Å (*upper panel*) and Mg IX 368 Å (*lower panel*) emission in an active region observed on 11 May 1996, 17:09 UT. This figure illustrates that the plasma at temperatures of 10^6 K can occupy different magnetic loops than the plasma at $1.6–2 \times 10^6$ K.

temperature for each source, with O V sources covering between 130–180 arc sec^2, and Mg IX covering 400–680 arc sec^2, while Ne VI and Mg VII areas have values in between those two limits. The increase in area comes mostly from the increase in the loop cross-section with temperature, although the apparent extent along the longer axis of the sources increases also.

The loops seen in Mg VI, VII, and IX lines appear to have a different orientation to the O V and Ne VI loops. The marked difference between Mg IX and O V/Ne VI suggests that the coronal emission comes from different magnetic loops than the $2–4 \times 10^5$ K emission.

The O III 599 Å line, with peak temperature 9×10^4 K, generally outlines similar structures as the O V line, particularly when the quiet-Sun network is observed. However, the relative O III intensity between individual spatial features can often be unrelated to the brightness of the O V line in those features. This is in contrast to the comparison of O IV, O V, and Ne VI lines, whose intensities follow similar

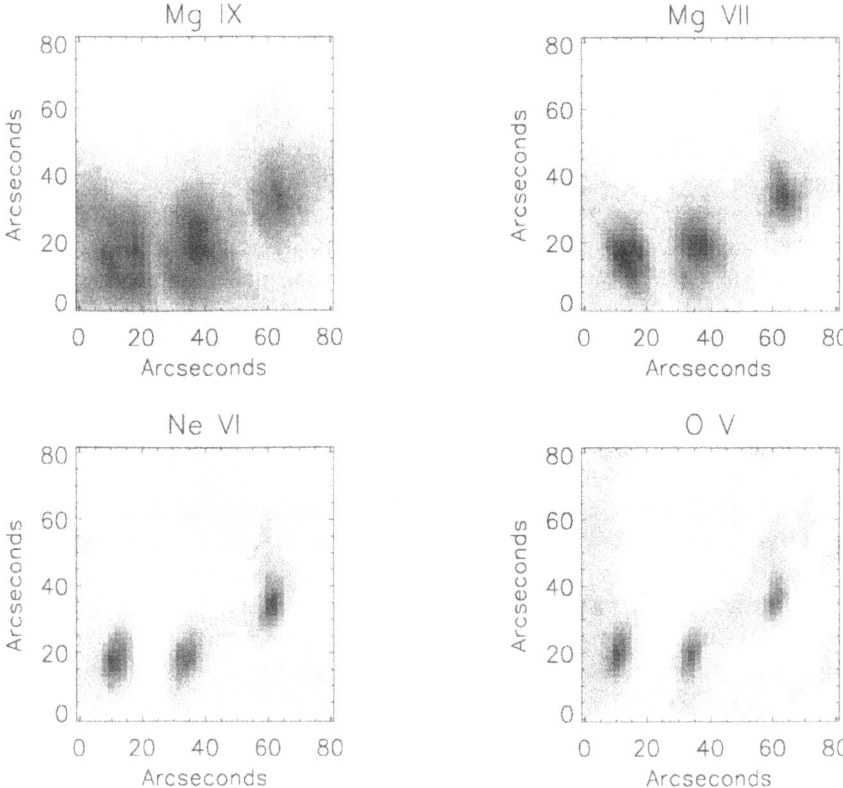

Figure 4. Comparison of three structures in the lines of ions of Mg IX (*upper left*), Mg VII (*upper right*), Ne VI (*lower left*), and O V (*lower right*) in the active region on 11 May 1996 at 17:09 UT. The size of the structures increases with temperature.

trend – if O V is bright, O IV and Ne VI lines are bright as well. In this active region, the spectral line of O III differs from the other lines in that the three sources from Figure 4 are barely seen in this line and their O III intensity is the same or lower than in other nearby features. For example, we find a small loop where the O III line intensity increases by a factor of two, while the O V line intensity there is lower by at least a factor of 1.5 than in the three sources.

The same active region was observed by CDS later that day at 23:42 UT. The O V intensity in sources 1 and 3 (numbered from left to right) decreased by over a factor of two and the two loops, while still seen, became comparable in brightness to the rest of the region. The peak O V intensity in the source 2 decreased only by about 10%, while the loop became more uniformly bright and its inclination to the N–S direction increased. The Mg IX intensity in loops 1 and 3 stays almost unchanged, decreasing by about 25% in the second loop. It is even clearer in these images that the Mg IX loops are oriented in different direction to the O V loops.

3.2. LIMB OBSERVATIONS

3.2.1. *19 June 1996*

The active region observed by CDS on 19 June 1996, 06:57 UT had emerged on the invisible side of the Sun and was seen for the first time when it appeared on the east limb. Figure 5 shows a comparison of loop structures at different temperatures. This active region has two distinct parts – a set of large loops extending up to 70 000 km above the limb, and much brighter, low-lying loops seen against the disk close to the limb. In the low temperature emission, two large loops are seen in the O IV 554 Å line (2×10^5 K) while there is a trace of two more loops in the Ne VI 563 Å line (4×10^5 K). The number of loops and their complexity increases for lines formed at higher temperatures. The contours of O IV and Ne VI loops overlap one another, but they appear offset relative to Mg IX loops (9.5×10^5 K), suggesting that the hotter loops at near coronal temperatures are spatially different from the cooler loops at the transition region temperatures.

The Si XII and Fe XVI emissions show a different, more complex set of loops, with significant emission above the tops of the Mg IX and Mg X loops. Si XII and Fe XVI images represent plasma at 1.6–2.5 million K. The high loops reach 70 000 km above the limb. A characteristic feature is the presence of emission along and above the tops of the cooler loops, and 'rays' extending even higher above. The 'rays' are suggestive of a magnetic field lines trapped under the top of one of the high loops and extending out far beyond the field of view.

Figure 6 shows a plot of intensity of the Fe XVI 361 Å line along the horizontal direction, at 100 arc sec above the bottom of the image in Figure 5. The tops of the high loops are seen as a small peak of intensity at the distance between 100 and 110 arc sec from the left of the image. The large peak of intensity corresponds to a long, low-lying loop. The ratio of Fe XVI 360 Å to Si XII 520 Å lines is a measure of temperature. Fe XVI is emitted at higher temperatures than Si XII line, and the ratio increases with temperature. The observed value of this ratio, shown as a dashed line in Figure 6, reaches a maximum value about 30 arc sec above the intensity peak of the high loops. This means that the temperature also increases above the peak intensity.

The density at the top of the large loops has been estimated from the ratio of Fe XIV 353/334 Å and Fe XII 364/338 Å lines to be $2–3 \times 10^9$ cm^{-3}, using a CHIANTI package for the theoretical intensities (Dere *et al.*, 1997).

To investigate the intense low lying structures on the disk, we superimposed the images in Fe XVI 361 Å and O IV 554 Å. It appears that the high temperature emission forms an elongated structure with the O IV emission in two footpoint ribbons. The density in this bright Fe XVI arcade, derived from the same lines as for the high loops, is $8–10 \times 10^9$ cm^{-3}. It is noted that higher density corresponds to the low-lying high-temperature emission on disk, not to the larger coronal loops.

Si XII and Fe XVI images look similar to images of post-flare loops frequently observed in X-ray wavelengths by the Soft X-ray Telescope on *Yohkoh*. While this

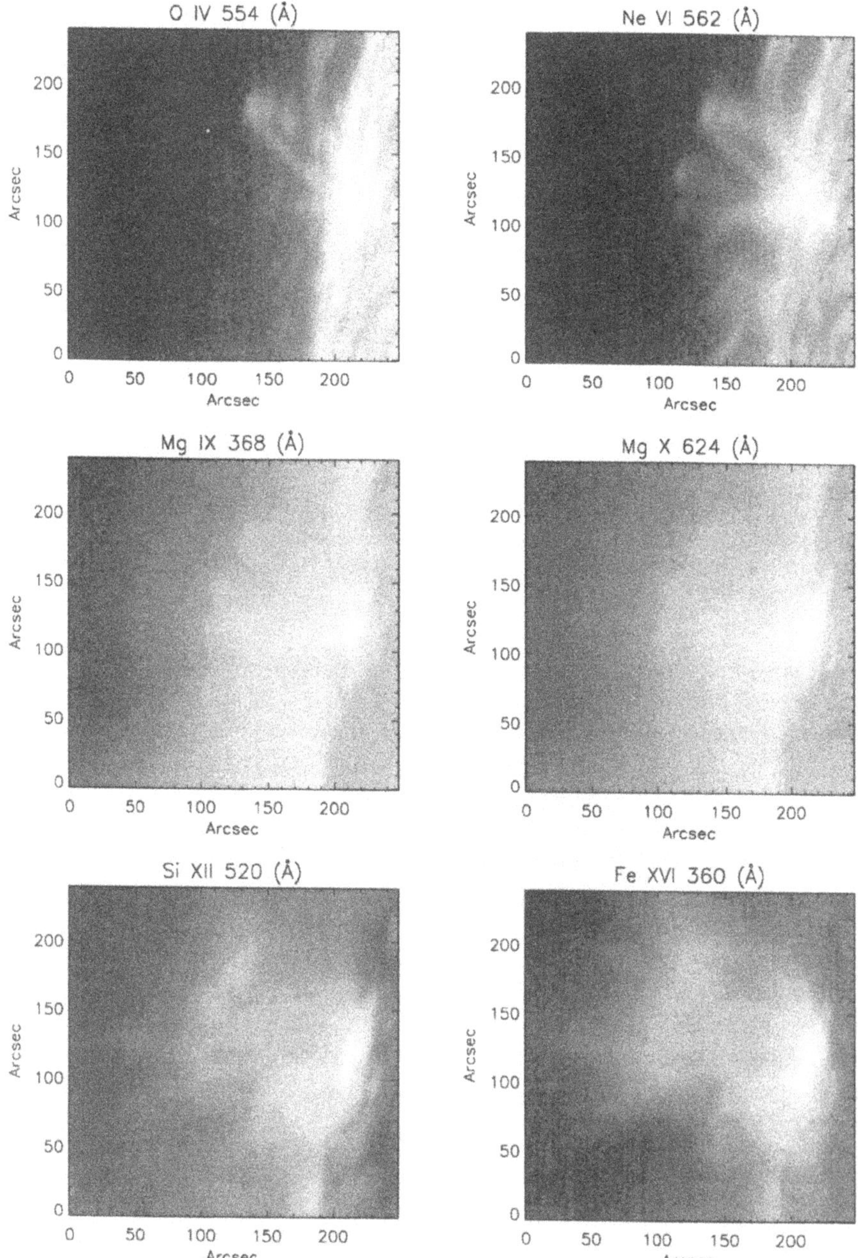

Figure 5. Images of an active region observed by the CDS on 19 June 1996, 06:57 UT, in the following lines (listed from *left to right*): O IV, Ne VI (*top panel*), Mg IX, Mg X (*middle panel*), Si XII, Fe XVI (*bottom panel*). The intensity scale is logarithmic to show the faint tops of the high loops. The dimensions of each panel are 4 × 4 arc min.

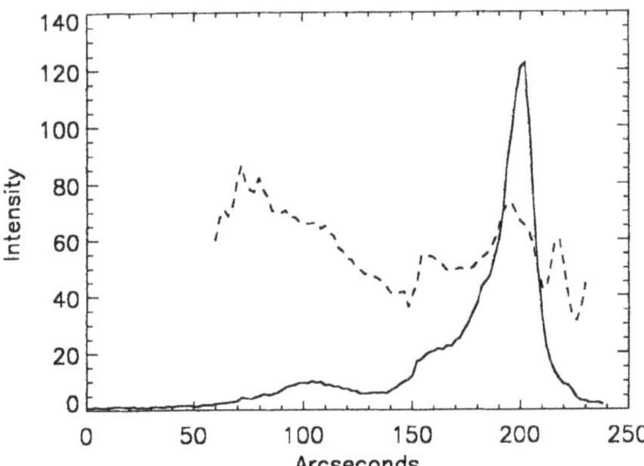

Figure 6. Spatial variation of intensity (continuous curve) of the Fe XVI 360 Å line in the active region shown in Figure 5. The X axis corresponds to the E–W direction, and the intensity is shown at 100 arc sec above the lower edge of the image, cutting across the top of the high loops and the bright loop on the disk. The 1σ statistical error is 2% of the intensity at the peak at 200″ and 6% of the intensity at the tops of high loops near 100″. The dashed curve shows the ratio of the Fe XVI 360 Å to Si XII 520 Å lines which increases with temperature. Both the intensity and the ratio have an arbitrary scale.

reminiscence of a post-flare loop system might be only a projection effect, there is a supporting evidence of increased activity in this active region before and after the CDS observation. First, the Large Angle Spectrometer and Coronagraph (LASCO) observed two coronal mass ejections (CME) in its C1 field of view from this active region before the CDS observation (Plunkett, 1996, private communication): a fast eruption on the 17 June, 02:03 UT and a large, slow eruption on the 18 June, from 09:00 to 24:00 UT. This makes it more plausible to suspect that a flare may have taken place in this active region. There were also two mass ejections seen by LASCO on the 19 June, after the CDS observation: a CME at 11:39 UT which was seen as an expanding arch at 1.6 solar radii, and a large CME starting at 17:00 UT. During this last eruption the tops of the high loops in Figure 5, which were also seen until then along the inner edge of the LASCO C1 field of view, disappeared. Subsequent CDS observations of this active region on the following days indeed reveal an active region which seems to have changed its structure – the high loops no longer being visible (see next section).

4. Bright Transition Region Sources

4.1. Relationship of Cool Loops to Sunspots

4.1.1. *20–26 June 1996, AR 7973*

The active region that was observed on the limb on the 19 June, discussed above, was later observed on the disk. CDS observations were made on 20, 23, 24, and 26 June.

Images taken on 20 June between 07:24 and 09:09 UT in Fe XVI and Mg IX lines show that the bright loop seen in the Fe XVI line in the 19 June data is still present. This data also reveals a very bright, compact source seen in O V image. The peak O V intensity of the source is 22 700 erg cm^{-2} s^{-1} $ster^{-1}$, while the total intensity is 2.44×10^{21} erg s^{-1} $ster^{-1}$. The source is also bright in other transition region lines (Ne V and Mg VII). However, it is not seen in the coronal lines.

White-light image from Mees Observatory taken at 21:35 UT shows a presence of a sunspot in this active region. The location of the sunspot was rotated back in time to 08:05 UT, when the bright source was observed. It was found that the bright O V source is located slightly below the sunspot, partially covering the penumbra. A more detailed analysis of the locations of the O V source relative to the sunspot will be made for observations on 23 June below.

Also on 20 June, from 20:43 UT, a 90-min movie of this active region was made using a wide slit ($90'' \times 240''$) in He I 584, O V 630, and Mg IX 368 Å lines. In this mode, a spectroheliogram from the whole $90'' \times 240''$ area is taken during one exposure without rastering. The duration of each exposure was only 3 s, and the consecutive exposures were taken every 32 s. The He I 584 image in Figure 7(a) reveals two sources of emission – one to the left of the image, elongated in the N–S direction (denoted 'A'), and the second one, more compact, located at the centre of the image (denoted 'B'). The He I lightcurves of these two sources are shown in Figure 8. The source 'A' shows intermittent variability – there are five peaks of intensity, two of which have an amplitude up to 10 percent of the total intensity, and the other three vary by only 3%. Source 'B' has a somewhat different behaviour – an overall decrease of intensity by 25% is seen in the first 30 min of observation, followed by a variability at 5% level and a larger increase towards the end of observation. The 1σ statistical error is only 0.2% for both light curves. The O V 630 Å image is much brighter in source 'B', and has a different shape than the He I source. Initially, it resembles two loops close to each other which merge into one source later. The area 'B' from Figure 7(a) has been further divided into boxes 'C' and 'D' in Figure 7(b). The O V lightcurve from the area 'C' shows oscillations with an amplitude of about 15% of the total intensity (Figure 9(a)). A lightcurve taken from the northern part of the source (from box 'D' in Figure 7(b)) shows even more variability - the largest 'pulse' of intensity reaches about 50% above the lowest intensity level (Figure 9(b)). The total duration of this burst is 5 min. The 1σ statistical error on intensity is 0.4% and 0.25% for the areas 'C' and 'D',

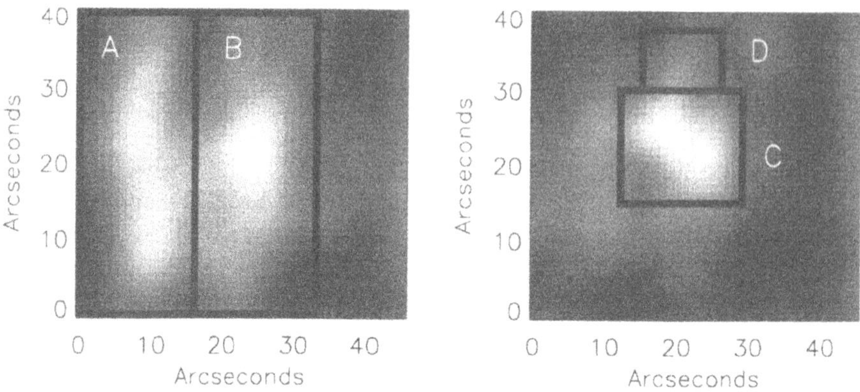

Figure 7. Images of a part of the active region taken with the wide 90″ × 240″ slit, on 20 June 1996 at 20:43 UT: (a) He I 584 Å (*left panel*), (b) O V 630 Å (*right panel*). A movie of this active region with 32-s cadence was recorded for 90 min. Boxes mark areas where lightcurves are shown in Figures 8 and 9.

respectively. In contrast to the O V data, the Mg IX image shows mostly source 'A'. Between 21:42 and 21:30 UT, the Mg IX intensity from this source follows the same trend as the He I intensity, but with a smaller amplitude less than 5%, and shows almost no variability afterwards.

Images taken three days later, on 23 June 14:25 UT, show a rather complex structure of this active region (Figure 10). There are several bright areas seen in high temperature lines. The bright source in the O V image, first noticed on 20 June, is still seen, located at E23 N10, in a place where the high temperature lines (Si XII, Fe XVI) have very little emission. Figure 10 shows the contours of the O V source on the Fe XVI and Mg IX images. The number of CCD pixels where the source intensity is above 50% of the peak intensity, is 33, which is equivalent to a 110 arc sec^2 area. The peak O V intensity and the total intensity of the O V source are 26 600 erg cm^{-2} s^{-1} ster^{-1} and 2.12×10^{22} erg s^{-1} ster^{-1}, respectively. A BBSO Hα image recorded on 23 June 1996 at 16:13 UT, shows a sunspot and distinct-looking bright structures around it. Similar structures are seen in the CDS image in the He I 584 line. The two images have been correlated to coalign them. After the coalignment, the O V contours are plotted on the Hα image (Figure 11). This shows that the O V bright source is located near the sunspot penumbra, partially overlaying the umbra.

The bright source is also seen in other transition region lines, up to Mg VII. For example, in the Fe XIV 353 Å window there is a Mg V 353.09 line (peak temperature of 3×10^5 K), which is bright in the same place as in the O V image. Again, O III emission behaves differently to the O V emission. The O III image shows a small loop in the location of the O V source, which is not brighter than other features nearby. For example, the O III line is nearly a factor two brighter at another location,

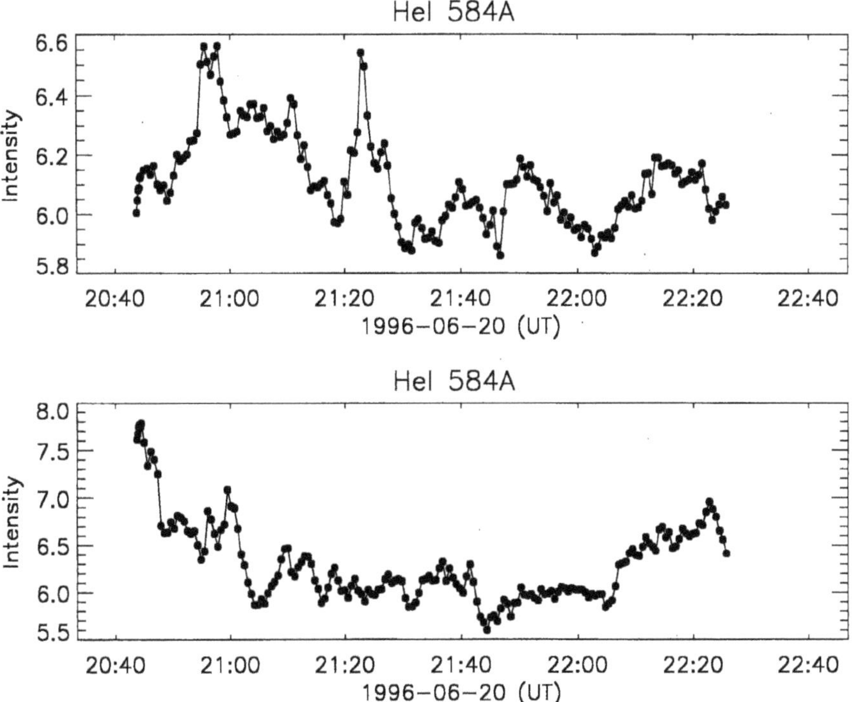

Figure 8. Time variation of the He I 584 Å line intensity from the areas marked by boxes in Figure 7: box 'A' (*upper panel*), box 'B' (*lower panel*). The 1σ statistical error is 0.2%.

almost unrelated to the active region, where the peak O V intensity drops by a factor three to 8600 erg cm^{-2} s^{-1} $ster^{-1}$.

The bright O V source is seen in the subsequent dataset obtained on 23 June, 19:51 UT. The same location is also very bright in Ne VI 562.8 Å and Mg VI (seen at 349.1 Å).

Observations of this AR continued through to 24 June, 19:07 UT and the bright O V source remains present, however, it resembles more the shape of a loop. The peak intensity is similar.

One more observation was made two days later on 26 June, 16:38 UT by which time two loops are seen in the bright O V source, and the peak O V intensity has decreased by about a factor three.

4.1.2. *26 June 1996, AR 7976*

Another active region was observed by CDS on 26 June at 21:59 UT. A $4' \times 4'$ raster, centered at E29.7 N14.3, was taken with a larger number of lines. This active region shows the same phenomenon of bright sources seen in the transition region lines, in particular O V and Ne VI, as the active region discussed in the previous

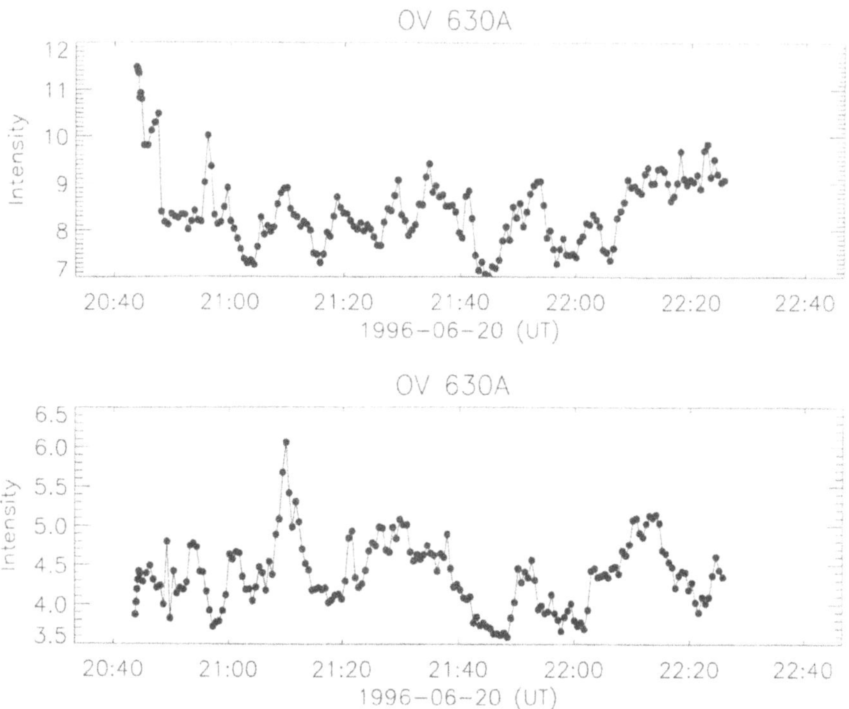

Figure 9. Time variation of the O V 630 Å line intensity from the areas marked by boxes in Figure 7: (a) box 'C' (*upper panel*), (b) box 'D' (*lower panel*). The 1σ statistical error is 0.4% and 0.25% in the upper and lower panel, respectively.

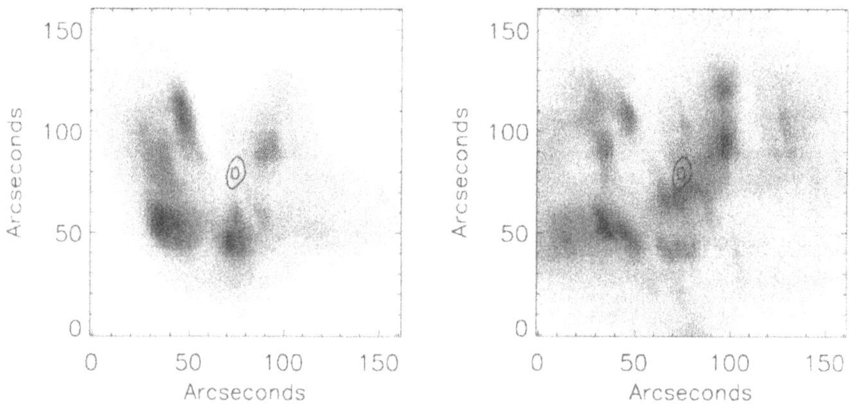

Figure 10. Images of the active region observed on 23 June 1996, 14:25 UT, in the lines of Fe XVI (*left panel*) and Mg IX (*right panel*). The contours show the location of the O V bright source. Contour levels are 50% and 90% of the peak OV intensity. The dimensions of the images are 2 × 2 arc min.

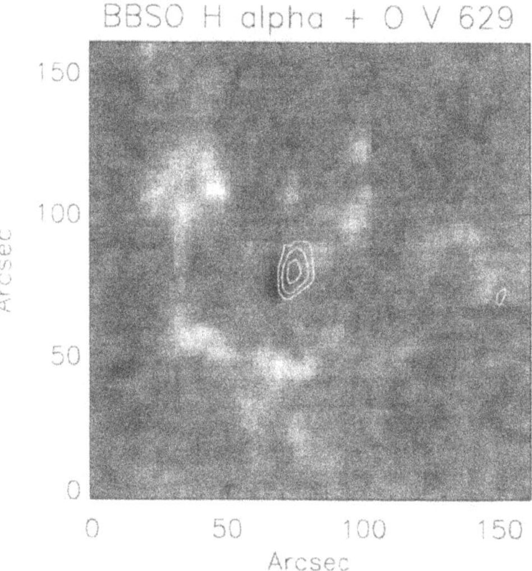

Figure 11. BBSO Hα image of the active region recorded on 23 June 1996 at 16:22 UT. The contours show the location of the O v bright source which partially overlaps the sunspot.

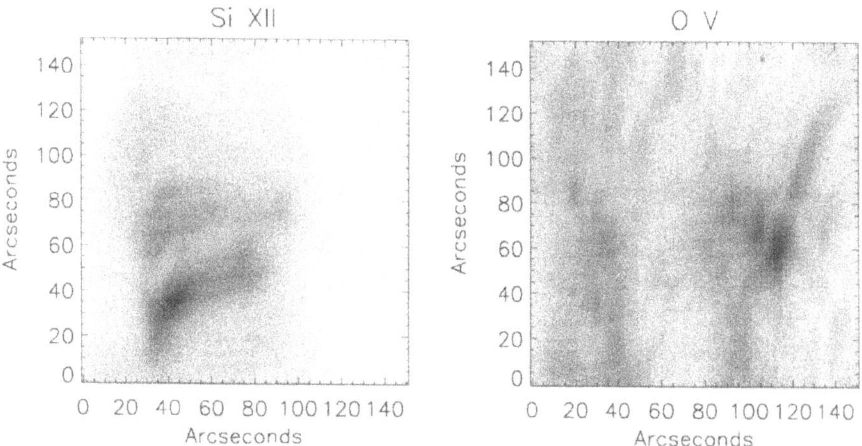

Figure 12. Si XII (*left panel*) and O v (*right panel*) images of an active region observed on 26 June 1996, 21:35 UT. A small sunspot is located near the upper end of the most intense O v source. The intensity scale of O v is nonlinear to better show the fainter emission in the surrounding area.

section. A bright source is evident in the O v image. Again, the source is not seen in the coronal lines. Instead, bright Fe XVI and Si XII emission in the shape of loops is coming out from an area to the left of the O v source (Figure 12). The Ca X and Mg IX images show thin loops crossing the intense O v source. The O III image has a much smaller area of emission than O v in the location of the intense source.

A white-light image from Mees observatory taken on 26 June, 21:35 UT shows a small sunspot in this active region. The sunspot diameter is about 12″. The coordinates of the sunspot at that time have been rotated to 22:34 UT when the O V bright source was observed. The O V source has an elongated shape, possibly a loop structure, the upper end of which appears to be rooted near the sunspot penumbra. The peak O V intensity is 34 900 erg cm^{-2} s^{-1} ster^{-1}, and the total O V intensity summed over 103 spatial pixels is 1.92×10^{22} erg s^{-1} ster^{-1}.

The KPNO magnetogram shows a strong magnetic field in this location, with two areas of opposite polarity - an area of one polarity on the left, and an area of opposite polarity on the right. The Si XII and Fe XVI loops connect the two polarities.

The emission in the He I 584 line is very weak at the location of the O V bright source and in the sunspot, and is much brighter at other locations. The O III 599 Å line is bright in the O V source, but has a similar level of intensity at four other locations. The Mg VII line is bright near this location, but not in precisely the same structure. This comparison between various lines indicates that the bright sources are tops of small magnetic loops with temperatures between 2×10^5 and 6×10^5 K.

4.1.3. *1, 2 August 1996 – AR 7981*

A large active region was observed by CDS on 1 and 2 August 1996. Images from the EUV Imaging Telescope (EIT) show that the active region extends about 5 arc min in the E–W and the N–S directions. There are four sunspots located on the right-hand side.

Data taken on 1 August at 21:37 UT shows the presence of two bright sources in O V images. The CDS image had 2×2 arc min^2 field of view centered at E3 S9.7, and was pointed towards the sunspot area on the right-hand side of the active region. The two O V sources are located close to each other – the source on the left is brighter and larger than the source on the right.

Another observation of this area was made on 2 August between 03:38 and 04:19 UT and consisted of a 4×4 arc min^2 raster in four lines (He I 584, O V 630, Mg IX 368 and Fe XVI 360 Å). Here, several bright O V sources are seen. A white light image from Mees Observatory, taken on 1 August, 21:35 UT, shows several sunspots. They are also seen in the BBSO Hα image on 2 August at 15:37 UT. The CDS images have been co-aligned with the Hα by correlating the He I 584 Å line with the Hα image. Figure 13(b) shows the BBSO Hα image, where the locations of three sunspots are marked with crosses. The same crosses are plotted on the O V image in Figure 13(a). The size of the crosses represents the sunspot's size. It can be seen that three bright O V sources are almost co-spatial with the sunspots.

The area around the largest sunspot was observed again on 2 August, 08:11 UT, with a 2×2 arc min^2 field of view, and using many more lines than in the previous observation. Figure 14(a) shows the bright transition region loops. The sunspot from the BBSO Hα data on 2 August, 15:37 UT was rotated to the time of 08:49 UT, when the brightest O V source was observed. The sunspot centre is marked with a cross on the Hα image (Figure 14(b)). The size of the cross is equal to the sunspot's

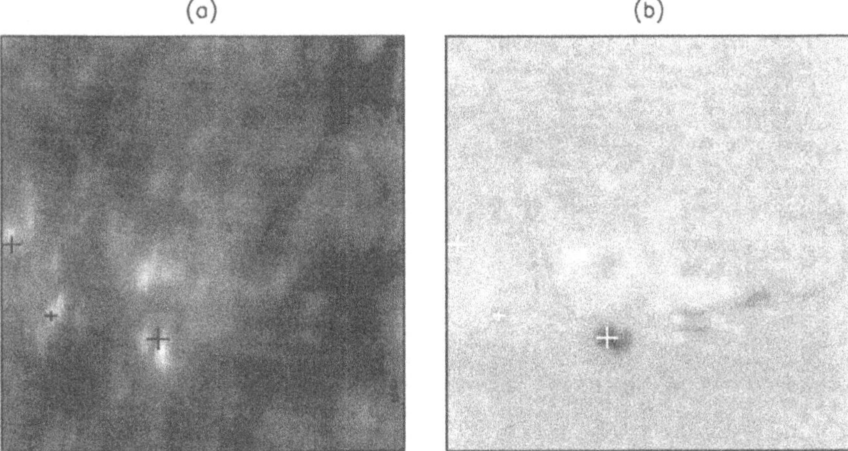

Figure 13. Images of the active region on 2 August 1996. (a) O V image taken between 03:38 and 04:19 UT (*left panel*). The crosses show the locations of the sunspots. (b) BBSO Hα image on 1 August at 15:01 UT (*right panel*). The crosses have been placed at the centre of sunspots. The dimensions of the images are 4 × 4 arc min.

diameter. A cross of the same size is placed on the CDS image, at a location derived from the raw CDS pointing values. The pointing uncertainty is 5″. When the bright Hα features above the sunspot were spatially correlated with similar features seen in some of the CDS lines, the sunspot centre moved 6″ below the cross, i.e., even closer to the bright O V source. The two bright O V sources are located on the outer edge of the penumbra.

The emission of two bright O V sources located next to the largest sunspot varies between the observations - the source on the left decreases in intensity, while the O V intensity of the right-hand side source increases to 4.4×10^{22} erg s^{-1} ster^{-1}. They are seen for over 12 hours on the 1 and 2 August. Other observations of this active region are discussed by Brynildsen *et al.* (1998).

The same active region returned at the end of August. CDS observation made on 28 August at 20:15 UT still shows a bright O V source near the sunspot. To find out whether this is the same source as seen on 1 and 2 August, we compared its location to the location of the sunspot. The source is found in the S–W part of the sunspot, covering the penumbra. That makes it possible that the 28 August source represents the same loops as the 1 and 2 August source. The peak and total O V intensities are 52 100 erg cm^{-2} s^{-1} ster^{-1} and 2.24×10^{22} erg s^{-1} ster^{-1}, respectively.

The datasets for the 1 and 2 August again illustrate the presence of coronal loops at different temperatures. Figure 15 compares the emission in Si XII and Mg IX lines and shows that the Si XII only comes from the upper part of the area displayed in Figure 15(a), while the Mg IX emission is also seen in the lower part of the area

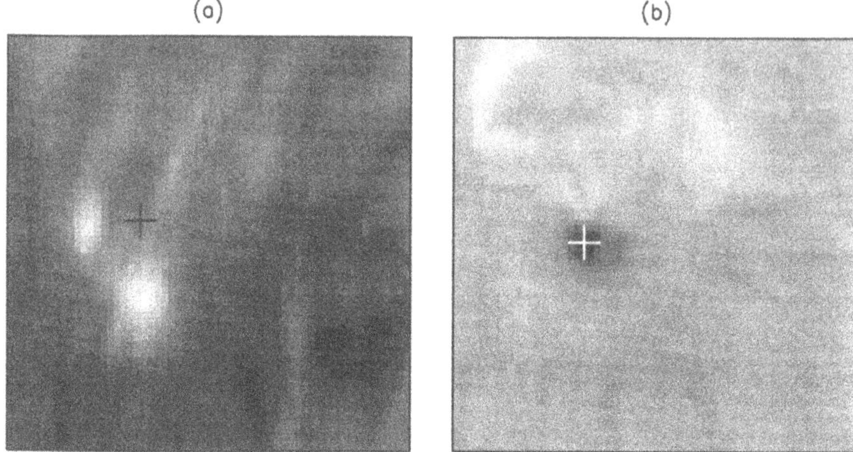

Figure 14. Images of the active region taken on 2 August 1996: (a) O V 630 Å at 08:11 UT (*left panel*), (b) BBSO Hα image on 2 August at 15:37 UT (*right panel*). The cross in the O V image shows the location of the sunspot when the raw CDS pointing values are used. The pointing uncertainty is 5″. A correlation technique gives the location of the sunspot 6″ below the cross. The cross in Hα image is placed at the centre of the sunspot. The dimensions of the images are 2 × 2 arc min.

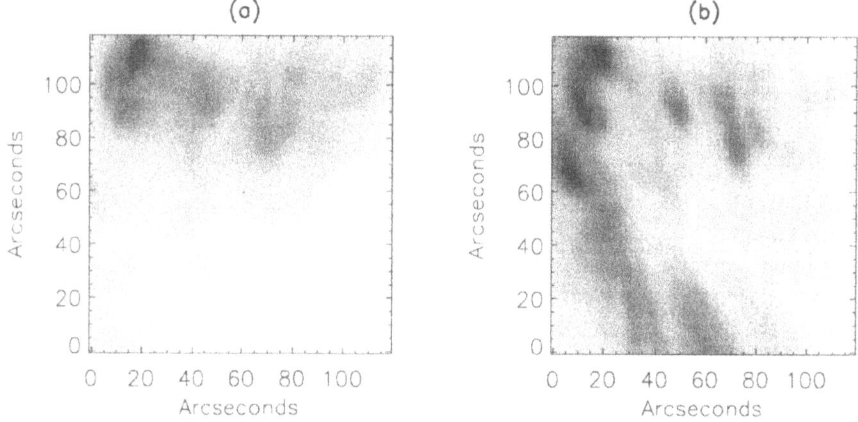

Figure 15. Comparison of (a) Si XII (*left panel*) and (b) Mg IX (*right panel*) emission in the active region on 2 August 1996, 08:11 UT. The loops seen in the bottom of Mg IX image have no corresponding emission in the Si XII line.

(Figure 15(b)), mostly in the form of long, straight lines which converge to the sunspot penumbra. The straight lines are legs of large loops that can be seen in their full length in EIT images. There is very little emission in coronal lines (Mg IX, Si XII) above the sunspot.

5. Discussion and Conclusions

We have examined five different sets of active region observations obtained with CDS between March and August 1996. We confirm that the transition region traditionally understood as the chromosphere-corona 'interface' in large magnetic loops is responsible for only a small part of the observed EUV emission in the so-called 'transition region' temperature range $8 \times 10^4 - 8 \times 10^5$ K. The majority of the EUV emission in this temperature range comes from loops and other unresolved structures which do not reach the coronal temperatures at all. The plasma with temperatures below 6×10^6 K exists independently of the coronal plasma, and undergoes changes with time which do not affect or do not depend on the changes of the coronal structures.

It seems that loops with different temperatures can co-exist within an active region, sometimes very close to each other, but often at different spatial locations. We have identified different classes of loops: some seen at all temperatures up to 2×10^6 K, others seen at temperatures of 10^6 K but not reaching 1.6×10^6 K, others seen at 4×10^5 K but not reaching the coronal temperatures. The hotter loops are seen higher above the limb. The footpoints of coronal loops are not particularly distinct in the lower temperature lines. The bright structures seen in the transition region temperatures $2-4 \times 10^5$ K (observed in spectral lines of O V and Ne VI) are most often smaller loops that do not reach coronal temperatures.

The cooler loops have smaller width and better defined cross-section than the hotter loops. This is particularly true of O V and Ne VI loops, but also often true for Mg IX. The Mg X loops can become a little 'fuzzy' at the edges, while Si XII and Fe XVI loops have the largest width at the loop top and most fuzzy edge. The cool loops often have much smaller size than the loops at coronal temperatures. It is not uncommon, however, that large loops at $2-4 \times 10^5$ K temperatures are seen at coronal heights.

We have found many bright sources in active regions at transition region temperatures: very bright, compact areas seen at 2×10^5 and 4×10^5 K (O V and Ne VI lines), sometimes extending in temperature up to 6×10^5 K (Mg VII line), but not reaching 9×10^5 K (Mg IX). The bright sources were seen to last for at least 12 hours, indeed one of these sources was observed to last for more than five days, and a source of similar intensity and location was still seen 26 days later. These intense compact sources are found near the regions of strong magnetic flux. Notably, three cases investigated in this paper have these transition region sources located very close to sunspots. The sources overlay at least the sunspot penumbra, and some can also be seen to cover part of the umbra.

We notice that many small structures seen at temperatures 9×10^4 K often do not follow the same intensity trends as seen among the spectral lines emitted at $2-6 \times 10^5$ K. Features that are brightest in the temperatures above 2×10^5 K may be undistinguishable from many other features when seen at the 9×10^4 K temperatures. Much brighter 9×10^4 K emission often comes from other locations.

Thus, the bright O V sources look rather inconspicuous in the O III images. This is even more obvious for the chromospheric emission at 2×10^4 seen in He I lines – there, the lack of He I emission at the location of some of the O V bright sources is even more striking.

We have explored various aspects of variability of active region emission lines and structures. We have discussed an active region where large scale restructuring of the magnetic field took place, accompanied by coronal mass ejections. In another active region, we observe heating taking place in a large coronal loop over one hour period, resulting in changes in the intensity distributions along the loop. Thirty-second cadence observations show oscillations of He I and O V intensities taking place on time scales of 5–10 min. Amplitudes of the order of 10–15% are more common, but intensity bursts with a 50% amplitude taking place in the legs of a magnetic loop have also been recorded.

This paper presents a mere glimpse at the wealth of active region observations which have already been accumulated by CDS. More quantitative analyses are in progress for specific active regions. Combining these CDS datasets with observations from other SOHO instruments, ground based and *Yohkoh* observations promises a rich harvest of results, increasing our understanding of the structure, heating and dynamics of solar active regions.

Acknowledgements

The CDS instrument was built and is operated by an international consortium led by the Rutherford Appleton Laboratory. We would like to thank the many members of the hardware groups at the Rutherford Appleton Laboratory and Mullard Space Science Laboratory in the United Kingdom, Max-Planck-Institut für Extraterrestrische Physik and Physicalisch-Technische Bundesanstalt in Germany, the Goddard Space Flight Center in the USA, and the University of Oslo, in Norway. This work was supported by the United Kingdom PPARC. SOHO is a project of international cooperation between ESA and NASA. We thank Haimin Wang, Big Bear Solar Observatory, for permission to use Hα images.

References

Arndt, M. B., Habbal, S. R., and Karovska, M.: 1994, *Solar Phys.* **150**, 165.

Balasubramaniam, K. S. and Simon, G. W. (eds): 1994, *Solar Active Region Evolution: Comparing Models and Observations*, Astronomical Society of the Pacific Conference Series, Vol. 68.

Brosius, J. W., Davila, J. M., Thomas, R. J., Saba, J. L. R., Hara, H., aand Monsignori-Fossi, B. C.: 1996, *Astrophys. J.*, submitted.

Brynildsen, N., Brekke, P., Fredvik, T., Haugan, S.V.H., Kjeldseth-Moe, O., Maltby, P., Rimmele, T., Harrison, R. A., Pike, C. D., Thompson, W., and Wilhelm, K.: 1998, *Solar Phys.*, in press.

Dere, K. P.: 1982, *Solar Phys.* **75**, 189.

Dere, K. P., Landi, E., Mason, H. E., Monsignori Fossi, B. C., and Young, P. R.: 1997, *Astron. Astrophys. Suppl.*, in press.

Doyle, J. G., Mason, H. E., and Vernazza, J. E.: 1985, *Astron. Astrophys.* **150**, 69.
Enome, S. and Hirayama, T. (eds): 1994, *Proceedings of Kofu Symposium*, NRO Report No. 360.
Feldman, U.: 1987, *Astrophys. J.* **320**, 426.
Feldman, U. and Laming, M.: 1994, *Astrophys. J.* **434**, 370.
Fludra, A., Doyle, G., Metcalf, T., Lemen, J. R., Phillips, K. J. H., Culhane, J. L., and Kosugi, T.: 1995, *Astron. Astrophys.* **303**, 914.
Habbal, S. R., Ronan, R., and Withbroe, G. L.: 1985, *Solar Phys.* **98**, 323.
Harrison, R. A. and 38 co-authors: 1995, *Solar Phys.* **162**, 233.
Harrison, R. A., Fludra, A., Pike, C. D., Payne, J., Thompson, W. T., Poland, A. I., Breeveld, E. R., Breeveld, A. A., Culhane, J. L., Kjeldseth-Moe, O., Huber, M. C. E., and Aschenbach, B.: 1997, *Solar Phys.* **170**, 123.
Neupert, W. M., Epstein, G. L., Thomas, R. J., and Thompson, W. T.: 1992, *Solar Phys.* **137**, 87.
Orrall, F. Q. (ed.): 1981, *Solar Active Regions*, Colorado University Press, Boulder.
Plunkett, S.: 1996, private communication.
Raymond, J. C. and Foukal, P. V.: 1982, *Astrophys. J.* **253**, 323.
Thomas, J. H. and Weiss, N. O. (eds.): 1992, *Sunspots: Theory and Observations*, NATO ASI Series, Kluwer Academic Publishers, Dordrecht, Holland.
Thomas, R. J. and Neupert, W. M.: 1994, *Astrophys. J. Suppl.* **91**, 461.
Uchida, Y., Kosugi, T., and Hudson, H. S. (eds): 1996, *Magnetodynamic Phenomena in the Solar Atmosphere – Prototypes of Stellar Magnetic Activity*, Kluwer Academic Publishers, Dordrecht, Holland.

HIGH-VELOCITY FLOWS IN AN ACTIVE REGION LOOP SYSTEM OBSERVED WITH THE CORONAL DIAGNOSTIC SPECTROMETER (CDS) ON SOHO

P. BREKKE[1], O. KJELDSETH-MOE[1], and R. A. HARRISON[2]
[1]*Institute of Theoretical Astrophysics, University of Oslo, Oslo, Norway*
[2]*Space Science Department, Rutherford Appleton Laboratory, Chilton, Didcot, Oxfordshire OX11 0QX, U.K.*

(Received 14 March 1997; accepted 25 April 1997)

Abstract. EUV spectra of coronal loops above an active region show clear evidence of strong dynamical activity. We present an example where the O v 629 Å line, formed at 240 000 K, is shifted from its reference position corresponding to line-of-sight velocities greater than 50 km s^{-1} with the shift extending over a large fraction of a loop. The observations were made with the Coronal Diagnostic Spectrometer (CDS) on the Solar and Heliospheric Observatory (SOHO), and are from active region NOAA 7981 on the east solar limb on 27 July 1996. An animation has been prepared showing the variation of the shift or flow velocity along the loop. This animation is to be found on the enclosed CD-ROM and gives a clear impression of the dynamical condition present in the loop. The appearance of the loop system in different lines formed over a range in temperature as well as the observed dynamics indicates that loops at different temperatures are not closely co-located. Finally, the results are discussed and related to mechanisms that may cause line shifts.

1. Introduction

This paper describes an observation, made with the Coronal Diagnostic Spectrometer (CDS) on the Solar and Heliospheric Observatory (SOHO), showing line shifts in the O v 629 Å line emitted from a loop in active region NOAA 7981. If the shifts are interpreted as gas flows, they mean that velocities of more than 50 km s^{-1} occur in the plasma at 240 000 K located in one leg of the loop. After describing the observation we compare it to model calculations of processes that may produce line shifts, such as syphon flows and waves. However, we do not attempt to model the observation in this paper. Instead our purpose is

– to emphasize that the solar transition region and corona may be very dynamic, and

– to show how CDS can reveal and illustrate the dynamical state of the solar atmosphere.

The latter point is clearly demonstrated in the animation on the CD-ROM accompanying this issue of *Solar Physics*, showing the variation of the shift in the O v 629 Å line along the loop.

Solar Physics **175:** 511–521, 1997.

2. Observations

The target for the observation is active region NOAA 7981 on the east solar limb on 27 July 1996. Between 10:00 UT and 10:52 UT on this date CDS ran an observing sequence pointed at the active region to observe the loop systems on the limb in 14 different spectral lines from ions with maximum equilibrium abundances at temperatures between ~10 000 K (He I 584 Å) and 2.7 MK (Fe XVI 360 Å). The observations were made with the normal incidence spectrometer (NIS) on CDS. For a description of this spectrometer and the CDS instrument in general we refer to Harrison *et al.* (1995).

The normal incidence spectrometer is stigmatic, providing a focussed image along the length of its entrance slit, and covers two wavelength ranges, 310–380 Å and 520–630 Å, simultaneously. Various slit widths are available and we use a $4'' \times 240''$ slit in this particular observation.

The spectral resolution of CDS-NIS is of particular interest for this investigation. According to Harrison *et al.* (1995) it amounts to values of $\lambda/\delta\lambda$ ranging from 3635 at the short wavelength end at 308 Å, to 4500 at 633 Å, the longest wavelength for NIS. With this resolution it is possible to measure Doppler shifts with an accuracy of approximately ± 10 km s^{-1} for lines observed with a good signal-to-noise ratio. With regard to the angular resolution, the size of the point spread function (FWHM) is $1.2''$ and $1.5''$, respectively, in the directions of the wavelength dispersion and perpendicular to it, while the half energy width is always less than $5''$. This makes it easy to distinguish individual loops above active regions if the density of loops is not too high, causing them to overlap.

Figure 1 shows images of the active region loops above the solar limb in 6 of the 14 lines. These images were made by rastering across the area outside the limb, moving the slit parallel to its long axis in steps of $4''$. Since the width of the entrance slit of the spectrometer was also $4''$ the raster completely covers its full area. The simultaneous images are built up from 60 exposures, each with an exposure time of 45 s. Thus, the size of the full image is $240'' \times 240''$ corresponding to the CDS field of view. The raster images are centred at $x = -1065$, $y = -136$, where the coordinates are approximately in arc seconds and refer to a coordinate system with its origin at sun center and x- and y-axes in the east–west and north –south directions. As a result the image field is located almost entirely outside the solar limb. The disc is present in the upper right corner of the images, but is easy to distinguish only in the cool He I and O V lines.

The intensity images are constructed from the integrated intensities in the lines. It should, however, be pointed out that the data retain the full spectral information, allowing us to derive line shifts and line widths at any point, in addition to the intensity.

Since the exposure time for each of the 60 exposures was 45 s, it takes 53 min to build up the full set of images, including overhead for each exposure. This is a significant point. The images are not simultaneous over the full field of view.

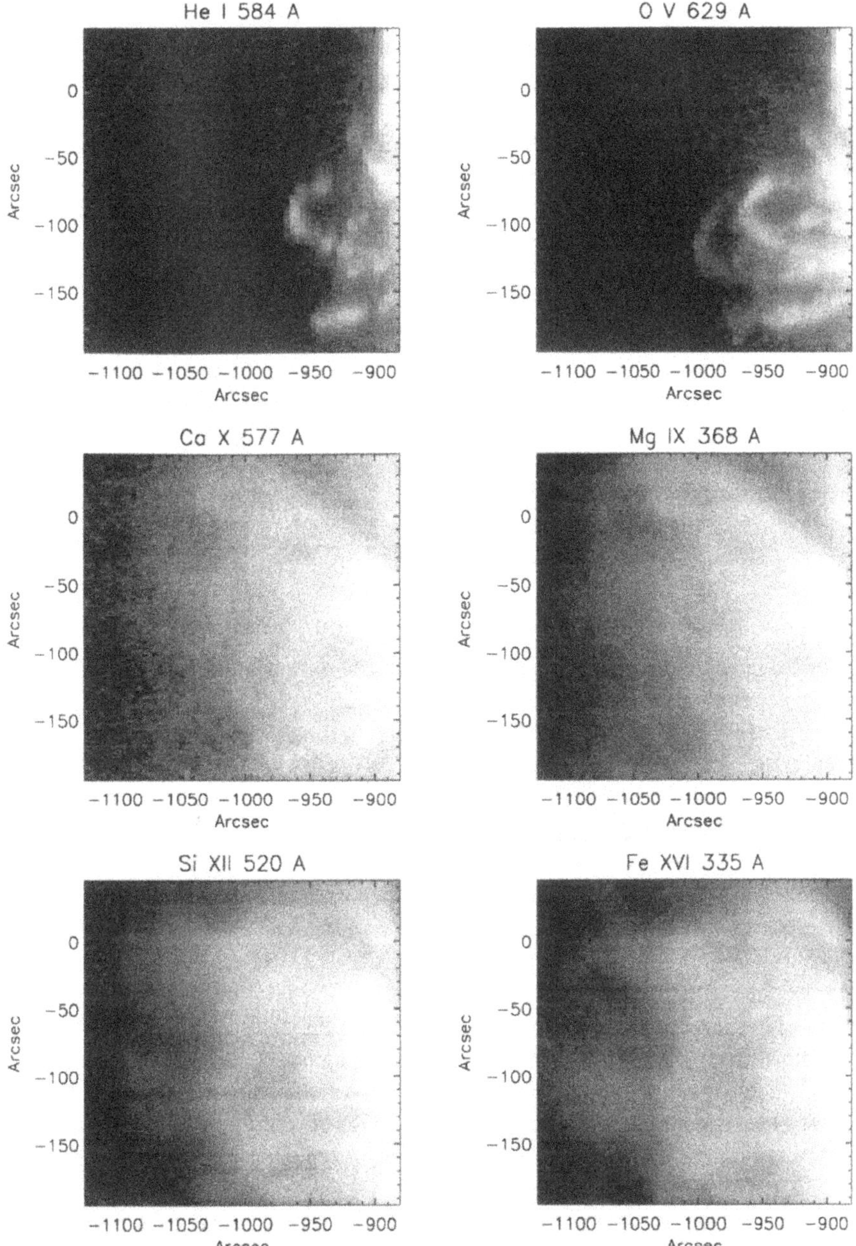

Figure 1. Monochromatic images in different lines, ordered from top left with increasing line-formation temperature. The observations were made on 27 July 1996, starting at 10:00 UT.

However, all points along one slit are observed simultaneously. This means that we have simultaneous observations of the north and south legs of the loops, i.e., along the vertical direction of the slit in the images. The various heights above the limb of the loop system are, however, observed at different times. For the cool loops in helium and oxygen the entire height of the loops are observed in about 20 min, while it takes 35 min to cover the hot coronal loops extending to higher altitudes.

As was mentioned above, the loop system was observed in 14 different lines. The six we have selected for Figure 1 are representative and cover three distinct temperature regions. The He I 584 Å line is emitted at ~10 000 K, while O V 629 Å is formed at 240 000 K, a temperature normally considered to be in the middle transition region. Ca X 577 Å and Mg IX 368 Å are formed at lower coronal temperatures of 740 000 K and 950 000 K, respectively, while the hot coronal lines, Si XII 520 Å and Fe XVI 335 Å, come from gas at temperatures of 1.9 MK and 2.7 MK. The temperature values refer to equilibrium ionisation temperatures, as compiled and calculated by Monsignori-Fossi (1995, private communication). Original sources for these calculations include data from Arnaud and Raymond (1992) for iron, and from Landini and Monsignori-Fossi (1991) for other ions.

3. Results

In this section we briefly describe how the appearance of the loop system depends on the temperature of the emitting gas. But mainly we address the observations of prominent shifts in the O V 629 Å line, indicating the presence of gas motions with high velocity in an active region loop with plasma at a temperature of 240 000 K.

3.1. Different Loops at Different Temperatures?

The images of the loop system of the active region at different temperatures show that the extent of the system is a function of temperature. With increasing temperature one sees more loops at progressively higher altitudes. Furthermore, we do not see the same loops at different temperatures. The cool loops below 300 000 K do not have hot coronal counterparts at 1.9–2.7 MK, and *vice versa*.

This property is not uncommon in active region loops. One notable example in the CDS observations from 22 March 1996 is described by Fludra *et al.* (1997). Their observations show that active region loops on the disc are clearly different structures in images from Mg IX 386 Å and Mg X 625 Å. These lines are formed only 150 000 K apart, at 950 000 K and 1.1 MK, respectively.

Our data from 27 July as well as many other observations with CDS in addition to the one from 22 March, clearly suggest that loops at transition region and coronal temperatures frequently occupy distinctly different locations. Thus, loops are usually not structures with a cool core surrounded by progressively hotter gas. A more typical picture may be of structures which are quasi-isothermal along an

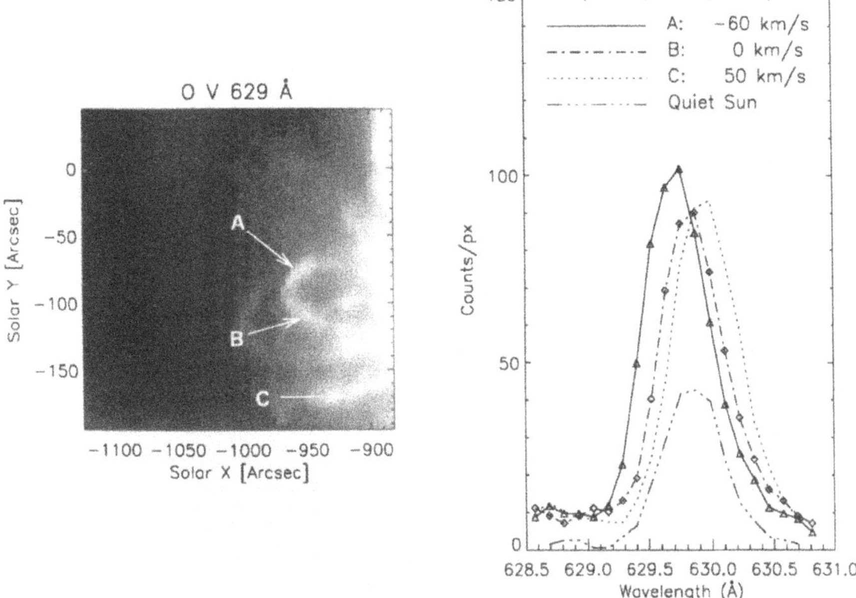

Figure 2. Active region loop system above the east limb observed in O V 629 Å on 27 July 1996 at 10:00 UT. The line profiles from three different spatial locations (A, B, and C) are displayed in the right panel together with a quiet-Sun disk-center profile.

extended portion of their length, existing side by side with other similar loop strands at different temperatures.

3.2. LINE SHIFTS AND POSSIBLE FLOWS

It turns out that strong line shifts are observed in the O V 629 Å line in part of the loop system. This is illustrated in Figure 2, where the left panel shows the monochromatic image of the O V line. Three spatial positions have been marked and the line profiles from these locations are displayed in the panel to the right, together with the average profile of the O V line from an extended quiet-region area at Sun center.

We note that the line shift at location A corresponds to a gas motion toward the observer with a velocity of 60 km s^{-1} relative to the quiet Sun. This is not the point with the highest blue shift in the loop. A neighbouring point at the same height has a shift corresponding to 66 km s^{-1}. At the top of the loop, in location B, there is no shift, and hence no motion along the line of sight. Location C is in a different loop from locations A and B. Here again we observe clear shifts, corresponding to motions of 50 km s^{-1} in the opposite direction, away from the observer.

3.2.1. *Animation of the Line Shifts*

To better understand what we are seeing, and particularly to discuss whether the shifts may correspond to flows of gas along magnetic field lines, one may map out the shift all along the loop where A and B are located. This has been done and the result is displayed in the animation in the CD-ROM attached to this issue of *Solar Physics*. The animation gives a strong impression of the possible dynamical condition present in the loop.

In the animation we are tracing one of the loops in the O V 629 Å line. The moving point in the O V image of the loop, to the left in the animation, marks the positions. On the right hand side the lower panel gives the observed line profile in the O V 629 Å line for the position where the spot is located, while the upper panel shows the corresponding shift in velocity units, km s^{-1}. Negative velocities correspond to blue shifts, or gas motion towards the observer. The line shifts have been derived from a fit to a Doppler profile. The uncertainty in the measured shifts amounts to $\sim$10 km s^{-1}. The dotted line in the lower right panel is the averaged line profile from an area at quiet-Sun center and is used as the wavelength reference. It should be noted that this reference line may itself be red-shifted by $\sim$5 km s^{-1} relative to the chromosphere or to an absolute reference system (see, e.g., Achour *et al.*, 1995; or Brekke, Hassler, and Wilhelm, 1997).

The marked positions start in the northern part of the loop, proceed through the top point and go down the southern leg. At low altitudes in the northern leg there are significant red shifts amounting to $\sim$20 km s^{-1}. However, it is doubtful whether these shifts relate to the loop we are studying. Thus, there seem to be low-lying, strongly emitting structures in the foreground separate from the rest of the loop. At higher altitudes the north leg radiates weakly in the O V line and no significant velocity is measured. However, at altitudes around 40 Mm above the solar limb, the radiation from the loop increases markedly and we suddenly enter a region with strong blue shifts, corresponding to velocities of 40 to 55 km s^{-1}, for an extended section of the loop. This part, with practically constant high line shift, persists almost to the top of the loop, where the shift suddenly drops to zero and remains at this value through the south leg.

It may be noted that the high blue shift starts just below the region with strong emission in the northern leg of the loop. At the top there is almost a gap in the loop, where the velocity also decreases suddenly. The emission increases again as one goes to lower altitudes, but emission in the south leg never becomes as intense as in the upper part of the north leg. The results of the animation are summarized in Figure 3. The upper panel shows the loop, with coordinates in megameters. The solar limb has been drawn into this picture for reference. The middle panel shows the intensities along the loop in arbitrary units, and the lower panel shows the shifts in km s^{-1}. The error bars give the measuring accuracy of $\pm$10 km s^{-1}.

Finally, it should be pointed out that we have looked for shifts in lines other than O V 629 Å in this loop system. We have not been able to distinguish any shifts above 20 km s^{-1}. This may not be surprising for the hotter lines. At coronal

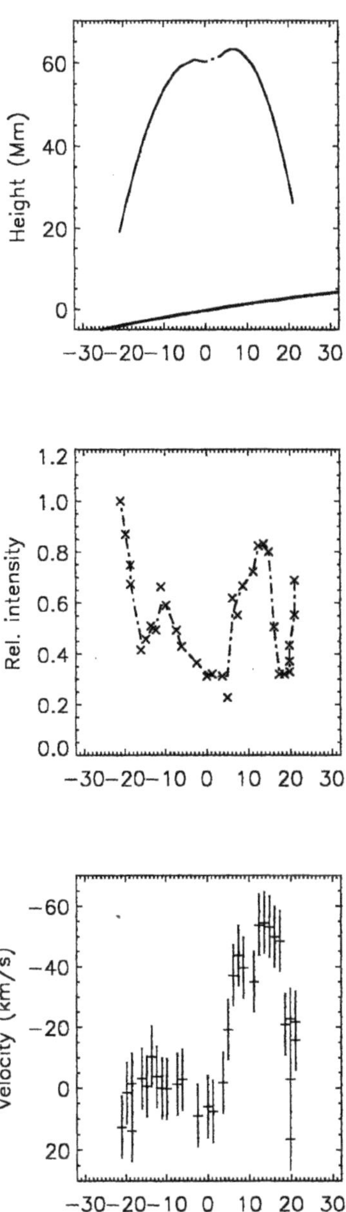

Figure 3. Upper panel: the loop, with coordinates in megameters. The solar limb has been drawn into this picture for reference. *Middle panel*: the intensity along the loop in arbitrary units. *Lower panel*: line shifts in km s^{-1}. Error bars are drawn for the measuring accuracy of ± 10 km s^{-1}.

temperatures, the density of loops is high and we have much more overlapping of loops. It may then be impossible to distinguish any regions with a clearly defined shift, since we will be looking through several such regions.

4. Discussion

In this section we shall touch on some of the problems of interpretation and compare the observations to calculations of mechanisms causing shifts in emission lines emitted from loops. We will not attempt to cover the literature on the subject comprehensively, but will discuss briefly two broad categories of explanations:

- The shifts are caused by gas flows, either going up one loop leg and down the other, or going first up and then down in the same leg.
- The shifts are caused by Alfvén waves or magnetosonic waves running along the magnetic field in the loop.

We may term these two model categories the flow models and the wave models and ask whether they are capable of causing shifts (or asymmetries being interpreted as shifts) as large as those we observe.

4.1. Gas Flows Along the Loop

Bray *et al.* (1991) and Mariska (1992) have reviewed the first class of models. It is possible to set up flows in loops if a strong asymmetry in the heating is introduced, with energy being deposited in a small part of the loop (see, e.g., Boris and Mariska, 1982; Mariska and Boris, 1983; McClymont and Craig, 1987; Mariska, 1988; Klimchuk and Mariska, 1988; McClymont, 1989; Spadaro, Antiochos, and Mariska, 1991). The most promising results come from flows in cool loops where velocities of 20–40 km s^{-1} are reached. Downflows of spicular material (see, e.g., Pneuman and Kopp, 1978; Athay and Holzer, 1982; Athay, 1987) or episodic heating followed by radiative cooling (Sturrock *et al.*, 1990; Raymond, 1990) may also cause flows to be set up, but usually with lesser velocities.

The loop we have observed is located at the solar limb. It is not possible to discern clearly its geometrical properties, but several uncertain points may be raised:

(1) The extent of the loop in the east–west direction as compared with its observed extent in the north-south direction, i.e., its azimuth angle.

(2) The tilt of the plane through the loop with the solar vertical direction.

(3) Whether the structure we have traced is a single loop or a composite of two or more loops.

The two first questions bear on the relation between the line-of-sight velocity as derived from the shifts to the real velocity of the gas in the loop. From the geometrical considerations it seems likely that the real velocities in the loop would be considerably larger than those measured from the shifts. Thus, while flow velocities of 50 km s^{-1} are sub-sonic at the temperature of formation for the O V

629 Å line, i.e., at 240 000 K, the real velocities are likely to exceed the sonic speed of 80 km s^{-1} at this temperature. It should also be pointed out that we cannot from our observations decide whether the flows are streaming upward or downward. This would depend on the loop geometry, particularly on the azimuth and the tilt of the loop.

A further difficulty is that the shifts are seen only in part of the loop. At least for gas moving up one leg and coming down the other one would expect to see the flows continue in the other leg. This difficulty can be overcome if we are really seeing two loops with markedly different temperatures in the two halves. The second half of the tracing in the animation would then be in the cool portion of a different loop, while the physically connected loop might continue at a much higher temperature. The gas velocities might then be unobservable in the south half of the loop owing to overlapping structures at coronal temperatures. There may be some evidence for this scenario. If we look at the loop picture in Figure 2 we note the gap at the top, already mentioned above. The considerable fuzziness at the top of the loop, and the difficulty with matching the north and south leg at the loop summit (see Figure 3), may be a further indication.

For flows going first up and then down in the same leg the observed velocity configuration does not appear likely, with high and nearly constant velocity in the upper part of the loop. Furthermore, the flowing gas apparently exists at the top of the loop and there would be no reason for it to stop abruptly there and not continue into the other leg.

It is uncertain to what degree the quoted calculations of flows may be applied to the situation we observe. Nevertheless, it is at present difficult to see how shifts of the size we are observing can be explained by models of flows in loops, similar to those quoted above. Such models will have to deliver much higher speeds to appear as reasonable explanations for the shifts reported here.

4.2. DISTURBANCES OR WAVES RUNNING ALONG THE LOOP

Acoustic waves generated by disturbances in the corona have been proposed by Hansteen (1993) as an explanation for the net red shifts seen in transition region lines. These waves do not result in sufficient shifts to explain our observations. Hansteen and Maltby (1992) and Hansteen, Maltby and Malagoli (1996) proposed to extend the same mechanism to MHD disturbances, which would become nonlinear as they propagated along magnetic field lines towards lower temperatures. With the Alfvén speed being much higher than the sound speed larger disturbances might result.

This idea has recently been pursued by Wikstøl, Judge, and Hansteen (1997). They have calculated the time dependent line profiles of transition region lines from gas in a coronal flux tube for downward running magneto-acoustic waves generated by disturbances at the top of loops, and from upward running sound waves. The disturbances are highly nonlinear and quickly steepen to form shocks.

The results for the magneto-acoustic waves are of particular interest. In this case the calculated gas velocities along the loop reach 110 km s^{-1}, while perpendicular to the magnetic field even higher velocities of 400 km s^{-1} are obtained. The loop observed in NOAA 7981 is more likely to be viewed from the side than with any large line-of-sight component along the loop. Thus, the high perpendicular velocities are very relevant for this case.

As with the flow models, the calculations of Wikstøl, Judge, and Hansteen (1997) do not directly apply to the observed case. However, their mechanism seems promising, since it appears that it may easily give line shifts as large as those observed. The wave model suffers from the same weakness as noted for the flow models when it comes to explaining why we see strong shifts only in part of the northern loop leg. Explanations along similar lines as discussed above for the flow models may be put forward, but they do not appear to be entirely satisfactory. Still, the wave mechanism seems considerably more promising than regular flows.

5. Conclusion

All the coronal instruments on SOHO have clearly demonstrated that the upper solar atmosphere is much more dynamical than was earlier expected. A reason for this emerging dynamical picture is that SOHO allows us to observe for long periods of time at a high time cadence. This makes it possible to study the general dynamical state of the Sun through its changing appearance over a range of spatial scales. Thus, we are able to catch dynamical events to a degree unsurpassed with earlier instruments.

In addition to observing time sequences at high cadence, the spectrometers on SOHO, CDS and SUMER add the dimension of spectral information. This gives us the opportunity to study directly the gas velocities, and thus the processes that causes the highly dynamical state. A previous set of observations with CDS emphasising the dynamical state of the solar atmosphere was presented by Brekke *et al.* (1997). The observations reported in this paper give an even clearer example of the highly dynamical state of the solar atmosphere, as well as a demonstration of the possibilities to study and effectively display the dynamics using CDS.

Acknowledgements

We would like to thank our many colleagues in the CDS consortium in the U.K., Germany, U.S.A., and Norway, who have made CDS into such an excellent instrument and provided operational and data reduction software support. This work was supported by the The Research Council of Norway. SOHO is a project of international cooperation between ESA and NASA.

References

Achour, H., Brekke, P., Kjeldseth-Moe, O., and Maltby, P.: 1995, *Astrophys. J.* **453**, 945.
Arnaud, M. and Raymond, J.: 1992, *Astrophys. J.* **398**, 394.
Athay, R. G.: 1987, *Nature* **327**, 685.
Athay, R. G. and Holzer, T. E.: 1982, *Astrophys. J.* **255**, 743.
Boris, J. P. and Mariska, J. T.: 1982, *Astrophys. J.* **258**, L49.
Bray, R. J, Cram, L. E., Durrant, C. J., and Loughhead, R. E.: 1991, *Plasma Loops in the Solar Corona*, Cambridge University Press, Cambridge.
Brekke, P., Hassler, D. M., and Wilhelm, K.: 1997, *Solar Phys.* **175**, 349 (this issue).
Brekke, P., Kjeldseth-Moe, O., Brynildsen, N., Maltby, P., Haugan, S. V. H., Harrison, R. A., Thompson, W. T., and Pike, C. D.: 1997, *Solar Phys.* **170**, 163.
Fludra, A., Brekke, P., Harrison, R. A., Mason, H.E., Pike, C. D., Thompson, W., and Young, P. R.: 1997, *Solar Phys.* **175**, 487 (this issue).
Hansteen, V.: 1993, *Astrophys. J.* **402**, 741.
Hansteen, V. and Maltby, P.: 1992, *Comm. Astrophys.* **16**, 137.
Hansteen, V. H., Maltby, P., and Malagoli, A.: 1996, in R. D. Bentley and J. T. Mariska (eds), *Magnetic Reconnection in the Solar Atmosphere*, PASP Conference Series, in press.
Harrison, R. A. *et al.*: 1995, *Solar Phys.* **162**, 233.
Klimchuk, J. A., and Mariska, J. T.: 1988, *Astrophys. J.* **328**, 334.
Landini, M. and Monsignori-Fossi, B.: 1991, *Astron. Astrophys. Suppl. Series* **91**, 183.
Mariska, J. T.: 1988, *Astrophys. J.* **334**, 489.
Mariska, J. T.: 1992, *The Solar Transition Region*, Cambridge University Press, Cambridge.
Mariska, J. T. and Boris, J. P.: 1983, *Astrophys. J.* **267**, 409
McClymont, A. N.: 1989, *Astrophys. J.* **347**, L47.
McClymont, A. N. and Craig, I. J. D.: 1987, *Astrophys. J.* **312**, 402.
Pneuman, G. W. and Kopp, R. A.: 1978, *Solar Phys.* **57**, 49.
Raymond, J. C.: 1990, *Astrophys. J.* **365**, 387.
Spadaro, D., Antiochos, S. K., and Mariska, J. T.: 1991, *Astrophys. J.* **382**, 338.
Sturrock, P. A., Dixon, W. W., Klimchuk, J. A., and Antiochos, S. K.: 1990, *Astrophys. J.* **350**, L31.
Wikstøl, Ø., Judge, P. G., and Hansteen, V. H.: 1997, *Astrophys. J.*, in press.

THE Mg/Ne ABUNDANCE RATIO IN A RECENTLY EMERGED FLUX REGION OBSERVED BY CDS

P. R. YOUNG and H. E. MASON
Department of Applied Mathematics and Theoretical Physics Silver Street, Cambridge CB3 9EW, U.K.

(Received 13 March 1997; accepted 14 May 1997)

Abstract. Evidence for the existence of the FIP-effect in the transition region is presented here based on recent observations from the Coronal Diagnostic Spectrometer (CDS) on-board the Solar and Heliospheric Observatory (SOHO). Observations of an emerging flux region in lines of Mg V–VII and Ne VI–VII reveal differences in the relative Mg/Ne abundance of a factor of 9.2 between two transition region brightenings separated by less than 1 arc min on the Sun. The lower abundance ratio is approximately equal to the photospheric Mg/Ne value and is associated with a small loop-like feature in the central, hottest part of the active region. The higher abundance ratio is found in spike-like structures at the edge of the active region. A density diagnostic of O IV is used to derive an electron number density of $10^{11.3}$ cm^{-3} for the low Mg/Ne brightening, while a Mg VII diagnostic gives a density of $10^{9.2}$ cm for the high Mg/Ne brightening.

1. Introduction

It is now generally accepted that, in the corona, elements with a low first ionisation potential (FIP) are around four times more abundant relative to elements with a high FIP than they are in the photosphere (Meyer 1985a, b). The dividing line between low FIP and high FIP elements is usually taken to be around 10 eV. The process responsible for this fractionation of elements could provide a valuable insight into the production and acceleration of the coronal plasma. The elements neon and magnesium are particularly useful for studying this phenomena as their FIPs are 21.6 and 7.6 eV, respectively. Several Ne and Mg lines fall into the extreme ultraviolet (EUV) wavelength range and have been extensively studied with the NRL-S082A slitless spectrograph on *Skylab*, see, e.g., Widing, Feldman, and Bhatia (1986), Widing and Feldman (1989, 1992, 1993), Feldman and Widing (1990, 1993), and Feldman, Widing, and Lund (1990). The variation of elemental abundances for different solar features at first seemed somewhat confused and was attributed to whether the magnetic field structures were open or closed.

A different pattern now seems to be emerging, based on the recent studies of the Mg VI/Ne VI ratio by Sheeley (1995, 1996). In the latter paper, it is suggested that a photospheric Mg VI/Ne VI ratio is always found at chromospheric heights, but it is only in strong enhancements such as two-ribbon chromospheric flares that the lines become strong enough to be measurable. Such flares are not necessarily associated with emerging flux. The coronal values of the Mg VI/Ne VI ratio are always present in 'long, spiky structures that extend outward from strong fields' (Sheeley, 1996).

Solar Physics **175:** 523–539, 1997.

Table I
A list of the NIS studies used on 6 June 1996 to observe AR 7968

Study/version number	Exposure time (s)	Area covered (arc min^2)	Slit (arc sec^2)	Duration (min)	FITS file
LARGE_T/v9	15	4×4	2×240	38	s3072
TEDE12/v4	50	1×2	4×240	14	s3073
TEDE12/v4	50	1×2	4×240	14	s3074
TRANREG/v10	8.5	2×2	2×240	12	s3075
TEDE12/v4	50	1×2	4×240	14	s3076
TEDE12/v4	50	1×2	4×240	14	s3077

The main limitation of the NRL-S082A observations used to derive these results was the feature of overlapping images, which severely restricts the availability of suitable active region datasets. The Coronal Diagnostic Spectrometer (CDS) instrument on the Solar and Heliospheric Observatory (SOHO) permits the unambiguous study of the relationship between these Ne and Mg enriched structures in active regions. In particular, this paper illustrates the powerful diagnostic capability of the Normal Incidence Spectrometer component of CDS (Section 2) applied to the analysis of transition region brightenings in an active region observed on 6 June 1996. We derive the Mg/Ne relative elemental abundance (Section 9) and the transition region electron densities (Section 8) for a small, compact loop and the base of a diffuse coronal structure. The analysis methods and limitations are discussed in Sections 5 and 7. Finally, in Section 10, we discuss the nature of the very intense transition region brightenings seen in this active region.

2. The Instrument

The Normal Incidence Spectrometer (NIS) of CDS is described in detail in Harrison *et al.* (1995); we note here only those aspects relevant to the observations discussed in this paper. A full discussion of the first results from CDS is given in Harrison *et al.* (1997).

NIS creates two spectrally-dispersed images of the entrance slit of the instrument on a CCD detector. The two wavelength bands are 307–379 Å (NIS 1) and 514–632 Å (NIS 2) with spectral resolutions of around 0.32 and 0.54 Å, respectively. The full length of the slit projects to around 4 arc min on the Sun, and spatial resolution along the slit is 2–3 arc sec. To build up images of a region on the Sun, a mirror between the telescope and the slit can be rotated such that 4 arc min of the Sun can be seen for any one pointing of the instrument. For spectroscopic work, two NIS slits are provided: 2×240 arc sec^2 and 4×240 arc sec^2, providing spatial resolutions of 2 and 4 arc sec, respectively.

To cover a 4×4 arc min^2 area with the 2×240 arc sec^2 slit taking both full NIS 1 and NIS 2 bands generates a huge amount of data that would take around 10 hr to telemeter to Earth. To optimize the observations, it is necessary to pre-select parameters such as the area to be covered, the slit and the spectral windows. A set of such parameters is called a *study* and the studies used in the current work are LARGE_T, TEDE12, and TRANREG (Table I). The unique feature of these CDS studies is to allow the simultaneous extraction of both spatial and spectral information, thus enabling quantitative spectroscopic diagnostic analyses to be carried out.

3. An Overview of the CDS Observations

The details of the CDS observing sequence used to study the active region can be found in Table I. The LARGE_T raster was used to provide an overview of the region in several lines formed at different temperatures – some images from this raster are presented in Figure 1. TEDE12 is a study designed to look at many different lines, yielding diagnostics of temperature, density and also abundances. As the spatial extent of TEDE12 is only 1×2 arc min^2, the raster was run twice to cover both halves of the active region. TRANREG takes some of the strongest transition region lines with a short exposure, providing a 'snapshot' of the active region.

The active region studied here emerged on 3 June 1996 to the solar-west of a large decaying active region, and was assigned the identification AR 7968. The NIS observations were performed on 6 June 1996, between 22:00 UT and 24:00 UT when the active region was close to disk centre. A Kitt Peak magnetogram image taken at 15:15 UT is shown in Figure 1, together with four images from the LARGE_T raster. The coordinates shown for each image are in arcseconds from the centre of the Sun, which is assigned $(0, 0)$. During the 7 hours between the taking of the magnetogram image and the running of LARGE_T, the active region rotated approximately 70 arc sec.

All of the ions discussed in this work are formed over a narrow temperature region and so we identify a single temperature with each ion that corresponds to the temperature of maximum ionisation, $T_{\max}$. This quantity is derived for the iron ions from Arnaud and Raymond (1992) and for all other ions from Arnaud and Rothenflug (1985). The images in Figure 1 are from ion lines formed at different temperatures (given in the caption of Figure 1). The coolest line is O V 629.7 Å, formed in the transition region, where we see four footpoints. The rightmost footpoint appears to coincide with the strong magnetic spot seen in the magnetogram image. Weakly-emitting, spike-like structures can be seen emerging from this region in Ca X, analogous to the structures described by Sheeley (1996). There is little high temperature (Si XII) emission, however, instead, the high temperatures are seen over the remaining three O V footpoints, which correspond to the

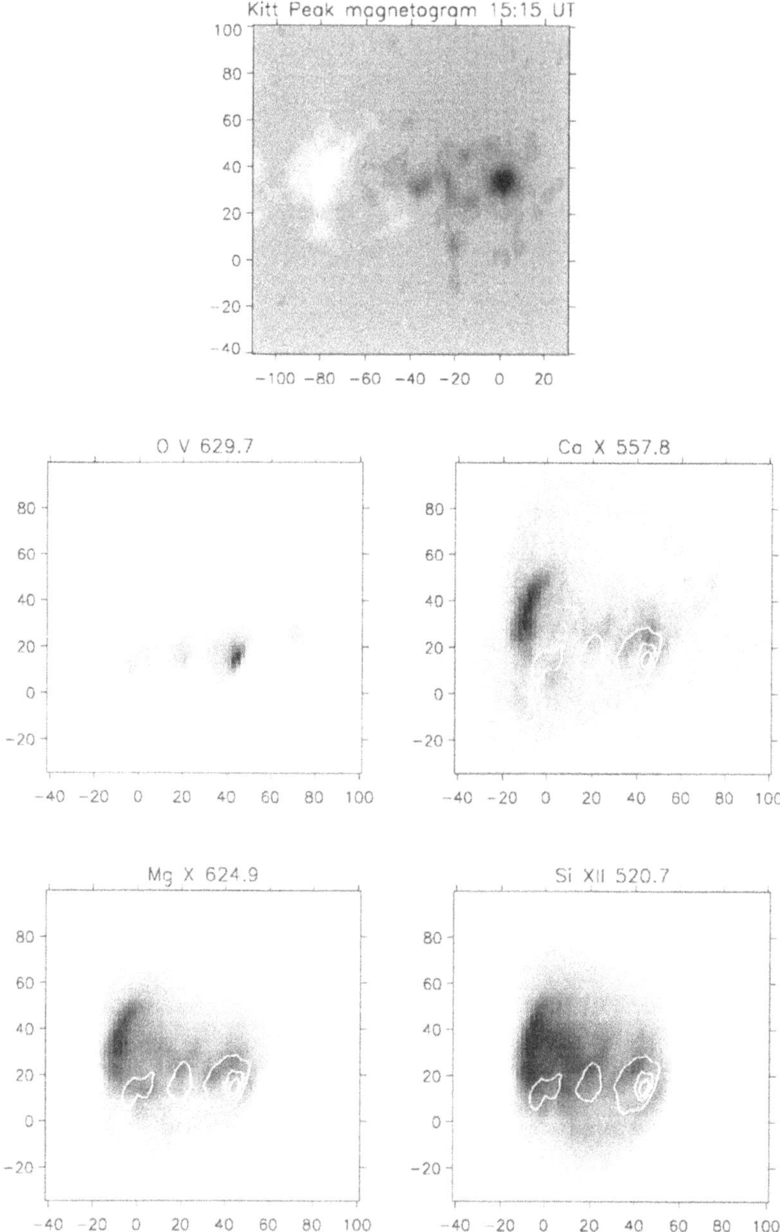

Figure 1. Images of the active region. The uppermost image is a Kitt Peak magnetogram taken between 15:15 and 16:10 UT, while the lower four images are from the LARGE_T raster, each taken simultaneously between 22:00 and 22:40 UT. The four lines have $\log T_{max}$'s of 5.5, 5.8, 6.2, and 6.3, respectively. The contours on the Ca X, Mg X, and Si XII images show the O V emission. The x and y axes give solar coordinates in arc sec with $(0, 0)$ being the centre of the Sun. The active region rotated by aound 70 arc sec betwen the taking of the magnetogram image and the NIS images, and so the displayed images cover approximately the same spatial region.

Table II

Details of the EUV emission lines studied in this work. The different components to the lines are specified in descending order of importance. T_{max} is the temperature of maximum ionisation of the ion.

Ion	log T_{max}	CDS wavelengths (Å)	Band of NIS	Transition: Configurations	Transition: Terms	CHIANTI wavelengths (Å)
Mg V	5.4	353.1	1	$2p^4 - 2s2p^5$	$^3P_2 - ^3P_2$	353.09
					$^3P_1 - ^3P_1$	353.30
Mg VI	5.6	349.2	1	$2s^22p^3 - 2s2p^4$	$^2D_{5/2} - ^2D_{5/2}$	349.17
					$^2D_{3/2} - ^2D_{3/2}$	349.14
					$^2D_{5/2} - ^2D_{3/2}$	349.12
					$^2D_{3/2} - ^2D_{5/2}$	349.19
Mg VII	5.8	319.0	1	$2s^22p^2 - 2s2p^3$	$^1D_2 - ^1D_2$	319.03
		367.7	1	$2s^22p^2 - 2s2p^3$	$^3P_2 - ^3P_2$	367.67
					$^3P_2 - ^3P_1$	367.68
Ne VI	5.6	562.8	2	$2s^22p - 2s2p^2$	$^2P_{3/2} - ^2D_{5/2}$	562.80
					$^2P_{3/2} - ^2D_{3/2}$	562.71
Ne VII	5.7	561.7	2	$2s2p - 2p^2$	$^3P_2 - ^3P_2$	561.73
					$^3P_1 - ^3P_1$	561.38
O IV	5.2	553.3	2	$2s^22p - 2s2p^2$	$^2P_{1/2} - ^2P_{3/2}$	553.33
		625.9	2	$2S2p^2 - 2p^3$	$^4P_{5/2} - ^4S_{3/2}$	625.85

fractured magnetic fields seen in the magnetogram image. Considerable structure can be seen in the coronal lines (Mg X and Si XII) in this area with small, apparently loop-like structures with length-scales of around 10–40 arc sec. Fludra *et al.* (1997) display images from a limb active region which displays compact low-lying, high temperature structures in addition to the usual larger coronal loops, and we suggest that the small loop structures seen in the Figure 1 images are low-lying (height ~ 20 000 km), connecting the magnetic fragments that can be seen in the magnetogram image.

We aim in this paper to compare transition region density and abundance diagnostics in the two right-most footpoints seen in the O V image of Figure 1. We will use emission lines of Mg V–VII, Ne VI–VII, and O IV; detailed information for which can be folmd in Table II.

4. Mg VI and Ne VI Features Seen in the TEDE12 Data

The Mg VI 349.2 Å and Ne VI 562.8 Å lines were observed with the TEDE12 study, and are of interest because their temperatures of maximum ionisation, T_{max}, are the same (Table II) so implying that they are emitted from essentially the same plasma elements and hence images in the two lines should be similar.

The images from the two TEDE12 rasters s3073 and s3074 are shown in Figure 2, where they are displayed in two halves as TEDE12 only covered a 1×2 arc min^2 region and so was run in adjacent positions to cover the whole of the active region. The image derived in putting together the two halves corresponds to the region displayed in Figure 1.

It is clear that, although similar structures can be seen in the two images – as one would expect, there are two brightenings (seen on the right-hand side of s3073 and the left-hand side of s3074) that are considerably more intense in Ne VI than they are in Mg VI. We will henceforth concentrate on the s3074 data where, of the two brightenings seen in the Mg VI image we label the left-hand one A and the right-hand one B – see Figure 3. The strong Ne VI brightening corresponds to A, while B coincides with the base of the Ca X spike-like structures.

5. Line Analysis

To quantitatively estimate the difference in abundance between the two brightenings it is necessary to fit line profiles to the magnesium and neon lines. To improve the photon statistics for the weaker lines.we sum over several pixels in the areas around the two brightenings – the areas used are indicated in Figure 3. For the area chosen for feature B, the total counts in the lines vary from around 4500 in the Mg V 353.1 Å and Ne VII 561.7 Å lines to almost 30 000 counts in the Ne VI 562.8 Å line. For feature A the counts range from 1000 counts in the Mg V 353.1 Å line to around 100 000 counts in the Ne VI 562.8 Å line.

Clearly, with this level of photon counts the data are not photon-limited and the largest source of errors stem from the line-fitting process and the estimate of the continuum level in the data. The lines were fitted as Gaussians using the ezmgn.pro routine which is available as part of the CDS software. The largest difficulties in the line fitting came with feature A as this lies in the region of strong coronal emission and so the magnesium lines become more difficult to fit – this can be seen in Figure 4 where we compare the spectra from the two brightenings.

6. Atomic Data

Emission lines of Mg V–VII, Ne VI–VII, and O IV are interpreted here, and atomic data for these ions are taken from the CHIANTI atomic database that is described in Dere *et al.* (1997), where detailed information on the sources of atomic data can be found. We give brief details of these sources below.

Mg V is a member of the oxygen iso-electronic sequence for which there is very little published atomic data available. We use unpublished work of A. K. Bhatia for both electron collision strengths and transition probabilities. For Mg VI we use the published work of Bhatia and Mason (1980) for collision strengths and transition

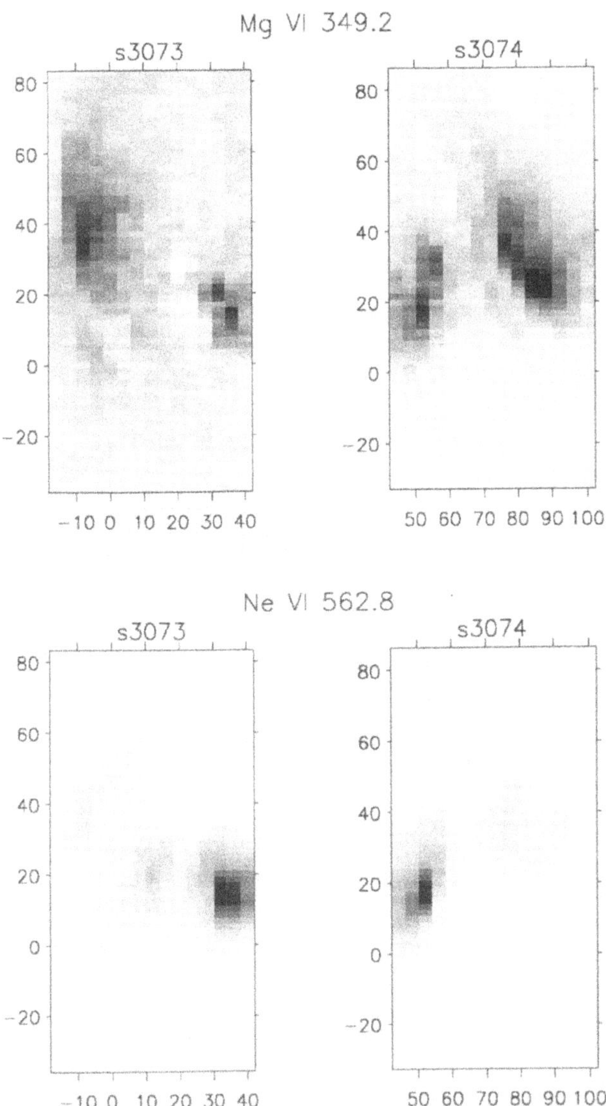

Figure 2. Images of the active region in the Mg VI 349.2 Å and Ne VI 562.8 Å lines. The x and y axes give solar coordinates in arc sec with (0, 0) being the centre of the Sun. The features A and B that are discussed in the text are shown in the s3074 image at approximate positions (52, 17) and (85, 22), respectively.

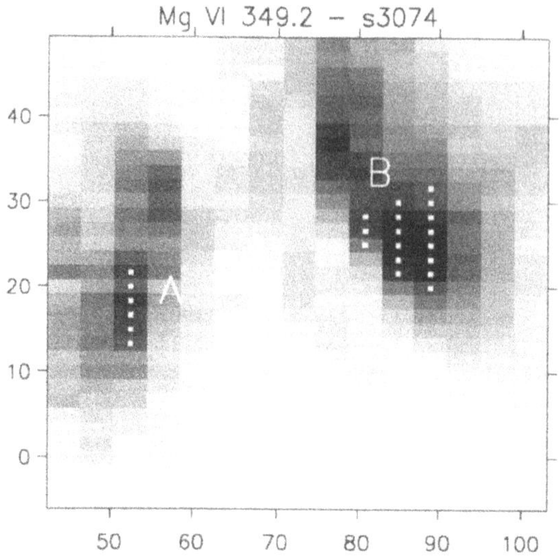

Figure 3. Blow-up of the s3074 Mg VI 349.2 Å image showing the pixels summed over for features A and B. The pixels are indicated by small white squares.

Table III
Line intensities for the magnesium and neon ions in the brightenings A and B

Ion	Line (Å)	Line intensities (ergs cm^{-2} s^{-1} sr^{-1})	
		A	B
Mg V	353.1	500 ± 200	580 ± 90
Mg VI	349.2	790 ± 140	740 ± 60
Mg VII	367.7	860 ± 200	600 ± 70
Ne VI	562.8	3200 ± 150	360 ± 30
Ne VII	561.7	630 ± 50	54 ± 9

probabilities. The recent collisional calculations of Bhatia and Doschek (1995) are used for Mg VII.

O IV and Ne VI are both members of the boron sequence and extensive calculations for these ions have been presented by Zhang and co-workers in a series of papers for more details see Dere *et al.* (1997). Ne VII is a member of the beryllium sequence and collisional and radiative data have been calculated by Zhang and Sampson (1992).

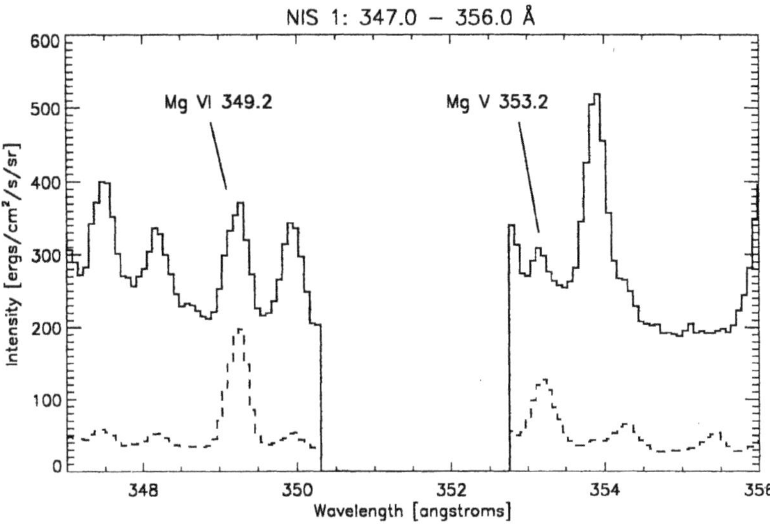

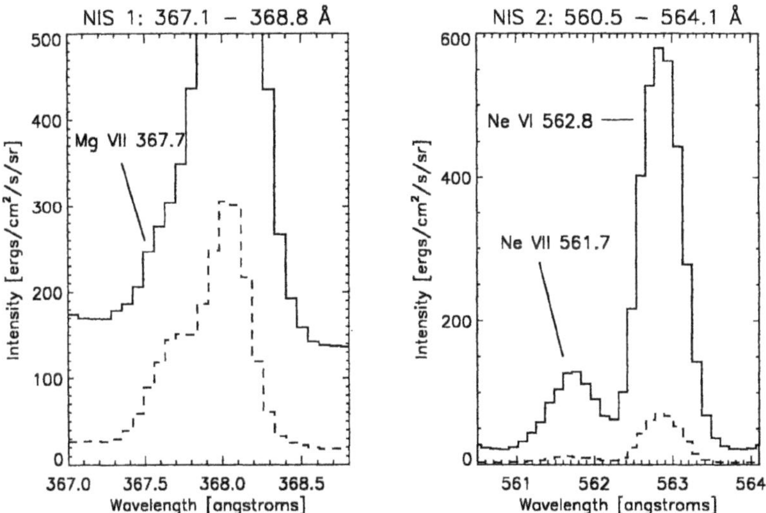

Figure 4. Comparisons of the neon and magnesium lines between the two features. The complete line corresponds to feature A while the broken line corresponds to feature B. Note that the magnesium lines have similar magnitudes in both A and B while the neon lines are clearly stronger in A. The intensity scales are the same for both A and B – the enhanced continuum seen in A is a real feature of the data.

7. Deducing Relative Abundances from the NIS Line Intensities

The use of emission lines to study ion relative abundances has been discussed extensively by previous authors (cf., Widing and Feldman, 1989; Meyer, 1993; Mason, 1995), and we detail below the method used here.

The expected intensity of a line emitted by an ion in a plasma of temperature T and with electron number density N_e is

$$I_{th} \sim \varepsilon(N_e)\mathcal{F}(T)\mathcal{A}\,, \qquad (1)$$

where $\mathcal{F}(T)$ the ionisation fraction and $\mathcal{A}$ the abundance of the element. ε represents the emissivity of a unit volume of the emitting ions, given here as

$$\varepsilon = \frac{n_j A_{ji}}{\lambda_{ji}}\,,$$

for the transition $j \rightarrow i$, where n_j is the population of the upper level relative to the total population within the ion, A_{ji} is the radiative decay rate for the transition and λ_{ji}; is the wavelength. Both ε and $\mathcal{F}$ are actually functions of both N_e and T, however ε is relatively insensitive to temperature for the ions considered here, while present ionisation balance calculations do not account for variations of $\mathcal{F}$ with density.

The observed intensity of a line consists of the sum of the emissions from all the plasma elements in the line of sight, with each element having its own density and temperature. For each plasma element we have a theoretical estimate of the emission from Equation (1) and so our estimate of the observed intensity is

$$I_{\rm obs} \sim \int_V I_{th}\,{\rm d}V\,,$$

where V is the volume of plasma along the line of sight. As it is impossible to accurately describe the detailed density and temperature structure in V, a derivation of the relative abundance requires some simplifications.

One method is to take emission lines from many different ions and construct an emission measure curve which attempts to describe the distribution of plasma with temperature. In matching up emission measure curves for different elements one arrives at estimates of the element abundances (see, e.g., Pottasch, 1964).

If two ions have $\mathcal{F}(T)$ functions that are very similar then it is possible to make the assumption that the emitting ions are in equal ratio for all plasma elements, i.e., that the $\mathcal{F}(T)$ functions are identical save for a scaling factor. The Mg VI and Ne VI ions provide just such a case and we will utilise this approximation here. However, to demonstrate that the observed effects are also seen in neighbouring ions we will also give line intensities from Mg V, Mg VII, and Ne VII.

From Equation (1), if we assume the $\mathcal{F}(T)$ functions for Mg VI and Ne VI are in the ratio Q, then the observed ratio of the 349.2 Å and 562.8 Å lines corresponds to

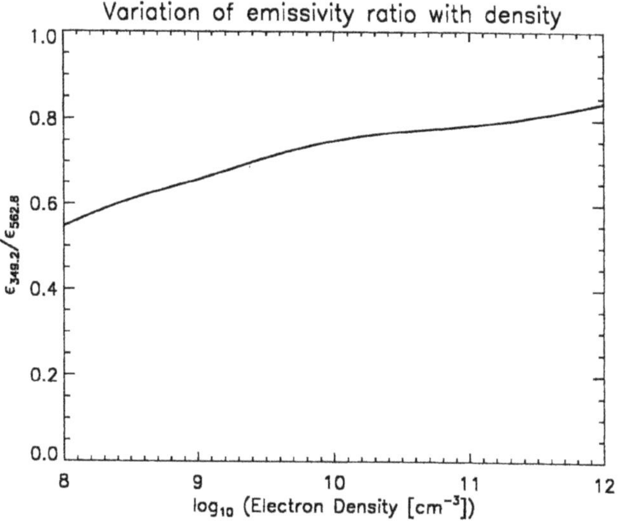

Figure 5. Variation of the ratio of the theoretical emissivities of the Mg VI 349.2 Å and Ne VI 562.8 Å lines as derived from the atomic data contained in the CHIANTI database.

$$R_{\text{obs}} = \frac{I_{349.2}}{I_{562.8}} = \frac{\varepsilon_{349.2}(N_e)}{\varepsilon_{562.8}(N_e)} Q \frac{A(\text{Mg})}{A(\text{Ne})} . \tag{2}$$

The ratio of the emissivities of the two lines can be calculated using the CHIANTI atomic emission code and is displayed in Figure 5. Clearly this function is a slowly-varying function of density and so the derived abundance is relatively insensitive to the density values chosen, however we use the values derived from the Mg VII and O IV density diagnostics discussed in the section below.

8. Electron Densities in the Brightenings

Although the principal density diagnostics in the NIS wavebands are only appropriate for the corona, there are two transition region ions that can yield densities under the extreme conditions presented here in the two brightenings. Mg VII emits four lines in the NIS 1 bandpass at wavelengths 319.0, 363.7, 365.2, and 367.7, and a useful density diagnostic can be formed by taking the 319.0 line relative to any of the latter three lines (Mason and Bhatia, 1978). Normally in active regions, however, the 319.0 Å line is blended with a Ni XV line (see, e.g., the SERTS-89 spectrum of Thomas and Neupert, 1994), which is formed at $\log T = 6.4$ and, as discussed by Young, Landi, and Thomas (1997), it is not possible to accurately de-convolve this blend. Additional difficulties with the NIS stem from partial blendings of the 365.2 Å and 367.7 Å lines with other lines*, while 363.7 Å is very

* The 365.2 Å line with Si XI 365.4 Å, and the 367.7 Å line with the strong Mg IX 368.1 Å line.

Table IV

Line intensities for the density sensitive O IV and Mg VII line pairs in the transition region brightenings

Feature	Ion	Line	Intensity (ergs cm^{-2} s^{-1} sr^{-1})
A	O IV	625.9	580 ± 90
		553.3	2610 ± 300
B	Mg VII	319.0	296 ± 40
		367.7	675 ± 80

Table V

Electron densities derived from the O IV and Mg VII density diagnostics

Feature	Ion	Ratio	Observed value	$\mathrm{Log}_{10}N_e$
A	O IV	625.9/553.3	0.22 ± 0.04	$11.3^{+0.2}_{-0.2}$
B	Mg VII	319.0/367.7	0.44 ± 0.08	$9.2^{+0.1}_{-0.2}$

weak. In feature B, however, there is very little high temperature coronal emission (see Figure 1) and so we can ignore the contribution of Ni XV to the 319.0 Å line, while the 367.7 Å line becomes of sufficient strength relative to the Mg IX 368.1 Å line to make it easily measurable (Figure 6).

The region over which the Mg VII lines have been summed is slightly different to the feature B shown in Figure 3, on account of the Mg VII lines being formed slightly higher up the spikelike structure. We display the averaged profiles in Figure 6 and give the fitted intensities in Table IV. Figure 8 shows the expected variation of the 319.0/367.7 line ratio with electron density and the density derived from the observations is given in Table V.

An O IV density diagnostic can be formed by taking one of the triplet of lines at 624.6 Å, 625.1 Å, and 625.9 Å relative to any of the multiplet of lines found at 553.3 Å, 554.1 Å, 554.5 Å, and 555.3 Å. The strongest lines in these two multiplets are 625.9 Å and 554.5 Å; however the 625.9 Å is normally very weak and so difficult to measure accurately close to the strong Mg X 624.9 Å line. In feature A, though, we find large counts in all of the oxygen lines with little enhancement in the coronal lines and so for the first time in solar spectroscopy we are able to use the 625.9 Å line in a density diagnostic – Figure 7 shows the 625.9 Å profile. From the fit to the line we reduce the intensity by 15% to allow for blending with the O IV 625.1 Å line, and we display this corrected intensity in Table IV.

The strong 554.5 Å line is saturated in feature A – hence the unusual profile seen at this location in Figure 7 and so we instead use the 553.3 Å line which is around five times weaker and so not saturated. As Figure 7 shows, this line is

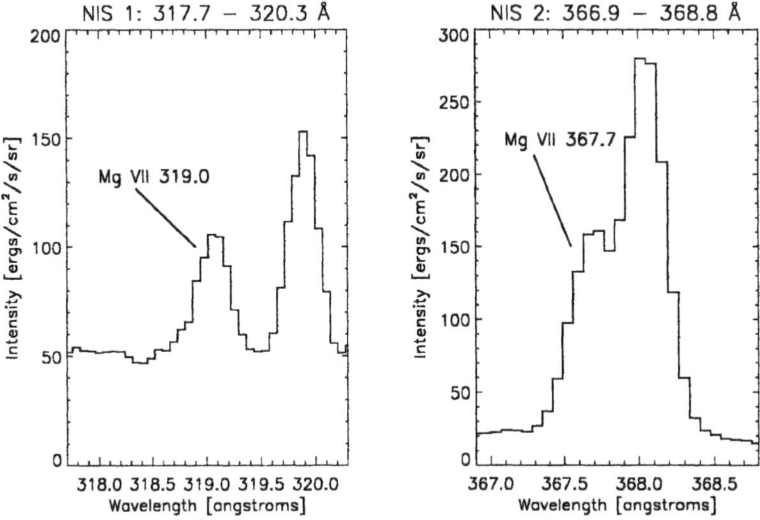

Figure 6. Profiles of the Mg VII 319.0 Å and 367.7 Å lines in brightening B. Note that the difference in continuum levels for the lines is due to a correction applied to the data by the NIS calibration routines. Essentially the instrument is less sensitive in the 319 Å region by around a factor two.

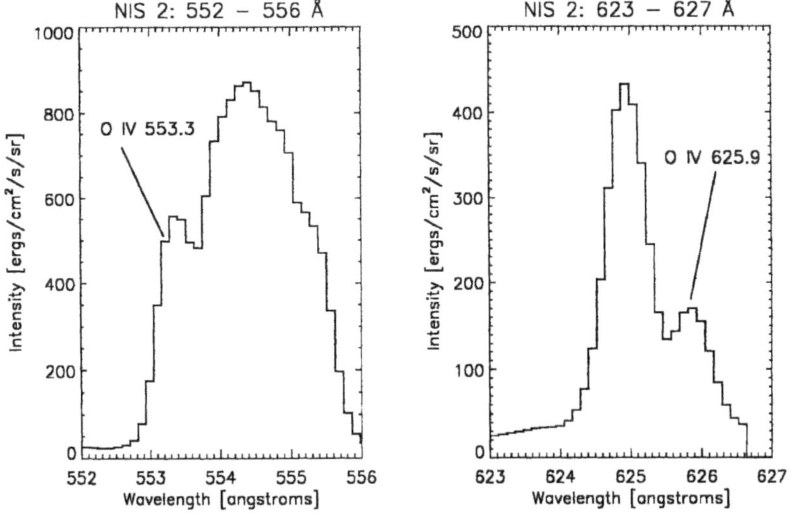

Figure 7. Display of the two O IV lines that were fitted to give the line intensities presented in Table IV. The unusual profile at around 554.5 Å belongs to the O IV 554.5 Å line which was saturated in this data. The strong line to the blue-side of O IV 625.9 Å is Mg X 624.9 Å.

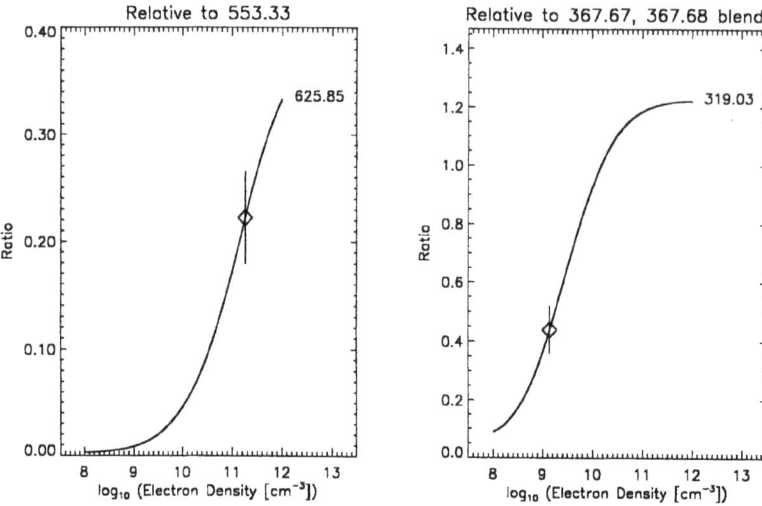

Figure 8. Theoretical variation of the O IV 625.9/553.33 (*left*) and Mg VII 319.0/367.7 (*right*) ratios with density, predicted from data in the CHIANTI atomic database. The observed values of the ratios are displaved with error bars.

blended slightly with the stronger 554.5 Å line but an intensity can be estimated by assuming a fixed width for the line*. The line intensities for the 553.3 Å and 625.9 Å lines are displayed in Table IV while the predicted density is shown in Table V. The theoretical variation of the 625.9/553.3 ratio with density is shown in Figure 8.

Although the Mg VI and Ne VI ions are formed at a different temperature to O IV and Mg VII (and so different densities may apply), we note that their $T_{\rm max}$'s are not far apart and so one would only expect small corrections to the derived densities. As the emissivity ratio shown in Figure 5 is relatively insensitive to density, we shall assume the O IV and Mg VII densities apply to Mg VI and Ne VI for both brightenings.

9. Relative Abundances in the Brightenings

Using Equation (2) we are now in a position to estimate the relative Mg/Ne abundances in the two brightenings. From Arnaud and Rothenflug (1985), we can estimate the Mg VI/Ne VI ion ratio, Q, as 1.22 by comparing ionisation fractions at the $T_{\rm max}$ of the ions. For feature A the density of $10^{11.3}$ cm^{-3} implies that the ratio of the emissivities is 0.78 (see Figure 5), and so with the observed ratio of 0.25 ± 0.05 for the 349.2 Å and 562.8 Å lines we have a relative abundance of

* The widths of the lines seen in NIS spectra are entirely instrumental and are typically 0.32 Å and 0.54 Å for NIS 1 and NIS 2, respectively.

$(0.25 \pm 0.05)/(0.78 \times 1.22) = 0.26 \pm 0.05$. For feature B, we have an emissivity ratio of 0.70 and an observed ratio of 2.0 ± 0.24, giving an estimated relative abundance of $2.0 \pm 0.24/(0.70 \times 1.22) = 2.4 \pm 0.3$.

The logarithmic photospheric abundance of magnesium is 7.58 ± 0.05 (Anders and Grevesse, 1989) on a scale where the hydrogen abundance is 12.0. The photospheric abundance of neon is 8.08 ± 0.06 (Grevesse, Noels, and Sauval, 1992), so giving a photospheric Mg/Ne relative abundance of 0.32 ± 0.06. Our results thus suggest that feature A is displaying photospheric abundances, while feature B shows magnesium enhanced over neon by a factor of 9.2.

10. Intense Transition Region Brightenings

The event A is an example of a particularly intense transition region brightening. Such brightenings have been described in Mason *et al.* (1997) and Fludra *et al.* (1997), but event A represents one of the few times when a diagnostic study has captured such an event. We note below some of the properties that have been inferred from this particular sequence.

(1) Very large counts in many of the transition region lines. The O V 629.7 Å line reaches a maximum value of around 57 000 ergs cm^{-2} s^{-1} sr^{-1} in feature A compared to typical quiet-Sun values of 1000 ergs cm^{-2} s^{-1} sr^{-1}, i.e., an enhancement of a factor of 50.

(2) Strongly enhanced continuum in both NIS 1 and NIS 2. The continuum level typically lies at around 0–20 counts and is at a fairly constant level in both bands. In the brightenings seen here the continuum rises to levels of 150–200 counts. The NIS continuum is thought to be due to scattered light, the principal component of which is Lα, which is likely to be considerably enhanced in such a brightening.

(3) The brightening does not appear at all in coronal lines, indeed the strong O V brightening in Figure 1 is not evident in Ca X 557.8 Å which is formed at temperatures of around $\log T = 5.8$.

(4) This paper suggests that such brightenings display photospheric abundances and that the electron density is around 10^{11} cm^{-3}.

It is not possible to estimate the lifetime of feature A, although the second set of TEDE12 runs in Table I show that both of the strong Ne VI brightenings displayed in Figure 2 are present at a similar intensity level, implying a lifetime of at least 40 minutes. A similar diagnostic study run 17 hours later on the active region did not reveal any such strong brightenings.

11. Discussion and Conclusions

In this paper we have used EUV emission line diagnostics observed by CDS to investigate the properties of two transition region brightenings seen in a newly emerged active region on 6 June 1996. The properties are as follows:

(1) Brightening A appears to be associated with a low-lying loop and is very intense in many of the transition region lines. It displays a photospheric Mg/Ne abundance of 0.26 ± 0.05 and has a density of around 10^{11} cm^{-3}.

(2) Brightening B lies at the base of larger loops that extend outwards from the active region. These loops show a larger Mg/Ne abundance of 2.4 ± 0.3, nine times larger than for feature A. A density of $10^{9.2}$ cm^{-3} was deduced from a Mg VII density diagnostic.

It is to be noted that, as suggested by the O V image in Figure 1, the emission lines of the oxygen ions O III, O IV, and O V are also stronger in feature A than in feature B. The FIP of oxygen is 13.6 eV and it is classed as a 'high-FIP' element (Feldman, 1992), and so is consistent with the result found for neon.

The absolute intensities of the emission lines studied here scale approximately as N_e^2, yet the ratio of the Ne VI 562.8 Å line, for example, in the two brightenings is only 8.8, while the difference in densities for the brightenings would imply an expected ratio of around 1000 (assuming that neon has the same absolute abundance in A and B). Clearly from this we can infer that feature A occupies a considerably smaller volume than feature B, although one can only speculate whether this is due to filamentary structure within A, or whether A merely has a very small vertical (i.e., parallel to the line of sight) extent.

The observations of Sheeley (1996) that high Mg/Ne ratios are associated with spikelike structures extending outwards from active regions, while low Mg/Ne ratios are found in very intense transition region enhancements in emerging flux regions are confirmed here. We note also that strong spikelike features are a common feature of CDS active region observations, particularly in the regularly-observed Mg IX 368.1 Å line. The ability of CDS to clearly separate the different lines and provide clean line profiles is used here to derive accurate estimates of the relative abundances and also electron densities. In particular CDS is able to fully resolve the O IV 625.9 Å line for the first time, from which transition region densities can be estimated. This work has shown that it is possible to extend the study of the variation of the Ne/Mg relative abundance to CDS data and future work will investigate how the ratio varies in various quiet and active structures on the Sun.

Acknowledgements

PRY and HEM acknowledge the financial support of PPARC, and thank the members of the CDS team, in particular R. A. Harrison, C. D. Pike, A. Fludra, and W. T. Thompson for help received in the planning and taking of observations. SOHO is a project of international cooperation between ESA and NASA. NSO/Kitt Peak data used here are produced cooperatively by NSF/NOAO, NASA/GSFC, and NOAA/SEL.

References

Anders, E. and Grevesse, N.: 1989, *Geochim. Cosmochim.* **53**, 197.
Arnaud, M. and Raymond, J. C.: 1992, *Astrophys. J.* **398**, 394.
Arnaud, M. and Rothenflug, R.: 1985, *Astron. Astrophys. Suppl. Ser.* **60**, 425.
Bhatia, A. K. and Doschek, G. A.: 1995, *Atom. Dat. Nucl. Dat. Tables* **60**, 145.
Bhatia, A. K. and Mason, H. E.: 1980, *Monthly Notices Roy. Astron. Soc.* **190**, 925.
Dere, K. P, Landi, E., Mason, H. E., Monsignori-Fossi, B. F., and Young, P. R.: 1997, *Astron. Astrophys. Suppl.*, in press.
Feldman, U.: 1992, *Phys. Scripta* **46**, 202.
Feldman, U. and Widing, K. G.: 1990, *Astrophys. J.* **363**, 292.
Feldman, U., Widing, K. G., and Lund, P. A: 1990, *Astrophys. J.* **364**, L21.
Feldman, U. and Widing, K. G.: 1993, *Astrophys. J.* **414**, 381.
Fludra, A., Harrison, R. A, Brekke, P., Mason, H. E., Pike, C. D., Thompson, W. T., and Young, P. R.: 1997, *Solar Phys.* **175**, 487 (this issue).
Grevesse, N., Noels, A., and Sauval, A. J.: 1992, in 'Coronal Streamers, Coronal Loops and Coronal and Solar Wind Composition', *Proc. of the First SOHO Workshop*, ESA SP-348, p. 305.
Harrison, R. A., Fludra, A., Pike, C. D., Payne, J., Thompson, W. T., Poland, A. I., Breeveld, E. R., Breeveld, A. A., Culhane, J. L., Kjeldseth-Moe, O., Huber, M. C. E., and Aschenbach, B.: 1997, *Solar Phys.* **170**, 123.
Harrison, R. A., Sawyer E. C., Carter, M. K., Cruise, A. M., Cutler, R. M., Fludra, A., Hayes, R. W., Kent, B. J., Lang, J., Parker, D. J., Payne, J., Pike, C. D., Peskett, S. C., Richards, A. G., Culhane, J. L., Norman, K., Breeveld, A. A., Breeveld, E. R., Al Janabi, K. F., McCalden, A. J., Parkinson, J. H., Self, D. G., Thomas, P. D., Poland, A. I., Thomas, R. J., Thompson, W. T., Kjeldseth-Moe, O., Brekke, P., Karud, J., Maltby, P., Aschenbach, B., Brauninger, H., Kuhne, M., Hollandt, J., Siegmund, O. H. W., Huber, M. C. E., Gabriel, A. H., Mason, H. E., and Bromage, B. J. I.: 1995, *Solar Phys.* **162**, 233.
Mason, H. E.: 1995, *Adv. Space Res.* **15**, 53.
Mason, H. E. and Bhatia, A. K.: 1978, *Monthly Notices Roy. Astron. Soc.* **184**, 423.
Mason, H. E., Young, P. R., Pike, C. D., Harrison, R. A., Fludra, A., Bromage, B. J. I., and Del Zanna, G.: 1997, *Solar Phys.* **170**, 143.
Meyer, J.-P.: 1985a, *Astrophys. J. Suppl.* **57**, 151.
Meyer, J.-P.: 1985b, *Astrophys. J. Suppl.* **57**, 173.
Meyer, J.-P.: 1993, *Adv. Space Res.* **13**, 377.
Pottasch, S. R.: 1964, *Space Sci. Rev.* **3**, 816.
Sheeley, N. R.: 1995, *Astrophys. J.* **440**, 884.
Sheeley, N. R.: 1996, *Astrophys. J.* **469**, 423.
Thomas, R. J. and Neupert, W. M.: 1994, *Astrophys. J. Suppl.* **91**, 461.
Widing, K. G., Feldman, U., and Bhatia, A. K.: 1986, *Astrophys. J.* **308**, 982.
Widing, K. G. and Feldman, U.: 1989, *Astrophys. J.* **344**, 1046.
Widing, K. G. and Feldman, U.: 1992, *Astrophys. J.* **392**, 715.
Widing, K. G. and Feldman, U. 1993, *Astrophys. J.* **416**, 392.
Young, P. R., Landi, E., and Thomas, R. J.: 1997, *Astron. Astrophys.*, submitted.
Zhang, H. L. and Sampson, D. H.: 1992, *Atom. Dat. Nucl. Dat. Tables* **52**, 143.

THE RELATIONSHIP BETWEEN UV AND X-RAY ACTIVE REGION STRUCTURES

S. A. MATTHEWS and L. K. HARRA-MURNION
Mullard Space Science Laboratory, University College London, Holmbury St. Mary, Surrey, RH5 6NT, U.K.

(Received 12 March 1997; accepted 19 June 1997)

Abstract. *Yohkoh* and the Coronal Diagnostic Spectrometer (CDS) on the Solar and Heliospheric Observatory (SOHO) jointly observed two brightenings in active region NOAA 7981 on 6 August 1996. Combining the UV data from CDS with information from the high time resolution coronal images obtained with the Soft X-ray Telescope (SXT) on *Yohkoh*, provides us with important information on the relationship between the transition region and corona. Our observations show that cool plasma ($T_e = 2.2 \times 10^5$ K) can lie at the same altitude as the hot coronal plasma ($T_e = 1$–4×10^6 K). The lower temperature structure is not formed from the cooling of the hotter coronal loop. We are also able to observe a low temperature cut-off of $T_e = 1$–4×10^6 K for a loop which repeatedly brightened over the period of approximately one day.

1. Introduction

The relationship between structures in the transition region and corona is not well understood, and this knowledge is vital for a complete understanding of the heating of active region plasma. There have been many previous studies of active regions in the UV and X-ray using the XUV spectroheliograph on-board *Skylab*. Cheng, Smith, and Tanberg-Hanssen (1980) discuss the X-ray emission at 2–3×10^6 K and UV emission at 6×10^5 K from active region loops and the spatial relationship between loops at different temperatures. They found that while hot coronal loops are low-lying and closely packed, the cool loops are fewer in number and large with smaller aspect ratios. A further detailed study of an active region by Cheng (1980) found that the spatial distribution of XUV structures in the region fell into three groups. Structures with temperatures in the range $\sim 2 \times 10^6$ K were found to be predominantly small, compact loops while those with transition region temperatures (1×10^5–2×10^5 K) were typified by large Ne VII and Mg IX loops. The third group encompasses chromospheric emission in He II and Hα which is seen as ribbons of emission over areas of strong vertical photospheric magnetic fields of opposite polarity. He concluded that there was no spatial correspondence between emission at different temperatures but that observations of emission at temperatures of less than 5×10^5 K were inconclusive. Dere (1982) observed two active regions using the XUV spectroheliogram on *Skylab* and studied the relationship between coronal and transition region structures. He found that the transition region emission exists in loops or in segments of loops whereas the coronal emission is always found in loops. Transition region loops and coronal

Solar Physics **175:** 541–551, 1997.

loops were found not to exist coaxially. Feldman and Laming (1994) conclude that in their study, the cold and hot loops often appear to occupy the same location, and frequently the cool and hot loops emerge in different locations. They do not observe the cool plasma being part of a large structure which extends into the hotter corona.

A number of recent rocket observations carried out in coordination with *Yohkoh* (e.g., Solar EUV Rocket Telescope (SERTS) and High Resolution Telescope and Spectrograph (HRTS)) have investigated the structure from the low transition region to corona. Strong and Bruner (1996) summarize the results of the various rocket experiments and find that while the cool plasma outlines the coronal legs, the hot plasma shows the apex of the coronal loops. However, the low temperature plasma is observed to extend into the corona. C IV emission in the form of loops was observed by Athay *et al.* (1983) with the Solar Maximum Mission (SMM). A study of these loops showed that their scale heights were more consistent with coronal temperatures than with the 1×10^5 K emission expected, and it was concluded that they were the result of rapid cooling of coronal loops (Strong and Bruner, 1996). Brosius *et al.* (1997) present a detailed study of 2 active regions observed with SERTS and the Soft X-ray Telescope (SXT) on *Yohkoh*. They find that the emission from Fe XV and Fe XVI closely follows the soft X-ray emission observed by SXT. The emission from the cooler He II and Mg IX emission tends to exist at the loop footpoints, as well as at structures removed from the hot loops.

The Coronal Diagnostic Spectrometer (CDS) on SOHO enables us to observe different temperature structures ranging from $10^4 \rightarrow > 10^6$ K simultaneously. The first results of active regions observed by CDS are described by Harrison *et al.* (1997) and Brekke *et al.* (1997). The latter paper illustrates how even a small difference in line formation temperature (e.g., Mg IX at 9.5×10^5 K and Mg X at 1.1×10^6 K) results in different loop structures. Hara (1996) described observations from the Norikura Solar Observatory looking at visible active region emission from the coronal red line (Fe X), green line (Fe XIV), and yellow line (Ca XV) which are formed at $\approx 1 \times 10^6$ K, $\approx 2 \times 10^6$ K and $\approx 3.5 \times 10^6$ K, respectively. Again, even with this small temperature difference the loop structures are quite different.

In this paper observations from the Coronal Diagnostic Spectrometer (CDS) on SOHO and the Soft X-ray Telescope (SXT) on *Yohkoh* are used to study the correlation between 2.2×10^5 K and $1-3 \times 10^6$ K plasma during small brightenings observed in active region NOAA 7981. One small brightening rose above the quiescent GOES 1–8 Å level of A7 to A9 ($A7 = 7 \times 10^{-8}$ W m^2) over a period of ≈ 1 hour. We investigate the morphological relationship between the transition region plasma at 2.4×10^5 K to the hot coronal plasma $>10^6$ K.

2. Observations

CDS consists of two spectrometers (the Normal Incidence Spectrometer (NIS) and the Grazing Incidence Spectrometer (GIS)) covering the wavelength range from 151–785 Å, and is described in detail by Harrison *et al.* (1995). The observations presented here were obtained by the NIS which is stigmatic and covers two wavelength ranges; 308–381 Å and 513–633 Å.

The SXT instrument on *Yohkoh* is a grazing incidence soft X-ray imaging telescope, sensitive to X-rays in the range 0.25 – 4.0 keV, and is described in detail by Tsuneta *et al.* (1991). The BCS, also on *Yohkoh* consists of four bent crystals covering the wavelength ranges around the resonance lines of He-like sulphur, He-like calcium, He-like iron and H-like iron. For these observations only the S XV emission was visible ($T_e \approx 5 \times 10^6$ K). A detailed description is given by Culhane *et al.* (1991).

Between 6–8 August 1996 NOAA active region 7981 was observed jointly by SXT and CDS–NIS. On 6 August, a number of brightenings located at S09 W58, were observed by both instruments. The CDS–NIS was used to observe the region formed over widely different temperatures; O V at 629.62 Å with $T_{\max} = 2.2 \times 10^5$ K, the Fe XIV at 334.172 Å with $T_{\max} = 1.8 \times 10^6$ K, and Fe XVI at 335.33 Å with $T_{\max} = 2.2 \times 10^6$ K. The temperatures are given by the $G(T_e)$ functions derived by the Atomic Data and Atomic Structure Package (Summers, 1994) using the ionization balance derived by Arnaud and Rothenflug (1985). A region of approximately 180 × 180 arc sec was scanned using the 2 × 240 arc sec slit available on CDS–NIS with a 16-s exposure time, producing an image which took approximately 30 min to accumulate. However, since we know that the CDS–NIS instrument builds up images from right to left, and we know that each pixel in the x-direction is just width of the slit, we can identify the time during this 30-min period that various sections of the image were produced with an uncertainty of approximately the exposure time, giving us somewhat improved time resolution over 30 min. The same active region was observed by the Soft X-ray Telescope (SXT), through the AlMgMn (Dagwood) and thin aluminium filters, and the Bragg Crystal Spectrometer (BCS), in the S XV channel, on *Yohkoh*.

In preparing the CDS–NIS data for comparison with the SXT images the data first had the appropriate calibration factors applied and then a correction for the CCD bias. The pointing information was also updated and cosmic ray events were removed. It is also important with CDS–NIS data to ensure that the background component is subtracted, particularly for weaker lines. As a result of small grating irregularities in the CDS–NIS, the line emission is superimposed on a scattered light component along the slit structure. For each spectral line, we averaged over the wavelength range of the line to form an image. After data corrections had been made the CDS–NIS images were co-aligned with SXT images at corresponding times. This was achieved using the pointing information for both instruments and by rescaling the CDS–NIS pixels to the size of the SXT pixels. This alignment will

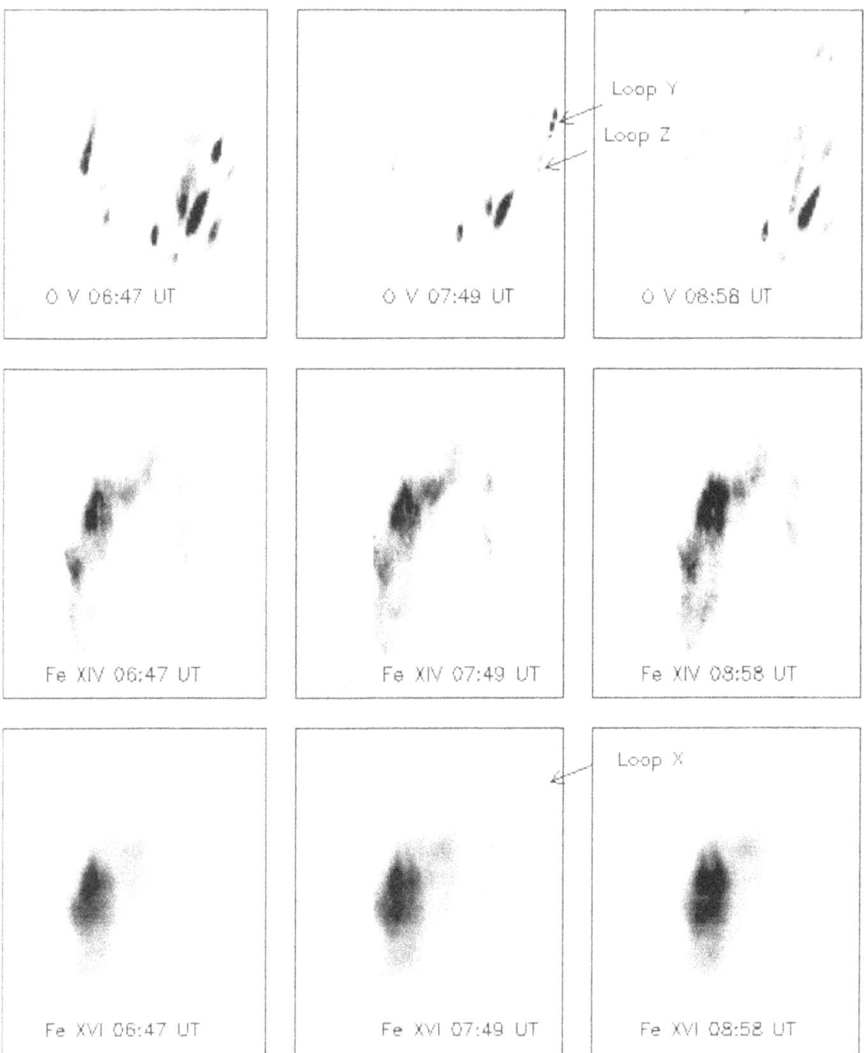

Figure 1. The top 3 images show O V images with time from left to right, the middle panel shows Fe XIV, and the bottom panel shows Fe XVI. The middle image for all the ions shows the time which the A9 brightening in GOES occurred. The size of each image is 182 × 179 arc sec. The times shown are the median times of the raster. The images are rastered from right to left.

of course be subject to errors in the determination of the pointing for both CDS and SXT. For CDS the error in the pointing determination is 5″ r.m.s. for both the x and y directions combined. The SXT data was first decompressed and corrections were made for the dark current, pin hole leak and vignetting.

The temporal evolution of the active region as seen in the O V, Fe XIV and Fe XVI lines of the CDS–NIS can be seen in Figure 1, showing an image both before and after the brightenings of interest. In this figure we can see the appearance of a loop-like structure in the central image of O V. This feature appears during the time-scale of a brightening observed in a similar location with SXT. The other important feature that these images show is the existence of diffuse Fe XVI emission in the top right of all 3 images which is clearly absent in the Fe XIV images.

For the time period of interest, 2 separate loops have been studied, called loop X and loop Y. These are shown in Figure 2 and will be discussed separately. Figure 2 also shows the GOES plot for the interval showing the time variation of the main brightening and the SXT light curves of loop X and loop Y. The CDS–NIS raster covering the time period of both of these events began at 07:36 UT. Rastering with CDS–NIS is executed from right to left.

Loop X

At approximately 07:30 UT on 6 August 1996 a small flare-like brightening of GOES class A9 began in loop X (see Figure 2). Using the filter ratios of thin aluminium and Dagwood, the temperature of loop X was determined to peak at $\approx 5 \times 10^6$ K. The temperature of this loop was also calculated using a synthetic spectral fit to the S XV data, and found to peak at 5×10^6 K. Figure 3 shows CDS–NIS images of the active region in Fe XVI (Figure 3(a)), Fe XVI overlaid with Fe XIV contours (Figure 3(b)), O V (Figure 3(c)), O V co-aligned with SXT (Figure 3(d)), SXT dagwood (Figure 3(e)) and SXT dagwood with Fe XVI contours (Figure 3(f)). In Fe XVI we observe diffuse emission in the region of loop X as observed with SXT, while a similar inspection of the Fe XIV shows that the same region (Figure 3(b)) is devoid of similar emission. This situation can be seen to be the case throughout all of our observations. In particular the three images in Figure 1 show this to be the case. Loop X consists of loop emission in Fe XVI with hotter SXT emission lying above that. No O V emission exists near the vicinity of the loop. Since loop X appears to continually brighten for the period of ≈ 1 day, the temperature does not fall significantly enabling cooler emission to be seen. A lower limit on the temperature of loop X can then be determined by the absence of Fe XIV emission in this region throughout the observations to be 1.8×10^6 K. There is also a small patch of emission (loop Z) which appears to be within loop X which has silmutaneously Fe XIV and Fe XVI emission. This may be the site of a small loop interacting with loop X.

Loop Y

Loop Y is smaller in size and intensity compared with loop X and behaves in a completely different manner. Using the filter ratio described above, the temperature of loop Y peaks at a value of 3.5×10^6 K. At the time of the brightening in loop Y observed with SXT (~07:38 UT), a transient loop feature is visible in O V as can be seen in the central image of Figure 1 and also in Figure 3(c). Figure 3(d) illustrates

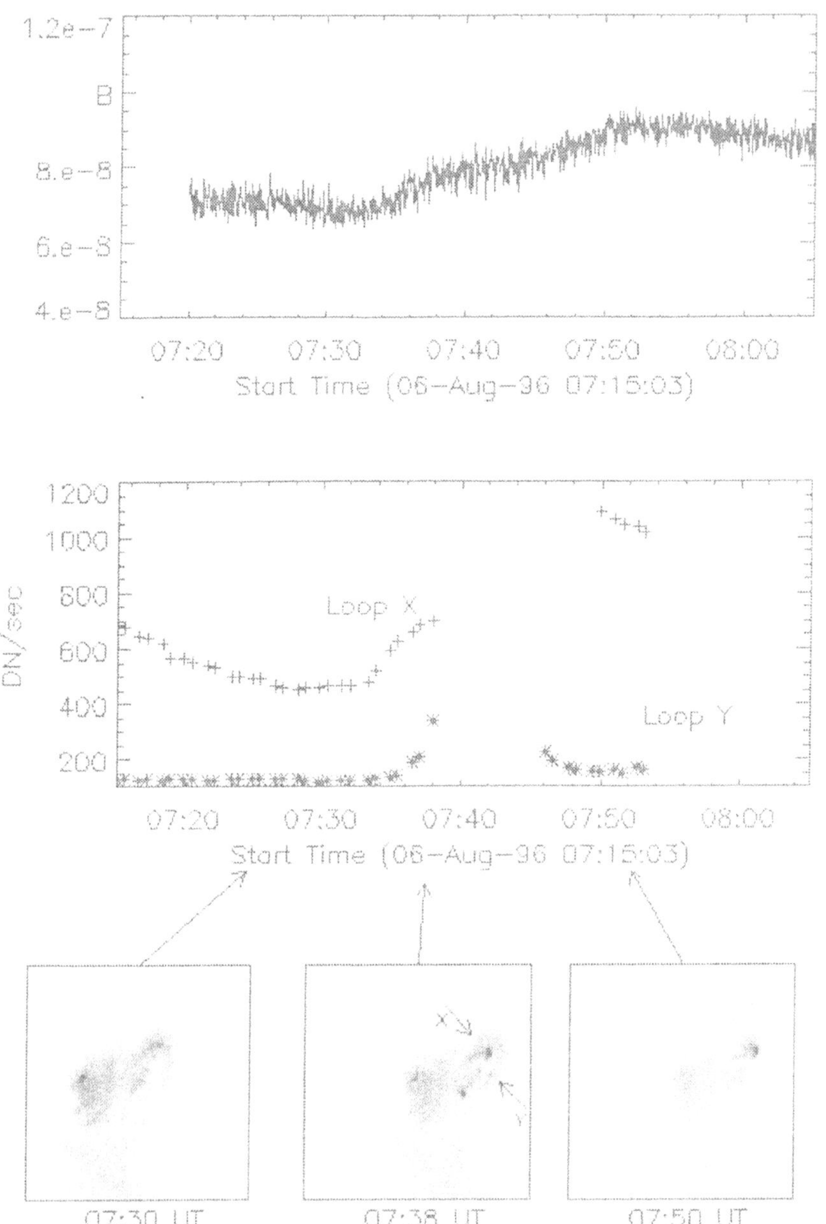

Figure 2. The top panel shows the GOES plot illustrating the A9 brightening which was observed by *Yohkoh* and SOHO. The middle panel shows the light curves of loop X (crosses) and loop Y (stars) separately. These were obtained using the Dagwood filter. The gap in the middle is due to points in which the images suffered from saturation. SXT images are shown at three times during the brightening. Loop X is the main brightening. The size of each image is 157 × 157 arc sec.

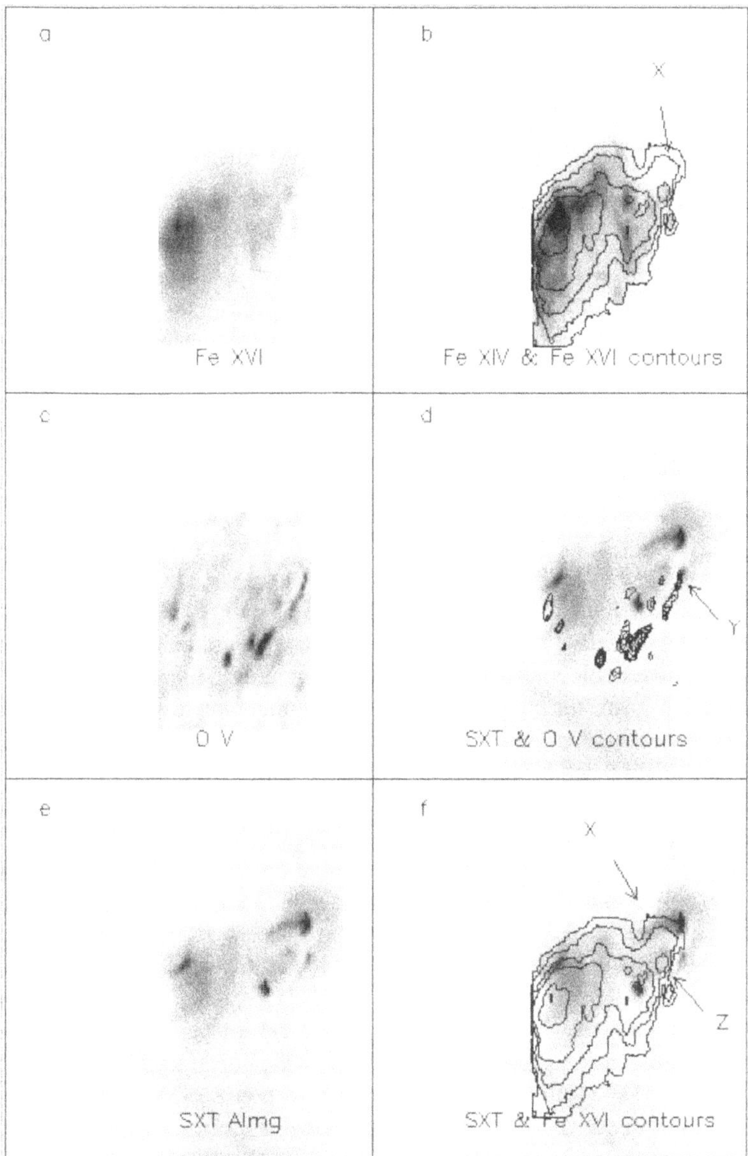

Figure 3. (a) Fe XVI image at 07:49 UT (b) Fe XIV image with Fe XVI contours overlaid. It is clear that where the main brightening occurs (loop X) the emission is seen in Fe XVI, but is not in Fe XIV giving a lower limit to the temperature of the structure. The raster occurs from right to left, hence this part of the image occurred from 07:36 UT $\rightarrow$ 07:39 UT. (c) O V image at 07:49 UT. (d) The O V image is co-aligned with the SXT image taking account of position and different pixel sizes (1 SXT pixel $= 2.5 \times 2.5$ arc sec, 1 CDS–NIS pixel $= 2.03 \times 1.68$ arc sec). The O V is shown by contours. Loop Y is seen in both SXT and O V. (e) SXT image (Dagwood filter) at 07:49 UT. (f) SXT image with Fe XVI contours overlaid. Loop X is observed in both SXT and Fe XVI. There is also a patch of emission labelled loop Z which seems to be a separate loop structure from loop Y and brightens in Fe XIV and Fe XVI.

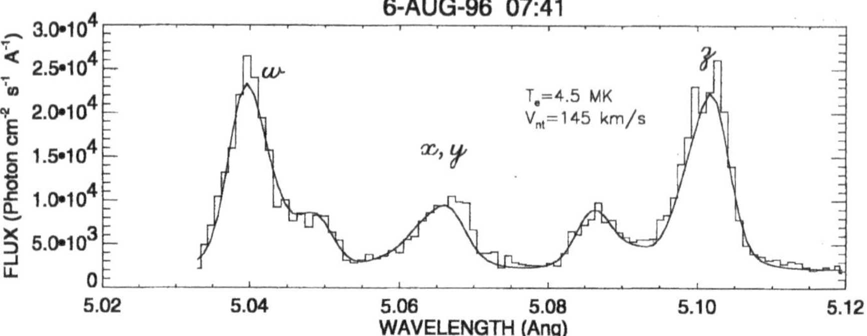

Figure 4. A synthetic fit to a S XV spectrum which has been accumulated for 440 secs. The resonance line w ($1s^2\,{}^1S_0 - 1s2p\,{}^1P_1$), intercombination lines x and y ($1s^2\,{}^1S_0 - 1s2p\,{}^3P_2$ and $1s^2\,{}^1S_0 - 1s2p\,{}^3P_1$, respectively) and the forbidden line z ($1s^2\,{}^1S_0 - 1s2s\,{}^3S_1$) are shown. The fit is obtained using the fit_bsc routine in the *Yohkoh* software using the atomic data as described in Harra-Murnion *et al.*, 1996.

that the cool O V loop Y lies on top of the hot SXT loop to within one pixel. Since loop Y is situated close to the limb at S09 W58, we would expect that a difference in height between the cool loop and hot loop would be apparent as a shift in the images. In this case, the cool and hot loops are coincident to within $\approx 3 \times 10^3$ km, accounting for location. The cool O V loop exists both in the previous raster (median time = 07:20 UT), and in the following raster (median time = 08:25 UT). In SXT loop Y can be seen to be brightest at 07:38:05 UT. The time taken for the loop to cool from its peak temperature to O V temperature ($\approx 2.2 \times 10^5$ K) by radiation and conduction can be calculated via a method detailed by Švestka (1987) to be approximately 5 min. Since loop Y is brightest in O V at ~07:37 UT, this together with its existence in preceeding and suceeding rasters, provides evidence against the possibility that the emission in O V is due to cooling of the SXT loop. Note that although the full time taken to produce the whole CDS–NIS raster is ~30 min it is possible to determine when specific portions of the images were produced within this 30-min time frame. Hence the brightening which occurs must be due to energy release within that cool structure, rather than a response to coronal energy input.

BCS observations in S XV indicate nonthermal velocities around the time of the flare-like brightening (~07:46 UT) of approximately 140 km s^{-1}. As discussed by Mariska (1994), the S XV spectral lines are broadened in the y direction by 1.2 km s^{-1} per arcsec. In our case if we look at the spatial extent of the SXT images and include all counts above the 25% level we obtain broadening a broadening of 44 km s^{-1}, which when subtracted gives a value for the nonthermal velocity of ~90 km s^{-1}. The values of the nonthermal velocity shows no deviation from that which we observe for much larger solar flares (e.g., Khan *et al.*, 1995), indicating that the same processes may also be acting in these small A class events. The values are certainly above those found in non-flaring active regions as discussed by Sterling (1997). He found values of 44 km s^{-1} in non-flaring active regions.

3. Discussion and Conclusions

Many models of the upper solar atmosphere have assumed the existence of loop structures which extend from the chromosphere ($T_e \approx 10^4$ K) up into the corona where temperatures often exceed 10^6 K. In this scenario one would expect that a series of observations made at the solar limb in a number of lines spanning the range of temperatures from the chromosphere to the corona could be superposed to illustrate one continuous structure. Thus when observing in coronal lines we would expect to see loop-like structures and in transition region and chromospheric lines, loop legs and footpoints.

Feldman and Laming (1994) used observations of the solar limb in Ne VII (564.22 Å, $T_e \approx 5 \times 10^5$ K) and Mg IX (368.07 Å, $T_e \approx 10^6$ K) to illustrate that the brightness in the region associated with transition region temperatures is independent of the brightness in higher temperature regions. They propose that these observations indicate that thermal continuity between the corona and chromosphere is not being maintained. Their observations of three active regions also indicated an inconsistency with predicted values of footpoint intensities. Indeed, Feldman and Laming (1994) consistently observe whole loop structures in transition region lines where footpoint emission is predicted, indicating that the cooler plasmas are not heated by conduction from above.

Our observations show clear loop structures visible in O V rather than loop leg or footpoint emission corresponding to the apex of loops seen in hotter lines, such as Fe XIV and Fe XVI (e.g., Dere, 1982; Strong and Bruner, 1996). Mariska (1992) and Dowdy, Rabin, and Moore (1986) suggest that only a fraction of the transition region actually connects to the corona and that most of it is actually contained in smaller loops of different temperatures which fail to reach coronal temperatures. In particular we note the existence of co-spatial emission of loop Y in SXT temperature plasmas ($\sim 3.5 \times 10^6$ K), as well as in regions where Fe XIV and O V lines are formed. The fact that loop Y is visible in O V in both the previous raster (median time 07:20 UT) and also in the following raster (median time 08:25 UT) strongly suggests that this loop cannot be explained by rapid cooling of the SXT loop. The brightening in the SXT loop reaches its maximum intensity at 07:38:05 UT, while the O V brightening appears at $\sim$ 07:37 UT, providing further support that the O V structure is maintained by some kind of in situ heating. We see no rise in the GOES flux prior to this event which could indicate a previous event cooling. Also apparent is the absence of cooler transition region emission in the region of loop X. Both of these results are consistent with the findings of Feldman and Laming (1994) in which they find that hot and cold loops often occupy the same spatial location, whilst in other cases hot and cold loops can be seen to emerge from different locations. Cool loops are often observed to exist but are in general referred to as post-flare loops (e.g., Schmieder *et al.*, 1996). They are generally large in size and follow a solar flare. In this paper we observe a different class of cool loop – one which exists alone, and which is not a by-product of post-flare cooling.

The possibility that the brightening in the O V loop is due to in situ heating may have important implications in terms of active region heating by transient brightening and microflares. Shimizu (1995) has suggested that transient brightenings in active region loops may contribute to active region heating, and in a detailed study of one active region he estimated the total energy released by these X-ray transient brightenings and flares by calculating both the total energy loss and the scaled energy loss. A power law distribution of the total energy loss was adopted and the distribution was assumed to extend from 10^{32} ergs down to 10^{24} ergs. Using this method Shimizu calculated the energy released to be a factor of nearly 5 smaller than the energy required to heat the active region. One possibility to account for the 'missing' energy is that smaller energy releases are produced by lower temperature plasma. Porter, Toomre and Gebbie (1984) found that UV brightenings in active regions occur far more frequently than hard X-ray microflares, and that activity may sometimes persist for hours. Also, Porter, Fontenla, and Simnett (1995) have found some evidence to suggest that an enhanced, steeper population does exist for low temperature emission, suggesting that if the flare process extends to sufficiently small events then microflares could be an important source of energy input to active regions. The brightening which we observe in O V in loop Y, apparently due to in situ heating, could then be contributing to the overall energy budget of the active region.

In the image of the active region shown in Fe XVI in Figure 3(f) we can see the presence of diffuse emission lying just beneath loop X as observed with SXT. Corresponding emission in Fe XIV shown in Figure 3(b), can be seen to be absent. This implies the presence of a very definite temperature cut-off, with plasma of differing temperatures being confined to separate structures.

In conclusion, for this active region the observations seem to provide no evidence to support the idea of continuous loop structures extending through the upper solar atmosphere. There also does not appear to be a connection between higher temperature coronal plasmas and the lower temperature transition region structures; we see no response in O V to the brightening in loop X. We have observed transition region plasma at 2.2×10^5 K co-spatial and at the same altitude as 3.5×10^6 K plasma, which cannot be the result of cooling of a hotter loop. This structure can be seen to brighten during our observations as the result of energy input which cannot be of coronal origin. In the same active region we also observe a high temperature coronal loop for which there is no coincident transition region emission. We conclude that two different types of cool loops exist – those which have cooled from coronal temperatures and those which exist and are maintained by in situ heating. A more detailed study of a similar region at the solar limb, with higher time resolution CDS–NIS observations, is under way and will provide us with the opportunity to investigate further the contention that UV brightenings contribute to active region heating and to make more accurate estimations of the altitudes of different temperature loops.

SAM and LKHM are grateful to PPARC for postdoctoral funding. SOHO is a project of international co-operation between ESA and NASA. *Yohkoh* is a mission of the Japanese Institute for Space and Astronautical Science. We thank J. L. Culhane, K. J. H. Phillips and R.A. Harrison for helpful discussions and proof-reading the manuscript. We thank an anonymous referee for improving the clarity of the paper.

References

Arnaud, M. and Rothenflug, R.: 1985, *Astron. Astrophys. Suppl. Ser.* **60**, 425.
Athay, R. G., Gurman, J. B., Henze, W., and Shine, R. A.: 1983, *Astrophys. J.* **265**, 519.
Brekke, P., Kjeldseth-Moe, O., Brynildsen, N., Maltby, P., Haugan, S. V. H., Harrison, R. A., Thompson, W. T., and Pike, C. D.: 1997, *Solar Phys.* **170**, 163.
Brosius, J. W., Davila, J. M., Thomas, R. J., Saba, J. L., Hara, H., and Monsignori-Fossi, B. C.: 1997, *Astrophys. J.*, submitted.
Cheng, C.-C., 1980, *Astrophys. J.* **238**, 743.
Cheng, C.-C., Smith, J. B., and Tandberg-Hanssen, E.: 1980, *Solar Phys.* **67**, 259.
Culhane, J. L. *et al.*: 1991, *Solar Phys.* **136**, 89.
Dere, K. P.: 1982, *Solar Phys.* **75**, 189.
Dowdy, J. F., Rabin, D., and Moore, R. L.: 1986, *Solar Phys.* **105**, 35.
Feldman, U. and Laming, J. M.: 1994, *Astrophys. J.* **434**, 370.
Hara, H.: 1996, PhD thesis, National Astronomical Observatory Japan.
Harra-Murnion *et al.*: 1996, *Astron. Astrophys.* **306**, 670.
Harrison, R. A. *et al.*: 1995, *Solar Phys.* **162**, 233.
Harrison, R. A., Fludra, A., Pike, C. D., Payne, J., Thompson, W. T., Poland, A. I., Breeveld, E. R., Breeveld, A. A., Culhane, J. L., Kjeldseth-Moe, O., Huber, M. C. E., and Aschenbach, B.: 1997, *Solar Phys.* **170**, 123.
Khan, J. I., Harra-Murnion, L. K., Hudson, H. S., Lemen, J. R., and Sterling, A. C.: 1995, *Astrophys. J.* **452**, L153.
Mariska, J. T.: 1992, *The Solar Transition Region*, Cambridge University Press, Cambridge.
Mariska, J. T.: 1994, *Astrophys. J.* **434**, 756.
Porter, J. G., Toomre, J., and Gebbie, K. B.: 1984, *Astrophys. J.* **283**, 879.
Porter, J. G., Fontenla, J. M., and Simnett G. M.: 1995, *Astrophys. J.* **438**, 472.
Schmieder, B., Heinzel, P., van Driel-Gesztelyi, L, and Lemen, J. R.: 1996, *Solar Phys.* **165**, 303.
Shimizu, T.: 1995, *Publ. Astron. Soc. Pacific* **47**, 251.
Sterling, A. C.: 1997, *Astrophys. J.* **478**, 807.
Summers, H. P.: 1994, *Atomic Data and Analysis Structure User Manual*, JET-IR(94)06.
Strong, K. T. and Bruner, M. E.: 1996, *Adv. Space Res.* **17** (4/5), 179.
Švestka, Z.: 1987, *Solar Phys.* **108**, 411.
Tsuneta, S. *et al.*: 1991, *Solar Phys.* **136**, 37.

SOHO CDS–NIS IN-FLIGHT INTENSITY CALIBRATION USING A PLASMA DIAGNOSTIC METHOD

E. LANDI[1], M. LANDINI[1], C. D. PIKE[2] and H. E. MASON[3]
[1]*Dipartimento di Astronomia e Scienza dello Spazio, Università di Firenze, Largo E. Fermi 5, 50124 Firenze, Italy*
[2]*Rutherford Appleton Laboratory, Chilton, Didcot, Oxfordshire OX11 0QX, U.K.*
[3]*Department of Applied Mathematics and Theoretical Physics, Silver Street, Cambridge, CB3 9EW, U.K.*

(Received 27 February 1997; accepted 20 May 1997)

Abstract. The internal intensity calibration of the Coronal Diagnostic Spectrometer (CDS) – Normal Incidence Spectrometer (NIS) is studied using the Arcetri diagnostic method. A large number of spectral lines observed by the CDS–NIS 1 and NIS 2 windows in a solar active region is analysed in order to determine the intensity calibration curve for both channels.

The plasma diagnostic method developed in Arcetri allows the measurement of the correction factors to the preliminary CDS–NIS internal intensity calibration curves and the determination of the relative calibration between NIS 1 and NIS 2. A further correction factor of approximately three is found to be necessary for a correct intercalibration of the two wavelength windows. Also the NIS 2 second-order sensitivity is measured. The Arcetri diagnostic method proves to be a powerful tool for intensity calibration studies.

1. Introduction

The Coronal Diagnostic Spectrometer (CDS) on board the Solar and Heliospheric Observatory (SOHO) is a pair of grazing and normal incidence spectrographs, which produce spectra of selected regions of the solar surface in six spectral windows of the extreme ultraviolet from 150 Å to 785 Å (Harrison *et al.*, 1995).

This spectral region is extremely rich in emission lines from a large number of highly ionized ions of the most abundant elements and its study represents a unique tool for a detailed diagnostic of temperature, density and chemical composition of the solar transition region and corona. Moreover it is a unique laboratory for testing atomic physics models and theoretical calculations of collision rates and transition probabilities. Initial results from CDS are reported in Harrison *et al.* (1997) and Mason *et al.* (1997).

Intensity calibration is a fundamental piece of information for any scientific use of the spectrograph, but getting such a result is not an easy undertaking in the extreme ultraviolet spectral range. A pre-flight calibration procedure was performed (Bromage *et al.*, 1996) for a limited number of wavelengths and a preliminary in-flight internal intensity calibration for NIS 1 and NIS 2 has been made with intensity ratios of some spectral lines observed in quiet Sun and active regions.

The large number of emission lines that are observed in the solar spectrum, coupled to the large amount of atomic data now available (CHIANTI – Dere *et al.*

Solar Physics **175:** 553–570, 1997.

Table I
Lines used for the preliminary intensity calibration for the CDS NIS 1 and NIS 2 spectral windows

Spectral window	Ion	λ (Å)
NIS 1	Si VIII	316.21, 319.84
NIS 1	Mg VIII	315.04, 339.00
NIS 1	Fe XI	352.68, 369.16
NIS 1	Fe XII	346.86, 352.11, 364.47
NIS 1	Fe XVI	335.41, 360.77
NIS 2	O III	525.80, 599.60
NIS 2	Ca X	557.77, 574.01
NIS 2	Mg X	609.79, 624.94

(1997), The Arcetri Spectral Code – Landi and Landini (1997a), ADAS – Summers *et al.* (1996)) and a new temperature and density diagnostic technique (Landi and Landini, 1997b), allow a rather detailed in-flight check of the current calibration curve and verification of the detector performance.

In the present work we have applied this new diagnostic technique to the intensities of lines emitted by a solar active region observed with the NIS spectrograph, and we have derived the correction factors to be applied to the preliminary intensity calibration of the instrument.

In Section 2 we describe briefly the preliminary intensity calibration, and in Section 3 we present the method used to verify the intensity calibration curve together with a brief description of the adopted theoretical data necessary for the analysis. In Section 4 we describe the data measured by the Normal Incidence Spectrograph. The results of the analysis are discussed in Section 5. They concern the internal calibration within each of the two spectral bands, NIS 1 and NIS 2, the relative calibration between NIS 1 and NIS 2, and the NIS 2 second-order efficiency.

2. Preliminary Intensity Calibration

A first estimate of the internal relative intensity calibration for NIS 1 and NIS 2 wavelength bands was obtained by 'bootstrapping' measurements of predicted and observed intensity ratios for electron density insensitive line pairs, well separated in wavelength.

The spectral lines used for NIS 1 and 2 are listed in Table I. The resulting relative intensity calibration curves are shown in Figure 1 (NIS 1) and Figure 2 (NIS 2).

The observed intensity ratios were derived from data obtained with a standard spectral atlas observing sequence (NISAT) when observing both quiet Sun and an active region. The theoretical intensity ratios were taken from the CHIANTI

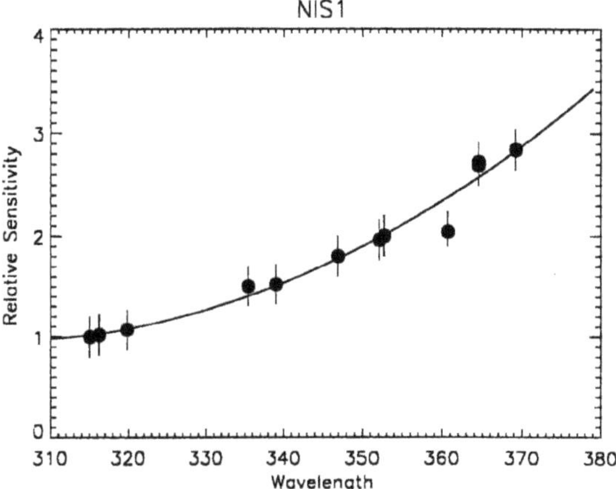

Figure 1. Preliminary NIS 1 in-flight intensity calibration.

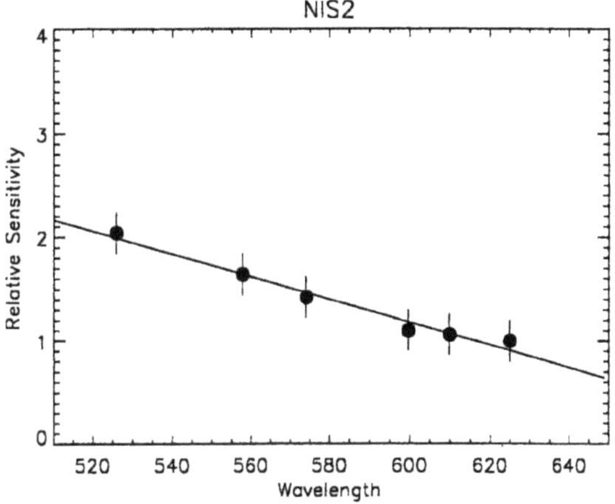

Figure 2. Preliminary NIS 2 in-flight intensity calibration.

package (Dere *et al.*, 1997). This simple method has a number of limitations and the available ratios are sparse, particularly for NIS 2. Nevertheless it appears to define well the large-scale characteristics of the internal calibration and is broadly consistent with the laboratory measures of the grating sensitivities.

3. The Theoretical Method

The method used for this calibration study is a byproduct of the temperature and density diagnostic procedure shown in Landi and Landini (1997b). Here we present a very brief outline of the theoretical method, while further details may be found in the quoted paper.

The intensity emitted by a thin plasma in a line is given by

$$I_{ij} = \frac{1}{4\pi} \int_h N_j(X^{+m}) A_{ij} \, \mathrm{d}h \quad \text{ph cm}^{-2}\,\text{s}^{-1}\,\text{st}^{-1} \,.$$

We can define the *contribution function* as follows:

$$G_{ij}(T, N_e) = \frac{N_j(X^{+m})}{N(X^{+m})} \frac{N(X^{+m})}{N(X)} \frac{N(X)}{N(H)} \frac{N(H)}{N_e} \frac{A_{ij}}{N_e} \,.$$

It is possible to express $G_{ij}(T, N_e)$ as the product of two distinct functions depending respectively on electron density and electron temperature, and on temperature alone:

$$G_{ij}(T, N_e) = f_{ij}(N_e, T) g(T) \,.$$

The $g(T)$ function is mainly due to the ionization equilibrium and is the same for all the lines of the same ion while $f(N_e, T)$ is determined mainly by the population of the upper level.

We define also the *differential emission measure* as

$$\text{DEM} = \varphi(T) = N_e^2 \frac{\mathrm{d}h}{\mathrm{d}T} \,,$$

so that the intensity of a line can be rewritten as

$$I_{ij} = \frac{1}{4\pi} \int_h G_{ij}(T, N_e) N_e^2 \, \mathrm{d}h = \frac{1}{4\pi} \int_h f_{ij}(T, N_e) g(T) \varphi(T) \, \mathrm{d}T \,.$$

If we choose the temperature T_{eff} as

$$\log T_{\text{eff}} = \frac{\int g(T) \varphi(T) \log T \, \mathrm{d}T}{\int g(T) \varphi(T) \, \mathrm{d}T} \,,$$

we can define the function $L_{ij}(N_e)$ as the ratio between the observed line intensity and the $G_{ij}(T_{\text{eff}} N_e)$ calculated as a function of N_e at the temperature T_{eff}:

$$L_{ij}(N_e) = \frac{I_{\text{obs}}}{G_{ij}(T_{\text{eff}}, N_e)} \,.$$

The diagnostic method relies on the observation that if we plot all the L-functions of the same ion calculated with the line fluxes derived from a well-calibrated spectrum versus the electron density, all the curves meet in a common point $(N_e^*, \; L(N_e^*))$. Moreover the L-functions of density independent lines must overlap. An example of this behaviour is shown in Figure 3. It is important to note also that unsplit multiplets of the same ion share this common behaviour, if we define their L-function as

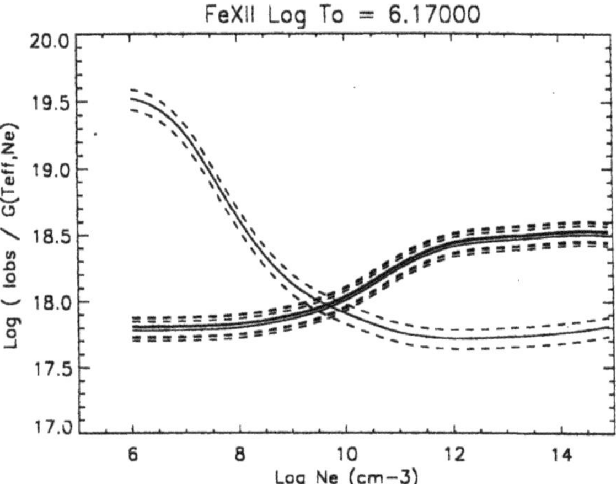

Figure 3. Fe XII L-functions of lines 338.28, 346.85, 352.11, 364.47 Å as a function of electron density N_e. The dashed curves indicate the experimental uncertainties of each L-function. The decreasing L-function belongs to line 338.28 Å while the L-functions of the other lines are nearly coincident, as expected.

$$L_{\text{mult}}(N_e) = \frac{I_{\text{mult}}}{\sum_{i=0}^{n} G_i(T_{\text{eff}}, N_e)},$$

where n is the number of lines contributing to the multiplet. The reasons for this are explained in Landi and Landini (1997a).

When this procedure is applied to a set of observed lines it allows us to measure the electron density of the emitting plasma.

It is possible that some L-functions do not cross the meeting point; in this case there are several possible explanations of this behaviour: blending from some other line of some other ion, problems in the atomic transition probabilities adopted for calculating the contribution function of the line, misidentification of the line and intensity calibration problems. We are concerned with this last point.

To calculate the L-functions of a given ion it is necessary to evaluate the effective temperature T_{eff}. This quantity is strictly tied to the Differential Emission Measure of the source, so the first step of the procedure is to determine this function. For this reason we selected the brightest and best theoretically known density independent lines, and using them we evaluated the Differential Emission Measure distribution along the observed line of sight. Several different procedures are available and the method we adopted in the present work is also reported in Landi and Landini (1997b). Then, for each ion, the second step is to use the DEM distribution to compute the effective temperature T_{eff} for each line and the third step is the display of the L-functions at that effective temperature versus electron density in order to verify line overlapping (for density independent lines) or proper crossing (for density-dependent ones). Thus, we are able to calculate the intensity correction factor versus the wavelength for all the lines whose L-functions do not meet the common crossing point and that are not affected by any blending or atomic physics problem. For this reason a careful study is also first performed to select the lines for which blending either does not occur or can be corrected, and for which we have the best available theoretical transition probabilities.

For each spectral window we start with the ion that contributes the largest number of lines and calibration corrections are plotted versus wavelength; each successive ion is corrected using the previous ions, until all the useful lines have been used.

Lines of the same ion which occur in the two NIS 1 and NIS 2 windows are used for intercalibration between the two windows and the second-order lines available in the NIS 2 section are used to evaluate the ratio of the second-order to first-order efficiency.

For this kind of study a large quantity of high-quality atomic data is required. In recent times some extensive databases have been created for matching the increasing need of theoretical transition probabilities, such as ADAS (Summers *et al.*, 1996) and CHIANTI (Dere *et al.*, 1997). In the present study we make use of the Arcetri Spectral Code (Landi and Landini, 1997a) which is an update of the older version (Monsignori Fossi and Landini, 1995) using the CHIANTI database and additional data for ions not yet included in the CHIANTI database.

4. The Observations

The CDS spectrometer on SOHO is composed of two different spectrometers, namely the Grazing Incidence Spectrometer (GIS) and the Normal Incidence Spectrometer (NIS). The GIS is equipped with a concave reflection grating, four microchannel plate and spiral anode detectors covering four spectral ranges: 151–221, 256–341, 393–492, and 659–785 Å. In this study we have limited our attention only to the NIS wavelength bands, covering the ranges 307–379 and 513–633 Å. Full details of the CDS spectrometer can be found in Harrison *et al.* (1995).

Both the CDS spectrometers were calibrated before launch. Details on the pre-flight calibration are reported in Bromage *et al.* (1996).

The observations we have analysed are of a weak active region on the disk observed as part of the NIS atlas observing program (NISATS) on 9 April 1996: data file s524r00. The observations consisted of ten exposures each stepped by 2 arc sec in longitude relative to the previous.

The slit used for the observations is the $2'' \times 240''$, so the total area covered by the spectrometer is $20'' \times 240''$. The exposure time for each position was 50 s. This exposure time is suitable both to achieve a satisfactory number of counts for all the lines neeessary for this study and to avoid saturation for the brightest lines of the spectrum.

The ten observed spectra were calibrated using the standard procedures and routines provided in the CDS software. The average of all ten exposures was taken and the result averaged over the spatial dimension. The units used for the present study are erg cm^{-2} s^{-1} arc sec^{-2}.

The observed spectra presented several very strong lines and a host of weaker lines, which have proved important for our purposes. The observed intensities have been obtained with a Gaussian fit to the data, using a program developed at Rutherford Appleton Laboratory by Jim Lang and David Brooks. The program allowed the fitting of blended lines. The s524 spectra were fitted by the authors also using a different set of routines developed by D. M. Zarro, now available in the CDS software. The results were in excellent agreement, providing an independent check on any subjective bias or program dependent features of the spectral line fitting procedures.

The values obtained are reported in Table II, which also identifies the lines used for the DEM study. Table III lists other lines which have been useful for the analysis. The tables include wavelength, identification, intensity and a note on whether the particular line has been used for DEM evaluation, NIS 1/NIS 2 intercalibration, NIS 1 and NIS 2 calibration or NIS 2 I–II-order calibration.

For most of the lines the number of observed counts is so high that the signal-to-noise ratio is very high and the errors are estimated at a few percent. We felt that the experimental uncertanties of the measured fluxes were much lower than the uncertanties of the theoretical data. In fact, there are several sources of uncertainty in the calculation of the theoretical Contribution Functions of the lines. These include the element abundances, the ionization equilibrium abundances of ions and the radiative and collisional transition probabilities for the level population calculations. We believe that all these factors contribute to limiting the accuracy of any diagnostic studies, and that it is important to take them into account when the experimental uncertainties fall to very low values, such as those provided by the CDS spectrometers.

For this reason we have set a standard error to all the observed fluxes, and used these values in all the subsequent analysis. We estimate that a reasonable uncertainty for the theoretical data is 15%. For cases where the observed uncertainty of some very weak lines is greater than this value, the original error has been maintained. This is the case for the Mg V ($\approx$ 30% experimental uncertainty) and Fe XVI 2 $\times$ 262.98 Å ($\approx$ 20%) lines.

5. Results

5.1. The Differential Emission Measure

The first step for the L-function study is the calculation of the differential emission measure, which is necessary to evaluate the T_{eff} for each ion considered in the study.

Table II

Observed intensities for the s524 active region. The observed intensity is expressed in 10^{-9} erg cm^{-2} s^{-1} arc sec^{-2}. Stars mark the lines used for the different purposes discussed in the text.

Ion	λ (Å)	I	DEM	NIS 1–NIS 2	NIS 1	NIS 2	NIS 2 IIor
Mg VIII	311.73	218.2	⋆		⋆		
Mg VIII	313.77	416.2	⋆		⋆		
Mg VIII	315.02	1046.4	⋆		⋆		
Mg VIII	317.04	259.9	⋆		⋆		
Al X	332.83	585.8	⋆	⋆			
Fe XVI	335.45	3757.8	⋆		⋆		
Mg VIII	339.05	215.9	⋆		⋆		
Fe XI	341.19	328.0	⋆	⋆			
Mg V	351.21	25.5	⋆				
Mg V	353.14	47.3	⋆				
Mg V	354.20	20.7	⋆				
Fe XI	356.59	121.9	⋆	⋆			
Ne V	359.42	138.5	⋆	⋆			
Fe XVI	360.75	1495.6	⋆		⋆		
Mg IX	368.03	4776.5	⋆				
O II	373.73	23.7	⋆	⋆			
O III	374.01	75.5	⋆	⋆			
O III	374.40	34.0	⋆	⋆			
Si XII	520.45	23.7	⋆				
O III	525.62	72.1	⋆	⋆		⋆	
Fe XVI	2 × 262.98	2.76	⋆	⋆			⋆
Al XI	549.98	129.8	⋆				
O IV	553.29	105.8	⋆			⋆	
O IV	554.02	76.6	⋆			⋆	
O IV	554.48	553.6	⋆			⋆	
O IV	555.26	117.0	⋆			⋆	
Ne VI	558.57	56.0	⋆				
Ne VII	560.04	4.92	⋆				
Ne VI	562.77	91.3	⋆				
Ne VII	564.61	13.9	⋆				
Ne V	569.78	32.2	⋆	⋆			
Ne V	572.29	45.7	⋆	⋆			
O III	597.78	14.1	⋆	⋆			
O III	599.55	186.3	⋆	⋆	⋆		
Si XI	2 × 303.32	140.0	⋆	⋆		⋆	
He II	2 × 303.78	1094.7	⋆				
O IV	608.35	114.4	⋆			⋆	
Mg X	609.73	1866.7	⋆	⋆		⋆	
Mg X	624.87	916.6	⋆	⋆		⋆	
O V	629.67	2331.7	⋆				

Table III

Observed intensities for the s524 active region. The observed intensity is expressed in 10^{-9} erg cm^{-2} s^{-1} arc sec^{-2}. Stars mark the lines used for the different purposes discussed in the text.

Ion	λ (Å)	I	DEM	NIS 1–NIS 2	NIS 1	NIS 2	NIS 2 IIor
Si VIII	314.32	345.5			⋆		
Si VIII	316.21	689.0			⋆		
Si VIII	319.85	1039.9			⋆		
Fe XV	327.06	102.3		⋆			⋆
Fe XIV	334.23	1543.6		⋆			
Fe XII	338.32	288.8		⋆			
Fe XII	346.89	489.3		⋆			
Si X	347.44	1440.3		⋆			⋆
Fe XII	352.13	879.3		⋆			
Fe XI	352.69	929.1		⋆			
Fe XIV	353.85	496.6		⋆			⋆
Si X	356.04	1087.5		⋆			⋆
Fe XII	364.44	1312.5			⋆		
Fe XI	369.14	243.3			⋆		
Fe XIV	2 × 257.40	3.69		⋆			⋆
Fe XIV	2 × 264.78	6.67		⋆			⋆
Ne IV	541.02	17.0		⋆			⋆
Ne IV	541.99	23.8		⋆			⋆
Ne IV	543.84	45.83		⋆			⋆
Fe XIV	2 × 274.21	12.6		⋆			⋆
Ca X	557.72	150.4				⋆	
Al XI	568.24	144.6		⋆			⋆
Ca X	573.99	87.4			⋆		
Si XI	580.88	170.5		⋆			⋆
Si XI	604.08	17.1		⋆			⋆

The differential emission measure determination allows a check on the intensity calibration curve of each spectral window and on the intercalibration between NIS 1 and NIS 2. Moreover it is possible to check the I–II-order calibration of both the spectral windows since some second-order lines are observed. A detailed discussion is provided on each of these items.

The derived differential emission measure is shown in Figure 4. The shape of the differential emission measure curve shows three peaks around $\log T = 5.4, 5.8,$ and 6.2. Several different effects may create this result. Three distinct structures may be seen along the line of sight: maybe three loops or systems of similar loops. This is possible since active regions contain a large number of loops, and the spectrum we are analysing has been emitted by a solar region $20'' \times 240''$ wide.

We can also explain the shape of the DEM by observing that the ions determining these peaks are: O III, O IV, and O V giving rise to the first peak; Ne V+Mg V, Ne VI, and Ne VII for the second peak; Mg VIII, Mg IX, Mg X, Al X, Al XI, Fe XI for the third one. This fact may cast some doubt both on the accuracy of the values for the ionisation equilibrium abundances of the oxygen, neon, and magnesium ions adopted in this study (Arnaud and Rothenflug, 1985; Arnaud and Raymond, 1992, for the iron ions) and to the abundances of the elements we have included in the theoretical calculation of the DEM itself (Feldman *et al.*, 1992).

The analysis has also outlined some other features:

– Ne V NIS 1 and NIS 2 lines disagree, showing that the NIS 2 fluxes are smaller than the NIS 1 by a factor 2.5–5. Also O III NIS 1 and NIS 2 lines show a similar behavior. Moreover the NIS 1 lines of Al X and Fe XI, whose effective temperature $T_{\rm eff}$ is very similar, never agree with those of Mg X (NIS 2), the latter being weaker than the former by a factor 1.5–3.5. We think that here there is some problem in the NIS 1–NIS 2 intercalibration.

– The NIS 2 second-order lines are in great disagreement with all the first-order lines as expected since no account is taken in the current calibration used for second-order lines.

– Ne V and Mg V have a very similar $T_{\rm eff}$ so their NIS 1 lines are expected to agree. Nevertheless the intensities of the weak Mg V lines are in better agreement with the NIS 2 Ne V lines instead of the NIS 1 line, in spite of the supposed difference between the NIS 1 and NIS 2 intensities. It is possible that difficulties arise in the measured intensities of Mg V, which are very near to some strong iron lines, or in the relative Ne/Mg abundance. Variations in the relative Ne/Mg abundance in another active region observed by CDS are discussed by Young and Mason (1997).

5.2. NIS 1–NIS 2 INTERCALIBRATION

As already pointed out in the DEM section, there seems to be a rather large discrepancy between the NIS 1 and the NIS 2 intensities, the former being much greater than the latter by a factor in the range 2.5–5. The strongest diagnostics for this behavior are O III and Ne V, since their lines have been observed in both channels. Their L-functions should be nearly identical for any value of the electron density since these lines are fairly density insensitive.

Several comments are needed. The O III NIS 1 lines ($2s^2 2p^2\,^3P - 2s^2 2p3s\,^3P$) are all blended with lines coming from other ions. Line 373.81 Å includes a second-order strongly density dependent Fe XII line at 186.88 Å, which could provide a significant contribution. Young *et al.* (1997) found that in the active region observed by SERTS in 1989 (Thomas and Neupert, 1994) Fe XII provides ≈ 58% of the total intensity. We think that in the present active region the Fe XII contribution is smaller than that found by Young *et al.* (1997), since this active region is weaker and the electron density is smaller than the values found by Young *et al.* (1997):

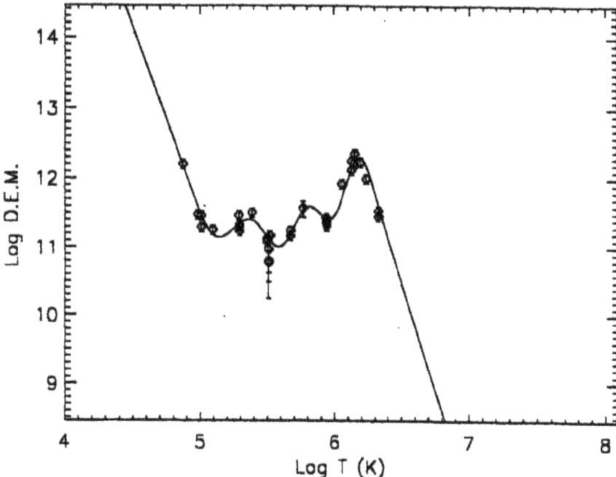

Figure 4. Differential emission measure of the s524 Solar active region. NIS 1 intensities have been corrected allowing for NIS 1 - NIS 2 intercalibration correction factor (see Section 5.2).

those authors report $N_e \approx 10^{10}$ cm^{-3} while the intensity ratio 352.11/338.27 (corrected for the NIS 1 internal calibration evaluated in the present paper) provide $N_e = 10^{9.5 \pm 0.1}$ cm^{-3}.

All the other lines of the multiplet (observed at 374.01 Å and 374.4 Å) are blended with N III $2s^2 2p\,^2P - 2s^2 3d\,^2D$ predicted by the Arcetri Spectral Code and the CHIANTI database at 374.20, 374.434, 374.442 Å. Nevertheless the synthetic spectrum obtained using the differential emission measure found in the present study shows that the contribution of the N III lines should be smaller than the O III intensity. This leaves the 374.01 Å line intensity nearly free of N III contribution (smaller than 10%), while the contribution to the 374.4 line should be smaller than 25%. We are then able to calculate the Fe XII contribution to 373.81 Å to be around 30% of the total observed intensity. This discussion shows that the N III and Fe XII blends are not able to contribute enough intensity to the O III lines to explain the difference between the NIS 1 and NIS 2 observed intensities.

Two general problems affect the observed Ne V 359.38 Å, 569.8 Å, and 572.3 Å lines used for NIS 1–NIS 2 intercalibration. The first of these problems is due to their collisional transition probabilities between configurations $2s^2 2p^2 - 2s2p^3$ included in the Arcetri Spectral Code and CHIANTI. We do not consider that any uncertanties in the atomic data used for Ne V are sufficient to explain the discrepancies between the Ne V NIS 1 and NIS 2 fluxes.

The other problem with Ne V lines is that their $T_{\rm eff}$ is slightly different [$\log T_{\rm eff}(359) = 5.52$, $\log T_{\rm eff}(568\text{–}572$ multiplet$) = 5.48$], and this affects the values of the L-functions of these lines. Nevertheless the effect of this difference is again too small to generate the large discrepancy between the NIS 1 and NIS 2 fluxes. A similar conclusion can be drawn also for the O III lines, which also show some relative temperature dependence. Since no blending occurs to the 359.38 Å line, any departure from this behaviour has been ascribed to intensity calibration problems.

Another check on the NIS 1–NIS 2 intercalibration can be obtained using lines belonging to ions whose Teff are very similar, since the observed intensities of their lines should be compatible with the same DEM. In the s524r00 spectrum this condition should be met by NIS 1 Al XI 332.8 Å, Fe XI 341.1 Å and 356.6 Å and Mg X 609.7 Å and 624.9 Å. Nevertheless the DEM analysis has shown that the Mg X NIS 2 fluxes are always lower than expected, though the required correction is smaller than the O III and Ne V one. It is important to note that this last measurement may be affected by possible element abundance problems, since the studied lines belong to ions of different elements. Therefore the Al–Fe–Mg proposed correction factor should be taken with caution.

Taken together with the conclusions drawn in the DEM and NIS 2 I–II-order sections (discussed later), this suggests that the current NIS 1 intensity calibration is approximately a factor of three too great compared to NIS 2.

The results obtained from this kind of analysis are shown in Figure 5. The reported L-function ratios are the correction factors to be applied to NIS 1 intensities in order to obtain the correct NIS 1–NIS 2 intercalibration.

5.3. NIS 1 INTERNAL CALIBRATION

The active region data in file s524 provide useful spectral lines for the calibration of the NIS 1 spectral window between 313 and 370 Å. Lines Fe XII 352.11 and Fe XI 352.67 are taken as reference, and their correction factor has been set to 1. The intensity of the Fe XVI 335.40 line includes also contribution from the far weaker Mg VIII 335.25 Å and Fe XII 335.33 Å lines: studying the L-functions of these two ions it has been possible to evaluate their contribution to the total intensity around 9.8% of the total emitted power and to determine the Fe XVI correction factor. Fe XII 364.46 Å line is blended with a far weaker Si XI line at 364.49 Å, whose contribution is less than 5%.

The derived correction factors are plotted in Figure 6. It is possible to note that there is no gross variation proposed to the current NIS 1 intensity calibration, nevertheless it appears that the short wavelength intensities need to be lowered by less than 20%. No line is available in the range 320–330 Å. Si VIII 314.34 Å and Mg VIII 313.74 Å are in marginal disagreement with the other nearby Mg VIII and Si VIII lines, but this is probably due to some fitting problem in this crowded region of the spectrum.

A straight line can fit the reported correction factors and provides a simple and useful way of correcting the NIS 1 intensities. The best-fit parameters for this straight line are

$$\text{Corr. fact.} = a + b\lambda \quad (\lambda \text{ in Å}) ,$$

$$a = -0.64 \pm 0.22 , \qquad b = (4.6 \pm 1.4) \times 10^{-3} \text{ Å}^{-1} .$$

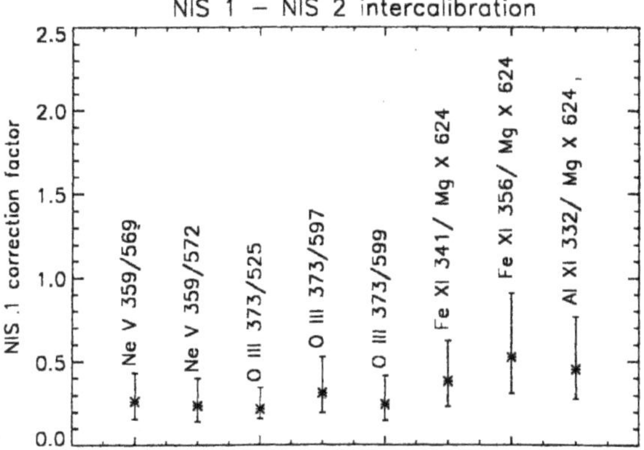

Figure 5. Correction factor to NIS 1 intensity relative to NIS 2 derived from the DEM analysis of s524 active region spectral atlas.

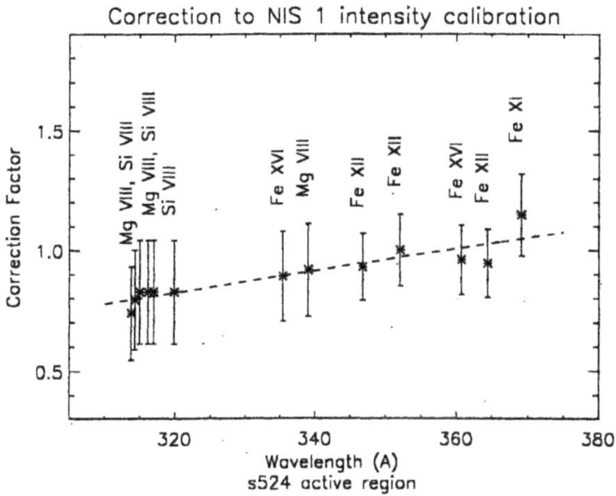

Figure 6. Correction factor to NIS 1 intensities as a function of wavelength. The dashed line shows the best linear fitting.

5.4. NIS 2 INTERNAL CALIBRATION

Checking the internal NIS 2 calibration is more difficult owing to the paucity of useful lines. Five lines have been taken into account. The correction factor of Mg X 624.95 has been obtained by comparing the 624.95 Å and 609.79 Å L-functions, the latter of which has been corrected for a blending contribution of O IV 609.82 Å. The correction factor of the O III 525.79 Å line is a weak function of temperature, due to the fact that the O III 525.79 Å and 599.59 Å lines show a different dependence on electron temperature, so their L-functions should be calculated at slightly different $T_{\rm eff}$. Nevertheless the uncertainty due to such

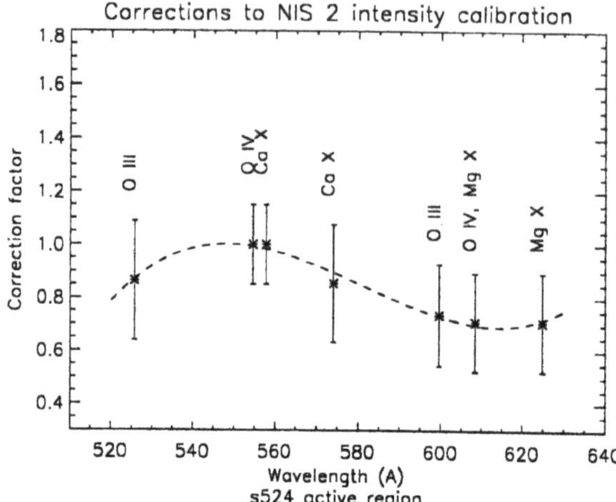

Figure 7. Correction factor to NIS 2 intensities as a function of wavelength. The dashed line represents the cubic polynomial that best fits the measured correction factors. Also the two Ca X lines have been included in the plot, though the correction factor of line 574.01 Å should be nearer to unity (see text).

behaviour (a few percent) is negligible. There are four O IV lines in the range 553.3–555.3 Å which could be useful, but since their wavelength difference is too small to show intensity calibration effects and their L-functions are in excellent agreement, only one of them has been considered for this study. Of interest could be the Ca X 557.77 Å and 574.01 Å features. Unfortunately the latter line is blended with C III 574.28 Å, and it is neither possible to separate them in the fit nor estimate the two contributions to the total intensity.

The O IV 554.51 line has been taken for reference (correction factor set to 1). The resulting correction factors versus wavelength are plotted in Figure 7. Also the correction factors of the two Ca X lines have been reported, though the 574.01 Å line should be considered with caution.

A cubic polynomial fit appears to be the most appropriate for calculating the NIS 2 correction factor as a function of wavelength, and the result is plotted as a dashed line in Figure 7. The adopted polynomial is

$$\text{Corr. fact.} = a + b\lambda + c\lambda^2 + d\lambda^3 \quad (\lambda \text{ in Å}),$$

$$a = -417.398\,, \qquad b = 2.17592\ \text{Å}^{-1}\,,$$

$$c = -3.76113 \times 10^{-3}\ \text{Å}^{-2}\,, \quad d = 2.16006 \times 10^{-6}\ \text{Å}^{-3}\,.$$

5.5. NIS 2 SECOND-ORDER CALIBRATION

The NIS 2 I–II-order calibration has been made possible by the presence of several second-order lines which can be compared to NIS 1 and NIS 2 first-order lines.

The lines which have been used for this calibration are Si XI 303.32, Fe XVI 262.98, Fe XIV 257.40, 264.78, and 274.21, Fe XV 284.16 Å. Also the strong He II 303.78 has been observed, but owing to the lack of other He II first-order lines and of ions whose T_{eff} is close to the He II value, it has not been possible to study its behaviour.

It is important to notice that most of these lines are typical active-region lines which cannot be observed as second-order lines in quiet-Sun conditions. The I–II-order calibration of NIS 2 is therefore possible only in active region data.

The Arcetri spectral code predicts two first-order lines blending Fe XV 284.16 Å: Al XI 568.12 Å and Ne V 568.41 Å. Therefore, it has been necessary to evaluate the flux of these two lines before using the Fe XV feature for calibration studies.

The Fe XVI 262.98 Å line is quite weak and has been observed in the longward wing of the strong O III 525.79 Å. It is possible that this line has been badly fitted and its measured intensity should be considered with caution.

The Fe XIV line 257.40 Å is quite weak and the experimental error on its measured intensity is greater than 15%. Young *et al.* (1997) found some problems in the Fe XIV intensities observed in the SERTS-89 spectrum, showing that Fe XIV lines could be separated into two groups: shortward of 274 Å and longward of 274 Å. Lines belonging to the same group were consistent with each other but not with the lines of the other group. In the s524 spectrum we confirm this tendency, since the correction factor provided by Fe XIV 257 and 264 Å lines disagree with the value provided by line 274.2 Å. Remembering the Young *et al.* results, we have adopted as Fe XIV correction factor the one provided by line 274.2 Å. It is interesting to note that the behaviour of the Fe XIV II-order lines closely resembles the one described by Young, Landi, and Thomas (1997) for the same lines. Fe XIV line at 270.52 was also expected to be observed as NIS 2 second-order line, but it blends with the first-order Ne IV 541.02 Å. Using the other two Ne IV lines observed at 541.99 and 543.84 Å, it is possible to determine the Fe XIV contribution to be $\approx$ 30% of the total intensity. This value creates excellent agreement between the 257 and 264 Å and 270.5 lines, and disagreement with the other Fe XIV lines, again in accordance with the results of Young, Landi, and Thomas (1997).

Another line, Si X 271.99 is expected to blend with Ne IV 543.88, and could be used for this study. Nevertheless it is not easy to use this line, both because of its weakness and because of the dominating Ne IV line. It is possible to determine the Si X contribution to be $\approx$ 20% of the total flux. The deriving correction factor should be taken with caution, though it is in agreement with the other values.

Only Si XI 303.32 Å has been compared to NIS 2 spectral lines, while all the other II-order lines belong to ions whose I-order lines fall in the NIS 1 wavelength range. This fact causes the correction factor for the NIS 2 I–II-order calibration to be dependent on the NIS 1–NIS 2 relative intensity calibration. The problems in the relative NIS 1–NIS 2 intensity calibration discussed in the present work lead the correction factor derived from only NIS 2 lines to be smaller than the value obtained with the other lines. The difference is due to the correction factor

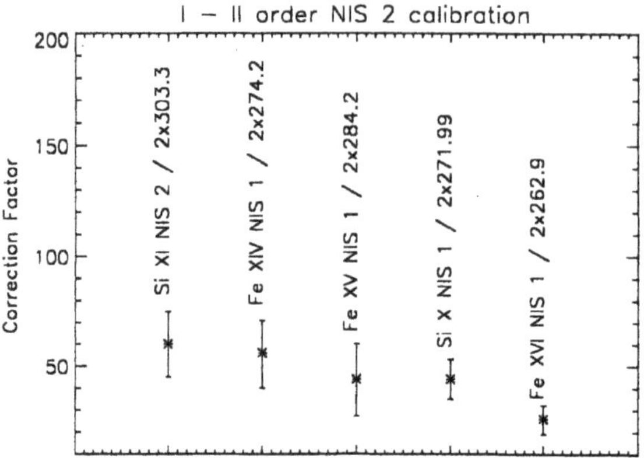

Figure 8. Correction factor to NIS 2 II-order lines. The fluxes measured by NIS 1 have been decreased by a factor 3 in order to consider the NIS 1–NIS 2 intercalibration correction.

necessary to intercalibrate the two CDS spectral windows. Thus, the NIS 2 I–II-order calibration can also be used as a check for NIS 1–NIS 2 intercalibration.

The result derived with the aforementioned lines is shown in Figure 8. The Si XI correction factor is much lower than the values provided by the Fe XIV and Fe XV lines, which agree reasonably with each other. Using the NIS 1–NIS 2 intercalibration correction of a factor 3 most of the difference can be compensated. Fe XVI line shows a lower correction factor, which can be ascribed to the problems in the fitting of this line. We are able to estimate the NIS 2 I–II-order correction factor as 55^{+15}_{-10} which means that the II-order sensitivity of the CDS–NIS 2 spectrometer is $(1.8 \pm 0.4) \times 10^{-2}$ relative to the first-order value.

6. Conclusions

In this work the intensity calibration of the CDS normal incidence spectrometer is discussed. A brief description of the preliminary in-flight calibration of both the NIS 1 and NIS 2 spectral windows is given, then results of the intensity calibration of both the spectral windows obtained with a new diagnostic method are described.

This new method has proved to be very useful for both the NIS 1 and NIS 2 internal intensity calibration and for the relative calibration between the two wavelength bands. While the preliminary in-flight internal calibration of the two windows has proved to be quite reliable and only small corrections were necessary, the relative calibration of the two spectral windows has shown a rather large discrepancy which had not been taken into account in the previous calibration works. Moreover it has been possible to determine the NIS 2 I–II-order intensity calibration using some second-order lines observed in the NIS 2 spectral range.

A similar study has been performed by the authors using another stronger active region spectrum (s4309) and the same procedures in order to check the results obtained with the present study. The corrections measured with the new data show excellent agreement with those found with the s524 spectrum. This adds confidence to the correction factors determined in the present work.

The adopted method for the analysis is recommended for further work on intensity calibration for the grazing incidence spectrometer of CDS (GIS), both for the internal calibration of the four spectral windows which compose this spectrograph and their relative calibration, and for relative calibration with the NIS spectral windows.

It should be noted that the results of the laboratory calibration of the NIS absolute intensity calibration (Bromage *et al.*, 1996) are currently incorporated into the CDS/NIS calibration software together with the relative calibration within each waveband determined from in-flight data as described in Section 2.

The absolute intensities derived for the observed spectral lines in NIS 1 and NIS 2 are therefore directly dependent upon the laboratory results. The fact that we find in this analysis that it is necessary to correct the absolute intensities of NIS 1 lines by a factor of three is tantamount to requiring a large adjustment to the laboratory results. Such a correction is much larger than the error estimates given for the laboratory calibration. One possible source of this discrepancy could be that the in-flight measures use full illumination of the telescope and spectrograph aperture, something which was not possible in the laboratory, even though attempts were made in the analysis to compensate for this effect. Alternatively, of course, there could have been a real change in the instrument efficiencies since the pre-flight data were taken. Further analysis of the laboratory data will appear in Lang and Kent (1997).

A single measure in the laboratory of the NIS 2 second-order efficiency gave a ratio of 25 relative to first order. This should be compared to the value of 55 derived here, noting however that our value is dependent upon the correctness of the NIS 2 over NIS 1 calibration mentioned above.

Acknowledgements

We are grateful to all the SOHO and in particular CDS team members who have made this solar mission such a success. H. E. Mason acknowledges the financial support of the U.K.–PPARC.

References

Arnaud, M. and Raymond, J. C.: 1992, *Astrophys. J.* **398**, 394.
Arnaud, M. and Rothenflug, R.: 1985, *Astron. Astrophys. Suppl. Ser.* **60**, 425.
Bromage, B. J. I., Breeveld, A. A., Kent, B. J., Pike, C. D., and Harrison, R. A.: 1996, *UCLan Report CFA/96/09.*

Dere, K. P., Landi, E., Mason, H. E., Monsignori Fossi, B. C., and Young, P. R.: 1997, *Astron. Astrophys. Suppl. Ser.*, in press.
Feldman, U., Mandelbaum P., Seely J. L., Doschek G. A., and Gursky, H.: 1992, *Astrophys. J. Suppl. Ser.* **81**, 387.
Harrison, R. A., Sawyer, E. C., Carter, M. K., Cruise, A. M., Culter, R. M., Fludra, A., Hayes, R. W., Kent, B. J., Lang, J., Parker, D. J., Payne, J., Pike, C. D., Peskett, S. C., Richards, A. G., Culhane, J. L., Norman, K., Breeveld, A. A., Breeveld, E. R., Al Janabi, K. F., McCalden, A. J., Parkinson, J. H., Self, D. G., Thomas, P. D., Poland, A. I., Thomas, R. J., Thompson, W. T., Kjeldseth-Moe, O., Brekke, P., Karud, J., Maltby, P., Aschenbach, B., Bauninger, H., Kuhne, M., Hollandt, J., Siegmund, O. H. W., Huber, M. C. E., Gabriel, A. H., Mason, H. E., and Bromage, B. J. I.: 1995, *Solar Phys.* **162**, 233.
Harrison, R. A., Fludra, A., Pike, C. D., Payne, J., Thompson, W. T., Poland, A., Breeveld, E. R., Breeveld, A. A., Culhane, J. L., Kjeldseth-Moe, O., Huber, M. C. E., and Aschenbach, B.: 1997, *Solar Phys.* **170**, 123.
Landi, E. and Landini, M.: 1997a, *The Arcetri Spectral Code*, in preparation.
Landi, E. and Landini, M.: 1997b, *Astron. Astrophys.*, in press.
Lang, J. and Kent, B. J.: 1997, 0RAL Technical report, in preparation.
Mason, H. E., Young, P. R., Pike, C. D., Harrison, R. A., Fludra, A., Bromage, B. J. I. J., and Del Zanna, G.: 1997, *Solar Phys.* **170**, 143.
Monsignori Fossi, B. C. and Landini, M.: 1995, in S. Bowyer and R. F. Malina (eds), *Astrophysics in the Extreme Ultraviolet*, p. 543.
Summers, H. P., Brooks, D. H., Hammond, T. J., Lanzafame, A. C., and Lang, J.: 1996, RAL Technical Report RAL-TR-96-017, March 1996.
Thomas, R. J. and Neupert, W. M.: 1994, *Astrophys. J. Suppl. Ser.* **91**, 461.
Young, P. R. and Mason, H. E.: 1997, *Solar Phys.* **175**, 523 (this issue).
Young, P. R., Landi, E., and Thomas, R. J.: 1997, *Astron. Astrophys. Suppl. Ser.*, submitted.

EIT OBSERVATIONS OF THE EXTREME ULTRAVIOLET SUN

D. MOSES[1], F. CLETTE[2], J.-P. DELABOUDINIÈRE[3], G. E. ARTZNER[3], M. BOUGNET[3], J. BRUNAUD[3], C. CARABETIAN[3], A. H. GABRIEL[3], J. F. HOCHEDEZ[3], F. MILLIER[3], X. Y. SONG[3], B. AU[4], K. P. DERE[4], R. A. HOWARD[4], R. KREPLIN[4], D. J. MICHELS[4], J. M. DEFISE[5], C. JAMAR[5], P. ROCHUS[5], J. P. CHAUVINEAU[6], J. P. MARIOGE[6], R. C. CATURA[7], J. R. LEMEN[7], L. SHING[7], R. A. STERN[7], J. B. GURMAN[8], W. M. NEUPERT[8,a], J. NEWMARK[8,b], B. THOMPSON[8,b], A. MAUCHERAT[9], F. PORTIER-FOZZANI[9], D. BERGHMANS[10], P. CUGNON[10], E. L. VAN DESSEL[10] and J. R. GABRYL[10]

[1]*Naval Research Laboratory, Washington, DC 20375, U.S.A.*
[2]*Observatoire Royal de Belgique, Brussels, Belgium*
[3]*Institut d'Astrophysique Spatiale, Université Paris XI, 91405 Orsay Cedex, France*
[4]*Naval Research Laboratory, Washington, DC 20375, U.S.A.*
[5]*Centre Spatial de Liège, B4031, Liège, Belgium*
[6]*Institut d'Optique Théorique et Appliquée, 91403 Orsay, France*
[7]*Lockheed Palo Alto Research Laboratory, Palo Alto, CA 94304, U.S.A.*
[8]*NASA/Goddard Space Flight Center, Greenbelt, MD, U.S.A.*
[9]*Laboratoire d'Astronomie Spatiale, Marseille, France*
[10]*Observatoire Royal de Belgique, B-1180 Brussels, Belgium*

(Received 24 March 1997; accepted 17 June 1997)

Abstract. The Extreme Ultraviolet Imaging Telescope (EIT) on board the SOHO spacecraft has been operational since 2 January 1996. EIT observes the Sun over a 45 × 45 arc min field of view in four emission line groups: Fe IX, X, Fe XII, Fe XV, and He II. A post-launch determination of the instrument flatfield, the instrument scattering function, and the instrument aging were necessary for the reduction and analysis of the data. The observed structures and their evolution in each of the four EUV bandpasses are characteristic of the peak emission temperature of the line(s) chosen for that bandpass. Reports on the initial results of a variety of analysis projects demonstrate the range of investigations now underway: EIT provides new observations of the corona in the temperature range of 1 to 2 MK. Temperature studies of the large-scale coronal features extend previous coronagraph work with low-noise temperature maps. Temperatures of radial, extended, plume-like structures in both the polar coronal hole and in a low latitude decaying active region were found to be cooler than the surrounding material. Active region loops were investigated in detail and found to be isothermal for the low loops but hottest at the loop tops for the large loops.

Variability of solar EUV structures, as observed in the EIT time sequences, is pervasive and leads to a re-evaluation of the meaning of the term 'quiet Sun'. Intensity fluctuations in a high cadence sequence of coronal and chromospheric images correspond to a Kolmogorov turbulence spectrum. This can be interpreted in terms of a mixed stochastic or periodic driving of the transition region and the base of the corona. No signature of the photospheric and chromospheric waves is found in spatially averaged power spectra, indicating that these waves do not propagate to the upper atmosphere or are channeled through narrow local magnetic structures covering a small fraction of the solar surface. Polar coronal hole observing campaigns have identified an outflow process with the discovery of transient Fe XII jets. Coronal mass ejection observing campaigns have identified the beginning of a CME in an Fe XII sequence with a near simultaneous filament eruption (seen in absorption), formation of a coronal void and the initiation of a bright outward-moving shell as well as the coronal manifestation of a 'Moreton wave'.

[a] Present address is SEL/NOAA 325 Broadway, Boulder, CO 80303, U.S.A.
[b] Space Applications Corp.

Solar Physics **175:** 571–599, 1997.

Table I
EIT bandpasses

Wavelength	Ion	Peak temperature	Observational objective
304 Å	He II	8.0×10^4 K	chromospheric network; coronal holes
171 Å	Fe IX, X	1.3×10^6 K	corona/transition region boundary; structures inside coronal holes
195 Å	Fe XII	1.6×10^6 K	quiet corona outside coronal holes
284 Å	Fe XV	2.0×10^6 K	active regions

1. Introduction

EIT is a new approach for imaging the EUV Sun on an extended spaceflight platform (Delaboudinière *et al.*, 1995). Normal incidence multilayer optics are used to image the Sun on an EUV sensitive 1024×1024 format charge coupled device (CCD) camera. With a telescope focal length of 1.652 m and a pixel size of 21 μm $\times$ 21 μm, the plate scale of the EIT images is 2.6 arc sec per pixel. The optics are divided into quadrants of matched multilayer pairs that are tuned to bandpasses selected for observation of the chromosphere and corona at all phases of the solar cycle. The quadrants are isolated by a rotating mask that permits only one bandpass to view the Sun at a time.

Images from the synoptic program are presented in Figure 1. The full details in these images are inadequately represented in the reduced journal prints, so full resolution digital images are included in the CD-ROM accompanying this volume. The in-flight observations of solar structures match the predictions of Table I (Delaboudinière *et al.*, 1995):

– The Fe IX, X image is generally less structured than the other coronal channels. Diffuse, unresolved emission is present over most of the quiet Sun, including coronal hole regions. Open field structures are seen in this bandpass as well as closed structures. Active region structures, as seen in this channel, are more confined to the core of the active region. Active region loops are narrow structures that in many cases are only filled at the base. These partially filled loops are similar to the 'spiky' Ne VII and Mg IX active region structures observed by *Skylab* (Sheeley, 1980; Webb, 1981). The primary difference is that the Fe IX, X structures do not appear to become narrower with height. A region of supressed emission is frequently observed surrounding an active region. X-ray bright points (XBP) are the most numerous in this channel (by 10%). Detection of faint structures in the Fe IX, X channel is more difficult than in the other coronal channels due to the relative intensity of the quiet corona (including coronal holes).

– The Fe XII image is more dominated by the closed field regions of the quiet-Sun. All but the hottest active region loops are visible in this wavelength. As in the Fe IX, X channel, some active region loops appear to only be partially filled. The appearance of some active regions in Fe XII is that the loops diverge more

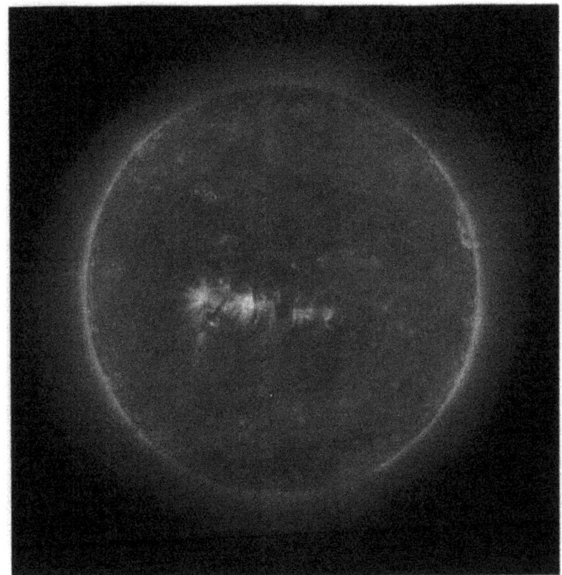

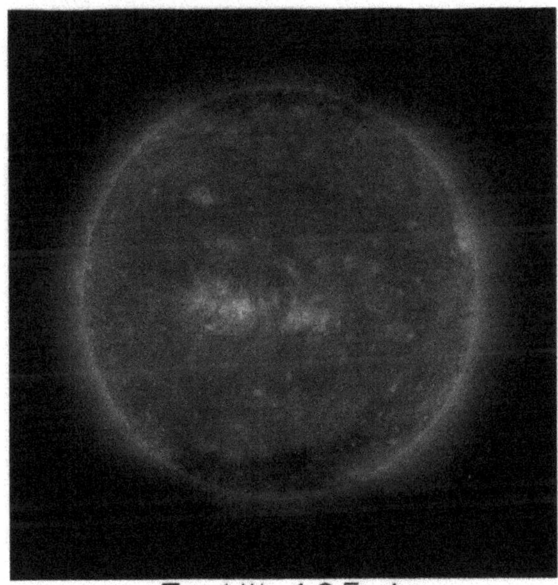

Figure 1a.

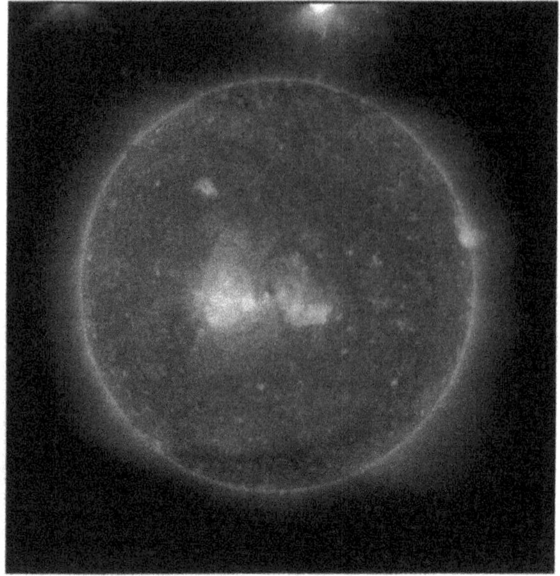

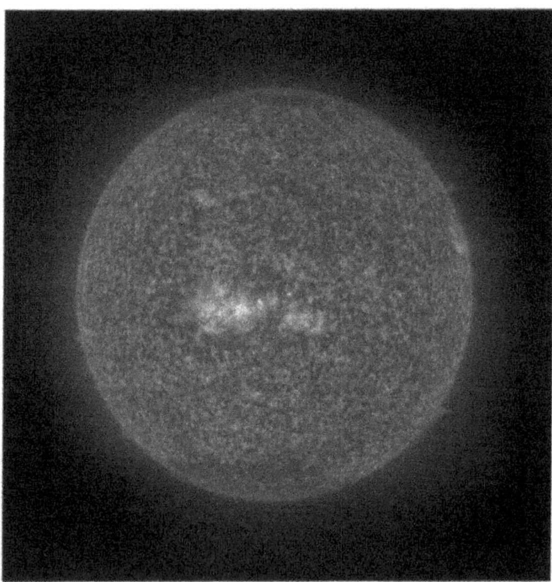

Figure 1b.

Figure 1. EIT synoptic images for 12 May 1996. Solar north is oriented up in this figure.

than higher-temperature active region loops (Fe XV and *Yohkoh* SXT). However, application of the technique of Klimchuk *et al.* (1992) on an isolated loop resulted in the same constant cross section for that loop as found at higher temperatures (Klimchuk, Moses, and Portier, 1997). Coronal holes are most easily visible in this channel because of the contrast between the emission intensity of the open field region and the closed field quiet Sun. Similarly, XBPs are most easily identified in this channel because of the contrast between the emission associated with the XBP magnetic bipole and the emission associated with the relatively weaker quiet-Sun magnetic regions or the coronal hole open field regions. Because of the high contrast and the high sensitivity to quiet Sun structures of the Fe XII channel, it is frequently used for movies of coronal evolution.

– The Fe XV image is dominated by the hot loops (e.g., active region loops). Of all the EIT observations, these most closely compare with the *Yohkoh* SXT images. Outside the active regions, this emission is obvious in the X-ray bright points, regions of erupting filaments and large-scale height loop structures. The large-scale height structures are best visible above the limb. In quiet Sun regions, there is also low level structure in Fe XV that maps out some of the internetwork field structure. The quiet-Sun Fe XV structure is not leakage of the He II into this channel as it does not form a complete network structure and cannot be removed by a normalized subtraction of a He II image. Instead, it is the high temperature component of the more general, diffuse emission seen in the other coronal channels. Also apparent in this channel is a visible light leak at the north edge of the field of view which is discussed in Section 3.

– The He II image is dominated by the transition region network structure. Bright network elements are visible throughout the region of closed loop structures. The coronal holes are marked by a reduction in the He II network elements. Active regions are generally bright in the He II image with a blending of Si XI emission. A region of supressed emission is frequently observed immediately around the active region enhancements. Macrospicules are apparent in the open field regions and along some large-scale closed regions. They are apparently aligned with the direction of the magnetic field. Prominences are always present. Filaments are slightly more difficult to identify because the contrast of the absorption features is frequently comparable to the contrast of the network structure.

Intercomparison of the images from the different coronal channels shows that, with increasing temperature, the plasma is more structured. As will be discussed in section 2.2, with increasing temperature for all channels, the emission exhibits greater variability with solar activity (time).

Detailed comparison of quiet Sun regions of coronal images and the He II images (e.g., Figure 11 in Section 3.2) shows that there is a transition region enhancement for every enhancement of coronal emission, but not the converse. This is in agreement with previous comparisons of X-ray bright points and He I 10830 Å emission (e.g., Moses *et al.*, 1994). With the availability of near-continuous photospheric magnetic field data from SOHO, the assumption that the relationship between the coronal counterparts of transition region features and the connectivity of the magnetic field can now be addressed.

A comparison of the appearance of prominences and filaments in the four EIT channels can be made from Figure 2. On the equator, the prominence is aligned primarily perpendicular to the line of sight. This structure can be seen in absorption in the coronal channel due to a combination of H I, He I and He II continuum absorption (Cheng, Smith, and Tandberg-Hanssen, 1980; Foukal, 1978; Schmahl and Orrall, 1979). The prominence in the south east can be seen to extend onto the disk as a polar crown filament in the He II as well as in the coronal channels. There is a coronal void around this prominence which reflects the temperature structure of the corona and is not due to absorption. The amount of absorption in the coronal lines depends upon the observed wavelength, the overall geometry of the observation, the relative elemental abundance, the density and the ionization balance of the cool material and thus can be used as a diagnostic of the plasma parameters of prominences.

2. Instrumentation

To move beyond a descriptive analysis of the data, several instrument issues must be addressed. These issues were either impossible to resolve pre-launch or their importance was not appreciated before the initial analysis began. This work has taken a large part of the EIT consortium effort during the first year of operations.

2.1. Imaging

The images obtained in-flight by EIT are consistent with the assertion by Delaboudinière *et al.* (1995) that the instrument resolution is limited by the plate scale of the CCD. Analysis of the off-disk images raises interesting questions about the extended distribution of He II in the solar atmosphere (Delaboudinière, 1998). In order to quantify these observations, the wings of the point response function (PRF) must be known in the EUV to better than the dynamic range of the CCD (1:5000). The PRF has been evaluated with wavefronts obtained with Zygo interferometer measurements of the telescope in each quadrant. The wavefronts obtained at 6330 Å were converted into optical path differences to obtain the complex pupil function. The Fourier transform of the complex pupil function gave the point spread functions of the optics in the four EUV wavelengths. The system PRF was obtained after convolution with the pixel size. Figure 3 is a plot of 2 perpendicular cuts in the PRF determined by the wavefront analysis for the 284 Å quadrant. (N.B. The quadrant structure of the telescope entrance pupil produces an asymmetric PRF.) To verify the Zygo measurements with measurements in the EUV, modifications were made in the IAS calibration facility in an attempt to produce a point source in the EUV with large angle scattering less than the level to be measured. The

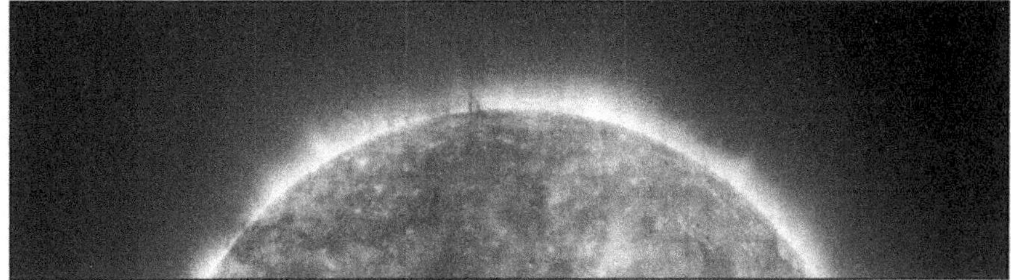

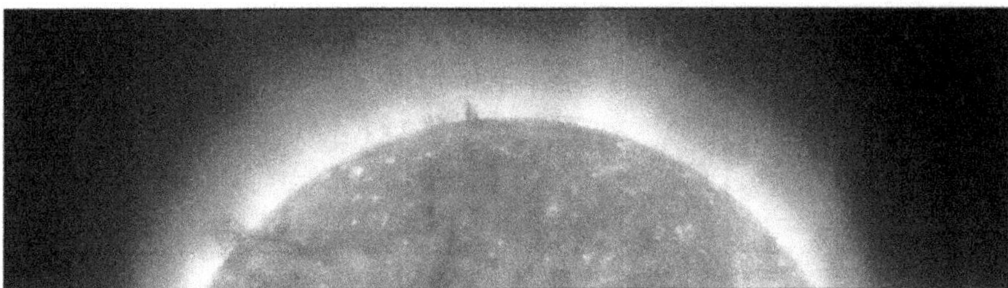

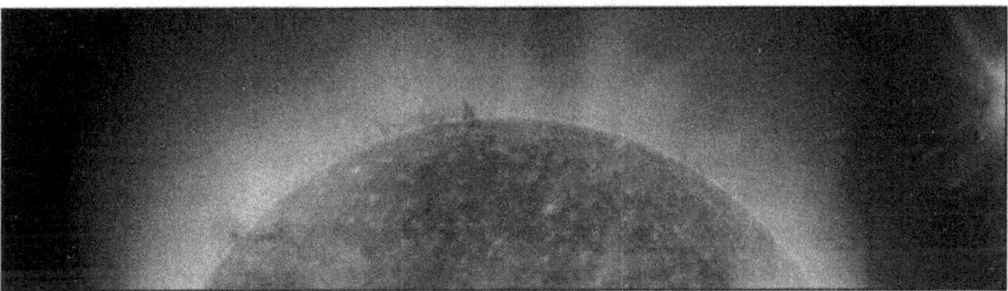

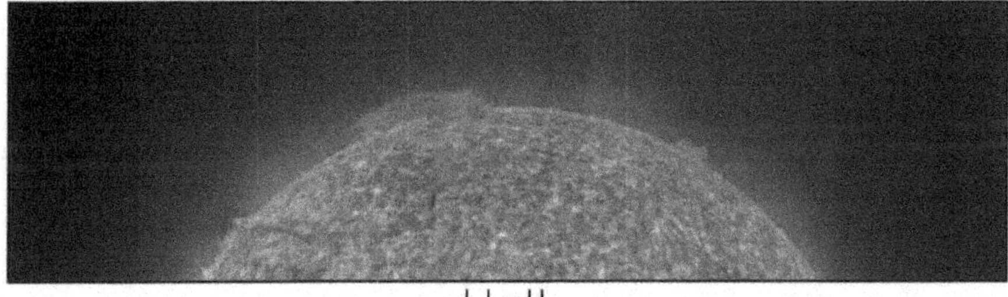

Figure 2. H I, He I, and He II absorption features in the EUV coronal images of the east limb on 28 February 1997.

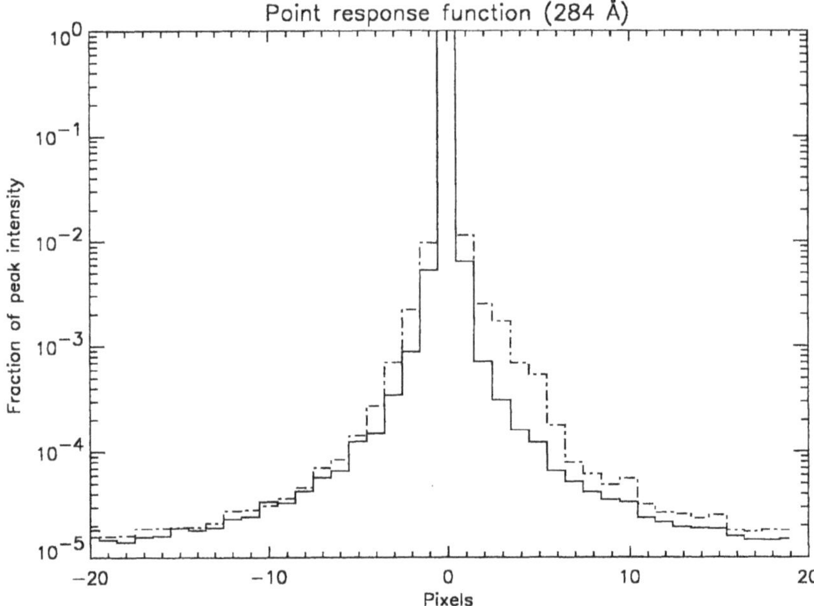

Figure 3. EIT extended point response function for the 284 Å quadrant from a wavefront analysis. The two lines represent perpendicular cuts through the peak of the response.

spare EIT optics were tested in this configuration as part of the EIT Calroc sounding rocket program. These measurements confirmed the Zygo wavefront analysis result that the far wings (1 arc min) of the PRF are suppressed by nearly 5 orders of magnitude relative to the core. However, vibration in the extended IAS facility prevented reliable measurement in the EUV near the core of the PRF. Observations of active regions in Fe XV just behind the solar limb verify that the EUV scatter measurements at IAS overestimate the spread of the PRF near the core.

2.2. EUV RESPONSE

Optimized exposure times have been derived from the inflight images (Table II). The variation of the instrument throughput is monitored by the total flux in a full field image for each bandpass (Figure 4). The utility of this monitor is determined by the variability of the solar flux in each waveband. The intrinsic solar variability increases with the temperature of the dominant emission line in the bandpass. Thus the monitor is very good for the He II and the Fe IX, X channels where the solar variability is low, while it degrades for the Fe XII channel and is poor for the Fe XV channel where changes in the instrumental response are masked by solar changes. Since the extremes of the wavelength regime are 171 Å (Fe IX, X) and 304 Å

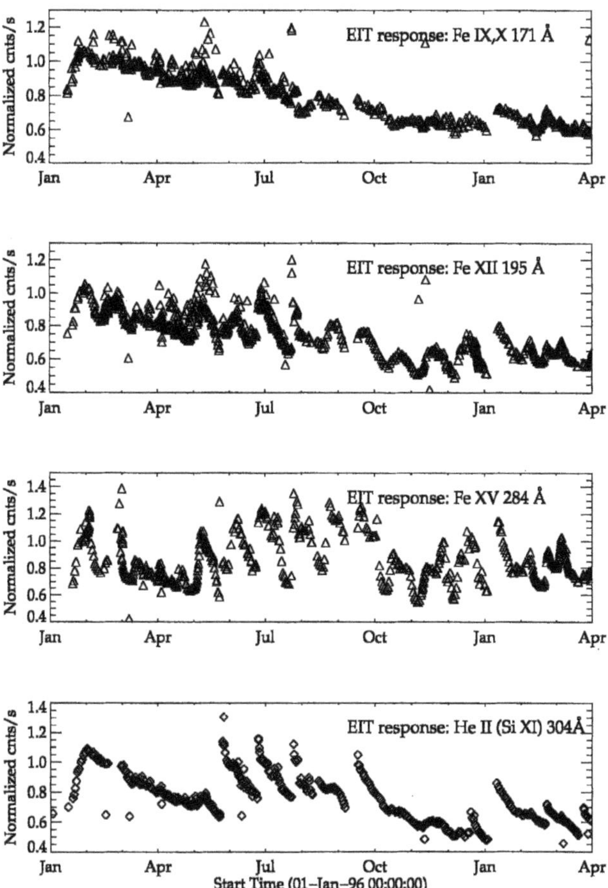

Figure 4. EIT response: total flux in each channel.

Table II
Nominal exposure times

Channel	Quiet Sun	Low activity	High activity
He II	32 s	25 s	18 s
Fe XI, X	7 s	5.5 s	2 s
Fe XII	12 s	9 s	4 s
Fe XV	102 s	62 s	47 s

(He II), this technique allows the full wavelength dependence of the degradation to be monitored.

Since the 304 Å radiation has the lowest penetration depth (range) of the wavelengths observed, the He II channel is the most sensitive to degradation asso-

Table III
EIT CCD temperature cycle history

Action	Date and Time	Duration
Heater on	1996 23 May 19:21	45 hr
Heater on	1996 23 June 19:10	24 hr
Heater on	1996 23 July 15:48	2.5 hr
Heater on	1996 5 Aug 18:50	15 hr
Heater on	1996 9 Aug 15:05	71.5 hr
Heater on	1996 6 Sep 19:49	236 hr
Heater on	1996 19 Dec 13:14	2.5 hr
Heater on	1997 3 January 21:50	156 hr
Heater on	1997 20 February 21:57	13 hr

ciated with evolution of surfaces and best illustrates the operative processes. The initial increase in response over the first month of observations is attributed to an overall cleanup of the optical surfaces and detector from the contamination associated with the launch and early spacecraft operations. Subsequently the response declined exponentially with an e-folding time of 210 days. In order to reverse the decline, the CCD was heated to 18C on 23 May 1996. Initial recovery was to the highest throughput observed in flight. Subsequent response declines have a shorter e-folding time (85 days) than the initial February to May interval. Successive heat cycles were conducted according to the requirements of the observing schedules and in exploration of the causes of the decline (Table III). An instrument failure occurred on 29 July that left the shutter open in the 304 Å channel for 7 hours. The total instrument response immediately declined 10 to 20% following this exposure. The response degradation was not uniform. Instead the degradation was patterned in proportion to the EUV exposure.

The degradation process consists of several components which are difficult to separate in detail. The two basic processes contributing to the degradation are (1) the absorption of EUV before it interacts with the CCD by a surface contaminant and (2) the reduction of charge collection efficiency (CCE) in the CCD due to EUV induced device damage.

The condensation of contaminant on the CCD is expected since the detector is one of the coldest surfaces in the instrument. In anticipation of in-flight contamination, (1) a cleanliness program was maintained in the construction of the instrument and (2) heaters for the CCD were included in the design. Accumulation of contamination at a constant rate will result in an exponential decline in response. If no polymerization of an organic component is involved, then the condensate will evaporate rapidly during temperature cycling. The straylight baffling around the CCD unfortunately reduces the vacuum conductance out of the vicinity of the CCD so that further outgassing during a temperature cycle is not feasible.

Reduction of the CCE is probably caused by the charging of the native backside oxide and the generation of interface trap sites. Since each interacting EUV photon generates more than one electron, the CCE can be determined by the comparison of the observed signal to the photon shot noise (Janesick, Klassen, and Elliott, 1987). The limitations of this approach are significant, but the degradation of the device for the 29 July event has been demonstrated to be totally CCE degradation. The proportion of the cumulative degradation from ordinary operations which can be attributed to CCE degradation is discussed in Defise *et al.* (1997). Although self annealing is observed, the rate of self annealing is dependent on the device temperature and possibly the level of prior damage. The self-annealing process during temperature cycling is at best 100 times slower than the condensate reversal.

2.3. Flat-Field Corrections

Spatial variations of the instrument response across the field of view must be corrected before performing almost any quantitative analysis of EIT images. These non-uniformities are produced by several independent contributions. Some of these flat-field components were known before SOHO launch, like the optical vignetting, the CCD flat-field and the grid pattern produced by the focal aluminum filter. However, the grid pattern could only be accurately measured after launch, using in-flight solar images. Moreover, during the course of the mission, two unpredictable artifacts appeared: a light leak affecting mainly the Fe XV images and a radiation induced non-uniform degradation of the CCD response. The characteristics of these last components can only be indirectly deduced from the affected solar images themselves, which limits the accuracy of any reconstruction due to the incomplete or distorted information available from flight data. Consequently, the ambitious 1% target for the accuracy of each flat-field component will not be met for some of them.

2.3.1. *The CCD Flat-Field*

A CCD flat-field calibration at EIT's working wavelengths was obtained on the ground with a monochromatic beam of synchrotron radiation using the method developed by Kuhn, Lin, and Loranz (1991). This analysis has been extended from the initial work reported in Delaboudinière *et al.* (1995). Figure 5 shows large-scale trends of the CCD flat field. Of particular interest is the steady gain enhancement towards the sensor edges, that reaches an amplitude of 10 to 15%. This last feature went unnoticed in a first partial analysis (Delaboudinière *et al.*, 1995). The accurate determination of these large-scale variations required special care, as the Kuhn, Lin and Loranz algorithm can produce false linear trends when intensities in the input images are improperly scaled. The accuracy of calculated large-scale variations was verified, but any residual edge-to-edge slope cannot be determined to better than about 10%. At smaller spatial scales, the accuracy of the results is close to 1%.

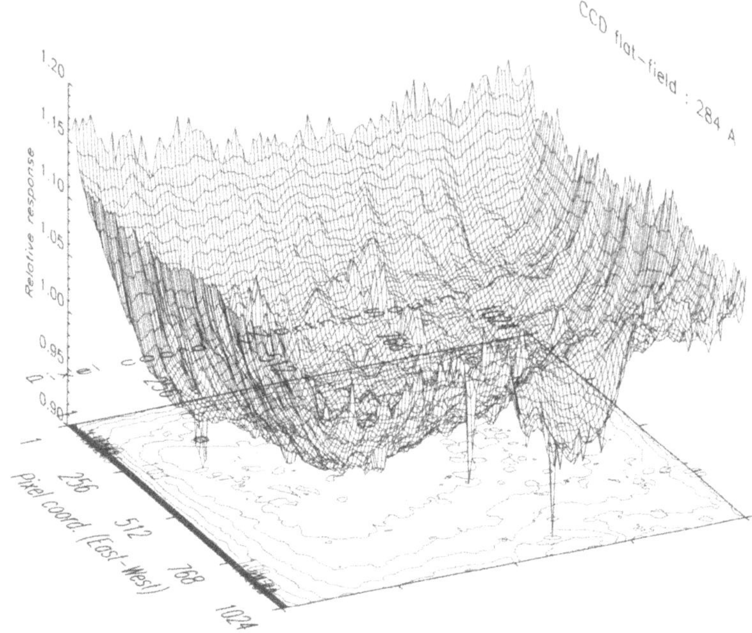

Figure 5. Three-dimensional plot of the CCD flat-field, showing the amplitude of the large-scale gain variations. The original map was smoothed over a 24 × 24-pixel domain to suppress local fluctuations.

Although the flat-field shows little change with wavelength, a systematic 10% increase in the amplitude of the defects can be measured from 171 Å to 304 Å. A comparison of the ground-based flat-fields with those deduced from the in-flight solar images shows a very good agreement over most of the sensor. However, significant discrepancies are observed inside some of the larger blemishes. Contaminants apparently interact preferentially with the surface defects. This could be interpreted in one of two ways: (1) the defects act as condensation nuclei or seeds for microcrystal formation or (2) the electronic properties of the defects are activated by chemical interaction with the condensate. Further analysis is needed to reconcile both measurements over the whole CCD within the 1% accuracy of the calculation. However, if we exclude the radiation induced aging described below, no change of the intrinsic CCD flat-field was detected in in-flight images during the first year in space.

2.3.2. *The Al-Filter Grid Pattern*

One of the redundant aluminum filters used to block long-wavelength radiation (mainly visible light) is located 14.5 mm in front of the CCD detector. Such Al filters are mechanically supported by a nickel grid (Delaboudinière *et al.*, 1995) which casts a shadow on the CCD. The finite extent of the asymmetrical entrance pupil of the telescope produces a smoothed out-of-focus pattern. The resulting periodic intensity modulation, with an amplitude of about 15%, proved to be the main flat-field artifact. Another set of filters is located on a filter wheel, 35 mm in front of the CCD. When an additional Al filter is selected on the filter wheel, it produces an additional grid pattern, with different characteristics, as it is located at a larger distance from the focal plane.

In order to extract the grid pattern from the solar images themselves, we relied on the fact that the pattern is purely periodic, while the spatial distribution of the intensity on the EUV Sun is essentially random (Note that the Sun acts as a noise source in this problem). The extraction process consists of the following steps:

– Full-field full-resolution daily synoptic images are averaged over a period of several weeks (rotational averaging of activity features). The sensor noise is also reduced by this averaging.

– The average image is then submitted to a non-linear filtering: a running 2-dimensional median filter is applied to 5×5 pixel sets, with a 21-pixel spacing between sampled pixels in both dimensions. As this 21-pixel interval is a close approximation of the grid period, the filter selectively amplifies the grid signal relative to all other spatial fluctuations in the image, i.e., it acts as a morphological filter.

– The Fourier transform of the resulting image reveals about 30 peaks, corresponding to the harmonics of the grid pattern. Two numerical filters are then applied in the frequency domain : suppression of the low-frequency f^{-1} noise background (subtraction of a smoothed average radial profile) and rejection of high-frequency noise in regions where no harmonics are present (square domains with cosine edge tapers).

– An inverse transform is finally applied to the filtered spectrum to reconstruct a cleaned image of the grid modulation.

The resulting pattern has the regularity and uniformity expected from the grid specifications (mesh period: 440 μm, mesh width: 40 μm). The grid is detected up to the very corners of the field of view, despite the very low signal levels in that part of the solar images. The grid corrections, which are now integrated in the EIT image display software, are accurate to about 1% over most of the field of view.

Diffraction calculations indicate a slight wavelength dependency, with a 5% increase of the modulation amplitude between 171 and 304 Å. However, the 90° rotation of the pattern for each EIT bandpass, following the changing orientation of the entrance pupil, produces most of the observed differences. Comparative measurements do not reveal significant temporal changes in the grid pattern since the beginning of the mission.

2.3.3. *Filter Pinholes and Visible-Light Leak*

Two light leaks were discovered early in the mission. They affect mainly the Fe XV and also to a lesser extent (50 : 1) the Fe IX, X images in the 'Clear' filter wheel position. Both leaks are located near the northern edge of the images and consist a bright featureless core surrounded by fainter halos containing sharp rays and

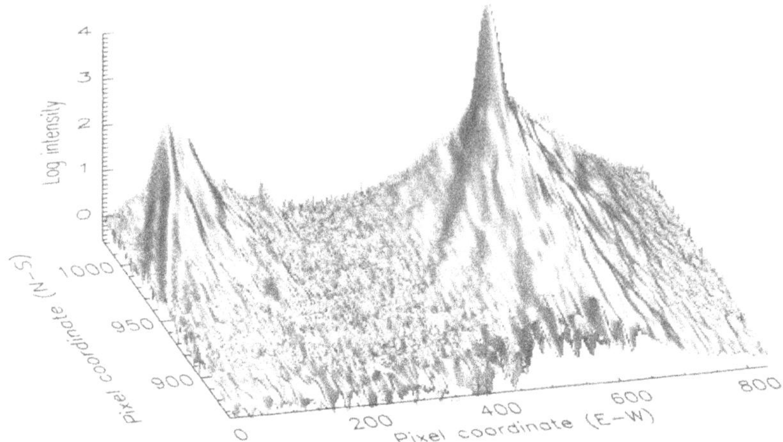

Figure 6. Two-dimensional distribution of the stray-light intensity produced by the light leaks in the 832×160 upper left area of the field of view, at 284 Å. To compensate the large intensity range between the core and faint extensions, several stepped exposures were combined and a logarithmic intensity scale is used. The intensities are scaled here according to the raw sensor output signal for a 10 s exposure.

extending up to 150 pixels from the core (Figure 6). No leak is detected in the Fe XII and He II images.

These leaks can be attributed to minute pinholes at the edge of the focal Al filter, and the light passing through these holes has been unambiguously identified as visible light reaching the CCD by study of the statistical fluctuations (photon shot noise dominated) of the signal. The visible light pattern is superimposed on the EUV solar image. The presence of visible light reaching the focal filter further implies the existence of visible light leaks in the entrance Al filters. Considering the large visible/EUV flux ratio and the limited intensity reaching the focal plane, both filter defects must affect only a small fraction of the filter area. These degradations probably occurred during SOHO's launch. The relative intensity of those leaks (50 : 1) at each of the two wavelengths, i.e., at each sector position, allow us to deduce that the tear in the front filter is located in the Fe XV sector near the edge adjacent to the neighboring Fe IX, X sector.

Being very bright and located in the dim and fairly uniform polar corona, this additive component can be extracted with a high accuracy from the Fe XV images. No significant changes in the pattern were observed in the course of the mission.

2.3.4. *Spatial Distribution of Radiation Induced Aging*

A degradation of the CCD response is induced by the radiation dose received by the detector, as explained in Section 2.2. As the average EUV intensity in the coronal and transition region images is highly non-uniform, the calculated flat-field is distorted by strong residuals of local solar activity features, especially outside the disk area. Figure 7 illustrates the distribution of the radiation damage as it appeared in mid-December 1996. While the response remains largely unaffected far from the solar limb, the degradation is strong (50% in this case) over the solar disk, in the form of a negative imprint of an average solar disk. A slight limb darkening can be recognized as well as a single grid pattern, which coincides with the 304 Å (He II) grid pattern. All these properties indicate that most of the degradation was primarily produced by the long-wavelength radiation at 304 Å, with significantly less contribution from the exposures made in all three coronal lines. (The Fe XII and Fe IX, X exposures are at shorter wavelengths, and the total dose in the Fe XV is low in comparison to the other channels). Moreover, the localized damage due to the active region located near the S–E limb at the time of the 29 July 1996 shutter failure can also be recognized. The largest damages are found within this region, where the response falls to about 20% of its average value on the rest of the disk. Finally, as the radiation damage is averaged over long periods, the changing apparent solar radius produces a smoothed limb profile. As the integrated damage always lags behind the actual solar radius, this can produce a spurious limb brightening or darkening in uncorrected images.

The successive CCD bakeouts carried out since late May 1996 have shown that the damage can be partly cured by this method (cf., Section 2.2). However, the degree of recovery is variable over the field of view and also from one bakeout to the next. Indeed, from the limited information derived from the few bakeouts executed so far, we can already deduce that the CCD response is affected by different radiation-induced mechanisms having different properties. The magnitude of each effect also depends on the time span separating successive bakeouts and their response to bakeout is strongly variable. Some components disappear completely in a few hours (e.g., small off-limb effects), while others remain relatively unaffected (e.g., active region blemish).

Two techniques are now available to monitor changes in the spatial evolution of the radiation damage:

– Reconstruction of the EUV response from sets of solar images. An adaptive median filtering is applied, which again makes use of the rotational averaging effect. This the only way to extract the true EUV response. However, the accuracy of the reconstruction is low, due to the permanent changes in the solar images.

– Calibration lamp images. In this case, a diffuse illumination is provided by a lamp located next to the CCD. This illumination proved to be very stable with time, but it falls in the visible and near-infrared wavelength ranges. The CCD response can be very different at these long wavelengths, and comparisons with EUV results have even shown that some components affecting the EUV response escape detection in the calibration lamp images.

Considering their respective drawbacks, both approaches must be combined to improve the accuracy of the reconstruction. However, in view of the continuous evolution of this aging effect, it might be difficult to reach an accuracy better than 5%. For instance, preliminary EUV – visible comparisons also hint at the super-

Figure 7. Distribution of the radiation induced aging of EIT's CCD sensor on 19 December 1996. This image was obtained by taking the ratio between calibration lamp images made in December 1996 and June 1996, when the response was still almost uniform.

position of different factors affecting differently the detector sensitivity in either wavelength range. Further analysis is in progress and should help in the identification of the underlying physical mechanisms (opacification of surface contaminants or degradation of the CCD charge collection efficiency: see Section 2.2). The ultimate goal of this effort would be to build a consistent model of the radiation damage. This fundamental undertaking will rely heavily on a permanent monitoring of the accumulated radiation dose and of the corresponding changes in the uniformity of the CCD response, which will be pursued until the end of the mission.

3. Observations

The versatility of the EIT instrument and the organization of the SOHO Experiment Operations Facility (EOF) allow a diverse range of scientific investigations. The

EIT synoptic program consists of one full frame full resolution image in each waveband each day. In addition, EIT supports SOHO Joint Operations Programs (JOPs) involving the coordinated observation of targets with the other SOHO instruments and outside observational facilities. EIT is deeply involved in the SOHO JOP program because of the context provided by a full disk EUV image. The results of the JOP participation are extremely diverse and are outside the scope of this paper. Beyond the synoptic program and JOPs, special EIT observing programs are conducted.

In this section we present reports of some of the new results from the EIT science investigations. The temperature structure of the corona is the subject of the first two sections. From the synopic program, the large-scale thermal structure of the corona in the temperature regime of 1 to 2 Mk is described in Section 3.1. In order to investigate in detail the thermal structure with height of an active region, a special observing sequence was executed for an active region on the limb (Section 3.2).

The variability of the upper chromosphere and corona is well observed with EIT image sequences and is the subject of the last three sections. In Section 3.3, movies in the He II and Fe XII channels show the dynamics of the EUV Sun. An analysis of the variability in these channels during a special high-cadence observing program is presented in this section. Sequences dedicated to identifying outflows and eruptions are the subject of Sections 3.4 and 3.5. Programs searching for evolution of coronal plumes have recorded dynamic events in the polar coronal holes which are labeled 'EUV jets'. Programs searching for the initiation of coronal mass ejections have identified these events in the EIT images.

3.1. Coronal Temperature Diagnostics

Although many observations and much analysis exists on coronal structures at temperatures up to about 1 MK and above 2 MK, principally from EUV and soft X-ray observations by *Skylab*, SMM and *Yohkoh*, less attention has been directed toward the EUV structure of the corona between 1 and 1.5 MK where soft X-ray imagers have relatively little sensitivity. Initial observations from EIT confirm that this is an uniquely interesting temperature regime, particularly during the interval of solar minimum activity.

The three EIT coronal imaging channels were designed for optimum coverage of the temperature range of 1.0 to 2.5 MK. Each coronal bandpass was chosen to be dominated by a different ionization state of Fe under specific conditions (e.g., non-flaring structures in the Fe XII channel), so that abundance variations would not be a factor in the relative fluxes observed in each channel. The assignment of a plasma temperature value to a given set of fluxes observed in the three channels in a given pixel is, mathematically, a poorly constrained inversion problem requiring a set of assumptions, including the cospatiality of material being observed in optically thin lines. However, with sufficient attention to the limitations of this approach, the thermal distribution and evolution of the corona in this temperature regime can be explored.

3.1.1. *Morphology of Line Ratio Maps*

Before quantitative analysis is presented, it is instructive to investigate the morphology of maps of the ratio of the Fe XII and Fe IX, X channels as a guide to the overall temperature distribution of 1 to 2 MK plasma in the corona. These two spectral bands have peak sensitivity at 1.2 MK and 1.5 MK, respectively, sufficiently close that cospatiality of at least some features can be expected. Even in the circumstance that the emission regions are not cospatial in the two channels, the relative intensity of the channels indicates the relative thermal distribution in the corona. The thermal distribution of coronal plasma in the 1 to 2 MK regime is presented for 12 May 1996 in Figure 8. Note the residual grid pattern in Figure 8. The aluminum filter support grid shadow discussed in Section 2.3 would dominate the ratio image without correction. This demonstrates the importance of the instrument work described in Section 2 even for qualitative analysis.

This ratio map generally outlines magnetically closed field regions. The diffuse, unresolved coronal regions in the Fe XII and Fe IX, X channels are cooler than the structures associated with stronger field regions and are thus also dark in the ratio image. For this reason, the ratio maps are sometimes used for enhanced contrast coronal displays for easy comparison with other data. X-ray bright points are associated with compact bipolar regions and are identified in the ratio image as bright compact structures. Active regions are indicated as the hottest structures observed on the disk by their intensity in the ratio image.

The coronal hole regions are clearly defined as cool, open field regions by the dark regions in the ratio image both on and off the disk. Polar plumes are identified in the ratio images as the darkest and thus the coolest structures in coronal holes. Plume-like structures have been observed outside the polar holes (see below for an example in a decaying active region) and they are some of the coolest extended coronal structures observed by EIT.

Closed off-limb regions in the ratio map are predominantly indicative of large-scale structures because the increased line of sight reduces the contribution of any given compact structure. In general, the extended, closed-field structures seen off-limb in the ratio images are hotter than the corresponding on-disk structures because the line of sight is primarily integrating through the region of higher temperature plasma which has a greater scale height. Extended structures outlined by the ratio maps include both high-latitude and low-latitude structures.

3.1.2. *Coronagraph Comparisons*

The hot high-latitude structures observed in the EIT temperature maps correspond to the persistent closed loop structures seen in the LASCO C1 green line (Fe XIV) coronagraph images as reported by Schwenn *et al.* (1997). These structures are of particular interest because of their role in determining the characteristics of the

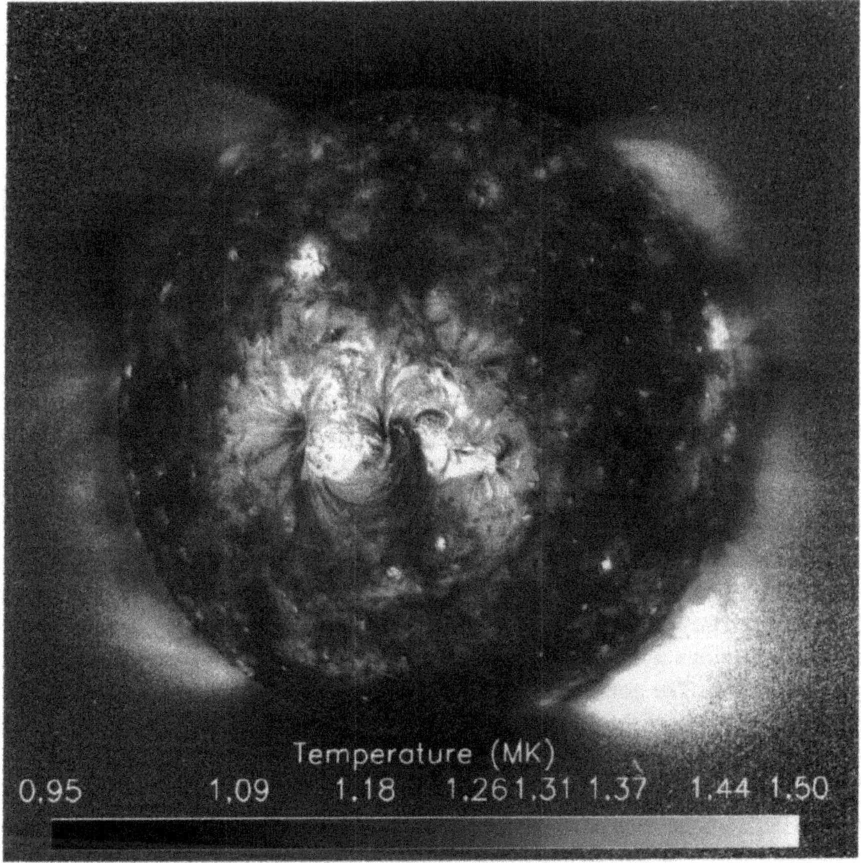

Figure 8. Temperature map derived from ratio of Fe XII and Fe IX, X from 12 May 1996

heliospheric current sheet. This correspondence directly links these high latitude structures with the large-scale high-temperature zones observed by Guhathakurta and Fisher (1994) using ground based green line (Fe XIV) and red line (Fe X) ratios. With the improved signal to noise ratio of the EIT temperature maps and the capability to follow a structure as it rotates on or off the solar disk, the EIT observations provide a valuable enhancement to the work with coronagraphs on this topic.

The *high-latitude* off-limb large-scale structures observed by EIT show a coupling of emission intensity and temperature. Comparable structures observed in the study by Guhathakurta and Fisher show the same coupling throughout most of the last solar cycle. With the low-noise EIT temperature maps, a temperature enhancement of these high-latitude structures at the boundary with the polar hole can frequently be observed. This can be seen in the both southern polar hole boundaries and the westward northern polar hole boundary of Figure 8. These temperature enhancements are also identified as regions of enhanced emission. As such, this temperature enhancement is better interpreted as a high-latitude concentration of closed-field flux elements than as some example of increased heating in the region of transition from closed to open field magnetic geometry.

Also, as found in the Guhathakurta and Fisher study, the *low-latitude* large-scale closed field structures observed by EIT do not show as direct coupling of emission intensity and temperature. This is particularly true for the case of structures associated with an active region surrounded by a weak magnetic field region. Due to the higher density of the active region loops, this is observed on the disk as well as on the limb. With the on-disk view of these structures, their magnetic origins can be determined. The region of enhanced temperature associated with the active region on the disk in Figure 8 is clearly much larger than the region of enhanced emission associated with the active region in the Fe IX, X and Fe XII images. A close examination of the Fe IX, X images shows, in comparison to the intensity of the overall quiet-Sun emission in this line, an extended region of suppressed emission surrounding the active region. Interestingly, there is also a kind of 'moat' around the active region in He II suggestive of the Hα circumfaculaire described by Bumba and Howard (1965). Apparently, flux elements in an active region which have diffused far enough from the neutral line of the active region bipole form large-scale connections with opposite polarity flux elements in the surrounding mixed polarity region. The consequence is a region of large-scale structures beyond the temperature range of the Fe IX, X images.

3.1.3. *Quantitative Analysis*

For an initial estimation of the emission measure and temperature of the structures observed by EIT, the preflight calibration described in Delaboudinière *et al.* (1995) has been combined with the CHIANTI solar spectrum (Newmark *et al.*, 1996). The CHIANTI (Dere *et al.*, 1997) database is maintained on the world wide web for easy access by the astrophysical spectroscopy community:

– http://wwwsolar.nrl.navy.mil/chianti.html.

With the limitations in applicability of the pre-flight EIT calibration described in Section 2 above, the results of this calculation must be considered preliminary until the conclusion of inflight intercalibrations involving the EIT Calroc and SEM CALSO–3 sounding rocket programs and the SOHO CDS, SUMER, CELIAS/SEM and UVCS. At this stage of analysis, it is preferable to just consider ratios of bandpasses. The temperature sensitivity for a unit emission measure of ratios of the three coronal channels is presented in Figure 9. Although all of the curves become double valued at higher temperatures, the response of the instrument to these higher temperatures is so low (Delaboudinière *et al.*, 1995) that these regimes become important only for hot regions of very high emission (e.g., flares).

In the images of 12 May 1996 from Figure 8 above, one can identify several interesting regions in the Fe IX, X and Fe XII images that are straightforward to

Table IV
Plasma parameter calculations

Feature	Counts s^{-1} (195 Å)	Fe XII/Fe IX, X	Temp × 10^6 K	EM (cm^{-5})
Plume (base)	50	0.55	1.35	7×10^{27}
Coronal hole	20	0.67	1.4	2.7×10^{27}
XBP	175	2.0	1.7	3.7×10^{28}

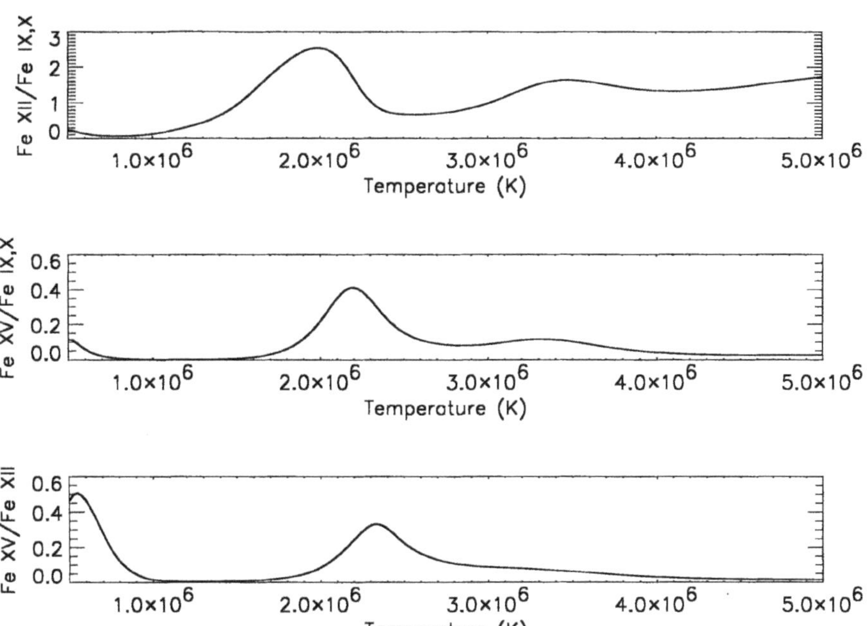

Figure 9. Temperature response of image ratios: Fe XII/Fe IX, X (*top*), Fe XV/Fe IX, X (*middle*), and Fe XV/Fe XII (*bottom*).

interpret. Since the emission is a slowly varying function of electron density, a value of 1×10^9 cm^3 is chosen for all the calculations. Values of temperature and emission measure for the polar coronal hole, the base of a plume in the coronal hole and an X-ray bright point are presented in Table IV. Analysis of other structures and other line ratios is not as straightforward.

3.1.4. *Temperature Structure of Active Region Loops*

An active region at latitude S 07° and heliographic longitude 260° was observed by EIT from May 1996 to September 1996 (Neupert *et al.*, 1997). By the end of this period it had decayed so that only a neutral line with a quiescent filament bisecting somewhat enhanced magnetic fields of opposite polarity remained. Figure 10 shows the appearance of this region in Fe IX, X and Fe XII as it approached the west limb

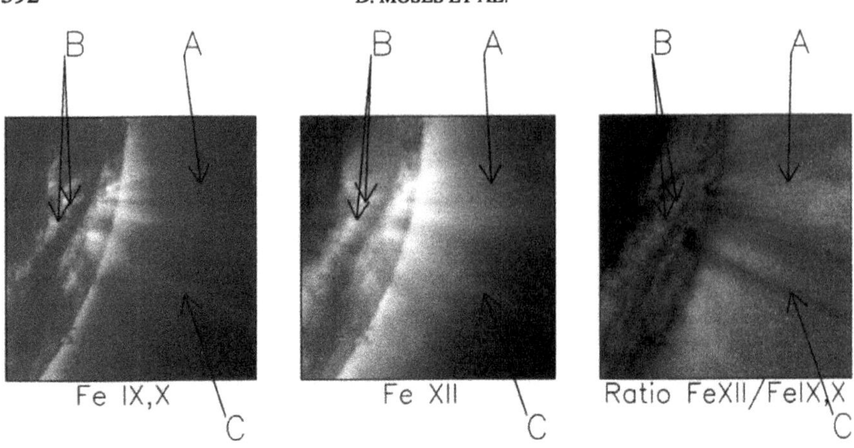

Figure 10. Fe IX, X, Fe XII and ratio images for 30 September 1996. Large-scale loops are identified as 'A', small scale loops are identified as 'B' and an extended feature is identified as 'C'.

on 30 September. A complex system of high (200 000 km) loops were present both over the major neutral line, and over leading and trailing portions of the region, where they may have connected to distant weak field regions of mixed polarities. A group of such features, observed most strongly in Fe IX, X and Fe XII, but to a lesser extent in Fe XV (not shown) is indicated as 'A' in the Figure. Smaller loop-like features (several indicated as 'B') were present at lower heights in the central portion of the region. A more nearly radial feature (designated as 'C'), for which only one footpoint was observed, may have been either a very high loop or the base of an open field region extending into the outer corona.

3.1.5. *Low Coronal Loops (Height Appreciably Less Than the Separation of Loop Footpoints and the Pressure Scale Height)*

Low loops (less than 10 000–20 000 km) have been observed in their entirety in a one or both spectral bands ('B' in the figure), suggesting that such features are nearly isothermal and in the range of 1.2–1.7 MK. Similar features, although at higher temperatures ($T_e > 2$ MK), have been well-observed by *Skylab*, SMM, and *Yohkoh*.

3.1.6. *Large-Scale Loops (Heights to 200 000 km and Comparable to or Greater than Footpoint Separation)*

EIT was able to track several features to heights well in excess of 100 000 km on the limb. The diminution in brightness with height for feature 'A' implies a density scale height of approximately 60 000 km. This result is comparable to the scale height expected at these temperatures. In 'A' the ratio of Fe XII to Fe IX, X increases to a maximum at a height of 190 000 km, suggesting the presence of coronal loops hottest ($T_e = 1.7$ MK) at their tops. Fe XV was also present, although more diffuse, so hotter plasmas may have been intermingled with the loops at 1.2–1.7 MK.

Table V
General characteristics of the 28 December 1996 image sequences (T: total duration, N: number of images, ΔT: average sampling time)

Sector	Begin	End	T	N	ΔT
He II	$16^h10^m21^s$	$19^h20^m56^s$	$3^h10^m35^s$	173	66.483 s
Fe XII	$20^h53^m26^s$	$21^h59^m08^s$	$1^h05^m42^s$	55	70.516 s

3.1.7. *Extended Features*

By comparison with large closed loop feature discussed above, the nearly radial feature denoted by 'C' may have been the base of a magnetically open field region. For this feature the Fe XII/Fe IX, X ratio was less than in the surrounding corona. Diminished emission of Fe XV (not shown) in this feature confirms the relatively low coronal temperature. The pressure scale height in this structure is in the range of 150 000 to 200 000 km. This result is significantly higher than the scale height expected at these temperatures. (N.B. This temperature structure is in contrast to the apparent positive gradient of the plumes in the polar coronal hole.)

3.2. Intensity Oscillations in the Transition Region and Corona

The EIT's ability to produce high-cadence images covering a wide field of view has brought the picture of a highly dynamical quiet corona, with small-amplitude variations of the local emission at all spatial and time scales. This small-scale activity is observed everywhere and can thus contribute significantly to the energy transfer from the photospheric and chromospheric levels into the corona through the transition region. Movies of the variability of the He II and Fe XII emission are presented in the CD-ROM.

3.2.1. *Image Data*

Subimages (320 × 320 pixel) of the central part of the solar disc were sampled every minute on average (unequal sampling) for a total duration of about 3 hours in the He II line and 1 hour in the Fe XII line (Table V). The intensity averaged over the whole data set, with compensation for solar differential rotation, is shown for both sequences in Figure 11. For this first analysis, these corrected images were submitted to a one-dimensional discrete Fourier transform for unequally spaced data (Deeming, 1975) along the time dimension, after an initial preprocessing.

3.2.2. *Initial Results*

The dynamics of the solar plasma is expected to vary according to the local state of the plasma, e.g., with magnetic field strength, density and temperature. Changes in these parameters are traced by spatial variations of the EUV brightness, allowing one to distinguish the dark cell interiors and bright cell boundaries in the transition region network, or dim loop patterns and bright points in the corona. In the time-

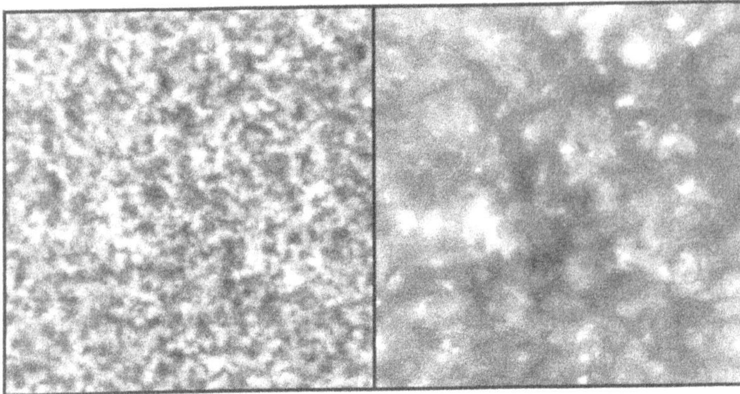

Figure 11. Spatial distribution of the average intensity in corrected and projected images of the He II image sequence (*left*) and the Fe XII image sequence (*right*).

averaged intensity maps (Figure 11), the observed contrast and sharpness of small-scale features (network, faint coronal loops) implies a relative stability of the medium-scale magnetic patterns over timescales of the order of the integration time, i.e., several hours.

Taking advantage of this property to improve the signal/noise ratio in the power spectra, individual spectra computed for single pixel locations were grouped by intensity ranges according to the local time-averaged intensity, which reflects the local plasma state. The resulting spectra are dominated by a smooth f^{-1} 'red noise' spectrum. They show only minor changes over the full intensity range, as well in the He II line as in the Fe XII line. The dynamical behavior thus seems largely independent of the local plasma state, and the amplitude of the fluctuations is simply proportional to the local average intensity. This invariance allows us to average out the spectra over the whole field-of-view, further reducing the uncertainty in the power estimates (Figure 12). In these global spectra, the noise background in the 0–2.5 mHz range is accurately reproduced by a $f^{-1.55\pm0.1}$ dependency in He II and $f^{-1.7\pm0.15}$ in Fe XII. These exponent values are close to the $-\frac{5}{3}$ value of the Kolmogorov turbulence spectrum (Landau and Lifchitz, 1971). Such an energy distribution corresponds to a fully developed turbulence (inertial regime) or to an intermittency mechanism in a non-linear system (Schuster, 1984). Spatial variation in the velocities exhibited by the C IV line formed at a temperature similar to the He II were studied by Dere (1989). He found that the power in the velocities varied with the wave number k as $k^{1.5}$ or $k^{5/3}$ as expected for the inertial regime of a turbulent plasma.

For the Fe XII coronal images, the power is falling off more rapidly with frequency, and beyond about 3 mHz the f^{-1} component merges with a uniform white-noise background, indicating the presence of a second noise contribution. Assuming that the fluctuations of the exposure times due to the shutter mechanism

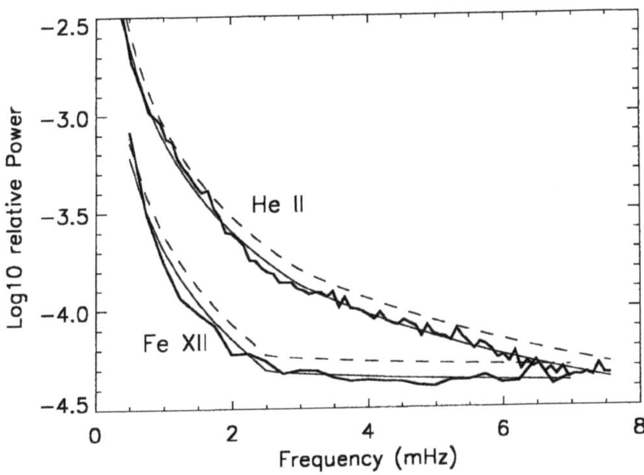

Figure 12. Average power spectra computed for the He II (304 Å) and Fe XII (195 Å) image sequences (full-field average). The thin lines represent the fitted f^{-1} profile, while the dashed lines show the 99.9 confidence levels.

are purely random, the associated 0.4 s standard deviation would correspond to a uniform noise background at a power of 5×10^{-5}. This is in good agreement with the high frequency part of the Fe XII power spectrum, indicating that this component has an instrumental origin. The same calculation shows that this background is 10 times lower in the He II power spectrum, which is one order of magnitude below the observed background. In this case, the noise background is produced by the Sun over the whole frequency range.

Using chi-squared statistics based on the dispersion of power estimates, the confidence levels (Groth, 1975) were also computed for the power estimates (see Figure 12). No significant local power excess is found, in particular within the 5–min and 3-min period ranges of photospheric and chromospheric oscillations. These oscillations are thus either fully reflected at the base of the corona or strongly attenuated, with residual amplitudes below our detection level. For the present data sets, this level equals 0.4% in relative amplitude. We must also stress that our preliminary study is based on a statistical approach and does not inform us about the spatial coherence of the detected waves. If waves originating in the low atmosphere can only propagate up to the corona along localized vertical magnetic field structures, their signature might elude detection in our spatially averaged spectra. A full 3-dimensional transform as well as the separate analysis of individual features, like single loops or bright points, is thus required and will be applied to the data in a subsequent analysis. Obviously, image sequences produced by EIT will deliver their full content of unique dynamical information only by applying such detailed analysis schemes involving jointly the spatial and temporal dimensions.

3.3. Dynamics of the Polar Coronal Hole

From previous space observations (e.g., Bohlin, Sheeley, and Tousey, 1975; Ahmad and Webb, 1978; Walker *et al.*, 1993) it was anticipated that EIT observations of structures within the polar coronal hole would contribute to the study of the dynamics of the polar coronal hole. A report by DeForest *et al.* (1997) on coordinated SOHO studies of polar plumes justifies this anticipation and can be found in this volume. The discovery (Gurman *et al.*, 1996) of EUV jetlike events in EIT Fe XII image sequences opens a new perspective on the study of polar coronal hole dynamics. These EUV jet events could be the 'building blocks' of dynamic phenomena in coronal holes.

Figure 13 shows a time series of images taken during the brightest EUV jet event observed so far. This jet begins in a bright, low-lying looplike feature some 1.3×10^4 km long and extends to the edge of the field of view, over 16.5×10^4 km above the surface, some 7 min later. During this time, the leading edge of the jet rises at an apparently constant speed of 400 km s^{-1}. Measurement of some 15 similar events yields a distribution of average speeds of 100–400 km s^{-1} and maximum observed lengths of 2 to 16.5×10^4 km. A movie showing an EUV jet eruption is included in the companion CD-ROM.

The EUV jetlike events are similar in appearance to the soft X-ray jets observed with the *Yohkoh* Soft X-ray Telescope (Shibata *et al.*, 1992). While the soft X-ray jets occurred predominantly in active regions and flaring, soft X-ray bright points (Shimojo *et al.*, 1996), the EUV jets appear to be confined to the polar coronal holes. The average speed of the EUV jets in the current (limited) sample, 325 km s^{-1}, is higher than that found for 100 soft X-ray jets by Shimojo *et al.* (1996). The maximum observed height of the EUV jets is about half that of the X-ray jets of Shimojo *et al.*'s study. Thus, the EUV jets appear to be a new regime of jet-like phenomena.

The EUV jets are not well observed in the Fe IX, X or the Fe XV channels. The contrast in EIT Fe IX, X images between structures in the coronal hole and the coronal hole itself limits the detection of faint structures. Exposure times in the Fe XV channel for quiet-corona structures are much longer than the other two coronal channels. Fe XV exposures long enough to show the EUV jets would blur these relatively faint features over several pixels, thus raising the brightness threshold for detection.

The footpoints of the microjets constitute a diverse collection of features: low-lying, bright loops as in Figure 13; unresolved structures that vanish as the microjet fades; supergranule-sized areas of emission that eject the microjet helically; and one case (so far) of what appears to be a thin flux tube breaking loose from the solar surface at one end while remaining fixed at the other. All these footpoints appear to have in common is that they occur within a coronal hole, where the magnetic field is diverging, and even small instabilities or eruptions are unlikely to be constrained magnetically.

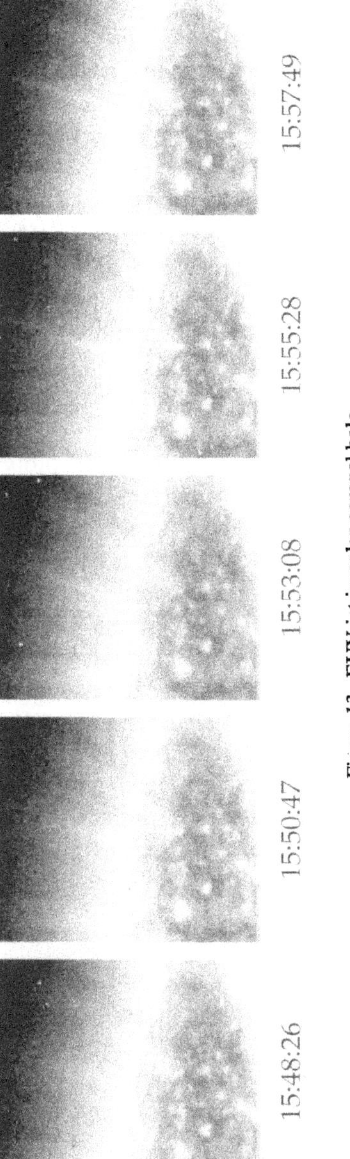

Figure 13. EUV jet in polar coronal hole.

3.4. CME WATCH

With the availability of flexibility in the SOHO telemetry resources beginning in December 1996, it has been possible to initiate sustained high cadence full disk imaging to search for large-scale coronal transients. The instrument is left in a single waveband in order to maximize temporal coverage of phenomena in a single temperature regime. To improve cadence in some circumstance, the resolution is reduced by on-chip binning to improve the cadence while supporting other programs.

During the first CME watch, a CME was observed in Fe XII on 23 December 1996 with full resolution from initiation (Dere *et al.*, 1997). The site of initiation was compact and simultaneously contained a filament eruption, a coronal void, and an outward propagating shock front. This front became the leading edge of the CME observed in LASCO. The evolution of these features is seen in the CME movie on the accompanying CD-ROM.

Associated with this CME was a disturbance tentatively identified as a coronal manifestation of a "Moreton wave". At the time of the initiation of the CME, the event began as a localized increase in emission propagating across the solar disk. The average increase over the entire wavefront began at greater than 10%, decreasing with each successive image. However, individual regions exhibited more than 100% increase in emission.

Based on the five 'coronal Moreton wave' events observed thus far, typical speeds range from 250–400 km s^{-1}, and the wavefronts which are initially propagating radially begin to break up and travel in separate fronts. The magnitude of the wave shows little correlation with the peak thermal X-ray flux of the associated event; in fact, the majority of EIT observations close to flare events do not show any evidence of these waves. The model of a weak fast-mode MHD shock developed by Uchida (e.g., Uchida, 1974) explains several features of the EIT observations, including the eventual localization of the wavefronts, the preferential direction of propagation (away from active regions) and the lack of correspondance with the peak thermal X-ray flux of the flare. A movie of a coronal Moreton wave is included in the CD-ROM.

Acknowledgements

NRL received support from the Office of Naval Research and NASA grants NDPRS-86759E and NDPRS-09930F. The French laboratories were supported by CNRS and CNES. The Belgian laboratories were supported by SPPS via PRODEX. LMPARL recieved support through NASA contract NAS5–32627 and the Lockheed Independent Research Program. We wish to acknowledge the assistance of Drs L. Acton, G. E. Brueckner, J. Cook, L. Floyd, N. R. Sheeley, O. C. St. Cyr, K. Waljeski, and Y. M. Wang. We are particularly indebted to the operations staff including L. Allen, R. Doney, E. Einfalt, E. Larduinat, J.-P. Olive, S. Passwaters, K. Schenk, H. Schweitzer, S. Stezelberger, and D. Wang. Finally we would like to thank the LASCO team for cooperation in overall operations and the SUMER team for the critical telemetry bandwidth increases they have occasionally allowed EIT.

References

Ahmad, I. A. and Webb, D. F.: 1978, *Solar Phys.* **58**, 323.

Bohlin, J. D., Sheeley, N. R., and Tousey, R.: 1975, in M. J. Rycroft (ed.), *Space Research*, p. 651.

Bumba, V. and Howard, R.: 1965, *Astrophys. J.* **141**, 1492.

Cheng, C. C., Smith, J. B., and Tandberg-Hanssen, E.: 1980, *Solar Phys.* **45**, 393.

Deeming, T. J.: 1975, *Astrophys. Space Sci.* **36**, 137.

Defise, J.-M. *et al.*: 1997, *SPIE* **3114**, in preparation.

DeForest *et al.*: 1997, *Solar Phys.* **175**, 393 (this issue).

Delaboudinière, J.-P.:1998, *Solar Phys.*, submitted.

Delaboudinière, J.-P., Artzner, G. E., Brunaud, J., Gabriel, A. H., Hochedez, J.-F., Millier, F., Song, X.-Y., Au, B., Dere, K. P., Howard, R. A., Kreplin, R., Michels, D. J., Moses, J. D., Defise, J.-M., Jamar, C., Rochus, P., Chauvineau, J.-P., Marioge, J.-P., Catura, R. C., Lemen, J. R., Shing, L., Stern, R. A., Gurman, J. B., Neupert, W. M., Maucherat, A., Clette, F., Cugnon, P., Van Dessel, E. L. *et al.*: 1995, *Solar Phys.* **162**, 291.

Dere, K. P.: 1989, *Astrophys. J.* **340**, 599.

Dere, K. P. *et al.*: 1997, *Solar Phys.* **175**, 601 (this issue).

Foukal, P.: 1978, *Astrophys. J.* **223**, 1046.

Groth, E. J.: 1975, *Astrophys. J. Suppl.*, No. 286, **29**, 289.

Guhathakurta, M. and Fisher, R.R.: 1994, *Solar Phys.* **152**, 81.

Gurman, J. B. *et al.*: 1996, *EOS* **77**, (46), F557.

Janesick, J. R., Klassen, K. P., and Elliot, T.: 1987, *Opt. Eng.* **26** (10), 972.

Klimchuk, J. A., Moses, D., and Portier-Fozzani, F.: 1997, in preparation.

Klimchuk, J. A., Lemen, J. R., Feldman, U., Tsuneta, S., and Uchida, Y.: 1992, *Publ. Astron. Soc. Japan* **44**, L181.

Kuhn, J. R., Lin, H., and Loranz, D.: 1991, *Publ. Astron. Soc. Pacific* **103**, 1097.

Landau, L. and Lifchitz, E.: 1971, *Mécanique des fluides*, Editions de Moscou.

Moses, D. *et al.*: 1994, *Astrophys. J.* **430**, 913.

Neupert W. M. *et al.*: 1997, *EOS* **78** (17), S256.

Newmark, J.: 1996, *EOS* **77** (46), F557.

Schmahl, E. J. and Orrall, F. Q.: 1979, *Astrophys. J.* **231**, L41.

Schuster, H. G.: 1984, *Deterministic Chaos: An Introduction*, Physik-Verlag, Weinheim.

Schwenn, R. *et al.*: 1997, *Solar Phys.* **175**, 667 (this issue).

Sheeley, N. R.: 1980, *Solar Phys.* **66**, 79.

Shibata, K., Ishido, Y., Acton, L. W., Strong, K.T., Hirayama, T., Uchida, Y.: 1992, *Publ. Astron. Soc. Pacific* **44**, L173.

Shimojo, M., Hashimoto, S., Shibata, K., Hirayama, T., Hudson, H., and Acton, L. W.: 1996, *Publ. Astron. Soc. Pacific* **48**, 123.

Uchida, Y.: 1974, *Solar Phys.* **39**, 431.

Walker, A. B. C., DeForest, C. E., Hoover, R. B., and Barbee, T. D. W.: 1993, *Solar Phys.* **148**, 239.

Webb, D. F.: 1981, in F. Q. Orrall (ed.), *Solar Active Regions*, Colorado Associated University Press, Boulder, p. 165.

EIT AND LASCO OBSERVATIONS OF THE INITIATION OF A CORONAL MASS EJECTION

K. P. DERE[1], G. E. BRUECKNER[1], R. A. HOWARD[1], M. J. KOOMEN[1,a], C. M. KORENDYKE[1], R. W. KREPLIN[1,b], D. J. MICHELS[1], J. D. MOSES[1], N. E. MOULTON[1,c], D. G. SOCKER[1], O. C. ST. CYR[1,d], J. P. DELABOUDINIÈRE[2], G. E. ARTZNER[2], J. BRUNAUD[2], A. H. GABRIEL[2], J. F. HOCHEDEZ[2], F. MILLIER[2], X. Y. SONG[2], J. P. CHAUVINEAU[3], J. P. MARIOGE[3], J. M. DEFISE[4], C. JAMAR[4], P. ROCHUS[4], R. C. CATURA[5], J. R. LEMEN[5], J. B. GURMAN[6], W. NEUPERT[6], F. CLETTE[7], P. CUGNON[7], E. L. VAN DESSEL[7], P. L. LAMY[8], A. LLEBARIA[8], R. SCHWENN[9] and G. M. SIMNETT[10]

[1]*E.O. Hulburt Center for Space Research, Naval Research Laboratory Washington D.C., 20375-5320 U.S.A.*

[2]*Institut d'Astrophysique Spatial, Université Paris XI, 91405 Orsay, France*

[3]*Institut d'Optique Théorique et Appliquée, 91403 Orsay, France*

[4]*Centre Spatial de Liège, Liège, Belgium*

[5]*Lockheed Martin Palo Alto Research Laboratory, Palo Alto, CA 94304 U.S.A.*

[6]*NASA Goddard Space Flight Center, Greenbelt, MD 20771 U.S.A.*

[7]*Observatoire Royal de Belgique, Brussels, Belgium*

[8]*Laboratoire d'Astronomie Spatiale, Marseille, France*

[9]*Max-Planck-Institut für Aeronomie, Lindau, Germany*

[10]*Space Research Group, School of Physics and Space Research University of Birmingham, Birmingham, U.K.*

(Received 18 March 1997; accepted 18 June 1997)

Abstract. We present the first observations of the initiation of a coronal mass ejection (CME) seen on the disk of the Sun. Observations with the EIT experiment on SOHO show that the CME began in a small volume and was initially associated with slow motions of prominence material and a small brightening at one end of the prominence. Shortly afterward, the prominence was accelerated to about 100 km s^{-1} and was preceded by a bright loop-like structure, which surrounded an emission void, that traveled out into the corona at a velocity of 200–400 km s^{-1}. These three components, the prominence, the dark void, and the bright loops are typical of CMEs when seen at distance in the corona and here are shown to be present at the earliest stages of the CME. The event was later observed to traverse the LASCO coronagraphs fields of view from 1.1 to 30 $R_\odot$. Of particular interest is the fact that this large-scale event, spanning as much as 70 deg in latitude, originated in a volume with dimensions of roughly 35″ (2.5×10^4 km). Further, a disturbance that propagated across the disk and a chain of activity near the limb may also be associated with this event as well as a considerable degree of activity near the west limb.

1. Introduction

Coronal mass ejections (CMEs) are often seen as spectacular eruptions of matter from the Sun which propagate outward through the heliosphere and often interact

[a] Sachs Freeman Assoc.
[b] Universities Space Research Assoc.
[c] Allied Signal Corp.
[d] Computational Physics Inc.

Solar Physics **175:** 601–612, 1997.

with the Earth's magnetosphere. The nature and cause of CMEs is a fundamental, unsolved problem in solar physics. Often they are associated with prominence eruptions and/or solar flares. The classical CME consists of 3 parts: a bright coronal loop, a dark void, and the eruptive prominence (Kahler, 1987). Other reviews of CMEs include those by Wagner (1984) and Hundhausen (1987). In the past, essentially all coronal mass ejections have been observed by coronagraphs that occult the emissions from the disk of the Sun in order observe the much weaker photospheric light that has been Thomson-scattered by coronal electrons. Consequently, it has not been possible to observe the earliest stages of a CME, aside from the eruptions of the prominence material which can be observed in Hα. We present here the first observations of all 3 components of a coronal mass ejection during its initiation phase.

2. Observations

The Solar and Heliospheric Observatory (SOHO), a joint European Space Agency (ESA) and National Aeronautics and Space Administration (NASA) effort, was launched late in 1995 and has provided unprecedented observations of the Sun and heliosphere. In this paper, we present observations made by 2 of the instruments on SOHO, the Extreme Ultraviolet Imaging Telescope (EIT) and the Large Angle Spectrometric Coronagraph (LASCO). A detailed description of these instruments is provided by Delaboudinière *et al.* (1995) and Brueckner *et al.* (1995).

Briefly, the EIT consists of a Ritchey–Chrétien telescope with EUV reflecting multilayer coatings, several filters and a CCD detector. Different multilayer coatings are placed on 4 quadrants of both the primary and secondary mirrors in order that any 1 of 4 wavelengths ranges can be observed at a given time, while providing a good registration of images obtained in the four wavelengths. The four wavelengths and dominant emitters are 171 Å (Fe IX and Fe X), 195 Å (Fe XII), 284 Å (Fe XV) and 304 Å (He II). The EIT observations that we will discuss were obtained with the 195 Å Fe XII channel. Fe XII is emitted most efficiently at temperatures near 1.5×10^6 K. A strong Fe XXIV line at 192 Å could also contribute to the 195 Å channel signal if 1.5×10^7 K plasmas, as can occur in solar flares, are in the line of sight. Another contributor to the observed pattern of emission at 195 Å is the presence of absorbing features. Absorption of coronal emissions is often noticed in all 3 of the coronal channels at locations of prominences and macrospicules, both seen in He II. The mechanism for this absorption is probably photo-ionization of He I at wavelengths below 504 Å and of He II at wavelengths below 228 Å. This turns out to be a particularly useful feature since, in this case, it allows one to see both the prominence and coronal structures in the CME.

The LASCO instrument consists of an internally occulted coronagraph, C1, and 2 externally occulted coronagraphs, C2 and C3. The C1 coronagraph is able to observe the corona to within about 1.1 $R_\odot$ with a field of view that extends to about 3 $R_\odot$. Observations with C1 are made through a tunable, narrow-band Fabry–Pérot filter. The passband of the Fabry–Pérot filter is around 0.8 Å wide. The set of observations made with the C1 coronagraph were made at line-center of the Fe XIV λ5303 line and off-band at 5309 Å. The field of view of C2 extends from about 1.5 $R_\odot$, the edge of the occulter, to 6 $R_\odot$. Similarly, the C3 coronagraph has a field of view from 4 to 30 $R_\odot$. The signal in C2 and C3 consists of broad-band visible photospheric light scattered primarily by electrons in the corona.

During the period from 22 December 1996 02:00 UT until 28 December at 16:00 UT, the LASCO and EIT experiments were able to obtain a unique set of observations. For most operations, the combined LASCO and EIT telemetry rate is 5.2 Kbs^{-1}. During this period, a large fraction of the SOHO experiment telemetry was dedicated to EIT and LASCO and resulted in a combined telemetry rate of about 25 Kbs^{-1}. Consequently, it was possible to obtain and downlink 1 EIT Fe XII image every 12 min, 1 C1 Fe XIV image every 24 min, 1 C2 image every 24 min and 1 C3 image every 50 min.

3. The 23 December 1996 Coronal Mass Ejection

The 23 December 1996 coronal mass ejection originated in NOAA active region 8005. This active region first appeared roughly one solar rotation before the CME studied here. On 24 November 1996, it was first evident as a small brightening in the EIT Fe XII images and early on 25 November it was seen as a small bright region in the EIT He II image. The new magnetic flux emerged into a region of enhanced diffuse flux that was the remnant of a older active region. The orientation of the new flux with respect to the pre-existing flux was such that new field lines connecting the new and old regions could be expected to form. Feynman and Martin (1995) have suggested that these conditions are favorable for the production of CMEs. When viewed earlier near disk center, the Observatory of Paris-Meudon Hα spectroheliograms showed a filament through its center. *Solar Geophysical Data* indicates the observation of several small sunspots on 16 and 17 December. By the time of the 23 December CME, the region was decaying in EUV intensity and the sunspots were no longer observed. The region did not reappear on the next solar rotation. The joint USAF/NOAA report of solar and geophysical activity indicated a small filament eruption near 14S 67W beginning at 20:16 UT on 23 December. This disappearing filament was located near region 8005 (13S, 67W) and was associated with a small flare (B2/sf) which peaked at 20:53 UT.

The first sign of the CME in the EIT 195 Å images was at 20:20 UT when a small brightening and the motion of a short segment of a prominence were observed. The length of the section of the prominence involved was approximately 35″ (2.5 × 10^4 km). The southern end of this segment was displaced from its position at 20:03 UT by about 15" (1.1 × 10^4 km, indicating a velocity of about 11 km s^{-1}. The prominence was seen as an absorption feature. Attached to the

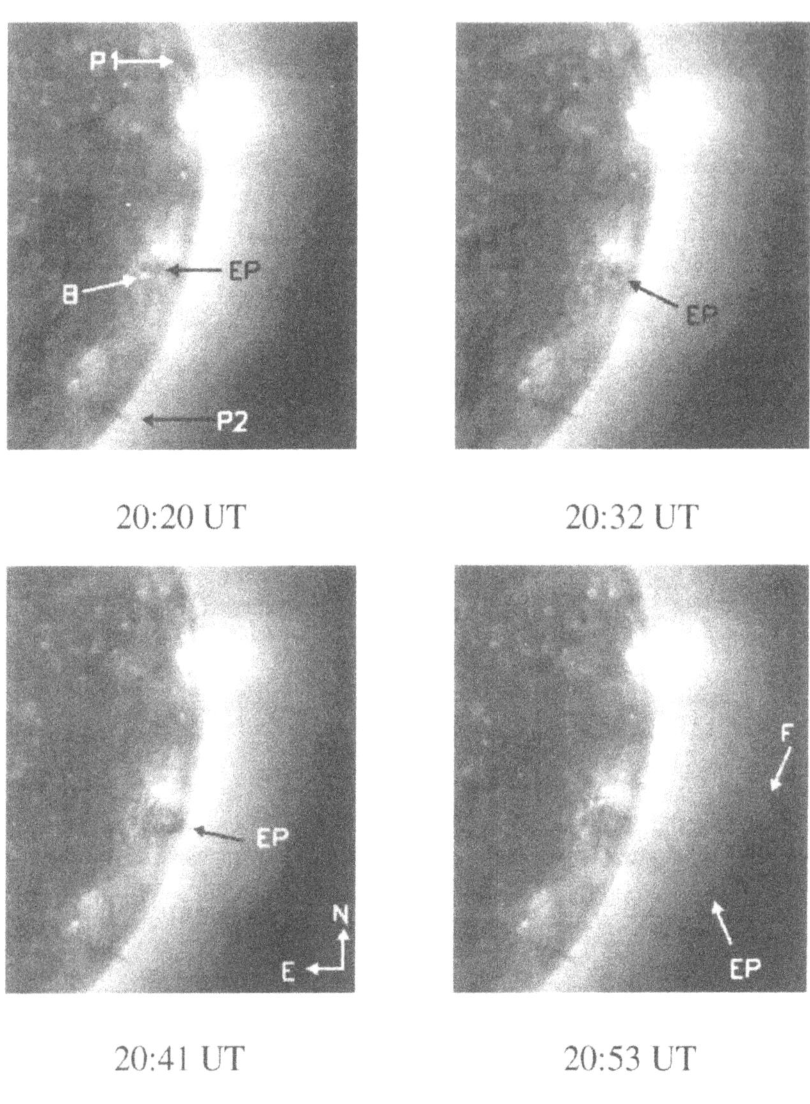

Figure 1. EIT images of the initiation of a CME. The letters in the figures are used to indicate the following structures: EP, the eruptive prominence; B, the flare-like brightening; P1 and P2, prominences; F, the outer front of the CME. Solar north and east are as indicated in the images at 20:41 UT.

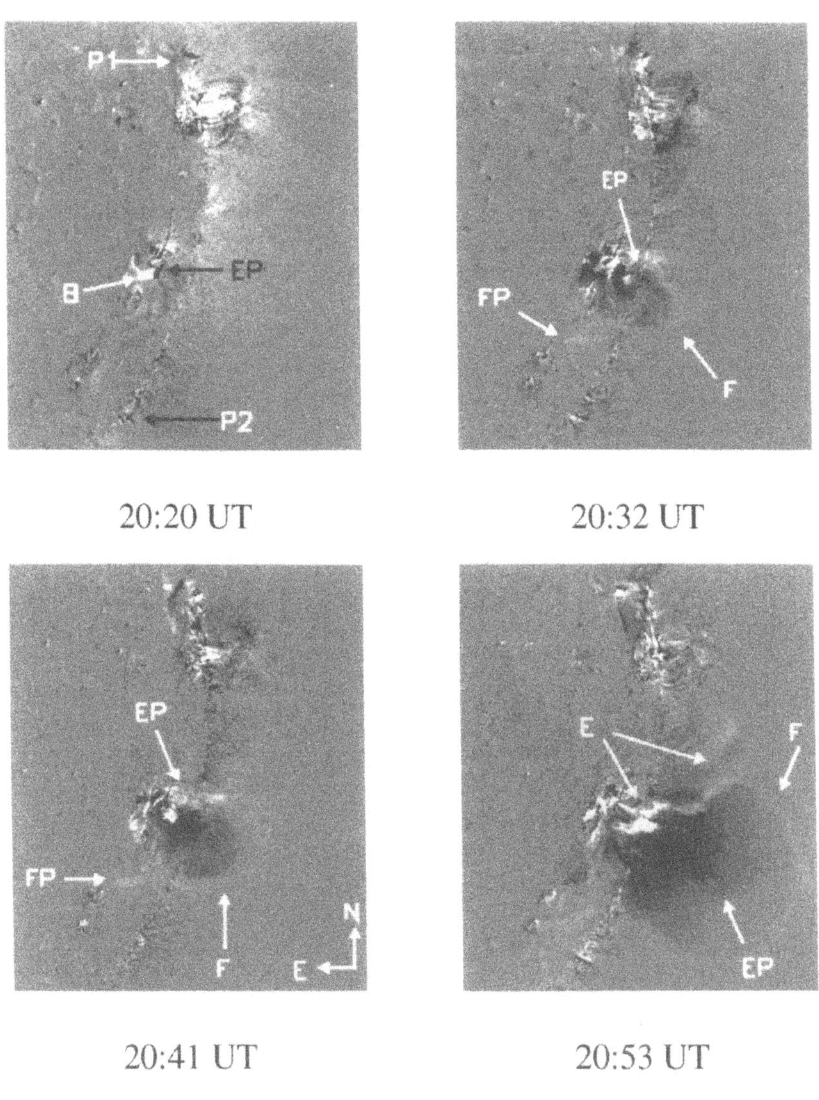

Figure 2. EIT running difference images of the initiation of a CME. The letters in the figures are used to indicate the following structures: EP, the eruptive prominence; B, the flare-like brightening; P1 and P2, prominences; F, the outer front of the CME; FP, the suspected footpoint of one side of the CME; E, a secondary set of ejecta. Solar north and east are as indicated in the images at 20:41 UT.

southern end of the prominence segment is a curved bright linear structure. Near the middle of this bright structure is a much brighter knot of emission. These features can be seen in Figure 1 and Figure 2. Figure 1 shows 4 EIT 195 Å images at the times indicated. Figure 2 shows running difference images of the EIT 195 Å emission. This bright knot of emission is probably the the site of the X-ray flare observed by the GOES detectors. In each case the image at a given time is the difference from the preceding image, all of which are shown in Figure 1 except for the image at 20:03 UT. Increases in intensity at a given position are indicated by white and decreases by black. Figure 1 best displays the prominence eruption while the difference images in Figure 2 best show the weak coronal emissions associated with the CME.

From the running difference images in Figure 2, as early as 20:32 UT, a weak but relatively bright loop-like structure, surrounding a relatively dim area, can be seen and is evident in the next two frames as it expands outward. Under the assumption that these structures are ejected radially from the Sun, their velocity is on the order of 200 km s^{-1}.

The exact geometry of the bright loop-like structure is not clear. This observational problem is related to the question of whether the loop-like structure is a compressional wave that propagates outward through the corona or whether it is the expulsion of coronal field and plasma (Sime and Hundhausen, 1987; Low and Hundhausen, 1987). One clue for distinguishing between the two scenarios would be to examine the footpoints of the apparent loops. The difference images at 20:32 UT and 20:41 UT show a relatively bright stationary elongated structure that appears to be the southern footpoint of the expanding CME loop. This footpoint is denoted by 'FP' in Figure 2. This same location is also consistent with being the location of the footpoint of the loop seen at 20:53 UT. It does not appear possible to identify a northern footpoint of the CME from these images. Just to the north of the prominence seen at 20:20 UT, a set of coronal loops span the neutral line from which the prominence apparently originated. These loops seem to be unaffected by the nearby CME, at least between 20:20 and 20:53 UT. These facts suggest that the bright loop-like structure remains connected to fixed footpoints for a period of time.

As seen in Figure 2, the eruptive prominence starts out as a fairly simple linear structure at 20:20 UT. By 20:32 UT it is considerably more complex. At 20:41 UT it appears as a collection of twisted, kinked filaments. By 20:53 UT it has been extended outward while material continues to be ejected from the base of the CME. In the EIT running difference image at 20:53 UT, bright ejecta, labeled 'E' in Figure 2, can be seen. These would constitute a new phase of material ejection following the initial prominence eruption. This new material is also bright in Fe XII indicating that it is at coronal temperatures as opposed to the cooler temperatures of the prominence material. This new material also seems to be connected to the active region to the north.

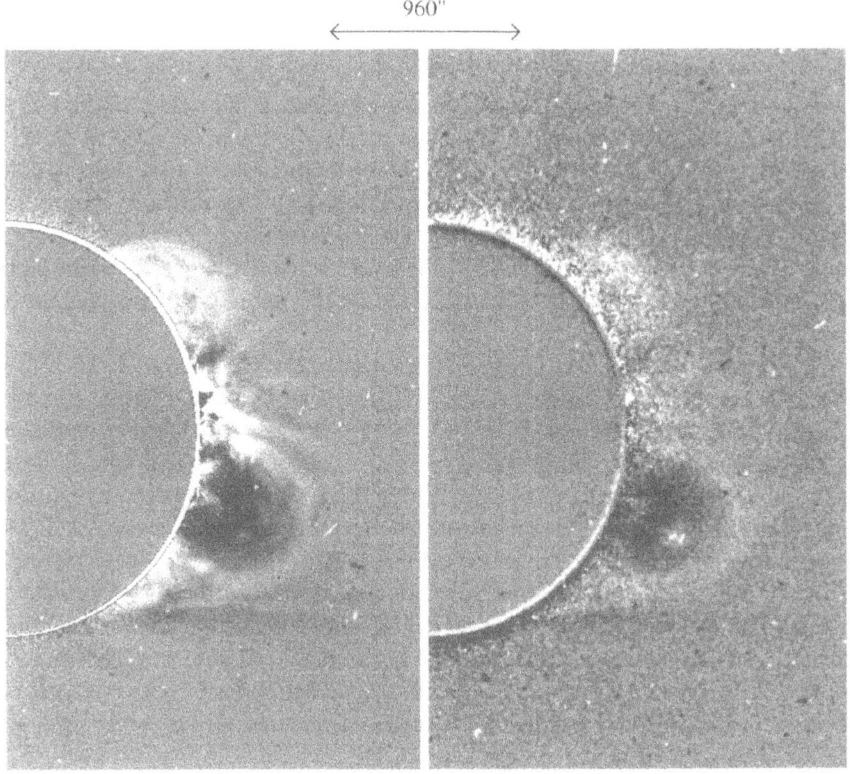

Figure 3. Images of the CME in the LASCO C1 coronagraph.

There is evidence that a weak large-scale wave passes across the solar disk. Although the signature of this wave is not especially distinct, it appears to originate near the site of the CME and travel outward as a slight brightening. At 21:22 UT the wave is seen to the northeast of the CME site at a distance of 3×10^5 km, indicating a velocity of 80 km s^{-1}, if it is assumed that it was initiated at the same time as the CME.

Beginning at 21:10 UT, the EIT images, not presented here, show the formation of a system of postflare loops, as are often observed following an eruptive prominence or CME (Sheeley *et al.*, 1975; Kahler, 1977; Webb, Krieger, and Rust, 1976; Tsuneta *et al.*, 1992). The pre-event images show a dark linear structure that would appear to mark the neutral line dividing the 2 magnetic polarities of this active region. The post-flare loops do not form along this neutral line but to the east of it.

As the CME traveled outward, it was observed several minutes later in the C1 coronagraph, as show in Figure 3. Both panels were obtained by subtracting an earlier image obtained around 18:30 UT from an image at the stated time at the same wavelength. The image on the left was obtained at line center of the Fe XIX 5303 Å coronal emission line. The image on the right was obtained at 5309 Å where the signal is due to scattering of photospheric light by free electrons, the K corona. Both C1 images show a distinct bright loop-like structure enclosing a region of decreased intensity. In addition, the K corona image shows an additional bright interior structure that is probably the prominence. The signature of the CME was not seen in the preceding C1 image at 20:38 UT, nor in the subsequent image at 21:28 UT.

The leading edge of this CME first appears in the C2 field at 21:16 UT, as shown in Figure 4. The images shown in Figure 4 are running difference images wherein the image at the stated time has been differenced from the immediately preceding image. There was a gradual swelling of the existing white-light streamer (particularly on the equatorward side) during the several hours prior to that time. The prominence material first appears in the next C2 image, taken at 21:37 UT. The angular span of this event as measured in C2 was 70 deg, centered on position angle 250. The images of the CME in C2 do not appear to have the same loop-like or bubble-like appearance as the images in the EIT or C1. There is evidence of a bright concave-outward structure appearing late in the mass ejection, and this has been taken to be a signature of disconnecting magnetic field lines by some researchers (Simnett *et al.*, 1997, for example). The original streamer seen in white-light was significantly reduced in size following the CME.

In order to show the development of the CME through the fields of view of the EIT, and C1 and C2 coronagraphs, we have traced the outline of the leading edge of the CME as seen in each of the images presented here. This composite is shown in Figure 5 where the circle shows the disk of the Sun. The CME started out with a fairly circular geometry but became more elongated with time when seen later in the C2 field of view. The height of the CME as a function of time is plotted in Figure 6. The values for the height assume that the CME is moving radially outward from the site where it was first seen on the disk. The velocities in the EIT and C1 fields of view are on the order of 200 km s^{-1} and 400 km s^{-1} in the C2 field of view.

3.1. Accompanying Phenomena

A number of nearly simultaneous events were observed in the EIT along the SW limb that may or may not be related to the central CME discussed above. While it is not possible to demonstrate any direct connection between these events, they appear to form a sequence. To the north (82W, 13N), a prominence (P1 in Figure 1) appears around 19:30 UT on 23 December with increasing size and apparent motion along the disk. A second period of growth of the prominence commences around

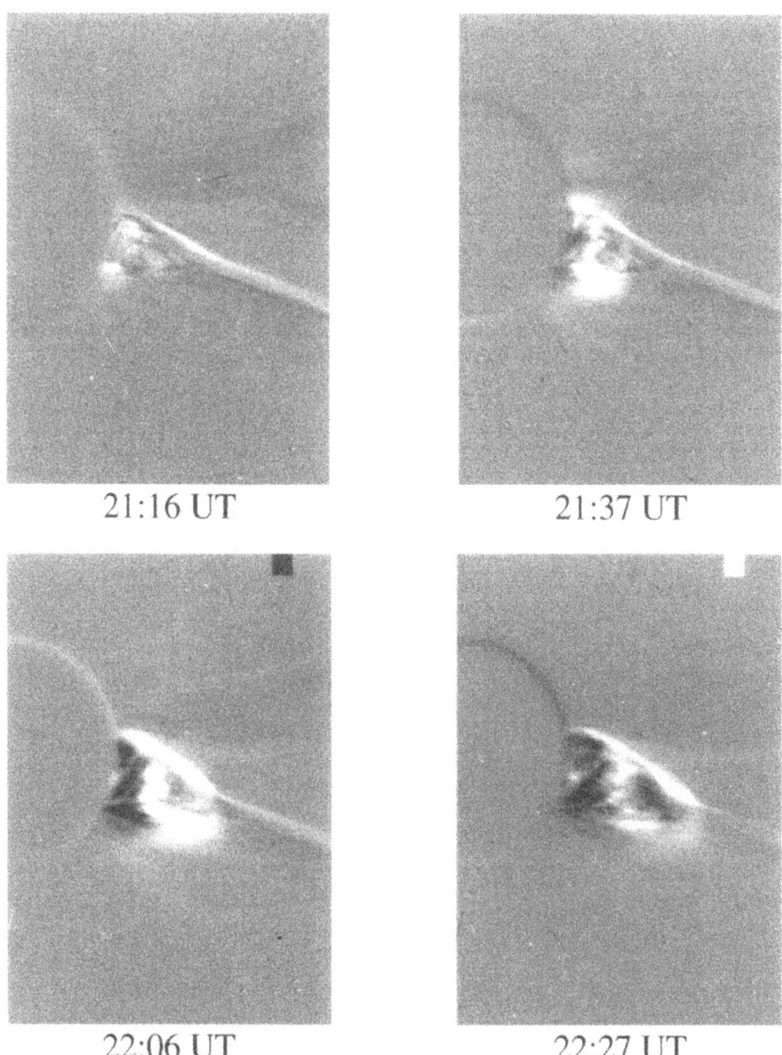

Figure 4. Images of the CME in the LASCO C2 coronagraph.

00:41 UT on 24 December. An ejection of coronal material appears above this site at around 05:14 UT on 24 December. The prominence and the coronal ejection appear to be associated with NOAA active region 8004 which, when viewed closer to disk

Figure 5. The leading edge of the 23 December 1996 CME as seen in the EIT, the LASCO C1 and C2 coronagraphs. The numbers can be associated with the time and telescope as follows: 4, 20:53 UT (EIT); 5, 21:04 UT (C1); 6, 2116 (C2); 7, 21:37 UT (C2); 8, 22:06 UT (C2); 9, 22:27 UT (C2); 10, 22:56 UT (C2). The lower lines refer to the front 'F' seen in the first 4 panels of Figure 2.

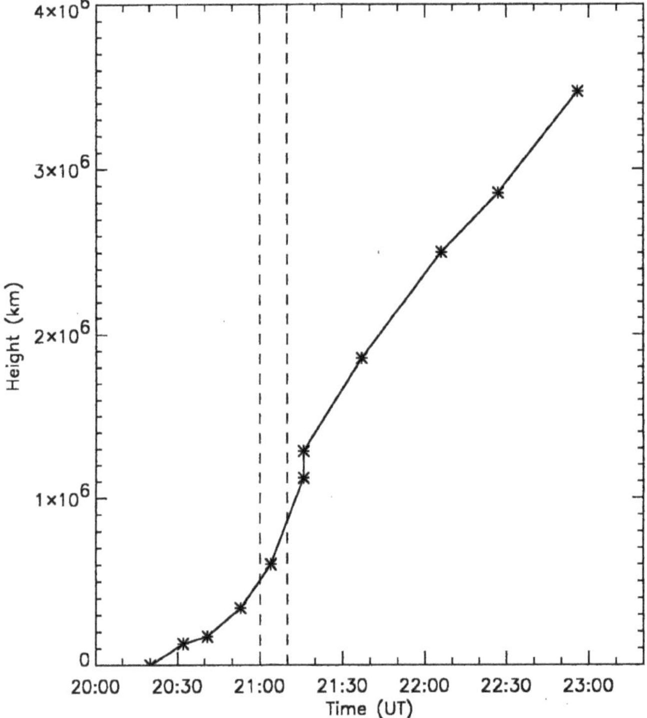

Figure 6. Height of the leading edge as a function of time. The C1 observations are inside the dashed lines, EIT to the left of it and C2 to the right.

center, showed an active region prominence through its center. On 21 December, a new active region (NOAA 8007) emerged to the east of NOAA 8004.

To the south, as seen in Figure 1, an elongated, nearly vertical prominence at 37S 90W began to stretch further out into the corona and at 00:31 UT on 24 December the prominence (P2 in Figure 1) was ejected outward into the corona. At 01:21 UT on 24 December the outward motion of large-scale loop-like coronal structures above the prominence began. The southern end of these large coronal loops appeared to be near 61S 47W. At around 01:21 UT on 24 December, the motion of prominence material near this footpoint began. These prominence activities continue through 03:13 UT on 24 December.

4. Discussion

These observations obtained with the EIT and LASCO experiments on SOHO provide a unique view of the earliest phases of a coronal mass ejection. To the best of our knowledge, similar observations have never been obtained in the past. The primary signature of CMEs in EUV and X-ray images is the creation of post-flare loop systems following the eruption of CMEs (Sheeley *et al.*, 1975; Kahler, 1977; Webb, Krieger, and Rust, 1976; Tsuneta *et al.*, 1992). More recently, transient voids of coronal X-ray emitting material are understood to signal the launch of a CME (Hudson, Lemen, and Webb, 1997).

Coronal mass ejections are clearly large-scale phenomena. At the Sun they are often related to streamers which they can completely disrupt and, once launched, can traverse the heliosphere at least to 1 AU. In the present case where the CME has been observed by EIT and LASCO from its earliest phases, the initiation begins in a very small volume. The initiation of a large-scale CME in an small-scale eruption is a key fact that must be included in our understanding of the creation of CMEs.

The EIT and LASCO observations show that from essentially the very beginning, the 3 main common components of a CME are present: the bright loop-like outer shell, the coronal void and the eruptive prominence. This would indicate that the basic physical state of the CME is completely determined at its initiation and that during its transit of the heliosphere, the development of the CME is just the evolution of this physical state. More observations of the initiations of CME, particularly with better time cadences, would be important for understanding this important phenomena.

References

Brueckner, G. E. *et al.*: 1995, *Solar Phys.* **162**, 357.

Delaboudinière, J.-P. *et al.*: 1995, *Solar Phys.* **162**, 291.

Feynman, J. and Martin, S. F.: 1995, *J. Geophys. Res.* **100**, 3355.

Hudson, H., Lemen, J., and Webb, D.: 1997, in B. Bentley and J. Mariska (eds), 'A Signature of CME Onset in Soft X-Rays', *Magnetic Reconnection in the Solar Atmosphere*, ASP Conf. Ser. 111, p. 379.

Hundhausen, A.: 1987, in V. J. Pizzo, T. Holzer, and D. G. Sime (eds), 'The Origin and Propagation of Coronal Mass Ejections', *Proceedings of the Sixth International Solar Wind Conference*, NCAR TN-306, p. 181.

Kahler, S.: 1977, *Astrophys. J.* **214**, 891.

Kahler, S.: 1987, in V. J. Pizzo, T. Holzer, and D. G. Sime (eds), 'Observations of Coronal Mass Ejections Near the Sun', *Proceedings of the Sixth International Solar Wind Conference*, NCAR TN-306, p. 215.

Low, B. C. and Hundhausen, A. J.: 1987, *J. Geophys. Res.* **92**, 2221.

Sheeley, N. R., Jr., Bohlin, J. D., Brueckner, G. E., Purcell, J. D., Scherrer, V. E., Tousey, R., Smith, J. B., Jr., Speich, D. M., Tandberg-Hanssen, E., Wilson, R. M., De Loach, A. C., Hoover, R. B., and McGuire, J. P.: 1975, *Solar Phys.* **45**, 377.

Sime, D. G. and Hundhausen, A. J.: 1987, *J. Geophys. Res.* **92**, 1049.

Simnett, G. M. *et al.*: 1997, *Solar Phys.* **175**, 685 (this issue).

Tsuneta, S., Takahashi, T., Acton, L., Bruner, M. E., Harvey, K. L., and Ogawara, Y.: 1992, *Publ. Astron. Soc. Japan* **44**, L211.

Wagner, W. J.: 1984, *Ann. Rev. Astron. Astrophys.* **22**, 267.

Webb, D. F., Krieger, A. S., and Rust, D. M.: 1976, *Solar Phys.* **48**, 159.

FIRST RESULTS FROM THE SOHO ULTRAVIOLET CORONAGRAPH SPECTROMETER

J. L. KOHL[1], G. NOCI[2], E. ANTONUCCI[3], G. TONDELLO[4], M. C. E. HUBER[5], L. D. GARDNER[1], P. NICOLOSI[4], L. STRACHAN[1], S. FINESCHI[1], J. C. RAYMOND[1], M. ROMOLI[2], D. SPADARO[6], A. PANASYUK[1], O. H. W. SIEGMUND[7], C. BENNA[9], A. CIARAVELLA[5,*], S. R. CRANMER[1], S. GIORDANO[9], M. KAROVSKA[1], R. MARTIN[10], J. MICHELS[1], A. MODIGLIANI[2], G. NALETTO[4], C. PERNECHELE[4], G. POLETTO[8] and P. L. SMITH[1]

[1]*Harvard-Smithsonian Center for Astrophysics, Cambridge, MA 02138, U.S.A.*
[2]*Università di Firenze, I-50125 Firenze, Italy*
[3]*Osservatorio Astronomico di Torino, I-10025 Pino Torinese, Italy*
[4]*Università di Padova, I-35131 Padova, Italy*
[5]*Space Science Dept., ESA/ESTEC, NL-2200 AG, Noordwijk, The Netherlands*
[6]*Osservatorio Astrofisico di Catania, I-95125 Catania, Italy*
[7]*Space Sciences Laboratory, University of California, Berkeley, CA 94720, U.S.A.*
[8]*Osservatorio Astrofisico di Arcetri, I-50125 Firenze, Italy*
[9]*Università di Torino, I-10125 Torino, Italy*
[10]*Institut d'Astrophysique Spatiale, F-91405 Orsay, France*

(Received 14 March 1997; accepted 18 June 1997)

Abstract. The SOHO Ultraviolet Coronagraph Spectrometer (UVCS/SOHO) is being used to observe the extended solar corona from 1.25 to 10 $R_\odot$ from Sun center. Initial observations of polar coronal holes and equatorial streamers are described. The observations include measurements of spectral line profiles for H I Lα and Lβ, O VI 1032 Å and 1037 Å, Mg X 625 Å, Fe XII 1242 Å and several others. Intensities for Mg X 610 Å, Si XII 499 Å, and 520 Å, S X 1196 Å, and 22 others have been observed. Preliminary results for derived H^0, O^{5+}, Mg^{9+}, and Fe^{11+} velocity distributions and initial indications of outflow velocities for O^{5+} are described. In streamers, the H^0 velocity distribution along the line of sight (specified by the value at e^{-1}, along the line of sight) decreases from a maximum value of about 180 km s^{-1} at 2 $R_\odot$ to about 140 km s^{-1} at 8 $R_\odot$. The value for O^{5+} increases with height reaching a value of 150 km s^{-1} at 4.7 $R_\odot$. In polar coronal holes, the O^{5+} velocity at e^{-1} is about equal to that of H^0 at 1.7 $R_\odot$ and significantly larger at 2.1 $R_\odot$. The O^{5+} in both streamers and coronal holes were found to have anisotropic velocity distributions with the smaller values in the radial direction.

1. Introduction

The Ultraviolet Coronagraph Spectrometer on the Solar and Heliospheric Observatory (UVCS/SOHO) is being used for spectroscopic observations of the extended solar corona. The instrument is designed to obtain the data for diagnostic techniques aimed at determining empirical values for densities, velocity distributions, and outflow velocities of hydrogen, electrons, and several minor ions.

The velocity distribution of neutral hydrogen in the corona tends to reflect the proton distribution as well, because the characteristic lifetime for a hydrogen atom at coronal densities is significantly shorter than typical coronal expansion times

[*] Also Osservatorio Astronomico di Palermo. Palermo, Italy.

Solar Physics **175:** 613–644, 1997.

(Withbroe *et al.*, 1982). For example, at $r = 2.5\ R_{\odot}$ in a low-density coronal hole, a newly created H I atom will move about $0.02\ R_{\odot}$, before it is ionized. The correspondence between neutral hydrogen velocities and protons has been discussed by Leer (1988), Olsen and Leer (1996), and Allen, Habbal, and Hu (1996).

This paper describes examples of UVCS/SOHO observations during the first year of operations, and the derivation of particle velocity distributions and outflow velocities in the extended corona. The results are derived from direct application of spectroscopic diagnostic techniques. For example, velocity distributions are derived directly from the observed line profiles. In the future, more accurate values are expected to be obtained through the development of self-consistent empirical models of the observed coronal structures. The accuracy and confidence in the values presented here are discussed in the text.

The primary goals of the UVCS/SOHO observations are to provide detailed empirical descriptions of the extended solar corona, and to use the descriptions to gain information about the physical processes controlling the large-scale and small-scale features of the extended corona and the acceleration and composition of the solar wind near the Sun.

Although it is premature at this stage to claim any definitive information in that regard, the observations have revealed several unexpected results. The most surprising are the velocity distributions in polar coronal holes for particles of different mass. The velocities deviate significantly from those of a thermal distribution. The O^{5+} line-of-sight (LOS) velocities above $2.1\ R_{\odot}$ are larger than those of a thermal distribution at 1×10^{8} K and the O^{5+} velocities are much larger than the H^{0} velocities. This difference in the velocity distributions of H^{0} and O^{5+} suggest that a large fraction of the O^{5+} velocity distribution is not caused by transverse wave motions. There is also evidence that the O^{5+} velocity distribution in the solar radial direction is much narrower than in the LOS direction. These two results suggest that a significant fraction of the O^{5+} velocity distribution may be caused by ion cyclotron resonant acceleration of O^{5+} ions by high-frequency MHD waves (McKenzie, Banaszkiewicz, and Axford, 1995). However, there is no independent confirmation of the existence of such waves although several plausible processes for producing them have been proposed. It is not clear from the data if any of the proton velocity distribution along the LOS is caused by ion cyclotron resonance. However, the H^{0} velocities along the LOS are larger than would be expected for a thermal distribution at the expected electron temperature (Ko *et al.*, 1997). The values for the proton velocities at e^{-1} ($v_{1/e}$) approach those predicted by McKenzie, Banaszkiewicz, and Axford (1995) for their ion-cyclotron resonance model.

Outflow velocities for O^{5+} reach values of 130 to 230 km s^{-1} in coronal holes within heliocentric distances of $3.0\ R_{\odot}$. Clear evidence for acceleration in the extended corona is observed. The particles do not appear to enter the extended corona at supersonic velocities. This information further constrains theoretical models aimed at identifying the dominant heating and acceleration processes.

UVCS/SOHO measurements of LOS velocity distributions near the base of equatorial streamers are much closer to those expected in a thermal distribution at the expected electron temperature, but there is a tendency for the more massive particles to have excess velocities. This behavior is more pronounced in regions of lower density. Future work should investigate the possibility that ion-cyclotron resonance is important in equatorial streamers as well as in coronal holes. The proton and O^{5+} outflow velocities in equatorial streamers appear to reach a value between 175 and 205 km s^{-1} near heliocentric heights of 7 $R_{\odot}$.

The UVCS results are not limited to the information presented here. A determination of chemical abundances in streamers indicates significant departures from photospheric values (Raymond *et al.*, 1997). There have been at least two spectroscopic observations of CMEs, one on 7 June 1996 and one on 23 December 1996. The latter observation indicates intensity increases by a factor of 500 in the H I Lyman lines and the presence of ions at low stages of ionization such as C III. Large spectral line shifts indicate particle flow velocities above 100 km s^{-1} along the LOS. There have also been observations of Comet SOHO 6 and 8 in H I Lα, observations of several stars at wavelengths shortward of the Hubble Space Telescope range that have not been seen since the Copernicus Mission, and observations of H I Lα in the ram direction of the solar system moving through the local interstellar cloud and of H I Lα and He I 584 Å in the wake direction.

2. The UVCS Instrument

The UVCS/SOHO has been described previously by Kohl *et al.* (1995) and, therefore, only a brief descripton is provided here. UVCS consists of a telescope spectrometer unit (TSU) and a remote electronics unit (REU). The TSU is a triple telescope with external and internal occultation and a high-resolution spectrometer assembly. Three spherical telescope mirrors focus co-registered images of the extended corona onto the three entrance slits of the spectrometer assembly. The spectrometer assembly consists of three channels:

– The Lα channel is a toric grating spectrometer that is optimized for line profile measurements of H I 1216 Å and may also be used for other spectral lines in the 1145 Å to 1287 Å spectral range. An extended wavelength range (1100 Å to 1361 Å) may be observed by rotating the grating.

– The O VI channel is a toric grating spectrometer that is optimized for measurements of the O VI lines at 1032 Å and 1037 Å and may also be used for other spectral lines in the 984 Å to 1080 Å (first-order) and 492 Å to 540 Å (second-order) spectral range. An extended wavelength range in first and second order (937 Å to 1126 Å and 469 Å to 563 Å, respectively) may be observed by rotating the grating. The O VI channel includes a convex mirror between the grating and the O VI detector to focus the H I 1216 Å (first-order) and Mg X 610 Å and 625 Å (second-order) radiations onto the O VI detector.

– The white-light channel (WLC) is a visible light polarimeter that measures polarized radiance in the 4500–6000 Å wavelength band.

The field of view (FOV) of UVCS is provided in Kohl *et al.* (1995). The instantaneous FOV is the portion of the solar image that passes through the entrance slits of the ultraviolet channels. The slit heights correspond to 40 arc min in the direction parallel to the limb tangent. The WLC has a 14 arc sec × 14 arc sec spatial field at the center of the UV instantaneous FOV. Internal mirror motions are used to step the instantaneous FOV from 1.2 to 10 $R_{\odot}$. Offset pointing adjustments with the UVCS pointing mechanism can be used to extend the FOV down to the solar limb and onto the disk or up to 12 $R_{\odot}$. UVCS can rotate its FOV about Sun-center in order to observe the full corona.

3. Diagnostic Techniques for the Extended Corona

Several diagnostic techniques were used to provide the information contained in this paper. The basic techniques have been described previously (Kohl and Withbroe, 1982; Withbroe *et al.*, 1982, 1985; Noci, Kohl, and Withbroe, 1987; van de Hulst, 1950), and therefore will be discussed only briefly here.

Hydrogen and ion velocity distributions are determined from measured spectral line profiles of H I Lα, O VI 1032 Å and other spectral lines. The shape of the profile depends on the velocity distribution of the particles emitting or scattering the detected photons. In general, spectral lines of the extended solar corona are formed by both collisional excitation followed by spontaneous radiative decay and by resonant scattering of light from deeper in the solar atmosphere. In the case of resonant scattering, the line shape is affected by the line profile of the incident light for non-90° scattering. However, the line shape is dominated by Doppler shifts that can be the result of several different types of particle motions (e.g., microscopic motions such as thermal motions or ion/cyclotron resonance induced motions, transverse wave motions, and macroscopic motions such as bulk outflow along a LOS through an extended object). The profiles can also be affected by variations in conditions along the LOS. There may be local sites with different densities and/or different local velocity distributions (e.g., foreground and background structures). There may be local non-Maxwellian velocity distributions. The LOS may pass through several flux tubes (i.e., regions of dissimilar magnetic field, electron temperature, bulk outflow, and MHD waves).

Contributions along the LOS can be affected by the angular dependence of resonant scattering. UVCS provides synoptic observations that can be used in the future to help determine the contributions to the profiles from structures along the LOS.

For near 90-deg scattering, which tends to be the case for observations above the limb, the angular dependence of resonant scattering and the line profile of the incident light have only a small effect on the observed profiles. Equations describing the angular dependence of resonant scattering are provided in Withbroe *et al.* (1982). A model code, which utilizes those equations, was used to simulate observable spectral lines. It was found that for objects located in the plane of the sky or objects distributed along the LOS as if they are spherically symmetric, and for precisions of $\pm 10\%$, the observed line profile can be converted to a velocity distribution using the usual formula

$$\frac{\Delta\lambda}{\lambda_0} = \frac{V}{c}, \quad (1)$$

where $\Delta\lambda$ is the wavelength relative to the line center, λ_0, and c is the speed of light. This is the case for polar coronal holes and the equatorial streamers near solar minimum that are described in this paper.

Outflow velocities are determined by the Doppler dimming method. There are several ways of determining outflow velocities from Doppler dimming. Spectral lines of different ions can have different sensitivities to outflow velocity because of their different spectral line widths. The relevant line width is that determined by the velocity distribution in the direction of the incoming light to the corona. The curves in Figure 1 are for line widths corresponding to Maxwellian distributions at three different temperatures.

A detailed analysis requires an empirical model which uses as inputs, measured electron temperatures and electron densities. This information is used in the model to calculate the ionization balance terms R and predict the observed EUV intensity as a function of the outflow velocity. The model must also include the details of the resonant scattering process and any collisionally excited contribution to the spectral line.

In this paper, Doppler dimming and brightening of O VI lines are presented. The Doppler dimming of the O VI line at 1037.613 Å is particularly useful because pumping by C II 1037.018 Å extends its velocity sensitive range to include values from about 90 km s^{-1} to about 300 km s^{-1} (Noci, Kohl, and Withbroe, 1987). Pumping by C II occurs when the O^{5+} outflow velocity is large enough to shift the spectral line profile of the chromospheric C II light onto the coronal O VI scattering profile in the reference frame of the outflowing O^{5+}. Figure 1 shows the effect.

Because Doppler dimming only affects the resonantly scattered component of spectral lines and O VI lines have both resonantly scattered and collisionally excited contributions to the observed intensities, it is necessary to account for both contributions in deriving outflow velocities.

Noci, Kohl, and Withbroe (1987) describe a method whereby, the intensity ratio of the O VI 1032 Å and 1037 Å lines can be used to determine the outflow velocity. Their method can be understood in the following way. In the absence of resonant scattering, the line ratio is 2:1 because of the factor of 2 difference in the collision strengths. This is observed to be the case, for example, in the transition region. The ratio for resonant scattering is 4:1; a factor of two for the strength of the absorption and another factor of two because of the difference in the

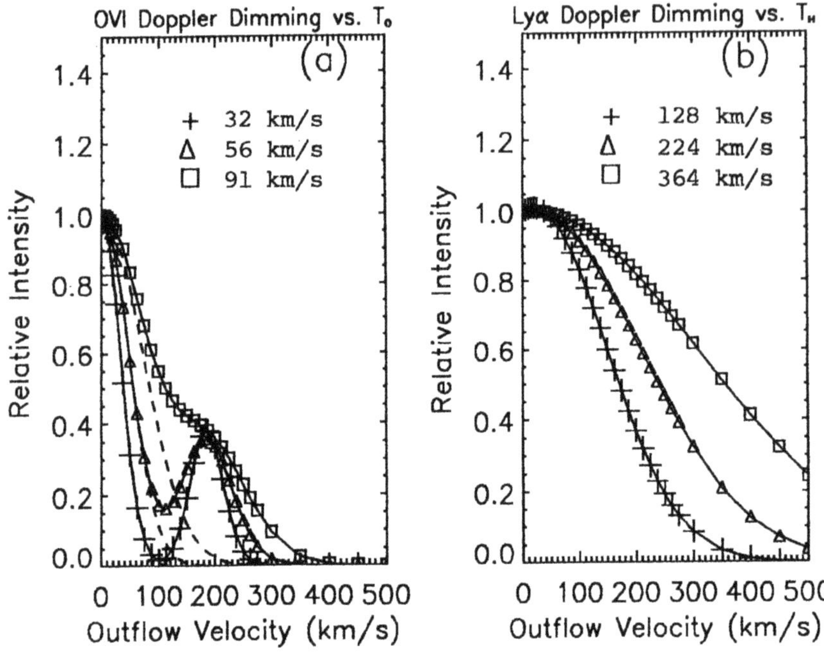

Figure 1. Doppler dimming curves for the resonantly scattered components of O VI 1032 Å and 1037 Å (a) and H I 1216 Å (b). In the plot on the left (a), the O VI 1037 curves are shown as solid lines and points. The curves for O VI 1032 Å follow those for O VI 1037 Å at the smallest velocities and then follow the dashed curves. The intensities are normalized to the value for zero outflow velocity. Curves are provided for three different values of $v_{1/e}$, in the direction of the incoming radiation to the corona from lower in the solar atmosphere. The curves include pumping of the coronal O VI 1037.613 Å line by chromospheric C II 1037.018 Å. The curves are for a single point in the plane of the sky.

intensity of the incident light from the transition region. Kohl and Withbroe (1982) have shown that, in the absence of C II 1037.018 Å pumping of the O VI 1037.613 line, the ratio determines the relative contributions of collisional excitation and resonant scattering to the observed intensity. It can be seen from Figure 1 that for radial velocity distributions with $v_{1/e}$ less than 56 km s^{-1}, resonant scattering of O VI 1032 Å becomes very small when the outflow velocity increases above about 125 km s^{-1} and the transition region O VI 1032 Å radiation is shifted off the coronal absorption profile. In that case, the line ratio would become 2 since the observed light would be only collisionally excited. However, if the outflow increases above 125 km s^{-1}, the C II line will begin to pump the O VI 1037 line and the ratio will drop below 2. The ratio would continue to decrease until the velocity reaches about 180 km s^{-1} when the C II line is centered on the O VI 1037 line. As the C II light is shifted further to the red, the resonantly scattered contribution begins to decrease and the line ratio approaches 2 at approximately 300 km s^{-1}.

The ratio is affected by the change in electron density with height in the corona since the collisional component scales as the square and the resonantly scattered component linearly. The ratio is affected by the velocity distribution in the direction of the incoming radiation to be scattered in the corona (see Figure 1). A detailed model is needed to derive precise values for the outflow velocities, but the general behavior of the outflow with height can be seen from the measured ratio.

4. Preliminary Identification of Spectral Lines Observed by UVCS/SOHO

The spectral lines detectable with UVCS/SOHO depend strongly on the observed coronal structure and the heliographic height. The strongest lines such as H I 1215.67 Å and 1025.72 Å, O VI 1031.96 Å and 1037.61 Å, and Mg X 609.76 Å and 624.95 Å are observed in polar coronal holes. Tables I and II list observed wavelengths and preliminary identifications for spectral lines observed in active region streamers on the east limb of the Sun on 23 and 24 July 1996 at about 1.5 $R_{\odot}$. A preliminary wavelength scale was used for the measured values. Identifications with low confidence are indicated by a question mark. Forbidden lines are indicated with double brackets and intercombination lines with closing brackets. Several spectral features that are present in the observations but currently unassigned, are not included in the tables. The intensity scale is in arbitrary units. The intensity values are approximately proportional to the observed values of integrated spectral radiance after background subtraction. Table I lists spectral lines detected with the UVCS/SOHO O VI detector. Lines with first order wavelengths longward of 1180 Å were observed via the path in the UVCS/SOHO O VI channel that includes an auxiliary convex mirror. Table II lists spectral lines detected with the UVCS/SOHO Lα Channel. Only first-order lines are detectable in the latter channel because of a magnesium floride window.

5. UVCS/SOHO Measurements of Coronal Spectral Line Profiles

UVCS has observed spectral line profiles of H I 1216 Å and 1026 Å, O VI 1032 Å and 1037 Å, Fe XII 1242 Å, S X 1196 Å, and several other lines in streamers, and of H I 1216 Å, O VI 1032 Å, and 1037 Å and Mg X 625 Å in polar coronal holes. The observed spectral line profiles are discussed in Sections 6 and 7.

The profiles have been corrected for an instrument characteristic that effecively smears the wavelengths such that a broad line wing with an amplitude of about 1 –3% of the line peak is produced. The zero integral functions are used to correct this effect by redistributing the photons in the artificial line wing. The zero integral functions were derived from observations of the narrow interplanetary H I Lα line. The functions for the Lα channel and the O VI channel are similar, but not identical. Convolution of the zero integral function with the observed line profile yields a

Table I
Preliminary identifications for observed spectral lines in the O VI channel

Wavelength[a] 1st order	Wavelength[a] 2nd order	Int.[b]	$\lambda_{\rm ID}$	Identification
949.97		7.1	950.15	H I Lδ, [Si IX] $2p2\ ^3P_1 - 2p2\ ^1S_0$
972.56		7.0	972.54	H I Lγ
974.06	487.03	9.3	487.04	[Fe XIII] $3s^23p^2\ ^3P_2 - 3s3p^3\ ^5S_1$
981.10	490.55	9.7	490.20	[Ar XIII] $2p^2\ ^3P_2 - 2s2p^3\ ^5S_2$
982.60	491.30	9.0	491.44	[S XIII] $2s^2\ ^1S_0 - 2s2p\ ^3P_1$
993.26	496.21	13.8	496.25	[Ar XIV] $2s^22p\ ^2P_{3/2} - 2s2p^2\ ^4P_{1/2}$
998.77	499.39	485	499.37	Si XII $2s\ ^2S_{1/2} - 2p\ ^2P_{3/2}$
1018.88		1.3	1018.60	[Ar XII] $2p^3\ ^4S_{3/2} - 2p^3\ ^2D_{5/2}$
1020.04	510.02	11.1	510.08	[Fe XIII] $3s^23p^2\ ^3P_2 - 3s3p^3\ ^5S_2$
1025.73		50.1	1025.72	H I Lβ
1028.10		1.8	1028.04	[Fe X] $3d\ ^4D_{7/2} - 3d\ ^4F_{7/2}$
1031.96		819	1031.91	O VI $2s\ ^2S_{1/2} - 2p\ ^2P_{3/2}$
1037.50		286	1037.61	O VI $2s\ ^2S_{1/2} - 2p\ ^2P_{1/2}$
1041.26	520.63	242	520.66	Si XII $2s\ ^2S_{1/2} - 2p\ ^2P_{1/2}$
1054.87		1.5	1054.90	[Ar XII] $2p^3\ ^4S_{3/2} - 2p^3\ ^2D_{3/2}$
1070.63	535.32	10.4	535.32	[S XII] $2s^22p\ ^2P_{3/2} - 2s2p^2\ ^4P_{3/2}$
1100.14	550.07	50.0	550.01	Al XI $2s\ ^2S_{1/2} - 2p\ ^2P_{3/2}$
1115.52	557.76	21.8	557.74	Ca X $3s\ ^2S_{1/2} - 3p\ ^2P_{3/2}$
1196.19		24.2	1196.25	[S X] $2p^3\ ^4S_{3/2} - 2p^3\ ^2D_{5/2}$
1215.67		18100	1215.67	H I Lα
1219.71	609.86	1530	609.76	Mg X $2s\ ^2S_{1/2} - 2p\ ^2P_{3/2}$
1238.57		13.0	1238.82	N V $2s\ ^2S_{1/2} - 2p\ ^2P_{3/2}$
1241.80		47.5	1242.03	[Fe XII] $3p^3\ ^4S_{3/2} - 3p^3\ ^2P_{3/2}$
1249.90	624.95	520	624.95	Mg X $2s\ ^2S_{1/2} - 2p\ ^2P_{1/2}$
1260.42	629.73	226	629.73	O V $2s^2\ ^1S_0 - 2s2p\ ^1P_1$

[a] In Å.
[b] In arbitrary units.

correction that, when subtracted from the observed profile, provides an improved approximation of the coronal profile. After about three iterations, convergence is obtained.

The results of this corrective procedure were checked by comparing the corrected line profile of chromospheric H I Lα with observations from other instruments (Gouttebroze *et al.*, 1978; Wilhelm, 1996) and by comparing UVCS/SOHO observations of H I Lα in streamers with typical streamer profiles observed with UVCS/Spartan 201 (see Strachan *et al.*, 1994). An example of an uncorrected line profile together with the same profile after correction is provided in Figure 2.

Table II
Preliminary identifications for observed spectral lines in the Lα channel

Wavelength[a]	λ_{ID}	Int.[b]	Identification
1196.25	1196.25	28.1	[S X] $2p^3\,{}^4S_{3/2} - 2p^3\,{}^2D_{5/2}$
1215.67	1215.67	18680	H I Lα
1238.56	1238.82	12.9	N V $2s\,{}^2s_{1/2} - 2p\,{}^2P_{3/2}$
1241.69	1242.03	60.1	[Fe XII] $3p^3\,{}^4S_{3/2} - 3p^3\,{}^2p_{3/2}$
1277.17	1277.23	2.3	[Ni XIII] $3p^3\,{}^3P_1 - {}^1S_0$
1349.89	1349.38	28.8	[Fe XII] $3p^3\,{}^4S_{3/2} - 3p^3\,{}^2P_{1/2}$

[a] In Å.
[b] In arbitrary units.

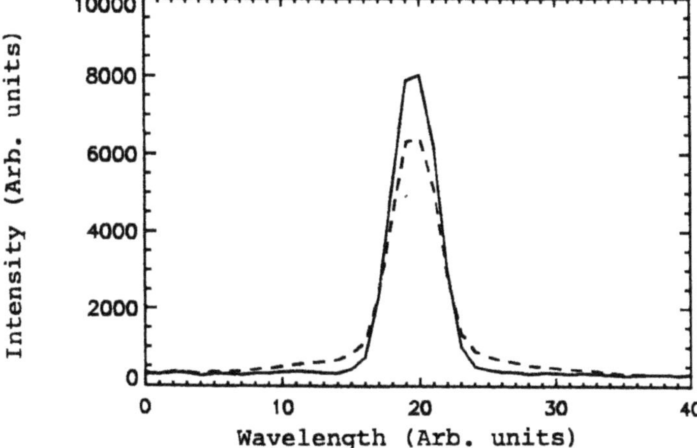

Figure 2. Example of profiles corrected (solid) and uncorrected (dashed) for an instrument effect that removes photons from the profile core and creates an artificial line wing.

The corrected profiles are normally curve-fit with a standard algorithm that minimizes the following equation for χ^2:

$$\chi^2 \equiv \sum_{i=1}^{N} \left(\frac{y_i - y(x_i; a_1 \ldots a_M)}{\sigma_i} \right)^2 , \tag{2}$$

where σ_i are the standard deviations for each data point (x_i, y_i), N is the number of data points used for the fit and M is the number of free parameters. The standard deviations are assumed to equal $n^{1/2}$ where n is the number of counts for each data point. The resulting minimum value of χ^2 is used to estimate the goodness-of-fit. It is customary to use the reduced χ^2 for this purpose which is the minimum χ^2 divided by the degrees of freedom $\nu = N - M$. Tables providing the probability of exceeding $\nu^{-1}\chi^2_{\rm min}$ in are provided in Bevington (1969). For the case of the spectral line profile curve-fits discussed in this paper which typically have about 18 deg of freedom. A $\nu^{-1}\chi^2_{\rm min}$ value of about 1.4 corresponds to a probability for exceeding that value of 10%. This value has been adopted in this paper as an indicator of an acceptable curve-fit. Larger values for $\nu^{-1}\chi^2_{\rm min}$ might be caused by small deviations from the ideal expression due, for example, to the angular dependence of resonant scattering (Withbroe *et al.*, 1982), and by departures from the assumed distribution of standard deviations or by instrument effects. Several such cases are discussed in Sections 6 and 7.

In addition to the above instrument characteristics, the observed spectral lines are broadened by the usual optical point spread function of the spectrometer. Laboratory measurements of the effective instrument profile were performed at both the complete instrument level and for the spectrometer alone. The results are described in Kohl *et al.* (1995), Pernechele *et al.* (1995), and Gardner *et al.* (1996). In-flight measurements have confirmed the laboratory results for the Lα channel and refined the value of the laboratory upper limit for the O VI channel. Since the UVCS/SOHO telescope is highly vignetted and the exposed telescope area depends on the heliographic height of the observation, the spectral resolution can be a function of the observed height. This effect has been determined to be rather small and is ignored in this paper. The spectral resolution also depends on the slit width used for each observation and the quantization error of the detector. The FWHM of the instrument function for each observation is provided along with the other descriptions of the observations in Sections 6 and 7.

The observed line profiles can have contributions from instrument stray light plus F corona and, in the case of H I Lα, interplanetary hydrogen. These contributions are negligible in streamers at most of the heliocentric heights discussed here, but can be significant in coronal holes. In Section 7, constrained curve fits are described for removing these backgrounds.

6. Measurements of Velocity Distributions for H^0, O^{5+}, and Fe^{11+} in Streamers

Figures 3–5 provide several examples of spectral line profiles from an equatorial region of the extended solar corona observed on 12 October 1996 where the wavelength scale is converted to LOS velocity using Equation (1). This structure appears to be a helmet streamer in H I Lα and H I Lβ, but it has a bifurcated appearance in the O VI and Fe XII lines (see Figure 6).

In general, there are four quantities that could affect the H I Lα and O VI 1032 Å intensities in the streamer. They are the electron temperature, the electron density, the elemental abundance, and Doppler dimming. Detailed modeling is expected to help identify the quantities and processes responsible for the observed intensities of Figure 6 (Noci *et al.*, 1997; Raymond *et al.*, 1997).

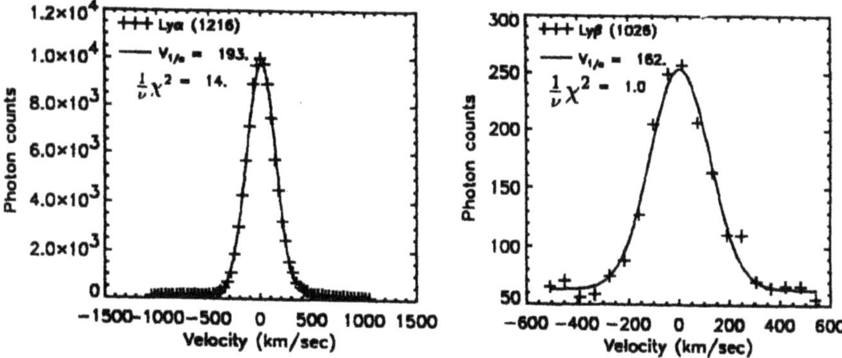

Figure 3. UVCS 12 October 1996 observations of H I Lα and Lβ profiles for an equatorial streamer above the west limb (position angle = 270°) centered at a heliocentric height of 2.1 $R_\odot$. The spatial resolution elements for Lα and Lβ are 0.23 × 8.4 arc min and 0.23 × 21 arc min, the FWHM of the instrument line profiles are 0.28 Å and 0.23 Å, respectively, and the integration time is 3600 s. Curve fits for a single Gaussian plus a constant background are shown. The data points are shown by crosses and the fitted curve by a solid line. The fully corrected best fit $v_{1/e}$, and the $\nu^{-1}\chi^2$ are provided in the figure.

The profiles have been corrected for the artificial line wing effect described in Section 5. Each of the corrected profiles are curve-fit to a single Gaussian plus a constant background. This provides a value for the velocity ($v_{1/e}$) corresponding to the Doppler half width, which is corrected for the finite instrumental resolution and provided in each figure. The FWHM of the instrumental profiles for the instrument configuration used for the observation are also provided. Several types of motions which may contribute to the measured velocity distribution are listed in Section 3.

The H I Lα and Lβ profiles at a projected heliographic height of 2.1 $R_\odot$ are provided in Figure 3. The H I Lα profile is for the upper bright region of the streamer as seen in Figure 6. Profiles for other regions are similar. To improve statistics, Lβ is for a slice across the entire streamer. In the case of H I Lβ, the value of $\nu^{-1}\chi^2 = 1.0$ indicates a good fit within the limitations of the rather low statistical precision of the data. In the case of Lα, which has much higher statistical precision, $\nu^{-1}\chi^2 = 14$.

Figure 7 provides values for the elements of $\nu^{-1}\chi^2$ for the Lα profile. It can be seen that the largest contributions come from near the base of the profile. The fit to the center portion of the profile is very good and indicates high confidence in the derived $v_{1/e}$.

The most likely candidates for explaining the poorer fit at the base are the following: (1) departures from a Maxwellian velocity distribution along the LOS, (2) a region along the LOS with a broader velocity distribution than that of the dominant contribution from the streamer, and (3) a small inaccuracy in the function used to remove the artificial line wings induced by the instrument or some other instrument effect. Less likely explanations include foreground and background

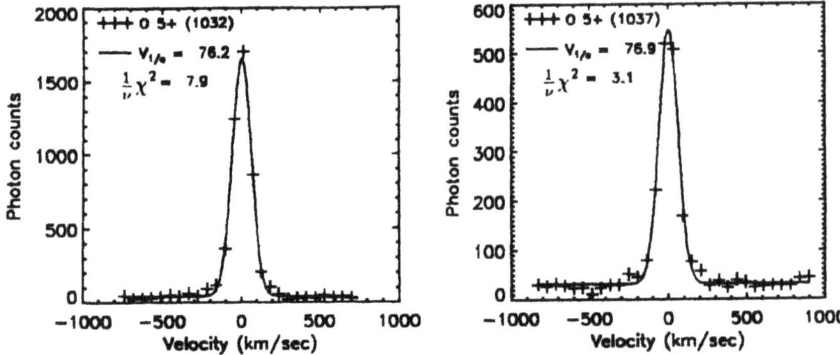

Figure 4. UVCS 12 October 1996 observations of O VI 1032 Å and 1037 Å profiles for an equatorial streamer above the west limb (position angle = 270°) centered at a heliocentric height of 2.1 $R_\odot$. The spatial elements for both lines are 0.23 × 9.3 arc min, the FWHM of the instrument line profiles are 0.23 Å and the exposure time is 3600 s. Curve fits for a single Gaussian plus a constant are shown. The data points are shown by crosses and the fitted curve by a solid line. The fully corrected best fit $v_{1/e}$ and the $\nu^{-1}\chi^2$ are provided in the figure.

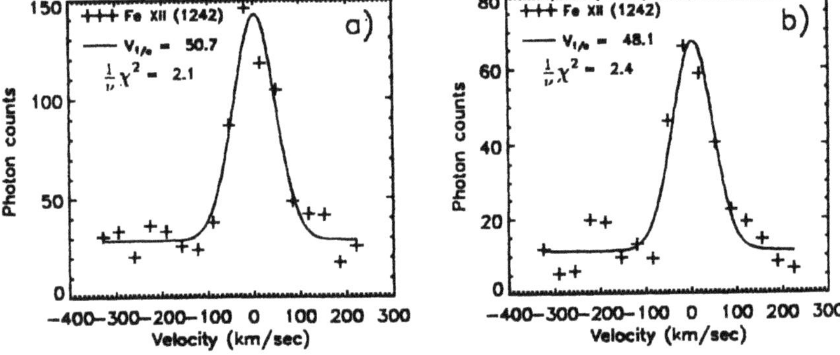

Figure 5. UVCS 12 October 1996 observations of Fe XII 1242 Å for an equatorial streamer above the west limb (position angle = 270°) centered at a heliocentric height of 2.1 $R_\odot$. The spatial resolution element across the entire streamer (a) is 21 arc min, and the spatial element across the upper bright region (b) is 8.4 arc= min. The FWHM of the instrumental profile is 0.18 Å, and the integration time is 3600 s. Curve fits for a single Gaussian plus a constant are shown. The data points are shown by crosses and the fitted curve by a solid line. The fully corrected best fit $v_{1/e}$ and the $\nu^{-1}\chi^2$ are provided in the figure.

structures with LOS outflows of about 240 km s^{-1}, but this requires extremely divergent magnetic field geometries or much larger outflow velocities than are indicated by the Doppler dimming results provided in Sections 8 and 9.

The difference in $\nu^{-1}\chi^2$ for the two lines is believed to result, primarily, from the larger values of σ_i for H I Lβ. Similar values for $\nu^{-1}\chi^2$ can be obtained for Lα if a subset of the data, corresponding to a much smaller exposure time, is used.

The velocity distributions derived from the two lines correspond to $v_{1/e}$ of 193 km s^{-1} and 162 km s^{-1}, respectively for Lα and Lβ.

The uncertainty is considerably smaller for Lα compared to Lβ because of the larger number of counts. There is a tendency for Lβ to be narrower than Lα in streamers (see Figure 8). The difference might have to do with line formation mechanisms. A significant fraction of Lβ is collisionally excited. That intensity is proportional to the square of the electron density (see discussion in Section 3). The smaller $v_{1/e}$ for Lβ might be explained by smaller $v_{1/e}$ in the most dense regions of streamers along the LOS.

O VI 1032 and 1037 profiles for the upper bright region in the streamer are provided in Figure 4. The statistical precision is at an intermediate level and the $\nu^{-1}\chi^2$ values indicate a departure from a single Gaussian at the precision level of the data. Values for the r.m.s. velocity (i.e., the second moment of the distribution) are equal to $2^{-1/2}$ of $v_{1/e}$ as expected for a Gaussian distribution. Hence, the O^{5+} LOS velocity distribution approximates a single Gaussian.

The velocity distributions derived from the two O VI lines are very similar. They correspond to $v_{1/e}$ of 76 km s^{-1} and 77 km s^{-1}, respectively, for the 1032 and 1037 lines. The two values are equal within the uncertainties.

The Fe XII 1242 Å profile for the upper bright region of the streamer is provided in Figure 5(b). An Fe XII profile with better statistics and integrated over the entire segment of the slit across the streamer is provided in Figure 5(a). The two profiles have equal widths to within the uncertainties. Some departures from a single Gaussian within the precision of the data is indicated by the values of $\nu^{-1}\chi^2$ although the relationship between $v_{1/e}$ and $v_{\rm r.m.s.}$ indicates a Gaussian shape.

The best-fit values of $v_{1/e}$ for H I 1216 Å, O VI 1032 Å, S X 1196 Å, and Fe XII 1242 Å for several heliographic heights in the 12 October 1996 streamer are provided in Figure 8. The profiles used in Figure 8 are integrated over a slice of the corona that extends across the entire structure and is centered on the designated height (see Figure 6).

In the case of H I 1216 Å, it is found that $v_{1/e}$ reaches a maximum value of about 195 km s^{-1} at a heliographic height of 2.1 $R_\odot$ and falls to about 150 km s^{-1} at 5.0 $R_\odot$. The O VI values are smaller (about 70 km s^{-1} at 2.1 $R_\odot$) and increase with height to about 150 km s^{-1} at 5 $R_\odot$. The Fe XII values are about 50 km s^{-1} at 2.1 $R_\odot$. S X 1196 Å values are also shown. Figure 8 includes values of $v_{1/e}$ for H I Lα and O VI 1032 Å in an equatorial streamer observed on 5 February 1997. The profiles for the two observations are similar but not identical. The 5 February 1997 observations indicate that the H I Lα values of $v_{1/e}$ continue to decrease slowly for heights up to 8 $R_\odot$.

Although the velocity distributions of coronal particles of different mass observed by UVCS never correspond to those expected in a thermal distribution, the velocities at the base of streamers come the closest. If we define the kinetic temnerature as one that describes all of the observed particle motions, the velocities at 1.5 $R_\odot$

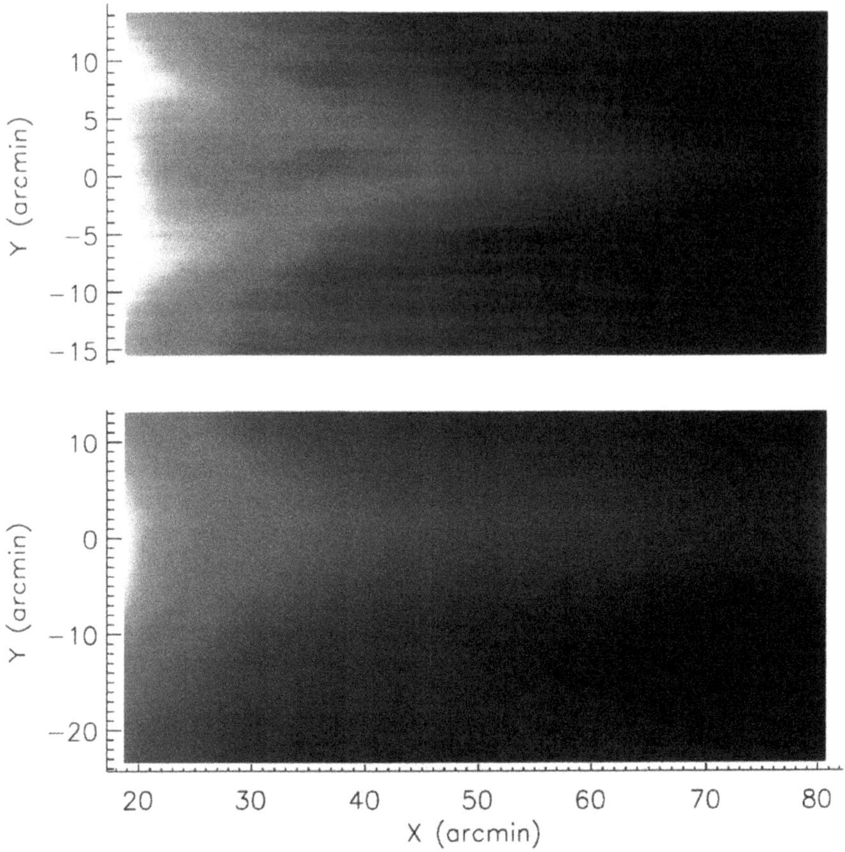

Figure 6. Images of the UVCS 12 October 1996 observation of an equatorial streamer above the west limb in H I Lα and O VI 1032 Å. The images result from an interpolation between strips of the corona corresponding to the UVCS entrance slit. The region has a strikingly different appearance in the two spectral lines. Fe XII images resemble those of O VI 1032 Å. The spatial scales are solar coordinates in arc min where the origin is at the center of the disk, the x axis is approximately in the heliographic west direction and the positive y axis is the projected heliographic north direction.

correspond to 2.0 × 10^6 K for H^0, 3.2 × 10^6 K for O^{5+} and 5.3 × 10^6 K for Fe^{11+}. The electron temperature derived from the charge state balance of hydrogen is 2.4 × 10^6 K (assuming the proton outflow velocity is less than 80 km s^{-1} and no Doppler dimming) which is similar to that derived from the Lα profile. At 2.4 × 10^6 K, O^{5+} would have $v_{1/e}$ = 50 km s^{-1} which is within 10 km s^{-1} of the observed value. At the same temperature, Fe^{11+} would have 27 km s^{-1} which is within 13 km s^{-1} of the observed value. The kinetic temperatures corresponding to the observed values of $v_{1/e}$ at 3 $R_\odot$ have larger deviations from a thermal distribution.

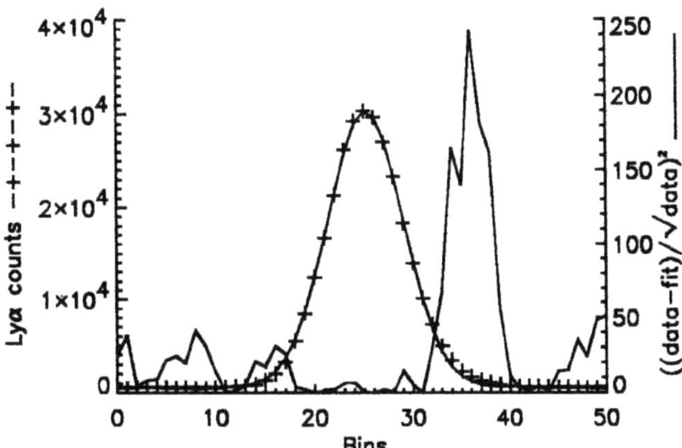

Figure 7. Chi-squared elements (solid line) and H I Lα spectral line profiles (solid line with crosses) for equatorial streamer of 12 October 1996 versus detector bins in the wavelength dispersion direction.

Figure 9 provides values of $v_{1/e}$ for O VI 1032 Å in three regions of the equatorial streamer. The region with the smallest O VI intensity also has smaller oxygen $v_{1/e}$ than the brighter regions.

Based on the values of polarized radiance from the WLC (see Section 8) and the intensity of Lα, it is believed that the streamer axis, which is dark in O VI, has a higher electron and proton density than the bright region.

It is interesting that the observed velocities come closest to a thermal distribution in the higher density regions (i.e., lower heights and streamer axis). In the lower density regions, the velocities of the more massive particles more strongly exceed those of a thermal equilibrium plasma at the observed hydrogen temperature. A plausible explanation for this behavior is that some process is preferentially accelerating particles of higher mass, and collisions are driving the velocities toward thermal equilibrium. The data suggest that such a process may be occurring throughout the streamer, but collisions are effectively thermalizing the plasma in the higher density regions. The velocity distributions in lower density coronal holes (see Section 7) may be providing even stronger evidence of such a process.

Values of reduced $\nu^{-1}\chi^2_{\rm min}$ for the equatorial streamer profiles are provided in each of the figures that show profiles. It can be seen that the values for H I Lα are rather large, while the curve-fits in Figure 3 appear to be reasonably good. The high statistical accuracy of the points causes $\nu^{-1}\chi^2_{\rm min}$ to be very sensitive to small differences between the observations and an ideal Gaussian curve. Data with poorer statistics such as the Fe XII 1242 Å line in Figure 5 may have similar departures from an ideal Gaussian, but the larger values of σ_i result in a much smaller value of $\nu^{-1}\chi^2_{\rm min}$.

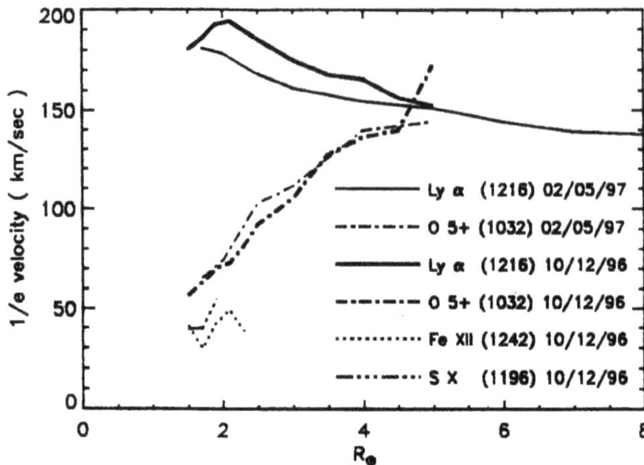

Figure 8. Plot of $v_{1/e}$ vs height for 12 October 1996 observation of a slice across the entire equatorial streamer. The profiles are dominated by the brightest regions. Values of $v_{1/e}$ are provided for H I Lα, O VI 1032 Å, S X 1196 Å, and Fe XII 1242 Å. Values for a similar streamer observed on 5 February 1997 are also provided.

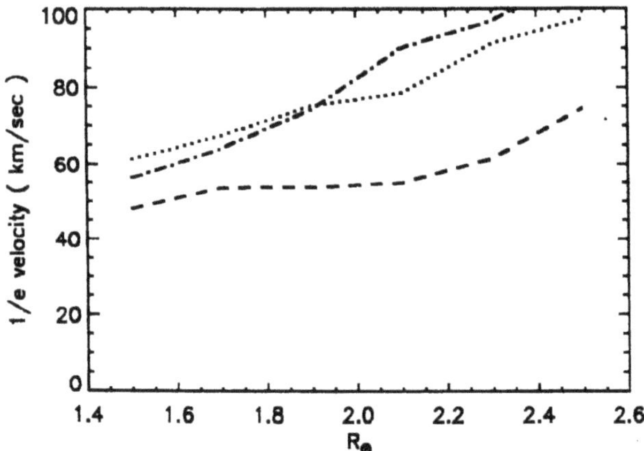

Figure 9. Velocity at e^{-1} of the peak of the LOS velocity distribution for O^{5+} versus heliocentric height for regions of the 12 October 1996 equatorial streamer. The curves for the two bright regions in O VI 1032 Å are shown as dots and as dot-dashes and the dark region curve is shown as dashes.

7. Measurements of Velocity Distributions for H^0, O^{5+}, and Mg^{9+} in Polar Coronal Holes

Figure 10 is an example of a H I Lα line profile in a polar coronal hole where the wavelength scale is converted to LOS velocity using Equation (1). The profile was measured at a central heliographic height of 3.0 $R_\odot$ on 11 May 1996 by the

UVCS/SOHO instrument. A curve-fit to a single Gaussian plus a constant is shown. The value of $\nu^{-1}\chi^2_{\rm min}$ is 7.6. The Gaussian that minimizes χ^2 clearly fails to fit the peak of the line and the base of the line. Such profiles are typical of those observed in polar coronal holes by UVCS/SOHO.

The integrated intensity of the observed line is 1.9×10^8 photons s^{-1} cm^{-2} sr^{-1}. The expected interplanetary intensity is about 3×10^7 photons s^{-1} cm^{-2} sr^{-1} and is broadened by the instrument profile FWHM = 0.28 Å. Laboratory measurements of instrumental stray light for observations at 2.7 $R_\odot$ determined that the equivalent stray light intensity for Lα is about 1×10^{-8} of the intensity of Lα at Sun center or about 4.5×10^7 photons s^{-1} cm^{-2} sr^{-1}. It has the profile of chromospheric H I Lα (Gouttebroze *et al.*, 1978) convolved with the instrument profile. In-flight determinations of a combination of stray light and the ultraviolet F corona based on detection of Si III 1206 and C III 977 (ions normally not expected in the corona) are consistent with laboratory values (Gardner *et al.*, 1996). A constrained curve fit was used to account for the stray light plus F corona and interplanetary Lα in the observation and to fit the coronal component to a Gaussian curve. For the curve fits, the line shapes of the stray light plus F corona and interplanetary Lα and the intensity of the stray light plus F corona were held fixed. All other quantities were allowed to vary. The result is shown in Figure 10(a). The coronal component is well fit to a single Gaussian curve with a velocity corresponding to the Doppler half width $v_{1/e}$ = 230 km s^{-1}. Any other coronal contributions to the observed profile would be at the level of the uncertainties in the stray light plus F corona and interplanetary Lα intensities. A plasma in thermodynamic equilibrium with the expected electron temperature of coronal holes at this height (about 1×10^6 K) would have a hydrogen $v_{1/e}$ of 129 km s^{-1}.

These H I Lα profiles observed by UVCS/SOHO are reminiscent of H I Lα profiles observed in coronal holes in 1993 by the Ultraviolet Coronal Spectrometer on the Spartan 201 spacecraft (Kohl, Strachan, and Gardner, 1996). That paper described curve-fits of the profiles to two Gaussians, in the case of the south polar coronal hole, and attributed part of the more narrow component to the tip of an obvious streamer along the LOS. The remainder of the narrow component was attributed to either less obvious streamers or to sites along the LOS with smaller values of $v_{1/e}$. Another alternative explanation for the non-Gaussian shape is local non-Maxwellian velocity distributions along the LOS.

The derived coronal component at 3.0 $R_\odot$ from UVCS/SOHO has $v_{1/e}$ = 230 km s^{-1}. This value is within 20% of the average broad component at 2.5 $R_\odot$ from Spartan 201 which has $v_{1/e}$ = 288 km s^{-1}. The Spartan narrow components are a factor of 10 brighter than the expected stray light levels. UVCS/SOHO synoptic images clearly show that streamers are outside the LOS for polar coronal hole observations at 3.0 $R_\odot$ in the June 1996 time frame. This suggests that the broad components observed by SOHO are formed in the coronal holes.

Figures 11–13 provide several examples of spectral line profiles from the north polar coronal hole observed by UVCS/SOHO in June 1996. Figure 12 provides

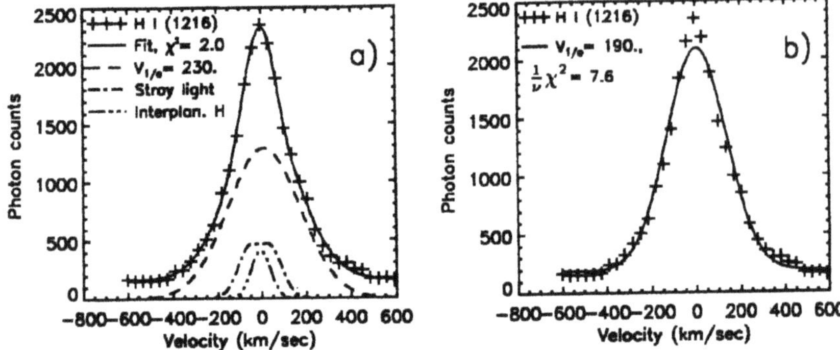

Figure 10. UVCS 11 May 1996 observations of H I Lα profiles for the south polar coronal hole (position angle = 180°) centered at a heliocentric height of 3.0 $R_\odot$. The spatial resolution element is 0.23 × 38.5 arc min, the FWHM of the instrument line profile is 0.28 Å, and the integration time is 10 800 s. Computer fits for a single Gaussian plus a constant (b) and three Gaussians plus a constant (a) are shown. Constrained fits were used to account for interplanetary Lα and stray light plus F corona. The data points are shown as crosses, the coronal profile by dashes and the fitted profile by a solid line. The fully corrected best fit $v_{1/e}$ and the $\nu^{-1}\chi^2_{\rm min}$ are provided on the figure.

profiles of O VI 1032 Å and 1037 Å with the slit centered at 2.1 $R_\odot$. The low intensity of the O VI emission at that height accounts for the fairly small number of accumulated counts. The spatial element used for the integration includes both polar plumes and inter-plume regions.

The 1032 profile has a complex shape that appears to consist of one Gaussian, which accounts for most of the line intensity, plus a very narrow component. The width of the narrow component is indistinguishable from the instrument profile FWHM. The ratio of the 1032 narrow component intensity to the disk intensity corresponds to that expected from in-flight measurement of instrument stray light plus F corona. Hence, it is believed that the narrow spikes on the profiles at this height are due to stray light. Observations of a polar coronal hole on 7 April 1996 with about the same observational parameters, were similar.

Figure 12 shows curve-fits to the 1032 and 1037 lines. A background of about 200 counts has been subtracted from the data. The $\nu^{-1}\chi^2_{\rm min}$ of 2.7 is considered acceptable. The Doppler half widths of the broad components of the 1032 and 1037 profiles correspond to $v_{1/e}$ of about 485 km s^{-1}.

The variations of $v_{1/e}$ with heliocentric height for H I Lα, O VI 1032 Å, and 1037 Å are provided in Figure 14 for observations of the north polar coronal hole from 7 April 1996 and the June 1996 data. The $v_{1/e}$ for both the April and June 1996 observations at 2.1 $R_\odot$ are much larger than expected. For example, the $v_{1/e}$ of O^{5+} for a plasma in thermodynamic equilibrium at the expected electron temperature of about 1 × 10^6 K is 32 km s^{-1}.

There are several plausible explanations for the large LOS velocities. It is unlikely that outflow velocities make a large contribution. Expected outflow velo-

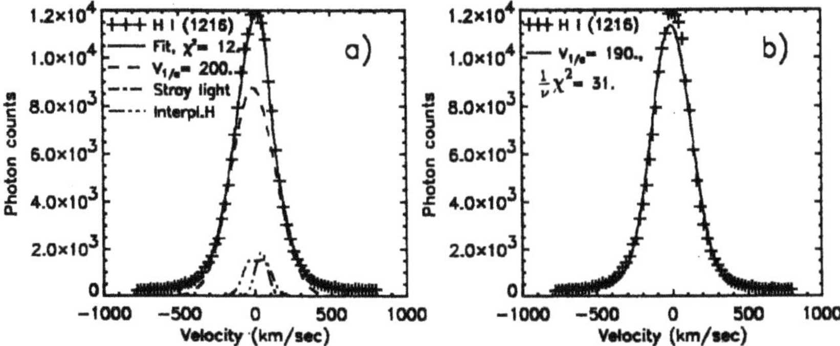

Figure 11. UVCS/SOHO 21 June 1996 observations of the H I Lα profile for the north polar coronal hole (position angle = 0.0°) centered at a heliocentric height of 2.1 $R_{\odot}$. The spatial resolution element is 0.35 × 21.7 arc min, the FWHM of the instrument line profile is 0.23 Å and the integration time is 31 800 s. Computer fits to a single Gaussian plus a constant (b) and three Gaussians plus a constant (a) are shown. Constrained curve fits were used to account for interplanetary Lα and stray light plus the F corona. The data points are shown as crosses the coronal profile by dashes and the fitted profile by a solid line. The best fit $v_{1/e}$ and the $\nu^{-1}\chi^2_{\min}$ are provided in the figure.

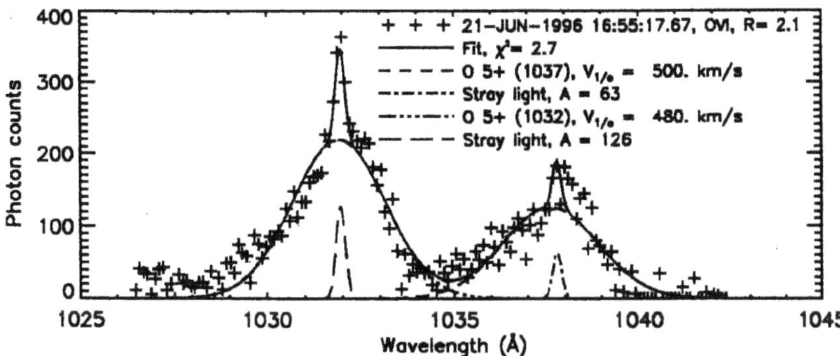

Figure 12. UVCS/SOHO 21 June 1996 observations of O VI 1032 Å and 1037 Å profiles for the north polar coronal hole (Position angle = 0.0°) centered at a heliocentric height of 2.1 $R_{\odot}$. The spatial resolution element is 0.35 × 20.3 arc min, the FWHM of the instrument line profile is 0.25 Å, and the integration time is 31 800 s. Computer fits for two Gaussians plus a constant are shown for each of the two lines. Constrained curve fits were used to account for conbributions from stray light plus the F corona. The data points are shown as crosses, coronal profiles by dashes and the fitted profile by a solid line. The best fit $v_{1/e}$ and $\nu^{-1}\chi^2_{\min}$ are provided in the figure.

cities at 2.1 $R_{\odot}$ (see Strachan *et al.*, 1993) are about 150 km s^{-1} to 250 km s^{-1} and the component along the LOS would be significantly smaller and would not be expected to account for very much of the observed line width. If other macroscopic motions such as turbulence are responsible for the large $v_{1/e}$, they might be expected to affect H^0 as well, but the H^0 have smaller velocities than those of O^{5+} above 2 $R_{\odot}$ (see Figure 14).

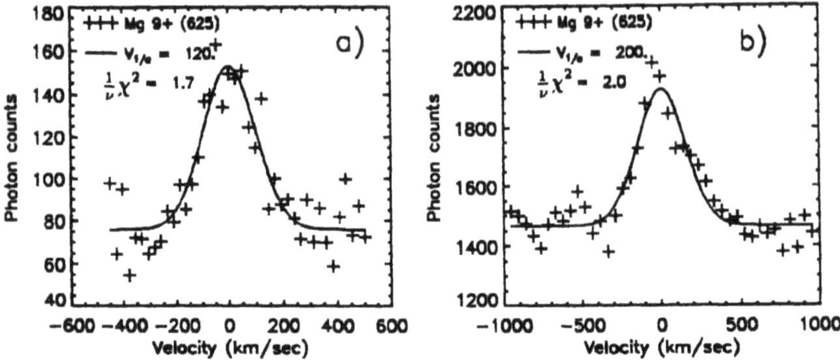

Figure 13. UVCS/SOHO observations of Mg X 625 Å for the north polar coronal hole (position angle = 0.0°). An observation centered at 1.5 $R_\odot$ was made on 17 June 1996 (a) and one centered at 1.75 $R_\odot$ was made on 19 June 1996 (b). The spatial resolution elements are 0.70 × 7.35 arc min (a) and 0.70 × 21.7 arc min (b). The instrumental FWHM are 0.22 Å (a) and 0.23 Å (b). The data points are shown as crosses and the curve fit to a single Gaussian plus a constant is shown as a solid line. The best fit $v_{1/e}$ and the $\nu^{-1}\chi^2_{\rm min}$ are provided in the figure.

Microscopic velocities might be due, in part, to the remnant of the thermal distribution from lower heights and they might also be due, in part, to ion-cyclotron resonant acceleration by high-frequency MHD waves (McKenzie, Banaszkiewicz, and Axford, 1995). There might also be a contribution to the observed velocity distribution from transverse wave motions (Withbroe *et al.*, 1985).

Figure 11 provides a profile of H I Lα in the coronal region where the O VI profiles of Figure 12 were observed and during the same period of time. The profile has been curve-fit to a single Gaussian plus a constant background. The $\nu^{-1}\chi^2_{\rm min}$ value is 31, indicating a low confidence in the goodness-of-fit. The high precision data indicate a deviation from a single Gaussian near the base of the profile. The profile was also curve fit in the same manner described earlier for the 11 May 1996 observation (see Figure 10). The stray light plus F corona levels for observations at 2.1 $R_\odot$ are somewhat higher than for the observations at 3.0 $R_\odot$. The curve fit provides a reasonable fit to the constrained values for the stray light plus F corona, interplanetary Lα and a single Gaussian component for the coronal profile. The value of $\nu^{-1}\chi^2$ for this high precision data reflects rather small departures from the fitted curve that may be due to a variety of causes. The derived value for $v_{1/e} = 200$ km s^{-1}.

Figure 13 provides profiles of Mg X 625 Å observed on 17 and 19 June 1996 at heliocentric heights centered at 1.5 and 1.75 $R_\odot$, respectively, in the north coronal hole. Single Gaussian fits yielded $v_{1/e}$ of 130 and 210 km s^{-1}, respectively, for the two heights.

Polar plumes are apparent in both the O VI and H I intensities at 1.7 $R_\odot$ (see Figure 15). The plumes are brightest near the center of the slit projection, which has

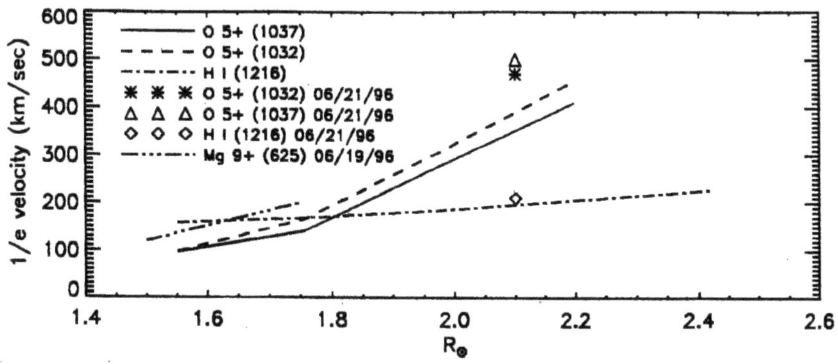

Figure 14. UVCS/SOHO measurements of $v_{1/e}$ versus heliocentric height for line-of-sight velocity distributions derived from H I Lα, O VI 1032 Å and 1037 Å and Mg X 625 Å for polar coronal holes observed on 7 April 1996 and June 1996. The values of $v_{1/e}$ are for single Gaussian fits to the line after removal of stray light, F corona and interplanetary H I Lα.

a higher latitude and a smaller heliocentric height than the regions corresponding to the extremities of the slit.

Figure 16 provides spectral line profiles for O VI 1032 in a region of the north polar coronal hole observed on 7 June 1996. Profiles are shown for a concentration of polar plumes at a heliocentric height of 1.7 $R_\odot$ and a region between two concentrations of plumes. The profiles were curve-fit to two Gaussians plus a constant background. The stray light plus F corona and interplanetary Lα are a very small contribution to the profile. Neither profile in Figure 16 can be well fit to a single Gaussian curve. The two Gaussian fits reveal that the narrow component is a much larger fraction of the total line intensity for the concentration of plumes than for the region between plume concentrations. The widths of the narrow components are comparable to the instrument profile FWHM and are, currently, very uncertain.

Table III provides values of $v_{1/e}$ for the broad components and the broad component fractions of the line intensity for detector rows (see Figure 15) corresponding to plume concentrations (rows 27–31 and 36–39), between plume concentrations (rows 23–27 and rows 32–35) and a region well outside plume concentrations (rows 5–17). The narrow components of both O VI lines are much stronger in the plume concentrations than in the other regions. These data suggest that polar plumes are the source regions of the narrow components observed at 1.7 $R_\odot$, and regions of the coronal hole outside the plumes are the source regions of the broad components for the observation of Figure 16. Stray light is expected to account for less than 5% of the narrow component of Figure 16(b). Observations from other time periods in 1996 are similar to Figure 16, but the enhancement of the narrow component is smaller. There is a tendency for the LOS velocity distribution in plume concentrations to be more narrow than in regions away from plume concentrations. The O VI 1032 image for 31 May 1996 makes it clear that streamers

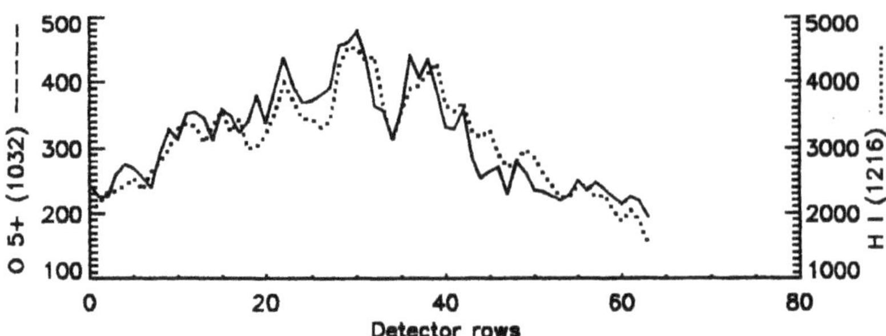

Figure 15. Variation of O VI 1032 Å (solid) and H I Lα (dots) intensities versus position along the instantaneous field of view, which corresponds to the UVCS/SOHO entrance slit, for an observation on 7 June 1996. The row numbers of the abscissa refer to spatial bins of 28 arc sec. The slit was oriented such that it was perpendicular to the radial line passing through the north solar pole and intersected it at 1.7 $R_\odot$ and row 30. The peaks in intensity apparently correspond to concentrations of polar plumes.

are outside the LOS for polar observations centered at 1.7 $R_\odot$ and so streamers are not the origin of the narrow O VI components for the observations of 7 June 1996. Observations at higher heights (see Figure 12) have much smaller narrow components that may be entirely due to stray disk light plus F corona.

Figure 14 is a plot of $v_{1/e}$ versus heliographic height for the north polar coronal hole as observed on 7 April 1996. It also includes data for the June 1996 observations. The data plotted are for single Gaussian coronal profiles which were obtained with the constrained curve fit procedure described earlier. Differences in the values for O VI 1032 and 1037 are believed to fall within the uncertainty limits. In general, the O^{5+} values of $v_{1/e}$ are smaller than those for H^0 at 1.5 $R_\odot$, comparable to H^0 at about 1.7 $R_\odot$ and significantly larger than those for H^0 at 2.1 $R_\odot$. The Mg X 625 Å line has $v_{1/e}$ between those of O^{5+} and H^0 at 1.5 $R_\odot$ and comparable or slightly larger values than both at 1.75 $R_\odot$.

It is interesting to compare the H^0 and O^{5+} values for $v_{1/e}$ at 2.1 $R_\odot$. The H^0 value of 200 km s^{-1} is considerably smaller than the O^{5+} value of 485 km s^{-1} for the June 1996 data. The O^{5+} values for 7 April 1996 are slightly smaller than those for June 1996, but still significantly larger than the H^0 value.

Recent calculations by Olsen and Leer (1996) and by Allen, Habbal, and Hu (1996) indicate that the H^0 velocities at 2.5 $R_\odot$ should be representative of the proton motions. Since transverse wave motion should be nearly the same for O^{5+} and protons, the observed velocities for H^0 and O^{5+} suggest that something other than transverse wave motion is contributing, significantly, to the O^{5+} velocity distribution. The explanation might be that the larger $v_{1/e}$ for O^{5+} is a result of ion-cyclotron resonant acceleration of O^{5+} ions about the coronal magnetic field lines (McKenzie, Banaszkiewicz, and Axford, 1995). More experimental and theoretical work is needed to resolve this issue.

Table III
UVCS/SOHO O VI 1032 and 1037 Å parameters for different concentrations of plumes

λ (Å)	Rows	I_B/I_{tot}	$v_{1/e}$ (km s^{-1})
1032	27–31 and 36–39	0.55	280
1037	36–39	0.47	370
1032	23–27 and 32–35	0.85	230
1037	32–35	0.719	290
1032	5–17	0.74	310
1037	5–17	0.86	330

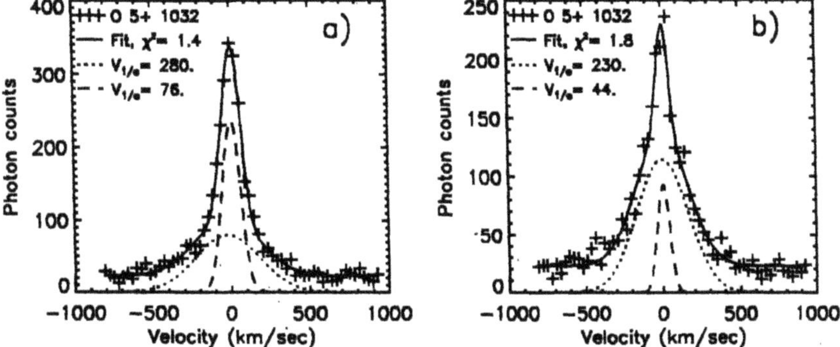

Figure 16. UVCS/SOHO 7 June 1996 observations of O VI 1032 Å profiles for regions in the brightest polar plume concentrations (a) and between plume concentrations (b). The observations are for the north polar coronal hole and combine rows 27–31 and 36–39 (a) and rows 23–27 and 32–35 (b) (see Figure 15). Row 30 corresponds to a heliocentric height of 1.7 $R_{\odot}$, the slit width corresponds to 8.7 arc sec, the FWHM of the instrument profile is 0.16 Å and the integration time is 16 000 s. The data points are shown by crosses, the fitted components by dots and dashes and the fitted profile by a solid line. The best fit $v_{1/e}$ and the $\nu^{-1}\chi^2_{\min}$ are provided in the figure.

8. Preliminary Doppler Dimming Measurements in Helmet Streamers

The outflow velocity of O^{5+} in the equatorial streamer above the west limb of the Sun on 5 February 1997 can be estimated from the intensity ratio of O VI 1032 Å to O VI 1037 Å (see discussion in Section 3). Figure 17 is a plot of that ratio versus heliocentric height along a nearly straight line near the axis of the streamer. The 5 February 1997 streamer is similar in appearance to the one shown in Figure 6. Observations of the former include larger heights.

As discussed in Section 3, the resonantly scattered components have a ratio of 4:1 for velocities less than about 100 km s^{-1} and the collisional components have a ratio of 2:1 at all heights. Ratios above 2 indicate that the O^{5+} outflow velocity is less than about 100 km s^{-1} (see Figure 1) where we assume for the moment, that $v_{1/e}$ in the direction of the incoming radiation is less than 56 km s^{-1}. The

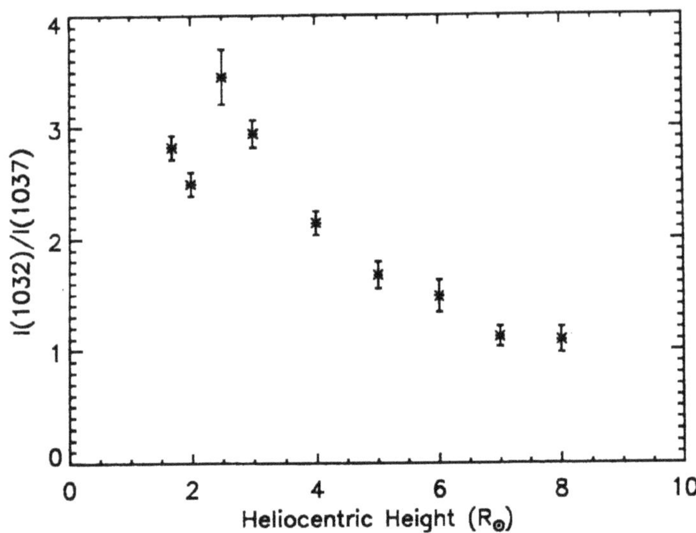

Figure 17. Observed intensity ratio for O VI 1032 Å to O VI 1037 Å near the axis of an equatorial streamer above the west solar limb on 5 February 1997. The error bars are for 1 standard deviation in the ratio.

plot of Figure 17 indicates that the ratio is above 2 for heights less than 4 $R_{\odot}$. The increase in the ratio from 1.7 $R_{\odot}$ to 2.5 $R_{\odot}$ is probably due (at least in part) to a decrease in electron density with height. UVCS/WLC measurements of polarized radiance, which is proportional to electron density, indicate that the electron density decreases with height in the streamer. The intensity of the collisional components scale as the square of the electron density and the resonantly scattered components scale linearly with electron density. The observed increase in the ratio from about 2.5 to 3.5 indicates that the O^{5+} outflow velocity remains below 100 km s^{-1} and the collisional component decreases with height as expected.

Since the collisional component will continue to decrease with height, the observed decrease in the ratio from about 3.5 to 2.0 between 2.5 $R_{\odot}$ and 4.0 $R_{\odot}$ indicates that Doppler dimming of the resonantly scattered components occurs at those heights since, otherwise, the resonant components would become more dominant.

At 4 $R_{\odot}$ the ratio becomes approximately 2. The Doppler dimming of both O VI lines is the same for velocities less than about 100 km s^{-1}. Hence, a ratio of 2 indicates that the resonantly scattered components of both lines are negligible. This occurs when the O^{5+} outflow velocity reaches about 100 km s^{-1}.

The observed decrease in the ratio below 2 indicates that the resonantly scattered component of the O VI 1037 line is larger than that of O VI 1032. This is expected to occur when the O^{5+} outflow velocity is large enough for the chromospheric C II 1037.018 Å radiation to be Doppler-shifted enough to overlap with the coronal absorption line profile of O VI 1037.613 Å. The precise outflow velocity required to allow this pumping depends on the spectral width of the chromospheric emergent intensity and the width of the coronal absorption profile.

Figure 18 is a plot of the O VI line ratio versus O^{5+} outflow velocity for an electron density at 7 $R_\odot$ of 4.4×10^4 cm^{-3}. This value is derived from measurements of visible polarized radiance assuming a spherically-symmetric corona. This approach yields a lower limit for the electron density at 7 $R_\odot$ in the 5 February 1997 streamer. Curves for seven different coronal absorption line widths are provided. These line widths depend on the velocity distribution of the coronal O^{5+} in the direction of the incoming C II chromospheric radiation. In the figure, this velocity distribution is designated by the value of $v_{1/e}$ for the approximately radial direction. The curves of Figure 18 take into account both the collisional and resonantly scattered components of the two lines for the above density. Higher electron densities will raise the curves to higher line ratios for the same kinetic temperature. This occurs because the collisional components, which have a 2:1 ratio, will be relatively brighter for the higher densities. The O VI line ratio is not sensitive to the electron temperature or to the elemental abundance.

The line ratio at 7 $R_\odot$ is 1.1 which corresponds to 56 km s^{-1} for $v_{1/e}$ in the direction of the incoming radiation. This is an upper limit on $v_{1/e}$ for that direction since the electron density used here is a lower limit and higher electron densities raise the curves of Figure 18.

Although we do not have a measurement for $v_{1/e}$ of O^{5+} for the LOS direction in the streamer at 7 $R_\odot$, the upper limit derived above for the incoming radiation direction is significantly less than the LOS values for the highest heights provided in Figure 8.

The O^{5+} outflow velocity at 7 $R_\odot$ derived from Figure 18 is between 175 and 205 km s^{-1}. The discussion above about the O VI ratio at lower heights assumed the Doppler dimming curves of Figure 1 for the case where $v_{1/e}$ is less than 56 km s^{-1} for the velocity distribution in the incoming radiation direction. Hence, the O^{5+} outflow velocity values discussed earlier for the lower heights in the streamer are approximately correct.

It is important to recognize that the derived O^{5+} outflow velocity is not necessarily equal to that of the protons. Doppler dimming of H I Lα can be used to estimate the proton outflow velocity.

Although the approach outlined above provides a reasonably accurate estimate of the O^{5+} outflow velocities in the observed equatorial streamer, more precise values could be obtained with a self-consistent empirical model of the O VI intensity ratio and the absolute O VI intensities as well as an analysis of Doppler dimming in H I Lα. That model would require independent determinations of the electron temperature and electron density. It is also desirable to estimate the LOS contributions to the observed intensities. The UVCS synoptic data can be used for that purpose.

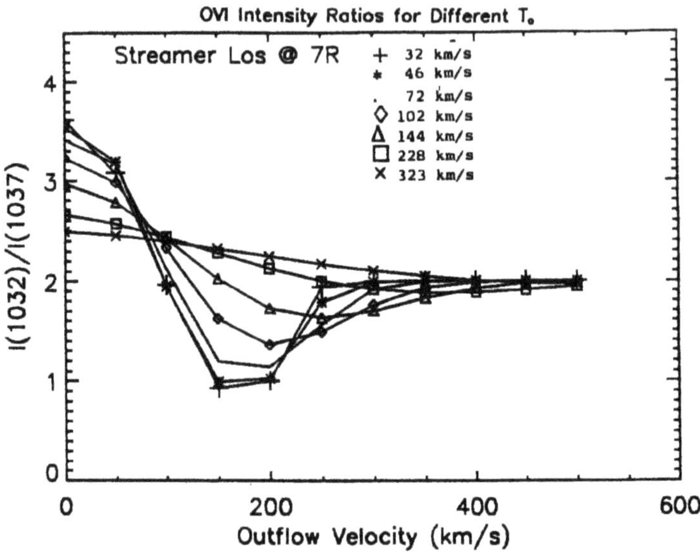

Figure 18. Modeled intensity ratio for O VI 1032 Å to O VI 1037 Å versus outflow velocity for a spherically-symmetric corona with an electron density that reproduces the observed visible polarized radiance at a heliocentric height of 7 $R_{\odot}$ for an equatorial streamer above the west solar limb on 5 February 1997. Seven curves are provided for different velocity distributions for the coronal O^{5+} in the direction of incoming C II radiation from the chromosphere. The curves are labled with the value of $v_{1/e}$ for the velocity distribution in the direction of the incoming radiation.

9. Preliminary Doppler Dimming Measurements in Polar Coronal Holes

Outflow velocities for O^{5+} can be estimated for polar coronal holes in the same manner used for the streamer in Section 8, Figure 19 is a plot of the O VI 1032/1037 ratio for the north polar coronal hole which was observed during the month of January 1997. The data used for the line ratios are standard UVCS synoptic observations for heliocentric heights between 1.5 and 2.5 $R_{\odot}$ and a series of special observations at 3 $R_{\odot}$. Stray light plus F corona and detector backgrounds have been removed from the data. The statistical error bars are vanishingly small on the plot.

The ratio is about 2.5 at 1.5 and 1.75 $R_{\odot}$. This ratio implies that the O VI 1032 line has a resonantly scattered component. If we assume that the Doppler dimming curve in Figure 1 (for $v_{1/e}$ less than 91 km s^{-1} in the direction of the incoming radiation) is applicable, then the O^{5+} outflow velocity is less than 150 km s^{-1}. However, if the velocity distribution in the direction of the incoming O VI transition region radiation is as wide as the observed line widths in the LOS direction (see Section 7), then the appropriate Doppler dimming curve would be much different.

The ratio reaches 2 at 2 $R_{\odot}$. If we again assume that Figure 1 is applicable, then the O^{5+} outflow velocity is expected to be between 100 km s^{-1} and 125 km s^{-1} for radial values of $v_{1/e}$ from 32 km s^{-1} to 91 km s^{-1}.

The ratio decreases from 2 to 0.7 over a range of heights from 2 to 3 $R_\odot$. The only known explanation for the ratio falling below two is pumping of O VI 1037.613 Å by the C II 1037.018 Å line.

Figure 20 is a plot of the O VI line ratio versus O^{5+} outflow velocity for a spherically-symmetric electron density with 6×10^4 cm^{-3} at 3 $R_\odot$. This model is consistent with observed visible polarized radiance values for 3 $R_\odot$ in the January 1997 polar coronal hole. Curves for seven different coronal absorption line widths are provided. These line widths depend on the velocity distribution of the coronal O^{5+} in the direction of the incoming C II chromospheric radiation. In the figure, this velocity distribution is designated by the value of $v_{1/e}$ for the approximately radial direction. The curves of Figure 20 take into account both the collisional and resonantly scattered components of the two lines for the above density. The ratio is not sensitive to the electron temperature or to the elemental abundance.

The line ratio at 3 $R_\odot$ is 0.70 which corresponds to $v_{1/e}$ less than 91 km s^{-1} in the direction of the incoming radiation.

The value for $v_{1/e}$ in the LOS direction for O^{5+} in the January coronal hole at 3 $R_\odot$ is about 600 km s^{-1}. The upper limit derived above for the $v_{1/e}$ in the direction of the incoming radiation is significantly less. Since the direction of the incoming radiation is approximately parallel to the expected magnetic field direction in polar coronal holes, and the LOS direction is approximately perpendicular to the field, it appears that $v_{1/e}$ parallel to the magnetic field is significantly smaller than $v_{1/e}$ in the perpendicular direction.

The O^{5+} outflow velocity at 3 $R_\odot$ derived from Figure 20 is between 130 and 230 km s^{-1}. The discussion above about the O VI ratio at lower heights referred to the Doppler dimming curves of Figure 1 where the value of $v_{1/e}$ in the direction of the incoming radiation was assumed to be less than 91 km s^{-1}. Hence, the discussion of the O^{5+} outflow velocities at lower heights should be reasonably accurate.

It is important to recognize that the derived O^{5+} outflow velocity is not necessarily equal to that of the protons. Doppler dimming of H I Lα can be used to estimate the proton velocity.

The estimates of outflow velocity from the above procedure treat the coronal hole as a section of a spherically-symmetric source. However, it is apparent from the observations that polar coronal holes contain polar plumes that extend up to several solar radii. It could be seen from Section 7 that the O VI spectral line widths in plumes at 1.7 $R_\odot$ appear to be much smaller than those away from plume concentrations. The density in plumes is believed to be considerably larger than in the interplume regions. The outflow velocity in plumes is unknown at the present time.

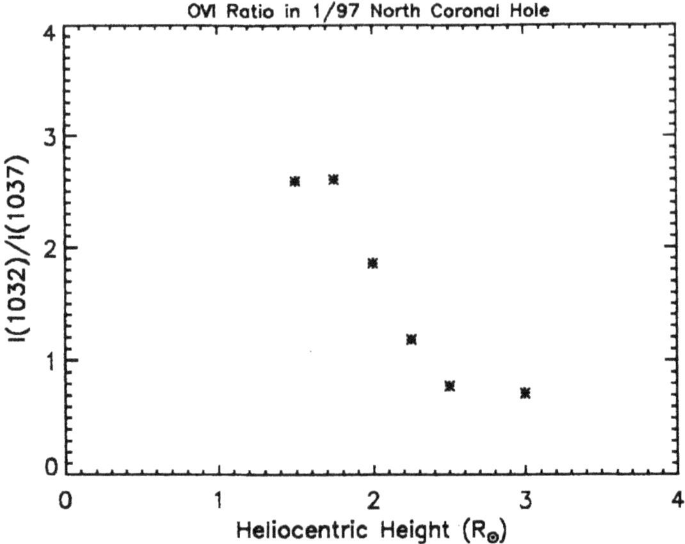

Figure 19. Observed intensity ratio for O VI 1032 Å to O VI 1037 Å near the axis of the north polar coronal hole as observed during the month of January 1997. The statistical error bars are too small to be disernible from the data points.

10. Summary

This paper provides examples of UVCS/SOHO spectroscopic observations of the extended solar corona during the first year of operations. Values are presented for velocity distributions of protons and minor ions and for O^{5+} outflow velocities in equatorial streamers and polar coronal holes. The results presented here are obtained from direct application of spectroscopic diagnostic techniques. More precise values can be obtained by developing a self-consistent empirical model that specifies densities, velocity distributions, and outflow velocities for the primary particles and several minor ions, and also specifies abundances relative to hydrogen. The models should specify local values for plasma parameters in three dimensions. For structures that are stable for a few days, daily UVCS synoptic observations, which began on 10 April 1996, can be used to help determine the three dimensional structure of the observed corona.

Measurements of spectral line profiles for H I Lα and Lβ, O VI 1032 Å and 1037 Å, S X 1196 Å, and Fe XII 1242 Å in equatorial streamers provide measurements of LOS values for the velocity distribution of the associated ions (specified by $v_{1/e}$). At heliocentric heights of 1.5 $R_\odot$ in equatorial streamers where the density is the highest, the $v_{1/e}$ along the LOS approach those of a thermal distribution at about 2×10^6 K. This value is in approximate agreement with the electron temperature derived from the ionization balance of hydrogen. Departures from thermal values

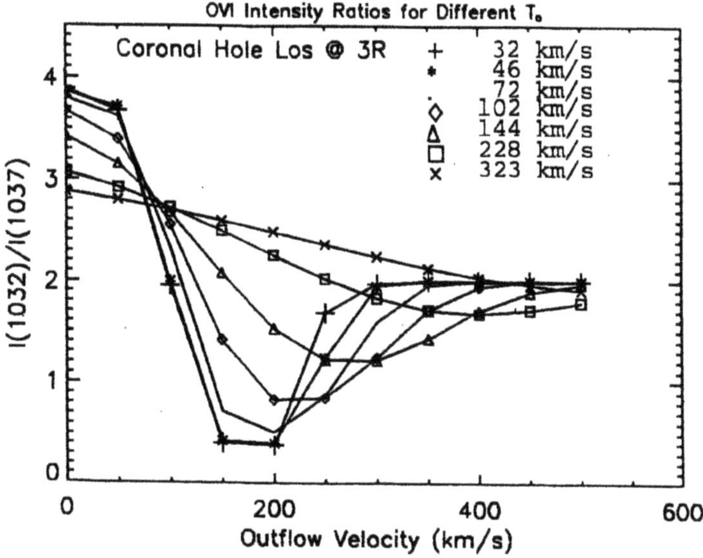

Figure 20. Modeled intensity ratio for O VI 1032 Å to O VI 1037 Å versus outflow velocity for a spherically-symmetric corona with an electron density that reproduces the observed visible polarized radiance at a heliocentric height of 3 $R_\odot$ for the north polar coronal hole observed during the month of January 1997. Seven curves are provided for different velocity distributions for the coronal O^{5+} in the direction of incoming C II radiation from the chromosphere. The curves are labled with a value of $v_{1/e}$ for the velocity distribution in the direction of the incoming radiation.

become larger with increasing mass. The $v_{1/e}$ for H^0 decrease from a maximum value of about 180 km s^{-1} at about 2 $R_\odot$ to about 140 km s^{-1} at 8 $R_\odot$. The $v_{1/e}$ for O^{5+} increase with height and reach values of 150 km s^{-1} at 4.7 solar radii which are approximately equal to the H^0 values. The observed velocities come closest to a thermal distribution in the highest density regions of equatorial streamers. In lower density structures, the $v_{1/e}$ show larger departures from equilibrium values with increasing mass. It appears that some process is preferentially accelerating particles of higher mass, and collisions are driving the velocities toward thermal equilibrium.

Measurements of the intensity ratio of O VI 1032 Å and 1037 Å have been used to estimate outflow velocities of O^{5+} ions in equatorial streamers. The O^{5+} outflow velocity is found to be less than 100 km s^{-1} below 3.5 $R_\odot$ and reaches a value of about 100 km s^{-1} at 4 $R_\odot$. The outflow velocity reaches a value of 175 to 205 km s^{-1} at 7 $R_\odot$.

The O VI line ratio was also used to estimate the $v_{1/e}$ for O^{5+} in the direction of the incoming radiation from the chromosphere. It is found that the $v_{1/e}$ in that direction at 7 $R_\odot$ in the observed equatorial streamer is less than 56 km s^{-1}. The

O^{5+} values of $v_{1/e}$ for the LOS direction at 7 $R_\odot$ have not been measured but the value at 5 solar radii is 150 km s^{-1}.

UVCS observations of coronal holes at 1.7 $R_\odot$ in H I Lα, O VI 1032 Å and O VI 1037 Å indicate bright ray-like structures that are likely to be concentrations of polar plumes along the LOS.

Both UVCS/SOHO and Spartan 201 observations indicate that a significant fraction of the protons along the LOS in coronal holes have $v_{1/e}$ larger than expected for a Maxwellian velocity distribution at the expected electron temperature of about 1×10^6 K. UVCS/SOHO observations, after removal of stray light plus F corona and interplanetary Lα, yield good constrained curve fits to single Gaussians for the coronal line profiles of H I Lα and O VI 1032 and 1037 at heights above 2.0 $R_\odot$.

If one compares the $v_{1/e}$ of O^{5+} and H^0 for various heights in polar coronal holes, one finds that the O^{5+} $v_{1/e}$ are smaller than those of H^0 at 1.5 $R_\odot$, are comparable at about 1.7 $R_\odot$ and become significantly larger than those of H^0 at 2.1 $R_\odot$. Mg^{9+} $v_{1/e}$ are larger than those of both H^0 and O^{5+} at 1.7 $R_\odot$.

Measurements of the intensity ratio of O VI 1032 Å and 1037 Å have been used to estimate outflow velocities of O^{5+} ions in polar coronal holes. The O^{5+} outflow velocity is found to be less than 150 km s^{-1} below 1.7 $R_\odot$ and reaches a value between 100 km s^{-1} and 125 km s^{-1} at 2 $R_\odot$. The outflow velocity reaches a value between 130 and 230 km s^{-1} at 3 $R_\odot$.

The O VI line ratio was also used to estimate the $v_{1/e}$ in the direction of the incoming radiation to coronal O^{5+} from the chromosphere. It is found that the $v_{1/e}$ in that direction at 3 $R_\odot$ in the observed polar coronal hole is less than 91 km s^{-1}. The O^{5+} values of $v_{1/e}$ for the LOS direction at 3 $R_\odot$ have been determined to be about 600 km s^{-1}. Since the direction of the incoming radiation is approximately parallel to the expected magnetic field direction in polar coronal holes, and the LOS direction is approximately perpendicular to the field direction, it appears that $v_{1/e}$ parallel to the magnetic field is significantly smaller than $v_{1/e}$ in the perpendicular direction.

The differences of the $v_{1/e}$ with mass in coronal holes suggests that very significant fractions of the O^{5+} velocities are not caused directly by transverse wave motions. The outflow velocities are expected to be too small to contribute significantly to the $v_{1/e}$. It appears that some microscopic motion (perhaps ion-cyclotron resonant acceleration) is the dominant process responsible for the observed difference between the $v_{1/e}$ for O^{5+} and H^0.

There is a tendency for the O^{5+} LOS velocity distribution in plume concentrations to be more narrow than the distribution in regions away from plume concentrations.

There are several types of observations that are not reported in this paper, but are in process. They include direct measurements of electron temperatures derived from measurements of the spectral line profile of coronal electron scattered H I Lα from the chromosphere, and determinations of proton outflow velocities derived from measurements of visible polarized radiance and Doppler dimming of resonantly scattered H I Lα. UVCS determinations of minor ion abundances in streamers are presented elsewhere in this volume (Raymond *et al.*, 1997). UVCS has observed several CMEs in the extended corona where the intensities of the H I Lyman series and ions at low stages of ionization are enhanced by two to three orders of magnitude over their usual values. There have also been observations of a comet in H I Lα, observations of several stars at wavelengths shortward of the Hubble Space Telescope range and observations of interplanetary H I 1216 Å and He I 584 Å.

UVCS/SOHO has also made many coordinated observations with other instruments on SOHO, with instruments on other satellites and with ground-based observatories. Among the most important of those are coordinated observations with SOHO's CELIAS, LASCO, SUMER, SWAN, MDI, and CDS. Coordinated observations of particular interest also include solar wind properties with *Ulysses*, CMEs with *Yohkoh*, and measurements of solar wind velocities and properties with radio scintillation measurements through the corona from astronomical sources and man-made satellites.

Acknowledgements

The authors wish to acknowledge the contributions of A. van Ballegooijen to the development of the UVCS instrument and the work of Brenda Bernard in the administration of the UVCS investigation. This work is supported by the National Aeronautics and Space Administration under grant NAG5–3192 to the Smithsonian Astrophysical Observatory, by Agenzia Spaziale Italiana and by Swiss funding through ESA's PRODEX programs, national funds and the Swiss Federal Institute of Technology Zürich.

References

Allen, L. A., Habbal, S. R., and Hu, Y. Q.: 1996, *J. Geophys. Res*, submitted.

Bevington, P. R.: 1969, *Data Reduction and Error Analysis for the Physical Sciences*, McGraw-Hill Book Co., New York, p. 313.

Gardner, L. D., Kohl, J. L., Daigneau, P. S., Dennis, E. F., Fineschi, S., Michels, J., Nystrom, G. U., Panasyuk, A., Raymond, J. C., Reisenfeld, D.= J., Smith, P. I,., Strachan, L., Jr., Suleiman, R., Noci, G. G., Romoli, M., Ciaravella, A., Modigliani, A., Huber, M. C. E., Antonucci, E., Benna, C., Giordano, S., Tondello, G., Nicolosi, P., Naletto, G., Pernechele, C., Spadaro, D., Siegmund, O. H. W., Allegra, A., Carosso, P. A., and Jhabvala, M. D.: 1996, 'Stray Light, Radiometric, and Spectral Characterization of UVCS/SOHO: in R. E. Huffman and C. G. Stergis (eds), Laboratory Calibration and Flight Performance', *Ultraviolet Atmospheric and Space Remote Sensing: Methods and Instrumentation*, SPIE 2831, p. 2.

Gouttebroze, P., Lemaire, P., Vial, J. C., and Artzner, G.: 1978, *Astrophys. J.* **225**, 655.

Ko, Y., Fisk, L. A., Geiss, J., Gloeckler, G., and Guhathakurta, M.: 1997, *Solar Phys.* **171**, 345.

Kohl, J. L. and Withbroe, G. L.: 1982, *Astrophys. J.* **256**, 263.

Kohl, J. L., Strachan, L., and Gardner, L. D.: 1996, *Astrophys. J.* **465**, L141.

Kohl, J. L., Esser, R., Gardner, L. D., Habbal, S., Daigneau, P. S., Dennis, E. F., Nystrom, G. U., Panasyuk, A., Raymond, J. C., Smith, P. L., Strachan, L., van Ballegooijen, A. A., Noci, G., Fineschi, S., Romoli, M., Ciaravella, A., Modigliani, A., Huber, M. C. E., Antonucci, E., Benna, C., Giordano, S., Tondello, G., Nicolosi, P., Naletto, G., Pernechele, C., Spadaro, D., Poletto, G., Livi, S., von der Luhe, O., Geiss, J., Timothy, J. G., Gloeckler, G., Allegra, A., Basile, G., Brusa, R., Wood, B., Siegmund, O. H. W., Fowler, W., Fisher, R., and Jhabvala, M.: 1995, *Solar Phys.* **162**, 313.

McKenzie, J. F., Banaszkiewicz, M., and Axford, W. I.: 1995, *Astron. Astrophys.* **303**, L45.

Leer, E.: 1988, 'Solar Wind Conference' in V. J. Pizzo, T. E. Holzer, and D. G. Sime (eds), *Proc. 6th Int. Solar Wind Conference*, Boulder, CO, p. 89.

Noci, G., Kohl, J. L., and Withbroe, G. L.: 1987, *Astrophys. J.* **315**, 706.

Noci, G., Kohl, J. L., Antonucci, E., Tondello, G., Huber, M. C. E., Fineschi, S., Gardner, L. D., Naletto, G., Nicolosi, P., Raymond, J. C., Romoli, M., Spadaro, D., van Ballegooijen, A. A., Siegmund, O. H. W., Benna, C., Ciaravella, A., Giordano, S., Michels, J., Modigliani, A., Panasyuk, A., Pernechele, C., Poletto, G., Smith, P. L., and Strachan, L.: 1997, *Advances in Space Research, Cospar Proceedings*, in press.

Olsen, E. L. and Leer, E.: 1996, *Astrophys. J.* **462**, 982.

Pernechele, C., Naletto, G., Nicolosi, P., Poletto, L., and Tondello, G.: 1995, *Proc. SPIE* **2517**.

Raymond, J. C., Kohl, J. L., Noci, G., Antonucci, E., Tondello, G., Huber, M. C. E., Gardner, L. D., Nicolosi, P., Fineschi, S., Romoli, M., Spadaro, D., Siegmund, O. H. W., Benna, C., Ciaravella, A., Cranmer, S., Giordano, S., Karovska, M., Martin, R., Michels, J., Modigliani, A., Naletto, M. G., Panasyuk, A., Pernechele, C., Poletto, G., Smith, P. L., Suleiman, R. M., and Strachan, L.: 1997, *Solar Phys.*, in press.

Strachan, L., Kohl, J. L., Weiser, H., and Withbroe, G. L.: 1993, *Astrophys. J.* **412**, 410.

Strachan, L., Gardner, I. D., Hassler, D. M., and Kohl, J. L.: 1994, *Space Sci. Rev.* **70**, 263.

van de Hulst, H. C.: 1950, *Bull. Astron. Inst. Neth.* **11**, 135.

Wilhelm, K.: 1996, personal communication.

Withbroe, G. L., Kohl, J. L., Weiser, H., and Munro, R. H.: 1982, *Space Sci. Rev.* **33**, 17.

Withbroe, G. L., Kohl, J. L., Weiser, H., and Munro, R. H.: 1985, *Astrophys. J.* **297**, 324.

COMPOSITION OF CORONAL STREAMERS FROM THE SOHO ULTRAVIOLET CORONAGRAPH SPECTROMETER

J. C. RAYMOND[1], J. L. KOHL[1], G. NOCI[2], E. ANTONUCCI[3], G. TONDELLO[4], M. C. E. HUBER[5], L. D. GARDNER[1], P. NICOLOSI[4], S. FINESCHI[1], M. ROMOLI[2], D. SPADARO[6], O. H. W. SIEGMUND[7], C. BENNA[8], A. CIARAVELLA[1,9], S. CRANMER[1], S. GIORDANO[8], M. KAROVSKA[1], R. MARTIN[10], J. MICHELS[1], A. MODIGLIANI[2], G. NALETTO[4,7], A. PANASYUK[1], C. PERNECHELE[4], G. POLETTO[11], PETER L. SMITH[1], R. M. SULEIMAN[1] and L. STRACHAN[1]

[1]*Harvard-Smithsonian Center for Astrophysics, Cambridge, MA 02138, U.S.A.*
[2]*Università di Firenze, I-50125 Firenze, Italy*
[3]*Osservatorio Astronomico di Torino, I-10025 Pino Torinese, Italy*
[4]*Università di Padova, I-35131 Padova, Italy*
[5]*Space Science Dept., ESA/ESTEC, NL-2200 AG, Noordwijk, The Netherlands*
[6]*Osservatorio Astrofisico di Catania, I-95125 Catania, Italy*
[7]*Space Sciences Laboratory, University of California, Berkeley, CA 94720, U.S.A.*
[8]*Università di Torino, I-10125 Torino, Italy*
[9]*European Space Agency*
[10]*Institut d'Astrophysique Spatiale, F-91405 Orsay, France*
[11]*Osservatorio Astrofisico di Arcetri, I-50125 Firenze, Italy*

(Received 14 March 1997; accepted 28 May 1997)

Abstract. The Ultraviolet Coronagraph Spectrometer on the SOHO satellite covers the 940–1350 Å range as well as the 470–630 Å range in second order. It has detected coronal emission lines of H, N, O, Mg, Al, Si, S, Ar, Ca, Fe, and Ni, particularly in coronal streamers. Resonance scattering of emission lines from the solar disk dominates the intensities of a few lines, but electron collisional excitation produces most of the lines observed. Resonance, intercombination and forbidden lines are seen, and their relative line intensities are diagnostics for the ionization state and elemental abundances of the coronal gas.

The elemental composition of the solar corona and solar wind vary, with the abundance of each element related to the ionization potential of its neutral atom (First Ionization Potential–FIP). It is often difficult to obtain absolute abundances, rather than abundances relative to O or Si. In this paper, we study the ionization state of the gas in two coronal streamers, and we determine the absolute abundances of oxygen and other elements in the streamers. The ionization state is close to that of a $\log T = 6.2$ plasma. The abundances vary among, and even within, streamers. The helium abundance is lower than photospheric, and the FIP effect is present. In the core of a quiescent equatorial streamer, oxygen and other high-FIP elements are depleted by an order of magnitude compared with photospheric abundances, while they are depleted by only a factor of 3 along the edges of the streamer. The abundances along the edges of the streamer ('legs') resemble elemental abundances measured in the slow solar wind, supporting the identification of streamers as the source of that wind component.

1. Introduction

Coronal streamers have been studied for many years, yet relatively little is known about them beyond their general morphology and average density. Even these are subject to controversy based on differing assumptions about 3D structure and

Solar Physics **175:** 645–665, 1997.

projection effects. Streamers overlie active regions and, at solar minimum, a large, stable streamer straddles the equatorial magnetic neutral line. The magnetic field confines the streamer material and a current sheet lies along its axis (e.g., Koutchmy and Livshits, 1992). The slow speed solar wind, which shows large excursions in speed and elemental abundances, is believed to originate in streamers (e.g., Gosling *et al.*, 1981; Feldman *et al.*, 1981), though the relative contributions of streamers and coronal mass ejections are still uncertain (e.g., Gosling, 1996). Thus streamers are the key to understanding the physical causes of variations in the wind speed, density and composition.

Elemental abundances vary by about a factor of four from place to place in the solar transition region and corona. The pattern is generally believed to follow the First Ionization Potential (FIP) effect seen in the solar wind and cosmic rays (e.g., Geiss *et al.*, 1995; DuVernois and Thayer, 1996). Elements whose neutral atoms have ionization potential larger than about 10 eV are less abundant that those whose neutral atoms are more easily ionized. This is strikingly seen in measurements of the relative intensities of Mg VI and Ne VI lines in various solar features (e.g., Widing and Feldman, 1992). While several theoretical explanations have been proposed (e.g., Fontenla, Avrett, and Loeser, 1993; Peter 1996; Wang 1996), no consensus has emerged. It is especially important to determine observationally whether low-FIP elements are enhanced relative to H, or high-FIP elements are depleted, as the abundances are nearly all always determined relative to O or Si, rather than to hydrogen. It is also important to measure the helium abundance, because helium is variable and strongly depleted in the solar wind, and no measurements of its abundance in the corona are available. The coincidence of low-He regions in the solar wind with field reversals thought to be associated with streamers was reported by Gosling *et al.* (1981).

Ultraviolet emission lines can add enormously to our empirical knowledge of streamers. Their profiles reveal ion and electron temperatures. Line intensities provide additional information about the composition, density and ionization state of the streamer plasma. The Doppler dimming technique (Kohl and Withbroe, 1982) gives the plasma flow speed.

Here we present a list of emission lines detected in spectra of a quiescent equatorial streamer and an active region streamer obtained with the Ultraviolet Coronagraph Spectrometer (UVCS) aboard the SOHO spacecraft, and we discuss the implications of the line intensities for the ionization state and elemental abundances. Over 30 spectral lines are detected, over a brightness range of 10 000. They include resonance, intercombination and forbidden lines of ions ranging from H I to Si XII and Fe XIII]. Section 2 describes the streamers observed and the instrumental configuration, Section 3 describes the data reduction and presents the emission lines, along with the reasons for some of the identifications. In Section 4, we analyze the relative intensities to determine the ionization state of the streamer plasma and the abundances of low, intermediate and high-FIP elements, with emphasis on the absolute abundances.

2. Observations

The Ultraviolet Coronagraph Spectrometer (UVCS) aboard the SOHO spacecraft observes the solar corona between about 1.4 and 11 $R_\odot$. It is described by Kohl *et al.* (1996). It has two ultraviolet spectral channels. The OVI channel covers the range 945–1123 Å (473–561 Å in second order), and the LYA channel covers 1160–1350 Å. The OVI channel includes a 'redundant' mirror designed for observation of the Lα line with the OVI detector. It allows observations in the range 1160–1270 Å (580–635 Å in second order).

Figure 1 shows images of the corona obtained on 26 July 1996 in the light of Lα and O VI $\lambda\lambda$1032, 1037. They were obtained by augmenting the daily UVCS synoptic scans with additional spectra. The images shown were obtained by interpolating among the heights actually observed (1.4, 1.5, 1.6, 1.7, 1.85, 2.0, 2.25, 2.5 3.0, 3.5, and 4.0 $R_\odot$ at the equator) to show the overall streamer morphology. The east limb clearly shows a complex of streamers arising from several active regions. The west limb shows a much simpler quiescent equatorial streamer, centrally bright in Lα and edge-brightened in O VI.

We have performed long integrations at a variety of grating positions on the two streamers in order to detect faint emission lines. The first example is the quiescent streamer on the west limb, typical of the equatorial streamers seen at solar minimum. The structure was stable over several months in the daily UVCS synoptic scans in Lα and O VI. Such streamers generally lie above the equatorial magnetic neutral line (e.g., Koutchmy and Livshits, 1992). As seen in Figure 1, it was bright in O VI at its edges, while Lα was centrally peaked (cf., Noci *et al.*, 1997). We observed it for 3 hours at a height of 1.5 $R_\odot$ on 25 July 1996 at PA = −90° E of N with OVI channel grating positions of 50 000, 140 000, and 225 000 to cover most of the UVCS spectral range. Slit widths were 100μ (28″) in the OVI channel and 50μ (14″) in the LYA channel. Wide slits provide larger count rates, but degrade spectral resolution. The slit widths chosen are a compromise between spectral resolution and the the ability to detect faint lines. For the LYA channel the spectral resolution was about 0.3 Å, while for O VI it is about 0.4 Å. The memory size of the image processor buffers limits the number of pixel bins per image, so the exposures were binned by 2 pixels in the spectral direction and 10 in the spatial direction (70″). Thus we sacrificed spatial resolution for coverage of the entire detector in order to avoid missing unexpected spectral lines. In order to improve the count levels of the faintest lines, we further binned the data along the slit during the ground processing. Figure 2 shows the brightness in O VI λ1032 and Lα along the slit. We extracted the region from 188″ to 608″ south of PA = 90° as typical of the O VI-bright edge and the region from 235″ N to 53″ S as representative of the O VI-faint center of the streamer.

The second streamer was an active region streamer complex observed on 23 and 24 July 1996 at a height of 1.7 $R_\odot$ and PA = 9°. Similar slit widths and detector masks were used, and again we covered a wide range of grating angles, extending

Figure 1. Images of the corona in the light of Lα (a) and O VI (b) on 26 July 1996. The west limb shows a quiescent streamer at the equator, while the east limb shows a complex of active region streamers. The radius of the central circle without data is 1.5 solar radii.

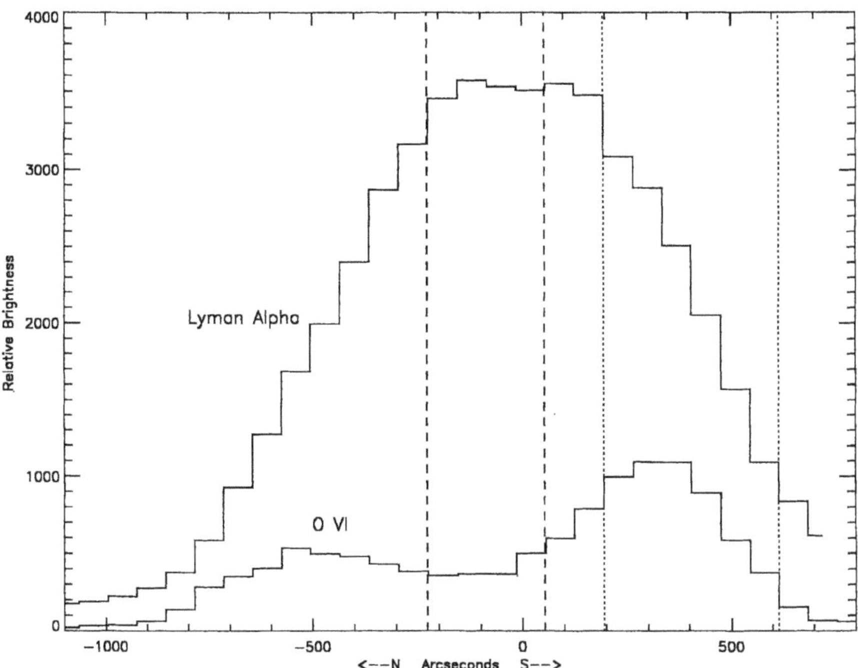

Figure 2. Relative brightness along the spectrograph slit in the Lα and O VI lines for the west limb observation. Dotted and dashed lines show the regions extracted for the 'Leg' and 'Center' spectra, respectively.

to the extremes of the grating range to pick up the [Fe XII] λ1349 line, the [Si VIII], [Si IX], Lδ blend at 950 Å, and the O V λ630 line. A total of 14 grating positions were observed for 75 min each over the two days to completely cover the redundant spectral range and to remove any ambiguities as to direct vs redundant wavelength identifications. The position of the grating mechanism slightly affects the telescope pointing, so we introduced compensating shifts to keep the pointing at 1.7 $R_\odot$. We extracted the spectrum of the band 1000″ wide centered on PA = 95°.

3. Data Reduction

The UVCS Display and Analysis Software was used to flatfield, to remove image distortion, to wavelength calibrate and to radiometrically calibrate the data. The radiometric calibration was obtained from laboratory measurements at SAO. We have verified this calibration by comparing spectra of several stars with archival IUE spectra and, at wavelengths below 1160 Å, the Voyager spectrum of τ Tau. At long wavelengths in the LYA channel, we applied a correction based on comparison of stellar spectra (in particular ρ Leo) with IUE archival spectra. We believe the calibration uncertainty to be less than 15% absolute for first-order lines. The second-order response of the OVI channel has yet to be fully cross-calibrated with the other SOHO instruments, and larger uncertainties may be present. Based on laboratory measurements of the grating and detector response, we take the efficiency to be 23% that at 1032 Å. We expect a decline in sensitivity toward short wavelengths in the second order response due to the SiC telescope mirror and the Ir coating on the diffraction grating. The only checks we have are the doublet ratios Si XII $I(499)/I(521)$ and Fe XIII] $I(487)/I(510)$. The former ratio is intrinsically 2:1, and the Fe XIII] ratio is 1:1.66 according to the transition probabilities of Huang (1985) and Ellis and Martinson (1984), while the ratios of count rates are 1.69:1 and 1:2.25. (We assume that the 974 Å feature is entirely due to Fe XIII]. The contribution from Ne VII $2s2p^1P - 2p^2\,{}^1D$ should be small.) Thus we apply tentative correction factors of 1.0, 1.08, 1.18, 1.30, and 1.33 to the intensities of the lines at 521, 510, 499, 490, and 487, respectively. In comparing the lines of Mg X and Si XII, we must span a larger wavelength range in the second order, and we must include the 'redundant' mirror efficiency. The redundant mirror reflectivity is known for first order lines, as we can compare with intensities measured in the LYA channel. For now we must assume constant efficiency between 521 and 609 Å and a redundant mirror efficiency equal to that at first-order wavelengths, 29%.

There is some error introduced in combining spectra taken a day apart for the active region streamers, and in combining the OVI and LYA channel data when the slits do not exactly coincide. We have scaled all the OVI channel data to the OVI λ1032 line at a grating position of 160 000, and we have scaled the LYA channel data to the intensity of Lα obtained in the OVI channel. This has been checked with the Lα, [Fe XII] and [S X] lines when they are seen in both channels. They agree to within roughly 10% for the 25 July data, but there are larger discrepancies between the 23 and 24 July spectra, in the sense that the higher ionization lines increase more from one day to the next than does O VI. This is evident in the difference between the [Fe XII] and [S X] intensities measured in the LYA channel and the redundant channel. Also, the Mg X doublet ratio appears to be almost 3:1, but the λ609 line was measured on 24 July, and the λ625 line on 23 July. Comparison of the ratios to the Si XII lines suggests that the ratio may be greater than 2:1, indicating a scattering component. For the abundance analysis, we will take the λ625 intensity as being less vulnerable to contamination by a scattering component.

Figure 3 shows some sample spectra from the OVI and LYA channels. It is difficult to show the complete OVI spectrum in a simple way because of the overlap of direct and redundant spectral ranges. The Mg X and (heavily vignetted) Lα lines in Figure 3(c) are examples of lines from the 'redundant' mirror. For each grating position we measured line intensities by direct summation, using eye estimates of the neighboring background. Measurements which are problematic due to direct-redundant blends or to a wire of the ion repulsion grid (sharp dips in Figures 3(b) and 3(c)) partially occulting the line are rejected, and the other measurements are averaged. Tables I–III present the lines and their intensities. We

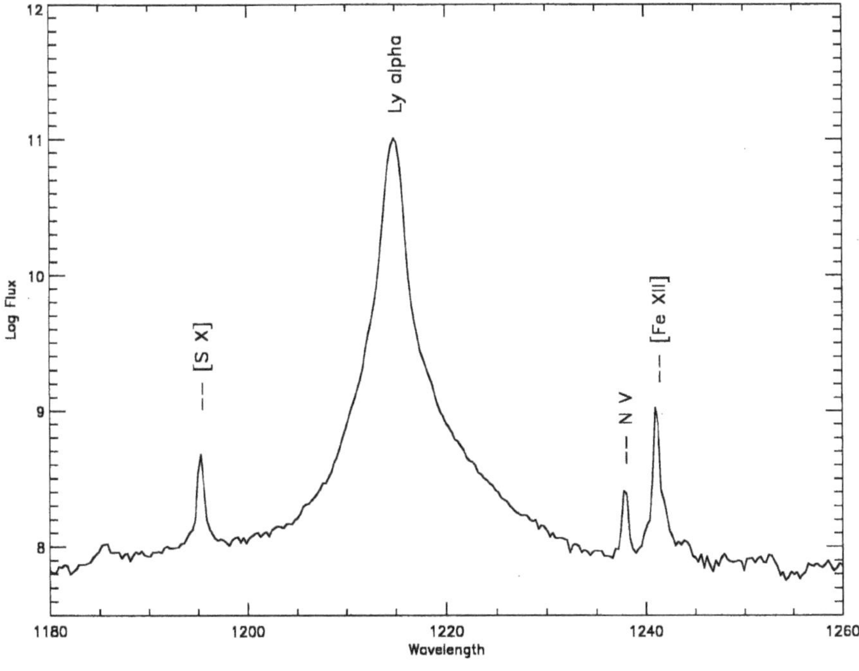

Figure 3a.

exclude known artifacts due to grating ghosts and chromospheric lines seen at some positions when the telescope internal occulter did not completely reject stray light.

3.1. Line Identifications

We have identified coronal emission lines based on published solar UV spectra and on predictions based upon atomic rates. In particular, we have relied on the spectra of Vernazza and Reeves (1978), Dere (1978), and Feldman and Doschek (1991). Uri Feldman has kindly provided a preprint of the line list from SUMER that contains several important lines (Feldman *et al.*, 1997). For the stronger lines, the identifications are obvious – Lα, Lβ, Lγ, the O VI, Mg X, and Si XII doublets, and the lines of [S X], [Fe XII], Al XI, and Ca X. Some of the weaker features are also straightforward due to their doublet structure – [Ar XII] $\lambda\lambda$1018, 1054 (Bhatia, Seeley, and Feldman 1989) and Fe XIII] $\lambda\lambda$ 487, 510 (Träbert, Hutton, and Martinson, 1987; Jupén, Isler, and Träbert 1993). We identify the feature at 950 Å as a blend of [Si IX] 950.15 $2s2p^2\,{}^3P_1 -{}^1S_0$, [Si VIII] $2s2p^3\,{}^4S_{3/2} - {}^2P_{1/2}$ at 949.22, and Lδ (Feldman *et al.*, 1997). Feldman *et al.* also identify the line at 1028 Å as [Fe X] and the line at 1277 Å as [Ni XIII].

Some other lines are problematic. We identify the feature at 1071 Å as the S XI XII] $2s^22p\,{}^2P_{3/2} - 2s2p^2\,{}^4P_{3/2}$ transition in second order, though Feldman *et al.*

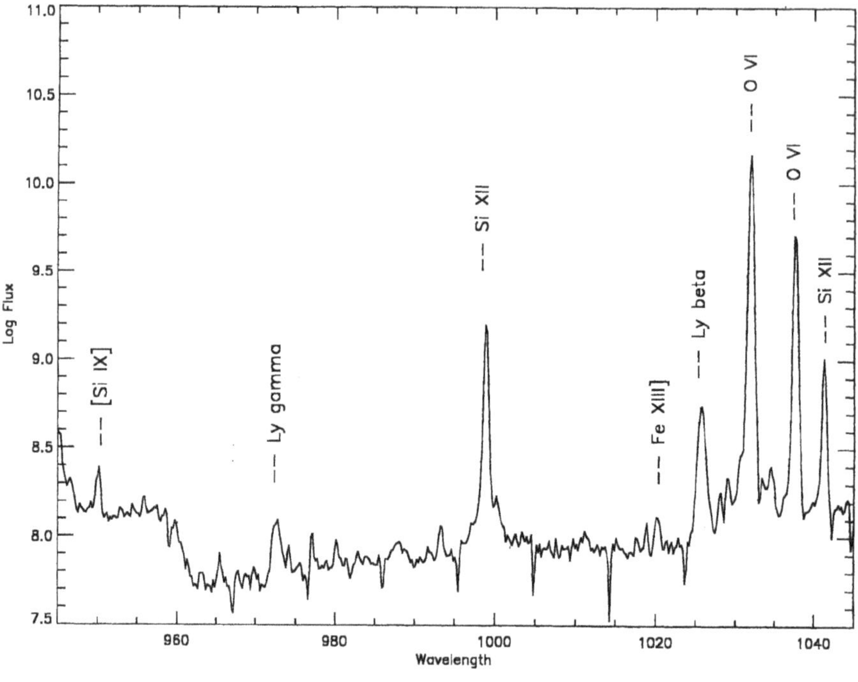

Figure 3b.

(1997) place this line at 538.72 (1078 Å second order). The 1078 Å feature in our spectrum is dominated by a grating ghost, but the unidentified line at 539.07 Å in the flare spectrum of Dere (1978) may also contribute. However, there are three other lines of the S XII] multiplet with similar intensities according to the collision strengths and A values of Zhang and Sampson (1994a). The energy levels of Kelly (1982) imply that if the 3/2–3/2 line lies at 539.07 Å, the brightest S XII] line will be at 553.39 Å, but we see nothing at 1107 Å. Placing the 3/2–3/2 line at 535.32 Å puts the remaining lines of the S XII] multiplet at 549.44, 516.52, and 500.01 Å (1/2–3/2, 5/2–3/2, and 3/2–1/2, respectively). These are buried beneath the Al XI, O VI, and Si XII lines. Therefore, we take 1071 to be the second order S XII] line. We note, however, that the 1071 Å feature is somewhat brighter than expected, and so this identification is still suspect.

There are several other unidentified lines. There are features at λλ 987.58, 1011.08, 1033.03, 1052.96, and 1079.35 which are about 10^{-4} as strong as Lα. We believe them to be grating ghosts of Lα, based on line width and brightness changes observed in the Coronal Mass Ejection of 23 December 1996. Some of these features, especially the one at 1079 Å, may mask real coronal features. We also see a number of other features near the detection limit in some spectra at 955.6, 965.8, 991.8, 1017.6, and 1064.3 Å. These may include the unidentified

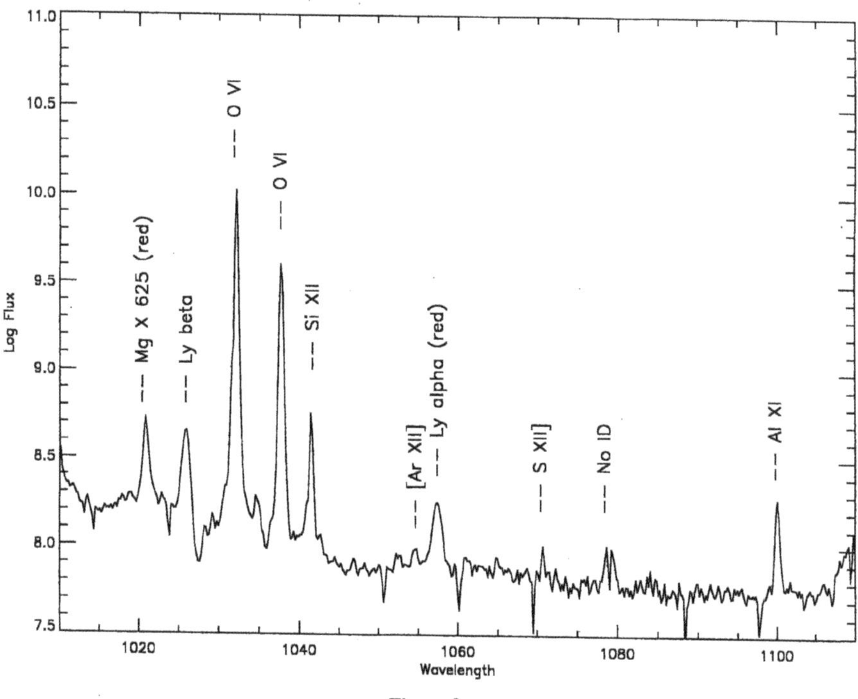

Figure 3c.

Figure 3. Sample UVCS spectra of the active region streamer complex. Units are photons cm^{-2} s^{-1} sr^{-1}, but no correction has be made for second order sensitivity or redundant mirror efficiency. (a) LYA channel spectrum. (b) OVI channel spectrum at grating position = 235 000. (c) OVI channel spectrum at grating position = 70 000.

high temperature lines at 955.89 and 965.19 in the SUMER list and the S XIII] line at 491.4 Å. We are certain that these are not detector artifacts, as they appear at different grating positions. We believe that they are not ghosts created by the diffraction grating. We have identified features at λλ 1029.11, 1034.68 in the OVI channel and 1185.94, 1245.15 in the LYA channel as ghosts (the latter pair having been seen in the laboratory). None of the other unidentified lines has the expected partner at an equal distance on the other side of O VI λ1032 expected for grating ghosts.

It is also worthwhile to note some lines that are not detected. The He II λ1085 line should be the strongest coronal He line in our wavelength range, and it is not seen. The only neon line would be Ne IX λ1248, though Feldman *et al.*, report a Ne VII line at 973.35. The former is a higher ionization state than those we observe, while the latter is lower than most we detect. The [Ni XIV] line detected by SUMER at 1034.09 Å is too badly blended with O VI and its ghost to be visible in our spectra. The [S X] line at 1212.93 is probably present, but it is too much dominated by Lα to be usefully measured.

Table I
OVI channel direct lines: 10^7 photons cm^{-2} s^{-1} sr^{-1}

λ_{meas}	λ_{ID}	Center	Leg	Active	Identification
949.97	950.15			7.1	[Si VIII], [Si IX], $L\delta$
955.80				1.1	No ID
972.56	972.54	22.7	10.7	7.0	Ly γ
974.06	487.06	–	9.2	9.3	[Fe XIII] $3s^2 3p^2\ ^3P_2 - 3s3p^3\ ^5S_1$
981.10	490.20	–	–	9.7	[Ar XIII] $2p^2\ ^3P_2 - 2s2p^3\ ^5S_2$
982.60	491.44	–	–	–	[S XIII] $2s^2\ ^1S_0 - 2s2p\ ^3P_1$
991.71		2.6	4.2	-	No ID
993.26	496.25	–	11.8	13.8	[Ar XIV] $2s^2 2p\ ^2P_{3/2} - 2s2p^2\ ^4P_{1/2}$
998.77	499.37	151	155	485	Si XII $2s\ ^2S_{1/2} - 2p\ ^2P_{3/2}$
1018.88	1018.60	–	1.9	1.3	[Ar XII] $2p^3\ ^4S_{3/2} - 2p^3\ ^2D_{5/2}$
1020.04	510.05	7.4	7.4	11.1	[Fe XIII] $3s^2 3p^2\ ^3P_2 - 3s3p^3\ ^5S_2$
1025.73	1025.72	114	59.4	50.1	$L\beta$
1028.10	1028.04	4.2	6.7	1.8	Fe X $3d^4 D_{7/2} - 3d^4 F_{7/2}$
1031.96	1031.91	373	819	889	O VI $2s\ ^2S_{1/2} - 2p\ ^2P_{3/2}$
1037.50	1037.61	139	286	310	O VI $2s\ ^2S_{1/2} - 2p\ ^2P_{1/2}$
1041.26	520.66	84.0	99.8	242.	Si XII $2s\ ^2S_{1/2} - 2p\ ^2P_{1/2}$
1054.87	1054.90	–	–	1.5	[Ar XII] $2p^3\ ^4S_{3/2} - 2p^3\ ^2D_{3/2}$
1070.63	535.32	13.6	16.6	10.4	[S XII] $2s^2 2p\ ^2P_{3/2} - 2s2p^2\ ^4P_{3/2}$
1100.14	550.01	36.3	24.9	50.0	Al XI $2s\ ^2S_{1/2} - 2p\ ^2P_{3/2}$
1115.52	557.74	–	20.1	21.8	Ca X $3s\ ^2S_{1/2} - 3p\ ^2P_{3/2}$

Table II
OVI redundant channel lines
Intensities in 10^7 photons cm^{-2} s^{-1} sr^{-1}

λ_{meas}	λ_{ID}	Center	Leg	Active	Identification
1196.19	1196.25	22.9	44.9	24.2	[S X] $2p^3\ ^4S_{3/2} - 2p^3\ ^2D_{5/2}$
1215.67	1215.67			18100	$L\alpha$
1219.71	609.76			1530	Mg X $2s\ ^2S_{1/2} - 2p\ ^2P_{3/2}$
1238.57	1238.82			13.0	N V $2s\ ^2S_{1/2} - 2p\ ^2P_{3/2}$
1241.80	1242.03	88.2		47.5	[Fe XII] $3p^3\ ^4S_{3/2} - 3p^3\ ^2P_{3/2}$
1249.90	624.95	257	205	520	Mg X $2s\ ^2S_{1/2} - 2p\ ^2P_{1/2}$
1260.00	629.73			226	O V $2s^2\ ^1S_0 - 2s2p\ ^3P_1$

4. Atomic Data

4.1. Ionization and Recombination

The determination of abundances is closely tied to the determination of the ionization state of the streamer, because we generally observe only one or two ions of each element. At the relatively high density and low speed of a streamer, coronal

Table III
Ly α Channel Lines
Intensities in 10^7 photons cm^{-2} s^{-1} sr^{-1}

λ_{meas}	λ_{ID}	Center	Leg	Active	Identification
1196.25	1196.25	20.4	38.2	28.1	[S X] $2p^3\ {}^4S_{3/2} - 2p^3\ {}^2D_{5/2}$
1215.67	1215.67	44640.	28550	18680	Lα
1238.56	1238.82	22.2	14.9	12.6	N V $2s\ {}^2S_{1/2} - 2p\ {}^2P_{3/2}$
1241.69	1242.03	78.8	78.5	60.1	[Fe XII] $3p^3\ {}^4S_{3/2} - 3p^3\ {}^2P_{3/2}$
1277.17	1277.23	–	–	2.27	[Ni XIII] $3p^4\ {}^3P_1 -{}^1S_0$
1349.89	1349.38	–	–	28.8	[Fe XII] $3p^3\ {}^4S_{3/2} - 3p^3\ {}^2P_{1/2}$

ionization equilibrium should be an excellent approximation, barring an important role for impulsive heating by, for instance, reconnection or shocks (cf., Raymond, 1990; Feldman *et al.*, 1992) or non-Maxwellian electron distributions (Scudder, 1994; Anderson, Raymond, and van Ballegooijen, 1996). At densities of 10^8 cm^{-3} (Koutchmy and Livshits, 1992) we can ignore the density sensitivity of dielectronic recombination and the population of metastable levels for the ionization stages we observe.

For hydrogen we compute the neutral fraction from the ionization rate of Scholz and Walters (1991) and the recombination rate of Hummer (1994). For helium, we use the computations of the Raymond and Smith (1977; updated version) spectral code, which agree extremely well with the He^+ fraction computed by Arnaud and Rothenflug (1985) at the temperatures tabulated in that paper. For other elements we adopt Arnaud and Rothenflug (1985) except for iron, where we use Arnaud and Raymond (1992). An exception is calcium. The concentration of Ca X depends strongly upon the Ca XI dielectronic recombination rate, which has been modified since the work of Arnaud and Rothenflug (e.g., Romanik, 1996), so for this ion we use the current version of the Raymond and Smith (1977) code. It agrees with Romanik's recombination rate to 10%. For most other ions, ionization state calculations agree with one another to about 0.1 dex in $\log T$. However, the recent calculation by Romanik (1996) shows some systematic differences, such as a 50% higher peak concentration for Fe XII. The ionic concentrations are likely to be reliable to better than 20% (H, He, N, O), 35% (Mg, Al, Si, S, Ar) or 50% (Ca, Fe, Ni), but there is no good way to verify these uncertainty estimates. An attempt to sort out the errors in ionization state, excitation rates, and instrumental calibration for SERTS observations of an active region (Brickhouse, Raymond, and Smith, 1995) could not reliably discriminate between the Arnaud and Rothenflug (1985) and Arnaud and Raymond (1992) calculations for iron.

4.2. Excitation Rates

For the small flow velocities which we expect at low altitudes in streamers, Doppler dimming should not be important. Velocity estimates based on Doppler dimming show that streamer material reaches 100 km s^{-1} only at about 5 $R_\odot$ (Kohl *et al.*, 1997), so speeds above 30 km s^{-1} at 1.5 $R_\odot$ are unlikely. Thus the photoexcitation rates are given by the absorption cross section (proportional to the oscillator strength) and the chromospheric illumination. We adopt the oscillator strengths of Weise, Smith and Glennon (1966). The disk intensities of Lα, Lβ, Lγ, and O VI λ1032 are taken to be 524, 4.13, 0.867, and 1.94 $\times$ 10^{13} photons cm^{-2} s^{-1} sr^{-1} based on measurements by UVCS on 4 December 1996. These are similar to the average quiet-Sun fluxes of Vernazza and Reeves (1978), lying 5–30% lower. We will return below to the question of variations in the relative brightness of these lines in active regions. For the other ions, we expect photoexcitation to be a minor correction.

Collisional excitation rates were obtained from various sources. For hydrogen we use Scholz and Waters (1991) for $n = 2$ and Callaway and McDowell (1983) for $n = 3$. We scale with oscillator strength and excitation energy for $n = 4$ and above. We include cascades from higher levels and recombination to the excited states (Hummer, 1994), which account for about 20% of the Lyman line collisional rates. The totals should be good to 10% (Lα) to 20% (Lγ).

For He II λ 1085 (the Balmer γ line of He II) we use RMATRIX calculations by McLaughlin (1997) for excitation to $n = 3$, scaled to $n = 5$ with the oscillator strength and threshold energy. There is a great need for more accurate rates, but at present we have only an upper limit on the line intensity.

For Li-like ions we adopt the collision strengths of Merts *et al.* (1980) for N V or Zhang, Sampson, and Fontes (1990) for higher ions. These are probably accurate to about 20%. Rates for [Fe XII] were obtained from Keenan, Tayal, and Henry (1990) with the resonance contribution reduced in accord with recent results of A. Binello (private communication) and cascades from higher levels computed as in Withbroe and Raymond (1983). The excitation rates of [S X] and [Ar XII] were computed from the direct collision strengths and cascades from upper levels of Bhatia and Mason (1980) with estimates of the resonance contributions to the excitation rates from Laming (1996). Other important rates include O V (Berrington *et al.*, 1985) and Fe XIII] (Fawcett and Mason, 1989). For [Si IX], we use the collision strengths of Zhang and Sampson (1996), and for [Fe X] the collision strengths of Bhatia and Doschek (1995). For S XII], we used the collision strengths of Zhang and Sampson (1994a) and estimated the contribution of cascades from excitation to $n = 3$ levels from Zhang and Sampson (1994b). For [Ni XIII], we scaled the [Fe XI] collision strengths of Bhatia and Doschek (1996) with threshold energy.

Table IV
Radiative and collisional contributions to H I and O VI

Line	Center		Leg		Active	
	Rad	Col	Rad	Col	Rad	Col
Lα	0.989	0.011	0.993	0.007	0.988	0.012
Lβ	0.43	0.57	0.53	0.47	0.41	0.59
Lγ	0.19	0.81	0.23	0.77	0.38	0.62
1031.91	0.48	0.52	0.60	0.40	0.61	0.39
1037.61	0.32	0.68	0.43	0.57	0.43	0.57

5. Abundances

5.1. Collisional and Radiative Contributions

Most of the lines we observe are formed by collisional excitation, but the lines of H I and O VI have major contributions from resonantly scattered chromospheric emission lines. In order to derive abundances we must separate the contributions.

From the excitation rates described above, we predict the Lyman line decrements for the scattered and collisional components. The Lα : Lβ : Lγ ratios are 13020 : 14.3 : 1.0 for the radiative component, assuming that the integrals of the chromospheric emission line profiles times the coronal absorption line profiles of the Lyman lines are the same. Detailed calculations with observed chromospheric profiles and a coronal temperature of 1.6×10^6 K show that the the Lα/Lβ ratio actually varies from 3% higher to 3% lower as the outflow speed increases from 0 to 100 km s^{-1}, but in the absence of a reliable estimate of the speed we assume the ratio above. The collisional components of the Lyman lines are in the ratios 21.8 : 2.88 : 1.0 for $\log T = 6.2$. The radiative ratios are independent of temperature, and the collisional ratios depend very weakly on temperature. The striking difference between these sets of ratios results from the enormous intensity of the chromospheric Lα line. By comparing the observed Lα to Lβ ratios with these predictions, we find the relative collisional and radiative contributions, then use those to find the contributions to Lγ. The resulting fractions are shown in Table IV. We note, however, that a similar analysis based on the ratios of Lγ and Lβ gives a somewhat different breakdown. This could be due to calibration uncertainties at the shorter wavelengths, to an error in the collisional excitation rates, or to an incorrect ratio of chromospheric intensities. In particular, Vernazza and Reeves (1978) show the Lα flux increasing by a factor of 3.4 between their average quiet Sun spectrum and active region spectrum, while the Lβ, Lγ and O VI lines increase by factors of 8.1–8.6. Thus a significant active region contribution to the chromospheric fluxes (of order 5–10% of the hemisphere below the streamer) will affect the relative radiative excitation ratios and require a larger radiative fraction of Lβ.

Similarly, for the O VI lines the collisional contribution has a 2:1 ratio, while the radiative contribution is 4:1 (a factor of 2 from the oscillator strengths times a factor of 2 for the chromospheric intensities). At outflow speeds near 200 km s^{-1}, pumping by the C II line at 1037.02 Å modifies the radiative ratio (Noci, Kohl, and Withbroe, 1987 ; Spadaro and Ventura, 1993), but the speeds low in a streamer should be much too small for that to be important. Table IV shows the contributions for the O VI lines derived from the intensities in Table I.

5.2. Oxygen Abundance Relative to Hydrogen

Now that the collisional and radiative contributions are determined, we can obtain two independent estimates of the O:H ratio. Both will depend on the ratio of ion concentrations of O VI and H I, $C_{\mathrm{O\,VI}}/C_{\mathrm{H\,I}}$, and thus on the assumed temperature. Fortunately, the ratio of concentrations is constant to within ±16% for $\log T = 6.0$–6.3, because of the interplay between the O^{5+} ionization rate and the O^{6+} dielectronic recombination rate. We adopt the value at $\log T = 6.2$, 15 200. The abundance derived from the radiative contributions is

$$\frac{N_{\mathrm{O}}}{N_{\mathrm{H}}} = \frac{I_{\mathrm{rad}}(1032)}{I_{\mathrm{rad}}(\mathrm{L}\beta)} \frac{C_{\mathrm{H\,I}}}{C_{\mathrm{O\,VI}}} \frac{B_{\mathrm{L}\beta}}{B_{\mathrm{O\,VI}}} \frac{f_{\mathrm{L}\beta}}{f_{1032}} \frac{I_{\mathrm{disk}}(\mathrm{L}\beta)}{I_{\mathrm{disk}}(\mathrm{O\,VI})} \frac{\delta\nu_{\mathrm{O\,VI}}}{\delta\nu_{\mathrm{H\,I}}} . \tag{1}$$

The branching ratios, B, for absorbed photons to be reemitted in the same line are 0.88 for Lβ and 1.0 for O VI. The oscillator strengths, f, are 0.0791 and 0.131 for Lβ and O VI λ1032, respectively (Weise, Smith, and Glennon, 1966), and the ratio of measured disk intensities, I, is 2.13. The scattering cross section at line center is inversely proportional to the line width. If the radial components of the thermal velocities of O and H are in thermal equilibrium, the ratio of widths $\delta\nu$ is $\frac{1}{4}$. While this is not true in coronal holes, the line widths at small heights in streamers are consistent with thermal equilibrium between O and H, and we assume it here. A more detailed calculation would integrate the chromospheric emission line profiles times absorption profiles which include turbulent velocities and outflow, but lacking the outflow velocity information we take the ratio of line center opacities. Note that we have chosen to compare O VI to Lβ because the nearby wavelengths minimize calibration uncertainties both in the coronal measurements and in the disk intensities. The resulting oxygen abundances are shown in Table V.

We obtain a similar expression for the abundance from the collisional contributions:

$$\frac{N_{\mathrm{O}}}{N_{\mathrm{H}}} = \frac{I_{\mathrm{coll}}(1032)}{I_{\mathrm{coll}}(\mathrm{L}\beta)} \frac{C_{\mathrm{H\,I}}}{C_{\mathrm{O\,VI}}} \frac{B_{\mathrm{L}\beta}}{B_{\mathrm{O\,VI}}} \frac{q_{\mathrm{L}\beta}}{q_{1032}} \tag{2}$$

The variation of the excitation rates q with temperature is small above 10^6 K, and it is nearly the same for the two lines. The abundances derived from the collisional parts of the intensities of the three spectra are also shown in Table V.

Table V
Radiative and collisional contributions to H I and O VI

Line	Center		Leg		Active	
	Rad	Col	Rad	Col	Rad	Col
$L\beta$	49.0	65.0	31.5	27.9	0 20.5	29.5
O VI 1032	180.0	193.0	494.0	325.0	539.0	347.0
O/H $\times 10^4$	0.68	0.52	2.9	2.0	4.9	2.0

A notable feature of Table V is the apparent systematic difference in the abundances derived for the collisional and radiative contributions. This difference could actually be real, in that the radiative contribution varies as the density, while the collisional contribution goes as the square of the density. Thus the collisional contribution weights denser regions along the line of sight more heavily, and there could be an anti-correlation between density and abundance, just as there is a difference in the strength of the FIP effect between fast and slow solar wind (Geiss *et al.*, 1995). However, a modest error in our separation of the collisional and radiative components of $L\beta$ could cause the apparent difference between radiative and collisional abundances. As noted above, an active region contribution to the chromospheric fluxes would have caused an underestimate of the radiative portion of $L\beta$, and the radiative-collisional discrepancy would be greatest for the active region streamer. The $L\beta$ breakdown based on the $L\gamma$ to $L\beta$ ratios would actually imply slightly larger abundances for the collisional components than the radiative ones.

While the breakdown of $L\beta$ into radiative and collisional components is somewhat uncertain, the average of the abundances derived for the two components is robust. The striking result is the overall difference between the streamer center and legs. The active region streamer abundance is quite similar to the quiescent streamer legs. The simplest interpretation is that this difference is real. The only alternative is that the ionization state is much different, as would occur if the temperature in the streamer center is above 3×10^6 K. However, such a high temperature would imply very bright Si XII emission and enhancements of high temperature lines such as S XIII] λ491, which are ruled out by the spectra.

5.3. He and Ne Upper Limits

The brightest lines of helium and neon in our spectral range should be He II λ1085 (Balmer γ of He^+) and the $1s2s\ ^3S_1 - 1s2p\ ^3P_2$ transition of Ne IX at 1248 Å. Conservative upper limits on the fluxes are 1×10^7 photons cm^{-2} s^{-1} sr^{-1}. Comparison of the He IX upper limit to the collisional portion of $L\beta$ implies a nearly temperature-independent upper limit He/H ≤ 0.048, or 50% of the photospheric value for the case of the quiescent streamer center. The other spectra give less restrictive upper limits. Though none of the upper limits are very restrictive, they do rule out models in which rapid diffusion of helium in the transition region and low helium velocity in the solar wind lead to a build-up of helium abundance in the corona (e.g., Hansteen, Leer, and Holzer, 1996). The Ne upper limit is much more temperature-sensitive due to the very high excitation threshold for the λ1248 line. At $T_e = 10^{6.2}$, the upper limit ranges from 1.9 to 3.3 times the photospheric neon abundance, but if the temperature is taken to be $10^{6.3}$ the range of upper limits becomes 0.57 to 1.0.

5.4. Abundances of Other Elements

Finally, we turn to the other elements. The questions of abundance and ionization state are closely intertwined, so we compare predicted line ratios based on photospheric abundances at temperatures $\log T = 6.1, 6.2$, and 6.3 with the observations. We adopt the Feldman (1992) photospheric abundance set shown in Table VII, then use the ratios of observed to predicted intensities to infer abundances for the streamers. Table VI presents the ratios of line intensities to the collisional component of Lβ. Note that we assume that the oxygen abundances in Table V should be the same for collisional and radiative components, and that incorrect division of Lβ between collisional and radiative components causes the discrepancies. We have therefore adjusted the collisional and radiative components to force the abundances to be equal. For the comparison in Table VI, this amounts to reducing $I(\mathrm{L}\beta_c)$ by up to 30%. For Table VI we estimate that the [Si IX] contribution to the 950 Å blend is 3×10^7 photons cm^{-2} s^{-1} sr^{-1}, based on the expected Lδ/Lγ ratio and the assumption that [Si VIII] is relatively faint.

Both first- and second-order lines show that $\log T = 6.2$ or slightly lower provides the best overall match. In particular, the ratios Fe X: Fe XII: Fe XIII, and Si IX: Si XII are independent of abundances, and they indicate $\log T = 6.2$. The S XII] line suggests the presence of some higher temperature gas, and the O V line definitely requires a small amount of much cooler gas. If the S XIII] and Ar XIII] identifications are correct for the 982 and 983 Å features, a considerable amount of material hotter than $\log T = 6.3$ must be present. While it is important to explore the range of parameters, we will assume a single temperature gas near $\log T = 6.2$ (perhaps slightly less in the quiescent streamer center, slightly more in the active region streamers) and consider only the lines which are strong in that temperature range. We note that the widths of the Lyman lines imply ion temperatures close to $\log T = 6.2$ (Noci *et al.*, 1997), and that this temperature is typical of the values derived from white light observations and the assumption of hydrostatic equilibrium (Koutchmy and Livshits, 1992).

Table VII gives estimated abundances for the three spectra, along with abundance estimates for the solar photosphere (Feldman, 1992) and the solar corona (Meyer, 1985). The uncertainties are difficult to quantify. The helium upper limits and the oxygen abundances are quite insensitive to the assumed ionization temperature, and for iron we have three ionization states. Two ionization states of Si

Table VI
Predicted and observed intensities relative to $L\beta_{coll}$

	Predicted			Observed		
	6.1	6.2	6.3	Center	Leg	Active
1st order						
N V$_{1238}$	2.15	1.43	0.62	0.39	0.65	0.65
O VI$_{1032c}$	43.6	42.4	32.6	3.43	14.3	18.0
Si IX$_{950}$	0.45	0.14	0.012	–	–	0.16
S X$_{1196}$	6.27	6.53	2.81	0.38	1.68	1.46
Ar XII$_{1018}$	0.033	0.25	0.53	–	0.083	0.067
Fe X$_{1028}$	0.44	0.11	0.016	0.075	0.29	0.093
Fe XII$_{1240}$	4.22	3.91	0.88	1.40	3.44	3.11
Ni XIII$_{1277}$	0.002	0.020	0.050	–	–	0.12
2nd order						
Si XII$_{499}$	3.36	35.1	56.7	2.68	6.70	25.1
Al XI$_{550}$	2.72	4.47	3.15	0.66	1.09	2.59
S XII$_{535}$	0.007	0.078	0.255	0.24	0.73	0.54
Ca X$_{558}$	2.00	1.56	0.72	–	0.88	1.13
Fe XIII$_{510}$	0.31	0.72	0.39	0.13	0.32	0.58
Mg X$_{625}$	19.5	13.0	7.89	4.56	8.99	26.9

are observed, so we also have relatively high confidence in that abundance in the active region streamer, but only one is observed in the quiescent streamer. The other elements depend upon observations of just one ionization state. We consider Mg, which depends on the second order lines observed with the redundant mirror, Ca, for which we observe only a minor ionization state, and Ni, for which we observe a very weak line of a minor ionization state with poor atomic data, to be the least reliable abundance estimates. The nitrogen abundance is also uncertain, in that a small amount of cool gas, inferred from the O V line, could contribute a significant fraction of the N V flux. An estimate of the uncertainties in the derived abundances is 0.1 dex for He, O, Si and Fe, 0.2 dex for Mg, Al, S and Ar, and 0.3 dex for Ca and Ni. However, these estimates are mere guesses at uncertainties dominated by systematics. Figure 4 shows the abundances relative to Feldman's photospheric compilation.

6. Discussion

The most striking result from Table VII is the severe depletion of all elements compared with photospheric abundances in the center of the quiescent equatorial streamer. The high-FIP elements O, S, and Ar are depleted by an order of magnitude, and low-FIP elements by roughly a factor of 5. The depletion may be even more

Table VII
Elemental abundance estimates

Element	Photosphere	Corona	Center	Leg	Active
H	12.00	12.00	12.00	12.00	12.00
He	10.99	11.00	≤10.7	≤11.0	≤11.1
N	8.00	7.59	≤7.40	≤7.60	≤7.60
O	8.93	8.39	7.80	8.40	8.50
Ne	8.11	7.54	≤8.4	≤8.6	≤8.6
Mg	7.58	7.57	7.1	7.4	7.7
Al	6.47	6.44	5.7	5.9	6.3
Si	7.55	7.59	6.7	7.1	7.5
S	7.21	6.94	6.2	6.6	6.6
Ar	6.65	6.33	–	6.1	6.2
Ca	6.36	6.47	–	6.1	6.2
Fe	7.51	7.57	7.0	7.5	7.4
Ni	6.25	6.33	–	–	6.3

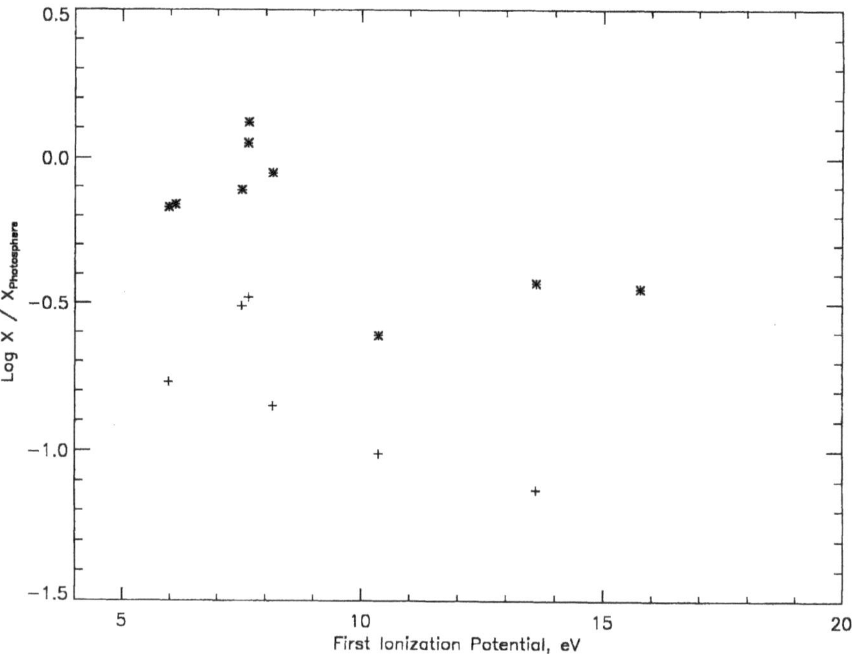

Figure 4. Abundances relative to the photospheric values of Feldman (1992) plotted against First Ionization Potential. Crosses refer to the center of the quiescent streamer and diamonds to active region streamer. The values for the quiescent streamer legs are close to those of the active region streamer.

severe than indicated in the table, because projection effects could account for some of the apparent heavy element emission. Assuming that the structure follows the magnetic neutral line around the solar equator, the UVCS line of sight passes through the streamer edges at least twice. The heavy element emission from the streamer center cannot be entirely leg emission seen in projection, because the line ratios (e.g., Si XII to O VI) differ in the two spectra, but the contribution of foreground and background emission to the streamer spectrum could be significant.

The relative intensities of the heavy element lines in the streamer center and leg spectra are quite similar, the hydrogen lines showing the largest differences. It is likely that the equatorial streamer core is a static region of closed magnetic field, so gravitational settling may be partly responsible for the low abundances. However, the FIP effect still seems to be present, so separation in the upper chromosphere probably operates as well. The simplest picture might be injection of material into both the core and the legs with the FIP abundance pattern. Material flows out through the legs into the slow solar wind in a time shorter than the gravitational settling time (a few days), while material in the streamer core is confined for longer periods. The similarity of the active region streamers to the quiescent streamer legs may indicate that the core region is absent, that material cycles quickly though the active corona, or that the core does not reach the height of our observation.

The second major result is the similarity of the abundances in the leg and active region streamer spectra to the abundances observed in the slow solar wind (e.g., Geiss *et al.*, 1995). This supports the overall association of streamers with the slow wind (Gosling *et al.*, 1981). The variability of slow solar wind abundances might be attributed to differences between streamers, but it could be that shear instabilities mix varying quantities of the highly depleted streamer core material into the expanding plasma at the streamer edge. Recent theoretical models have identified the streamer-coronal hole boundary as the source of much of the solar wind (e.g., Wang, 1994), but the abundances in streamer legs suggest that they produce only slow wind. Material may be ejected episodically, but if so it seems to originate in the streamer legs rather than the core.

The UVCS observations have several important implications for the interpretation of observations at other wavelengths. The drastic depletion of heavy elements in the streamer core implies that the X-ray emissivity of the gas is much lower than is generally assumed, so that the standard spectral analysis would severely underestimate the amount of hot gas, and might yield an incorrect temperature. Variable abundances should be considered in analyzing narrow-band EUV and X-ray images, as well. Heavy elements dominate the radiative cooling rate of coronal gas, so theoretical models will be affected.

At the most basic level, the difference between the spectra of streamer legs and core implies a real physical difference between these regions. The helmet shape cannot be a geometrical projection effect of a thin sheet seen edge-on, as suggested by Koutchmy and Livshits (1992). Finally, in answer to the question of whether high-FIP elements are depleted or low-FIP elements are enhanced in the corona (e.g., Feldman, 1992), we find that depletion of the high-FIP elements dominates. Table VII shows a slight overabundance of Mg in the active region streamer. There have been indications of over-abundance of 'extremely-low-FIP' elements compared with low-FIP elements (e.g., Falconer, Davila, and Thomas, 1997), but Al and Ca are extremely-low-FIP elements, and they appear to follow Fe and Si, which are close to photospheric values. Further studies of the variations among streamers, the second order sensitivity of UVCS, and the ionization state of streamer gas will be needed to determine whether the lowest-FIP elements are enhanced more than Fe and Si.

Important tasks for future UVCS observations will include studies of the density and velocity structure in streamers, the abundance variation from one streamer to another, and the evolution of a single streamer over time.

Acknowledgements

Support for this work is provided by NASA Grant NAG5-3192 to the Smithsonian Astrophysical Observatory, and by ASI. The authors thank the SOHO flight operations team, along with A. van Ballegooijen, U. Feldman, N. Brickhouse, J. Laming, and B. McLaughlin for advice and discussion of atomic rates and wavelengths.

References

Anderson, S. A., Raymond, J. C., and van Ballegooijen, A: 1996, *Astrophys. J.* **457**, 939.
Arnaud, M. and Raymond, J.: 1992, *Astrophys. J.* **398**, 394.
Arnaud, M. and Rothenflug, R: 1985, *Astron. Astrophys. Suppl.* **60**, 425.
Berrington, K. A., Burke, P. G., Dufton, P. L., and Kingston, A. E.: 1985, *ADNDT* **33**, 195.
Bhatia, A. K. and Doschek, G. A.: 1995, *ADNDT* **60**, 97.
Bhatia, A. K. and Doschek, G. A.: 1996, *ADNDT* **64**, 183.
Bhatia, A. K. and Mason, H. E.: 1980, *Monthly Notices Roy. Astron. Soc.* **190**, 925.
Bhatia, A. K., Seeley, J. F., and Feldman, U.: 1989, *ADNDT* **43**, 99.
Brickhouse, N. S., Raymond, J. C., and Smith, B.W.: 1995, *Astrophys. J. Suppl.* **97**, 551.
Callaway, J. and McDowell, M. R. C.: 1983, *Comm. Atmospheric Mol. Phys.* **13**, 19.
Dere, K. P.: 1978, *Astrophys. J.* **221**, 1062.
DuVernois, M. A. and Thayer, M. R.: 1996, *Astrophys. J.* **465**, 982.
Ellis, D. G. and Martinson I.: 1984, *Phys. Scripta* **30**, 255.
Falconer, D. A., Davila, J. M., and Thomas, R. J.: 1997, preprint.
Fawcett, B. C. and Mason, H.E.: 1989, *ADNDT* **43**, 245.
Feldman, U.: 1992, *Phys. Scripta* **46**, 202.
Feldman, U. and Doschek, G. A.: 1991, *Astrophys. J. Suppl.* **75**, 925.
Feldman, U., Behring, W. E., Curdt, W., Schüle, U., Wilhelm, K., and Lemaire, P.: 1997, preprint.
Feldman, U., Laming J. M., Mandelbaum, P., Goldstein, W. H., and Osterheld, A.: 1992, *Astrophys. J. Suppl.* **398**, 692.
Feldman, W. C., Asbridge, J. R., Bame, S. J., Fenimore, E. E., and Gosling, J.T.: 1981, *J. Geophys. Res.* **86**, 5408.
Fontenla, J. M., Avrett, E. H., and Loeser, R.: 1993, *Astrophys. J.* **406**, 319.
Geiss, J., Gloeckler, G., von Steiger, R., Balsiger, H., Fisk, L. A., Galvin, A. B., Ipavich, F. M., Livi, S., McKenzie, J. F., Ogilvie, K. W., and Wilken, B.: 1995, *Science* **268**, 1033.

Gosling, J. L.: 1996, *ARAA* **34**, 35.
Gosling, J. L., Borrini, G., Asbridge, J. R., Bame, S. J., Feldman, W. C., and Hansen, R. T.: 1981, *J. Geophys. Res.* **86**, 5438.
Hansteen, V. H., Leer, E., and Holzer, T. E.: 1996, *Astrophys. J.* **428**, 843.
Huang, K.-N.: 1985, *ADNDT* **32**, 503.
Hummer, D. G.: 1994, *Monthly Notices Roy. Astron. Soc.* **268**, 109.
Jupén, C., Isler, R. C., and Träbert, E.: 1993, *Monthly Notices Roy. Astron. Soc.* **264**, 627.
Keenan, F. P., Tayal, S. S., and Henry, R. J. W.: 1990, *Solar Phys.* **125**, 61.
Kelley, R. L.: 1982, *Atomic and Ionic Spectrum Lines Below 2000 Å, ORNL-5922*, Oak Ridge National Laboratory, Oak Ridge.
Kohl, J. L. and Withbroe, G. L.: 1982, *Astrophys. J.* **256**, 2630.
Kohl, J. L. *et al.*: 1996, *Solar Phys.* **162**, 313.
Kohl, J. L. *et al.*: 1997, *COSPAR Proceedings*, submitted.
Koutchmy, S. and Livshits, M.: 1992, *Space Sci. Rev.* **61**, 393.
Laming, J. M.: 1996, private communication.
McLaughlin, B.: 1996, private communication.
Merts, A. L., Mann, J. B., Robb, W. D., and Magee, N. H., Jr.: 1980, *Los Alamos Informal Report LA-8267.*,
Meyer, J.-P.: 1985, *Astrophys. J. Suppl.* **57**, 173.
Noci, G., Kohl, J. L., and Withbroe, G. L.: 1987, *Astrophys. J.* **315**, 706.
Noci, G. *et al.*: 1997, *COSPAR Proceedings*, submitted.
Peter, H.: 1996, *Astron. Astrophys.* **312**, L37.
Raymond, J. C.: 1990, *Astrophys. J.* **365**, 387.
Raymond, J. C. and Smith, B. W.: 1977, *Astrophys. J. Suppl.* **35**, 419.
Romanik, C. J.: 1996, Ph.D. Thesis, University of California, Berkeley.
Scholz, T. T. and Walters, H.R.J.: 1991, *Astrophys. J. Suppl.* **380**, 302.
Scudder, J. D.: 1994, *Astrophys. J.* **427**, 446.
Spadaro, D. and Ventura, R.: 1993, *Astron. Astrophys.* **276**, 571.
Träbert, E., Hutton, R., and Martinson, I.: 1987, *Monthly Notices Roy. Astron. Soc.* **227**, 27P.
Vernazza, J. E. and Reeves, E. M.: 1978, *Astrophys. J. Suppl.* **37**, 485.
Wang, Y.-M.: 1994, *Astrophys. J.* **437**, L67.
Wang, Y. M.: 1996, *Astrophys. J.* **464**, L91.
Weise, W. L., Smith, M. W., and Glennon, B. M.: 1966, *Atomic Transition Probabilities – Hydrogen through Neon, NSRDS 4, 1*, U.S. Government Printing Office, Washington D.C.
Widing, K. G. and Feldman, U.: 1992, *Astrophys. J.* **392**, 715.
Withbroe, G. L. and Raymond, J. C.: 1983, *Astrophys. J.* **285**, 347.
Zhang, H. L. and Sampson, D. H.: 1994a, *ADNDT* **56**, 41.
Zhang, H. L. and Sampson, D. H.: 1994b, *ADNDT* **58**, 255.
Zhang, H. L. and Sampson, D. H.: 1996, *ADNDT* **64**, 1.
Zhang, H. L., Sampson, D. H., and Fontes, C. J.: 1990, *ADNDT* **44**, 31.

FIRST VIEW OF THE EXTENDED GREEN-LINE EMISSION CORONA AT SOLAR ACTIVITY MINIMUM USING THE LASCO-C1 CORONAGRAPH ON SOHO

R. SCHWENN and B. INHESTER
Max Planck Institut für Aeronomie, Lindau, Germany

S. P. PLUNKETT
Space Research Group, School of Physics and Space Research, University of Birmingham, Birmingham, U.K.

A. EPPLE and B. PODLIPNIK
Max Planck Institut für Aeronomie, Lindau, Germany

D. K. BEDFORD, C. J. EYLES, G. M. SIMNETT and S. J. TAPPIN
Space Research Group, School of Physics and Space Research, University of Birmingham, Birmingham, U.K.

M. V. BOUT, P. L. LAMY and A. LLEBARIA
Laboratoire d'Astronomie Spatiale, Marseille, France

G. E. BRUECKNER,* K. P. DERE, R. A. HOWARD, M. J. KOOMEN, C. M. KORENDYKE, D. J. MICHELS, J. D. MOSES, N. E. MOULTON, S. E. PASWATERS, D. G. SOCKER, O. C. ST. CYR** and D. WANG
E.O. Hulburt Center for Space Research, Naval Research Laboratory, Washington D.C., U.S.A.

(Received 20 June 1996; accepted in final form 17 November 1996)

Abstract. The newly developed C1 coronagraph as part of the Large-Angle Spectroscopic Coronagraph (LASCO) on board the SOHO spacecraft has been operating since January 29, 1996. We present observations obtained in the first three months of operation. The green-line emission corona can be made visible throughout the instrument's full field of view, i.e., from 1.1 $R_\odot$ out to 3.2 $R_\odot$ (measured from Sun center). Quantitative evaluations based on calibrations cannot yet be performed, but some basic signatures show up even now: (1) There are often bright and apparently closed loop systems centered at latitudes of 30° to 45° in both hemispheres. Their helmet-like extensions are bent towards the equatorial plane. Farther out, they merge into one large equatorial 'streamer sheet' clearly discernible out to 32 $R_\odot$. (2) At mid latitudes a more diffuse pattern is usually visible, well separated from the high-latitude loops and with very pronounced variability. (3) All high-latitude structures remain stable on time scales of several days, and no signature of transient disruption of high-latitude streamers was observed in these early data. (4) Within the first 4 months of observation, only one single 'fast' feature was observed moving outward at a speed of 70 km s^{-1} close to the equator. Faster events may have escaped attention because of data gaps. (5) The centers of high-latitude loops are usually found at the positions of magnetic neutral lines in photospheric magnetograms. The large-scale streamer structure follows the magnetic pattern fairly precisely. Based on our observations we conclude that the shape and stability of the heliospheric current sheet at solar activity minimum are probably due to high-latitude streamers rather than to the near-equatorial activity belt.

* Principal Investigator (PI).
** Computational Physics Inc., Washington D.C., U.S.A.

Solar Physics **175:** 667–684, 1997.

1. Introduction

The well-known green coronal spectral line at 530.3 nm was discovered in 1869 by Young and Harkness during a total solar eclipse. The line's true spectral characteristic was revealed by Grotrian (1939) and Edlèn (1942) who found evidence for a high temperature state of the corona. In fact, the emissivity of the green line, which is due to a forbidden transition of the Fe XIV ion, peaks at a temperature of about 2×10^6 K (Burgess and Seaton, 1964; Esser *et al.*, 1995). For comparison, the 'red' coronal emission line at 637.4 nm from Fe X ions has its maximum emissivity at a temperature around 10^6 K. The green line is the brightest of all coronal emission lines in the visible spectral range.

Soon after the invention of the coronagraph by Lyot (1930), regular observations of the corona were performed from several high-altitude stations around the globe (e.g., Waldmeier, 1951, 1957; Billings, 1966; Leroy and Trellis, 1974). The green corona appeared to be well correlated with solar activity, in both its general brightness variations during the sunspot cycle and its shape. The striking absence of green emission above both polar regions at activity minimum led Waldmeier (1957) to use the German term 'Koronalöcher', i.e., coronal holes.

Coronal green-line observations from ground-based coronagraphs rarely exceed a distance range of 6 arcmin above the Sun's limb, i.e., 1.4 $R_\odot$ measured from Sun center. Then the faint coronal signal is swamped by the unavoidable sky background. The green emission line drop-off is particularly steep since its emissivity is proportional to almost the square of the coronal electron density (Guhathakurta, Fisher, and Altrock, 1993) which, in turn, decreases as $R_\odot^{-2}$ or even steeper (see, e.g., discussion in Billings, 1966). The reason is that in high-density regions, i.e., close to the Sun, collisional excitation dominates radiative excitation. The second mechanism gains relative importance farther out (Raju and Desai, 1993). On the other hand, the 'white' (or continuum) corona is due to Thomson scattering of photospheric light by coronal electrons, and naturally its intensity depends linearly on electron density. This difference in the generation mechanism is responsible for an enormous radial gradient in the line-to-continuum intensity ratio; from values of 10 to 80 (depending on coronal activity) at 1.1 $R_\odot$, it decreases steeply with radial distance. From about 1.4 $R_\odot$ the ratio stays about constant at a value of the order of 1 (Raju and Desai, 1993).

The innermost green corona (up to 1.4 $R_\odot$) is clearly visible in ground-based coronagraphs. Very highly resolved coronal structures close to the limb could thus be revealed (e.g., Picat *et al.*, 1973). Observations of emission line brightness distributions around the Sun's limb are being continued on a routine basis. For example, the National Solar Observatory on Sacramento Peak (USA) has been making daily scans at 1.15 $R_\odot$ (since 1994 also at 1.25, 1.35, and 1.45 $R_\odot$) in several spectral lines (Smartt, 1982).

Due to limitations caused by scattered light originating in the Earth's atmosphere, associations of emission line features with large-scale coronal structures as seen, e.g., during eclipses, have not been very successful. There is a distinct observational gap between close-to-the-limb coronagraphic measurements and what we know from eclipses (out to 15 $R_\odot$) and *in-situ* measurements (down to 60 $R_\odot$) about the large-scale coronal and heliospheric structure (Bird and Edenhofer, 1990, Schwenn, 1990). Space-borne coronagraphy (Koutchmy, 1988) could fill in only parts of this gap, providing a wealth of observations between about 2 and 10 $R_\odot$. Unfortunately, the external occulters used so far on all space-borne coronagraphs caused vignetting at the inner edge and did not allow useful observations inside about 2 $R_\odot$.

This problem has finally been overcome, as we will demonstrate. The images to be shown and discussed in this paper were obtained using the internally occulted C1 instrument of the Large-Angle Spectroscopic Coronagraph (LASCO) onboard the Solar and Heliospheric Observatory (SOHO). The LASCO system further includes two externally occulted coronagraphs (C2 and C3) such that a total range of 1.1 to 32 $R_\odot$ is covered without gaps and with appropriate spatial resolution. SOHO was launched on December 2nd, 1995 into the planned heliocentric orbit (Domingo *et al.*, 1995). The scientific data are still in a somewhat provisional state, in that quantitative evaluations based on calibrations have not been performed. Even so, some basic signatures show up clearly which are worth discussing and allow important conclusions to be made.

2. The C1 Coronagraph: a Short Review of the Instrument

We restrict ourselves to a few features of this newly developed instrument which are considered important in the context of this presentation (for more specific details, the reader is referred to the description by Brueckner *et al.*, 1995):

– The C1 coronagraph is an internally occulted system, based on the concept invented by Lyot (1930). Thus, the instrument does not suffer from diffraction-limited spatial resolution at the inner edge, as externally occulted coronagraphs usually do.

– The C1 system uses parabolic mirrors rather than lenses, as first proposed by Newkirk and Bohlin (1963) (see also Zirin, 1970). Recent advances in mirror polishing techniques have resulted in stray-light levels of the instrument as low as 10^{-8} $B_\odot$ at 3 $R_\odot$ and 10^{-7} $B_\odot$ at 1.5 $R_\odot$, as measured in the laboratory. Preliminary analysis of the inflight performance indicates that these values have not noticeably changed since the start.

– In the focal plane of the objective mirror there is a field mirror with a center hole (equivalent to 1.1 $R_\odot$) that lets the bright sunlight pass through. Thus, it acts as an internal occulter.

– The field of view forms an inscribed circle on a 6×6 $R_\odot$ square, which is covered by a 1024×1024 pixel cooled CCD. The pixel width is equivalent to

5.6 arc sec, resulting in a spatial resolution of about 11 arc sec. The diffraction limited spatial resolution is of the order of 4 arc sec.

– A tunable Fabry–Pérot interferometer spectral filter system allows narrow passband images of the solar corona to be obtained simultaneously over the entire field of view. Using different blocking filters we can select certain spectral regions. Each of these can be scanned in small spectral steps allowing line profile determinations (including the respective nearby continuum regions) at any point within the field of view.

The C1 instrument has been in operation since January 29, 1996. All systems work nominally. Several test and calibration sequences were run during the first few months. That included spectral calibrations in order to determine the in-flight characteristics of the spectrometer. Exposure times were selected such that optimum use is made of the CCD full-well capacity. Typical exposure times are in the range between 10 and 25 s.

3. Some Comments on Image Processing

Careful calibration measurements of the instrument have been performed both on the ground and in space. A detailed analysis allowing precise photometric studies is under way and will be published in a dedicated paper. Here we restrict ourselves to a qualitative study of large scale structures of the green emission line corona. We made sure that our conclusions are not affected by quantitative corrections such as flat-fielding, photometrics, etc.

Unprocessed direct images from the LASCO-C1 coronagraph do not show much coronal signal. A coronal signal only becomes clearly visible after subtracting a nearby continuum image from an image taken at line center. The reason for this is the strong radial gradient of the instrumental stray light, rather than its absolute level. This is illustrated in Figure 1 which shows radial cuts through two measured intensity distributions along a radial line in the solar equatorial plane. These are real count rates accumulated in a 10 s exposure time; only the constant camera offset bias has been taken out. One of the images was taken at the center of the green line at 530.3 nm, the other one in the nearby continuum at 530.9 nm. The passband difference was obtained by appropriately tuning the Fabry–Pérot interferometer; all other settings remained unchanged. Thus, the two images are completely equivalent in terms of bandwidth, sensitivity etc.

Each image contains contributions from instrumental stray light and the continuum (or 'white') corona, plus an additional contribution from the emission line corona for one of the images. Thus, the difference between the two curves as shown in the bottom part of the figure is due to the green emission line alone. All other contributions cancel out. For this particular example, there is a significant signal in the range up to about 2 $R_{\odot}$. The green-line intensity varies strongly, both with respect to position above the limb and with time.

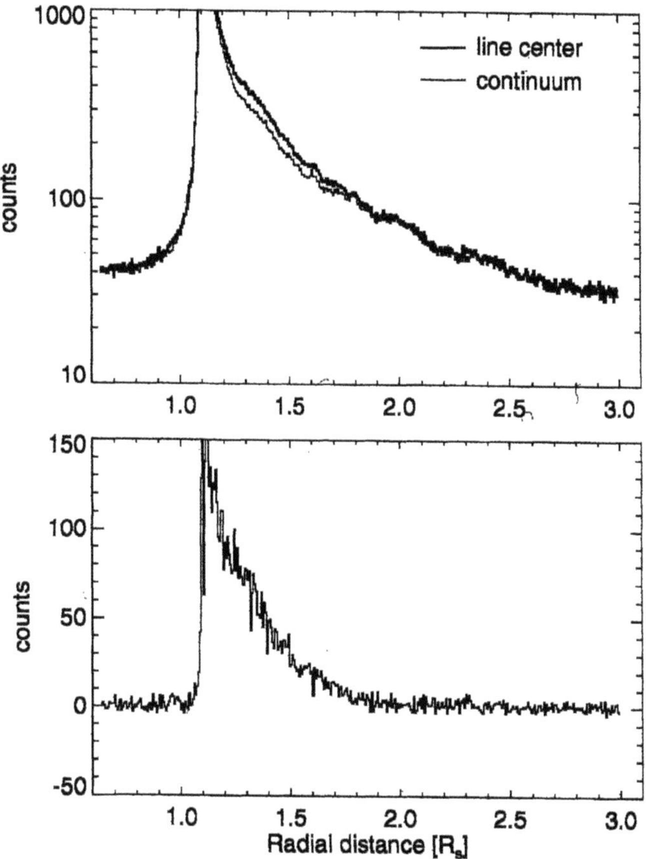

Figure 1. Radial cuts through two intensity distributions measured by LASCO-C1 on February 1, 1996 along a radial line in the solar equatorial plane (western part only). Count rates accumulated in a 10 second exposure time are plotted, with the constant camera offset removed. One of the images was taken at the center of the green line at 530.3 nm, the other one in the nearby continuum at 530.9 nm. The difference between the two curves is shown in the bottom panel.

Note especially the steep drop-off of the total signal from the occulter edge to the outer edge in Figure 1, which is very similar to the laboratory measurement using an artificial Sun (Figure 8 in Brueckner *et al.*, 1995). Over the range of the significant green line signal, the stray light varies radially by an order of magnitude. Therefore, any azimuthal contrast due to coronal structures is much less than the contrast due to the radial stray light profile. That is the reason why the corona is hard to discern in direct images taken with this instrument.

The instrument's stray light pattern is almost cylindrical, such that polar cuts through an image look very similar to equatorial ones.

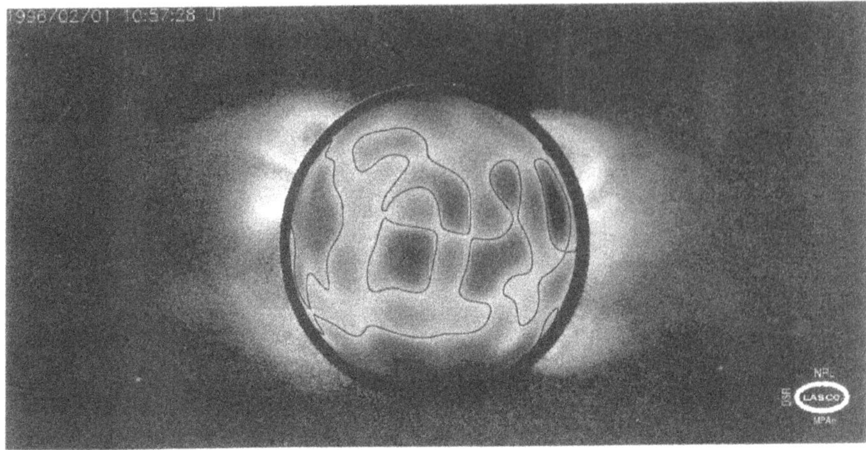

Figure 2. The green-line emission corona observed by the LASCO-C1 coronagraph on board the SOHO spacecraft on February 1, 1996 at 10:57 UT. The image results from a subtraction of two images, one taken at a wavelength of 530.3 nm (line center), the other one at 530.9 nm (continuum). The insert is a magnetogram kindly provided by the Wilcox Solar Observatory, Stanford.

The whole differenced green-line image used for Figure 1 is shown in Figure 2. Before discussing its content, a few technical comments should be given:

– For this particular image only half of the total field of view was used, in order to save telemetry.

– For effective transmission from the spacecraft to the ground, the data were compressed by the onboard computer using the lossless Rice algorithm.

– Image processing on the ground routinely includes analogous decompression, subtraction of a constant CCD bias, and a succeeding correction due to minor exposure time variations for individual images. Continuum images can then be subtracted from the line-center images, pixel by pixel.

– For the images shown in this paper no flat-field calibration was applied, mainly for technical reasons. The present study would not benefit from flat-fielding. Note that a radial neutral density filter is built into the optics in order to enhance the dynamic range. Its effect would be removed by flat-fielding and the large-scale structures reaching throughout the whole field of view would be more difficult to discern. Further corrections achieved by flat-fielding do not affect any part of the image by more than 10% and do not change the overall qualitative appearance. For further quantitative analysis careful photometric calibration procedures will be applied.

– The two exposures differenced here were taken on February 1 at 10:57 UT and at 19:26 UT, respectively. The long time separation does not have a significant effect on the result. We have checked this by differencing a sequence of continuum images acquired over several days.

– At times, a minor geometrical mismatch (a lateral shift by a few per cent of a pixel width) may occur between the two images to be subtracted. The very steep radial intensity gradients of the original images cause some brightening on one side and darkening on the opposite side in the subtracted images. This is corrected routinely using a computer program.

– All images contain some hundred randomly distributed 'bad' pixels caused by cosmic ray particles penetrating the CCD. They are removed by replacing them with an interpolation between the surrounding 'good' pixels.

– A linear color table is applied for image presentation. A noticeable coronal signal can be discerned from the noise as far out as to the 3.2 $R_\odot$ edge of the C1 field of view.

4. Signatures of the Green Corona at Solar Activity Minimum

Figure 2 is a very typical example of the appearance of the extended green corona at the present minimum phase of the solar activity cycle.

– We find bright apparently closed loop systems centered at latitudes of 30° to 45° in both hemispheres. The loop tops usually reach out to about 1.5 $R_\odot$, in some cases to well beyond 2 $R_\odot$.

– Their helmet-like outer extensions (we may call them 'streamers') are clearly bent towards the solar equator.

– Around the equator, there is a more diffuse bright pattern, clearly separated from the high-latitude loops.

– Towards the outer edge of the C1 field of view, the two streamers and the equatorial diffuse pattern on either side appear to merge to form one large equatorial streamer system.

– There is no detectable green-line emission above either of the poles. We are revisiting Waldmeier's 'Koronalöcher', the coronal holes.

– All visible edges of the polar coronal holes appear to be sharp and well-defined. We found that they agree in much detail with those seen in *Yohkoh*-SXT data and in SOHO-EIT data.

It is worth noting that in these difference images polar plumes can never be seen, in strong contrast to similar difference images using the red line. Apparently, the entire corona in coronal holes does not produce sufficient green-line emissivity. However, upon differencing line-center images taken several hours apart among themselves, signatures of polar plumes begin to appear. They are due to spatial motions of the plumes and their associated electron densities, probably caused by solar rotation, such that the continuum contribution to the line-center images is altered.

The insert in Figure 2 is the photospheric magnetogram for that same day, kindly provided by the Wilcox Solar Observatory in Stanford. It will be discussed in some detail in Section 6.

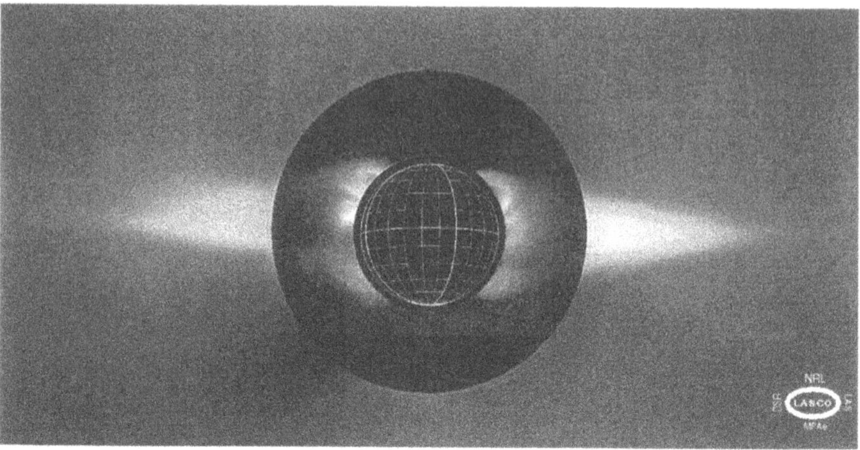

Figure 3. A composite image of a C1 green-line emission image taken on February 1, 1996 at 10:57 UT and a C2 white-light image taken at 02:21 on the same day. The C1 image (see Figure 2) has been scaled to the spatial resolution of the C2 image (11.4 arc sec pixel^{-1}) and extends to the inner edge of the C2 field of view at a radial distance of 2.2 $R_{\odot}$.

Figure 3 shows two merged images from our C1 and C2 instruments, taken almost simultaneously. This latter coronagraph is an externally occulted system. It images the white-light corona in three wideband spectral channels, covering a circular field of view from 2.2 to 6 $R_{\odot}$ (see Brueckner *et al.*, 1995). Figure 3 demonstrates how the inner coronal structure is continued outward: the bending of the high-latitude streamers toward the equatorial plane continues, and a single flat near-equatorial sheet forms. This sheet is also observed by the C3 instrument and can be discerned out to the edge of its field of view at 32 $R_{\odot}$.

5. Variations of the Green Corona with Solar Rotation

Early on, even during the commissioning phase, we started an observational program which would allow the green-line emission to be imaged with a fairly regular cadence. The data that were used for this analysis were mainly taken between March 3 and March 18, 1996, i.e., during about half a solar rotation. The time separation between images was usually 1 to 2 hr, with some longer gaps randomly distributed.

All images (about 200 total) were inspected for the present analysis. When the data are analyzed as a 'movie' of the time sequence, many faint features are revealed which otherwise might easily escape attention (see Movie 1 on CD-ROM).

We try to illustrate the basic signatures in the few images shown in Figure 4:

– The high-latitude arches remain rather stable on a time scale of days. They apparently corotate with the Sun without major variations.

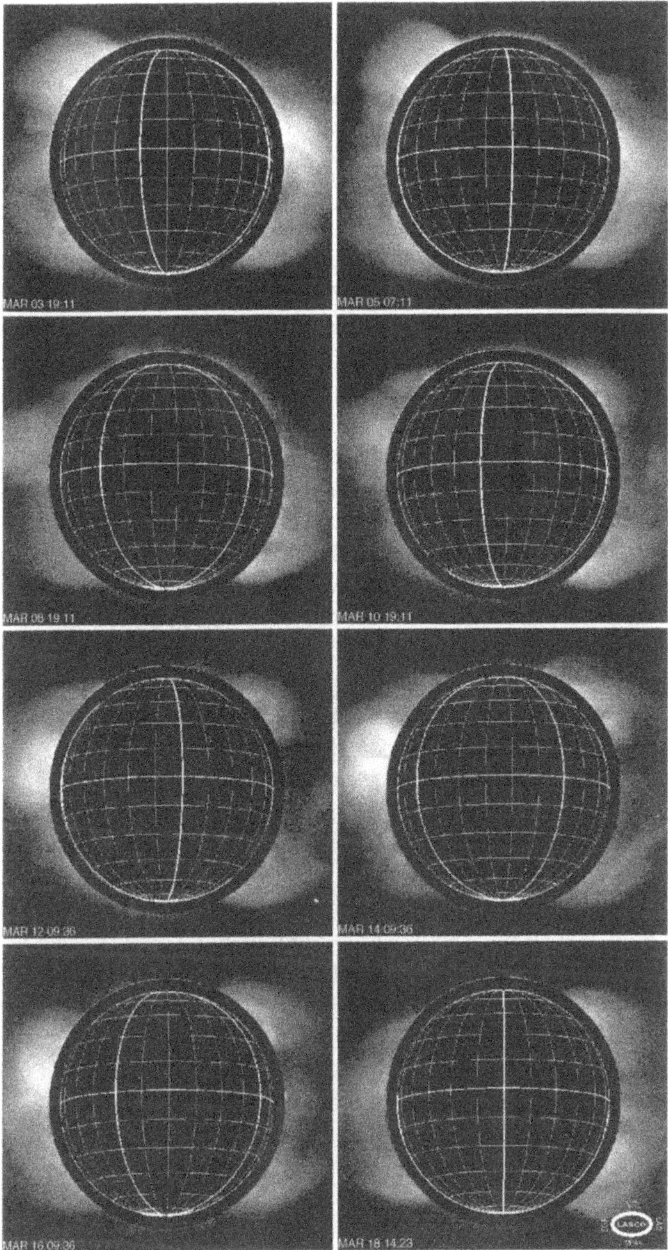

Figure 4. A set of images similar to Figure 2, taken between March 3, 1996 and March 18, 1996.

– The structural changes are fairly gradual. Our preliminary analysis of the March time interval has not yet produced evidence for the occurrence of a coronal reorganization following a mass ejection. A systematic search is underway and some events of this type have been noted in the more recent data.

– Gradual evolution eventually leads to different large-scale patterns. For example: one of the high-latitude streamers may be absent for a few days (e.g., on March 5, Figure 4(b)), and the polar coronal hole appears to extend to lower latitudes.

– The near-equatorial features are generally more variable. They may disappear completely for several days, while the prominent high-latitude streamers remain unchanged (e.g., around March 12 on the west limb, Figure 4(e)). Sometimes they look like typical streamers, while at other times they may turn brighter than the other features by an order of magnitude (Figure 4(f)). Apparently, these changes occur in close correlation with the underlying active regions.

– The very bright loop-like structure seen about March 14 (Figure 4(f)) was of particular interest. As this structure rotated into the field of view the loops were clearly visible beyond 1.5 $R_{\odot}$. They remained visible for several days until solar rotation moved them in front of the solar disk, i.e., out of the C1 field of view. There were some dynamical changes of the loops, but the structure remained stable without discernible disruptions. This region in the corona stayed 'active' for at least 3 solar rotations. Detailed associations with an evolving active region observed by several other instruments are under way.

– There are apparent motions of fine structures (outward or laterally) within the streamer features, with speeds rarely exceeding 0.1 $R_{\odot}$ per hour, i.e., 20 km s^{-1}. For the very slow motions, it is not easy to distinguish uniquely between evolutionary effects and apparent motions caused by corotation.

– Sometimes (e.g., on March 8) green features are seen above the poles, too. Close inspection suggests that these are just projections of the high-latitude streamers corotating with the Sun which are swept across the SOHO–Sun line.

The first direct signature of a fast dynamic event and its aftermath was not observed until May 16th. A moving loop-like feature could be seen in four successive frames taken between 15:07 and 17:32 UT. An apparent speed of 70 km s^{-1} (constant in the C1 field of view) was determined. The event occurred on the west limb, at a latitude of 15° south. Later on it was also detected in the C2 and C3 coronagraphs as part of a white-light mass ejection. This event and others identified later need more detailed study. The important point to be stressed here is the fact that these ejections occurred preferentially at low latitudes, between the high-latitude streamers which appeared not to be affected.

6. Correlation with Magnetic Topology

The structure of the corona is governed by the solar magnetic field. The magnetic topology is not just that of a simple magnetic dipole, not even at the comparatively quiet minimum phase. Higher order multipole components are always involved (see, e.g., Hoeksema, 1986; Bird, 1990). Green line images appear to be particularly well-suited to make this structure visible. A coronal image like Figure 2 leads to the immediate impression that there are several magnetic loop systems anchored at the Sun. As a consequence, there must be a series of polarity changes around the limb, apart from the global polarity switch from 'positive' at the northern coronal hole to 'negative' in the south. The radial extent of these multipole moments is much more limited than that of the overall magnetic dipole centered in the polar coronal holes. It has been shown that essentially ALL interplanetary magnetic field lines are rooted in coronal holes (Levine, 1982; Wang and Sheeley, 1992; Hoeksema, 1995).

We compared the green-line images with photospheric magnetic field data kindly provided by the Wilcox Solar Observatory in Stanford. Their magnetograms are assembled from daily observations of magnetic structures when they cross the solar central meridian. The published synoptic map for Carrington rotation 1905 is shown in Figure 5. It was used for computing the insert in Figure 2.

For a more detailed comparison of magnetograms with coronagraph images taken at a given time one has to take solar rotation into account. For example: west-limb features in Figure 2 observed on February 1 had passed central meridian 7 days earlier, i.e., around January 24–25. The magnetogram (Figure 5) shows at that time a neutral line at 30° north that appears to be centered beneath the Fe XIV loop on the west limb. The lower loop corresponds to the magnetic neutral line around 40° south. The equatorial, more diffuse feature in the coronal image does not have such a unique equivalent in the magnetogram. That may be due to a line-of-sight effect or to a temporal change during the 7-day corotation time. East-limb features in Figure 2 only reached central meridian around February 7–8. At that time we find a similar pattern in the magnetogram: 3 basically horizontal neutral lines which match the 3 bright loop systems at about the same latitudes.

It is mere coincidence that on this day the coronal patterns above both limbs, as well as their associated magnetograms, look very similar. In contrast, on March 5 (Figure 4(b)) we noted the absence of a high-latitude streamer on the north-west limb. The magnetogram shows on February 28 a 'hole' in the northern high-latitude neutral line, similar to the situation on January 31 (see Figure 5), i.e., one solar rotation earlier. Very rarely, the warps in the neutral line are such that there is only one neutral line over the whole latitude range on a given day (e.g., on January 21, Figure 5). The equivalent coronagraph data confirm this pattern; there is only one streamer at about the expected equivalent latitude, e.g., on March 10 (see Figure 4(d)) above the east limb.

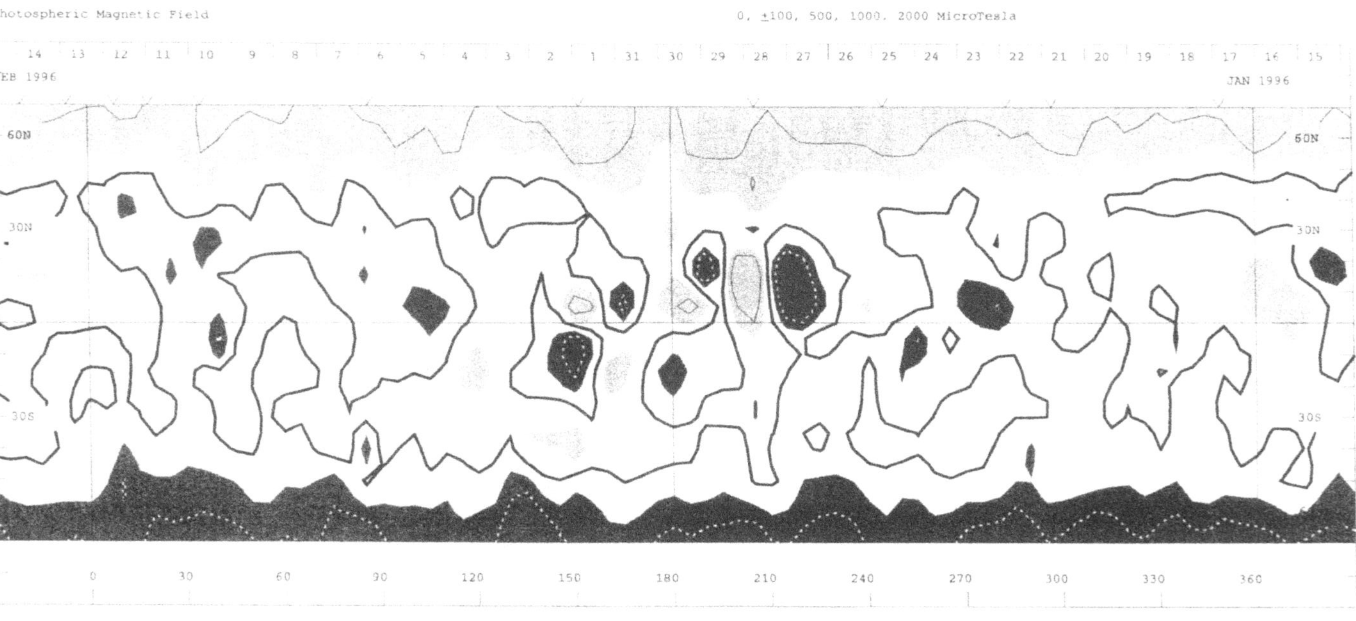

Figure 5. The photospheric magnetogram of Carrington rotation 1905 in January and February 1996, provided by the Wilcox Solar Observatory in Stanford. The solid lines denote neutral magnetic lines, i.e., they separate opposite magnetic polarities. In the hatched regions (the darker shade indicates 'negative' polarity) the field strength exceeds 50 mT.

Furthermore, we reconstructed the three-dimensional coronal magnetic field using the magnetogram of Figure 5. According to Wang and Sheeley (1995), the reconstructed magnetic field agrees best with interplanetary observations if the measured magnetic surface field is interpreted as the line-of-sight component of a purely radial photospheric field. In addition, we apply the correction proposed by Ulrich (1992) to account for the saturation of the observed line splitting at lower solar latitudes. For our model, we binned the surface field data into 30 steps in sine(latitude) and 72 equidistant steps in longitude and fitted it using spherical harmonics up to 18th order. The coronal magnetic field is then reconstructed from a potential field model, assuming that there is a spherical source surface at 3 $R_{\odot}$ and that there are no currents flowing inside (Altschuler and Newkirk, 1969; Schatten, Wilcox, and Ness, 1969). Figure 6 shows part of the field lines that result from this model. Only those field lines have been plotted which are closed, have a length of less than 1.5 $R_{\odot}$ and which emerge out of regions with a surface field strength larger than 0.2 G. For these field lines, the assumed location of the source surface is of little consequence. In comparison with the green line image of Figure 2, we note a striking similarity in the overall structures. The agreement could certainly be further improved using more realistic model assumptions, e.g., by allowing currents inside the source surface.

The magnetograms of the first 3 months of the SOHO mission show that neutral lines between 30° and 50° latitude in both hemispheres are often very persistent and stable on a time scale of several days, e.g., between longitudes 0° to 110° (perhaps even 150°) in the north or from 90° to 190° in the south (Figure 5). Our model calculation shows that this southern neutral line is the center of a long magnetic arch system that reaches all around the visible half of the Sun. There is certainly a close association with the so-called 'polar crown filaments', visible in Hα data at similar latitudes. Filaments (called prominences when observed above the limb) are regions of cool and dense material at coronal heights held against gravity by surrounding magnetic field structures. Shortly before activity minimum 'the polar prominences are oriented nearly parallel to the equator and sometimes form a nearly uninterrupted crown around the polar caps', as de Jager (1959) phrased it, at latitudes between 45° and 50°. They move towards the poles with increasing solar activity (Ananthakrishnan, 1952) and are usually very stable. The coronal streamers and their polar edges as we see them in LASCO-C1 data are also very stable. For example, both southern edges remained almost unchanged from at least March 3 to March 8 (Figure 4). This stability helps to avoid mixing of different structures along the line of sight of the coronagraph. That is the reason why the polar edges appear so 'sharp' in most images.

It is interesting to note in this context what the source-surface magnetic field calculated by the Stanford group (published in *Solar Geophysical Data*) looks like. For the time of rotation 1905 (see Figure 5) everything outside their assumed source surface at 2.5 $R_{\odot}$ is almost perfectly flat. The extrapolated current sheet does not extend beyond $\pm 10°$ in latitude through several solar rotations. That is why the

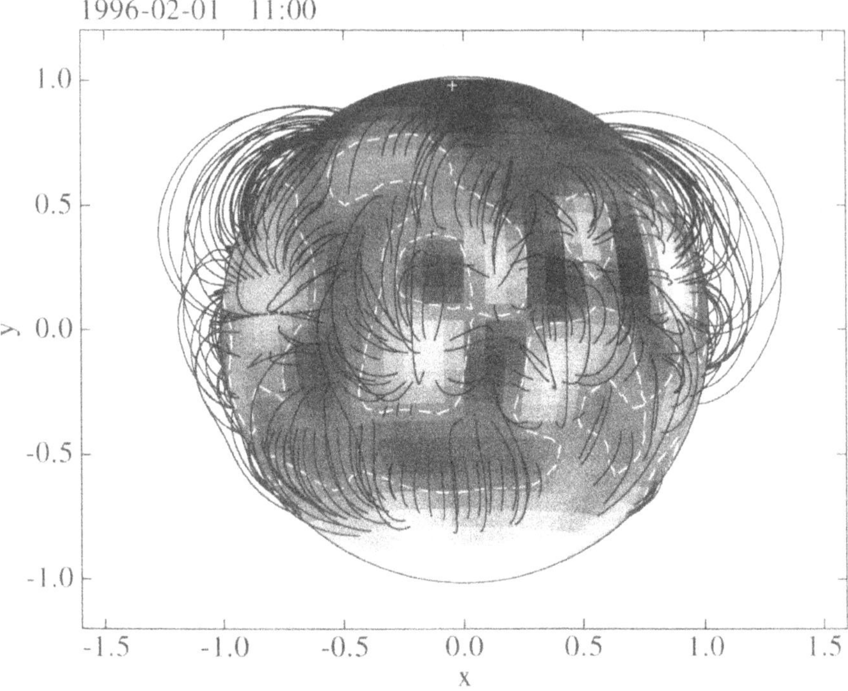

Figure 6. Reconstruction of the three-dimensional coronal magnetic field using the magnetogram of Figure 5. The measured magnetic surface field is interpreted as the line-of-sight component of a purely radial photospheric field. A correction taking into account saturation of the observed line splitting at lower solar latitudes has been applied. For further details see text.

equatorial streamer sheet in the LASCO-C2/C3 fields of view (see Figure 3) does not move up and down by much; the inclination of the solar dipole to the rotation axis is presently extremely small.

The near-equatorial features are more variable, both in space and time, as the magnetograms indicate. The dynamics of some remaining active regions cause the green-line corona to vary in intensity by a factor of 10 and more. This variability, in conjunction with the smaller scale and often inclined magnetic structures, may be the reason for the more diffuse appearance of this region.

7. Discussion

We now begin to put the LASCO observations in context with previous work. This turned out to be a difficult task. Immense data sets from various ground-based coronagraphs have been accumulated over the past 60 years. Unfortunately, they are hard to locate and to access. Because of observational constraints, these coronal observations did not extend far enough to shed light on the relationships between green-line coronal structures and those of the heliosphere. Most of the scientific work reported in the literature has concentrated on a very limited range close to the Sun. In the context of this paper, it appears appropriate to go back as far as to Waldmeier's diligently collected data sets from 1939 to 1954 which include the activity minimum in 1944 (Waldmeier, 1957). His very regular limb scans of the green corona show the latitudinal brightness variations through all phases of the activity cycle. In the minimum years, the brightness was concentrated in two well-separated belts centered at latitudes of 30°. He noticed also distinct features (he called them 'Nebenzonen' or 'Schultern' in German) at latitudes of 50° to 60°. It might well be that Waldmeier had already seen the bright footpoints of the high-latitude streamer loops now revealed by LASCO-C1. Leroy and Noëns (1983) used Pic du Midi coronagraph data from 1944 to 1974 for a more quantitative analysis and concluded that their 'data are not sufficient to settle definitely what happens near latitude 45° just before the minimum'.

In more recent studies Guhathakurta, Fisher, and Altrock (1993) and Guhathakurta and Fisher (1994) analyzed coronagraph data (limb scans at 1.15 $R_\odot$) obtained at the National Solar Observatory on Sacramento Peak between 1984 and 1992. They derived coronal temperatures from the green to red line intensity ratio. They found two zonal bands of high coronal temperature at latitudes around 50°. These bands tend to lie in regions above a magnetic neutral line and bright features in coronal polarization brightness 'which usually correspond to the centers of coronal helmet streamers'. LASCO-C1 confirms the persistence of high-latitude streamers above magnetic neutral lines.

Loucif and Koutchmy (1989) analyzed 19 eclipse pictures in order to study coronal variations with the solar cycle. At high solar activity, all streamers were found to be oriented radially, on the average, and not much lateral bending was observed. In contrast, the idealized drawings of the 1954, 1965, and 1976 activity minimum eclipses compare very well with our Figure 1, in that the high-latitude streamers are bent towards the equatorial plane such that they are finally parallel to it.

The formation of antiparallel magnetic fields beyond the closed loops requires the existence of current sheets (see, e.g., Koutchmy and Livshits, 1992). The idea, developed by Schulz (1973), of a large-scale warped azimuthal current sheet separating the whole heliosphere into halves with opposite polarity led Alfvén (1977) to propose the 'ballerina model' of the heliosphere (see also Schwenn, 1990; Mariani and Neubauer, 1990). At the passage of space probes through the current sheet, i.e., at magnetic sector boundaries, a series of discontinuities is often encountered. They were usually thought to be due to multiple crossings of a single, fluctuating current sheet (Behannon, Neubauer, and Barnstorf, 1981). Only recently, evidence was found for the possible existence of multiple heliospheric current sheets (Crooker *et al.*, 1993). These authors show sketches of multiple streamers which resemble our Figure 2. That would mean that signatures of higher order multipole moments are also convected out into the heliosphere by the evolving solar wind. We are not yet used to this new concept, maybe because of the well-known source-surface magnetic field models which always show just one current sheet at 2.5 $R_\odot$ at activity minimum. However, since large streamer systems coexist at different latitudes, different current sheets must also exist, unless they merge somewhere.

Having observed the coronal structures now for 3 months continuously, we conclude that at this minimum phase of the solar cycle there is always a near-equatorial 'streamer sheet' surrounding the Sun. Its initial width of about one solar diameter is determined by the position of high-latitude streamers, which in turn are located above polar crown filaments and their underlying magnetic neutral lines. The current sheets pertaining to the original streamers are embedded in this streamer belt.

We suggest that the flatness and stability of the distant heliospheric current sheet(s) is probably a result of the persistence of the high-latitude streamers. In other words, the heliospheric current sheet is associated with high-latitude phenomena rather than with streamers above the 'activity belt', as was implied in many descriptions of the ballerina model (see, e.g., Schwenn, 1990, and references therein). At times no streamers at all are visible above the equatorial regions for several days (see March 10–14 images, Figures 4(c–e)), while only the high-latitude streamers persist.

The transition from closed to open magnetic topology at the top of a streamer is not yet understood. In the LASCO-C1 observation sequence performed so far there is no evidence for major transient disruptions occurring on the streamer tops. In particular, the high-latitude streamers do not show any signature of 'blowout' events, as defined by Howard *et al.* (1985). Those which potentially may have occurred within the C1 field of view during data gaps did not change the remaining structure noticeably. Such processes, if they occur at all, can only be very minor and must be fairly fast. The only outstanding transient (which may have been a mass ejection event) seen in C1 green line images occurred at 15° south latitude. Further analysis pending, we can already state that transient streamer disruption within 3 $R_\odot$ is certainly not a very common process at this time of the activity cycle.

8. Summary

Despite the early phase of the SOHO mission and the LASCO data processing, some preliminary conclusions can be drawn with respect to the green-line emission corona at solar activity minimum:

(1) The LASCO-C1 instrument makes the coronal green-line emission visible to almost the edge of its field of view at 3.2 $R_\odot$.

(2) Bright, apparently closed, loop systems are almost permanently present, centered at latitudes of 30° to 45° in both hemispheres. Their helmet-like extensions

are bent towards the equator plane. Further out they merge into one large equatorial streamer sheet clearly discernible out to 32 $R_{\odot}$.

(3) At mid latitudes a more diffuse pattern is usually visible, well separated from the high-latitude loops and with a very pronounced variability.

(4) The polar coronal holes are discernible by the complete absence of green-line emission. Their borders toward the high-latitude streamers are usually very clear-cut.

(5) All high-latitude structures remain stable on time scales of several days.

(6) No signature of transient disruption of high-latitude streamers was observed in the first few months of operation.

(7) Within the first 4 months of observation, only one single 'fast' feature was observed moving outward at a speed of 70 km s^{-1} close to the equator. Faster events may have escaped attention because of data gaps, but this seems unlikely since no sudden morphological changes were detected.

(8) The centers of high-latitude loops are usually found at the positions of magnetic neutral lines in photospheric magnetograms. The large-scale streamer structure follows the magnetic pattern fairly exactly.

(9) The shape and stability of the heliospheric current sheet at solar activity minimum are probably due to high-latitude streamers rather than to the near-equatorial activity belt.

Acknowledgements

The LASCO instrument was constructed by a consortium of institutions: the Naval Research Laboratory (Washington D.C, U.S.A.), the Department of Space Research (University of Birmingham, U.K.), the Max-Planck-Institut für Aeronomie (Katlenburg-Lindau, Germany), and the Laboratoire d'Astronomie Spatiale (Marseille, France). The authors wish to thank again all members of their teams (see Brueckner *et al.*, 1995) who have made these observations possible. Our gratitude is also due to the various project teams who managed, fabricated and successfully launched the SOHO spacecraft. We sincerely thank the Flight Operations Team (FOT) at Goddard Space Flight Center (GSFC) which has done a really professional job in running this most complicated mission smoothly. We thank the Science Operations Coordinators at GSFC, Eliane Larduinat, Laura Allen and Piet Martens who have managed successfully to take care of so many different and often diverging interests, and finally to the SOHO project scientists Art Poland and Vincente Domingo and their assistants Luis Sanchez Duarte and Bernhard Fleck for their tireless efforts to cope with so many different scientists in order to achieve an optimum scientific success of the SOHO mission.

The NRL effort was supported by NASA under NDPR S-92385-D. The German effort was supported under grant 010C88024 by the Deutsche Agentur für Raumfahrtangelegenheiten (DARA). The French participation was financially supported by the Centre National D'Etudes Spatiales (CNES). The British contribution was supported by the Particle Physics and Astronomy Research Council in the United Kingdom.

References

Alfvén, H.: 1977, *J. Geoph. Res.* **91**, 13 679.

Altschuler, M. D. and Newkirk, G., Jr.: 1969, *Solar Phys.* **9**, 131.

Ananthakrishnan, R.: 1952, *Nature* **170**, 158.

Behannon, K. W., Neubauer, F. M., and Barnstorf, H.: 1981, *J. Geophys. Res.* **86**, 3273.

Billings, D. E.: 1966, *A Guide to the Solar Corona*, Academic Press, New York.

Bird, M. K. and Edenhofer, P.: 1990 in R. Schwenn and E. Marsch (eds.), *Physics of the Inner Heliosphere*, Springer-Verlag, Berlin, p. 13.

Brueckner, G. E. *et al.*: 1995, *Solar Phys.* **162**, 357.

Burgess, A. and Seaton, M.: 1964, *Monthly Notices Roy. Astron. Soc.* **127**, 355.

Crooker, N. U., Siscoe, G. L., Shodhan, S., Webb, D. F., Gosling, J. T., and Smith, E. J.: 1993, *J. Geophys. Res.* **98**, 9371.

De Jager, C.: 1959, in S. Flügge (ed.), *Handbuch der Physik*, Springer-Verlag, Berlin, p. 80.

Domingo, V., Fleck, B., and Poland, A. I.: 1995, *Solar Phys.* **162**, 1.

Edlèn, B.: 1942, *Z. Astrophys.* **22**, 30.

Esser, R., Brickhouse, N. S., Habbal, S. R., Altrock, R. C., and Hudson, H. S.: 1995, *J. Geophys. Res.* **100**, 19 829.

Grotrian, W.: 1939, *Naturwissenschaften* **34**, 87.

Guhathakurta, M. and Fisher, R. R.: 1994, *Solar Phys.* **152**, 181.

Guhathakurta, M., Fisher, R. R., and Altrock, R. C.: 1993, *Astrophys. J.* **414**, L145.

Hoeksema, J. T.: 1986, in R. G. Marsden (ed.), *The Sun and the Heliosphere in Three Dimensions*, D. Reidel Publ. Co., Dordrecht, Holland, p. 241.

Hoeksema, J. T.: 1995, *Space Sci. Rev.* **72**, 137.

Howard, R. A., Sheeley, N. R., Jr., Koomen, M. J., and Michels, D. J.: 1985, *J. Geophys. Res.* **90**, 8173.

Koutchmy, S.: 1988, *Space Sci. Rev.* **47**, 95.

Koutchmy, S. and Livshits, M.: 1992, *Space Sci. Rev.* **61**, 393.

Leroy, J.-L. and Noëns, J.-C.: 1983, *Astron. Astrophys.* **120**, L1.

Leroy, J.-L. and Trellis, M.: 1974, *Astron. Astrophys.* **35**, 283.

Levine, R. H.: 1982, *Solar Phys.* **79**, 203.

Loucif, M. L. and Koutchmy, S.: 1989, *Astron. Astrophys. Suppl. Ser.* **77**, 45.

Lyot, B.: 1930, *C.R. Acad. Sci. Paris* **191**, 834.

Mariani, F. and Neubauer, F. M.: 1990, in R. Schwenn and E. Marsch (eds.), *Physics of the Inner Heliosphere*, Springer-Verlag, Berlin, p. 18.

Newkirk, G. and Bohlin, J. D.: 1963, *Appl. Optics* **2**, 131.

Picat, J. P., Fort, B., Dantel, M., and Leroy, J.-L.: 1973, *Astron. Astrophys.* **24**, 259.

Raju, K. P. and Desai, J. P.: 1993, *Solar Phys.* **147**, 255.

Schatten, K. H., Wilcox, J. M., and Ness, N. F.: 1969, *Solar Phys.* **6**, 442.

Schulz, M.: 1973, *Astrophys. Space Sci.* **24**, 371.

Schwenn, R.: 1990, in R. Schwenn and E. Marsch (eds.), *Physics of the Inner Heliosphere*, Springer-Verlag, Berlin, p. 99.

Smartt, R. N.: 1982, *SPIE* **331**, 442.

Ulrich, R. K.: 1992, in M. S. Giampampa and J. A. Bookbinder (eds.), *Cool Stars, Stellar Systems and the Sun*, ASP, San Francisco.

Waldmeier, M.: 1951, *Die Sonnenkorona I*, Birkhäuser Verlag, Basel.

Waldmeier, M.: 1957, *Die Sonnenkorona II*, Birkhäuser Verlag, Basel.

Wang, Y.-M. and Sheeley, N. R., Jr.: 1992, *Astrophys. J.* **392**, 310.

Wang, Y.-M. and Sheeley, N. R., Jr.: 1995, *Astrophys. J.* **447**, L143.

Zirin, H.: 1970, *Sky Telescope* **218**, 215.

LASCO OBSERVATIONS OF DISCONNECTED MAGNETIC STRUCTURES OUT TO BEYOND 28 SOLAR RADII DURING CORONAL MASS EJECTIONS

G. M. SIMNETT, S. J. TAPPIN, S. P. PLUNKETT, D. K. BEDFORD and C. J. EYLES
School of Physics and Space Research, University of Birmingham, B15 2TT, U.K.

O. C. ST. CYR, R. A. HOWARD, G. E. BRUECKNER, D. J. MICHELS, J. D. MOSES, D. SOCKER, K. P. DERE, C. M. KORENDYKE, S. E. PASWATERS and D. WANG
E.O. Hulbert Center for Space Research, Naval Research Laboratory, Washington, D.C. 20375–5320, U.S.A.

R. SCHWENN
Max-Planck-Institut für Aeronomie, Katlenburg-Lindau, Germany

P. LAMY, A. LLEBARIA and M. V. BOUT
Laboratoire d'Astronomie Spatiale, Marseille, 13376, France

(Received 16 December, 1996; accepted 12 February, 1997)

Abstract. Two coronal mass ejections have been well observed by the LASCO coronagraphs to move out into the interplanetary medium as disconnected plasmoids. The first, on July 28, 1996, left the Sun above the west limb around 18:00 UT. As it moved out, a bright V-shaped structure was visible in the C2 coronagraph which moved into the field-of-view of C3 and could be observed out to beyond 28 solar radii. The derived average velocity in the plane of the sky was 110 ± 5 km s^{-1} out to 5 solar radii, and above 15 solar radii the velocity was 269 ± 10 km s^{-1}. Thus there is evidence of some acceleration around 6 solar radii. The second event occurred on November 5, 1996 and left the west limb around 04:00 UT. The event had an average velocity in the plane of the sky of $\sim$54 km s^{-1} below 4 $R_\odot$, and it accelerated rapidly around 5 $R_\odot$ up to 310 ± 10 km s^{-1}. In both events the rising plasmoid is connected back to the Sun by a straight, bright ray, which is probably a signature of a neutral sheet. In the November event there is evidence for multiple plasmoid ejections. The acceleration of the plasmoids around a projected altitude of 5 solar radii is probably a manifestation of the source surface of the solar wind.

1. Introduction

One of the essential requisites of solar coronal mass ejections is that they should not carry a significant net magnetic flux into the heliosphere. Not only is this not observed in the interplanetary medium, but it is clear that within the lifetime of the Sun, a steady state must have been established, within which current transient activity must fit. Therefore there must be a significant number of coronal mass ejections (CMEs) which disconnect from the Sun and propagate out as detached plasmoids. Such events so far have been difficult to detect, either from eclipse observations or from spaceborne coronagraphs over the last two decades. The first report of a disconnection event was by Illing and Hundhausen (1983) who interpreted observations from the Solar Maximum Mission (SMM). Webb and Cliver (1995) made a systematic search of both eclipse data and spaceborne coronagraph data up to

Solar Physics **175:** 685–698, 1997.

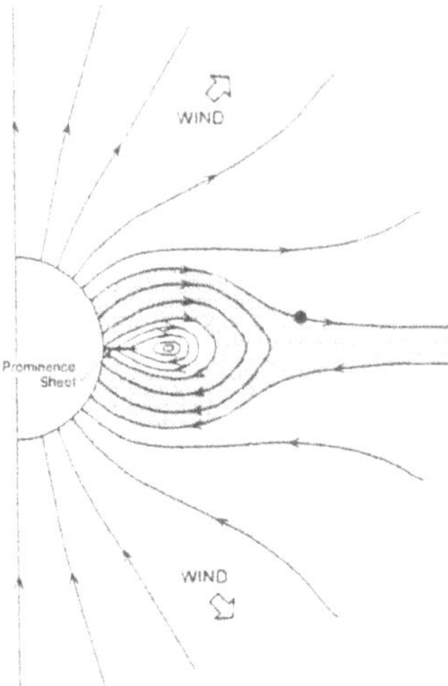

Figure 1. A model of the magnetic topology in a helmet streamer overlying a prominence. The lines of force in the high equatorial region enclose the equatorial interplanetary current sheet. Regions of high density are shaded. The closed magnetic field in the low-density cavity is normally in static equilibrium. The vertical prominence sheet is shown as a line extending out into the low density cavity. In this model the length runs in the azimuthal direction. (After Low, 1994.)

1980 for disconnection events. Although the data are not ideal for such a study, they nevertheless concluded that possibly over 10% of CMEs have disconnection features. St. Cyr and Burkepile (1990) suggested that this feature was only present in possibly 3% of all events.

McComas *et al.* (1989) had proposed a solution to the dilemma through the balancing of the newly-opened magnetic flux in CMEs via reconnection across the current sheet which extends above structures such as coronal helmet streamers. This would serve to detach a 'U'-shaped structure out into the corona. McComas *et al.* (1991) and McComas *et al.* (1992) presented observations of such events detected by SMM on July 27, 1988 and June 1 1989, respectively. However, their data were limited to the coronal region from 1.5 to $\gtrsim 5\ R_{\odot}$.

Low (1996) has reviewed the theoretical modelling for such events. In an ideal situation, the closed field regions in the corona may be thought of as a high density helmet dome, with a streamer emerging from the top, representing the neutral sheet extending into interplanetary space. Frequently there will be a prominence in the low corona, with a low density 'cavity' above it. Figure 1 (reproduced from Low (1994)) shows a schematic of this situation. In the data we present here there is evidence for the disconnection and ejection from the Sun of structures virtually identical to that depicted in Figure 1. Thus we have convincing evidence that not only do disconnection events occur, but they occur in a manner predicted by theory.

2. The Observations

The observations we report here were taken with the set of three nested coronagraphs (LASCO) on the Solar and Heliospheric Observatory (SOHO) in 1996. A full description of LASCO may be found in Brueckner *et al.* (1995). They cover the corona in three concentric bands, from 1.1–3.0 (C1), 1.5–6.0 (C2), and 3.7–30 (C3) $R_{\odot}$. The images we present in this paper (with one exception) are from the C2 and C3 coronagraphs; they are responding to photospheric light which is Thomson-scattered off free electrons in the corona. Therefore the intensity reflects excess mass in the corona, and does not indicate temperature. The C1 coronagraph contains a scanning Fabry–Pérot interferometer and for the November 5 event the onset of the CME was observed in the line of Fe X. The two disconnection events originated near the west limb of the Sun and the estimated onset times were ~18:00 UT on July 28 and ~04:00 UT on November 5. There were no active regions near the west limb on either occasion and we know of no chromospheric response to the ejections.

2.1. The July 28–29 Event

From the beginning of July 28, the equatorial streamer off the west limb gradually brightened. At the same time it swelled such that by 15:04 UT it was apparent that an eruption of some sort was probable. The C2 images showing this are presented in Figure 2.

The eruption first became visible in the C2 coronagraph at 18:56 UT when a bright feature appeared at ~2 $R_{\odot}$ about 5° below the equator. This rapidly expanded outwards and some sample images from C2 are shown in Figure 3. Note that the original bright equatorial streamer is still visible just to the north of the equator off the west limb. At 22:02 UT on July 28 the ejected mass appears pear-shaped and is embedded within the expanded helmet streamer. By 00:00 UT it has moved outwards, leaving behind a bright strip connecting back to the Sun, with a darker region to the north. In the image at 01:19 UT on July 29, some thin bright lines are just visible on the northern edge of the dark region. These are seen better in the image at 02:37 UT, which has had the contrast enhanced to increase their visibility. The thin bright line bifurcates at around 3 $R_{\odot}$ in this image, and the lines are reproduced in white in the upper part of the frame to aid the eye. The trailing edge of the ejected mass is just visible to the right of the frame.

The evolution of the ejection is now seen best in the C3 coronagraph. Four partial frames are shown in Figure 4, taken at 03:39, 04:08, 04:48, and 06:06 UT, respectively. These images are difference images, and each has had the pre-event coronal image subtracted from it. This emphasises transient behaviour, but can clearly produce artefacts if long-lived features such as coronal streamers move through the field of view due to solar rotation. The ejection now begins to look like a bright bubble on a stem, with the brightest feature very close to the top of the

Figure 2. The growth of the streamer on July 28. The images are from the C2 coronagraph, where the diameter of the occulting disc is 1.5 $R_\odot$. The times of the images are in the upper left of each frame. The west limb is to the right and north is at the top on all figures. The image is taken with a clear filter and polarizer. The appearence of the west equatorial streamer at 00:49 UT was similar to that over the previous 12 hours. The image at 15:04 UT shows not only a significant brightening, but also a marked swelling. Note that the appearance of the east limb is little changed.

stem. In this sequence another, fainter, ejection may be seen moving outwards at a projected angle around 20° south of the equator.

Unfortunately LASCO suffered a data gap between 06:06 and 14:00 UT, so when the event is next seen it has a rather different appearance, as may be seen in Figure 5. These are again differenced C3 images, so the very dark radial feature off the west limb is an artefact caused by rotation of the bright streamer between the

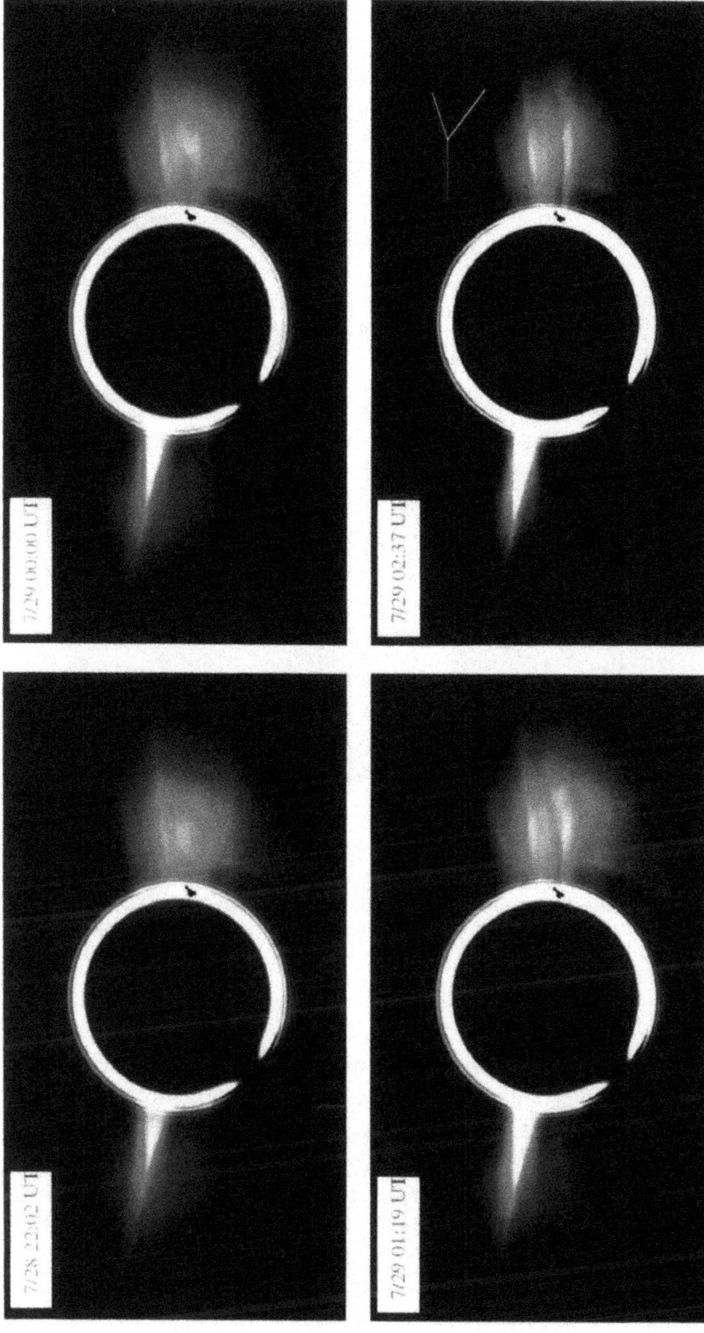

Figure 3. The evolution of the July 28–29 event as seen by the C2 coronagraph. The times of the images are given in the upper left of each frame and the sequence goes from left to right. (July 29, 22:02 UT) This image shows a bright, pear-shaped feature which is embedded within the expanded helmet streamer. (July 29, 00:00 UT and July 29, 01:19 UT) The mass has moved outwards, leaving a bright strip connecting back towards the Sun. (July 29, 02:37 UT) In this image the contrast has been enhanced to highlight the bifurcation of a thin bright sheet slightly to the north of the main ejection. The feature is reproduced in white in the upper part of the frame to aid the eye. The trailing edge of the ejected mass is just visible on the right.

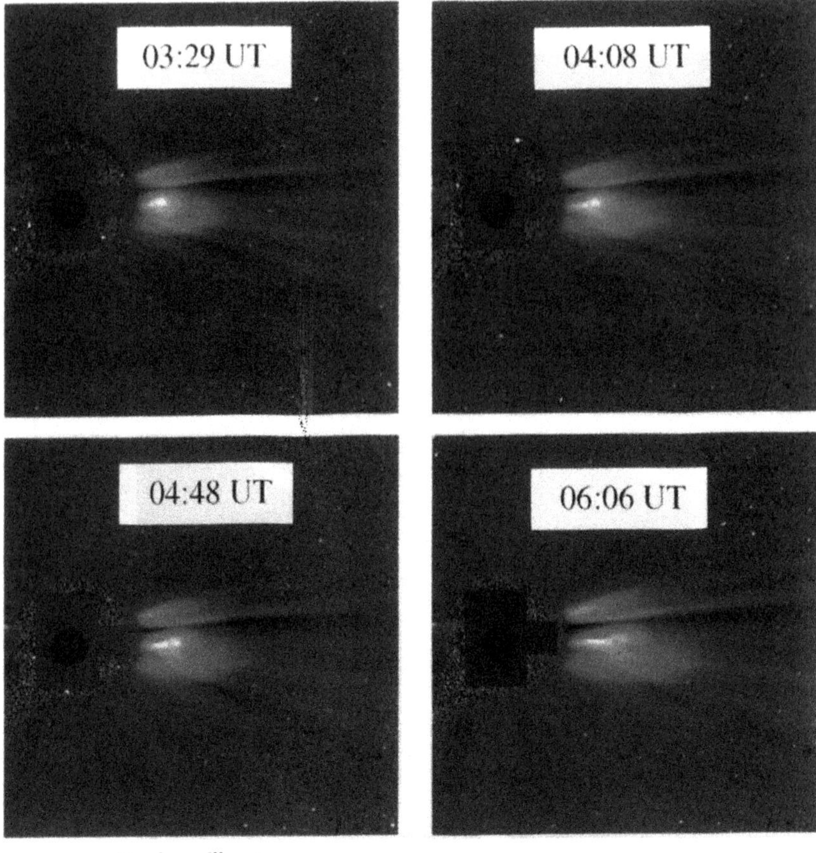

⟵ 20 solar radii ⟶

Figure 4. The ejection as seen in the C3 coronagraph, in four images taken at 03:29, 04:08, 04:48, and 06:06 UT, respectively. These images have all had a pre-event image subtracted, and therefore may contain some artefacts caused by rotation of the streamer. The appearance of the transient bright feature is not affected by this. This sequence shows the evolution of the height of the ejected mass with the bright, thin column linking back to the Sun.

image shown and the subtracted image. However, the important feature of these images is the dark ‘<’-shaped rarefaction which is visible in the centre of the image at 14:00 UT. This feature could be tracked all the way out to the edge of the field of view, although after around 25 $R_{\odot}$ it became very difficult to identify and could be seen best by viewing a moving sequence of images. The projected angle between the two dark lines in the ‘<’ is 75°. Not only does the rarefaction propagate all the way out to 30 $R_{\odot}$, but it maintains the same angle throughout.

It has recently been possible to supplement this paper with a CD-ROM movie to illustrate the dynamic phenomena taking place during this mass ejection. The

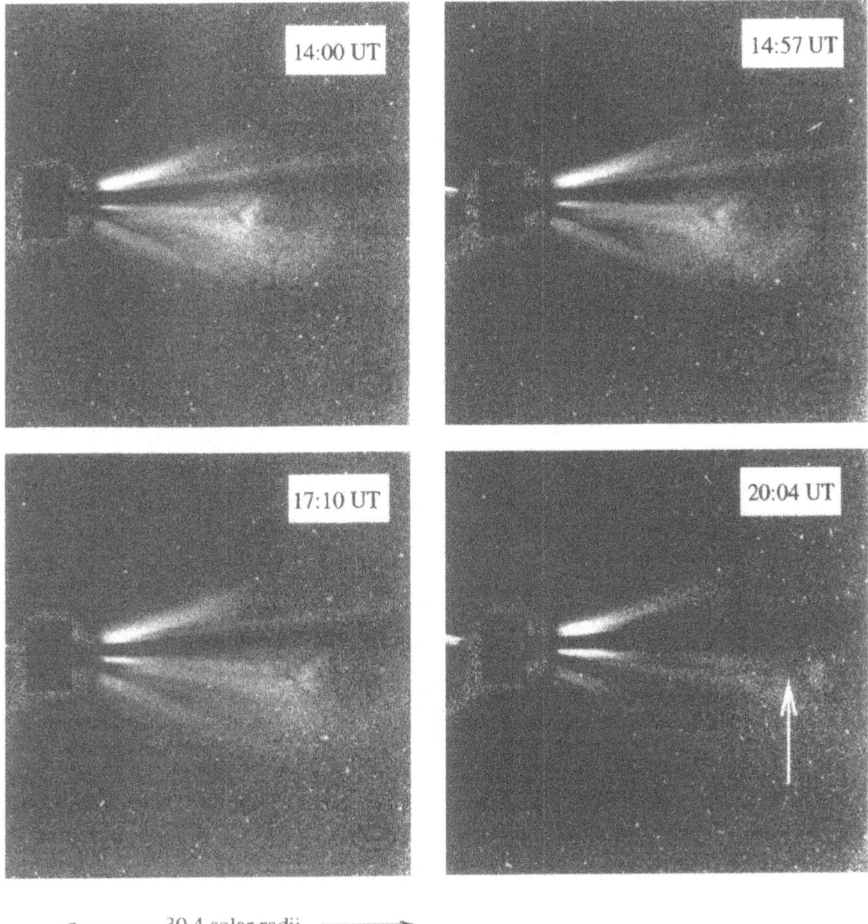

Figure 5. The motion of a '<'-shaped rarefaction which developed behind the ejected mass. It is visible in the centre of the image taken at 14:00 UT on July 29. By 17:10 UT it is still quite clear, but by 20:04 it is becoming difficult to identify. The white arrow points to the apex of the '<' in this last frame. The projected angle, in the plane of the sky, between the two dark lines in the '<' is 75°.

first movie covers the July 28–29 event, and it is in two parts. The first part shows a composite of the C2 and C3 fields of view. The second part shows just the C2 field of view. The main mass ejection is very clear, and in the frame at 03:17 UT on July 29 the beginning of the '<'-shaped rarefaction behind the main mass ejection can be seen. The second movie shows the C2 field-of-view for November 4 until 06:00 UT on November 6. This is the subject of the second event discussed below. There are two features of interest here. The first is that just prior to the mass ejection off the west limb, there is an event off the east limb. Second, from 19:00 UT on

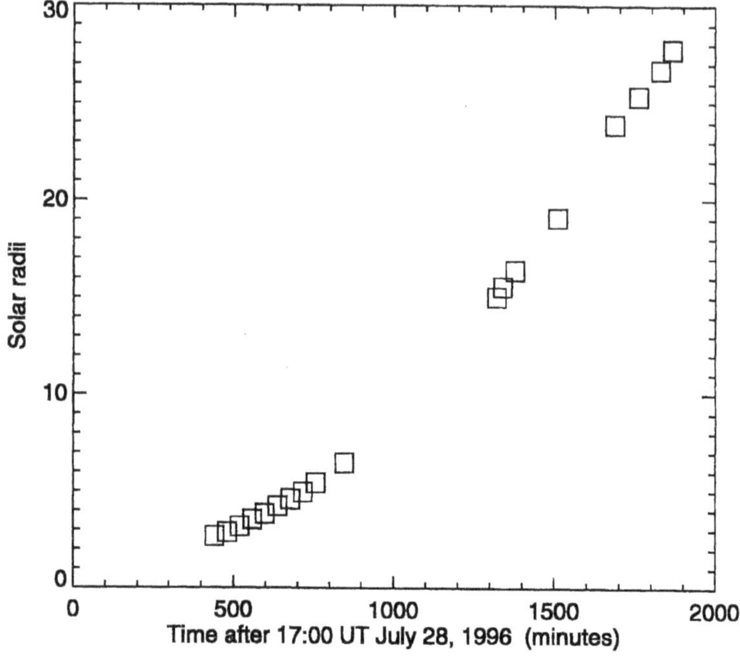

Figure 6. The height – time plot of the apex of the '<', clearly identified in Figure 5. The velocity in the lower part of the plot is 110 ± 5 km s^{-1} and in the outer region it is 269 ± 10 km s^{-1}.

November 5 the bifurcation, which we have argued is a manifestation of a current sheet caused by reconnecting magnetic fields, is clearly visible behind the main mass ejection. The movie representation of the events provides a much better visual impact than the still frames shown in the figures.

The apex of the '<' provides a useful reference for studying the evolution of the velocity of the ejected material. This is shown in Figure 6. Here it is clear that the event has undergone significant acceleration in the vicinity of 5 to 6 $R_\odot$. The velocity before the data gap is not quite constant, as there is a very slight acceleration between 2 and 5 $R_\odot$, but the average projected velocity is 110 ± 5 km s^{-1} In the region above 15 $R_\odot$ the velocity is 269 $\pm$10 km s^{-1}. It is very clear from Figure 6 that the '<' feature is moving out at a constant velocity, with neither acceleration nor deceleration.

2.2. The November 5 Event

The second event we discuss here occurred on November 5. During this period the LASCO/C1 coronagraph was making occasional line scans with the Fabry–Pérot interferometer in the line of Fe X, and Figure 7 shows an Fe X image taken at 04:30 UT. This image is obtained by dividing the image taken at the center of

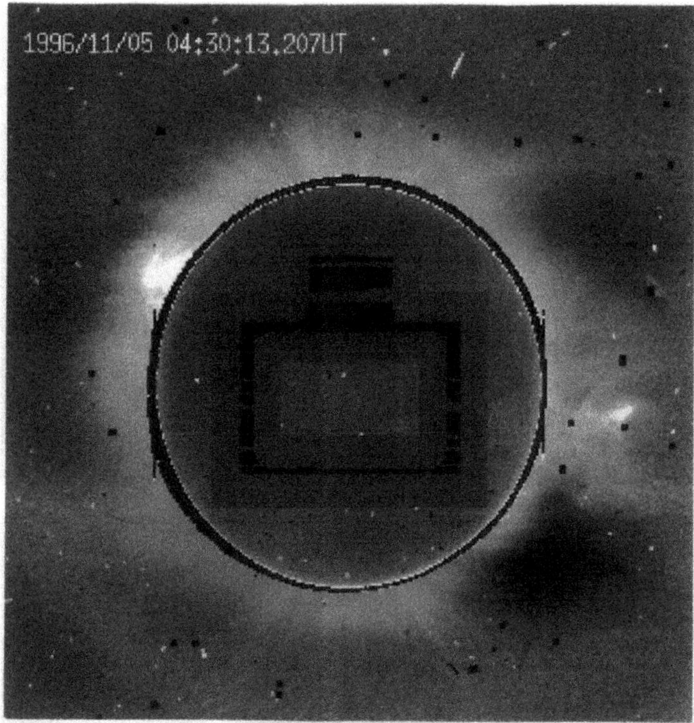

Figure 7. The onset of the event at 04:30 UT as seen in the spectral line of Fe X at λ637.4 nm. The image is made by taking an image in the centre of the line and dividing it by the average of the images taken 0.02nm either side of line centre. The diameter of the occulting disc is 1.1 $R_{\odot}$.

the line, at λ637.4 nm, by the average of images taken 0.02 nm either side of line centre. The bright patch approximately 0.2 $R_{\odot}$ above the west limb is believed to represent hot material in the corona. Unfortunately there was no further Fe X line scan made that day, so we do not know if this represents the onset of the mass ejection or merely a coronal site where high density/high temperature material is concentrated. However, examination of the remainder of this image shows no other detached coronal feature; only the region off the north-east limb is of comparable brightness, and this is clearly linked down to below the level of the occulting disc (1.1 $R_{\odot}$). Underneath the bright feature (towards the west limb) there is a hint of a bright arc which is concave outwards. If this is a real feature it is in the correct orientation to be the same structure as is seen later.

By around 10:00 UT there is a mass ejection seen moving above the west limb by the C2 coronagraph. However, the image clearly shows a bubble with a dense (bright) core. Figure 8 (upper frame) shows the C2 image at 12:30 UT, where the dense plasma ball is visible inside a somewhat less dense cloud of gas. Note the sharp boundary to the south of the event in what might be described as the halo

Figure 8. Upper panel: the appearance of the ejected mass at 12:30 UT on November 5. *Lower panel*: the sharp, linear features which were visible behind the ejected mass at 19:30 UT. The one in the center is immediately behind the main mass ejection and shows a bifurcation at around 4 $R_{\odot}$. There are two similar features at lower projected heights.

surrounding the main ejection; to the north it merges with the pre-existing streamer in this projected image. Figure 8 (lower frame) shows the image at 19:20 UT, when the ejected plasmoid has moved out towards the right edge of the C2 image. Close examination of this image reveals several narrow, bright rays, with bifurcations at increasing radial distance. The most obvious one in Figure 8 is right behind the main ejection, which is actually a double feature. There are two more visible, one almost exactly on the equator and the other about 25° north. These thin density

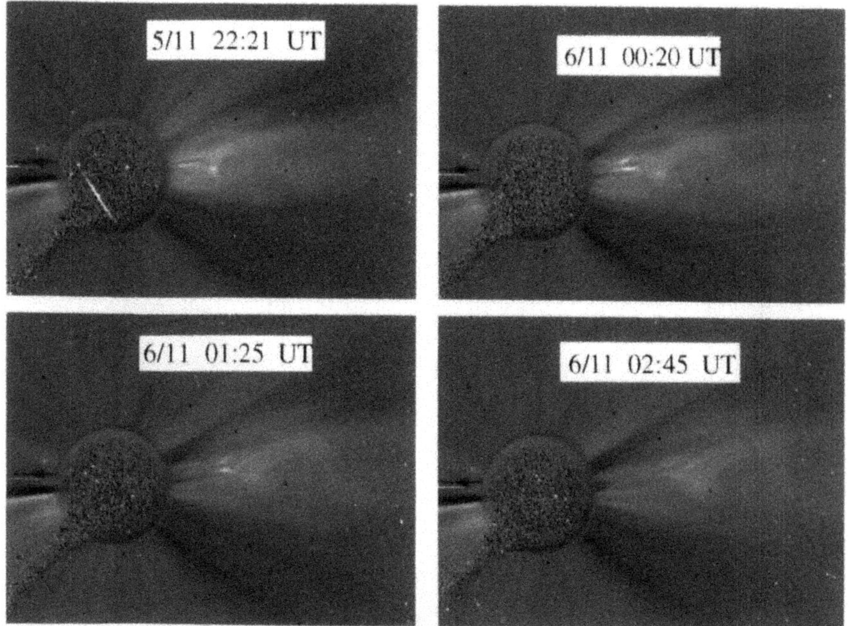

Figure 9. The evolution of the November 5–6 event around the time of the acceleration. The images are from the C3 coronagraph.

enhancements are interpreted as signatures of current sheets formed by magnetic reconnection in the structures associated with the CME; they are visible in the C2 images for several hours, and the 'Y' features move outwards, presumably tracking the main CME.

The CME continued to move outwards through the corona and underwent a significant acceleration around the end of November 5. Figure 9 shows a sequence of four C3 images which illustrate the evolution of the event from 22:21 UT to 02:45 UT (November 6). The image at 00:20 UT (November 6) is particularly interesting as it shows (a) the main bubble with a thin bright straight line linking back to the Sun, but also a second bright line to the north. This line is probably associated with a second ejected bubble which is behind the first bubble as seen from SOHO. Thus a given event may comprise at least two distinct ejected mass bubbles, or plasmoids.

The 'Y' shaped boundary in the bright rays gives the event a distinct feature which may be readily tracked. Figure 10 shows the height-time plot in the plane of the sky for this feature. The points above 10 $R_{\odot}$ lie on a straight line, corresponding to a projected velocity of 310 ± 10 km s^{-1}. For most of the passage through C2 the velocity is around 54 km s^{-1}, although there is some slight acceleration as may be seen by a careful examination of Figure 10. We have plotted the Fe X point from

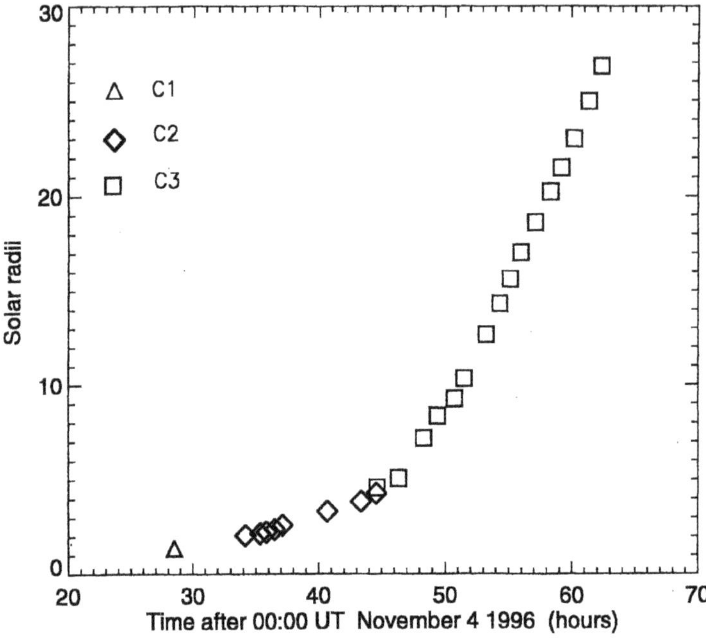

Figure 10. The height-time profile of the trailing edge of the mass ejection for the November 5 event. The different symbols indicate the coronagraph which has imaged the event at the indicated times.

C1 as well, and this falls on a reasonable extrapolation of the C2 points. The main acceleration of the event occurs around 5 $R_{\odot}$.

3. Discussion

The two events reported here are clear examples of disconnected plasmoids ejected from the Sun. The ejected mass has the appearance of a bubble of dense gas. Comparison of the west limb appearance of the July event (Figure 3, upper panels) with the November event (Figure 8, upper panel) illustrates this well. Moreover, from casual inspection of these images one could be excused for thinking that they were the same event. There have been many attempts to understand the underlying physics of a CME (e.g., Low, 1994). Furthermore, numerical modelling is now advanced to the point that many details for different types of CME can be addressed (e.g., Wang *et al.*, 1995). In fact, Wang *et al.* (1995) have described simulations which have somewhat similar properties to the events discussed here.

An unexpected observation is that of the thin bright rays, with bifurcations at the trailing edge of the ejected plasmoid. The July 28–29 event has one such feature, while the November 5 event has several. We interpret these features as current sheets, similar to those discussed by Petschek (1964). Plasma flows in towards the boundary layer induce magnetic reconnection, while at the same time matter is ejected outwards along the magnetic field lines. There is enhanced plasma pressure at the centre of the current sheet, which shows up in the C2 images as a bright feature due to a density enhancement. The bifurcation point is then interpreted as the local magnetic reconnection site. The November 5 event shows at least three current sheets which move outwards through the corona at approximately the same speed. This is interpreted as multiple, discrete mass ejections in the same general direction. If they were at widely separated longitudes, then we would expect the projected images to show some velocity dispersion. We believe this is an important facet of this event which should be taken into account by the modellers.

The other interesting feature of the July 28–29 event is the 'V'-shaped rarefaction which is visible moving out through the corona to beyond the edge of the field of view at $\sim 30\ R_{\odot}$. This rarefaction is already visible in some of the C2 images (not shown), so it forms relatively close to the Sun, and is probably a manifestation of a cavity in the magnetic field such as is shown in Figure 1. We believe that the observations of this feature are partly a fortuitous result of a favourable viewing angle, as the rarefaction we have identified must clearly be a three-dimensional structure. Thus if it were viewed at a different angle, it would certainly look very different, and probably would not be visible at all. Thus for this event, the velocity of the ejecta was probably very close to 90° to the line of sight, in which case the projected velocity of 269 km s^{-1} must be very close to the actual velocity.

The acceleration of the plasmoids in the region of 5–6 $R_{\odot}$ we believe is an indication of where they get accelerated by the solar wind, such that they then move outwards into the interplanetary medium at the solar wind velocity. The velocity of both ejecta in the region beyond 10 $R_{\odot}$ is consistent with the velocity of the slow solar wind, when projection effects are taken into account. Furthermore, this general acceleration profile may well represent the acceleration profile of the slow solar wind. In this case, the source surface at low latitudes of the solar wind would be around 6 $R_{\odot}$. If this were not to be the case, then some other agent must be invoked for accelerating the plasmoids at this radial distance.

It is evident from the behaviour of the equatorial streamer on July 28 that prior to the mass ejection there had been a steady input of mass to the corona. This was responsible for the streamer swelling shown in Figure 2. The triggering of the CME is presumably a result of increased pressure within the magnetic field in or adjacent to the streamer. The pressure increase can be due to enhanced density, increased temperature, or most likely a combination of both. The C2 coronagraph responds to light scattered off free electrons and therefore any increase in brightness, such as that shown in Figure 2, is caused by increased mass along the line of sight. Virtually the only source of this mass is the chromosphere and it is presumably the result of gentle evaporation.

If our interpretation of the bright rays visible in Figures 3 and 8 as current sheets is correct, the magnetic reconnection that leads to the current sheets will also accelerate charged particles. Coulomb losses of these particles will increase the gas temperature within the swelling structure, thus increasing the tendency towards instability (Simnett and Harrison, 1985).

Acknowledgements

We acknowledge the support of our colleagues in the LASCO consortium and the funding agencies which made LASCO possible.

References

Brueckner, G. E. and 14 co-authors: 1995, *Solar Phys.* **162**, 357.
Illing, R. M. E. and Hundhausen, A. J. : 1983, *J. Geophys. Res.* **88**, 10 210.
Low, B. C.: 1994, *Proceedings of the Third SOHO Workshop 'Solar Dynamic Phenomena and Solar Wind Consequences'*, ESA SP-373, p. 123.
Low, B. C.: 1996, *Solar Phys.* **167**, 217.
McComas, D. J., Gosling, J. T., Phillips, J. L., Bame, S. J., Luhmann, J. G., and Smith, E. J.: 1989, *J. Geophys. Res.* **94**, 6907.
McComas, D. J., Phillips, J. L., Hundhausen, A. J., and Burkepile, J. T.: 1991, *Geophys. Res. Letters* **18**, 73.
Petschek, H. E.: 1964, *The Physics of Solar Flares*, NASA SP-50, p. 425.
St. Cyr, O. C. and Burkepile, J. T.: 1990, NCAR Technical Report, NCAR TN-352+STR.
Simnett, G. M. and Harrison, R. A.: 1985, *Solar Phys.* **99**, 291.
Wang, A. H., Wu, S. T., Suess, S. T., and Poletto, G.: 1995, *Solar Phys.* **161**, 365.
Webb, D. F. and Cliver, E. W. : 1995, *J. Geophys. Res.* **100**, 5853.

THE RELATIONSHIP OF GREEN-LINE TRANSIENTS TO WHITE-LIGHT CORONAL MASS EJECTIONS

S. P. PLUNKETT[1,2], G. E. BRUECKNER[2], K. P. DERE[2], R. A. HOWARD[2], M. J. KOOMEN[2], C. M. KORENDYKE[2], D. J. MICHELS[2], J. D. MOSES[2], N. E. MOULTON[2], S. E. PASWATERS[2], O. C. ST. CYR[2], D. G. SOCKER[2], D. WANG[2], G. M. SIMNETT[3], D. K. BEDFORD[3], D. A. BIESECKER[3], C. J. EYLES[3], S. J. TAPPIN[3], R. SCHWENN[4], P. L. LAMY[5] and A. LLEBARIA[5]

[1]*Universities Space Research Association, Mail Code 682.3, NASA Goddard Space Flight Center, Greenbelt, MD 20771, U.S.A.*

[2]*E.O. Hulburt Center for Space Research, Naval Research Laboratory, Washington, DC 20375, U.S.A.*

[3]*School of Physics and Space Research, University of Birmingham, Edgbaston, Birmingham B15 2TT, U.K.*

[4]*Max Planck Institut für Aeronomie, D-37189 Katlenburg-Lindau, Germany*

[5]*Laboratoire d'Astronomie Spatiale, F-13376 Marseille Cedex 12, France*

(Received 18 March 1997; accepted 12 July 1997)

Abstract. We report observations by the Large Angle Spectrometric Coronagraph (LASCO) on the SOHO spacecraft of three coronal green-line transients that could be clearly associated with coronal mass ejections (CMEs) detected in Thomson-scattered white light. Two of these events, with speeds >25 km s^{-1}, may be classified as 'whip-like' transients. They are associated with the core of the white-light CMEs, identified with erupting prominence material, rather than with the leading edge of the CMEs. The third green-line transient has a markedly different appearance and is more gradual than the other two, with a projected outward speed <10 km s^{-1}. This event corresponds to the leading edge of a 'streamer blowout' type of CME. A dark void is left behind in the emission-line corona following each of the fast eruptions. Both fast emission-line transients start off as a loop structure rising up from close to the solar surface. We suggest that the driving mechanism for these events may be the emergence of new bipolar magnetic regions on the surface of the Sun, which destabilize the ambient corona and cause an eruption. The possible relationship of these events to recent X-ray observations of CMEs is briefly discussed.

1. Introduction

The green emission-line corona has been studied in detail from ground-based observatories, notably Sacramento Peak Observatory, for many years. The green coronal line, at a wavelength of 5303 Å, arises from a forbidden transition in the Fe XIV ion, and has a peak emissivity at a temperature of about 1.9×10^6 K (Mason, 1975; Esser *et al.*, 1995). Cinematography revealed transient events in the emission-line corona, in which the morphology of the structures emitting in the green line was observed to change dramatically in a short period of time. Dunn (1971) provides a review of the early observations of green-line transients, and attempts to divide these events into a number of classes, based on their morphology and speed. Correlation of green-line events with other forms of solar activity has been attempted by a number of authors. DeMastus, Wagner, and Robinson (1973)

Solar Physics **175:** 699–718, 1997.

studied the association of green-line events with Hα activity on the limb and with solar radio bursts. They concluded that 'the same mechanism which releases chromospheric and prominence material into the corona also produces the green-line transients' and that metric and/or centimetric radio bursts were often associated with 'non-trapped' green-line events. Wagner *et al.* (1974) reported a transient that was observed both by an Fe XIV coronagraph and a K-coronameter, in which a depletion of the K-corona was observed during a restructuring of the emission-line corona. In another event, Fisher (1977) observed a depletion of the inner corona in Fe XIV, with associated metric radio activity.

The relationship of green-line transients to coronal mass ejections (CMEs), observed in Thomson-scattered white light, has been difficult to establish, partly because of limitations imposed by the available instrumentation. Ground-based observations are restricted to a few tenths of a solar radius above the limb, and are limited in their operations by weather conditions. Space-borne coronagraphs have all been of the externally-occulted type, in which the inner edge of the field of view is limited by vignetting to about 2.0 $R_\odot$. Nonetheless, Wagner *et al.* (1983) were able to report the simultaneous detection of a CME in white light out to 10 $R_\odot$ (using the SMM Coronagraph/Polarimeter and the P78-1 coronagraph on Solwind) and an Fe X red-line (6374 Å) transient. However, this event was markedly different in both character and speed to the green-line transients. Wagner (1987) attempted to relate green-line transients and white-light CMEs observed with the Skylab/ATM S-052 coronagraph. He concluded that 'no report can be given today of a dramatic coronal transient whose morphology can be followed in time through cinematography in the Fe XIV coronal emission line and correspondingly in its progress through the outer white-light corona'.

There has been a great deal of interest in recent years in the detection of CMEs in the lower corona using soft X-ray emission, particularly since the launch of the *Yohkoh* satellite in 1991. Klimchuk *et al.* (1994) studied a sample of 29 eruptive X-ray events, and concluded that their characteristics were similar to white-light CMEs, although they could not identify a CME that was associated with any of the X-ray events they studied. Other studies (e.g., Hiei, Hundhausen, and Sime, 1993; Sime, Hiei, and Hundhausen, 1994) suggest that the X-ray structures are associated with an erupting prominence behind the leading edge of a CME, and are probably due to reconnection of field lines that are opened up during the eruption. Gopalswamy *et al.* (1997) claim to have observed the frontal loop of a CME in X-rays, although they had no simultaneous white-light observations to confirm their conclusions. One of the most interesting results of the X-ray studies has been the observation that the coronal X-ray emission frequently shows a marked dimming around the time of an eruption (e.g., Hudson, Acton, and Freeland, 1996; Hudson and Webb, 1997). The relationship of these dimming events to CMEs also remains unclear.

It is important for proper interpretation of the results quoted above, and the results to be presented in this paper, to remember that the green-line emission reveals the presence of significant quantities of Fe XIV ions, whereas observations in white light show the distribution of electron density in the corona. Thus, the green-line emission represents a small portion of the full differential emission measure distribution of the corona, whereas the white-light observations are insensitive to temperature and show the entire corona. This fact alone means that there may be many events that are clearly evident in white light, but may not show up at all in green-line observations. Similar cautionary remarks apply to comparisons of X-ray and white-light observations.

The Large Angle Spectrometric Coronagraph (LASCO) (Brueckner *et al.*, 1995) on the Solar and Heliospheric Observatory (SOHO) affords the opportunity to overcome many of the limitations imposed by earlier instrumentation. LASCO is a set of three nested coronagraphs, with overlapping and concentric fields of view. The inner coronagraph, C1, is an internally occulted mirror coronagraph, with a field of view covering 1.1–3.0 $R_\odot$ with essentially no vignetting. The middle and outer coronagraphs, C2 and C3, are externally occulted instruments, with fields of view extending from 2.0–6.0 $R_\odot$ and 3.7–32.0 $R_\odot$ respectively. C1 includes a Fabry–Pérot interferometer and a set of blocking filters that can be used to isolate selected coronal emission lines, including the Fe XIV green line, and to perform imaging spectroscopy of the inner corona. The outer coronagraphs image the corona in white light. This paper reports an analysis of two events observed by LASCO in both the green line and as white-light CMEs during the first year of operation in 1996 and relates the emission-line transients to the structures observed in the outer corona. The events discussed were not the only ones of their type observed during this time; however, they serve to illustrate very well the association between green-line transients and CMEs. Detailed analyses of other events will be published in future papers.

2. Observations

The normal LASCO mode of operation includes a 'synoptic' program of observations, in which C1 acquires images of the green-line emission corona at a fairly regular cadence of about one hour, while C2 and C3 record the white-light corona beyond 2 $R_\odot$ at a similar cadence. The observations reported here were all taken using this program. The C2 and C3 synoptic observations employed a limited field of view over the polar regions (out to 3.3 $R_\odot$ for C2 and 18 $R_\odot$ for C3) in order to save on telemetry bandwidth. Time-lapse movies are examined on a regular basis to search for, among other things, evidence of transient activity in the green-line data and CMEs in the white-light data.

2.1. THE 17 JUNE EVENT

Figure 1 shows a series of green-line images recorded on 17 June 1996, in which a 'loop-like' transient is clearly evident. The background continuum has been removed from these images by subtracting an image recorded at a wavelength close to that of the green line (5309.6 Å), but sufficiently far removed from line centre to avoid contamination by emission in the line itself. The continuum image used in this case was recorded at 08:54 UT, which is sufficiently close to the times of the on-line images shown in Figure 1 that the effects of solar rotation on the structures along the line of sight can safely be ignored. To allow as wide a dynamic range of intensities as possible to be recorded, the instrument has a built-in radial filter to reduce the intensity in the inner portion of the field of view (out to 1.3 $R_{\odot}$); the effects of this radial filter have been retained in the images shown in Figure 1. One further piece of information about the instrument design is necessary for correct interpretation of these images; the optical design of the C1 instrument is such that the effective wavelength transmitted by the Fabry–Pérot varies slightly with radial distance across the field of view. The wavelength shift from the inner to the outer edge of the field is about 0.4 Å. The images in Figure 1 were recorded with the Fabry–Pérot tuned to an effective wavelength of 5302.9 Å at 1.1 $R_{\odot}$. The fall-off in green-line signal at radial distances greater than about 1.5 $R_{\odot}$ can be attributed, at least in part, to the fact that the effective wavelength is slightly offset from line centre in these parts of the images. There is a strong Fraunhofer line in the solar spectrum just to the blue side of the Fe XIV line, so that if the effective wavelength shifts into this feature, the measured signal decreases sharply. More recent tests have demonstrated that, with appropriate tuning, the green-line emission can be detected almost to the edge of the C1 field of view, at 3.0 $R_{\odot}$.

On the day that the images in Figure 1 were obtained, there was an active region close to the east limb of the Sun. The coronal loop systems associated with this active region are clearly visible in the C1 green-line images (see, for example, the first panel in Figure 1). The active region emission is superimposed on the general quadrupolar nature of the green-line corona, with two bright extended streamers at latitudes of about ±40° (Schwenn *et al.*, 1997).

The C1 image at 01:30 UT on 17 June shows a small loop beginning to rise just above the occulter, at a latitude of about 10° N, right above the location of the active region. By the time of the next image at 02:03 UT, this loop (labelled as feature A in Figure 1) has expanded both vertically and laterally, now stretching over some 30° or more in latitude, and reaching to a height of about 1.6 $R_{\odot}$. It also appears to have swept up some of the material from the surrounding, diffuse corona visible in the earlier image, forming a cavity behind the expanding bright loop. At the same time, the northern leg of the loop system, at about 20° N, brightened noticeably just above the occulting disk. In the next image, at 02:47 UT, the closed loop structure is no longer visible, although close examination of this image on a computer screen reveals a faint structure at about 2.3 $R_{\odot}$ that does not show up on the printed version. The faintness of this feature is at least partly due to the shift in effective wavelength across the image, as discussed above. At this time, the bright legs of the loop are clearly evident, as they are in the last image in Figure 1, at 03:32 UT. These images give the appearance of a structure that has been forced open by

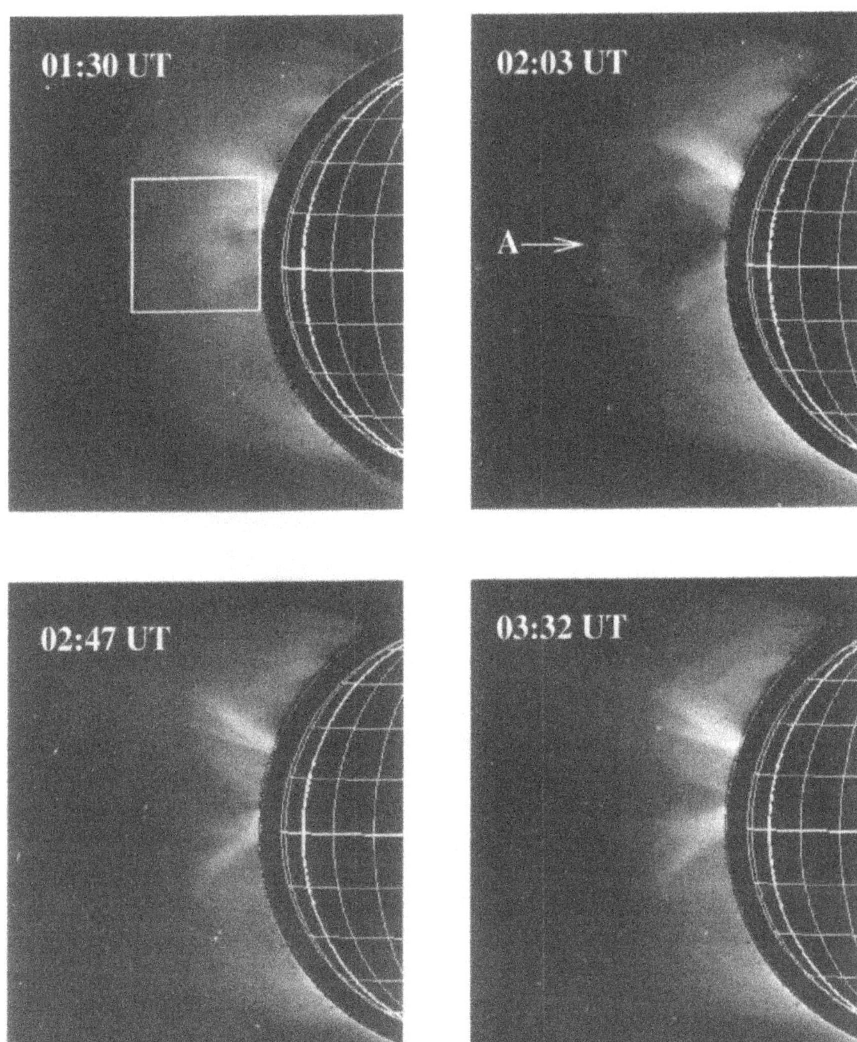

Figure 1. A sequence of green-line images recorded by the LASCO C1 coronagraph on 17 June 1996. An image taken in the nearby continuum at 08:54 UT has been subtracted from each of the images shown here. The black disk indicates the inner edge of the field of view at 1.1 $R_{\odot}$, while the heliographic grid indicates the position of the Sun behind the occulter. The white rectangle drawn on the first image marks the region used to generate the light curve shown in Figure 8.

the eruption of the loop system, and may be reconnecting to structures at higher latitudes, resulting in the brightening of the legs as the eruption proceeds. The whole process fits very well with the picture of the 'whip-like' events, classified by Dunn (1971) as either 'realignments' or 'accelerated expansions'. A movie of this

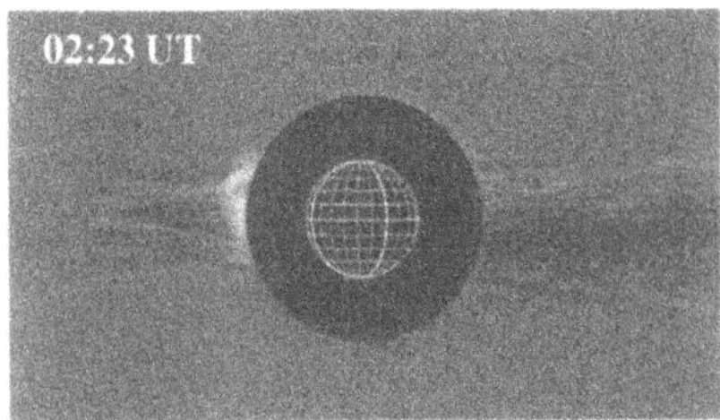

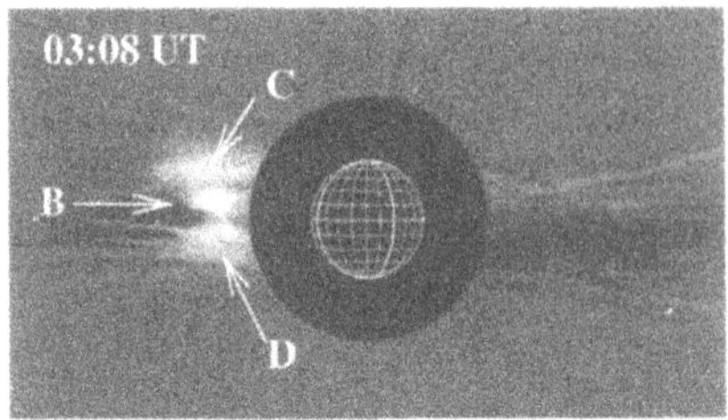

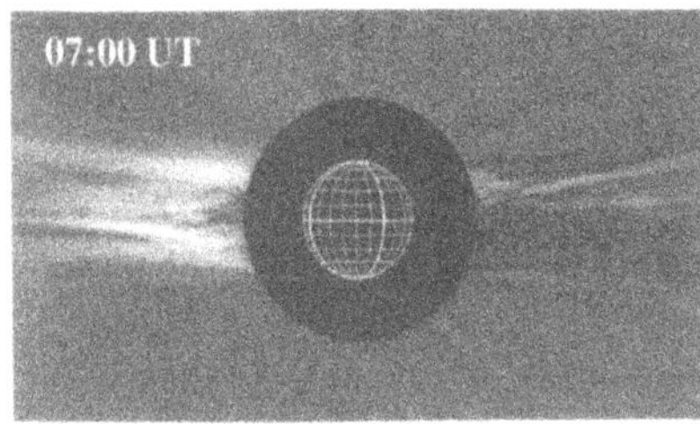

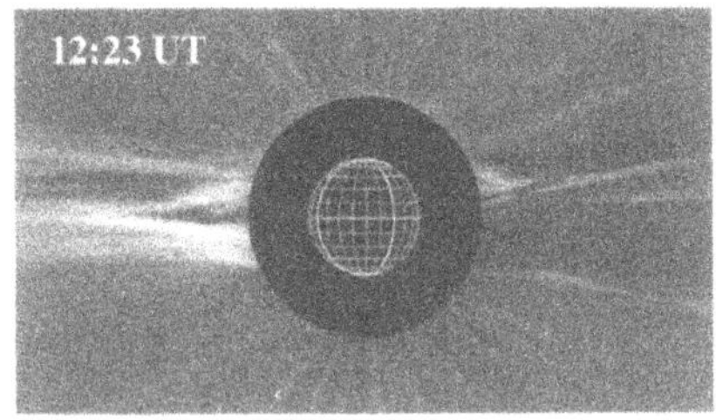

Figure 2. Development of a CME in the LASCO C2 coronagraph on 17 June, 1996. A base image made up of the average of three images recorded prior to the CME has been subtracted from each image shown here. The images have been scaled as the square root of the intensity, to enhance the visibility of faint features. The black circle marks the location of the occulting disk and the heliographic grid shows the position of the Sun. Labelled features are discussed in the text.

event, including the images shown in Figure 1, can be found on the accompanying CD-ROM.

The C2 and C3 coronagraphs detected a white-light CME on this same day, first visible as a bright mound extending right along the eastern edge of the C2 occulting disk at 2 $R_{\odot}$ in the image recorded at 02:23 UT. Figure 2 shows the development of this CME in C2 over a period of ten hours. The images presented here are difference images, with a base image made up of the average of the three images recorded immediately prior to the onset of the CME (23:15 UT on 16 June, 00:54 UT and 01:39 UT on 17 June) subtracted from each displayed image. This procedure effectively removes the quiescent white-light corona and allows the (relatively faint) CME to be seen. One must exercise caution in interpreting these difference images; for example, the dark patches visible on the west limb in the two latest images shown in Figure 2 are probably due to rotation of a bright structure in the period since the base image was taken, rather than any ejection of material from this region. Also, the intensity scaling that is necessary to make the features of interest more clearly visible may artificially enhance the apparent darkness of the voids in some images. This is particularly true in those cases where a square-root intensity scaling has been applied, as all negative differences are then shown as black.

From its first detection in C2 at 02:23 UT, the CME develops a rather complex structure, in which a bright 'blob' of material moves out just north of the equator (feature B in the image at 03:08 UT in Figure 2), accompanied by at least two jet-like features (labelled C and D) to the north and south of this blob and a general brightening of the surrounding streamers. There were no C2 images recorded between 03:41 UT and 06:05 UT, after which it becomes more difficult to trace individual features in the event. Nonetheless, the bright blob can be followed through part of the C3 field of view, out to about 11 $R_{\odot}$ at 16:55 UT. As the C2 image at 12:23 UT shows, the streamers continue to brighten out to several solar radii during this time.

2.2. The 18–19 October Events

On 18 October 1996, a very gradual green-line loop transient was observed by the C1 coronagraph off the southeast limb, at a latitude of about 50° S. Around this time, an extended filament channel that had persisted for several rotations, and gave rise to a number of large CMEs, reappeared on the southeast limb. Starting at about 16:00 UT, the loop expanded slowly outwards in the C1 field of view over a period of more than twelve hours, taking on an extended tongue-like shape as it did so. It also curved noticeably to the north of its original location as it expanded. This slowly-expanding loop was visible in green-line emission at a height of about 1.6 $R_{\odot}$ at 18:53 UT on 18 October, and rose to a height of 2.2 $R_{\odot}$ at 05:03 UT on 19 October, before disappearing from the C1 field of view. This loop was quite faint, and can only be clearly seen with the aid of contrast enhancement on a computer screen.

On the next day, 19 October 1996, another spectacular green-line transient was observed by LASCO at a location slightly to the north of the earlier event. This event was much faster than the one on 18 October and appeared much more as a whip-like eruption than a gradual expansion of a pre-existing, closed loop. Figure 3 shows a sequence of C1 green-line images over a period of six hours from 15:05 UT to 21:19 UT on 19 October. Again, the images taken at line centre have had a nearby continuum image, recorded at 10:23 UT, subtracted from them. Beginning at 15:05 UT, a loop-like structure is visible at a height of about 1.5 $R_{\odot}$, rising beneath the diffuse emission from the streamer overlying the filament channel. This loop, labelled A' in Figure 3, is more clearly visible in the image at 17:25 UT, where it has risen to a height of about 1.8 $R_{\odot}$, and much of the diffuse Fe XIV emitting material that was present in the earlier images has disappeared. In subsequent images, this bright loop front is no longer detectable, and one is left with a void in the emission-line corona in the images recorded at 19:15 UT and 21:19 UT. A C1 green-line movie from 18–19 October, showing both of the events discussed in this section, can be found on the accompanying CD-ROM.

Two white-light CMEs from approximately the same location as the green-line events were observed by the C2 and C3 coronagraphs on 19 October. The first event in C2 began with a brightening and expansion of the streamer off the southeast limb at 04:27 UT, followed by a loop-like CME that is first visible in the image

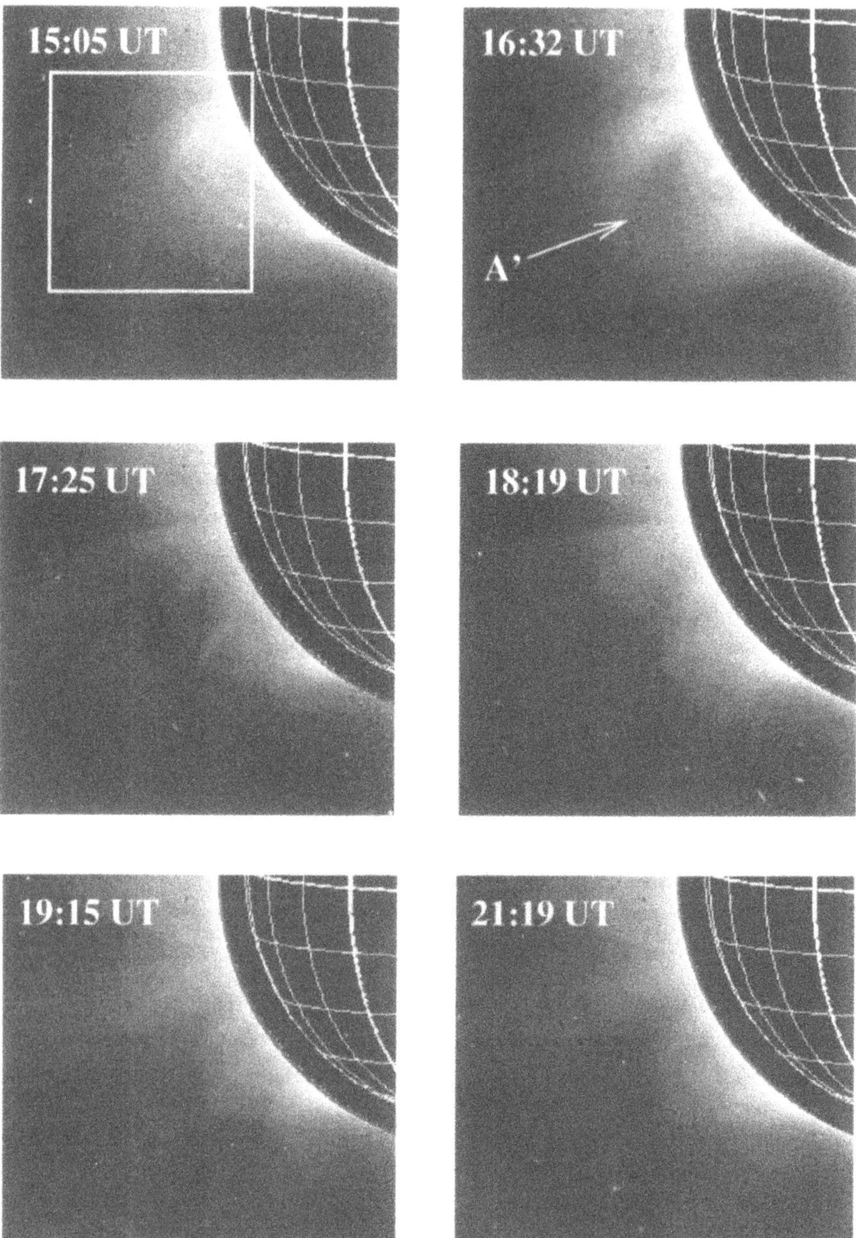

Figure 3. A sequence of green-line images recorded by the LASCO C1 coronagraph on 19 October 1996. An image taken in the nearby continuum at 10:23 UT has been subtracted from each of the images shown here. The black disk indicates the inner edge of the field of view at 1.1 $R_{\odot}$, while the heliographic grid indicates the position of the Sun behind the occulter. The white rectangle drawn on the first image marks the region used to generate the light curve shown in Figure 11.

recorded at 06:05 UT. The leading edge of this loop (feature B' in Figure 4) can be followed for almost 24 hours as it moves outwards through the fields of view of C2 and C3, to a projected radial distance of almost 30 $R_{\odot}$. The event has the appearance of a bright loop near the top of a streamer, and would probably be classified as a slow 'streamer blowout', adopting the terminology used by Howard *et al.* (1985) to classify CMEs detected with the SOLWIND coronagraph. There is some evidence (see, for example, the C3 image at 22:45 UT on 19 October shown in Figure 5) that the loop B' disconnects from the Sun to form a detached plasmoid, with a distinctive tear-drop shape; however, this aspect of the event will not be pursued here. Other examples of disconnected CMEs observed by LASCO have been reported by Simnett *et al.* (1997).

The second CME was much faster than the first, and is only visible in one frame in C2, at 18:00 UT, where it is marked as feature C′ in Figure 4. This event can be followed in C3 to 30 $R_{\odot}$, as an expanding loop structure with a rather diffuse front (see Figure 5). Note that by the time of the C3 image at 22:45 UT in Figure 5, this second event has overtaken the earlier slow streamer blowout, although some of the difference in apparent speeds may be due to projection effects. At 19:50 UT, another loop was visible in C2 rising along essentially the same path as C′ (this feature is labelled D′ in Figure 4), with a well defined edge that showed some evidence for a helical structure. This feature is just visible above the occulter in Figure 5, very close to the shadow of the supporting pylon, while the leading edge C′ is at a projected distance of about 14 $R_{\odot}$. We interpret the loop D′ to be prominence material rising into the cavity behind the leading edge of the CME. This interpretation is supported by the image shown in Figure 6, recorded by the Extreme Ultraviolet Imaging Telescope (EIT) on SOHO (Delaboudinière *et al.*, 1995). This image, taken at 19:06 UT in the light of the He II 304 Å line, shows a huge erupting prominence above the southeast limb, extending all the way to the edge of the EIT field of view at 1.4 $R_{\odot}$, and possibly to even greater heights. This is clearly consistent with the timing of the loop in C2 at 3.0 $R_{\odot}$ some 44 min later. The appearance of feature D′ in C2 was followed closely in time by a fainter, diffuse front propagating to the south and west of the Sun when viewed in projection on the sky. This diffuse front is visible upon close examination towards the bottom of the last two frames shown in Figure 4, where it is labelled E', and is also visible in the C3 images taken somewhat later. Its appearance is suggestive of a 'halo' event propagating close to the line of sight between the Sun and the spacecraft. The relationship of this halo feature to the earlier CMEs is unclear from the white-light imagery, although it is probably associated in some way with the faster of the two events and/or the prominence eruption. This aspect of the events will not be considered in any detail in this paper.

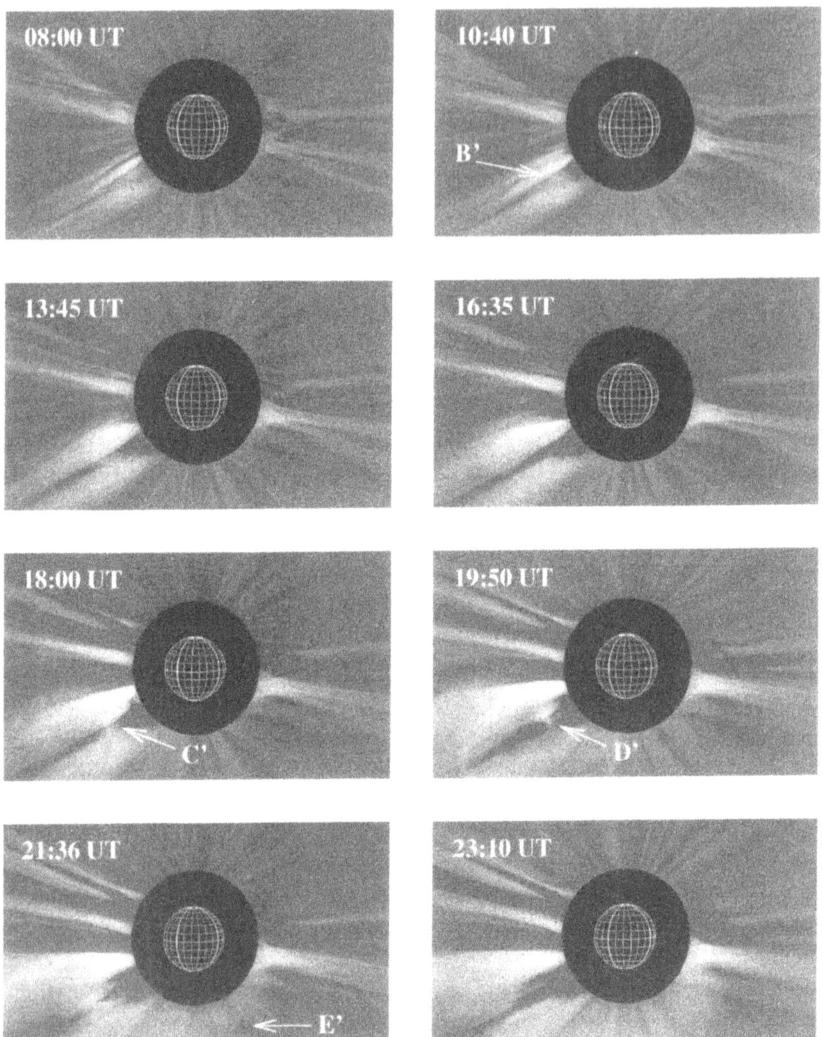

Figure 4. Development of CME activity in the LASCO C2 coronagraph on 19 October 1996. A base image made up of the average of three images recorded prior to the onset of the CME at 04:27 UT has been subtracted from each image shown here. The images have been scaled as the square root of the intensity, to enhance the visibility of faint features. The black circle marks the location of the occulting disk and the heliographic grid shows the position of the Sun. Labelled features are discussed in the text.

3. Height–Time Plots and Light Curves

The relationships between the various features in the events described above can best be understood by means of height–time plots. Figure 7 shows a plot of the

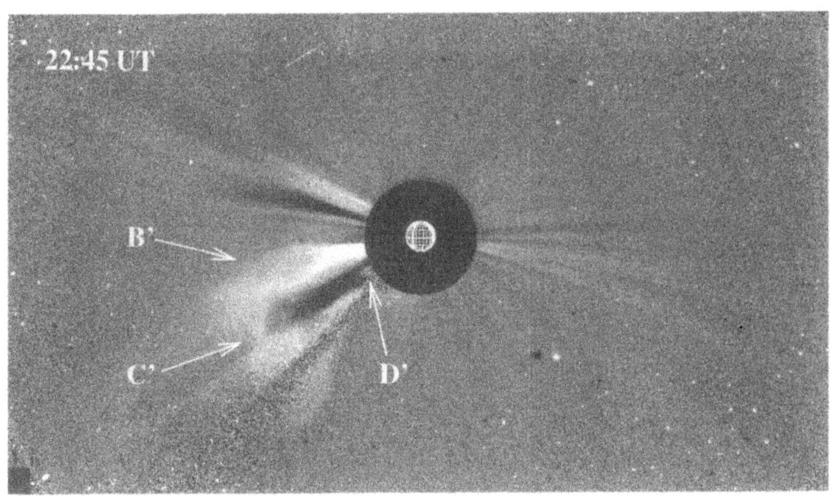

Figure 5. C3 white-light image recorded at 22:45 UT on 19 October 1996, showing two CMEs to the southeast. This image has been divided by a base image made from the average of three frames recorded prior to the CME. The pylon that supports the occulter is visible towards the bottom left of the image, where it blocks out part of the CME. The black circle marks the location of the occulting disk and the heliographic grid shows the position of the Sun. Labelled features are the same as in Figure 4.

projected height of the 17 June CME as a function of time. The C1 data points show the location of the leading edge of the bright loop visible in Figure 1, while the C2 and C3 data points show the location of the leading edge of the bright 'blob' near the equator (see the image at 03:08 UT in Figure 2). This latter feature is difficult to trace, because of the complicated structure of this event in white light; however, with appropriate intensity scaling of the images on a computer screen, it can be reliably located up to 17:00 UT. It is difficult to unambiguously quantify the errors in measuring the locations of individual features. In Figure 7, we have adopted a conservative estimate of the error as ± 10 pixels for each point. This corresponds to an error of ± 0.05 $R_\odot$ for C1, ± 0.1 $R_\odot$ for C2 and ± 0.5 $R_\odot$ for C3. Figure 7 clearly shows that the C1 loop rises beneath the white-light event in C2 and C3 (at 02:47 UT, the Fe XIV loop is still below the height of the white-light feature at 02:23 UT). The two sets of data can both be modelled by straight-line fits, indicating a rise at constant velocity in each case. There is no evidence for any significant acceleration of the CME out to about 11 $R_\odot$. The velocities deduced from a least-squares fit are approximately 180 km s^{-1} for C1 and 120 km s^{-1} for C2/C3. Extrapolation of the height–time plots back to the solar surface at 1.0 $R_\odot$ (under the assumption of constant velocity) suggests that the white-light CME left the surface at about 00:00 UT, while the 'launch time' for the green-line transient is estimated to be 01:22 UT. Assuming the Fe XIV loop continued to expand with the same velocity, it would have caught up with the leading edge of the CME at about

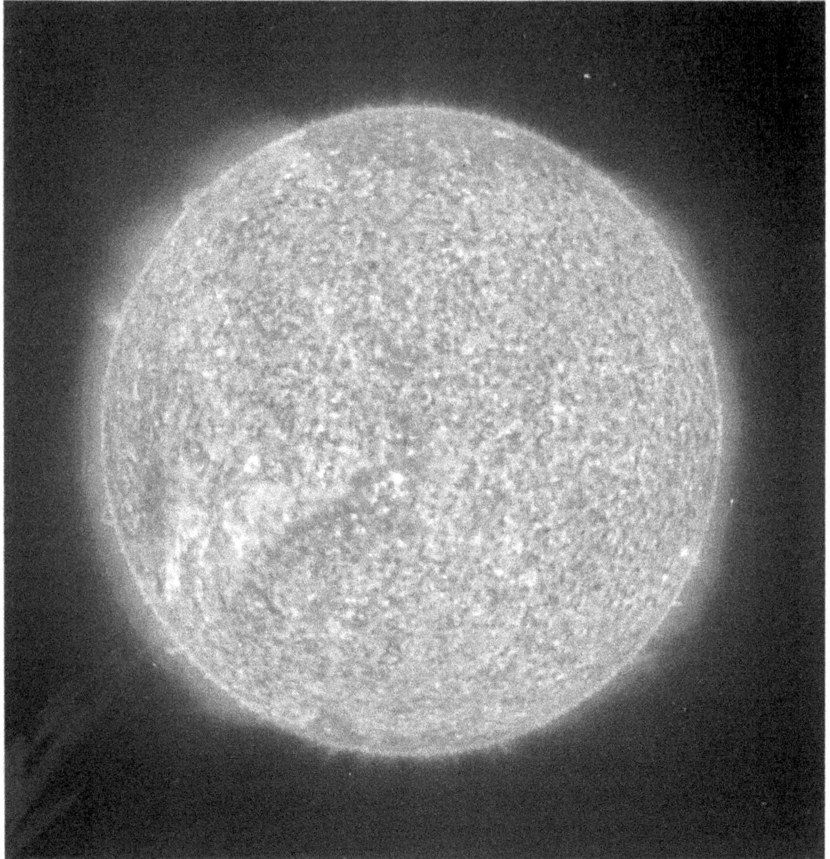

Figure 6. EIT image at a wavelength of 304 Å, recorded at 19:06 UT on 19 October 1996. A large erupting prominence is visible off the southeast limb.

04:13 UT. Although there is no evidence for a classical two-part structure in the CME (leading edge followed by prominence material), it is nevertheless tempting to associate the Fe XIV loop transient with a prominence eruption following the launch of the CME.

A light curve of the measured green-line intensity off the east limb on 16/17 June is shown in Figure 8. This light curve shows the total count rate in the region marked by the white rectangle in Figure 1. The general upward trend over a period of several hours can be attributed to the rotation of the active region, and its associated coronal loop systems, onto the limb. Between 01:30 UT and 02:03 UT on 17 June, there is a sharp decrease in the green-line intensity, corresponding to the time when the loop transient was observed in this region. This decrease in

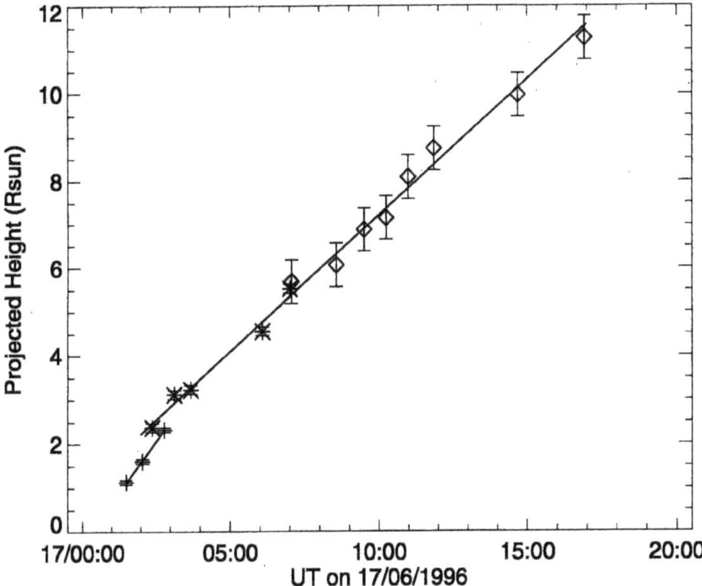

Figure 7. Height vs time plot for the CME of 17 June 1996. The crosses mark the position of the erupting loop seen in emission in the green line, while the asterisks and diamonds mark the position of the leading edge of the most prominent white-light feature in this complex event in C2 and C3, respectively. The straight lines show the best fits to both features under the assumption of constant velocity.

intensity could either be due to a net removal of mass from the region, or to a sudden change in the coronal plasma temperature. The appearance of the transient event in Figure 1 strongly suggests the former interpretation to be correct. It is not until some two hours later that the rotation of new structures into the field of view fills the void left behind by the loop eruption.

Figure 9 shows a height–time plot for several different features in the slow streamer blowout CME of 18–19 October. The leading edge can be followed right through the fields of view of all three coronagraphs, to almost 30 $R_\odot$. As before, an error of ± 10 pixels has been adopted for each data point. The CME is quite faint in the outer parts of the C3 field of view, so that the leading edge can only be reliably identified if the images are rebinned to a coarser spatial resolution. The error associated with these data points is then increased. It is clear from this plot that the earlier (and slower) of the two erupting loops seen in the C1 green-line images corresponds to the leading edge of this CME in white light. The speed of this feature in C1 is less than 10 km s^{-1}. The height–time plot for this feature in C1 and C2 can be modelled with a parabola, indicating a slow rise with a constant acceleration of about 1 m s^{-2} below 5 $R_\odot$. This is similar to the results obtained by Sheeley *et al.* (1997) from measurements of discrete 'blobs' in the low-latitude solar wind. There is a sharp break in the height–time plot around 5–6 $R_\odot$; the

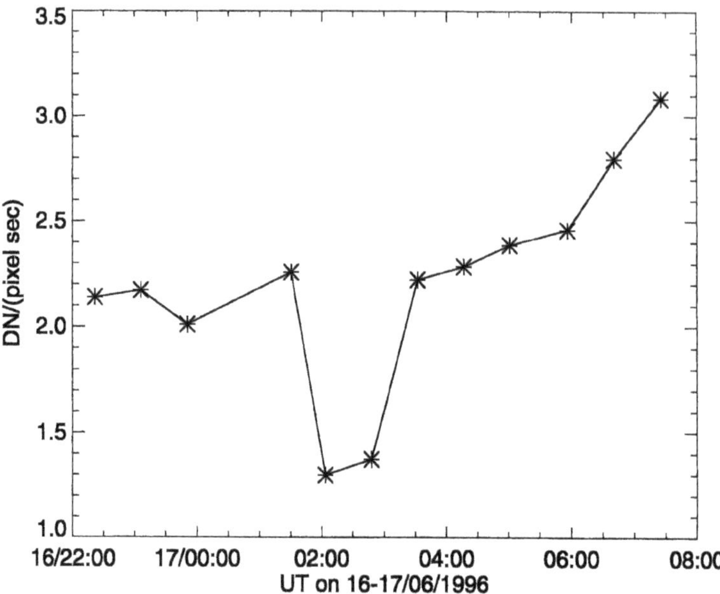

Figure 8. Light curve of the intensity of the green line off the east limb of the Sun on 17 June 1996. The region used for this light curve is indicated by a white rectangle in Figure 1.

data are best fit by a straight line above 6 $R_\odot$, giving a projected speed of about 330 km s^{-1}. This is close to where the event first begins to take on the tear-drop shape, indicating possible disconnection from the Sun. Thus there is a very rapid acceleration over a projected distance of about 1 $R_\odot$ around this height, with relatively little acceleration (or deceleration) above or below this location. The data are not consistent with a model in which the CME undergoes constant acceleration over the entire field of view. This rapid acceleration around 5–6 $R_\odot$ may indicate the location of the source surface of the solar wind, as suggested by Simnett *et al.* (1997). In this interpretation, the CME is accelerated by the same mechanism that accelerates the solar wind, and above this altitude it simply moves outwards at the ambient solar wind velocity. The sharp break is not the result of any systematic errors in going from C2 to C3; no such break is visible in the height–time profiles for many other events (e.g., Figures 7 and 10).

Figure 10 shows the height–time plot for the second, faster CME on 19 October. The white-light data in C2 and C3 are consistent with a model in which the leading edge moves radially outwards at an approximately constant speed of 490 km s^{-1} to a projected distance of at least 30 $R_\odot$. However, it should be noted that there is only one data point available in C2; thus, there could be some acceleration in the region below about 6 $R_\odot$ that was missed because of the cadence at which the images were recorded. Extrapolation of the constant-velocity fit back to the solar surface suggests the onset of the CME to be about 17:07 UT. Of course, the physical meaning of any

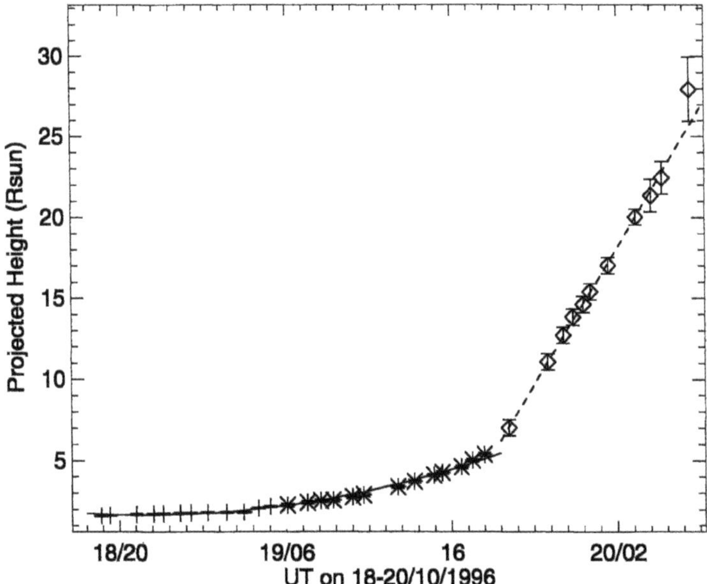

Figure 9. Height vs time plot for the slow streamer blowout CME of 18–19 October 1996. The crosses mark the position of the expanding loop seen in Fe XIV emission in C1, while the asterisks and diamonds mark the positions of the leading edges of the white-light CME in C2 and C3 respectively. The solid curve shows the best fit to the leading edge in C1 and C2, assuming constant acceleration, while the dashed line shows the best fit to the same feature in C3 under the assumption of constant velocity.

such extrapolation is questionable; there is no observational evidence that the CME begins at the solar surface rather than in the high corona. The fast erupting loop (speed about 25 km s^{-1}) seen in the green-line emission clearly lies well below the trajectory followed by this leading edge, but fits very well to the trajectory followed by the second part of the white-light CME, which we have identified above as being due to the eruption of the prominence off the southeast limb. This feature can only be followed in C3 to a radial distance of about 8 $R_{\odot}$, at which stage it falls within the shadow cast by the pylon used to support the occulter, and can no longer be clearly distinguished in the images. The C2 and C3 locations for this feature, and the Fe XIV loop positions in C1, can all be modelled by a parabolic height–time curve, with a best-fit constant acceleration of approximately 6 m s^{-2}. The prominence material rises more slowly than the leading edge in the C3 field of view, and there is no evidence that it catches up with the leading edge at any time.

In Figure 11, the light curve for the region off the southeast limb marked with a white rectangle in Figure 3 is shown. There is a sudden decrease in the intensity around 15:05 UT on 19 October which, apart from a slight increase around 18:00 UT, persists for many hours after the eruption from this region. As for the 17 June event, the C1 images suggest that this is due to a depletion of mass

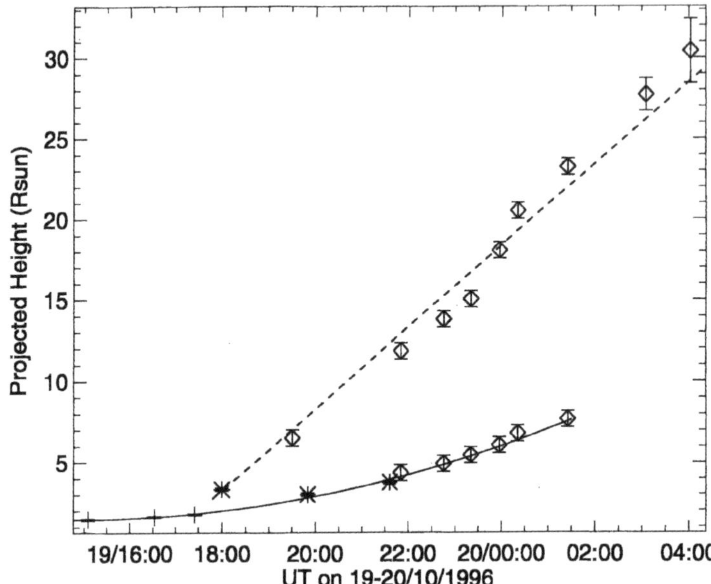

Figure 10. Height vs time plot for the fast, loop-like CME of 19 October 1996. The crosses mark the position of the erupting loop seen in Fe XIV emission in C1, while the asterisks and diamonds mark the positions of the leading edges of the white-light features (leading front and prominence core) in C2 and C3, respectively. The dashed line shows the best fit to the leading edge in C2 and C3, assuming constant velocity, while the solid curve shows the best fit for the prominence core, assuming constant acceleration, and that the Fe XIV loop in C1 and the prominence material in C2/C3 are the same feature.

from this part of the corona, leaving a void behind, rather to any change in the plasma temperature.

4. Conclusions

The observations presented in this paper show, for the first time, an unambiguous connection between coronal green-line transients and CMEs observed in Thomson-scattered white light in the outer corona. We have presented observations of two events with fast, 'whip-like' green-line transients associated with CMEs. In both cases, the fast green-line events trail the leading edge of the CME in white light. In another case, the green-line feature corresponding to the leading edge of the CME has also been identified. It is apparent that the 'whip-like' inner coronal transients are a manifestation of a prominence eruption, at least in the case of the second event on 19 October 1996. There are a number of other events in the LASCO database that also support this conclusion, although these events have not been discussed in this paper. These results provide definitive evidence for the conclusion first reached more than twenty years ago by DeMastus, Wagner, and Robinson (1973), that the

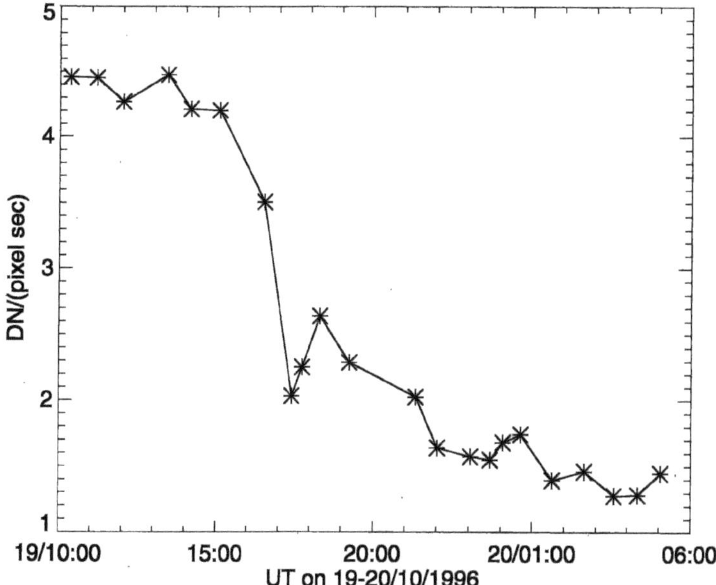

Figure 11. Light curve of the intensity of the green line off the southeast limb of the Sun on 19 October 1996. The region used for this light curve is indicated by a white rectangle in Figure 3.

green-line transients are produced by the same mechanism that causes prominence eruptions and/or sprays, in which significant amounts of chromospheric material are expelled into the corona. It would seem that the mechanism which drives these eruptions must heat at least some of this plasma to temperatures approaching 2×10^6 K, where it becomes visible in the green-line emission, or else coronal plasma along the trajectory followed by the erupting prominence gets swept up and expelled along with the colder prominence material. The morphology of the events studied, that of an erupting loop, shows no evidence for reconnection following the eruption in which the pre-existing streamers could be seen to reform. However, there is some indirect evidence for reconnection of open field lines to locations on the solar surface remote from the eruption site. For example, the appearance of the 'halo' event in C2 and C3 during the latter stages of the second 19 October CME suggests the formation of a coronal arcade propagating across the disk from the site of the prominence eruption. In the case of the 17 June event, there is some evidence that the bright legs left behind by the erupting loop reconnect to streamers at higher latitudes, which remain essentially undisturbed throughout the whole process of eruption.

The events discussed here are of two very different types: a fast 'bullet' type of eruption (17 June and 19 October) with a high initial speed and little or no acceleration of the leading edge, and a slow event with a long buildup followed by rapid acceleration in the outer corona (18–19 October). The driving mechanism is probably very different in the two types of event. In both of the fast green-line events, the Fe XIV eruption seems to be driven by the emergence of a loop-like structure from beneath the occulting disk, indicating the addition of new magnetic flux to the pre-existing coronal configuration. Thus it appears that these transients may be initiated by the emergence of new magnetic flux from the photosphere, as proposed by Steinolfson (1992). In this model, the new flux disturbs the quiescent state of the overlying corona, resulting in the expulsion of plasma and magnetic field that is visible as an erupting loop. Confirmation of this hypothesis is difficult, because conventional methods of generating magnetograms have very little sensitivity in areas near the edge of the disk, which is where the likely source of most lower coronal transients observed in coronagraphs must lie. The solution to the question of how, or if, such a process might be related to the initiation of a CME, the leading edge of which may precede the Fe XIV eruption by many hours, remains unclear at this time, as does the driving mechanism for the slower events, such as the one on 18 October.

The depletions of the green-line intensity in the lower corona in association with eruptive transients may be analogous to the X-ray dimming events reported by Hudson and Webb (1997). If this is the case, then it would appear that the X-ray dimmings may also be a signature of an eruptive prominence, rather than the leading edge of the CME itself.

The projected speeds of a few tens of km s^{-1} observed for some green-line transients are similar to the speeds of expansion of X-ray loops above active regions measured by Uchida *et al.* (1992). The question of whether these phenomena are related is beyond the scope of this paper, but it is worthwhile to note that Uchida *et al.* (1992) found that the X-ray loops expanded almost continuously and they did not associate them with CMEs.

In conclusion, the main results from the analysis presented in this paper are as follows:

(1) Green-line transients in the lower corona (below 2 $R_{\odot}$) have been clearly shown to be associated with CMEs observed in white light in the outer corona. For one event, we have identified the leading edge of a CME as a slowly-expanding loop feature in the green-line emission.

(2) The fast (speed >25 km s^{-1}) green-line events appear to be related to the prominence core of the CME, rather than the leading front observed in white light. These events are similar to the 'whip-like' events described by Dunn (1971).

(3) In both fast events that we have studied, a dimming of the green-line emission corona is observed at the time of eruption. This dimming is interpreted as being due to a net removal of mass from a large volume of the corona, leaving behind a dark void. The relationship of these dimmings to similar events observed in X-rays is unclear at present.

The question of what the 'typical' appearance of the leading edge of a CME might be in the lower corona remains unsolved by these observations, and can only be addressed by continued dedicated observations in the future. Clearly there is a need for further collaboration and detailed analyses of a greater number of events to relate the emissions in visible light, X-rays and EUV, and white-light observations at larger radial distances.

Acknowledgements

Much of this work was done while the lead author was at the University of Birmingham, where the analysis was made possible with the use of Starlink computing facilities. We are grateful to our colleagues in the LASCO/EIT consortium, who have all contributed to the success of these instruments. We thank especially the LASCO and EIT operations teams at GSFC, who continually work under stressful conditions to ensure that the scientific goals of these instruments are met. Work at NRL and USRA was supported under NASA contract NDPR S-92385-D. The British contribution was supported by the UK Particle Physics and Astronomy Research Council. Work in MPAe was supported by the Deutsche Agentur für Raumfahrtangelegenheiten (DARA). The French effort was supported by the Centre National d'Etudes Spatiales (CNES). SOHO is a mission of international cooperation between ESA and NASA.

References

Brueckner, G. E. and 14 co-authors: 1995, *Solar Phys.* **162**, 357.
Delaboudinière, J.-P. and 27 co-authors: 1995, *Solar Phys.* **162**, 291.
DeMastus, H. L., Wagner, W. J. and Robinson, R. D.: 1973, *Solar Phys.* **31**, 449.
Dunn, R. B.: 1971, in C. Macris (ed.), *Physics of the Solar Corona*, D. Reidel Publ. Co., Dordrecht, Holland, p. 114.
Esser, R., Brickhouse, N. S., Habbal, S. R., Altrock, R. C., and Hudson, H. S.: 1995, *J. Geophys. Res.* **100**, 19829.
Fisher, R. R.: 1977, *Solar Phys.* **55**, 135.
Gopalswamy, N., Hanaoka, Y., Kundu, M.R., Enome, S., Lemen, J. R., Akioka, M., and Lara, A.: 1997, *Astrophys. J.* **475**, 348.
Hiei, E., Hundhausen, A. J., and Sime, D. G.: 1993, *Geophys. Res. Letters* **20**, 2785.
Howard, R. A., Sheeley, N. R., Jr., Koomen, M. J., and Michels, D. J.: 1985, *J. Geophys. Res.* **90**, 8173.
Hudson, H. S., Acton, L. W., and Freeland, S. L.: 1996, *Astrophys. J.* **470**, 629.
Hudson, H. S. and Webb, D. F.: 1997, in N. U. Crooker, J. A. Jocelyn, and J. Feynman (eds.), *Coronal Mass Ejections: Causes and Consequences*, AGU Geophysical Monograph Series, AGU, Washington, DC, in press.
Klimchuk, J. A., Acton, L. W., Harvey, K. L., Hudson, H. S., Kluge, K. L., Sime, D. G., Strong, K. T., and Watanabe, T.: 1994, in Y. Uchida, T. Watanabe, K. Shibata and H. S. Hudson (eds), *X-ray Solar Physics from Yohkoh*, Universal Academy Press, New York, p. 181.
Mason, H. E.: 1975, *Monthly Notices Roy Astron. Soc.* **170**, 651.
Schwenn, R. and 23 co-authors: 1997, *Solar Phys.* **175**, 667 (this issue).
Sheeley, N. R., Jr. and 18 co-authors: 1997, *Astrophys. J.* **484**, 472.
Sime, D. G., Hiei, E., and Hundhausen, A.J.: 1994, in Y. Uchida, T. Watanabe, K. Shibata and H. S. Hudson (eds), *X-ray Solar Physics from Yohkoh*, Universal Academy Press, New York, p. 197.
Simnett, G. M. and 18 co-authors: 1997, *Solar Phys.* **175**, 685 (this issue).
Steinolfson, R. S.: 1992, *J. Geophys. Res.* **97**, 10811.

Uchida, Y., McAllister, A., Strong, K. T., Ogawara, Y., Shimizu, T., Matsumoto, R., and Hudson, H. S.: 1992, *Publ. Astron. Soc. Japan* **44**, L155.
Wagner, W. J.: 1987, *Solar Phys.* **114**, 81.
Wagner, W. J., Hansen, R. T., and Hansen, S. F.: 1974, *Solar Phys.* **34**, 453.
Wagner, W. J., Illing, R. M. E., Sawyer, C. B., House, L. L., Sheeley, N. R., Jr., Howard, R. A., Koomen, M. J., Michels, D. J., Smartt, R. N., and Dryer, M.: 1983, *Solar Phys.* **83**, 153.

MHD INTERPRETATION OF LASCO OBSERVATIONS OF A CORONAL MASS EJECTION AS A DISCONNECTED MAGNETIC STRUCTURE

S. T. WU and W. P. GUO
Center for Space Plasma and Aeronomic Research and Department of Mechanical and Aerospace Engineering, The University of Alabama in Huntsville, Huntsville, AL 35899, U.S.A.

M. D. ANDREWS
Hughes STX Corp., Naval Research Laboratory, Code 7665A, Washington, DC 20375, U.S.A.

G. E. BRUECKNER, R. A. HOWARD, M. J. KOOMEN, C. M. KORENDYKE, D. J. MICHELS, J. D. MOSES, D. G. SOCKER and K. P. DERE
H.O. Hulburt Center for Space Research, Naval Research Laboratory, Washington, DC 20375-5320, U.S.A.

P. L. LAMY, A. LLEBARIA and M. V. BOUT
Laboratoire d'Astronomie Spatiale, Marseille, France

R. SCHWENN
Max-Planck-Institut für Aeronomie, Lindau, Germany

G. M. SIMNETT, D. K. BEDFORD and C. J. EYLES
Space Research Group, School of Physics and Space Research, University of Birmingham, Birmingham, U.K.

(Received 4 February, 1997; accepted 10 March, 1997)

Abstract. We present a qualitative and quantitative comparison of a single coronal mass ejection (CME) as observed by LASCO (July 28–29, 1996) with the results of a three-dimensional axisymmetric time-dependent magnetohydrodynamic model of a flux rope interacting with a helmet streamer. The particular CME considered was selected based on the appearance of a distinct 'tear-drop' shape visible in animations generated from both the data and the model.

The CME event begins with the brightening of a pre-existing coronal streamer which evolves into a 'tear-drop' shaped loop followed by a Y-shaped structure. The brightening moves slowly outward with significant acceleration reaching velocities of $\sim$450 km s^{-1} at 30 $R_{\odot}$.

The observed CME characteristics are compared with the model results. On the basis of this comparison, we suggested that the observed features were caused by the evacuation of a flux rope in the closed field region of the helmet streamer (i.e., helmet dome). The flux rope manifests itself as the cavity of the quasi-static helmet streamer and the whole system becomes unstable when the flux rope reaches a threshold strength. The observed 'tear-drop' structure is due to the deformed flux rope. The leading edge of the flux rope interacts with the helmet dome to form the typical loop-like CME. The trailing edge of this flux rope interacts with the local bi-polar field to form the observed Y-shaped structure. The model results for the evolution of the magnetic-field configurations, velocity, and polarization brightness are directly compared with observations.

Animations have been generated from both the actual data and the model to illustrate the good agreement between the observation and the model. These animations can be found on the CD-ROM which accompanies this volume.

1. Introduction

In this paper, we present a qualitative and quantitative comparison of a coronal mass ejection (CME) as observed by LASCO with the results of a three-dimensional

Solar Physics **175:** 719–735, 1997.

axisymmetric magnetohydrodynamic (MHD) model of helmet streamer and flux-rope interactions. The particular CME (July 28–29, 1996) considered was selected based on the appearance on a distinct 'tear-drop' shape visible in animation generated from both the data and the model.

Features similar to this event were observed by the SMM coronagraph (Illing and Hundhausen, 1983; Burkepile and St. Cyr, 1993; Webb and Cliver, 1995). They considered that these 'V'- or 'U'-shaped events could be disconnected magnetic structures but did not present any specific suggestion on the cause of these events.

We present observation from the C2 and C3 coronagraphs on LASCO*. The C2 and C3 instruments are externally occulted Lyot (1930) coronagraphs. The C3 telescope design is similar to the NRL developed coronagraph which flew on OSO-7 (Koomen *et al.*, 1975) and SOLWIND. The C2 telescope is a similar instrument with a new type of occultor . Both instruments have 1024 × 1024 ccd cameras in the focal plane.

The C2 telescope images the region from approximately 2.0 to 6.0 $R_\odot$ with a resolution of ∼23″ and a pixel size of 11.4″. The C3 telescope images the region of approximately 3.7 to 32 $R_\odot$ with a resolution of ∼113″ and a pixel size of 56.5″. Each of the instruments has a number of optical filters. The data reported here were collected using the Orange filter (540–640 nm) for C2 and the Clear filter (400–850 nm) for C3. Additional information on the design and performance of the C2 and C3 coronagraphs is given by Brueckner *et al.* (1995). A more detailed description of this event is given by Simnett *et al.* in this issue. We briefly describe the essential features of this event which are pertinent to our physical interpretation.

The agreement of the observations and numerical simulations suggests that this type of event is due to the flux rope within the streamer becoming unstable. Increasing the magnetic field strength with the flux rope destabilizes the helmet streamer. The leading edge of the flux rope pushes out the helmet dome to create the loop-like CME (Wu, Guo, and Dryer, 1997). The trailing edge of the flux rope interacts with the bipolar magnetic field in the lower part of the streamer to form the observed Y-shaped magnetic structure as seen by LASCO as well as previous SMM coronagraph.

In this paper, we use the streamer and flux-rope model of Wu, Guo, and Dryer (1997) to quantitatively explain these LASCO observations as disconnected magnetic structures. A summary of the observations is given in Section 2. The description of the simulation model and method of treatment are given in Section 3. Simulation results are presented in Section 4. Finally, the discussions and concluding remarks are given in Section 5.

* LASCO is the Large Angle Spectroscopic Coronagraph which is one of the instruments on the Solar and Heliospheric Observatory (SOHO) which is a mission of international cooperation between NASA and ESA. The LASCO experiment was developed by a consortium of institutions from four countries: E.O. Hulburt Center for Space Research, Naval Research Laboratory, Washington, DC; Laboratoire d'Astronomie Spatiale, Marseille, France; Max-Planck-Institut für Aeronomie, Lindau, Germany; and Space Research Group, School of Physics and Space Research, University of Birmingham, Birmingham, U.K.

2. Observations

The observations reported here were made as part of our regular synoptic sequence on July 28–30, 1996. Reduced field-of-view, full-resolution images from C2 and C3 were taken with a cadence of approximately 45 min. There were several significant gaps in the data sequence with the most significant occurring from 06:56 to 13:56 UT on July 29 due to a problem with the flight software. All of the data collected on July 28–30, 1996 has been used in the data animation (see Appendix A) which are reproduced on the accompanying CD-ROM.

Figures 1(a) and 1(b) show the pre-event corona as seen in the C2 and C3 coronagraphs. On the west limb, the corona structure is very simple. There is a single bright streamer visible from the edge of the C2 field out to at least 20 $R_\odot$. This simple structure is typical of the quiet corona observed by LASCO. The figures presented here show only the corona above the west limb. The coronal structures above the east limb are more complicated with multiple streamers which are not included in this study. The east limb streamer structures were not significantly affected by the west limb CME.

The CME begins with a slow brightening of the lower, southern edge of the pre-existing streamer as illustrated in Figure 2(a). The streamer brightens and extends southward over a period of approximately 10 hours. The CME is first visible as a distinct structure in the C2 field of view at 18:56 UT on 28 July, 1996, Figure 2(b). A bright, relatively featureless blob is seen at $\sim$3 $R_\odot$ Figure 3(a). The structure moves outward to 4 $R_\odot$ and the 'blob' structure starts to become visible as a distinct structure. At 02:37 UT on 29 July, 1996 (Figure 3(b)), this blob is seen as a tear-drop-shaped loop with an extended tail. The CME extends over $\sim$2 $R_\odot$ in the east–west direction and $\sim$1.5 $R_\odot$ in the north–south direction. There is a bright knot within the loop that does propagate outward with the loop. It is not clear whether the loop and the bright knot are a connected structure. We may be seeing distinct structures that appear to be connected due to projection effects.

The development of the CME and motion through the field of view are not well observed in C3. The initial development of the CME as seen in C3 is similar to that seen in C2. However, the fully developed loop is not observed until after the data gap. Figure 4 shows the late stages of the event in C3 as a loop with a trailing inverted 'V' shape directed back toward the Sun. This structure is seen to move outward until it disappears from the C3 field of view at >32 $R_\odot$.

It is possible using this sequence of images to track the features as they move outward. While there are a number of features that can be tracked within the C2 or C3 field of view, the only two features which can be tracked across the time gap are the leading and trailing edges of the loop. The most accurately tracked feature is the back of the loop. The feature is seen at nine positions in C2 and five positions in C3. The structure can be tracked from $\sim$2.5 to 31 $R_\odot$. The leading edge is seen at eight positions in the C2 field of view but only two positions in the C3 field of view.

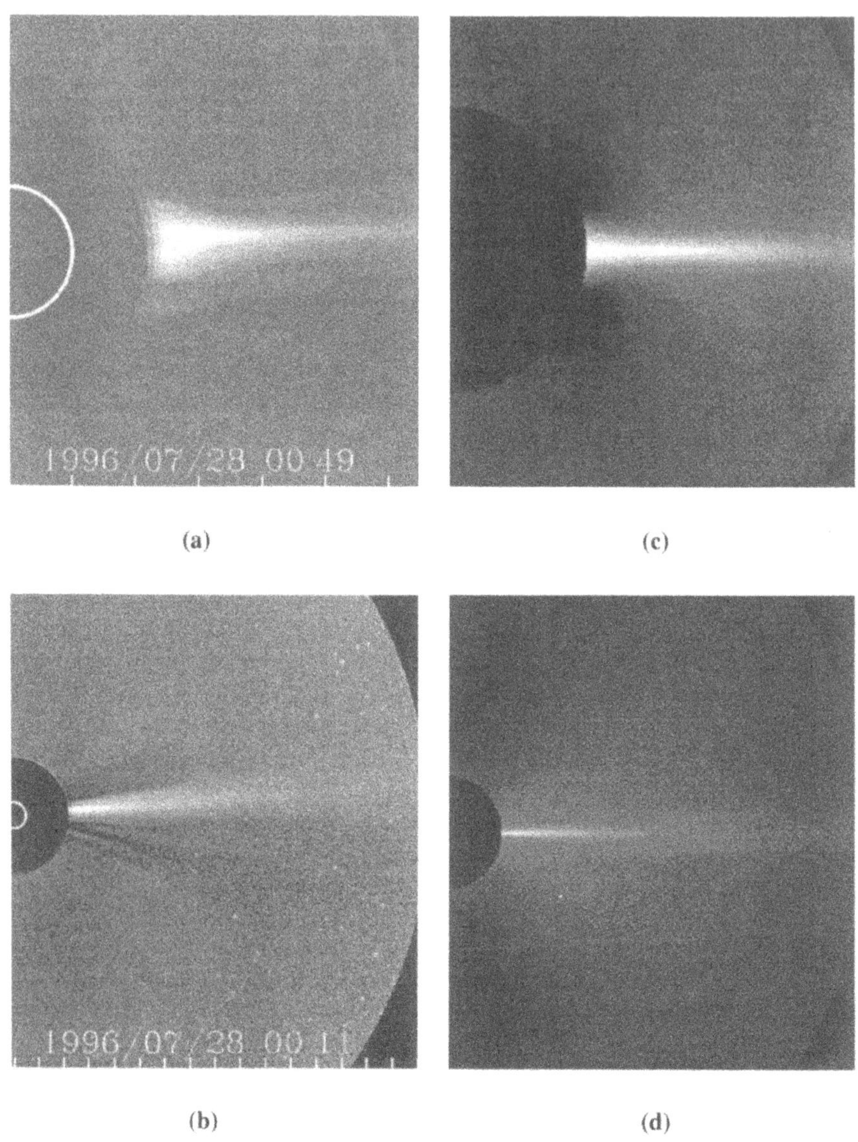

Figure 1. Comparison of the observed and the modeled pre-event corona. (a) LASCO C2 observation of the west limb at 00:49 UT on July 28, 1996. The bright half circle indicates the location of the solar surface. The radius of the occulting disk is 2.15 $R_{\odot}$. The bright structure in the middle is a single streamer. (b) LASCO C3 observation of the west limb at 00:11 UT on July 28, 1996. The radius of the occulting disk is 4.39 $R_{\odot}$. The bright structure is the extension of the same streamer observed in C2. (c) Model prediction of the area corresponding to C2 at $t = 0$ hr. The bright structure is produced by the model streamer. (d) Model prediction of the area corresponding to C3 at $t = 0$ hr. The ray-like structure is the indication of the current sheet above the model streamer.

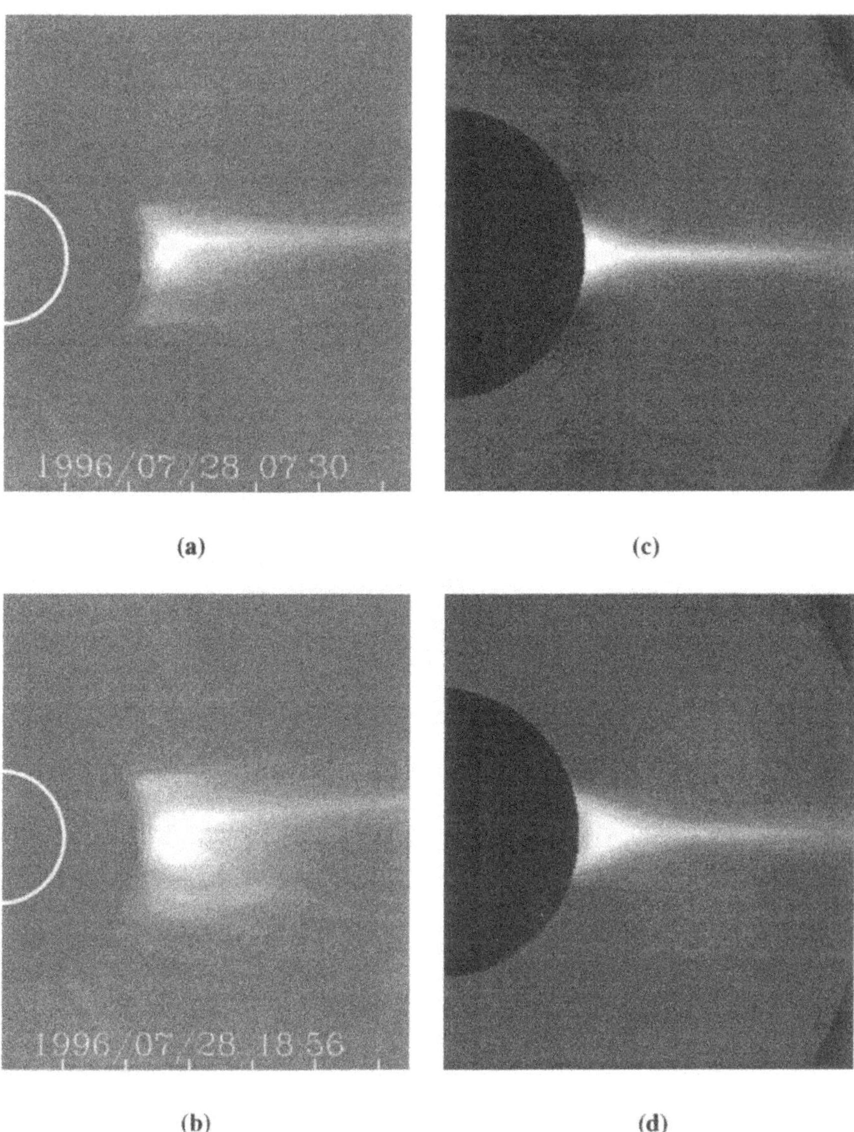

Figure 2. Early development of the event in the C2 field of view. (a) LASCO C2 observation at 07:30 UT on July 28, 1996. The streamer brightens as compared to Figure 1(a). (b) LASCO C2 observation at 18:56 UT on July 28, 1996. The distinct bright structure to the south of the streamer indicates the appearance of the CME. (c) Model prediction at $t = 2$ hr. (d) Model prediction at $t = 3$ hr.

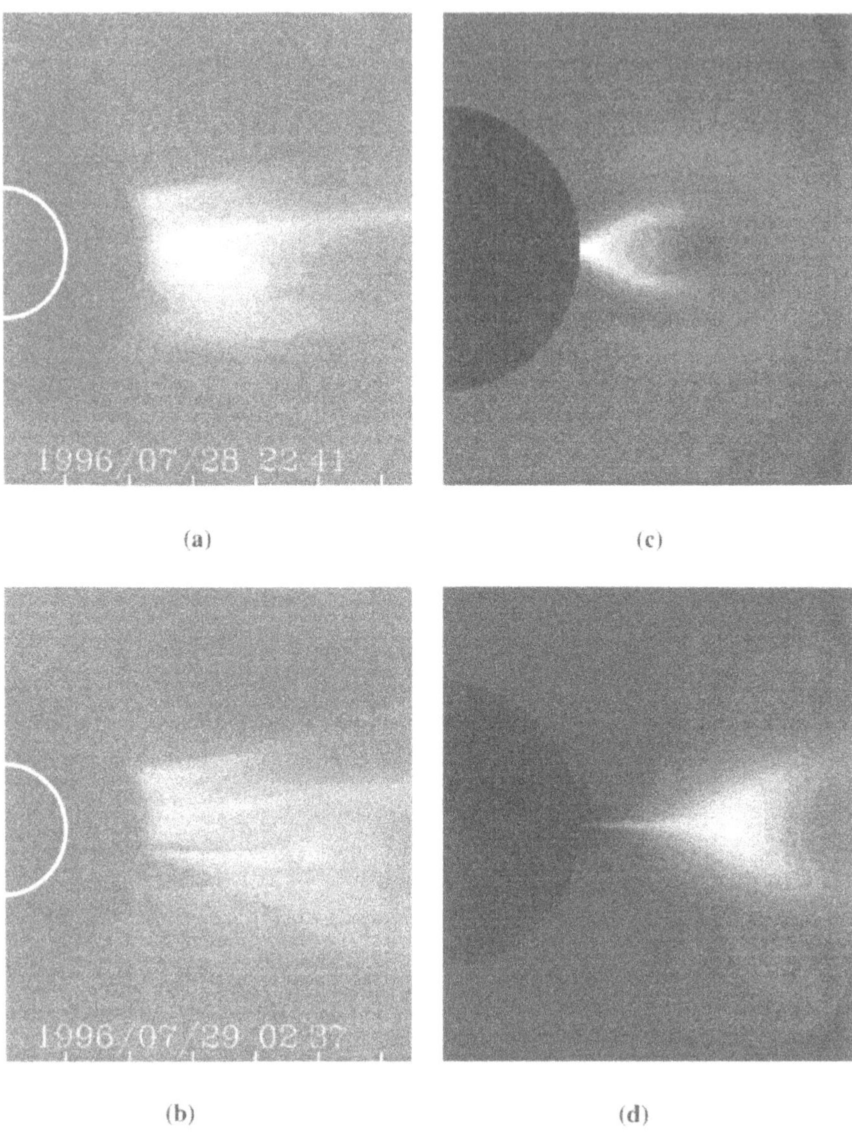

Figure 3. Evolution of the event in the C2 field of view. (a) LASCO C2 observation at 22:41 UT on July 28, 1996. The 'blob' structure becomes visible. (b) LASCO C2 observation at 02:37 UT on July 29, 1996. The 'blob' evolves to a tear-drop-shaped structure with an extended tail. (c) Model prediction at $t = 8$ hr. The 'blob' structure is the erupted flux rope. The inner part of the erupted flux rope is darker because of the original cavity in the initial streamer. The trailing edge of the flux rope is brighter. The dim loop-like structure in front of the flux rope is the CME loop which is formed by the compression of the coronal materials originally in the helmet dome. (d) Model prediction at $t = 11$ hr. Similar to Figure 3(b), the erupted flux rope deforms to the shape of tear-drop followed by a tail.

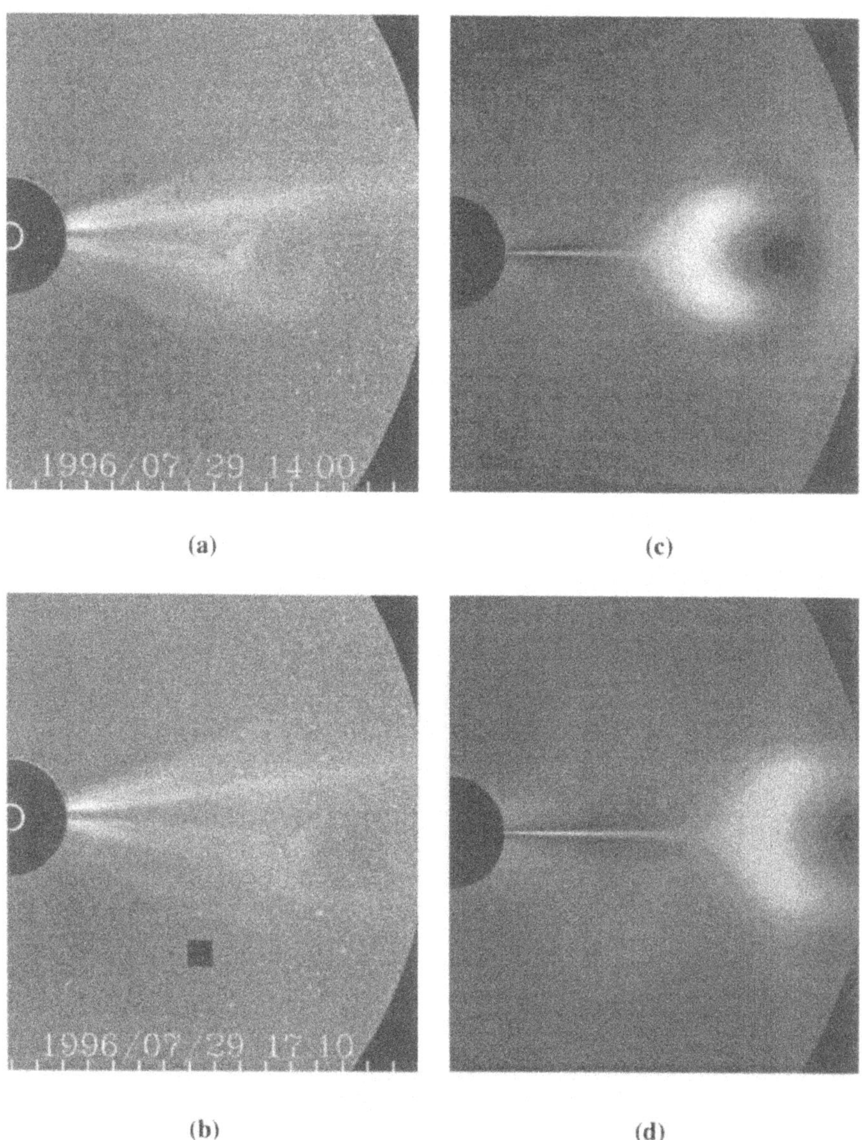

Figure 4. Propagation of the event in the C3 field of view. (a) LASCO C3 observation at 14:00 UT on July 29, 1996. The tear-drop-shaped loop and the following Y-shaped structure is more prominent. The inner part of the flux rope is darker. (b) LASCO C3 observation at 17:10 UT on July 29, 1996. The structure retains its overall shape as it propagates in the C3 field of view. (c) Model prediction at $t = 21$ hr. The model shows similar characteristics as compared to Figure 4(a). (d) Model prediction at $t = 24$ hr.

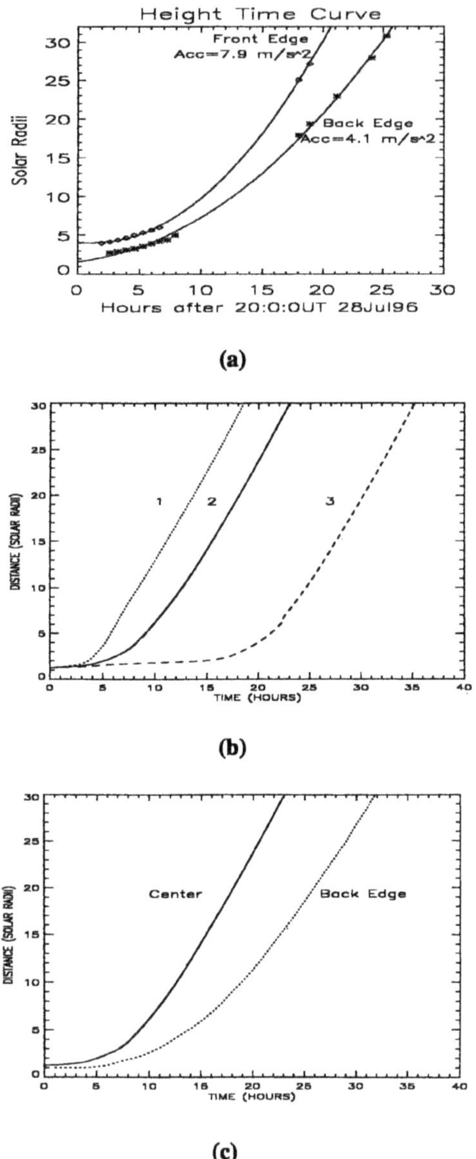

Figure 5. (a) The height of the front/back edge vs time curve of the observed tear-drop-shaped structure deduced from observations; (b) the height of the center of the flux rope vs time curve for the three cases in the model; and (c) the height of the center and back edge of the flux rope vs time for Case 2 of the model.

The height versus time data are plotted in Figure 5(a). These data have been fit with a second-order polynomial, $r(t) = r_0 + v_0 t + (1/2)at^2$, i.e., constant acceleration. Both sets of points are reasonably well fit by the curves. It does appear that the acceleration of the trailing edge of the loop is slower than the acceleration determined by the fit. The large data gap introduces large systematic uncertainties which makes it impossible to determine whether the assumption of constant acceleration is reasonable. The height curves do not extend down to the solar surface. This is consistent with our interpretation that the event begins above the solar limb. A constant acceleration for this type of event has been reported elsewhere (McComas *et al.*, 1991; Sheeley, 1997). The measured velocity, with large systematic errors, at 25–30 $R_\odot$ for both features is 400–500 km s^{-1}. The velocities predicted by a constant acceleration are within 50 km s^{-1} of the measured values.

3. Description of MHD Simulation Model and Method of Treatment

Many of the observed CME features resemble the recent magnetohydrodynamic simulation studies of streamer and flux-rope interactions given by Wu, Guo, and Dryer (1997). That study shows that when a flux rope inside the closed field region of the streamer (i.e., under the helmet dome) reaches a critical strength, it will destabilize the streamer. This flux rope will then escape from the closed field region of the streamer and push the helmet dome outward to form the classical observed loop-like CMEs (Burkepile and St. Cyr, 1993). The trailing edge of this flux rope interacts with the dipole magnetic field of the lower part of the streamer which forms these observed V-shaped disconnected magnetic structures as shown by Wu, Guo, and Dryer (1997). This model uses an initial high-density flux rope which does not reproduce the void region in front of the V-shaped structure.

Most recently, Wu and Guo (1997) constructed another self-consistent MHD model of a streamer interacting with a low-density, high magnetic-field flux rope. This model of a flux rope with cavity does reproduce the void region of the observed features. We shall use this model to interpret this event. The mathematical model, method of treatment, and physical parameters are given in the following sections.

3.1. Mathematical Model

The mathematical model employed for the present study is based on the model given by Wu *et al.* (1997) and Wu and Guo (1997), which is a set of basic non-dissipative (ideal) magnetohydrodynamic (MHD) equations in the system of spherical coordinates (r, θ, ϕ), where r is the radius, θ is the co-latitude, and ϕ is the longitude with the assumption of axisymmetry (i.e., $\partial/\partial\phi = 0$). These equations include the conservation laws of mass, momentum, and energy, together with the magnetic induction equation to govern the magnetic field and plasma flow nonlinear interactions. The model is outlined as follows:

(i) The computational domain is 1 to 35 $R_\odot$ and from the pole to the equator, N –S symmetry is assumed. The grid network is chosen to be 301 (radial direction) × 92 (meridional direction) which has been demonstrated to resolve the physical structures of interest to this study.

(ii) The numerical scheme used in this study is the three-dimensional axisymmetric combined difference scheme (Wu, Guo, and Dryer, 1997; Guo and Wu, 1997). The mass and energy conservation equations are solved by the upwind scheme, momentum and magnetic induction equation are solved by the Lax–Wendroff scheme, and a semi-implicit operator is included in the momentum equation which becomes effective in studying slow evolution cases.

(iii) The boundary conditions used in this study are: (a) the boundary condition at the pole and the equator are symmetric boundary conditions, (b) the outer boundary is determined by the extrapolation because the plasma flow at this boundary is supersonic and super-Alfvénic (Wu and Wang, 1987), and (c) the inner boundary is determined by the method of projected normal characteristics (Wu and Wang, 1987). This evolutionary boundary condition is used to assure the self-consistency of the numerical solution (Wu, Wang, and Guo, 1997).

3.2. METHOD OF TREATMENT

The initial condition is a helmet streamer containing a flux rope with cavity as theoretically suggested by Low (1994) and numerically simulated by Wu and Guo (1997). The detailed mathematical procedure is given by Guo and Wu (1997). We construct the pre-event helmet streamer solution by the relaxation method (Steinolfson, Suess, and Wu, 1982) using a full set of non-dissipative MHD equations with the uniform boundary condition at the solar surface as $n_0 = 3.2 \times 10^8$ cm^{-3}, $T_0 = 1.8 \times 10^6$ K. The magnetic field at the solar surface is dipole-like with $B_0 = 2.0$ G at the equator, which corresponds to plasma $\beta = 1.0$ at the equator. The resulting initial configuration, helmet streamer and flux rope with cavity is obtained by emerging a helical flux rope into the closed field region of the streamer. Figures 1(c) and 1(d) display the simulation solution for the initial polarization brightness. The polarization brightness is obtained by integration of the Thompson scattering of photospheric light by electrons using computed density along the line-of-sight (Billings, 1972). Figures 6(a) and 6(b) show the magnetic field lines and the velocity of the initial pre-event corona in the model.

As pointed out by Wang *et al.* (1995), the pre-event corona determines the simulated CME characteristics. While the observed pre-event corona (Figures 1(a) and 1(b)) is significantly more complicated, the model does closely resemble the pre-event corona as shown in Figure 1(c) and 1(d). The physical parameters used for this solution are shown in Table I, where the effective flux rope radius (see Figure 6(a)) is defined as the distance from the center of the flux rope to the surface of the Sun.

Table I
Physical parameters of the streamer and flux rope

a	effective flux-rope radius $= 0.26 R_{\odot}$
β_c	plasma β at the center of torus $= 0.40$
B_c	magnetic field strength at the center of the flux rope $= 1.16$ G

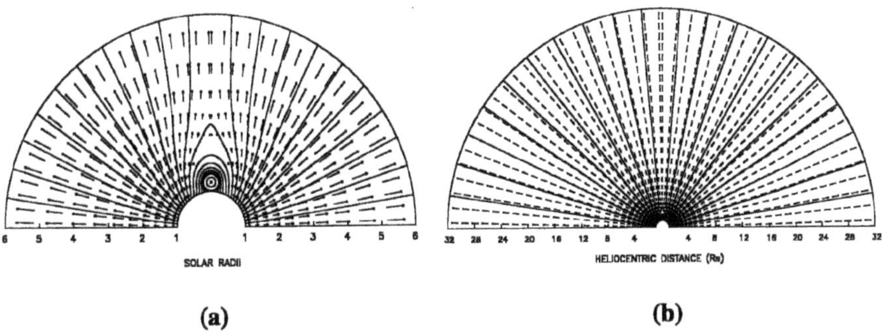

Figure 6. The magnetic field lines (indicated by solid lines) and velocity (indicated by arrowed line segments) of the pre-event coronal state in the model. (a) Magnetic field lines and velocity between 1–6 $R_{\odot}$. Note the existence of the flux rope in the closed field region of the streamer. (b) Magnetic field lines and velocity between 1–32 $R_{\odot}$. Note the existence of the current sheet above the streamer.

4. Numerical Results

The successful simulation of a physical event requires both a qualitative and quantitative agreement between model and observation. It is not sufficient to produce a model which 'looks like' the observed event. It is necessary to accurately predict the quantitative features of the event. In this particular event, the model must reproduce the observed 'tear-drop' magnetic structure followed by a 'Y' shaped features as given in the left panel of Figures 2–4 and the distances vs time curve as shown in Figure 5(a). In this case, the model not only reproduces the observed shapes, it also provides reasonable estimates of mass, energy, and velocity.

The numerical simulation was carried out for three sets of parameters of the flux rope to cover a range of distance-time curves of the propagation of these observed magnetic structure. These three sets of parameters are given in Table II.

Initially the streamer and flux-rope system is in equilibrium. To induce this configuration to become unstable, we simply increase the azimuthal component of the magnetic field (B_{ϕ}) and simultaneously decrease the density within the flux rope. Detailed procedures are discussed by Wu and Guo (1997) and Guo and Wu (1997). Physically, this corresponds to increasing the photospheric shear and draining the mass from the apex portion of the flux rope. The flux rope begins to be unstable and moves upward, the center of the flux rope as a function of time is plotted in Figure 5(b) for these three cases. In comparison to the observed height

Table II
The three sets of parameters for the description of the flux rope

Case	a	β_c	B_c
1	1.35 $R_{\odot}$	0.07	0.36 G
2	0.62 $R_{\odot}$	0.09	0.68 G
3	0.51 $R_{\odot}$	0.10	0.77 G

–time curves in Figure 5(a), it is clear that Case 2 closely matches the observed event. To further investigate the simulation in comparison with observations, we selected Case 2 for presentation.

Figure 5(c) shows the distance–time curve for the center and trailing edge of the flux rope for Case 2 which we identify as the observed Y-shaped magnetic structure. In comparing these two curves, it is easily recognized that the flux rope was deformed. The center moves faster than the trailing edge of the rope. Two important factors cause this phenomenon, the flux rope expands upward due to the nonlinear interaction with the streamer which causes magnetic buoyancy to operate (Wu and Guo, 1997); and gravitational pull drags the flux rope which causes the delayed movement of the trailing edge. These features can be clearly seen in the evolution of magnetic field configuration as shown in Figure 7. The calculated polarization brightness is shown in the right panels of Figures 2–4, respectively, for direct comparison with the observed brightness.

The MHD simulation also produces estimates of the total mass contained in this event and each mode of energy (i.e., kinetic, potential, thermal, and magnetic). The total mass is estimated to be 1.5×10^{16} g which consists of the mass of the streamer and the flux rope. To compute the energy for each mode, we have assumed that the length of the streamer–flux rope system is a constant 0.3 $R_{\odot}$. The energy partitioning within the computational domain is shown in Figure 8. The event is started by increasing the B_{ϕ} component and decreasing the density within the flux rope. This produces a peak in the magnetic energy. The magnetic energy decreases monotonically as it is redistributed into the other modes. The kinetic and thermal energy decrease at large times is a consequence of the mass moving out of the computational domain. The simulated velocity for this event is $\sim$360 km s^{-1} for the center and $\sim$330 km s^{-1} for the trailing edge of the flux rope at $\sim$30 $R_{\odot}$, respectively. The acceleration for the center of the flux rope takes place below $\sim$10 $R_{\odot}$, and then it propagates at almost constant speed. The acceleration process for the trailing edge consists of two stages: (i) the majority of the acceleration takes place below 10 $R_{\odot}$, and (ii) after 10 $R_{\odot}$, the acceleration decreases.

The agreement between the model and the observations is shown in Figures 1–4. The streamer–flux rope model (Wu and Guo, 1997) does reproduce both the configuration and physical measurements of the observed event.

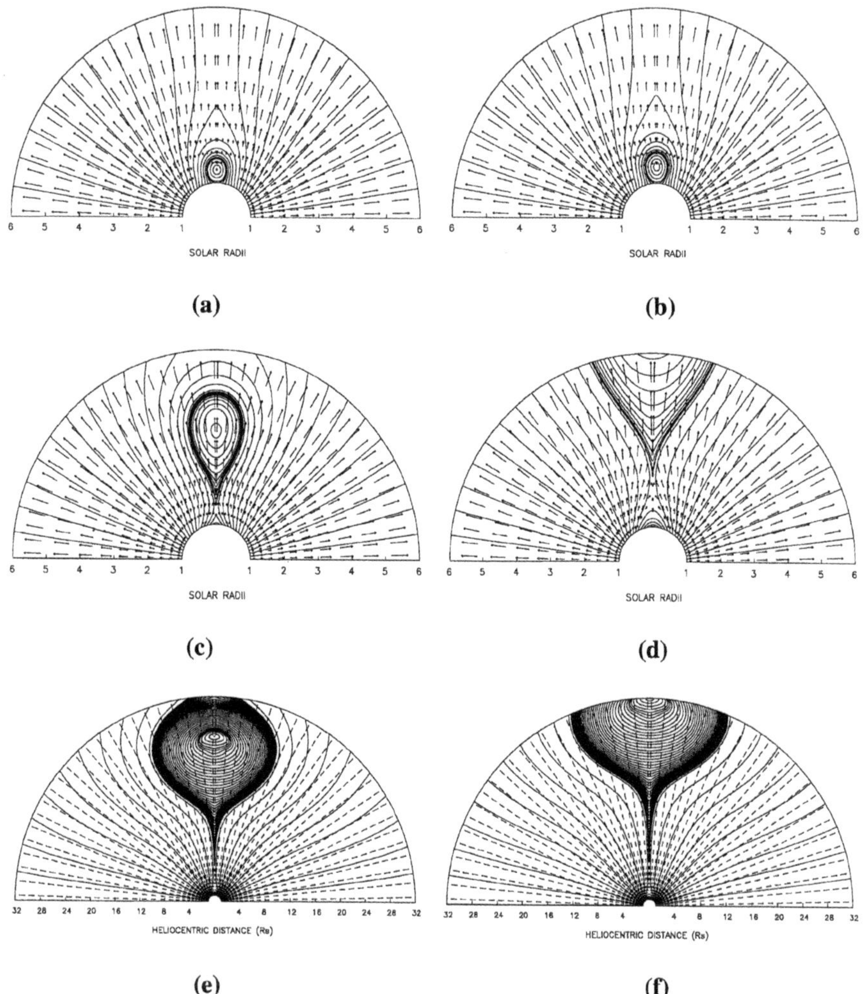

Figure 7. The evolution of the computed magnetic field lines and velocity in the model (a–f), at $t = 2$, 3, 8, 11, 21, and 24 hr, respectively. These figures are projections of the three-dimensional magnetic field lines and velocity on the solar meridional plane. The flux rope is actually a helical flux rope running in the azimuthal direction. Figure 6 of Wu, Guo, and Dryer (1997) showed a three-dimensional view of a similar configuration.

5. Discussion and Concluding Remarks

Illing and Hundhausen (1983) first suggested that events of this type were caused by magnetic disconnection during a coronal mass ejection. Cliver (1989) interpreted observations of the 1983 April 6 solar eclipse as a similar disconnected coronal mass ejection. He further noted that these types of disconnected magnetic features

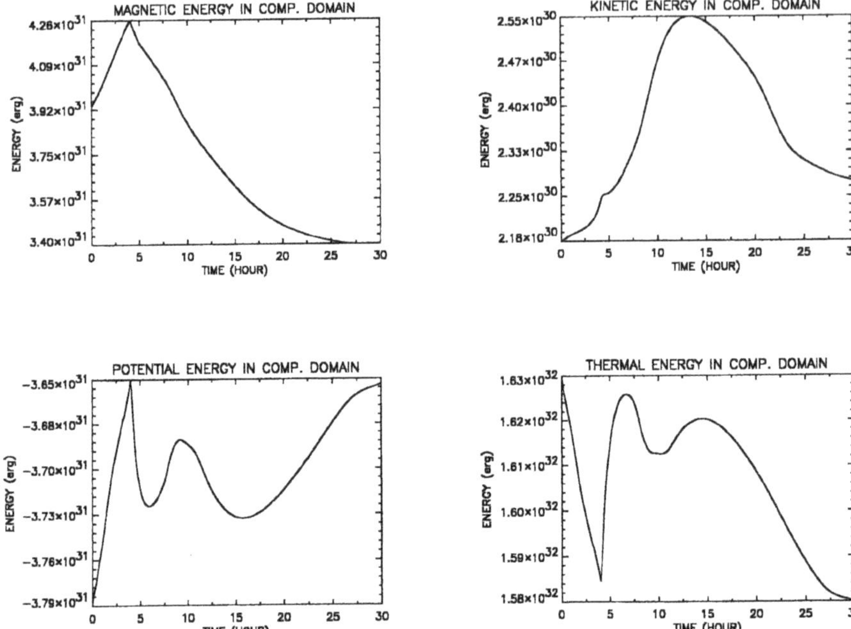

Figure 8. Time evolution of the magnetic, potential, kinetic, and thermal energy in the computational domain during the event. The kinetic and thermal energy decrease at large times is a consequence of the mass moving out of the computational domain.

were also observed during solar eclipse in 1960 and 1980. Recently, Webb and Cliver (1995) made a further study on evidence for magnetic disconnection in coronal mass ejections. Their study is based on considerations of the unobserved continuous buildup of the net interplanetary magnetic flux which implies that it is necessary to have some form of disconnection of the flux near the Sun. Most recently, Simnett *et al.* (1997) presented two CMEs which were well-observed by the LASCO coronagraphs. Because of the unique capability of the LASCO coronagraphs, they were able to follow these two events up to 30 $R_\odot$. It appears that both of these events exhibit magnetic disconnection structures in the coronal mass ejections.

In this paper we used the LASCO observations of July 28–29, 1996 together with the streamer and flux rope interaction model (Wu, Guo, and Dryer, 1997; Wu and Guo, 1997) to understand the physical processes of this type of event. The pre-event corona seems to be a helmet streamer as shown in Figures 1(a) and 1(b). Wu, Guo, and Wang (1995) and Wu, Guo, and Dryer (1997) presented a self-consistent MHD simulation model for steamer and flux-rope interaction in which the initial state of this model closely resembles observed configuration as shown in Figures 1(c) and 1(d).

As shown in Figures 2–4, the initial circular structure is deformed to a 'tear-drop' shape. The trailing edge of the 'tear-drop'-shaped magnetic structure forms a V-shaped configuration with a tail. This configuration is clearly shown by the model results. As the structure propagates outward, the V-shape configuration with the tail forms a complete Y-shaped configuration which is identical to the magnetic field configuration in the simulation (Figure 7). In summary of our investigation of the event on July 28–29, 1996, we conclude the following:

(i) The pre-event corona consists of a helmet steamer with a detached magnetic field structure inside the helmet dome. We interpret this physically as two magnetic flux systems. One is the helmet streamer and the other is an orthogonal flux rope.

(ii) This system will become unstable when the flux rope reaches a threshold strength. This initiates the observed event.

(iii) The flux rope expands upward while being pulled back by gravitational force. The density is redistributed within the streamer flux rope system to naturally produce the faint non-uniform structures observed.

(iv) The observed 'tear-drop'-shaped structure corresponds to the deformed flux rope.

(v) The Y-shaped structure (Figures 3 and 4) is formed due to the flux rope interacting with a local finite-size current sheet (i.e., the bipolar magnetic field of the streamer).

In short, we claim that this coronal mass ejection shows magnetic disconnection due to evacuation of a flux rope from the helmet streamer.

While the model and data presented herein are quite consistent, a final note is necessary. The present observation is a projection onto the plane of sky. The magnetic disconnection in the coronal mass ejection can occur under two different scenarios: (i) the flux-rope destabilization as presented in this study, and (ii) through the reconnection process within the current sheet in the streamer as presented by many investigators (Forbes, 1990; Martens and Kuin, 1989, Wang *et al.*, 1995). Stereo observations are required to resolve the full three-dimensional structure of these events.

Acknowledgements

The work performed by STW and WPG is supported by NRL through USRA N00014-C- 95-2058. MDA is supported at NRL by NASA contract DPR5-86760E.

Appendix A. Image Processing

This Appendix describes the image processing done on the LASCO images to generate the pictures shown in Figures 1–4 and each frame of the movie in the data animation. This is a four-step process.

First, the individual images are processed to remove small, bright features due to stars and/or cosmic-ray tracks in the detector. For each point in the image, the local median and standard deviation are calculated. Those pixels which are 7 standard deviations brighter than the median are replaced with the local median. This method effective removes most of the 'noise' due to stars and cosmic rays without significantly affecting the 'signal' from the corona. Star tracks can still be seen in the data animations.

The second step is to generate background and average coronal images. All of the images collected in the standard optical configuration over a two week period are processed to generate these images. There were over 200 standard images from each of C2 and C3 used in this processing. Each image has a standard background level subtracted and is corrected to a standard observation time. The minimum image is constructed by selecting the smallest observed value over all of the images at each pixel. That is for each pixel in the image plane the 200+ images are examined to select the smallest value at that individual pixel. This method does produce a relatively featureless background image, but does not remove all of the coronal structure. The average image is formed by an arithmetic mean over all of the images.

The third step is to correct the images for exposure time uncertainty. There is a small but significant uncertainty in the exposure time of the LASCO images. The images shown in this paper have been corrected to a standard exposure time using the average images. A mean is calculated for regions far from the Sun for each individual image and the average image. The ratio of these two means is used to normalize to the standard exposure time.

The final image processing step is to subtract the minimum image and apply a radial filter. The radial filter is calculated from the minimum by averaging over angles to produce a function of intensity versus radius. The filter is applied by dividing the difference image by the radial function. This image processing produces images which have had both the constant background level subtracted and the large radial variation of the corona signal removed. Finally, the images are byte-scaled to be displayed as 256 colors or gray levels.

References

Billings, D. E.: 1966, *A Guide to the Solar Corona*, Academic Press, San Diego, CA, p. 65.

Brueckner, G. E. *et al.*: 1995, *Solar Phys.* **162**, 357.

Burkepile, J. T. and St. Cyr, O. C.: 1993, NCAR/TN-369+STR, National Center for Atmospheric Research, Boulder, CO.

Cliver, E. W.: 1989, *Solar Phys.* **122**, 319.

Forbes, T. G.: 1990, *J. Geophys. Res.* **95**, 11 919.

Guo, W. P. and Wu, S. T.: 1997, *Astrophys. J.*, submitted.

Illing, R. M. E. and Hundhaussen, A. J.: 1983, *J. Geophys. Res.* **88**, 10 210.

Koomen, M. J., Detwiler, C. R., Bruckner, G. E., Cooper, H. W., and Tousey, R.: 1975, *Applied Optics* **14**, 743.

Low, B. C.: 1994, *Phys. Plasmas* **1**, 1684.

Low B. C. and Hundhaussen, J. R.: 1995, *Astrophys. J.* **443**, 818.
Lyot, G.: 1930, *C.R. Acad. Sci. Paris* **191**, 834.
Martens P. C. H. and Kuin, N. P. M.: 1989, *Solar Phys.* **122**.
McComas, D. J., Phillips, J. L., Hundhausen, A. J., and Burkepile, J. T.: 1991, *Geophys. Res. Letters* **18**, 73.
Sheely, N. R.: 1997, *Astrophys. J.*, submitted.
Simnett, G. M., Tappin, S. J., Plunkett, S. P., Bedrord, D. K., Eyles, C. J., St. Cyr, O. C., Howard, R. A., Brueckner, G. E., Michels, D. J., Moses, J. D., Socker, D., Dere, K. P., Korendyke, C. M., Paswaters, S. E., Wang, D., Schwenn, R., Lamy, P., Liebaria, A., and Bout, M. V.: 1997, *Solar Phys.* **175**, 685 (this issue).
Steinolfson, R. S., Suess, S. T., and Wu, S. T.: 1982, *Astrophys. J.* **255**, 730.
Wang, A. H., Wu, S. T., Suess, S. T., and Poletto, G.: 1995, *Solar Phys.* **161**, 365.
Webb, D. F. and Cliver, E. W.: 1995, *J. Geophys. Res.* **100** (A4), 5853.
Wu, S. T. and Wang, J. F.: 1987, *Comp. Method Appl. Mech. Engr.* **64**, 267.
Wu, S. T. and Guo, W. P.: 1997, in N. Crooker, J. Joselyn, and J. Feynman (eds.), *Coronal Mass Ejections: Causes and Consequences*, AGU Geophysical Monograph Series, American Geophys. Union, Washington, DC U.S.A., in press.
Wu, S. T., Guo, W. P., and Wang, J. F.: 1995, *Solar Phys.* **157**, 325.
Wu, S. T., Guo, W. P., and Dryer, M.: 1997, *Solar Phys.* **170**, 265.
Wu, S. T., Wang, A. H., and Guo, W. P.: 1997, *Astrophys. Space Sci.*, in press.

FIRST RESULTS FROM SWAN LYMAN α SOLAR WIND MAPPER ON SOHO

J. L. BERTAUX[1], E. QUÉMERAIS[1], R. LALLEMENT[1], E. KYRÖLÄ[2], W. SCHMIDT[2], T. SUMMANEN[2], J. P. GOUTAIL[1], M. BERTHÉ[1], J. COSTA[1] and T. HOLZER[3]
[1]*Service d'Aéronomie du CNRS, BP. 3, 91371, Verrières-le-Buisson, France*
[2]*Finnish Meteorological Institute, P.O. Box 503, SF 00101 Helsinki, Finland*
[3]*High Altitude Observatory, Boulder, Colorado, U.S.A.*

(Received 26 February 1997; accepted 18 June 1997)

Abstract. After one year of almost flawless operation on board the SOHO spacecraft poised at L1 Lagrange point, we report the main features of SWAN observations. SWAN is mainly dedicated to the monitoring of the latitude distribution of the solar wind by the Lα method. Maps of sky Lα emissions were recorded througout the year. The region of maximum emission, located in the upwind hemisphere, deviates strongly from the pattern that could be expected from a solar wind constant with latitude. It is divided into two lobes by a depression aligned with the solar equatorial plane called the Lyα groove already noted in 1976 *Prognoz* data. The north lobe is much brighter than the south lobe. These two characteristics can be explained qualitatively by an enhanced ionization along the neutral sheet where the slow solar wind is concentrated, which results from the higher low-latitude solar wind mass flux as measured by *Ulysses*. The groove is the direct imprint on the sky of the enhanced carving by the slow solar wind, at this time of solar minimum, when the tilt angle of the neutral sheet is small. The question is still pending to predict what will happen with the ascending phase of the solar cycle. Observations of comets are briefly mentioned, with the ability of SWAN to monitor the H_2O production of many comets. Operations of the instrument are briefly described, including some instrumental problems which could be solved by software modifications sent to the instrument.

1. Introduction

SWAN is the first instrument dedicated to the measurement of solar wind latitude distribution by the Lα method. As such, it can be qualified of 'solar wind mapper', though from an instrumental point of view, it could better be described as a full-sky Lα imager. In the present paper, we report the main features of the observations obtained during one year of nearly flawless operation of SWAN on the SOHO spacecraft, poised at L1 Lagrange point. The purpose of the paper is to give the reader an overview of in-flight performance and capabilities of the SWAN instrument, as well as the most significant scientific results in a qualitative manner. A more quantitative analysis concerning the solar wind distribution retrieval is deferred to another paper, soon to come (Kyrölä *et al.*, 1997), while the whole SWAN investigation was described in detail before the SOHO launch (Bertaux *et al.*, 1995).

The principle of the solar wind Lα method is the following: a flow of interstellar H neutral atoms is sweeping permanently across the solar system. These atoms are ionized in a minor proportion ($\simeq$ 20%) by the solar EUV flux, but mainly by charge-exchange with solar-wind protons. A cavity is carved in the flow, with a shape that reflects the 3D distribution of the solar wind mass flux. Actually only the H atoms which have not been ionized may be observed in Lα through resonance scattering of solar photons, and the 3D distribution of these H atoms reflects the solar wind distribution, in particular its latitude distribution (the longitude distribution is somewhat averaged out by the solar rotation). The fact that the solar wind latitude distribution would influence the Lα sky pattern was first suggested by Joselyn and Holzer (1975) ; it was first observed in 1974 by Mariner 10 UV spectrometer (Kumar and Broadfoot, 1979) and then fully confirmed in 1976 by *Prognoz* 5 and 6 Lα photometers (Lallement, Bertaux, and Kurt, 1985; Summanen, 1993). With the daily operation of SWAN from SOHO, this new method of studying the solar wind by Lα remote sensing has now achieved full maturity. This remote sensing method is superior to in situ measurements for the permanent monitoring of the latitude solar wind distribution, but it should be kept in mind that this is at the expense of a much coarser time resolution (several days), and a somewhat more complex retrieval algorithm than the plain classical in situ measurements.

From a simple survey of the Lα sky maps collected during one year by SWAN, three main characteristics are emerging, which are easy to explain after the fact, but were not fully expected by the SWAN team before the flight (except for the second one, the Lα groove).

First, there is a spectacular change of the Lα map shape when SOHO, following the Earth, is moving in the solar system from a downwind position at time of launch (2 December 1995) to an upwind position in early June 1996. This motion of 2 AU of the observatory along the line of the interstellar flow results in a great variation of the size of the bright Lα region (around the incoming flow at $\lambda = 252°, \beta = +6°$) and correspondingly the size of the dimmer downwind region, as illustrated by comparing maps of Figures 6 and 7. The orbit of the Earth has a size which is comparable to the characteristic size of the H Lα glow distribution, in particular the so-called MER feature (Maximum Emissivity Region).This behavior can be understood from the sketch of Figure 1.

The second striking feature is the so-called Lα groove, which is obvious in Figures 7 and 6, as a depression of Lα intensity aligned along the ecliptic in the upwind hemisphere. This groove, already detected in *Prognoz* data (Bertaux, Quémerais, and Lallement, 1996b), is due to enhanced solar wind ionization (and mass flux) along the interplanetary neutral sheet at a time of solar minimum (*Prognoz* and SWAN are about 2 solar cycles apart, or rather one single 22-year solar cycle, if one considers the polarity of the solar magnetic field). The third feature is connected with the fact that the groove divides the upwind hemisphere into two distinct regions, one in the north and one in the south (ecliptic hemispheres). Our previous mental image of a Lα map called for a bright hemisphere centered near the upwind direction and a dimmer hemisphere centered in the opposite downwind direction. The point of maximum emission brightness was to be found near the ecliptic, displaced from the upwind direction by the lateral parallax effect of the Earth, when moving from March to September (Figure 2). This was indeed the

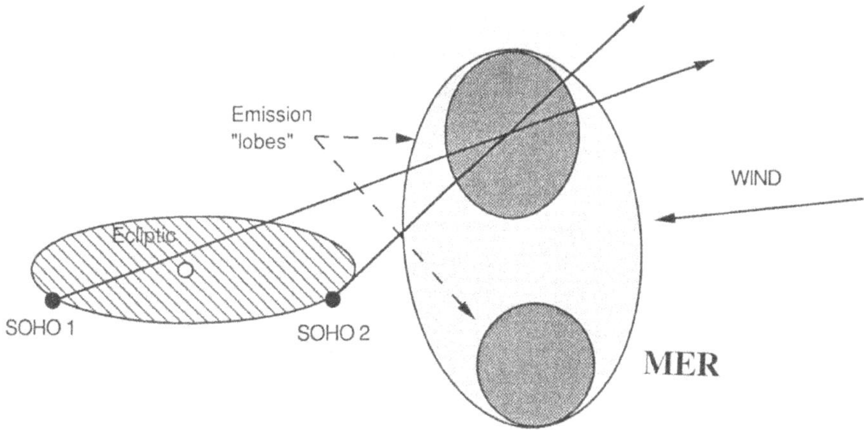

Figure 1. Sketch of the Maximum Emissivity Region (MER) divided into two lobes of emission by the enhanced ionization of solar wind in the solar equatorial plane. The latitude separation of the two lobes as seen from SOHO, as well as the overall angular size of the MER, changes during the year with the position of SOHO (*zoom* effect) from downwind position (SOHO 1) to upwind position (SOHO 2).

picture which emerged from the first partial Lα sky maps obtained in 1969 and 1970 by two Lα photometers on board OGO-5 (Bertaux and Blamont, 1971; Thomas and Krassa, 1972). The detection of the parallax effect (about 40°) was crucial to prove that the Lα emission was indeed coming from inside the solar system, and not from the galaxy as was previously suggested by several authors. Our mental image of Lα sky distribution has to be deeply modified now, at least for the present solar minimum conditions. There are two maxima of emission, one in the south, one in the north.They are widely separated from each other; they are moving in the sky as a function of Lagrange point L1 position along the year (SOHO is in a halo orbit around L1, located 1.5×10^6 km from Earth toward the Sun), not only because of the lateral parallax as detected in 1971, but also because of the zoom effect described above as the first feature.

2. Description of SWAN Instrument and Operations

SWAN has been described in detail in Bertaux *et al.* (1995) and only the most essential features are given here. SWAN is composed of two identical Sensor Units (SU) ; SU+Z mounted on the $+Z$ body face of SOHO can map slightly more than one hemisphere centered on $+Z$, while the second one is mounted on the $-Z$ face. Because of the attitude control of SOHO, which maintains $+X$ towards the center of the Sun, and the solar rotation axis in the plane XZ, the $+Z$ axis is near the north ecliptic pole (similarly the $-Z$ axis is nearly the south ecliptic pole), but not in a fixed celestial direction during the year. Full-sky maps are most often plotted

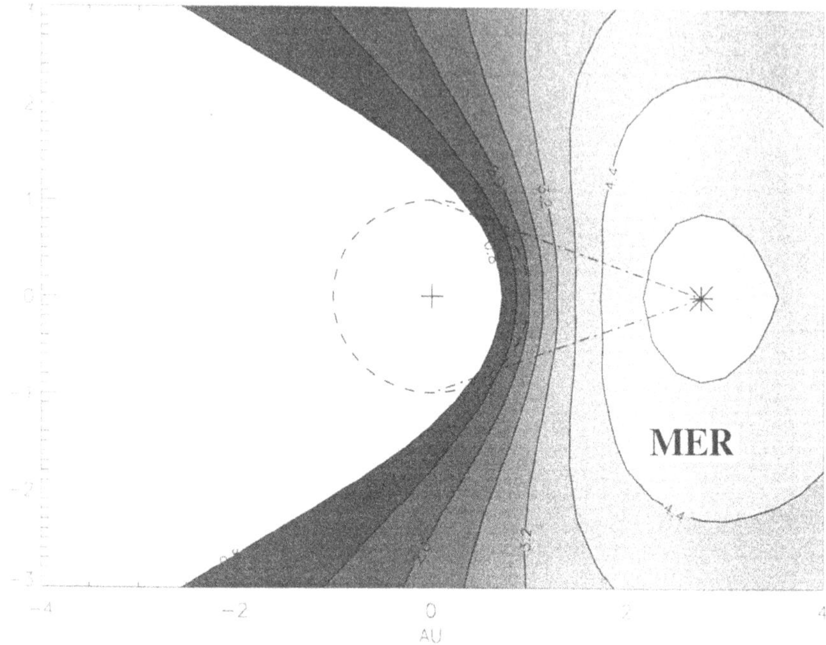

Figure 2. The parallax effect along the orbit of SOHO (dashed line). $L\alpha$ iso-emissivity contours for a typical model without solar anisotropies are represented (model values: lifetime of H atom at 1 AU TD = 1.2×10^6 s, $\mu = 1$). The Maximum Emissivity Region (MER) is defined as enclosed in the two highest isocontours. The maximum emission region (LOS integrated emissivity) is seen in different directions (dot-dashed lines) depending on the location along the orbit. The numbers are the emissivity values in $L\alpha$ photons s^{-1}.

in ecliptic coordinates (Figures 6 and 7), and the upper half is generated by SU+Z, while the lower half comes from SU$-Z$.

The instantaneous field of view of one sensor unit is a square of $5 \times 5°$ divided into 5×5 pixels of 1° square each. The central line of sight can be oriented toward any point of one hemisphere by a mechanical periscope system containing two mirrors at 45° incidence. While this system is oriented and maintained toward a specific position, UV photons are counted in each pixel of the multi-anode Hamamatsu detector for a typical duration of 13 to 45 s. The counting rate is typically 200 to 1000 counts s^{-1}. A motion of one or two of the mechanisms is then activated to measure the intensity distribution in another FOV of $5 \times 5°$. Covering the whole sky with a 1° resolution can therefore be accomplished in about one day.

In addition to this purely photometric mode, one hydrogen cell placed in the optical path can be activated to yield quantitative information on the spectral profile of the $L\alpha$ line, with a resolving power of $\simeq 3 \times 10^5$. One Electronic Unit (EU), contains a redundant microprocessor to control both sensor units and provides the electrical interface to the spacecraft.

One single observation is defined, for one sensor, by a rectangular area in the SWAN coordinate system inside two angular boundaries for both the inner mechanism (the one nearest to the detector) and the outer mechanism. Additional parameters to be defined are:

– The counting time during which photons are counted (13 to 45 s) while the mechanism is stopped.

– Whether the H cell is activated or not, at which power level, and for how long.

– The size of the elementary step by which both mechanisms are moved from 0.1° to tens of degrees ; one is kept first fixed (nominally the outer one) while the other one describes a full line between the boundaries ; then the other one is moved by one step and the first one is moved back step by step (one step per exposure).

– The level of high voltage to the detector.

– A time after which a following observation will automatically begin.

When the inner mechanism is kept fixed, moving the outer one provides a scan in the sky on a great circle passing through the Z SOHO axis direction (actually, half a circle centered on Z). When the outer mechanism is kept fixed, moving the inner one provides a scan along a small circle centered on the $+Z$ axis for sensor SU$+Z$, with an apex angle that depends on the position of the outer mechanism.

The first several months of observations were conducted from the SOHO-EOF (Experiment Operation Facility) control center at NASA/GSFC, during which the SWAN team exercised many modes of observations to build experience and confidence in the instrument and conducted several dedicated campaigns (i.e., comet C-1996 B2 Hyakutake, and spacecraft roll-maneuvers). Then a repetitive two-week program was implemented, which has been now (December 1996) reduced to a repetitive one-week program, as indicated in Figure 3. The program combines 4 full sky maps (one with the H cell), known comet observations, observations with one pixel of SU$+Z$ covered with a BaF2 filter to eliminate Lα emission, flat-field calibration and stellar observations. Due to memory limitations, such a program has to be loaded in three salvos of ground commands per week. These commands are generated at the institutes (Service d'Aéronomie or Finnish Meteorological Institute), and then sent to the SOHO-EOF (delayed commands) for a routing to SOHO and SWAN inside designated time windows. With such a scheme, SWAN has a duty cycle of about 95%. Some special operations were conducted either from GSFC and for some emergency procedures were conducted from MEDOC at IAS, Orsay, France.

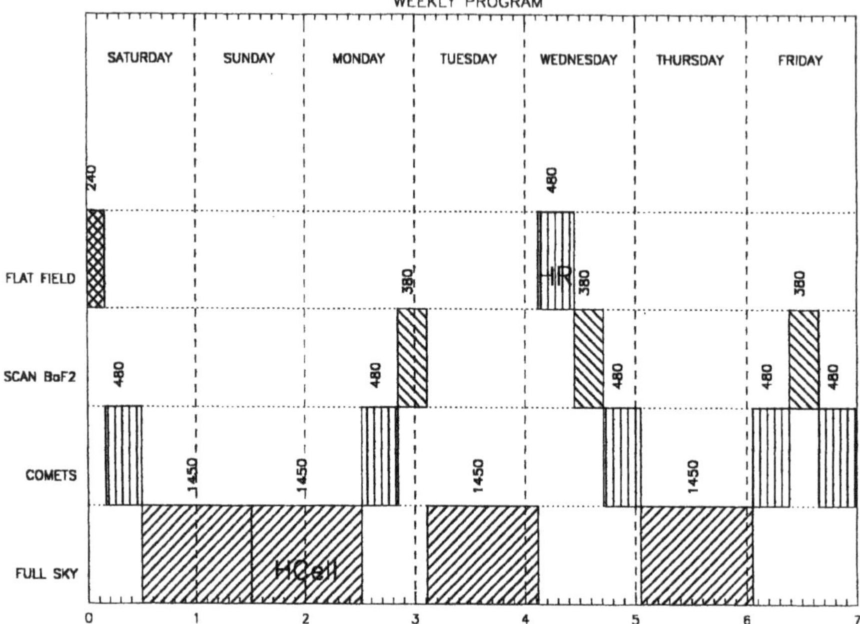

Figure 3. Typical SWAN weekly observation program. Time is shared between full-sky intensity maps, H-cell transmission maps, comet observations, flat fields, and long exposures for the BaF_2 pixel counting rate in fractions of the sky. The latter measurements will be used later for stellar decontamination. The duration of each observation is indicated (min).

3. Instrumental In-Flight Performances

The SWAN instrument was switched on for the first time in flight on 7 December 1995, five days after the SOHO launch. At that time, the HV of the detectors were not yet switched due to ESA fear of flash from incomplete outgassing. The one-shot pin-puller wax mechanisms serving as launch locks for the periscopes were both successfully operated, and the EU and SU were activated. After a 10-day period of commissioning and thermal stabilization of the periscopic mechanisms, the first light was seen by the instrument on 16 December. Since that day, the instrument has been in the scientific observation mode nearly all the time, in spite of some hardware problems or software errors that have been detected. Up to now, all the problems could be overcome by operational changes in the instrument operation, or flight software modifications, and have not impacted the scientific performances of the instrument. Here is a short description of the major incidents that have occurred during this first year of operation.

3.1. High voltage trip

In order to prevent a destructive high count rate in case of direct solar illumination of the photon counting detectors, the high-voltage power supply (HVPS) feeding the detector is equipped with a tripping circuit that shuts down the power in case of an overcurrent (one HVPS per SU). When the HVPS of SU+Z was switched on for the first time on 16 December 1996, the optical axis was oriented toward the $+Z$ axis and the level was progressively increased from its minimum value for which no photons are counted, because each of them triggers a pulse of electrons below the threshold of the PAD (Pulse Amplifier Discriminator) attached to each anode. The number of counts increased with the HV level. We were only half-way to the top when suddenly all counters read zero, making us believe for some hours that the sensor was permanently dead. Fortunately, it was the tripping circuit which had activated on its own, and a new switch-on command of the HVPS successfully reactivated the detector. It should be mentioned that on the ground, the trip-off circuit had never been triggered at such low levels and counting rates, and this difference of behavior between ground and space is still unexplained. The HV level of SU+Z was then kept at a moderate level, which still gives a hefty counting rate (see next section). However, even at this moderate level, there are still some occasional trips, with the consequence that the rest of the observation is lost. The same problem was also experienced with the HVPS of the SU$-Z$ sensor, which tripped for no obvious reason with a mean rate of one trip every 3 days. The original operational scenario required a ground telecommand to switch back on a tripped power supply, detected from a look at the data. This very constraining scheme was alleviated by adding a new piece in the flight software checking the status of the power supply every 20 s, and switching it back on automatically, in case of a trip. These random trip events could not be correlated to anything up to now and occasionally one or two exposures are lost.

3.2. Time Tagging of measurements (LOBT)

After 12.5 days of operation, the SWAN internal time tagging of each source packet went wrong. This could be traced to a software bug (difficult to detect on the ground, because it required more than 12 days of uninterrupted operation), which was taken care of by another software patch, and also by the daily updating of LOBT (local on-board time) from SOHO on-board time to SWAN, activated by flight operation team personnel at the SOHO-EOF.

3.3. Sticky mechanism

On the southward-looking sensor SU$-Z$, one of the two periscope mechanisms got stuck from time to time. It was known from previous ground thermal vacuum tests that this kind of mechanism is very sensitive to thermal gradients across the ball bearing caging. Thus, the power dissipation in each motor was adjusted by telecommand and the respective order of motion for each axis was inverted with a flight software modification, reducing the amount of rotation of the sticky mechanism. After this modification, all four mechanisms performed nominally, except for one specific incident. On 17 October 1996, this same mechanism was almost completely stuck, moving only over 1°. After several days of strenuous thinking and various attempts to unstick it by nerve stressing telecommanding, all unsucessful, it was realized that the motor commanding electronics might have been corrupted by a single event upset. Indeed after reset of this specific electronic board, the mechanism moved smoothly again.

3.4. Sensor unit digital voltage trip

During the first year of operations, each sensor unit digital voltage tripped once, resulting in red alerts in the housekeeping parameters being monitored by the flight operation team in Goddard. These two incidents seemed linked to a single event upset in the motor control electronics. After a reset of the faulty sensor, the instrument restarted nominally.

3.5. Main CPU reset

A reset of the instrument main computer occurred on 28 November 1996 at 18 UT. This reset was triggered by the internal watchdog device that restarts the instrument in case of a major software malfunction. This software error was very likely induced by a single event upset, likely related to a high intensity proton flux measured with the ERNE instrument at the same moment. The instrument restarted automatically after this problem, but the scientific operations could restart only after re-setting of some parameters via telecommand. It should be noted that each time the instrument is switched OFF or reset, for whatever reason, all software patches are lost and must be reloaded by ground command.

3.6. Flat field as a function of HV level

Flat field designates the fact that pixels of one detector have different sensitivities to Lα photons. Special observations in which the same square degree of the sky (void of stars) was measured with all the pixels successively were performed to yield the flat-field response of each detector. In Figure 4 are plotted the counting rate of all the pixels of SU+Z detector (as a function of the identification number of the pixel) when exposed to the same Lα intensity, for various HV levels. It is clear that the non-uniformity of the detector decreases with an increasing HV level, while the counting rate increases. This is indicative that it is the average gain of the microchannels in front of the various pixels which is not uniform (and therefore the counting ratio ρ_c) rather than variations of the quantum efficiency which is responsible for the non-uniformity.Therefore there are two reasons for desiring a high HV level, that had to be put in balance with the undesirable trip-off effect

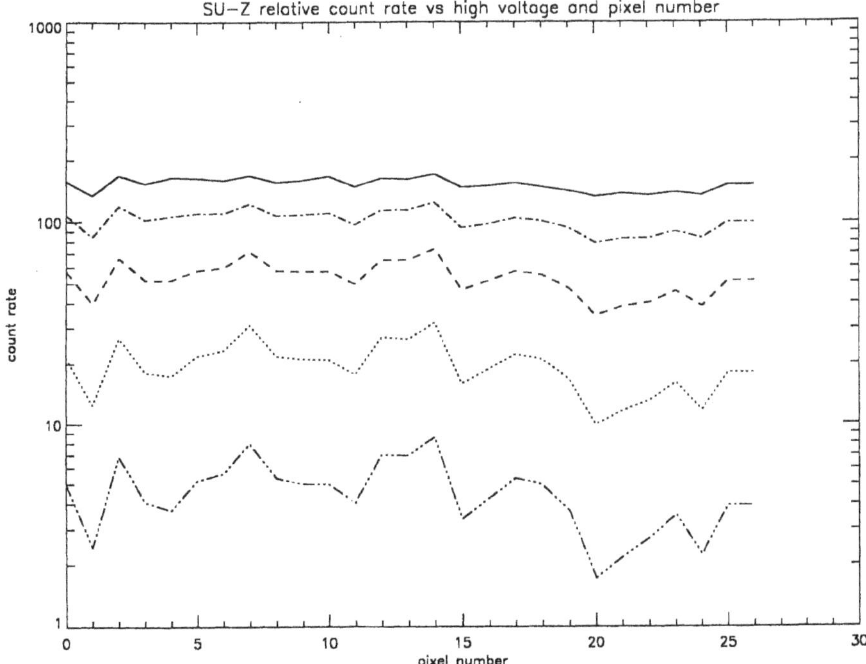

Figure 4. The counting rates for the 27 pixels for different levels of the high-voltage power supply, for sensor SU$-Z$. The HV levels are in reference digital units 100 to 140 by steps of 10 units and the counting rate increases with HV level. The direction of sight corresponds to a region with no stellar contamination. Results obtained for sensor SU$+Z$ are very similar.

mentioned above. All data presented in this paper are corrected for the flat field effect, which amounts, for the most used HV level, to about $\pm 15\%$. This flat field is monitored regularly; it has changed a little for SU$+Z$ detector, during the first three months, as well as its overall sensitivity.

3.7. Computation of the Line of Sight

The celestial coordinates of a line of sight of a given measurement are defined by several variables. First, the position of the two periscope mechanisms of the sensor unit must be known from analog encoder readings. This enables us to compute the coordinates in the so-called SOHO reference frame, where the X-axis is defined by the SOHO-to-Sun-center direction and, where the Z-axis is defined by the projection of the Sun rotation axis onto the plane perpendicular to this X-axis. Second, to compute the coordinates of the line of sight in a fixed reference frame (usually the ecliptic) we need only the date of the measurement and the position of SOHO in the solar system which then fully defines the transformation from the SOHO reference frame to the ecliptic reference frame. The position of the periscope mechanisms is given by encoders with a residual noise of the order of a fourth of a degree. When computing the star coordinates from full-sky scans, it soon appeared that the observed positions were offset by roughly 1 deg in the northern hemisphere and 3 deg in the southern hemisphere. A preliminary analysis of the errors observed for the star positions has enabled us to correct these pointing errors to within half a degree. In the future, a more thorough analysis should yield an even better pointing accuracy.

4. Absolute Photometric Sensitivity and Time Evolution

Several sources of information may be used to estimate the absolute photometric sensitivity of SWAN: ground calibrations, the observation of UV hot stars, and comparison with other instruments assumed to be well calibrated. The sensitivity at Lα to an extended source can be expressed as the number N_p, the counts per second for one square degree pixel for an intensity of 1 Rayleigh $=$ $10^6/4\pi$ photons cm^{-2} s^{-1} sr^{-1}. Because single photon pulse heights depend on the high voltage level, for one photon there will be only $\rho_c Q$ counts, where Q is the quantum efficiency of the CsI cathode ($\simeq$9%) and ρ_c is the counting ratio, which varies between 0 and 1 and increases with the HV level (the fraction of photon-generated pulses higher than the threshold). Therefore, N_p varies with the HV level, which should always be specified when the sensitivity of SWAN is specified. Most of the SWAN observations have been done with the same standard HV level (in digital units, 110 for SU$+Z$ and 130 for SU$-Z$).

4.1. Comparison with HST/GHRS

On 9 March 1996, the Goddard High Resolution Spectrometer on HST was used to measure the interplanetary Lα emission IP in the downwind direction (within a few degrees from downwind, and at a time when the Doppler shift between geocoronal emission and IP emission allows one to distinguish them with a high-resolution spectrum). According to the standard absolute calibration of GHRS, the IP emission was 200 ± 20 Rayleigh (J. T. Clarke, private communication), while SU$-Z$ counted 70 counts per second. Taking into account the fact that SU$+Z$ was more sensitive than SU$-Z$ by a factor 2.43 at that time, an estimate of N_p for this date can be given:

$$N_p\ (\mathrm{SU}+Z) = 0.84 \pm 0.08 \ \text{ at HV digital level} = 110\,,$$

$$N_p\ (\mathrm{SU}-Z) = 0.35 \pm 0.035 \ \text{ at HV digital level} = 130\,.$$

These numbers should be considered as a preliminary estimate, subject to change without advance notice.

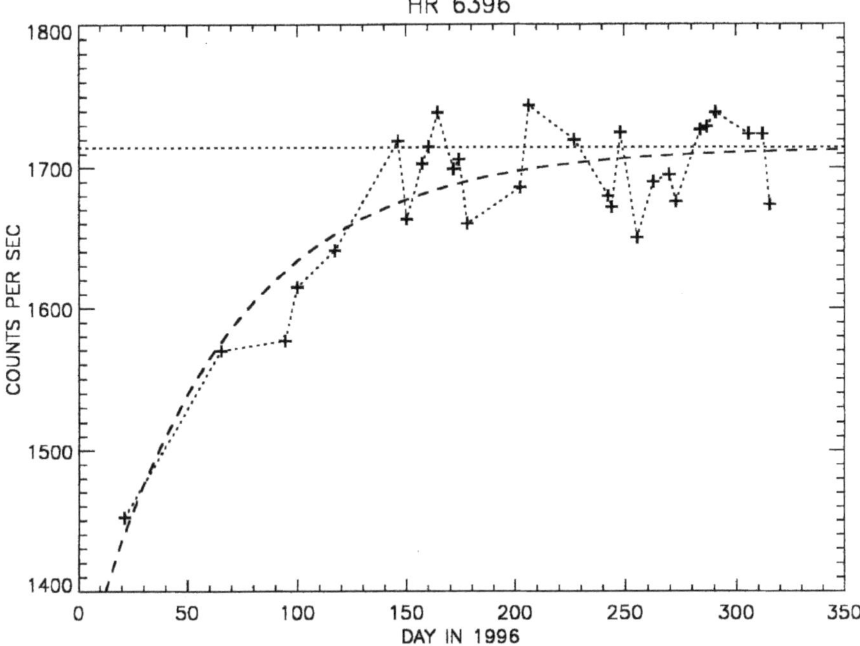

Figure 5. Time variation of the stellar count rate observed by the SU+Z sensor unit for star number 6396 of the HR catalog (ecliptic coordinates: 183.32 and 84.76 degrees). The time origin is 1996/001/00:00:00. The count rate shows an increase of 15% between DOY 1 and DOY 150 (day of year). The sensitivity of the sensor unit is stable after DOY 150.

4.2. EVOLUTION WITH TIME

An absolute calibration with stars is more complex, because SWAN is sensitive in the range 115–180 nm and it requires a good knowledge of the relative spectral sensitivity of SWAN. For the time being, the exercise has not yet been completed. However, numerous observations of the same stars during 1996 allowed us to monitor time variations of SWAN sensitivity. Surprisingly, it was found that the sensitivity of SU+Z sensor *increased* by about 15% during the first few month of observations, and then levelled off (Figure 5). In this figure, we have shown one good example of continuous observations of one star during 1996. The star selected here is number 6396 in the HR catalog (ecliptic longitude 183.32° and ecliptic latitude 84.76°). This is not a variable star and its high ecliptic latitude ensures that it is always observed by the north sensor during each of the full-sky maps. This means that these observations do not require any specific program of the instrument but are a normal output of full-sky observations.

To estimate a stellar count rate, we must remove the Lα background due to interplanetary hydrogen atoms in the line of sight. It is of the order of 500 counts per second per pixel. This is performed by imaging the 5 $\times$ 5 deg area of the sky

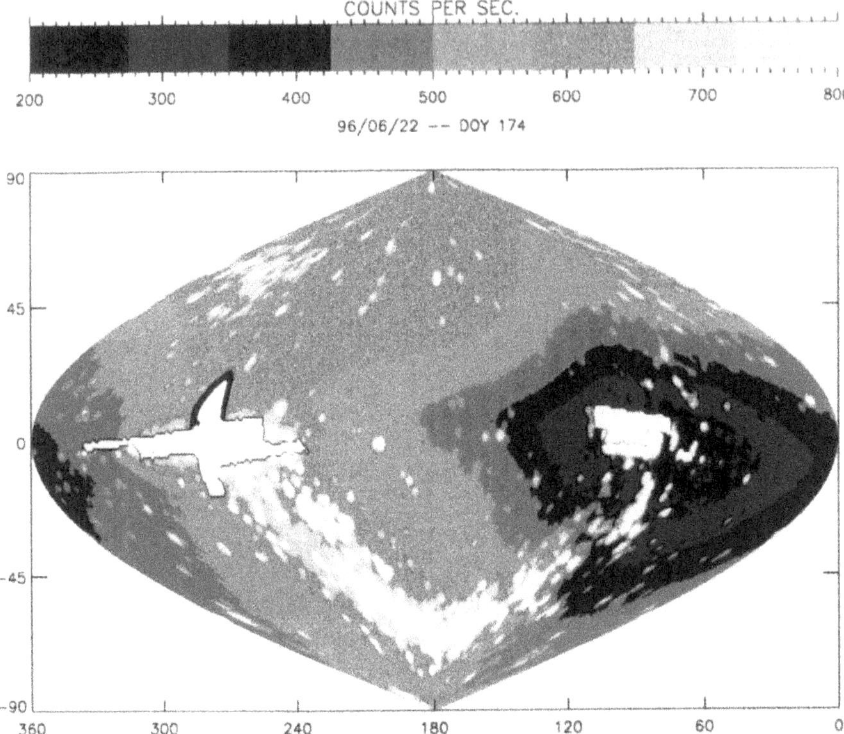

Figure 6. Color-coded Lα sky intensity maps (in ecliptic coordinates) recorded by SWAN on 22 June 1996 (day 174) when the Earth was on the upwind side. Iso-contours of the IPM Lα emission are visible at each change of color. They show the existence of a secondary minimum of emission along the ecliptic, more clearly visible in the 96/174 map on the left side. The minimum divides the region of maximum intensity into two bright areas, one north, one south and forms a groove in the sky. The Milky Way and isolated hot stars produced the great emissive circle (more conspicuous in the south) and the small white spots, respectively. White areas with roughly rectangular shapes do correspond to the absence of valid data due to stray light from the Sun or reflections of the solar light on the satellite in the anti-solar direction. The nearly triangular area is an obscuration of the sky by a thruster shield implemented for the protection of SWAN during thruster firings. The maximum intensity region moves in longitude and latitude as shown by its different locations in the two maps. In particular, when SOHO approaches the upwind part of the orbit (June map), the two maxima are seen at high latitude, which means that the spacecraft is passing between the two 'emission lobes' described in Figure 1.

surrounding the star and interpolating to the position of the star. The stellar count rate is then obtained as the sum of the count rates of the 25 pixels from which the interplanetary Lα background is removed. To minimize the effects of chromatic aberration, we have only used the measurements for which the peak emission of the star is centered on the 5×5 pixel array. Figure 5 shows the derived count rate as a function of time. The time origin of the plot is 1 January 1996 at 00:00 UT. The plot shows that the count rate has increased by 15% between the beginning of

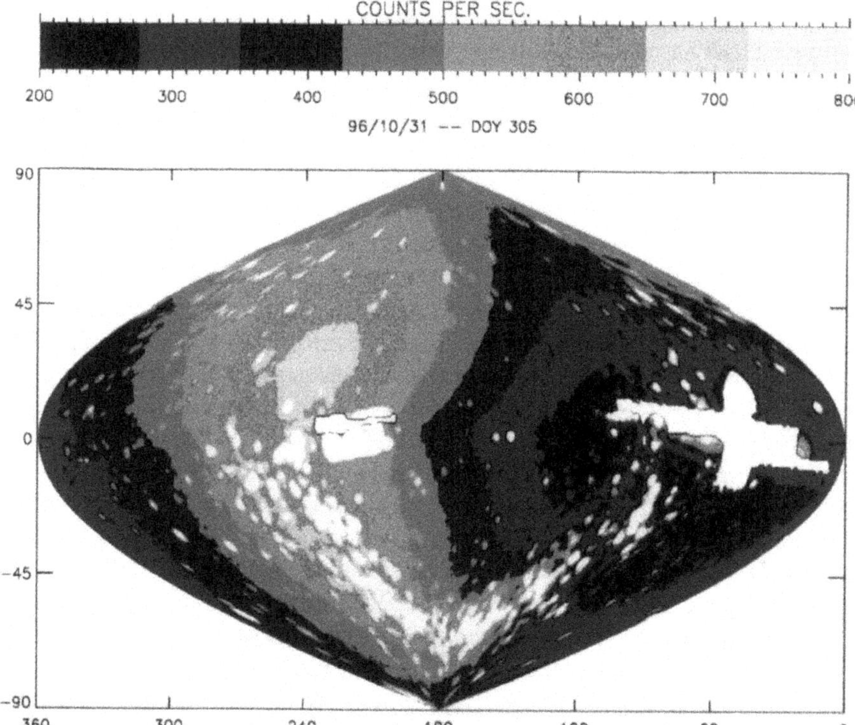

Figure 6. Same as Figure 6, for 31 October 1996 (day 305), when the Earth was on the downwind side.

1996 and DOY 150 (end of May). After that time, the count rate is stable. The same result is obtained when using other stars. On the other end, the analysis of the $-Z$ sensor shows that its sensitivity has not changed since the beginning of the mission, which is independently confirmed from measurements on the same region of the sky obtained with both sensors during special roll maneuvers of SOHO, when the attitude is rotated along the SOHO–Sun axis.

5. Description of Typical Sky Maps

The bulk of SWAN data, from which will be derived the distribution of solar wind, comes from the full-sky maps recorded three times a week in the photometric mode (without the H cell activated). Figure 6 represents, in ecliptic coordinates, the distribution of intensity recorded on 22 June 1996, and is typical of the many maps recorded up to now. The color code is adjusted to cover the range from 200 to 800 counts per second (per 1° pixel). The upper half of the map was recorded with SU$+Z$ detector, while the lower half was recorded with SU$-Z$. Actually, the line dividing the area covered by the two sensors is not strictly the ecliptic plane, because SOHO attitude is maintained with the Z axis parallel to the projection of the solar rotation axis on the ZY plane of the spacecraft, perpendicular to the SOHO–Sun X axis. Therefore the Z axis may be inclined to the ecliptic polar axis by up to 7.2°. Some artifacts must first be described briefly. The blank area around $\lambda = 90°$, $\beta = 0°$ (right side), is the region around the Sun. The upper half is blank because measurements are not recorded here, because the stray light scattered by the Sun shade placed in front of the SU+Z sensor opening is very intense and could damage the detector. On the contrary there are measurements recorded by the SU−Z, because the stray light generated by the Sun shade is smaller, due to the larger distance of the Sun shade from the sensor opening (about 2 meters). Still, the signal is rather intense and greater than 800 counts s^{-1}, the upper limit of the selected color code. Accordingly the lower half of the region around the Sun is also blank. On the left side there is another blank area along the ecliptic where measurements are avoided, around the anti-solar direction. This is because the glint of the Sun is reflected on some parts of the service module of the spacecraft, giving a very strong signal. Just above and below the left side of this blank area, there is a small increase of intensity which is probably due to a remnant of stray light from the glint, not totally eliminated by the entrance baffle of the sensor. Above this blank area can be seen a darker area with the shape of a shark fin. This is due to the obscuration of the sky by a piece of hardware which was attached to the spacecraft to protect the SWAN sensor from thruster firings during orbital maneuvers, and is the darkest area observed by SWAN. All the bright spots scattered in the sky are due to the UV emission of hot stars in the bandwidth of SWAN (115–180 nm). While the angular resolution of SWAN as defined by the optical quality at Lα is better than one degree (about 0.5°), bright stars are sometime seen over 2 to 3°; this is because of the chromatic aberration of the optical system, which includes the MgF_2 lenses of the H cell. In addition, in these plots the binning routine distributed each single pixel measurement in 4 adjacents bins, which has a tendency to spread somewhat the contamination of one star to the surrounding area. Most of the bright stars are seen along the galaxy, and contaminate the south ecliptic hemisphere more than the north hemisphere. All the rest is Lα emission from H atoms resonantly illuminated by the Sun, with a very small amount of dark current estimated at less than 1 count per second (not subtracted here). The counting rate plotted for the SU+Z sensor is the actual count rate (s^{-1}), while for the SU−Z sensor the counting rate was multiplied by 2.43, the ratio of sensitivities determined during special roll maneuvers of SOHO, in which the same part of the sky was successively observed with the two sensors. Since the counting time is 13 s, a typical intensity on the map of 500 counts s^{-1} corresponds to a statistical noise of 1.2% per 1 square degree bin of sky at least, since some bins have more than one measurement. The map of Figure 7 was recorded 31 October 1996, and contains the same artifacts as the first map, but displaced in the sky because the SOHO-produced artifacts are linked to the position of the Sun in the sky. Stars are of course seen at the same place in the sky. Color-coded with the same scale, both maps indicate a maximum of Lα intensity on the left side, which is the hemisphere from which comes the interstellar wind, and a minimum on the right hemisphere, where the wind is blowing, as expected from the effect of the solar ionization. However, there is a striking difference between the two maps in the relative area covered by the maximum region and the minimum region, as is evidenced for instance by the smaller area contained below the 500 counts s^{-1} isophote. This is not due to a time variation of the ionization, because the value of the maximum intensity is the same in the two maps, as well as the minimum intensities. Rather, this is due to the motion of the spacecraft which went downwind by nearly 2 AU from June to October, as is explained by the sketch of Figure 1. This is a kind of 'zoom effect' on the region of the minimum, or conversely there is a 'zoom effect' on the region of the maximum when travelling in the solar system from downwind to upwind along 2 AU only. This clearly indicates that the orbit of the Earth, which is mostly inside the ionization cavity void of H, approaches very near the Maximum Emissivity Region (MER) when being upwind and may be even embedded inside the Lα-illuminated volume of H. Therefore, a picture emerges in which, even for constant solar conditions and resulting constant H distribution, the sky seen from Earth in Lα experiences strong 'seasonal' changes. Of course, in order to detect possible time variations of the H distribution which would be linked to changes of the solar wind ionization, one will have first to model carefully this seasonal variation to eliminate it. A second spectacular feature is the small intensity depression almost aligned with the ecliptic plane, seen mostly on the left hemisphere from where comes the wind. This feature was first identified in *Prognoz* data and called a 'groove' (Bertaux, Quémerais, and Lallement, 1996b) because of its shape, and was interpreted as the direct sign of increased ionization by the solar wind in the neutral sheet. Indeed, *Ulysses* found a larger solar wind flux within $\pm 25°$ of the ecliptic (Goldstein *et al.*., 1996), though the solar wind is slower in this region (Figure 8). Because the charge-exchange cross-section is larger with a smaller solar-wind proton velocity, it also enhances this effect. Model calculations of the Lα pattern based on latitude variation of *Ulysses* solar wind measurements have indeed shown such a 'groove' pattern, separating the upper hemisphere into two regions: the north and the south (Summanen, Lallement, and Quémerais, 1997). This is indeed what is found in both maps presented here, the groove being much more pronounced on the June map than on the 31 October map, because of the proximity of the emission volume (or rather, the lack of H in the emission volume created by the enhanced ionization). Before these SWAN maps, the common view of the sky Lα pattern was that of a dipole, with a maximum of intensity around the upwind direction (near the ecliptic plane) and a minimum in the opposite direction, as predicted by a model in which the solar wind ionization would be constant with latitude (Figure 9). Actually, at this time of solar minimum (1996) we see two maxima, one in the north and one in the south. Figure 10 is a global average of the Lα intensity on a large series of maps collected by SWAN up to November 96. In

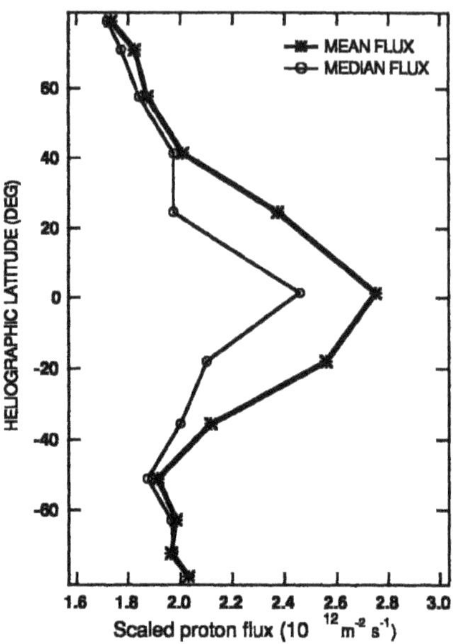

Figure 8. Solar wind proton fluxes measured by SWOOPS during the fast latitude scan of *Ulysses*, binned by solar rotation (from Phillips, 1994). Both the median and the average values are represented, to show that the average value is indeed larger than the median. The solar wind flux is enhanced by 40% between $-20°$ and $+20°$ of solar latitude, explaining the presence of the groove carved in the $L\alpha$ emission pattern of the upwind hemisphere.

this figure is displayed the position of the northern maximum detected on each of the maps (the position of the southern minimum is more delicate to determine due to the larger contamination of the south sky by galactic stars). A hand-drawn line connects the points in approximate chronological order, illustrating both changes in longitude (classical lateral parallax effect) and in latitude (due to the zoom effect, the maximum moves toward the pole when SOHO approaches the upwind position). The amount of change (about 120° of longitude, from 190° to 310°; and 30° in latitude, from 25° to 55°) is related to the distance and latitude of the point of the maximum emissivity region, in turn related to the latitude distribution of the ionization rate and solar wind flux. However, a detailed comparison with models is beyond the scope of this paper. It can be remarked in Figure 10 that only a fraction of the 'trajectory' of the maximum emission point is seen, with an apparent jump from March to June. As a matter of fact, the accumulation of points at 320° longitude corresponds to a group of hot stars whose contribution was not totally removed because they are not spatially resolved and produce a rather strong emission, larger than the brightest IPM emission seen at high latitude when SOHO is upwind. As a consequence, until a correction for this stellar emission is done, which is in progress, the high-latitude part of the maximum emission trajectory is missing.

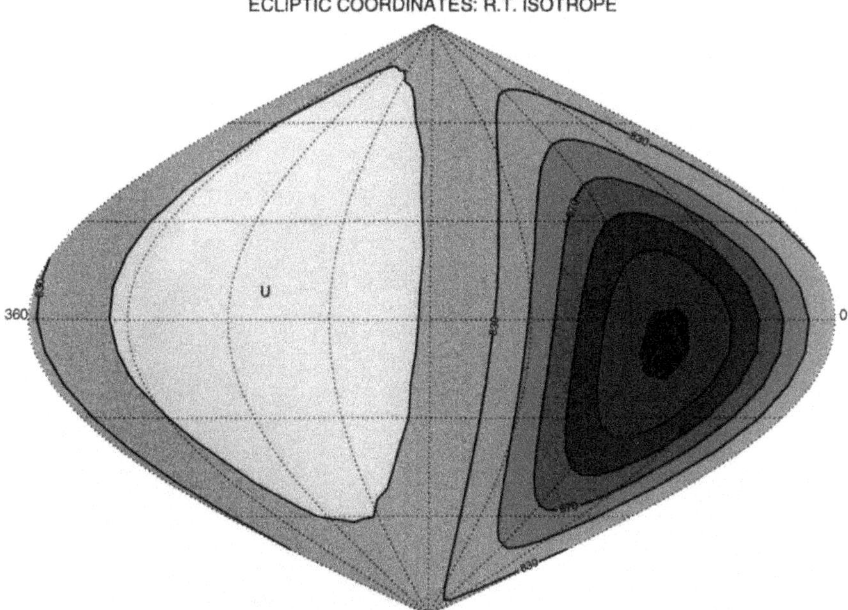

Figure 9. Model of Lα emission pattern produced by an isotropic solar wind and corresponding ionization rate. It shows only a smooth variation from the maximum region of the upwind hemisphere to the minimum region in the downwind hemisphere, while SWAN maps show a more complicated structure with two maxima separated by the groove. Parameters of the model: $T_D = 1.2 \times 10^6$ s, $\mu = 1$, H density at infinity 0.15 cm^{-3}, Lα excitation factor $g_0 = 1.35 \times 10^{-3}$ photons s^{-1} at 1 AU. Isophotes are in rayleighs and multiple scattering is fully included.

5.1. NORTH-SOUTH ASYMMETRY

The two maps presented here show clearly a north-south asymmetry, with the north ecliptic hemisphere being brighter in Lα than the south hemisphere. Though the two hemispheres are measured with two different detectors which have different calibration factors, we are sure that this is not an instrumental effect, because we had many opportunities to cross-calibrate the two detectors, in particular during the special roll maneuvers of the spacecraft conducted for this purpose, and for the purpose of other SOHO instruments. We have also some doubt that this could be the result of an asymmetry of illumination from the Sun, because the observed asymmetry has been a permanent feature since the beginning of the observations. Therefore, it must reflect a genuine asymmetry of the H distribution. At first, we thought that it might come from a smaller solar wind flux in the north than in the south, because indeed the *Ulysses* data showed a modest difference between the two

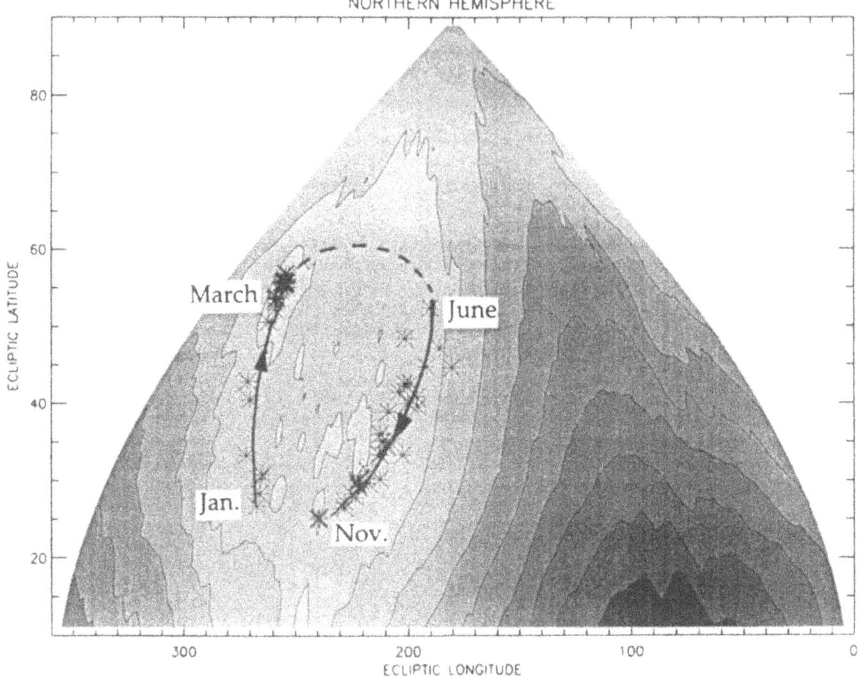

Figure 10. One-year evolution of the maximum intensity direction in the north hemisphere: the crosses represent the direction of maximum intensity, in ecliptic coordinates, as recorded by SWAN in the north, superimposed on a map of the averaged emission (mean value over almost the year). Bright isolated stars have been removed during the search for the maximum intensity. The maximum emission varies in longitude from 180 to 320 deg. This is the parallax effect described in Figure 2. Simultaneously, the latitude varies from a minimum value of about 20 degrees when SOHO was on the downwind side and the two bright regions (the two 'lobes') which characterize the upwind side were seen from more than 2 AU, to up to 60 deg when SOHO was passing between these two 'lobes' (see Figure 1). A hand-drawn line connects the points as a function of time.

hemispheres, at least at high latitudes. However, there is another possibility which comes from the fact that the interstellar wind vector is not strictly in the ecliptic, but at a latitude of +6°. If the interstellar wind were in the solar equatorial plane, and if the heliographic distribution of the solar wind were symmetric with respect to the equatorial plane, there would be a perfect symmetry of the H distribution and Lα emissivity that would be reflected as a symmetry in the intensity maps. In fact, it is not the ecliptic system which is relevant for solar-wind studies, but rather the heliographic system, inclined by 7.2° to the ecliptic. The ascending node of the heliographic equator is at 75° ecliptic longitude, almost coinciding with the downwind direction, and the angle of the wind vector to the solar equatorial plane is only 7°, or even 5° if the determination of the wind vector for interstellar helium atoms from *Ulysses* is considered (Witte, Banaszkiewicz, and Rosenbauer, 1996).

Still, through modelling we realized that even a small angle such as 5 to 7° will result in a non-symmetric Lα emission pattern for a symmetric solar-wind distribution. This can be explained by considering that H atoms coming in the upwind region of the north ecliptic will encounter less solar wind ionization than the ones with an equal angle to the wind vector in the south ecliptic, at least with the kind of solar wind distribution as shown by *Ulysses*. In conclusion, the marked north-south asymmetry could well result totally from the small, but non-zero, inclination of the interstellar wind vector to the solar equatorial plane.

5.2. SWAN MOVIE ON CD-ROM

The SWAN movie included in the CD-ROM is made from 55 individual full-sky images taken by SWAN over more than one year. The sequence of the movie is roughly one image per week, although the time between two images can go down to 5 days or be as large as 10 days. Each image is computed in ecliptic coordinates. One observation lasts 24 hours, which is the time necessary for each sensor unit to map half of the sky. The full sky image is made of two halves, each one being obtained by a different sensor unit. The cross-calibration factor used here is 2.85, the reference being set by the north sensor ($+Z$ side of spacecraft). The color code is defined up to 800 counts s^{-1}, which covers the range of intensity of the interplanetary Lα background. The sequence starts in April 1996 and ends in May 1997, just before the time when SOHO reaches its most upwind position.

6. Time Variation Due to Solar Illumination

It is clear from the discussion above that the main factor of Lα intensity variation with time (from map to map) is the actual position of SOHO in the solar system, with respect to the H emission volume. In Figure 11 is represented the variation of the counting rate in a fixed celestial position at $\lambda = 221°$, $\beta = 35°$, as a function of time. This direction was selected for not being contaminated by UV stars, while being relatively near the upwind direction. A part of the increase of the signal during the first 100 days may be due to the increase of sensitivity of the SU+Z as noted in Section 4. After day 100, when the sensitivity was stable, there is a general continuous increase of about 30% and a levelling-off after day 200, as a result of the motion of the spacecraft. On top of this general behavior, one can notice shorter time scale variations, with a period of 25 days, particularly visible between day 200 and day 300. This is clearly due to an increase of the solar illumination. We were able to check with our colleagues from the EIT instrument on SOHO that indeed there was at that time one single moderately active region on the Sun, passing in phase with the noted increase of sky backscattered Lα. A detailed analysis of such a correlation (beyond the scope of this paper) will allow us to establish the exact angular shape and size of the additional solar Lα produced by one single sunspot;

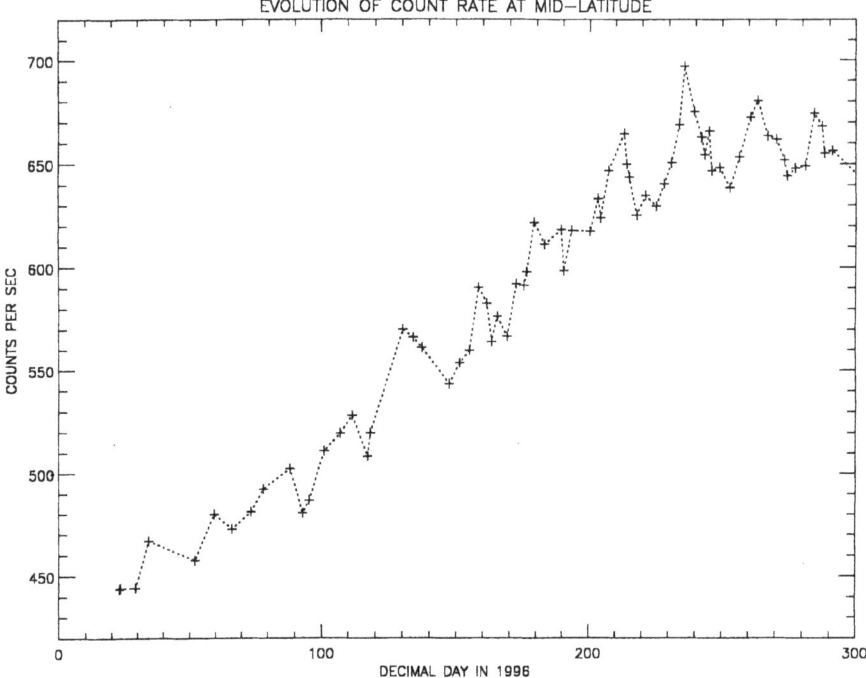

Figure 11. Temporal evolution of the count rate in a constant direction corresponding to $(l, b) = (221°, 35°)$ in ecliptic coordinates as a function of the day of the year (DOY) 1996. This direction has been chosen for being uncontaminated by stellar light. There is a significant increase at the beginning, due to a combination of two factors: the 15% sensitivity increase noted for stellar observations (Figure 5), and the increase linked to the motion of SOHO in the ecliptic plane towards denser H regions. In addition, there are peaks with a $\simeq$25 day periodicity due to solar rotation, clearly seen between days 200 and 300.

then, from the solar maps constructed from EIT, one will be able to produce a 2D model of the Lα solar illumination as a function of date, an important element when a detailed comparison of SWAN data with a model will be performed.

We note here a different approach adopted by Pryor *et al.*. (1996), the *Galileo* team dealing with the EUVS Lα data collected during the cruise to Jupiter. They have produced a sophisticated model of the solar illumination pattern based on the observed distribution on the solar disc of the equivalent width of the absorption line He I 1083 nm, known to be a proxy of the total solar Lα line. It will be interesting to compare the two lines of approach in the future. Their observations along great circles were interpreted with a model having an isotropic ionization rate, at variance with our contention. However, the *Galileo* data were collected in 1990, at a time of maximum solar activity, when the distribution of slow and fast solar wind may be expected to differ from the solar minimum conditions prevailing for SWAN data.

7. Hydrogen-Cell Results

Here we describe a different set of data obtained with the same instrument, but operated in a different mode of observation, allowing us to measure the velocity distribution of H atoms, with the high-resolution spectral analysis of the Lα emission line with a hydrogen absorption cell. One of the purposes of these measurements is to firmly establish the parameters of the interstellar flow, which need to be known as input parameters of the model of H distribution in the solar system, ionized by the solar wind.

7.1. Description of the Measurements

The two SWAN sensors are both equipped with a hydrogen cell in front of the detectors, i.e., a pyrex vessel filled with molecular hydrogen at a pressure of about a few hundred Pa. The cell is said to be 'activated' when one of the two filaments of tunsgsten inside it is heated electrically, and dissociates at its surface a fraction of the H_2 molecules to create atomic H. This temporarily produced atomic gas is able to resonantly scatter a fraction of the incident Lα light, provided the frequency of the photons as seen by the atoms in the cell is the resonance frequency. Photons scattered to the walls of the cell are lost from the light beam collected on the detector. This condition is fulfilled when the radial velocity of the H-cell atom (the atom velocity projected onto the line of sight) is equal to the radial velocity of one of the emitting interplanetary atoms. Since the thermal Doppler velocity spread inside the cell, where the temperature is close to 300 K, is smaller than the thermal broadening in the IPM, where the temperature is of the order of 10 000 K, the absorption line created by the cell is narrower (half-width 0.015 Å) than the IPM emission line (halfwidth 0.07 Å) and only a fraction of the light is absorbed, even when the Doppler shift between SOHO and the IPM is zero. When the Doppler shift is not zero, the H-cell absorption is even smaller and may go to zero.

All H-cell measurements are obtained by comparing the counting rate when the cell is OFF and the cell is ON, for the same direction of sight (the atomic H recombination into H_2 occurs within a fraction of a second). This allows us to measure both the intensity without any absorption by the cell, and the transmitted fraction of the light when the cell is ON. We will make use of the ratio $R = I_{ON}/I_{OFF}$ called the H-cell transmission (in previous works, it was also named 'reduction factor'), which is a function of the relative motion of the emitters and the absorbers, i.e., the relative motion of the IPM gas and SOHO, and of the temperature of the emitter (for a given H-cell regime or optical thickness). We may also refer to the absorption $A = 1 - R$.

7.2. Zero Doppler Shift Circles

Figure 12(a–c) shows SOHO at three different dates (22 January–25 April–7 July 1996) along its orbit. The vector characterizing the motion of the interstellar wind

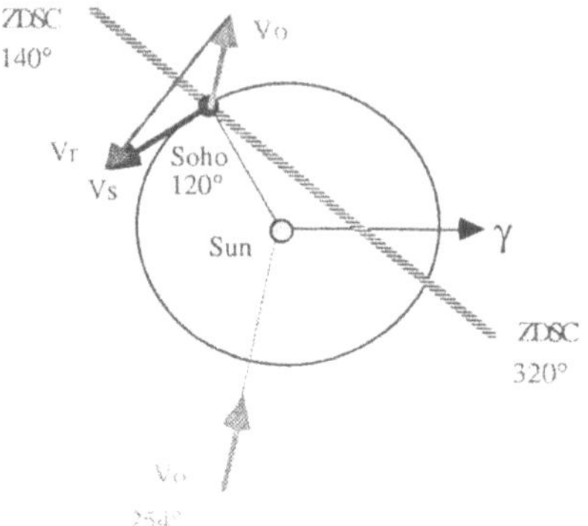

1996 January 22

Figure 12a.

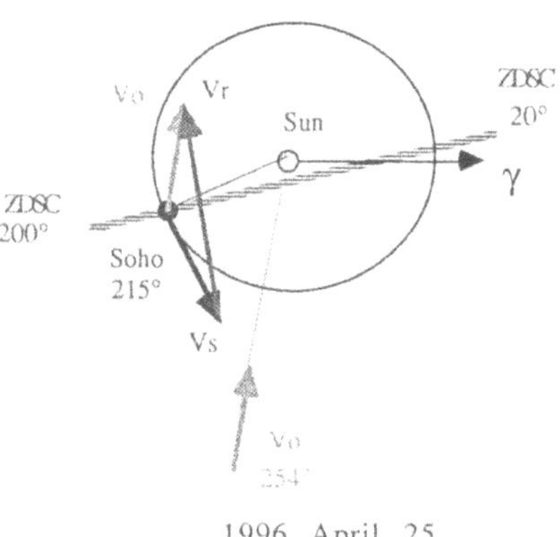

Figure 12b.

flow in the heliocentric frame is represented by the vector $\mathbf{V}_0$ (dashed light-grey vector), directed from $\lambda = 254°$, and $\beta = 7°$ ecliptic coordinates. For simplicity, we have drawn its projection onto the ecliptic plane, which is only slightly different from the true vector. This is, of course, a vector constant with time, while the SOHO heliocentric orbital motion, represented at the three different dates by the vector $\mathbf{V}_s$ (solid vector), is dependent on time. What counts for the H-cell absorption is the relative motion between the cell (SOHO) and the interstellar flow. This relative motion is the difference between the two heliocentric motions of the flow and the spacecraft, i.e., the difference between the two previous vector, and a vector $\mathbf{V}_r$ (dashed dark-grey vector) which depends on the location of SOHO. This relative motion vector $\mathbf{V}_r$ is varying all along the year both in direction and modulus (see Figure 12(a–c)). When the Earth (or SOHO) is moving towards the wind flow direction (approximately March 1st), the relative motion modulus is a maximum

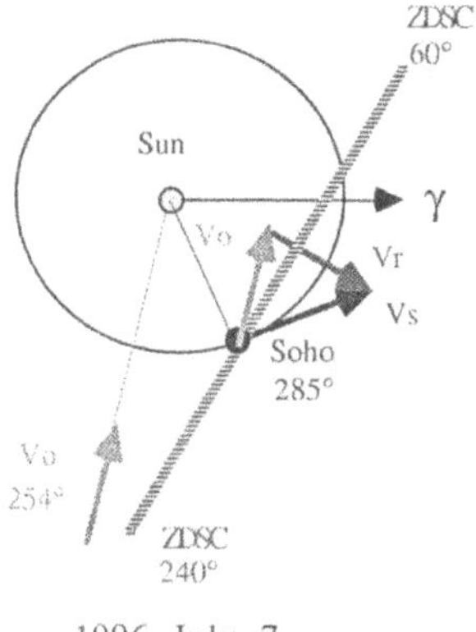

Figure 12c.

Figure 12. Schematic and first-order representation of the H-cell absorption pattern and of the Zero Doppler Shift Circle (ZDSC). All velocities are heliocentric. The absorption by the cell is centered and then a maximum when the relative motion between the emitter (the IPM gas with a mean velocity V_0, light-grey arrow) and the absorber (the atomic H in the cell with the SOHO velocity V_s, black arrow) is zero along the line of sight. This happens along a great circle perpendicular to the relative velocity vector $\mathbf{V}_r = V_s - V_0$ (dark-grey arrow). This circle projection is shown on the ecliptic plane and changes in direction along the Earth orbit. The modulus of $\mathbf{V}_r$ also changes from a maximum value when the IPM and SOHO are moving in opposite directions (March 6) to a minimum value six months later (6 September). The actual position of the ZDSC have been represented for the three dates corresponding to H-cell maps of Figures 13–15: 22 January (a), 25 April (b), 7 July (c). The small inclination of the wind vector to the ecliptic was neglected for clarity.

(about 50 km s^{-1}), and is close to the sum of the interstellar wind heliocentric velocity (about 20 km s^{-1}) and the Earth velocity (30 km s^{-1}). Six months later, the relative motion has decreased to about 10 km s^{-1}, the difference of the two velocities.

Consider Figure 12(a) (22 January). For this location, the relative motion between the IPM and SOHO is aligned with the $\lambda = 50°$; $\lambda = 230°$ axis. When the SWAN line of sight is along this axis, then for an H atom of the absorption cell, the light emitted by the IPM flow is either blue-shifted or red-shifted (for $\lambda = 230°$ and $\lambda = 50°$, respectively) by the modulus of the $\mathbf{V}_r$ vector (about 45 km s^{-1}). This is larger than the half sum of the Doppler widths of the IPM line and H-cell absorption width, which means that the cell absorption line is shifted out of the emission line, and there is zero absorption. On the contrary, when SWAN is looking towards a direction perpendicular to the vector $\mathbf{V}_r$, the IPM emission line is not Doppler-shifted for an observer linked to the instrument and the H-cell atoms do absorb at the center of the line. This is true for all directions perpendicular to $\mathbf{V}_r$, which defines a set of directions in a plane called the Zero Doppler Shift Circle (Bertaux and Lallement, 1984). The projection of the ZDSC onto the ecliptic is drawn in Figure 12(a–c) as the dashed axis elongated along the $\lambda = 140°$,

$\lambda = 320°$ axis. Like the vector $\mathbf{V}_r$, the ZDSC position depends on the location of SOHO along its orbit. On 7 July, its intersections with the ecliptic lie at $\lambda = 60°$, $\lambda = 240°$ (Figure 12(a–c)).

7.3. The Data

While an H cell has already been flown on the two Soviet spacecraft *Prognoz* 5 and 6 (Bertaux *et al.*, 1985), only a limited set of directions were measured and it was not possible to obtain an entire ZDSC. SWAN provides for the first time a series of complete ZDSC's, simply obtained by mapping the whole celestial sphere in the Cell ON/Cell OFF mode. The 22 January map is shown in Figure 13. What is plotted is the map of the ratio $I_{\mathrm{ON}}/I_{\mathrm{OFF}}$, or the fraction of light transmitted by the H cell for this location of SOHO. The map is in ecliptic coordinates. On most of the map, there is no absorption and the ratio is simply 1 (within the photon noise). There is a significant absorption only in a region forming a band along a great circle, which is the ZDSC for this date, and also in the geocorona around the Earth seen at $\lambda = 100°$, $\beta = 3°$. The geocorona is discussed in a later section. Along the central part of the ZDSC, the absorption is maximum and the H-cell transmission is about 0.8, which corresponds to 20% of the light being absorbed by the cell. In fact, the transmitted fraction is slightly different for the two sensors SU$-Z$ and SU$+Z$ because the optical thickness in the two cells is not exactly the same, and the ZDSC appears darker in the south hemisphere. It is clear that such a map gives crucial information on the relative velocity vector, and then on the interstellar wind motion.

The measured maximum absorption of 20% corresponds to our expectations for a centered absorption (no Doppler shift), an optical thickness τ of about 3 to 5 for a nominal H-cell regime we have used almost continuously since the launch, and an emission line temperature of about 8000 K.

On both sides of the ZDSC, the absorption decreases, since the absorption by the cell is no longer in the middle of the emission lines, and disappears when the Doppler shift between the emitters and the H-cell atoms is too large.

The data from April 25 are shown in Figure 14. Note that the ecliptic coordinates of the ZDSC have changed considerably. Each new ZDSC brings additional constraints on the motion of the interstellar gas. The width of the ZDSC is about the same as for 22 January. This is due to the similar modulus of the relative motion (see Figure 12(a–c)) as compared with the previous January data. On the contrary, for the 7 July data (Figure 15), the width of the ZDSC has substantially increased. This is because the modulus of the relative motion $\mathbf{V}_r$ between the gas and the spacecraft has decreased from about 45 km s^{-1} in January down to about 20 km s^{-1} in July. Indeed, the smaller is the relative motion modulus, the smaller is the Doppler shift for the same angle from the ZDSC (the Doppler shift for an angle α with the ZDSC varies as $\mathbf{V}_r \sin\alpha$), and the larger the absorption band around the ZDSC. The same Doppler shift corresponding to an angle $\alpha = 10°$ from the ZDSC

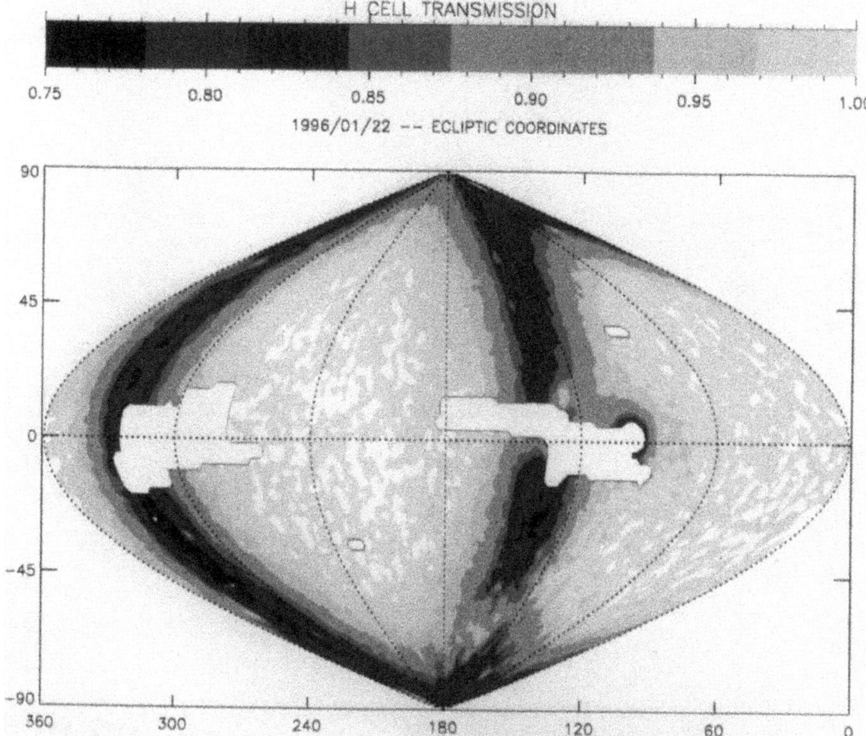

Figure 13. H-cell transmission ($R = I_{\rm ON}/I_{\rm OFF}$) maps at three different dates. The dates correspond to the positions in Figures 12(a–c). The evolution of the ZDSC is clearly visible. Its direction changes, but also its width. It is much broader in July because the modulus of the relative motion $\mathbf{V}_r$ (Figure 12(c)) between the IPM and SOHO is smaller. The dark region seen in July at 270° ecliptic longitude and at 90° in January is the terrestrial atomic H (the geocorona). Its narrow Lα emission is strongly absorbed by the cell because along this direction the Earth is nearly at rest with respect to SOHO.

which decreases the absorption almost down to zero in January here corresponds to α_2 such that $\sin\alpha = 45/20 \sin\alpha_2$ or $\alpha_2 = 23°$. This means that the absorption remains significant up to 23° away from the ZDSC center. For dates after August, the cell begins to absorb in all directions of the sky.

The ZDSC concept as described here was established by Bertaux and Lallement (1984) in the particular case of a thermal gas with a uniform temperature, which requires that the solar gravitation is exactly compensated by solar Lα radiation pressure ($\mu = 1$), flowing at a constant velocity through the solar system. As seen from the H-cell maps it represents rather well the absorption pattern to first order. However, the line of maximum absorption departs slightly from a perfect circle, due to the fact that the interstellar flow is not homogeneous, neither in velocity, direction nor temperature. Gravitation, solar radiation pressure, ionization and collisions modify the interstellar H flow according to the location in the heliosphere. Also,

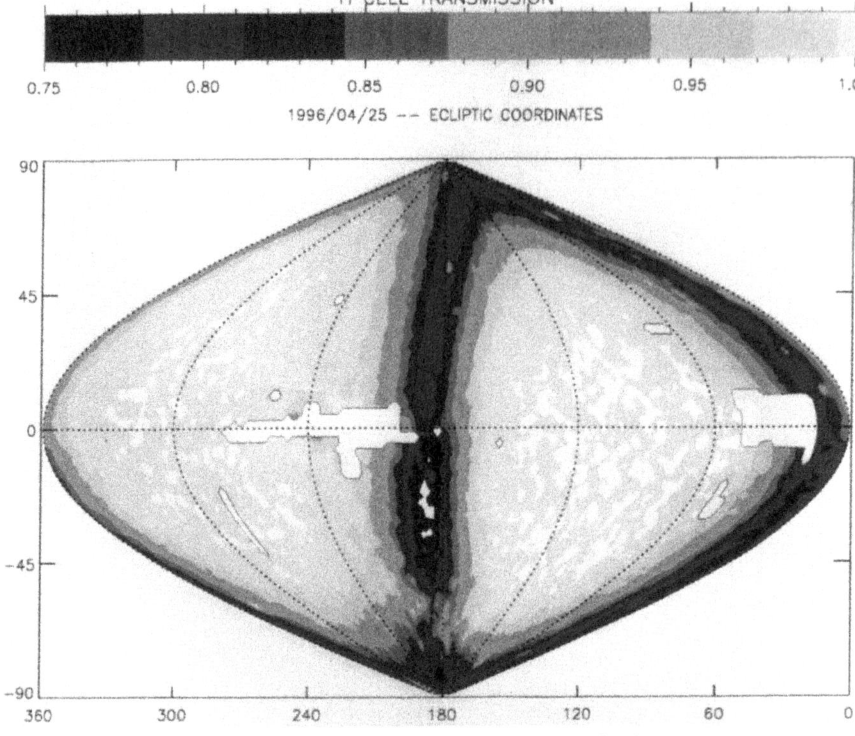

Figure 14. Same as Figure 13, for Earth location b.

the flow is almost certainly not Maxwellian, even at large distances from the Sun, due to the interaction with the plasma around the heliopause. Finally, radiative transfer effects which influence the shape of the emission line also depend on the line-of-sight geometry. Beyond a first-order analysis, a rather detailed and complicated modelling will be required to interpret fully the data.

7.4. In-Flight Performances of the H Cells

Hydrogen cells flying in SWAN were fabricated at Service d'Aéronomie with the same technology as used for *Prognoz* 5 and 6 cells, which includes the sealing of MgF_2 windows with AgCl. During the development of SWAN, numerous studies were conducted at Service d'Aéronomie in order to minimize the aging of the filaments, which lose their Tungsten material when they are switched on. Apparently this is the result of the presence of HCl inside the cell and cannot be avoided with the technology of sealing MgF_2 windows with AgCl. The presence of HCl inside the cell was identified from narrow absorption features detected during vacuum calibrations with the continuum spectrum of synchrotron radiation (see Figure 11 of Bertaux *et al.*., 1995). It can be noted that during the calibration of the *Prognoz*

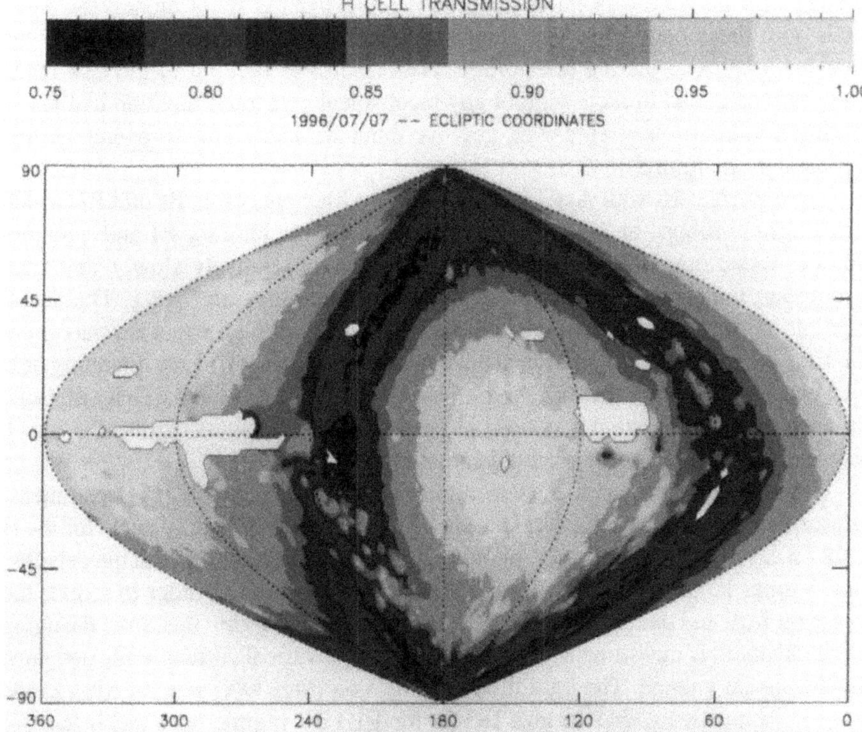

Figure 15. Same as Figure 13, for Earth location c.

Lα photometer in 1974, one absorption trough was also noted at 128.9 nm in the efficiency curve of the instrument, as indicated in Figure 4 of Bertaux *et al.* (1985). However, at that time we did not think that it was due to HCl gaseous absorption. The reason is that synchrotron radiation was not available and we used instead a series of atomic lines of various gases in a discharge lamp, and among those discrete wavelengths only one coincided with an HCl absorption.

For SWAN, when we discovered the presence of HCl in the cells we finally understood that it was the aging factor of the filaments (when heated, there is formation of tungsten chloride on the filament, which migrates in the cell and is reduced by H atoms and deposition of W atoms on the cell walls, as in high temperature halogen lamps; however, in the halogen lamp, the reverse process exists also because the wall is at a very high temperature, which is not the case for our H cell). In 1994, we procured from a UK company a batch of 6 new glass vessels with MgF_2 lenses sealed to the glass by another method, not using AgCl (direct sealing). Then we filled them with H_2 and processed them as usual at Service d'Aéronomie, and performed lifetime tests which indeed showed a much better behavior, with practically no aging. One of them was fully qualified in vibration and thermal testing, and we were very much tempted to replace in our flight models the AgCl cells with these new cells. The choice to make was excruciating, but finally we decided to stick to the old technology. The reason was that out of the new batch of 6, two cells had broken without any mechanical constraint and one had failed during a vibration test. However, now we think we know how to overcome this problem in the future.

The SWAN cells with AgCl sealing have two filaments each, F1 and F2, which are tungsten ribbons of 20×100 μm section, $\simeq$40 mm long for F1 and $\simeq$30 mm for F2. When operated, these filaments are thinning relatively slowly, and keep a constant temperature and H production rate while they are aging. The actual lifetime in flight was a great concern before launch, with an expected maximum lifetime of $\simeq$1000 hr of operation at the lowest regime called F1 Low. Planning of H cell observations were therefore optimized to match the expected SOHO lifetime. In each cell, one filament was used preferentially at its lowest regime (F1 Low), keeping the second one as a redundant back-up.

Until the end of November 1996, a total of 34 full-sky maps with H-cell measurements were obtained. Each map lasts 24 hours, with a 25% duty cycle for the H cell: on each position in the sky, photons are counted during 13 s with the cell ON, and 3 times longer with the cell OFF. This scheme is used in order to reduce the noise on I_{OFF} and therefore on the ratio $R = I_{\mathrm{ON}}/I_{\mathrm{OFF}}$. During this time, the aging of the filament is measured by the decrease of the electrical current while activated by a constant voltage. The total time of utilisation (time ON) was for SU+Z and SU$-Z$, respectively, 241 hr and 183 hr for F1 Low regime, and much less for the other higher regimes. Most of the time, the F1 Low regime is quite adequate to scan the IPM line shape by changing the angle α and therefore the Doppler shift, the so-called DASS method (Doppler angular spectral scanning, Bertaux and Lallement, 1984). However, around September, when SOHO has a small $\mathbf{V}_r$ with respect to the wind, one cannot scan the full line angularly, and furthermore one can no longer assume that the line shape is constant with the direction of sight, a pre-requisite for the DASS method. During this period, the IPM line shape was measured by using higher regimes of the H cell, with larger absorptions linewidths and resulting absorptions. These regimes were used only for a few hours, and resulted in an increased rate of aging for the cell of SU$-Z$ as shown by the F1 Low current behavior. It is less obvious for the SU+Z cell, but still noticeable.

The total current decrease 4.56×10^{-2} mA hr^{-1} was for SU+Z from 403 mA to 392 mA, This suggests an additional lifetime after 1996 of 6400 hr down to a current of 100 mA, or 21 year at the rate of 300 hr per year which is quite adequate for our scientific purpose. For SU$-Z$, the corresponding numbers are a loss of 22 mA in 183 hr, 0.12 mA hr^{-1}, and a lifetime of 2200 hr or 7.4 year. It should be noted however that the aging of this SU$-Z$ cell is mainly linked to the use of the higher regimes, which should be used cautiously. In addition, there is also the redundant filament which is kept as a back-up and rarely used up to now. Clearly, after nearly one year of use, the expected actual lifetime of the SWAN cells in flight is longer than what was hoped before flight, based on laboratory life tests. One possible reason is the absence of gravity, which on ground has a mechanical effect on the filaments, with a tendency to deform the filament and make the various spires approach each other, increasing locally the temperature of the filament.

8. The Geocorona

The geocorona is quite a conspicuous feature of sky maps, especially when the hydrogen cell is used, as seen in Figure 13 of the reduction factor. This is because geocoronal H, when extending in the exosphere, are transforming their thermal, kinetic energy, into gravitational energy. As a result, the temperature of geocoronal H decreases from 1000 K at the exobase level (500 km of altitude) to less than 300 K at 40 000 km of radial distance. The exospheric atoms are scattering solar Lα with a very narrow line width, which is completely absorbed by the H cell. From Figure 13 the region of the sky where the reduction factor of the combined emission sky+geocorona is significantly less than 1 extends over 20° of angular diameter, which, seen from the SOHO halo orbit at 1.5×10^6 km, corresponds to a diameter of 500 000 km. This is the largest extension ever seen of the geocorona by any spacecraft instrument up to now (except for unpublished observations by *Galileo* in 1990 and 1992 during Earth encounters; Pryor, private communication). One should mention that systematic observations of the geocorona are somewhat hampered by the very strong stray light which is coming from the anti-solar direction, due to solar light directly reflected from some parts of the Service Module of SOHO. Due to the particular halo orbit of SOHO around the L1 Lagrange point, the geocorona can be observed only during specific positions.

9. Observations of Comets with SWAN

The Lα emission is the most intense UV emission in comets: H atoms produced from photodissociation of H_2O, the major volatile component of a cometary nucleus, are directly exposed to the Sun during their expansion into a huge envelope of H: 2×10^7 km for comet Bennett (total extension), during the discovery observation of such an envelope (Bertaux and Blamont, 1970; Bertaux, Blamont, and Festou, 1973). During the year 1996, SWAN was able to detect the Lα emission of several comets and to monitor their H production rate: comets Honda–Mrkos –Padjusakova, Hyakutake 1, Szcepansky, Hyakutake 2, Tabur, Hale-Bopp. Some other comets were observed with no positive detection. SWAN can follow their Lα emission even near the Sun, which was the case for the unexpected and beautiful comet Hyakutake 2. This comet approached the Earth at only 0.11 AU on 26 March 1996, and offered a fantastic display in Lα (as well as to the naked-eye observers). Figure 16 is a map of the emission at that time, which was taken as one of a series

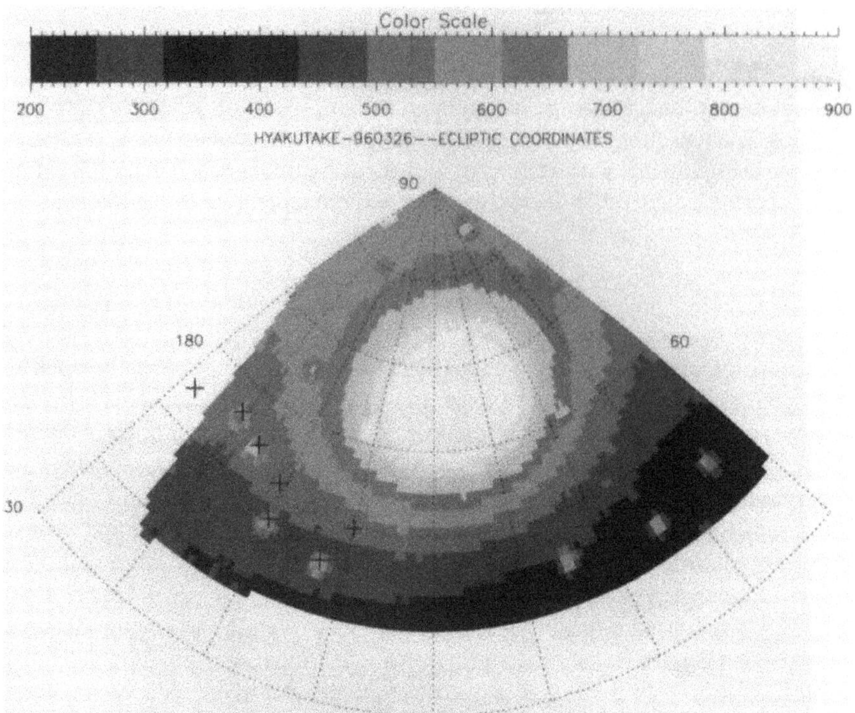

Figure 16. Lα emission from comet Hyakutake recorded by SWAN in March 1996, when the comet was closest to Earth at 0.11 AU. The stars of the 'Big Dipper' have been represented by crosses to give a better idea of the angular extent of the emission. Four of them are actually detected by SWAN.

of dedicated observations for this comet, started as soon as the comet discovery was reported in IAU circulars. These maps were made with a refined grid of 0.5° sampling. In Figure 16 the sky background has not been subtracted, but still the shape of the isophotes are dictated by the rounded shape of the huge cometary H cloud extending over more than 60° in the sky. Black crosses are fiducial marks for the position of the stars of the Big Dipper for comparison, to give an idea of the apparent angular size. The corresponding actual size of the H envelope was more than 20 million km in diameter at that time. After subtraction of the sky background, the comet Lα emission maps were analyzed and transformed into maps of the H column densities; then the production rate was derived from a comparison with a simple model and plotted as a function of time or radial distance to the Sun as in Figure 17. An analysis of the comet Hyakutake hydrogen envelope measured by SWAN is described in Bertaux *et al.* (1997), and concludes that the comet experienced a surge of activity around 21 March 1996, associated with the splitting of a chunk of about 200 m size which was completely disrupted and evaporated in about 10 days. The overall behavior of Comet Hyayutake B2 may be represented by a combination of a main source from the nucleus following the R^{-2} law and the complete evaporation of the fragment detached on 21 March, whose size may be estimated at 200 m, from the additional production rate on top of the R^{-2} law. Comet Wirtanen, the prime target of the ESA Rosetta mission, was detected with

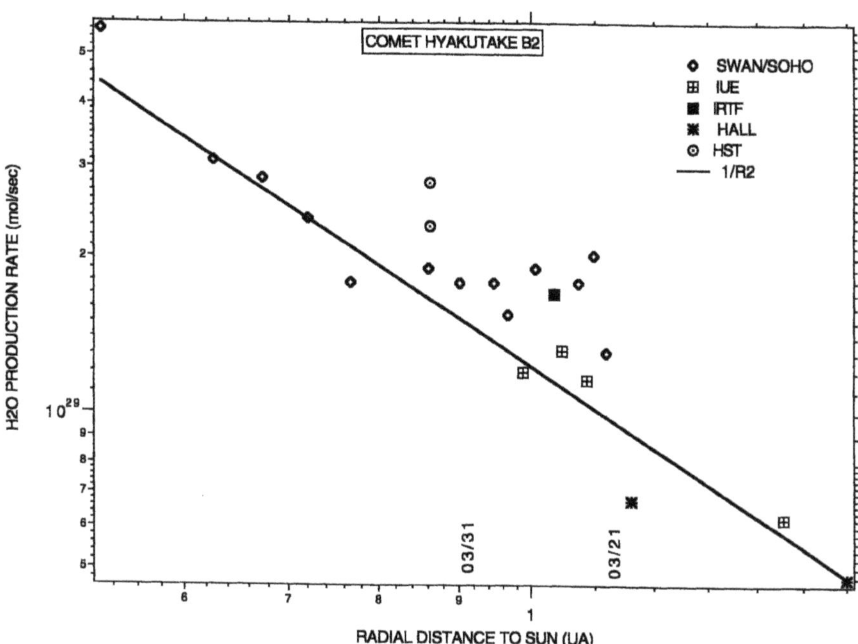

Figure 17. The SWAN production rate of water of comet Hyakutake derived from SWAN Lα images is compared to several other determinations, both from ground-based observatories (Hall and IRTF) and from space (IUE and HST). The agreement between the various methods and instruments is fair. The production rate is plotted as a function of distance to the sun and compared to an R^{-2} law. Time flows from right to left. There is a clear surge of activity appearing around 21 March, that we suggest is linked to the fragmentation of the nucleus observed at Pic du Midi and on HST images. Then the production rate levels off for two weeks, followed by a steady increase.The last point of the curve (16 April 1996) may indicate a second fragmentation of the nucleus.

SWAN on 10 February 1997, with an estimate H_2O rate of 5 to 10 × 10^{27} mol s^{-1}. The use of the absorption H cell will allow us to study with great refinement the velocity of cometary H atoms, when Doppler effect conditions are adequate (when the distance from the Earth to the comet goes through an extremum). This was the case for comet Hyakutake on 26 March, and around that date several maps were also obtained with the hydrogen cell. It is beyond the scope of the present paper to show a detailed analysis of these results; one difficulty is that the comet was in a region of the sky containing also the ZDSC, and one has to take care to disentangle the relative emissions of the comet and the sky and the different effects of the H cell on them. Let us only emphasize here that it is important to know the expansion velocity v of the atoms in order to deduce the production rate of H, Q(H), from an intensity measurement of Lα: indeed, the H number density $n(r)$ at distance r from the nucleus is expressed by

$$n(r) = Q/4\pi r^2 v \, ,$$

$$N(p) = Q/4\pi v p \, ,$$

where $N(p)$ is the column density at impact parameter p, which is proportional to the Lα emission rate (at least in the optically thin regime).

Up to now, the velocity has not been measured directly, and we assumed that it results from the residual energy left in H after photodissociation of H_2O and OH, with an estimate of 20 km s^{-1} and 8 km s^{-1}, respectively for H atoms produced from H_2O, and those produced from OH, to derive Figure 17. Of course a safer estimate of the production rate will be achieved by measuring directly v, which could also perhaps discriminate several populations of H atoms of different origin (i.e., organic material within the dust grains).

Because SWAN can observe relatively faint comets, and near the Sun when they cannot be observed from the ground, SWAN could detect some new comets before they are discovered from ground-based observations. This has not been the case up to now. There are two difficulties associated with such an exercise: the first is that the sky is full of stars which must not be confused with comets, and this requires a complete catalogue of stars which can be detected by SWAN above the sky background which is still not ready, and the second is that the data do not flow always in real time, requiring a delay of 2 to 3 days during normal operations.

10. Perspectives and Conclusions

After nearly one year of smooth operation in flight, SWAN has demonstrated its potential to achieve its main scientific objective: the monitoring of the heliographic latitude distribution of the solar wind by the mapping of the sky Lα emission. Even without a detailed modelling, maps are directly revealing a feature which was first noticed in *Prognoz* data: a groove, cut through the upwind hemisphere, roughly along the the ecliptic plane. This feature is a direct sign that the ionization rate by the solar wind is higher in the ecliptic plane. As indicated from *Ulysses* data and discussed in Bertaux, Lallement, and Quémerais (1996a), this is the result of the dichotomy of the solar wind expansion in two regimes: the slow solar wind remains concentrated near the solar neutral sheet, within about 20°, while the fast solar wind occupies the rest of space, above and below the neutral sheet. The enhanced ionization in the neutral sheet, or streamer belt, comes both from a larger solar wind mass flux, as indicated by *Ulysses* data, and from a larger charge-exchange cross-section with a smaller velocity (400 km s^{-1}).

It can be remarked that *Prognoz*, *Ulysses*, and SWAN data were collected at a time of solar minimum, when the inclination and the warpiness of the neutral sheet are minimal (Hoeksema, 1991). As a result, it is a quite favorable condition to carve a groove in the incoming flow of interstellar H atoms, since during a solar rotation the neutral sheet remains at the same low latitude in the upwind hemisphere. What will happen during the next cycle, during the ascending phase of the solar cycle will be quite interesting to monitor from SOHO. We might expect, when the inclination and warpiness of the neutral sheet will increase, to see the groove attenuated and disappear, because each point of the upwind hemisphere will experience an average mixture of fast and slow solar wind during a solar rotation.

Indeed, there are some indications both from OGO-5 data in 1969 (Bertaux and Blamont, 1971) and from *Galileo* (Pryor *et al.*, 1992, 1996) that the emission pattern in the upwind hemisphere is less complex at solar maximum, with only one maximum as mentioned by Summanen (1996). Still, the signature of solar wind anisotropies might exist, but rather in the downwind emission, as was first discovered. Detailed comparisons with models will be required for this exercise, and as was said before, the hydrogen cell results are important for this purpose. While parameters of the interstellar wind are not supposed to vary, one solar parameter, the Lα radiation pressure, connected to the disc integrated at line center, can be monitored with H-cell measurements, because of the bending of H atom trajectories in the solar system. We hope also to be able to distinguish in the H cell data the signature of the heliopause, which modifies the Maxwellian distribution of H atoms prevailing in the interstellar medium before the interaction with plasma around the heliopause. The prolongation of the SOHO mission, in addition to providing the monitoring of the solar wind in the ascending phase of the new solar cycle, would allow us to check the response of the geocorona to solar activity, and to monitor the H_2O production of many more comets in the years to come.

Acknowledgements

SOHO is a mission of international cooperation between ESA and NASA.The construction of the SWAN space instrument required the help of many people at various stages. We wish to thank particularly Jean-Pierre d'Allest, who as Directeur General made possible the participation of CNES to SWAN, once the instrument was selected by ESA; Vaino Kehla, Heikki Huomo, H. Hannula, E. Strommer at VTT (Finland), Risto Pellinen at FMI (Finland); E. Dimarellis as the first project manager of SWAN, C. and M. Bernard, G. Chrétiennot and M. Leclère at Service d'Aéronomie for their participation to H-cell fabrication and testing and C. Taulemesse for the design of the optics. We also wish to thank the ESA staff of GSFC for operations, and also Eliane Larduinat, Kim Tolbert, and Laura Allen at the Flight Operation Team for their full dedication to the operation of the SOHO instruments. Finally, we wish also to thank Cyril Pennanech for data and command management and Clara Veniard for her analysis of the stellar data and early comparison with IUE data. We would like also to express our thanks to John Phillips for communication of revised solar wind proton fluxes from SWOOPS-*Ulysses*, and John Clarke for the early communication of the HST Lα intensity measurement.

SWAN was financed in France by CNES with support from CNRS for staff salaries and in Finland by TEKES and the Finnish Meteorological Institute.

References

Bertaux, J. L. and Blamont, J. E.: 1970, *Compt. Rend. Astron. Soc.* **270**, 1581.
Bertaux, J. L. and Blamont, J. E.: 1971, *Astron. Astrophys.* **11**, 200.
Bertaux, J. L. and Lallement, R.: 1984, *Astron. Astrophys.* **140**, 230.
Bertaux, J. L., Blamont, J. E., and Festou, M.: 1973, *Astron. Astrophys.* **25**, 415.
Bertaux, J. L., Lallement, R., and Quémerais, E.: 1996a, *Space Sci. Rev.* **78**, 317.
Bertaux, J. L., Quémerais, E., and Lallement, R.: 1996b, *Geophys. Res. Letters* **23**, 3675.
Bertaux, J. L., Lallement, R., Kurt, V. G., and Mironova, E. N.: 1985, *Astron. Astrophys.* **150**, 1.
Bertaux, J. L., Kyrölá, E., Quémerais, E., Pellinen, R., Lallement, R., Schmidt, W., Berthé, M., Dimarellis, E., Goutail, J. P., Taulemesse, C., Bernard, C., Leppelmeier, G., Summanen, T., Hannula, H., Huomo, H., Kelha, V., Korpela, S., Leppälä, K., Strómmer, E., Torsti, J., Viherkanto, K., Hochedez, J. F., Chrétiennot, G., Peyroux, R., and Holzer, T.: 1995, *Solar Phys.* **162**, 403.
Bertaux, J. L., Costa, J., Quémerais, E., Lallement, R., Kyrölä, E., Schmidt, W., Summanen, T., Makkinen, T., Goukenleuque, C.: 1997, *Planetary Space Sci.*, in press.
Clarke, J. C.: 1996, private communication.
Goldstein, B. E., Neugebauer, M., Phillips, J. L., Bame, S., Gosling, J. T., McComas, D., Wang, Y. M., Sheely, N. R., and Suess, S. T.: 1996, *Astrophys. J.* **316**, 296.
Joselyn, J. A. and Holzer, T. E.: 1975, *J. Geophys. Res.* **80**, 903.
Hoeksema, J. T.: 1991, *Adv. Space Res.* **17** 15.
Kumar, S. and Broadfoot, A. L.: 1979, *Astrophys. J.* **228**, 302.
Kyrölä, E., Summanen, T., Schmidt, W., and Mákinen, T., Bertaux, J. L., Lallement, R., Quémerais, E., and Costa, J.: 1997, *J. Geophys. Res.*, submitted.
Lallement, R., Bertaux, J. L., and Kurt, V. G.: 1985, *J. Geophys. Res.* **90**, 1413.
Phillips, J. L.: 1994, private communication.
Pryor, W. R., Ajello, J. M., Barth, C. A. *et al.*: 1992, *Astrophys. J.* **394**, 363.
Pryor, W. R., Barth, C. A., Hord, C. W., Stewart, A. I. F., Simmons, K. E., Gebben, J. J., McClintock, W. E., Lineaweaver, S., Ajello, J. M., Toshika, W. K., Naviaux, K. L., Edberg, S. L., White, O. R., and Sandel, B. R.: 1996, *Geophys. Res. Letters* **23** (15), 1893.
Scherer, H. and Fahr, H.: 1995, *Solar Phys.* **161**, 383.
Summanen, T.: 1996, *Astron. Astrophys.* **314**, 663.
Summanen, T., Lallement, R., Bertaux, J. L., and Kyrölä, E.: 1993, *J. Geophys. Res.* **98** (A8), 13215.
Summanen, T., Lallement, R., and Quémerais, E.: 1997, *J. Geophys. Res.* **102** (A4), 7051.
Thomas, G. E. and Krassa, R. F.: 1971, *Astron. Astrophys.* **11**, 218.
Witte, M., Banaskiewicz, M., and Rosenbauer, H.: 1996, *Space Sci. Rev.* **78**, 289.

RECURRENCE OF ENERGETIC PARTICLE FLUX ANISOTROPY AS OBSERVED BY ERNE ON 9 JULY 1996

J. TORSTI, T. LAITINEN, R. VAINIO, L. KOCHAROV, A. ANTTILA and E. VALTONEN
Space Research Laboratory, Department of Physics, FIN-20014 Turku University, Finland

(Received 4 March 1997; accepted 28 May 1997)

Abstract. A solar energetic particle event was observed on 9 July 1996, by the ERNE sensors LED and HED on board the SOHO spacecraft. The arrival of the first protons in the energy range >20 MeV took place at 09:55 UT, 43 min after the maximum in the X-ray and Hα radiation of a flare located at S10 W30. The rise phase of the particle intensities at all energies was exceptionally rapid. At 12:50 UT, the intensities dropped in all energy channels. Simultaneously, the magnetic field instrument MFI on board WIND, not far from SOHO, detected a sharp and large change in the magnetic field direction. The analysis of the directional measurements of ERNE in the energy range 14–17 MeV shows the presence of a strong flux anisotropy during the whole period 10:10–12:50 UT. From 12:50 UT until about 16:00 UT, the directional analysis of the proton fluxes gives only a weak anisotropy at the limit of the sensor resolution. Later on, the flux anisotropy was found to recur, indicating a continuous injection of particles into the flux tubes connected to the SOHO spacecraft. These experimental results lead to strict limits on particle injection and transport models. The first period of the anisotropy and its recurrent phase cover 24 hours. This suggests an extended injection of particles. The strength and stability of the anisotropy indicate that, during these periods, SOHO was in an interplanetary sector where the particle transport was almost scatter-free. On the other hand, during the intermediate 3-hr period, we observed particles which traveled in a sector of diffusive transport or which were retarded by magnetic field disturbances not far from the observation site.

1. Introduction

On 9 July 1996, a solar flare occurred in the NOAA region 7978 on the western hemisphere of the Sun, S10 W30, not far from the root of the magnetic flux tube connected to the Earth and the SOHO* spacecraft. The maximum X-ray flux (class X2.6) as observed by the GOES 8 and 9 satellites was recorded at 09:12 UT (*Solar Geophysical Data*, 1996). The start and end of the X-ray flare were recorded at 09:01 UT and 09:49 UT, respectively. The soft X-ray e-folding time indicates the impulsive nature of the event. The optical importance of the flare was 1B. A type II radio burst was observed at 09:11-09:13 UT (*Solar Geophysical Data*, 1996). The LASCO observations on board SOHO showed that a large coronal mass ejection occurred simultaneously, probably in association with the flare. At 09:55 UT, the counting rates of the ERNE High Energy Detector (HED) started to rise at energies above 20 MeV nucl^{-1}. Consistent with velocity dispersion, the Low Energy Detector (LED) particle counting rates between 1.3 and 12 MeV nucl^{-1} started to rise about one hour later. This was the start of a solar energetic particle

* SOHO is an international co-operation project between ESA and NASA.

Solar Physics **175:** 771–784, 1997.

(SEP) event that continued for several tens of hours in interplanetary space close to the Earth, and was distinguishable in the proton energy range from the threshold of the ERNE sensors, 1.3 MeV, up to 50 MeV, where the transient flux no longer exceeded the cosmic-ray background.

Being a three-axis stabilized platform, the SOHO spacecraft offers a unique opportunity to measure directional particle distributions in detail within the view cones of the ERNE sensors. In this work, we present an analysis of the development of the energetic proton intensities measured by ERNE, concentrating on the angular distributions measured within the large view cone of HED. Anisotropy analysis is vital in studying the transport conditions in the interplanetary magnetic field (IMF) (Schulze, Richter, and Wibberenz, 1977), the injection scenario and the particle spectra in the source. Previous studies of particle anisotropies were made with particle telescopes on rotating spacecraft with a typical resolution of $\sim 45°$ in the directional measurements. Although the HED view cone is limited, the resolution of the trajectory measurements is 1–2 deg; hence, the real limits to the resolution in the measured angular distributions are set by the particle statistics gathered. The SEP event on 9 July 1996, being the largest event detected by ERNE until that time, provided the first opportunity to study angular distributions of high-energy protons near SOHO.

2. Instrument

2.1. General Description

The energetic particle instrument ERNE consists of two particle telescopes, HED and LED. Together, the telescopes cover the energy range from 1.3 MeV nucl^{-1} to about 110 MeV nucl^{-1} for protons and alpha particles, and a few times higher energies for heavier ions. The nuclear charge range is from hydrogen to zinc. In accelerator tests, the instrument was able to distinguish the isotopes up to neon in LED and up to iron in HED. A full description of the ERNE experiment is given in Torsti *et al.* (1995), but a short summary of both telescopes is given here. A schematic plot of the ERNE sensors is given in Figure 1.

LED consists of three silicon detector layers D1, D2, and AC. The first layer, D1, is divided into seven independent windows in a honeycomb structure, thus forming seven independent telescopes using the same stopping detector. Two of the front detectors are 20 μm, and five are 80 μm thick. The thickness of the stopping detector D2 is 1 mm. The two silicon detector layers are operated in anticoincidence with the third one of 0.5 mm thickness. The total half angle of the LED view cone is about 32°. The different viewing directions of the seven sub-telescopes provide a limited capability for directional intensity measurements. The total geometric factor of the two telescopes with the 20 μm front detector is 0.260 cm^2 sr, and that of the five with the 80 μm front detector is 0.655 cm^2 sr.

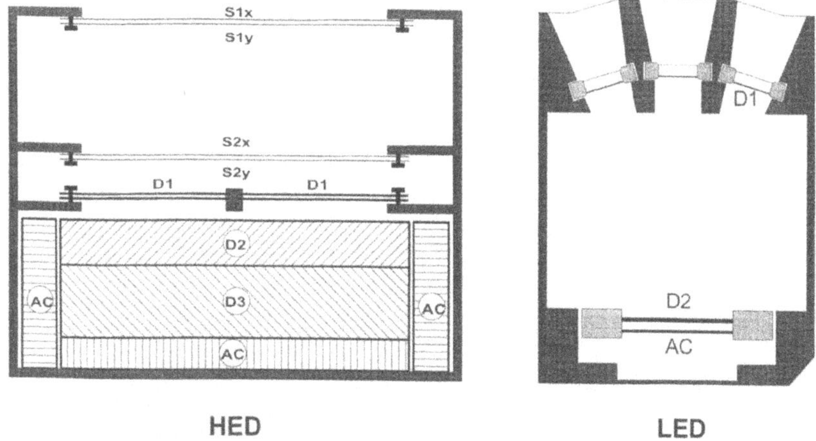

Figure 1. Schematic structure of HED and LED sensors.

The HED telescope structure is more complicated. It consists of four position-sensitive silicon strip detector layers, S1x, S1y, S2x, and S2y, each 300 μm thick, and of one 1 mm thick conventional silicon detector layer, D1. These are followed by two scintillator layers, the first one, D2, of CsI (8 mm) and the second one, D3, of BGO (15 mm). The scintillators are surrounded by plastic anticoincidence detectors (AC). The geometric factor of the telescope is dependent on the particle type and energy, being largest at about 1.3 times the threshold energy, ~ 36 cm^2 sr, for each ion species, and going down to ~ 25 cm^2 sr for the highest energies.

2.2. HED TRAJECTORY MEASUREMENT

In HED, the four position-sensitive strip detectors give two points of the particle trajectory through the detector stack with 1 mm accuracy in each layer. The S1x (S1y) and S2x (S2y) layers are separated by 31 mm. This provides an angular resolution of about 1–2° depending on the zenith angle of the trajectory. The area of the strip detectors is 70×70 mm^2 (with a cross-shaped inactive area of $2 \times 4 \times 70$ mm^2 in both pairs), giving a full opening angle of more than 140°. However, the acceptance is very low for the most oblique trajectories; hence, the effective opening angle of the telescope is about 120°. The detector zenith is pointing towards the nominal Archimedean spiral field direction: $\theta = 0°$ and $\phi = 315°$ in the polar geocentric solar ecliptic (GSE) coordinate system. In this work, we have analyzed trajectories based on the original pulse height data of HED. This enables the computation of pitch angle distributions with sufficient statistics using temporal resolutions ranging from 40 min during the event maximum on 9 July 1996, to several hours in the tail of the event on 10 July 1996.

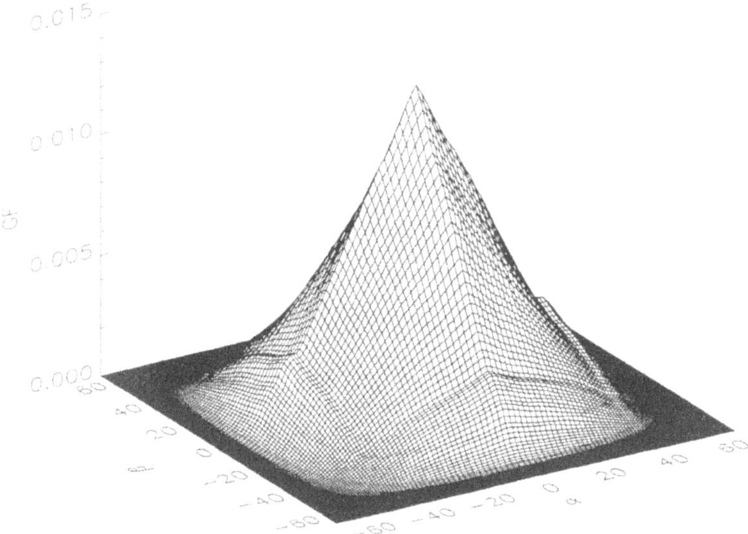

Figure 2. Differential geometric acceptance of HED as a function of the projection angles α and β in the proton energy range 14–17 MeV.

The position-sensitive strip detectors produce two pulse heights, one from each edge of the detector in a direction perpendicular to the strips. The strips are capacitively coupled together and the pulse heights are measured from the two ends of the capacitive chain. The position is a function of $X = RP/(LP + RP)$, where RP and LP are the pulse heights from the right- and the left-hand edges of the detector, respectively. In an ideal capacitive chain, X would give the coordinate in the units of the detector width measured from the left-hand edge, but as there is some charge attenuation in the chain, the value has to be corrected to give the actual coordinate. The attenuation correction is then made also for the measured total energy loss in the detector.

To penetrate to the fourth strip detector layer, a vertically incident proton needs about 12 MeV of energy. When measuring angular distributions, however, we also use particles with the most oblique trajectories. At 14 MeV, protons at all zenith angles are detected. Therefore, we selected protons in the energy range 14 –17 MeV for this anisotropy analysis. The total geometric factor of HED in this channel is 35 cm^2 sr and the angular response function is presented in Figure 2. The observed two-dimensional angular distributions (or directional intensities) are obtained by dividing the observed directional counting rates by the response function. Two examples of such distributions are given in Figure 3. The left-hand diagram represents the observed directional intensities of galactic cosmic-ray protons detected in the range 14–17 MeV during ten days, from 29 June to 8 July 1996. The plateau in the center area of the HED view cone shows that the cosmic-ray

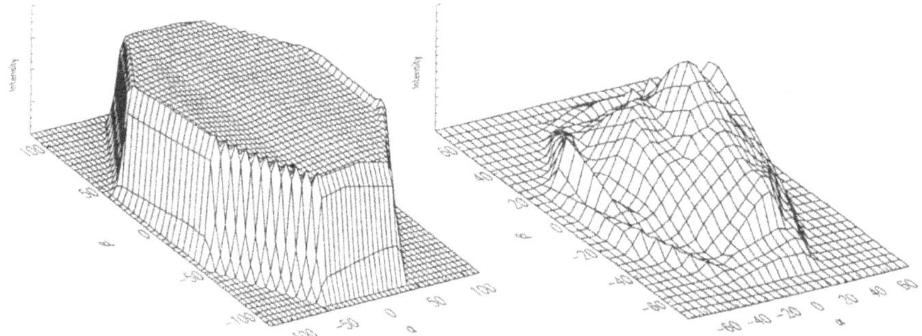

Figure 3. Directional distribution of the 14–17 MeV protons as measured by HED during the period 29 June–8 July 1996 (*left*), and during 10:14–10:55 UT on 9 July 1996 (*right*) during the rise phase of the SEP event.

flux is isotropic. The fluctuations of the intensities at the borders of the view cone reflect the fact that the geometric factor and, as a result, the statistics are smaller than in the center. The diagram on the right presents the proton flux detected during 10:14–10:55 UT on 9 July. In this case the directional distribution is clearly anisotropic. The maximum anisotropy points in the approximate direction of the interplanetary magnetic field at $\alpha \approx 15°$, $\beta \approx 0°$. The angular distribution of arriving protons is extremely narrow, with a half width of about 30°. Irregularities in the distribution are caused by statistical fluctuations.

3. Observations

3.1. Solar Wind Measurements

The solar wind had a relatively low speed during the SEP event on 9 July 1996, as measured by the MTOF proton monitor of the CELIAS instrument on board SOHO. It gradually decreased from $\sim$ 450 km s^{-1} at 00 UT on 9 July to $\sim$ 310 km s^{-1} at 24 UT on 10 July. During the same period, the density of solar wind protons increased from about 4 to almost 9 protons per cm^3. The thermal velocity of protons decreased from $\sim$ 50 km s^{-1} at 00 UT on 9 July in a few hours to $\sim$ 20 km s^{-1}, remaining approximately at this level during the rest of the time period.

The SOHO spacecraft does not carry a magnetometer. Fortunately, there is another spacecraft at 1 AU in the solar wind environment carrying a magnetic field instrument, namely the MFI instrument, on board the WIND spacecraft. The positions of SOHO and WIND at noon on 9 July 1996 in GSE coordinates measured in Earth radii were (248.7, −82.8, −15.1) and (210.7, 17.7, −8.2), respectively. At these locations and assuming the nominal Archimedean spiral field line orientation (45° tilt with respect to the Sun–Earth line), the two spacecraft would fly through the same IMF flux tubes with a time shift of 16 min in the frame corotating with

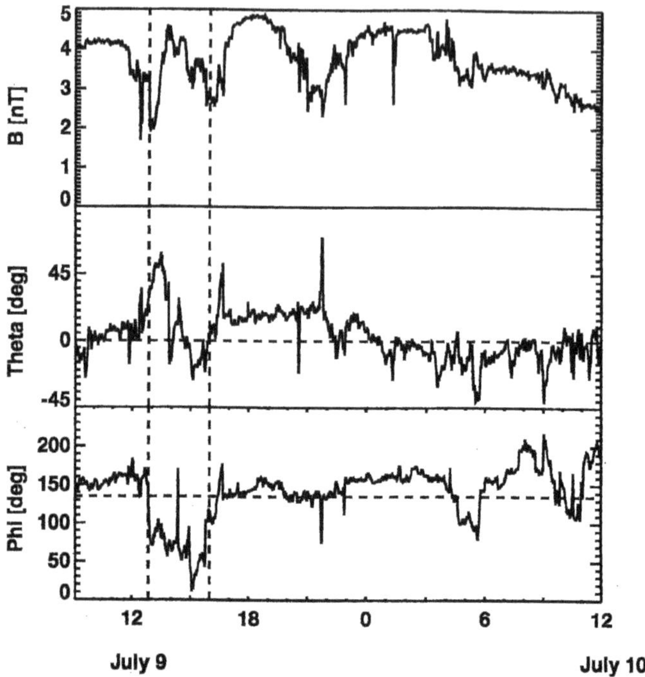

Figure 4. The WIND MFI measurements of the magnetic field in polar GSE coordinates on 9–10 July 1996. Horizontal dashed lines represent the direction of the ERNE sensor axis. Vertical lines limit the period of the disturbed magnetic field.

the Sun. The time shift depends on the field direction, and is longer if the field is less tilted with respect to the Sun-Earth line. We have plotted the magnetic field measurements of MFI in Figure 4 in polar GSE coordinates on 9–10 July 1996. It can be seen that a magnetic field disturbance passes WIND during the interval between 12:45 UT and 16:00 UT.

3.2. Energetic particle distributions

During the first hours of the particle event, strong and rapid variations in particle counting rates were observed by both ERNE telescopes. In Figure 5 we present the ERNE measurements of the proton fluxes in five energy channels during the 9 July SEP event along with the angle between the magnetic field, as measured by MFI, and the detector zenith of the ERNE sensors on board SOHO. For most of the time, the angle is below 30°, within the view cone of both ERNE sensors. About 4 hours after the X-ray event at the Sun (09:04 UT), simultaneously with the rapid changes in the IMF direction, a dramatic decrease in the proton counting rates was observed. The drop was most pronounced at low energies, almost two orders of magnitude at 1.6–3.0 MeV, and extremely rapid in this channel, 2-3 min. The drop

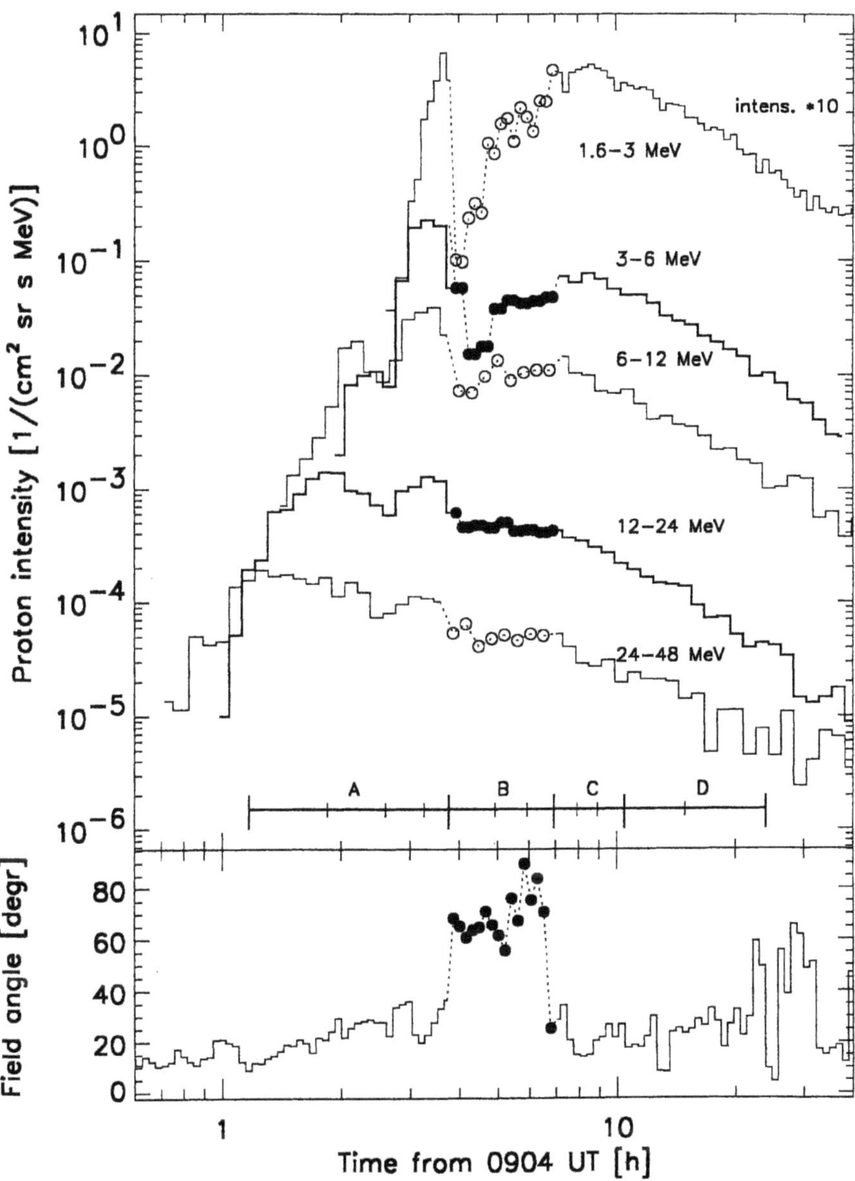

Figure 5. Intensity profiles of protons as observed by ERNE from 09:04 UT 9 July until 24:00 UT 10 July 1996. The energy channels 1.6–3 MeV, 3–6 MeV, and 6–12 MeV represent the ERNE/LED observations, and the channels 12–24 MeV and 24–48 MeV the ERNE/HED observations. The intensity in the lowest energy channel is multiplied by 10. The bottom panel gives the angle between the ERNE sensor axis and the interplanetary magnetic field observed by WIND/MFI. The intensities during the period 12:50–16:00 UT are presented as open and closed circles. The bar illustrates the time intervals (subintervals) A (A_1, A_2, A_3, A_4), B (B_1, B_2, B_3), C (C_1, C_2, C_3), and D (D_1, D_2) discussed in the text and shown in Table I.

is still detectable at 24–48 MeV within a factor of about 2. The magnetic field stays at large angles with respect to the ERNE sensor axis for about 3 hours. After this period, the field points again almost in the nominal direction, being also more stable than during the disturbance. During this time, after 16:00 UT on 9 July, the particle fluxes show a smooth power-law time decay.

ERNE HED detected over 3300 protons in the energy range 14–17 MeV during 9–10 July 1996. All the data were transmitted to Earth as pulse heights. These particles were used in the detailed analysis of the development of the angular distributions. Four main time intervals were selected for anisotropy analysis in the 14–17 MeV proton channel: (A) the event onset and flux maximum (9 July, 10:14 –12:50 UT) during which the magnetic field stayed within the LED view cone, (B) the passage of the magnetic field disturbance (9 July, 12:50–16:00 UT) when the magnetic field direction was outside the LED view cone, (C) the recovery of the quiet magnetic field (9 July, 16:00–19:30 UT), and (D) the decay phase of the SEP event (9 July, 19:30–10 July, 09:00 UT). The main intervals were divided into sub-intervals for which the two dimensional angular distributions were calculated. During each of the sub-intervals, axial symmetry was assumed with a fixed direction of the symmetry axis. We used two methods to obtain the axis: (i) we calculated the average field direction from the MFI measurements during the sub-interval, and (ii) we deduced the symmetry axis by visually searching the direction of maximum flux from the two-dimensional angular distribution. The first method was applicable during all sub-intervals, but it proved to be impossible to use the latter method during the main interval B, when the distribution was almost isotropic and the magnetic field direction was at the edge of, or even outside, the view cone of HED. The times and the symmetry axis directions of all the sub-intervals are given in Table I.

One-dimensional pitch-angle distributions were formed for each sub-interval by averaging the two-dimensional distributions over the phase angle around the symmetry direction. These distributions were then combined to form the high-resolution pitch-angle distributions with higher statistics for each main interval, weighting each sub-interval distribution with the duration of the sub-interval. The main interval pitch-angle distributions with statistical error limits are plotted in Figure 6.

Although the statistical fluctuations were large in the sub-interval distributions, a quantity characterizing the anisotropy of each distribution could still be calculated. We fitted each sub-interval pitch-angle distribution to an exponential anisotropy:

$$F(\mu) = N \exp(a\mu\,|\mu|^{1-q}) ,$$

where μ is the pitch-angle cosine and N and a are fitting parameters (when small enough, the parameter a is proportional to the first order anisotropy). In the focused transport model (e.g., Kunstmann, 1979), this function is the steady-state solution of the transport equation if the ratio of the particle mean free path to the focusing length is spatially constant and the parameter q is the spectral index of the magnetic

Table I
Intervals of the anisotropy measurements

	Day	Period (UT)	Number of protons	MFI opposite direction (GSE)		Symmetry axis (GSE)		Anisotropy characteristics
				lat. (deg)	long. (deg)	lat. (deg)	long. (deg)	
A	9 July	10:14–12:50	1337					Beam-like long-lasting anisotropy
A1	9 July	10:14–10:55	296	−4	331	5	329	
A2	9 July	10:55–11:40	409	−7	341	−7	326	
A3	9 July	11:40–12:20	337	−4	343	−9	324	
A4	9 July	12:20–12:50	295	−19	334	−14	320	
B	9 July	12:50–16:00	829					Quasi-isotropy
B1	9 July	12:50–14:00	324	−36	262			
B2	9 July	14:00–15:00	276	−5	250			
B3	9 July	15:00–16:00	229	13	235			
C	9 July	16:00–19:30	636					Recurrent long-lasting anisotropy
C1	9 July	16:00–17:00	214	−18	310	−13	307	
C2	9 July	17:00–18:00	194	−14	320	2	305	
C3	9 July	18:00–19:30	228	−18	330	14	303	
D	9 July	19:30–09:00	969					Long-lasting anisotropy
D1	9 July	19:30–24:00	450	−15	322	6	309	
D2	10 July	00:00–09:00	519	9	335	10	317	

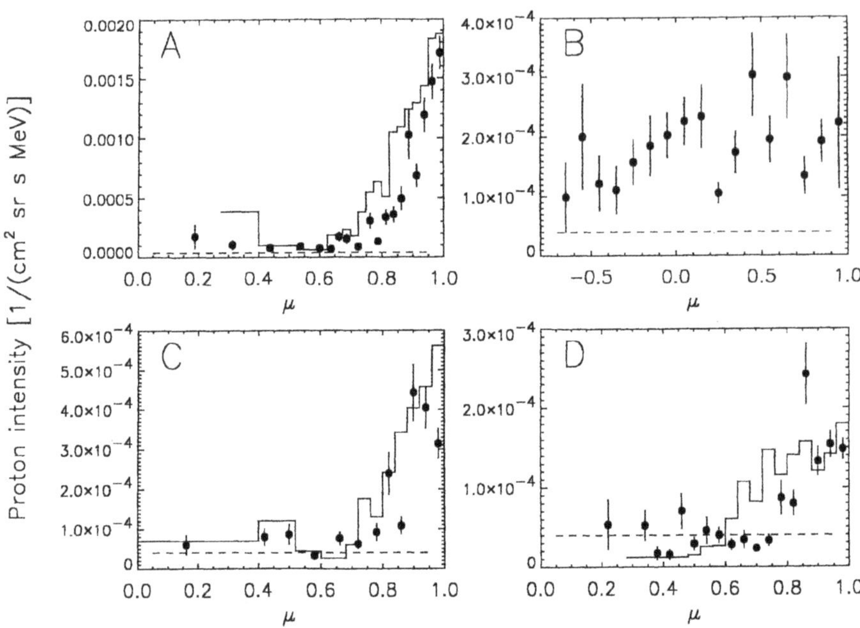

Figure 6. The pitch-angle distributions of the 14–17 MeV protons during four time intervals 9–10 July 1996: *A*: 9 July 10:14–12:50 UT, *B*: 9 July 12:50–16:00 UT, *C*: 9 July 16:00–19:30 UT, and *D*: 9 July 19:30 UT–10 July 09:00 UT. The distribution of the pitch-angle cosine, μ, is given both with respect to the average field direction measured by WIND/MFI (solid circles) and with respect to the symmetry axis estimated from the directional scatter diagram (histogram). The horizontal dashed line represents the isotropic background of galactic cosmic radiation.

field fluctuations. We adopted $q = 1$ in our calculations, corresponding to isotropic scattering. Using the observed angular distributions and assuming an exponential shape outside the sensor view cone, we estimated the average pitch angle, $\langle\mu\rangle$, in each sub-interval of Table I. The isotropic cosmic-ray background was subtracted. The average pitch angle is plotted as a function of time in Figure 7.

4. Results and Discussion

After the maximum of high-energy particle flux at SOHO, the MFI instrument on board WIND detected a magnetic field disturbance passing the spacecraft (Figure 4). The disturbance begins with a drop in the magnetic field magnitude from 4 to 3.3 nT. At 12:50 UT, the field magnitude suddenly drops to 2 nT, and the ϕ-angle of the magnetic field turns about 90° in 15 min, remaining below 90° for the next 3 hours. During this period, the field intensity goes from 2 nT to more than 4 nT and again back to 2.5 nT. The θ-angle experiences a rise from 20° at 12:50 UT to almost 60° at 13:30 UT. Then it falls to $-20°$ by 15:10 UT and starts to rise

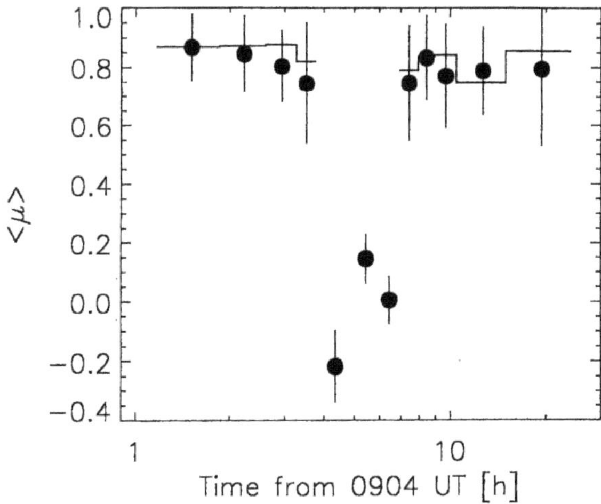

Figure 7. The time development of the average pitch-angle cosine, $\langle \mu \rangle$, of the angular distribution of protons. The axis of the proton beam was taken as the direction of the average field measured by WIND/MFI (solid circles), and the symmetry axis estimated from the directional scatter diagram (histogram).The energy range of protons is from 14 to 17 MeV.

again reaching 40° at 16:40 UT, after which it finally stabilizes at about 15°. A detailed study of the magnetic field fluctuations in the disturbed flux tube is beyond the scope of this paper, but we note that the disturbed region is characterized by large and rapid changes in the field direction, while the maximum magnetic field strength is approximately the same inside and outside the region. This seems to contradict the properties of magnetic flux ropes or magnetic clouds described by Marubashi (1986). For this reason, we treat the region simply as a different flux tube. Note that in this disturbed tube, the plasma number density is enhanced by $\approx 20\%$ in comparison with surrounding solar wind plasma. The energetic particle transport conditions inside the disturbed region are expected to be very different from those outside the region.

Looking at the counting rates at high energies (channels 12–24 and 24–48 MeV in Figure 5), we can see indications of a double-peaked structure of the event. During the period A1, the intensity of the 14–17 MeV protons rose to the maximum of the first peak, at 10:55 UT, in 45 min. The decline of the intensity to the minimum at 11:40 UT was again 45 min (period A2). The second rise to the maximum, period A3, took 40 min. The peak intensity was achieved at 12:20 UT. The second decline, period A4, ended at an intensity level which was about 30% from the maximum of the second peak.

We calculated the directional distributions during the four subintervals A1–A4, but were not able to find significant differences in the distributions. Therefore, we conclude that the interplanetary transport was identical. As shown in Figures 6 and 7, the anisotropy was strong during the main period A. The particle flux is concentrated in a narrow cone with a half width in the range 25°–40° during the whole period A. During this 3-hr period, there may be only a weak trend for the particle beam to become somewhat broader.

A dramatic change in the flux occurs during interval B, when the flux is almost isotropic. Even more surprising was perhaps the recovery of the anisotropic beam structure during interval C, after the passage of the magnetic disturbance, 6 hours after the onset of the event. The long-lasting high anisotropy gives direct evidence of continuous injection of energetic protons into interplanetary space, most likely from the interplanetary shock wave following the flare.

The directional anisotropy of the flux remains strong during phase D. At the end of the present analysis period, at 09:00 UT 10 July, the proton intensities of solar/interplanetary origin and of the galactic background are equal. During the last interval D2, i.e., 15–24 hours after the X-ray flare maximum, the average pitch angle is still in the range 25°–40° (Figure 7). This means that half of the solar protons pass the spacecraft in a cone with a half width of less than $\approx$ 30° from the field direction or symmetry axis.

The flux tubes of the interplanetary magnetic field are corotating with the Sun, meaning that charged particles stay in the spiral-shaped magnetic flux tubes in the rotating frame. In this frame, the spacecraft does not stay in a single flux tube, but flies through the tubes going from west to east as time goes on. This offers us a possibility to study latitudinal changes in the transport conditions. We have shown that during the solar energetic particle event on 9 July 1996, the angular distribution of 14–17 MeV protons evolved from a beam-like structure through an almost isotropic phase to a recurrent beam. This provides evidence of large differences in the transport conditions of the neighbouring flux tubes. The almost two orders of magnitude decrease at the low-energy proton channel suggests that in the disturbed magnetic region, the transport of particles is much slower than in the ambient flux tubes. The study of the plausible injection and transport model scenarios (Kocharov *et al.*, 1997) indicates that we observed the onset phase of the low-energy proton event twice. First it was seen in a flux tube with fast particle transport (undisturbed magnetic field), and later in a region of slow particle transport (rapid and strong disturbances in the field). The width of the disturbed flux tube is $\approx$ 0.05 AU. The boundary of the two regions was thin: the 1.6–3 MeV proton flux dropped almost two orders of magnitude in 2 min, giving a thickness of about 50 000 km consistent with the Larmor radius of the MeV-protons in IMF.

Another peculiarity of the 9 July event is the short delay time from the X-ray intensity maximum to the first proton intensity maximum. In the energy range >12 MeV the maximum was reached in less than two hours. Previous studies (e.g., Cane, McGuire, and von Rosenvinge, 1986, and references therein) divided X-ray flares into two groups depending on the decay time of the soft X-ray flux. On the basis of this definition, the 9 July flare was impulsive. Although the fast rise of proton intensity to the maximum is typical of well-connected impulsive flares, and thus of the present event, the particle transport of the 9 July event in interplanetary space was among the fastest detected at 1 AU (Cane, McGuire, and von Rosenvinge, 1986; Kallenrode, Cliver, and Wibberenz, 1992). Previous studies of interplanetary mean free paths of protons (Beeck and Wibberenz, 1986; Beeck *et al.*, 1987; Bieber *et al.*, 1994; Palmer, 1982; Kallenrode, 1993) have revealed two populations of ~ MeV particle mean free paths, one being ~ 0.1 AU, the other ~ 1.0 AU. Results of the detailed analysis of the event of 9 July 1996 (Kocharov *et al.*, 1997) show that the mean free path was more than 1 AU outside the above-described narrow magnetic region.

5. Conclusions

The interplanetary transport of energetic particles during most of the solar event on 9 July 1996, was extremely rapid. The high-energy protons arrived at the SOHO distance as a beam of particles. The half width of the proton beam was in the range 25°–40° with respect to the symmetry axis of the beam.

About 3 hours after the event onset, the spacecraft passed a strong interplanetary field disturbance, which caused the low-energy proton intensity to drop almost two orders of magnitude in 2–3 min, and changed the strong anisotropy in the 14–17 MeV energy range to almost isotropy.

The observed recurrence of a beam-like anisotropy six hours after the event onset suggests that the energetic particle injection in the SEP event on 9–10 July 1996 was extended, lasting until the end of the analysis period, i.e., 24 hours. Most of the time during the event, the SOHO spacecraft was in almost scatter-free regions of the interplanetary magnetic field. Between 12:50 and ~ 16 UT on 9 July 1996, the spacecraft flew through a disturbed magnetic flux tube, where the particle transport was much slower than in the neighbouring regions. The thickness of the boundary between the regions was of the order of the Larmor radius of the energetic particles. The particles seemed to retain the beam shape with a constant beam width until the end of the event. Therefore, from the particle transport point of view, the interplanetary sectors on both sides of the disturbed regions were identical.

The 9 July event was a small one. Therefore, the disturbance of the interplanetary medium due to the flux of energetic particles was presumably small. Because HED has a large view cone, it was able to follow the passage of a clean beam-shaped bundle of particles in an unchanged scatter-free space. When the anisotropy was strong, and the magnetic field direction was close to the direction of the ERNE axis, ERNE was able to sample in its view cone practically the total flux of the passing protons. We conclude that a beam mode of interplanetary transport was first detected during the event of 9 July 1996.

Acknowledgements

We are grateful to Dr R. P. Lepping for permission to use the magnetic field data of WIND/MFI. Thanks are extended to the SOHO/CELIAS and SOHO/LASCO teams for the solar wind and coronagraph data available in the SOHO archive. The Academy of Finland is thanked for financial support.

References

Beeck, J. and Wibberenz, G.: 1986, *Astrophys. J.* **311**, 437.

Beeck, J., Mason, G. M., Hamilton, D. C., Wibberenz, G., Kunow, H., Hovestadt, D., and Klecker, B.: 1987, *Astrophys. J.* **322**, 1052.

Bieber, J. W., Matthaeus, W. H., Smith, C. W., Wanner, W., Kallenrode, M.-B., and Wibberenz, G.:1994, *Astrophys. J.* **420**, 294.

Cane, H. V., McGuire, R. E., and von Rosenvinge, T. T.: 1986, *Astrophys. J.* **301**, 448.

Kallenrode, M.-B.:1993, *J. Geophys. Res.* **98**, 19037.

Kallenrode, M.-B., Cliver, E. W., and Wibberenz, G.: 1992, *Astrophys. J.* **391**, 370.

Kocharov, L., Torsti, J., Laitinen, T., and Vainio, R.: 1997, *Solar Phys.* **175**, 785 (this issue).

Kunstmann, J.: 1979, *Astrophys. J.* **229**, 812.

Marubashi, K.: 1986, *Adv. Space Res.* **6** (6), 335.

Palmer, I.D.: 1982, *Rev. Geophys. Space Phys.*, **20**, 335.

Schulze, B. M., Richter, A. K., and Wibberenz, G.: 1977, *Solar Phys.* **54**, 207.

Solar Geophysical Data: 1996, No.. 624, Pt. I.

Torsti, J., Valtonen, E., Lumme, M., Peltonen, P., Eronen, T., Louhola, M., Riihonen, E., Schultz, G., Teittinen, M., Ahola, K., Holmlund, C., Kelhä, V., Leppälä, K., Ruuska, P., and Strömmer, E.: 1995, *Solar Phys.* **162**, 505.

IMPLICATIONS OF PROTON ANISOTROPY DEVELOPMENT OBSERVED BY THE ERNE INSTRUMENT DURING THE 9 JULY 1996 SOLAR PARTICLE EVENT

L. G. KOCHAROV, J. TORSTI, T. LAITINEN and R. VAINIO
Space Research Laboratory, Department of Physics, FIN-20014 Turku University, Finland

(Received 4 March 1997; accepted 28 May 1997)

Abstract. We analysed the solar particle event following the 9 July 1996 solar flare. High-energy protons were detected by the ERNE instrument on board SOHO. Anisotropy of arriving protons revealed very peculiar non-monotonic development. A short period of almost isotropic distribution was imbedded into the prolonged period of beam-like distribution of 14–17 MeV protons. This implies the existence of a narrow magnetic channel with a much smaller mean free path than in the surrounding quiet solar wind plasma.

We used Monte Carlo simulations of interplanetary transport to fit the observed anisotropies and intensity–time profiles. Proton injection and transport parameters are estimated. The injection scenario is found to be very close to the scenario of the 24 May 1990 event, but the intensity and the interplanetary transport parameters are different. The extreme anisotropy observed implies prolonged injection of high-energy protons at the Sun and at the interplanetary shock front, and either a very large mean free path ($\geq$ 5 AU) outside the slow transport channel, or alternatively, a somewhat smaller mean free path ($\approx$ 2 AU) and enhanced focusing between the Sun and the Earth.

1. Introduction

The 10 MeV protons, travelling in the interplanetary medium, are scattered by short-scale fluctuations of the interplanetary magnetic field and focused by the large-scale component of the field (e.g., Toptyghin, 1983). These particles can be used as probes of the magnetic fields that they traversed before they were detected by a spacecraft (Wanner and Wibberenz, 1993). Being themselves a major phenomenon of the interplanetary space, they also carry information about their sources, i.e., the Sun and phenomena closely related to it such as interplanetary shocks.

The fluctuation conditions are not stationary. For this reason, differences in the proton mean free path between one event and another can amount to 2 orders of magnitude (Kunow *et al.*, 1991). The strong variation in the fluctuation level typically occurs on the angular scale 3–7° in heliolongitude (Wanner and Wibberenz, 1993). On the other hand, the large-scale component of the interplanetary field is also not stationary, implying variability of magnetic focusing. The Parker (1958) theory is a steady-state prediction that avoids discussion of interplanetary dynamic processes such as overtaking high-speed streams or falling behind slow-speed solar wind. However, such processes must change the magnetic field and affect the propagation of energetic particles. The effect of a distorted interplanetary field on the transport of energetic particles has already been employed to explain spacecraft observations (Tan *et al.*, 1992, Anderson *et al.*, 1995).

Analysis of the 9 July 1996 solar particle event observed by ERNE revealed complex behaviour depending on the proton energy channel selected (Torsti *et al.*, 1997). At high energies, $E > 12$ MeV, the shape of the intensity-time profile of the event was rather close to the profile of the 24 May 1990 event that we recently analysed (Torsti *et al.*, 1996). Both events demonstrated a double-peaked structure but the proton intensity in May 1990 was three orders of magnitude higher than in July 1996. At low energies, $E = 1.6–6$ MeV, a huge drop, of 1–2 orders of magnitude, was observed in ERNE count rates at $\approx$ 13:00 UT on 9 July 1996. This is a specific property of the event. Simultaneously, the extremely anisotropic flux of $\approx$ 15 MeV protons was replaced by a much more isotropic distribution. The $\approx$ 15 MeV proton beam reappeared at $\approx$ 16:00 UT. Simultaneously with the drop in the 1.6–6 MeV proton count rates and the vanishing of the $\approx$ 15 MeV proton beam, the magnetic field observed on board WIND abruptly changed its direction by $\approx 60°$ (Torsti *et al.*, 1997). During the period 13:00–16:00 UT, the magnetic field direction changed repeatedly. It became stable after 16:30 UT on 9 July 1996. These observations indicate that the interplanetary magnetic field and proton transport conditions changed at least twice during the event, at $\approx$ 13:00 UT and at $\approx$ 16:00 UT on 9 July 1996. For this reason, three periods in the proton count rate were selected for further analysis: Period 1 covering 10:14–12:50 UT, Period 2 covering 12:50–16:00 UT, and Period 3 covering 16:00–24:00 UT on 9 July 1996. Our study is aimed at the estimation of the injection and transport parameters. For this reason, it focuses on the 14–17 MeV proton anisotropy data and does not attempt to cover comprehensively all the energy bands observed.

2. Basic Model

Taking into account the similarity of the 24 May 1990 and the 9 July 1996 solar particle events, we attempted to fit observational data with two (prompt and delayed) injections of particles. In the same way as Torsti *et al.* (1996), we consider (i) solar injection of particles at the root of the magnetic line connected to the spacecraft and (ii) interplanetary shock wave injection of particles in the shock area that is magnetically connected to the observer for the prompt and delayed components, respectively. The intensity near the detector, $I(E, \mu, t)$, was obtained as a convolution of the injection profile, $q(t, E)$, and the interplanetary transport Green's function, $G(z_0, t - t_0, \mu)$:

$$I(E, \mu, t) = \int_{-\infty}^{t} q(t_0, E) \frac{G(z_0, t - t_0, \mu) V}{r_{\oplus}^2 \cos \psi(r_{\oplus})} \mathrm{d}t_0 \,, \tag{1}$$

where $q(t_0, E)$ is the production of accelerated particles per unit of heliocentric solid angle at time t_0 at spiral distance z_0 in the interplanetary medium; $\psi(r_{\oplus})$ is the tilt angle of the interplanetary magnetic field at the site of observations, $r = r_{\oplus}$; E, V, and $\mu = \cos(\theta)$ are proton energy, velocity and pitch angle cosine, respectively.

To calculate interplanetary transport functions, $G(z_0, t - t_0, \mu)$, and fit the data, some spatial dependence of the mean free path must be adopted. Kallenrode (1993) discussed a number of possibilities for the spatial dependence of the radial mean free path λ_r, and decided to use λ_r = constant. In such a case the dependence of the parallel mean free path on distance along the magnetic line is almost linear: $\lambda \sim z$ at $z > 0.7$ AU. For this reason, we fitted the data under the proposition that the parallel mean free path is linear in z at $z > 0.7$ AU and constant at $z < 0.7$ AU. This assumption on $\lambda(z)$ allows us to obtain reasonably good fits of the event. However, somewhat steeper spatial dependence could also be acceptable. The rigidity dependence of λ is not essential because protons in a single narrow energy band, 14–17 MeV, are studied. The Monte Carlo simulations of focused transport were used to obtain interplanetary transport functions (see Appendix). It is convenient to describe the pitch-angle scattering in terms of a parallel mean free path, λ, given by

$$\lambda = \frac{3V}{8} \int_{-1}^{1} \frac{1 - \mu^2}{\nu(\mu)} d\mu \,, \tag{2}$$

where $\nu(\mu)$ is the scattering frequency (e.g., Toptyghin, 1983). When defined in this way, the parallel mean free path may take any value, even exceeding 1 AU. The parallel mean free path at the Earth orbit, $\lambda_{\oplus}$, and the source function, $q(t, E)$, were adjusted to fit the anisotropies and intensity-time profiles observed. We use Parker's (1958, 1963) model of the interplanetary magnetic field at solar wind velocity 400 km s^{-1}. Throughout the paper, this standard model is designated as Model 1.

In order to perform fitting in a limited energy and temporal range, we used a simplified injection model. The source of the prompt component injection was positioned at the spiral distance of two solar radii from the centre of the Sun, $z_0 = 2R_S$. For the interplanetary shock injection, the source was fixed at $z_0 = 20R_S$ because we mainly consider the first 8 hours of the event when the shock was travelling not very far from this point. For both components, the injection time profiles are fitted by exponents:

$$q(t, E) = A \exp(\Phi(t)) \,, \tag{3}$$

where $\Phi(t)$ is a piecewise linear function of time, A is a normalization constant. We used

$$\Phi(t) = (D - 1/\tau)\,(t - t_m) \quad \text{at} \quad 0 < t < t_m \,;$$

$$\text{and} \quad \Phi(t) = -(t - t_m)/\tau \quad \text{at} \quad t > t_m \,, \tag{4}$$

Table I
Injection and transport parameters for Model 1

Component	Periods 1 and 3	Period 2			
	$\lambda_{\oplus}$	$\lambda_{\oplus}$	D	τ	t_m
	AU	AU	min^{-1}	min	min
Prompt	5.2	0.35	0.095	50	37
Delayed	5.2	0.35	0.05	230	132

where t_m, D, and τ are fitting parameters, $D\tau > 1$. These parameters are different for different injection components (see Table I). The 'zero' time is 09:04 UT on 9 July 1996.

3. Results and Discussion

We calculated 14–17 MeV proton intensity at the Earth–Sun L1 point as a function of pitch angle and time, $I(\mu, t) \sim \mathrm{d}^2N/\mathrm{d}\Omega\ \mathrm{d}t$. This intensity was integrated over Periods 1, 2, and 3 to obtain angular distributions, $\mathrm{d}N/\mathrm{d}\Omega$, for the periods. Then the angular distributions and the intensity-time profiles were compared with observations (see Figures 1 and 2). The choice of the integration periods was governed by two circumstances: (i) the limited count statistics available, and (ii) the preliminary analysis of magnetic field data which revealed the disturbed magnetic region observed during Period 2 (Torsti *et al.*, 1997). It is also important that the character of anisotropy is kept constant during each time period, so that the choice of the integration time does not affect our principal conclusions. To estimate the predicted count rate, the following procedure was applied. The angular distribution was extremely anisotropic during Periods 1 and 3, so that almost all protons were arriving inside the ERNE/HED view cone and contributing to the count rate. For these periods, the calculated intensity was simply integrated over pitch angles, and the intensity-time profile obtained, $\mathrm{d}N/\mathrm{d}t$, was superimposed on the observed count rate (see solid line in Figure 2). During Period 2, the proton distribution was almost isotropic, so that the fraction of protons contributing to the count rate was about $\frac{1}{4}$, and this fraction is only weakly dependent on model parameters. For this reason, to estimate the model count rate, it is sufficient to divide the angle integrated intensity by a factor ≈ 4. The corresponding curve is shown with the mainly dotted line in Figure 2. It is seen that, with the statistics available, the obtained fits are reasonably good at least as a first approximation. Thus, we can conclude that the mean free path was extremely high during Period 1, at least $\lambda_{\oplus} = 5$ AU, after which it abruptly dropped by more than factor 10 (Period 2), and finally, at the beginning of Period 3, the mean free path returned to its original extremely large value.

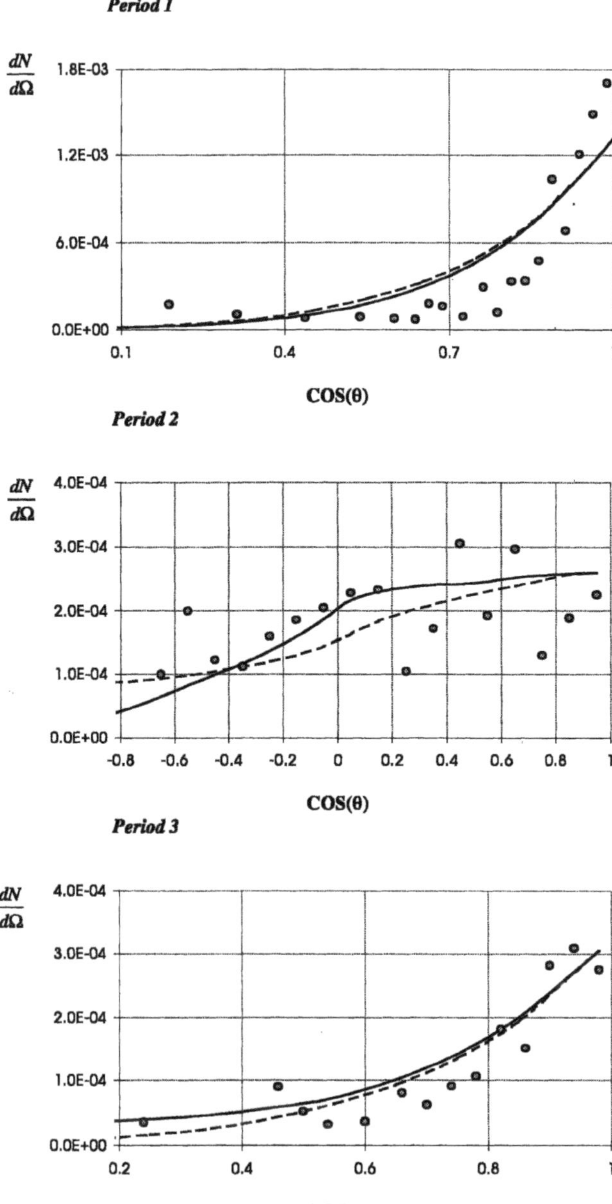

Figure 1. Illustration of fitting to the ERNE anisotropy measurements. The ERNE data are shown as points (see also Torsti *et al.*, 1997). Dashed lines correspond to 14–17 MeV proton distribution calculated in the framework of the standard transport model (Model 1). Solid lines correspond to the off-standard focusing model (Model 2). Period 1 covers 10:14–12:50 UT, Period 2 covers 12:50–16:00 UT, and Period 3 covers 16:00–24:00 UT on 9 July 1996.

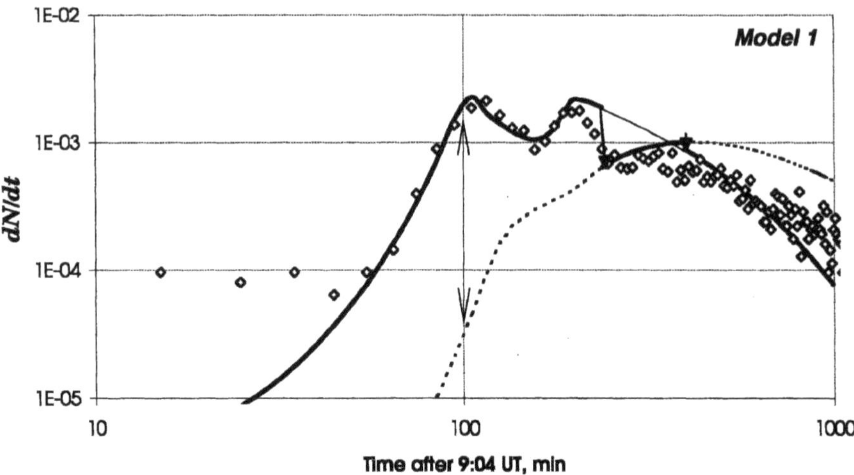

Figure 2. The 10-min-average count rates in the proton 14–17 MeV channel during the 9 July 1996 solar particle event as observed by ERNE (points) and the parameter fit according to Model 1 (curves). Solid line illustrates the intensity-time profile outside the slow transport channel. Mainly dotted line illustrates the intensity-time profile inside the slow transport channel. Periods observed by ERNE are shown with heavy solid portions of the curves and two heavy arrows (the double-headed arrow is discussed in the text).

In addition to the traditional transport model, we considered the possible effect of deviation of the interplanetary magnetic field from the standard Parker (1958, 1963) model (Model 1). Our alternative model (Model 2) does not attempt to reproduce the real interplanetary magnetic field but is aimed only at answering the question of whether or not a deviation of the real field from the standard Parker model can affect proton anisotropies observed by ERNE. In this attempt, the injection scenario and the parallel mean free path $\lambda(z)$ are proposed to be the same at all magnetic flux tubes traversed. To estimate the possible effect of off-standard focusing, we propose that the standard magnetic field, B_{P}, be distorted with the factor

$$\zeta = 1 + \zeta_0 + \zeta_1 \sin\left(\frac{\pi(z - z_s)}{2z_m}\right)$$

at

$$z < z_c \equiv 4z_m - \frac{2z_m}{\pi}\arcsin(\zeta_0/\zeta_1) + z_s\ ,$$

and

$$\zeta = 1 \quad \text{at} \quad z \geq z_c\ , \tag{5}$$

so that the magnetic field at point z is $B(z) = \zeta(z)B_{\mathrm{P}}(z)$. To limit the choice of the new transport parameters ($\lambda_{\oplus}$, ζ_0, ζ_1, z_s, and z_m), we imposed the following

Table II
Injection and transport parameters for Model 2

Component	D min^{-1}	τ min	t_m min
Prompt	0.11	35	32
Delayed	0.05	180	122

Period	$\lambda_\oplus$ AU	ζ_0	ζ_1	z_m AU	z_s AU
1 and 3	2.0	1.5	1.5	0.3	0.3
2	2.0	0.15	−0.75	0.6	0

limitations: (i) along the length of a magnetic tube, the ratio of maximum to minimum values of factor ζ is not more than 4, which is a typical value of variation of magnetic field strength as observed near the Earth orbit, (ii) the scale of the distortion, z_m, is about 0.5 AU, and the size of the distorted region, z_c, is 1–3 AU, and (iii) the large-scale magnetic field is not distorted near the Sun. We also took into account the fact that the expansion of a tube is likely to go on concurrently with the compression of some neighbouring tubes to keep the total magnetic flux roughly constant.

Table II presents a possible set of parameters fitting the 14–17 MeV proton observations. We have enhanced focusing between the Sun and the spacecraft at magnetic lines traversed during Periods 1 and 3. For Period 2, the magnetic field is decreased at $\approx$ 0.6 AU but increased at $\approx$ 1.8 AU from the Sun. The proton angular distributions obtained for Periods 1, 2, and 3 are depicted by solid lines in Figure 1. The intensity-time profiles are shown in Figure 3. It turns out that the 14–17 MeV proton observations can be fitted equally well in both transport models. In particular, almost the same angular distribution can be obtained in the standard model at $\lambda_\oplus \approx 5$ AU or in the off-standard focusing model at $\lambda_\oplus \approx 2$ AU. Employing only the 14–17 MeV energy band, we cannot select the most plausible model, because only a small portion of the intensity-time profile near the second peak was actually observed during Period 2. However, a very big difference between the model predictions really exists at the rise phase of the event (marked with double-headed arrows in Figures 2 and 3).

Around 12:00 UT, the rise phase of the event was observed at lower energies, $E \approx 4$ MeV. At about 12:50 UT, low energy channels revealed a huge drop of more than one order of magnitude (see Torsti *et al.*, 1997). This lends support for Model 1. In fact, in the 1.6–6 MeV proton energy band, the onset of the 9 July 1996 event was observed twice. On the first occasion, the onset was observed inside the fast propagation region during Period 1. Then SOHO entered the slow transport

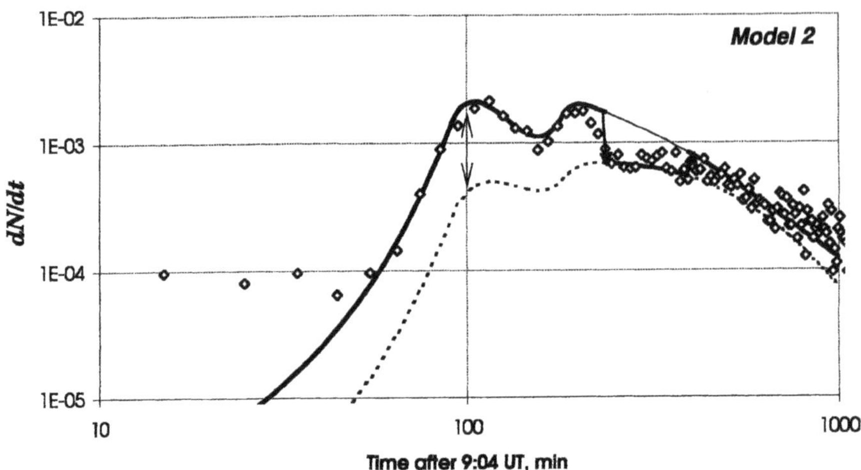

Figure 3. The same as in Figure 2 but for Model 2.

channel where the onset of the event in the 1.6–6 MeV band was observed again (see Figure 5 by Torsti *et al.*, 1997). In the 14–17 MeV band, reconstruction of the event according to Model 1 is shown in Figure 4. For these energies, the slow transport channel (STC) was traversed near the time of the maximum intensity of the delayed component (marked with STC in Figure 4). It can, of course, not be ruled out that some extra focusing further contributed to the sharpening of the proton angular distribution outside STC because, even at an unusually low scattering level ($\lambda_\oplus \approx 5$ AU) and standard focusing, the calculated distribution is still not sharp enough to reproduce precisely the extreme beam-like anisotropy observed by ERNE during Period 1.

4. Conclusions

(1) During the 9 July 1996 solar particle event, ERNE observations revealed that a narrow ($\approx 2°$ in heliolongitude) magnetic channel with special transport parameters (slow transport channel, STC) was imbedded in a wide region of fast propagation of high-energy protons.

(2) Inside STC, the magnetic field was disturbed, the proton angular distribution was almost isotropic, and the proton propagation was slow.

(3) Outside STC, the 14–17 MeV proton mean free path was extremely large, $\lambda_\oplus \geq 5$ AU, if focusing corresponded to the standard model of the interplanetary magnetic field. In contrast, the mean free path was less or about 0.35 AU inside STC.

(4) The same proton angular distribution may come from extremely weak scattering ($\lambda_\oplus \approx 5$ AU) in the standard interplanetary magnetic field or from a

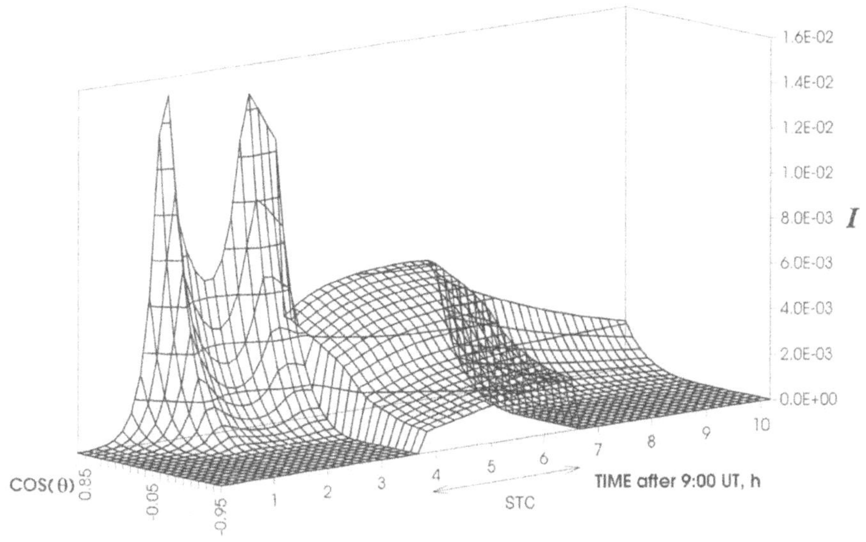

Figure 4. Reconstruction of the observed 14–17 MeV proton event in the pitch-angle cosine and time plane according to Model 1.

stronger scattering ($\lambda_\oplus \approx 2$ AU) at magnetic lines additionally compressed at ≈ 0.6 AU from the Sun. Proton anisotropies measured by ERNE are sensitive to the actual shape of the interplanetary magnetic field lines. A narrow beam-like proton distribution may be a signature of extra focusing and exceptionally weak scattering between the Sun and the spacecraft.

Acknowledgements

The Academy of Finland is thanked for financial support. SOHO is a mission of international co-operation between ESA and NASA.

Appendix. Monte Carlo Simulations of Interplanetary Transport

Calculations of the particle propagation are based on the model of focused transport (e.g., Toptyghin, 1983). The code employed is basically similar to that described by Torsti *et al.* (1996) but the scattering is dependent on particle pitch angle, i.e., we take into consideration the anisotropic scattering model. Thus, the transport equation for high-energy particles is in the form (Roelof, 1969; Earl, 1976):

$$\frac{\partial f}{\partial t} + \frac{\partial}{\partial z}\mu V f + \frac{\partial}{\partial \mu}\frac{V}{2L}(1-\mu^2)f - \frac{\partial}{\partial \mu}(1-\mu^2)\nu\frac{\partial f}{\partial \mu} = Q(z,\mu,t)\,, \quad \text{(A1)}$$

where $f = f(z, \mu, t)$ is the number of particles per unit of magnetic tube length and per unit of solid angle in velocity space; z is the coordinate of the observer (along the magnetic field line); μ is the cosine of pitch angle; V is the particle velocity; $L(z)$ is the focusing length, $1/L(z) = -(\mathrm{d}B/\mathrm{d}z)/B$; $B(z)$ is the magnetic field strength, $\nu = \nu(z, R)$ is the scattering frequency which determines the parallel mean free path, λ; R is the particle rigidity; $Q(z, \mu, t)$ is the source function. In the case of anisotropic scattering we adopt $\nu = \nu(\mu)$ as

$$\nu(\mu) = \frac{3V}{2(2-q)\,(4-q)\lambda_0}(\mu_0^2 + (1-\mu_0^2)\mu^2)^{(q-1)/2}\,, \tag{A2}$$

where q is the spectral index of MHD turbulence, the parameter λ_0 coincides with the mean free path if $\mu_0 = 0$. We adopted the values $q = 1.5$ and $\mu_0 = 0.04$.

In order to calculate numerical Green's functions, $G(z_0, t - t_0, \mu)$, we perform Monte Carlo simulations of particle transport. We consider an impulsive and isotropic injection of particles with energy E at time t_0 at the distance z_0 from the Sun. We calculate, per one Monte Carlo temporal step, Δt, the change of the particle pitch angle cosine as $\Delta\mu = (1-\mu^2)V\Delta t/(2L) + \eta$, and the change of the particle coordinate as $\Delta z = V\mu\Delta t$. The value of η, which is the change of the particle pitch angle cosine due to scattering, is calculated by means of the following method. The particle scattering can be considered with respect to the direction of the particle movement at the moment. In the isotropic scattering case, by means of such a 'turnable' axis, under the small-angle approximation, the scattering term of the Fokker–Planck equation can be written in the form

$$\frac{\nu}{\vartheta}\frac{\partial}{\partial\vartheta}\vartheta\frac{\partial f}{\partial\vartheta}\,. \tag{A3}$$

Hence, after each Monte Carlo step, the angular distribution (with respect to the 'turnable' axis) is

$$f(\vartheta, \phi, \Delta t)\,\mathrm{d}\Omega = \frac{1}{4\pi\nu\Delta t}\exp\left\{-\frac{\vartheta^2}{4\nu\Delta t}\right\}\vartheta\,\mathrm{d}\vartheta\,\mathrm{d}\phi\,, \tag{A4}$$

where ϑ is the angle of the particle scattering during a given Monte Carlo step; $\vartheta \in [0, \pi]$; ϕ is the azimuth angle of the particle turn about the axis, $\phi \in [0, 2\pi]$. Hence, the values of ϑ^2 and ϕ are determined from an exponential distribution and the uniform distribution, respectively. The new value of the pitch angle, θ, can be calculated by means of spherical geometry. In the case of anisotropic scattering, an additional streaming term in the phase space causes additional regular change in μ so that $\eta = \eta_0 + \eta_A$, where η_0 can be obtained in the same way as in the case of isotropic scattering and $\eta_A = (\mathrm{d}\nu/\mathrm{d}\mu)\,(1-\mu^2)\Delta t$. The actual Monte Carlo time step Δt depends on the current value of scattering frequency $\nu(\mu, z)$ to keep the scattering angle small in comparison with characteristic scales of the problem. The total number of traced particles was 10^5 for each Green's function calculated. The velocity of the solar wind is taken to be 400 km s^{-1}.

We have performed a number of simulations to verify the technique. In particular, the case of the constant mean free path has been considered. In this case, the focused transport equation was solved earlier by Ruffolo (1995) using the finite difference method. A good agreement between our intensity-time profile and the profile reported by Ruffolo has been obtained when no adiabatic deceleration is taken into account. On the other hand, provided that the ratio λ/L does not change with distance, an exact analytical expression may be obtained for the time integrated (steady-state) angular distribution of particles (for the steady-state distributions see, Roelof, 1969; Kunstmann, 1979; Earl, 1981). The comparison of such a distribution with the distribution obtained in Monte Carlo simulations also revealed a good agreement.

References

Anderson, K. A., Sommers, J., Lin, R. P., Pick, M., Chaizy, P., Murphy, N., Smith, E. J., and Phillips, J. L.: 1995, *J. Geophys. Res.* **100** (A1), 3.

Earl, J. A.: 1976, *Astrophys. J.* **205**, 900.

Earl, J. A.: 1981, *Astrophys. J.* **251**, 739.

Kallenrode, M.-B.: 1993, *J. Geophys. Res.* **98**, 19037.

Kunstmann, J. E.: 1979, *Astrophys. J.* **229**, 812.

Kunow, H., Wibberenz, G., Green, G., Müller-Mellin, R., and Kallenrode, M.-B.: 1991, in R. Schwenn and E. Marsch (eds), *Physics of the Inner Heliosphere II*, Springer-Verlag, New York, p. 243.

Parker, E. N.: 1958, *Astrophys. J.* **128**, 664.

Parker, E. N.: 1963, *Interplanetary Dynamical Processes*, Interscience, New York.

Roelof, E. C.: 1969, in H. Ogelman and J. R. Wayland (eds), *Lectures in High Energy Astrophysics*, NASA SP-199, p. 111.

Ruffolo, D.: 1995, *Astrophys. J.* **442**, 861.

Tan, L. C., Mason, G. M., Lee, M. A., Klecker, B., and Ipavich, F. M.: 1992, *J. Geophys. Res.* **97** (A2), 1597.

Toptyghin, I. N.: 1983, *Cosmic Rays in Interplanetary Magnetic Fields*, Nauka, Moscou.

Torsti, J., Kocharov, L. G., Vainio, R., Anttila, A., and Kovaltsov, G. A.: 1996 *Solar Phys.* **166**, 135.

Torsti, J., Laitinen, T., Vainio, R., Kocharov, L. G., Anttila, A., and Valtonen, E.: 1997, *Solar Phys.* **175**, 773 (this issue).

Wanner, W. and Wibberenz, G.: 1993, *J. Geophys. Res.* **98** (A3), 3513.

Solar Physics **175**, No. 2, 797–799

CD-ROM Viewing Instructions

This volume includes a CD-ROM which contains multimedia material (images, movies, and sounds) that are referred to in papers published in this issue. It can be accessed using any WWW browser by opening the URL:

file:/<your_CD-ROM_directory>/papers/index.htm

where <your_CD-ROM_> is the location of your system's CD-ROM drive. Some examples follow below. Images, movies, and sounds are in GIF or JPEG, MPEG and/or Quick Time, AIFF and AU formats, respectively. You may need additional software if your WWW browser does not support these formats.

Examples of how to open the Table of Contents of the CD-ROM
- -

– Windows (3.x and 95)

Launch your WWW browser of choice and then load the file named 'index.htm' under the 'papers' directory of the DOS drive that represents your CD-ROM. If your CD-ROM is 'D:', then

URL: file:/d:/papers/index.htm

or,

'Open File' menu and load d:\papers\index.htm

– Macintosh

Launch your WWW browser, select 'Open File', click on 'Desktop' and select the volume with the name of the CD-ROM: SOL_PHYS_1752

– Unix

URL: file:/cdrom/papers/index.htm

You may need to substitute the mounting point of the CD-Rom for '/cdrom' or for '/cdrom/SOL_PHYS_175_2', depending on the Unix version you are running.

Table of contents of the CD-ROM

GPSR Compliance
The European Union's (EU) General Product Safety Regulation (GPSR) is a set of rules that requires consumer products to be safe and our obligations to ensure this.

If you have any concerns about our products, you can contact us on

ProductSafety@springernature.com

In case Publisher is established outside the EU, the EU authorized representative is:

Springer Nature Customer Service Center GmbH
Europaplatz 3
69115 Heidelberg, Germany

www.ingramcontent.com/pod-product-compliance
Ingram Content Group UK Ltd.
Pitfield, Milton Keynes, MK11 3LW, UK
UKHW021903190726
13853UKWH00003B/1395

* 9 7 8 9 4 0 1 1 5 2 3 7 2 *